Photosynthesis: Structures, Mechanisms, and Applications

Harvey J.M. Hou
Mohammad Mahdi Najafpour
Gary F. Moore
Suleyman I. Allakhverdiev
Editors

Photosynthesis: Structures, Mechanisms, and Applications

Springer

Editors
Harvey J.M. Hou
Department of Physical Sciences
Alabama State University
Montgomery, AL, USA

Gary F. Moore
School of Molecular Sciences, Center
  for Applied Structural Discovery
  at the Biodesign Institute
Arizona State University
Tempe, Arizona, USA

Mohammad Mahdi Najafpour
Center for Research in Climate Change
  and Global Warming
Institute for Advanced Studies
  in Basic Sciences
Zanjan, Iran

Suleyman I. Allakhverdiev
Institute of Plant Physiology
Russian Academy of Sciences
Moscow, Russia

ISBN 978-3-319-48871-4   ISBN 978-3-319-48873-8 (eBook)
DOI 10.1007/978-3-319-48873-8

Library of Congress Control Number: 2017938427

© Springer International Publishing AG 2017
This work is subject to copyright. All rights are reserved by the Publisher, whether the whole or part of the material is concerned, specifically the rights of translation, reprinting, reuse of illustrations, recitation, broadcasting, reproduction on microfilms or in any other physical way, and transmission or information storage and retrieval, electronic adaptation, computer software, or by similar or dissimilar methodology now known or hereafter developed.
The use of general descriptive names, registered names, trademarks, service marks, etc. in this publication does not imply, even in the absence of a specific statement, that such names are exempt from the relevant protective laws and regulations and therefore free for general use.
The publisher, the authors and the editors are safe to assume that the advice and information in this book are believed to be true and accurate at the date of publication. Neither the publisher nor the authors or the editors give a warranty, express or implied, with respect to the material contained herein or for any errors or omissions that may have been made. The publisher remains neutral with regard to jurisdictional claims in published maps and institutional affiliations.

Printed on acid-free paper

This Springer imprint is published by Springer Nature
The registered company is Springer International Publishing AG
The registered company address is: Gewerbestrasse 11, 6330 Cham, Switzerland

Foreword

A huge energy crisis is before us, and, in my opinion, the field of photosynthesis is being borne again; we are indeed seeing its reincarnation in what we may call "sustainability." Those who would solve the energy crisis need all the information on different parts of the process of photosynthesis. They would want to know how to replicate it, how to mimic it, how to modify it, and how to exploit it to our benefit using all sorts of tools from various disciplines of science: biology, physics, chemistry, mathematics (for modeling), and engineering. This book *Photosynthesis: Structures, Mechanisms, and Applications*, edited by four international authorities, Harvey Hou (USA), Mahdi Najafpour (Iran), Gary Moore (USA), and Suleyman Allakhverdiev (Russia), is one of the first books in that direction. Harvey Hou is a leading scientist and pioneer in assessing the thermodynamic parameters of photosynthesis and in developing manganese/semiconductor catalysts for artificial photosynthesis. Mahdi Najafpour is an active and successful expert in exploring transition-metal compounds as water-oxidizing catalysts. Gary Moore is a prominent specialist in artificial photosynthesis and solar-to-fuel technology. Suleyman Allakhverdiev is an established world leader in the field of photosystem II and artificial photosynthesis. This book appears at a particularly appropriate time with a balanced combination of topics related to both basic and applied aspects of photosynthesis (see the Preface and the Table of Contents). It is in the right direction toward the goal of finding a solution to the energy crisis facing us all—because of growing world population and climate change. The book deals not only with possible ways and means to understand and improve natural photosynthesis, but it also includes possibilities around the corner (maybe by 2050) for what some of us call "artificial photosynthesis." For several highly innovative ideas being discussed where photosynthesis plays an important role, I refer the readers to a paper by Donald R. Ort (with 24 top international authorities): *Redesigning photosynthesis to sustainably meet global food and bioenergy demand*, Proceedings of the National Academy of Science USA 112 (28): 8529-8536. I strongly recommend this highly informative and innovative book to all the major libraries in the world as well as to all researchers engaged in solving problems related to bioenergy.

I end my remarks by citing the hope Jules Verne (1875) gave us—we should, of course, modify it after reading this remarkable book,

Photosynthesis: Structures, Mechanisms, and Applications: "I believe that one day water will one day be used as a fuel because the hydrogen and oxygen which constitute it, used separately or together, will furnish an inexhaustible source of heat and light. I therefore believe that, when coal (oil) deposits are oxidized, we will heat ourselves by means of water. Water is the fuel of the future." I am an optimist and have faith in the next generation of scientists to solve the problems of global warming and of both food and energy facing us today. For background on all aspects of photosynthesis, I refer the readers to now 42 volumes in the Springer series *Advances in Photosynthesis and Respiration Including Bioenergy and Related Processes* (see http://www.springer.com/series/5599).

University of Illinois at Urbana-Champaign Govindjee
Urbana, IL, USA
URL: http://www.life.illinois.edu/govindjee/

Preface

Photosynthesis powers our biosphere and offers an intellectual challenge to deepen our understanding of this globally important process. It is an inherently interdisciplinary field of research, involving times that span from geologic to sub-femtosecond scale. The process occurs in plants, algae, and cyanobacteria and has evolved over 3 billion years. Photosynthesis currently produces more than 100 billion tons of dry biomass annually, equating to a global energy storage rate of ~100 TW. However, addressing the environmental, socioeconomic, and geopolitical issues, associated with increasing global human energy consumption, will require the identification and development of new technologies capable of using renewable carbon-free or carbon-neutral energy sources. Among the renewable sources, solar is indeed quite promising as it alone is sufficient to meet global human energy demands well into the foreseeable future. In this context, photosynthesis inspires us to transform our current technological energy infrastructure to one that utilizes sunlight for the direct production of fuels and other industrially important chemicals.

Recent advances in understanding the detailed structures of photosynthetic systems have set the stage for dynamic functional studies. Likewise, sophisticated spectroscopic techniques have revealed important mechanistic details. In this book, we provide examples of such important advances and progress. The book contains chapters written by experts and leaders in their respective fields, representing an international collaborative effort including 52 leading experts and scientists in photosynthesis research from 27 laboratories in 13 countries including Bulgaria, Canada, China, Germany, India, Iran, Japan, New Zealand, Russia, Slovakia, Sweden, the UK, and the USA. We envision this book to serve as a resource for both young and established researchers as well as students in colleges and universities at the graduate as well as advanced undergraduate level.

Several new topics on photosynthesis are covered in this book; these include structural and mechanistic information on biological photosynthetic systems and on human engineered constructs that are inspired by their biological counterparts. The notion that biology can inspire technology is an embedded theme, and the chapters are organized in terms of the temporal sequence of events occurring in the process of photosynthesis. For example, chapters on the topic of light harvesting are followed by those on charge separation at the reaction centers, followed by charge stabilization and then chemical reactions and protection mechanisms. The book concludes with more specialized topics including artificial photosynthetic constructs and their potential applications in future technologies.

The opening chapter, by Mohammad Mahdi Najafpour, Harvey J. M. Hou, and Suleyman I. Allakhverdiev, provides an introduction and overview of natural and artificial photosynthesis. Chapter 2, by Pu Qian, and Chap. 3, by Erica Belgio and Alexander V. Ruban, evaluate the structure and function of native and recombinant light-harvesting systems from purple photosynthetic bacteria and higher plants. The next three Chaps. (4, 5, 6 and 7) focus on the charge transfer mechanisms of photosystem II and the oxygen-evolution complex: Chap. 4 is by Jessica Wiwczar and Gary W. Brudvig; Chap. 5 by Alice Haddy, Vonda Sheppard, Rachelle Johnson, and Eugene Chen; and Chap. 6 by Mahir D. Mamedov and Alexey Yu. Semenov. The structure, function, and organization of photosystem I are reviewed and discussed in Chap. 7 by Wu Xu and Yingchun Wang. The following seven Chaps. (8, 9, 10, 11, 12, 13 and 14) deal with the regulation and protection mechanisms of photosynthesis under diverse environmental conditions: Chap. 8 is by Seiji Akimoto and Makio Yokono; Chap. 9 by Leonid V. Kurepin, Alexander G. Ivanov, Mohammad Zaman, Richard P. Pharis, Vaughan Hurry, and Norman P.A. Hüner; Chap. 10 by Marek Zivcak, Katarina Olsovska, and Marian Brestic; Chap. 11 by Anjana Jajoo; Chap. 12 by Krishna Nath, James P. O'Donnell, and Yan Lu; Chap. 13 by Beth Szyszka, Alexander G. Ivanov, and Norman P.A. Hüner; and Chap. 14 by Amarendra N. Misra, Ranjeet Singh, Meena Misra, Radka Vladkova, Anelia Dobrikova, and Emilia L. Apostolova. Several stimulating new approaches on the topic of artificial photosynthesis are presented in the next five Chaps. (15, 16, 17, 18 and 19): Chap. 15 is by Mohammad Mahdi Najafpour, Saeideh Salimi, Małgorzata Hołyńska, Fahime Rahimi, Mojtaba Tavahodi, Tatsuya Tomo, and Suleyman I. Allakhverdiev; Chap. 16 by Mohammad Mahdi Najafpour, Seyed Esmael Balaghi, Moayad Hossaini Sadr, Behzad Soltani, Davood Jafarian Sedigh, and Suleyman I. Allakhverdiev; Chap. 17 by Harvey J.M. Hou; Chap. 18 by Art Van Der Est and Prashanth K. Poddutoori; and Chap. 19 by Babak Pashaei and Hashem Shahroosvand. The final chapter, Chap. 20, by Gary F. Moore, offers concluding remarks and a forward-looking perspective regarding research in natural as well as artificial photosynthesis.

We acknowledge all the authors for their contributions that have made this book possible. The book chapters were reviewed by Seiji Akimoto, Erica Belgio, Marian Brestic, Gary Brudvig, Alice Haddy, Anjana Jajoo, Leonid Kurepin, Krishna Nath, Pu Qian, Art Van Der Est, and Wu Xu. The complete book was reviewed by the four editors and revised by the authors. We are deeply thankful to Govindjee for his valuable advice and continued support. We are very grateful to André Tournois, Jacco Flipsen, and Mariska van der Stigchel of Springer for their help during the production of this book. Finally, the support from Alabama State University (to HJMH), the Institute for Advanced Studies in Basic Sciences (to MMN), Arizona State University (to GFM), and the Russian Science Foundation (grant no. 14-14-00039, to SIA) is greatly appreciated.

Montgomery, AL, USA	Harvey J.M. Hou
Zanjan, Iran	Mohammad Mahdi Najafpour
Tempe, AZ, USA	Gary F. Moore
Moscow, Russia	Suleyman I. Allakhverdiev

Contents

1. **Photosynthesis: Natural Nanomachines Toward Energy and Food Production** 1
   Mohammad Mahdi Najafpour, Harvey J.M. Hou, and Suleyman I. Allakhverdiev

2. **Structure and Function of the Reaction Centre – Light Harvesting 1 Core Complexes from Purple Photosynthetic Bacteria** 11
   Pu Qian

3. **Recombinant Light Harvesting Complexes: Views and Perspectives** 33
   Erica Belgio and Alexander V. Ruban

4. **Alternative Electron Acceptors for Photosystem II** 51
   Jessica Wiwczar and Gary W. Brudvig

5. **The Cl^- Requirement for Oxygen Evolution by Photosystem II Explored Using Enzyme Kinetics and EPR Spectroscopy** 67
   Alice Haddy, Vonda Sheppard, Rachelle Johnson, and Eugene Chen

6. **Vectorial Charge Transfer Reactions in the Protein-Pigment Complex of Photosystem II** 97
   Mahir D. Mamedov and Alexey Yu Semenov

7. **Function and Structure of Cyanobacterial Photosystem I** 111
   Wu Xu and Yingchun Wang

8. **How Light-Harvesting and Energy-Transfer Processes Are Modified Under Different Light Conditions: STUDIES by Time-Resolved Fluorescence Spectroscopy** 169
   Seiji Akimoto and Makio Yokono

9 **Interaction of Glycine Betaine and Plant Hormones: Protection of the Photosynthetic Apparatus During Abiotic Stress** 185
Leonid V. Kurepin, Alexander G. Ivanov, Mohammad Zaman, Richard P. Pharis, Vaughan Hurry, and Norman P.A. Hüner

10 **Photosynthetic Responses Under Harmful and Changing Environment: Practical Aspects in Crop Research** 203
Marek Zivcak, Katarina Olsovska, and Marian Brestic

11 **Effects of Environmental Pollutants Polycyclic Aromatic Hydrocarbons (PAH) on Photosynthetic Processes** 249
Anjana Jajoo

12 **Chlorophyll Fluorescence for High-Throughput Screening of Plants During Abiotic Stress, Aging, and Genetic Perturbation** 261
Krishna Nath, James P. O'Donnell, and Yan Lu

13 **Adaptation to Low Temperature in a Photoautotrophic Antarctic Psychrophile, *Chlamydomonas sp.* UWO 241** 275
Beth Szyszka, Alexander G. Ivanov, and Norman P.A. Hüner

14 **Nitric Oxide Mediated Effects on Chloroplasts** 305
Amarendra N. Misra, Ranjeet Singh, Meena Misra, Radka Vladkova, Anelia G. Dobrikova, and Emilia L. Apostolova

15 **Nanostructured Mn Oxide/Carboxylic Acid or Amine Functionalized Carbon Nanotubes as Water-Oxidizing Composites in Artificial Photosynthesis** 321
Mohammad Mahdi Najafpour, Saeideh Salimi, Małgorzata Hołyńska, Fahime Rahimi, Mojtaba Tavahodi, Tatsuya Tomo, and Suleyman I. Allakhverdiev

16 **Self-Healing in Nano-sized Manganese-Based Water-Oxidizing Catalysts** 333
Mohammad Mahdi Najafpour, Seyed Esmael Balaghi, Moayad Hossaini Sadr, Behzad Soltani, Davood Jafarian Sedigh, and Suleyman I. Allakhverdiev

17 **A Robust PS II Mimic: Using Manganese/Tungsten Oxide Nanostructures for Photo Water Splitting** 343
Harvey J.M. Hou

18 **Time-Resolved EPR in Artificial Photosynthesis** 359
Art van der Est and Prashanth K. Poddutoori

19 Artificial Photosynthesis Based on 1,10-Phenanthroline Complexes................................... 389
Babak Pashaei and Hashem Shahroosvand

20 Concluding Remarks and Future Perspectives: Looking Back and Moving Forward.................. 407
Gary F. Moore

Index... 415

Contributors

Seiji Akimoto Graduate School of Science, Kobe University, Kobe, Japan

Suleyman I. Allakhverdiev Controlled Photobiosynthesis Laboratory, Institute of Plant Physiology, Russian Academy of Sciences, Moscow, Russia

Institute of Basic Biological Problems, Russian Academy of Sciences, Pushchino, Moscow, Russia

Faculty of Biology, Department of Plant Physiology, M. V. Lomonosov Moscow State University, Moscow, Russia

Emilia L. Apostolova Institute of Biophysics and Biomedical Engineering, Bulgarian Academy of Sciences, Sofia, Bulgaria

Seyed Esmael Balaghi Faculty of Basic Sciences, Department of Chemistry, Azarbaijan Shahid Madani University, Tabriz, Iran

Erica Belgio School of Biological and Chemical Sciences, Queen Mary University of London, London, UK

Department of Autotrophic Microorganisms – ALGATECH, Institute of Microbiology ASCR, Třeboň, Czech Republic

Marian Brestic Department of Plant Physiology, Slovak University of Agriculture, Nitra, Slovakia

Gary W. Brudvig Department of Chemistry, Yale University, New Haven, CT, USA

Eugene Chen Department of Chemistry and Biochemistry, University of North Carolina at Greensboro, Greensboro, NC, USA

Anelia G. Dobrikova Institute of Biophysics and Biomedical Engineering, Bulgarian Academy of Sciences, Sofia, Bulgaria

Alice Haddy Department of Chemistry and Biochemistry, University of North Carolina at Greensboro, Greensboro, NC, USA

Małgorzata Hołyńska Fachbereich Chemie and Wissenschaftliches Zentrum für Materialwissenschaften (WZMW), Philipps-Universität Marburg, Marburg, Germany

Harvey J.M. Hou Department of Physical Sciences, Alabama State University, Montgomery, AL, USA

Norman P.A. Hüner Department of Biology and the Biotron Center for Experimental Climate Change Research, Western University, London, ON, UK

Vaughan Hurry Umeå University, Umeå, Sweden

Alexander G. Ivanov Department of Biology and the Biotron Center for Experimental Climate Change Research, Western University, London, ON, Canada

Institute of Biophysics and Biomedical Engineering, Bulgarian Academy of Sciences, Sofia, Bulgaria

Anjana Jajoo School of Life Science, Devi Ahilya University, Indore, MP, India

Rachelle Johnson Department of Chemistry and Biochemistry, University of North Carolina at Greensboro, Greensboro, NC, USA

Leonid V. Kurepin Department of Biology and the Biotron Center for Experimental Climate Change Research, Western University, London, ON, Canada

Department of Forest Genetics and Plant Physiology, Swedish University of Agricultural Sciences, Umeå, Sweden

Yan Lu Department of Biological Sciences, Western Michigan University, Kalamazoo, MI, USA

Mahir D. Mamedov A.N. Belozersky Institute of Physical-Chemical Biology, Lomonosov Moscow State University, Moscow, Russia

Amarendra N. Misra Centre for Life Sciences, School of Natural Sciences, Central University of Jharkhand, Ranchi, India

Meena Misra Centre for Life Sciences, School of Natural Sciences, Central University of Jharkhand, Ranchi, India

Gary F. Moore School of Molecular Sciences and the Biodesign Institute Center for Applied Structural Discovery (CASD), Arizona State University, Tempe, AZ, USA

Mohammad Mahdi Najafpour Department of Chemistry, Institute for Advanced Studies in Basic Sciences (IASBS), Zanjan, Iran

Center of Climate Change and Global Warming, Institute for Advanced Studies in Basic Sciences (IASBS), Zanjan, Iran

Krishna Nath Department of Biological Sciences, Western Michigan University, Kalamazoo, MI, USA

James P. O'Donnell Department of Biological Sciences, Western Michigan University, Kalamazoo, MI, USA

Katarina Olsovska Department of Plant Physiology, Slovak University of Agriculture, Nitra, Slovakia

Babak Pashaei Department of Chemistry, University of Zanjan, Zanjan, Iran

Richard P. Pharis Department of Biological Sciences, University of Calgary, Calgary, AB, Canada

Prashanth K. Poddutoori Department of Chemistry, Brock University, St. Catharines, ON, Canada

Pu Qian Department of Molecular Biology and Biotechnology, The University of Sheffield, Sheffield, UK

Fahime Rahimi Department of Chemistry, Institute for Advanced Studies in Basic Sciences (IASBS), Zanjan, Iran

Alexander V. Ruban School of Biological and Chemical Sciences, Queen Mary University of London, London, UK

Moayad Hossaini Sadr Faculty of Basic Sciences, Department of Chemistry, Azarbaijan Shahid Madani University, Tabriz, Iran

Saeideh Salimi Department of Chemistry, Institute for Advanced Studies in Basic Sciences (IASBS), Zanjan, Iran

Davood Jafarian Sedigh Department of Chemistry, Institute for Advanced Studies in Basic Sciences (IASBS), Zanjan, Iran

Alexey Yu Semenov A.N. Belozersky Institute of Physical-Chemical Biology, Lomonosov Moscow State University, Moscow, Russia

Hashem Shahroosvand Department of Chemistry, University of Zanjan, Zanjan, Iran

Vonda Sheppard Department of Chemistry and Biochemistry, University of North Carolina at Greensboro, Greensboro, NC, USA

Ranjeet Singh Centre for Life Sciences, School of Natural Sciences, Central University of Jharkhand, Ranchi, India

Biotechnology Department, Beej Sheetal Research Pvt. Ltd., Jalna, India

Behzad Soltani Faculty of Basic Sciences, Department of Chemistry, Azarbaijan Shahid Madani University, Tabriz, Iran

Beth Szyszka Department of Biology and the Biotron Center for Experimental Climate Change Research, Western University, London, ON, Canada

Mojtaba Tavahodi Department of Chemistry, Institute for Advanced Studies in Basic Sciences (IASBS), Zanjan, Iran

Tatsuya Tomo Faculty of Science, Department of Biology, Tokyo University of Science, Tokyo, Japan

Art Van Der Est Department of Chemistry, Brock University, St. Catharines, ON, Canada

Freiburg Institute of Advanced Studies (FRIAS), Albert-Ludwigs-Universität Freiburg, Freiburg, Germany

Radka Vladkova Institute of Biophysics and Biomedical Engineering, Bulgarian Academy of Sciences, Sofia, Bulgaria

Yingchun Wang State Key Laboratory of Molecular Developmental Biology, Institute of Genetics and Developmental Biology, Chinese Academy of Sciences, Beijing, China

Jessica Wiwczar Department of Molecular Biophysics and Biochemistry, Yale University, New Haven, CT, USA

Wu Xu Department of Chemistry, University of Louisiana at Lafayette, Lafayette, LA, USA

Makio Yokono Institute of Low Temperature Science, Hokkaido University, Sapporo, Japan

Mohammad Zaman Ballance Agri-Nutrients Limited, Tauranga, New Zealand

Marek Zivcak Department of Plant Physiology, Slovak University of Agriculture, Nitra, Slovakia

About the Editors

Harvey J.M. Hou, born in 1962 in China, is a professor in the Department of Physical Sciences at Alabama State University, Montgomery, Alabama, USA. He received his B.Sc. in physical chemistry in 1984 from Wuhan University and completed his Ph.D. in analytical chemistry in 1993 at Peking University (Beijing, China) with Xiaoxia Gao. He had his postdoctoral training at the Chinese Academy of Sciences with Peisung Tang and Tingyun Kuang, at Iowa State University with Parag Chitnis, and at Rockefeller University with David Mauzerall. Since 1995, he has served as a faculty member at the Chinese Academy of Sciences, Gonzaga University, the University of Massachusetts at Dartmouth, and Alabama State University. He began his research career in photosynthesis in 1993, working on photosystem II (PS II). In 1996, he visited the laboratory of Jacque Breton in France and studied the orientation of pigments in PS II. In the laboratory of Parag Chitnis, he examined the organization of PS I. Working with David Mauzerall, he systematically investigated the thermodynamics of electron transfer reactions in photosynthesis using pulsed photoacoustics. His work has uncovered a significant entropy change of reaction in PS I; further, he has demonstrated that the entropy change in PS I is dramatically different from that in PS II. Since he established his laboratory in 2002, he has maintained his long-term collaboration with David Mauzerall on the thermodynamics in cyanobacterial PS I and in heliobacteria. In 2006, he began collaboration with Gary Brudvig at Yale University and Dunwei Wang at Boston College on artificial photosynthesis and has developed a manganese/semiconductor system for solar energy storage. His research group has also investigated the responses of cyanobacteria and cranberry plants to environment. He co-chaired a symposium at the 15th International Congress of Photosynthesis Research, chaired the 28th Annual Eastern Regional Photosynthesis Conference, and co-organized the 38th Annual Midwest/Southeast Photosynthesis Meeting.

Mohammad Mahdi Najafpour received his Ph.D. in inorganic chemistry from Sharif University of Technology, Tehran, Iran, in 2009. Mahdi is a recipient of several awards and fellowships, notably the gold medal of the National Chemistry Olympiad in 2004, "Top Student" award of Sharif University of Technology in 2007, and fellowship of government for a research stay in Germany as a visiting scientist in 2009; he ranked first in the Khwarizmi Youth Festival in 2010, received Professor Ashtiani's award in 2012, was selected for the Third World Academy of Sciences (TWAS) young affiliateship in 2014, and awarded the Al-Biruni Award by the Academy of Sciences of Iran in 2015. Currently, he is an associate professor of chemistry at the Institute for Advanced Studies in Basic Sciences (IASBS) (Zanjan, Iran). As a nanobioinorganic chemist, Mahdi believes that, with learning strategies from natural systems, design of modern catalysts for all reactions using only earth-abundant, low-cost, and environmentally friendly metal ions is possible. Mahdi and his research group explore transition-metal compounds as water-oxidizing catalysts for artificial photosynthesis. He is the author of over 150 publications in these and other areas.

Gary F. Moore is an assistant professor in the School of Molecular Sciences and the Biodesign Institute Center for Applied Structural Discovery at Arizona State University. He is also guest faculty at Berkeley Lab. He obtained his B.Sc. (2004) from the Evergreen State College and his Ph.D. (2009) under Ana L. Moore from Arizona State University where he was a National Science Foundation fellow. Gary completed his postdoctoral training as a Camille and Henry Dreyfus Foundation energy fellow at Yale University (2009–2011) working with the research groups of Gary W. Brudvig, Robert H. Crabtree, Victor S. Batista, and Charles A. Schmuttenmaer. He began his independent research career as a principal investigator and staff scientist at Berkeley Lab (2011) working with the Joint Center for Artificial Photosynthesis before starting his current position at Arizona State University (2014). His group includes an interdisciplinary team of researchers that conduct use-inspired research with applications to photocatalysis, hard-to-soft matter interfaces, molecular electronics, chemical sensing, and proton-coupled electron transfer. He is a frequently invited speaker to international seminars, conferences, and workshops on the science and social policy of artificial photosynthesis and solar-to-fuel technologies.

About the Editors

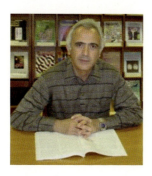

Suleyman I. Allakhverdiev is the head of the Controlled Photobiosynthesis Laboratory at the Institute of Plant Physiology of the Russian Academy of Sciences (RAS), Moscow; chief research scientist at the Institute of Basic Biological Problems RAS, Pushchino, Moscow Region; professor at M.V. Lomonosov Moscow State University, Moscow, Russia; head of the Bionanotechnology Laboratory at the Institute of Molecular Biology and Biotechnology of the Azerbaijan National Academy of Sciences, Baku, Azerbaijan; and invited adjunct professor at the Department of New Biology, Daegu Gyeongbuk Institute of Science and Technology (DGIST), Daegu, Republic of Korea. He is originally from Chaykend (Karagoyunly/ Dilichanderesi), Armenia, and he obtained both his B.S. and M.S. in physics from the Department of Physics, Azerbaijan State University, Baku. He obtained his Dr. Sci. degree (highest/top degree in science) in plant physiology and photobiochemistry from the Institute of Plant Physiology, RAS (2002, Moscow), and Ph.D. in physics and mathematics (biophysics) from the Institute of Biophysics, USSR (1984, Pushchino). His Ph.D. advisors were Academician Alexander A. Krasnovsky and Dr. Sci. Vyacheslav V. Klimov. He worked for many years (1990–2007) as visiting scientist at the National Institute for Basic Biology (with Prof. Norio Murata), Okazaki, Japan, and in the Département de Chimie-Biologie, Universite du Quebec at Trois Rivieres (with Prof. Robert Carpentier), Quebec, Canada (1988–1990). He has been a guest editor of many (more than 30) special issues in international peer-reviewed journals. At present, he is a member of the editorial board of more than 15 international journals. Besides being editor-in-chief of SOAJ *NanoPhotoBioSciences*, associate editor of the *International Journal of Hydrogen Energy*, section editor of *BBA Bioenergetics*, and associate editor of *Photosynthetica*, he also acts as a referee for major international journals and grant proposals. He has authored (or co-authored) more than 350 research papers, 6 patents, and 7 books. He has organized several (more than 10) international conferences on photosynthesis. His research interests include the structure and function of photosystem II, water-oxidizing complex, artificial photosynthesis, hydrogen photoproduction, catalytic conversion of solar energy, plants under environmental stress, and photoreceptor signaling.

Photosynthesis: Natural Nanomachines Toward Energy and Food Production

Mohammad Mahdi Najafpour, Harvey J.M. Hou, and Suleyman I. Allakhverdiev

Summary

Photosynthesis is of the most important reactions on Earth and it is a scientific field that is intrinsically interdisciplinary. This chapter provides a brief discussion on the importance of photosynthesis and potential future applications of photosynthesis as a blueprint for artificial photosynthetic systems.

Keywords

Plant • Chloroplast • Photosystem I • Photosystem II • Water oxidation • Water splitting • ATP synthase • Chlorophyll • Electron transport • S-state cycle • Artificial photosynthesis

Contents

1.1 Introduction 1
1.2 Organization and Function of Photosynthetic Nanomachines 2
1.3 Structures and Mechanisms of the Water Oxidation Nanomachines 5
1.4 Applications of Artificial Photosynthetic Nanomachines 7
References 8

1.1 Introduction

Nature uses unique and powerful strategies to capture the solar energy in an amazing process: Photosynthesis (from Greek, photo: light and synthesis: putting together). The phenomenon is the most important photobiological reaction on our planet and occurs in plants, algae and cyanobacteria. All creatures need energy to their life. Phototrophy is a process used by organisms to trap photons and store energy as chemical energy in the form of adenosine triphosphate (ATP), which transports energy within cells (Fenchel et al. 2012). There are three types of phototrophy: anoxygenic photosynthesis,

M.M. Najafpour (✉)
Department of Chemistry, Institute for Advanced Studies in Basic Sciences (IASBS), Zanjan 45137-66731, Iran

Center of Climate Change and Global Warming, Institute for Advanced Studies in Basic Sciences (IASBS), Zanjan 45137-66731, Iran
e-mail: mmnajafpour@iasbs.ac.ir

H.J.M. Hou
Department of Physical Sciences, Alabama State University, Montgomery, AL 36104, USA

S.I. Allakhverdiev
Controlled Photobiosynthesis Laboratory, Institute of Plant Physiology, Russian Academy of Sciences, Botanicheskaya Street 35, Moscow 127276, Russia

Institute of Basic Biological Problems, Russian Academy of Sciences, Pushchino, Moscow, Region, 142290, Russia

Faculty of Biology, Department of Plant Physiology, M. V. Lomonosov Moscow State University, Leninskie Gory 1-12, Moscow 119991, Russia

© Springer International Publishing AG 2017
H.J.M. Hou et al. (eds.), *Photosynthesis: Structures, Mechanisms, and Applications*,
DOI 10.1007/978-3-319-48873-8_1

oxygenic photosynthesis, and rhodopsin-based phototrophy (Fenchel et al. 2012). Photosynthesis usually converts carbon dioxide into different organic compounds using solar energy. Although, some anoxygenic photosynthetic organisms such as heliobacteria can't fix carbon photoautotrophically.

In anoxygenic photosynthesis used by green bacteria, phototrophic purple bacteria, and heliobacteria, light energy is captured and stored as ATP, without the production of oxygen (Bryant and Frigaard 2006; Blankenship et al. 1995; Feiler and Hauska 1995). This means water is not used as the primary electron donor. Such organisms use a single photosystem, which restricts them to cyclic electron flow only, and they are therefore unable to produce O_2 from the oxidization of H_2O. Anoxygenic phototrophs have photosynthetic pigments called bacteriochlorophylls. Bacteriochlorophyll a and b have maxima wavelength absorption at 775 nm and 790 nm, respectively.

In contrast, oxygenic photosynthesis in plants, algae and cyanobacteria results not only in the fixation of carbon dioxide (CO_2) from the atmosphere, but also release of molecular oxygen to the atmosphere (Blankenship 2013; Allen et al. 2011). The name of cyanobacteria is based on the bluish pigment phycocyanin, which is used to harvest light for photosynthesis in these cells. Plants, algae and cyanobacteria not only capture approximately 3000 EJ and produce more than 100 billion tons of dry biomass annually (Blankenship et al. 1995), but also they form oxygen, which is necessary for many organisms. The transition from anoxygenic to oxygenic photosynthesis in the eubacterial lineage was an important innovation in evolution and the vast majority of the scientific community is in agreement regarding the correlation between this event and the beginning of oxygen accumulation on Earth (Canfield 2005). The evidence from chemical markers (Kazmierczak and Altermann 2002; Summons et al. 1999), stromatolite fossils (Olson and Blankenship 2004), and microfossils (Schopf 1975) attest that cyanobacteria arose before 2.5 billion years ago (Konhauser et al. 2011; Gould et al. 2008). Before cyanobacteria, the atmosphere had much more carbon dioxide and organisms used hydrogen or hydrogen sulfide as sources of electron (Olson 2006), but the new oxygen-rich atmosphere was a revolution for complex life.

1.2 Organization and Function of Photosynthetic Nanomachines

In plants and algae, photosynthesis takes place in chloroplasts. Each plant cell contains about 10–100 chloroplasts (Fig. 1.1). The chloroplasts in modern plants are the descendants of these ancient symbiotic cyanobacteri (Gould et al. 2008). Chloroplast is one of three types of plastids, indicated by its high concentration of chlorophyll (Fig. 1.1a) (McFadden 2001). The plastids in plant cells are highly dynamics structures and divide during reproduction. Their behavior is strongly influenced and regulated by environmental factors like light color and intensity. They are composed of two inner and outer phospholipid membrane and an intermembrane space between them. Within the membrane is an aqueous fluid called the stroma, which contains stacks (grana) of thylakoids (Fig. 1.1b). The thylakoids are flattened disks, bounded by a membrane with a lumen or thylakoid space within it. The thylakoid membrane forms the photosystems (Fig. 1.1b, c, d). On the other hand, Cyanobacteria have an internal system of thylakoid membranes where the outer membrane, plasma membrane, and thylakoid membranes each have specialized roles in the cyanobacterial cell (McFadden 2001).

The process of photosynthesis starts with the absorption of light by pigments (Fig. 1.2) located in PSII (Blankenship 2013). Then, the energy is transferred to the reaction center, which are special chlorophyll (P_{680}^+). Energy transfer between the antenna pigments transports the excitation to the reaction center and is an important part of the photosynthetic process. Excitons trapped by a reaction center provide the energy for the primary photochemical reactions and causes the oxidation of its adjacent tyrosine (Yz) and

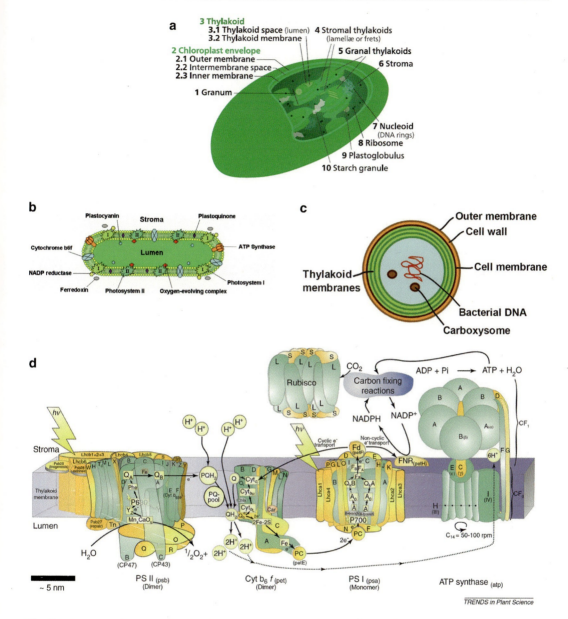

Fig. 1.1 Structures of a typical higher-plant chloroplast (**a**), thylakoid (**b**), thylakoids inside a cyanobacterium (**c**), and major protein complexes in a chloroplast (**d**). PSII; cytochrome b6f (Cyt b6f); PSI; ATP synthase; and Rubisco. Subunits are given single-letter names, omitting the three-letter prefix that denotes the complex of which each forms part. These prefixes are: psa for PSI; psb for PSII; pet (photosynthetic electron transport) for the cytochrome b6f complex and secondary electron carriers; atp for the ATP synthase; and rbc for Rubisco. Polypeptide subunits encoded in the chloroplast are coloured *green*; polypeptide subunits encoded in the nucleus are coloured *yellow*. Each major complex, such as PSI, PSII or ATP synthase, can be considered as a nano machine (see the scale in the image) (Reprinted with permission from Allen et al. 2011. Copyright (2011) by Elsevier)

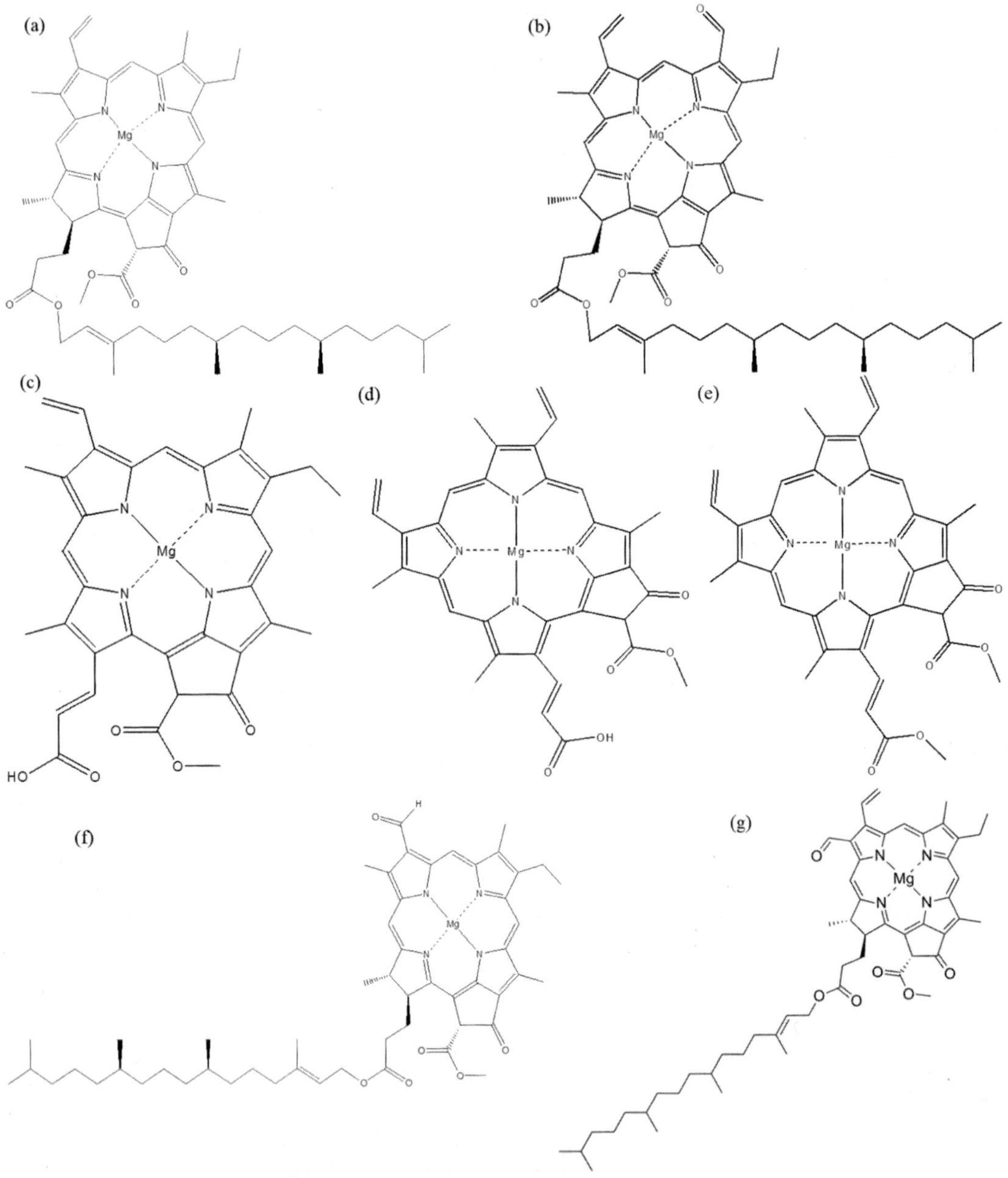

Fig. 1.2 Chemical structures of chlorophylls (**a**) chlorophyll a; (**b**) chlorophyll b; (**c**) chlorophyll c1; (**d**) chlorophyll c2; (**e**) chlorophyll c3; (**f**) chlorophyll d; (**g**) chlorophyll f

afterwards, the Yz^+ is reduced to Yz by the electron captured from the water oxidation complex (WOC) or the oxygen-evolving complex (OEC) (Ferreira et al. 2004; Umena et al. 2011). Two photosystems are connected by a series of intermediate carriers named plastoquinol, cytochrome b6f, and plastocyanin. Electrons transfer from P_{680} via Pheo QA to QB to give plastoquinol after two turnovers. Cytochrome b6f acts as an electron carrier mediates electron transfer between the two carriers plastoquinol and plastocyanin. The electron due the P_{680} excitation is carried to PSI to restore the function of chlorophyll P_{700}. The light absorption and electron

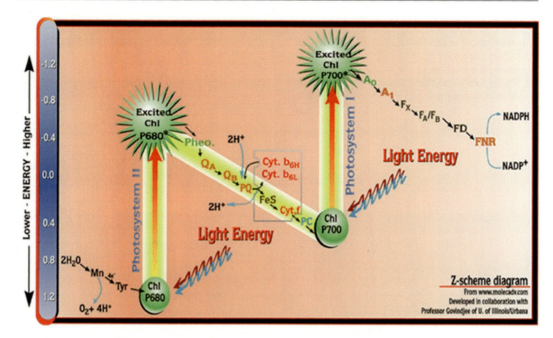

Fig. 1.3 Z-Scheme of electron transport in photosynthesis. Abbreviations used are (from *left* to the *right* of the diagram): Mn for a manganese complex containing 4 Mn atoms, bound to Photosystem II (PSII) reaction center; Tyr for a particular tyrosine in PSII; O2 for oxygen; H+ for protons; P680 for the reaction center chlorophyll (Chl) in PSII: it is the primary electron donor of PSII; Excited (Chl) P680 for P680* that has the energy of the photon of light; Pheo for pheophytin molecule (the primary electron acceptor of PSII; it is like a chlorophyll a molecule where magnesium (in its center) has been replaced by two "H"s); QA for a plastoquinone molecule tightly bound to PSII; QB for another plastoquinone molecule that is loosely bound to PSII; FeS for Rieske Iron Sulfur protein;Cyt. f for Cytochrome f; Cytb6 (L and H) for Cytochrome b6 (of Low and High Energy); PC for copper protein plastocyanin; P700 for the reaction center chlorophyll (Chl; actually a dimer, i.e., two molecules together) of PSI; it is the primary electron donor of PSI; Excited (Chl) P700 for P700* that has the energy of the photon of light; Ao for a special chlorophyll a molecule (the primary electron acceptor of PSI); A1 for a phylloquinone (Vitamin K) molecule; FX, FA, and FB are three separate Iron Sulfur Centers; FD for ferredoxin; and FNR for Ferredoxin NADP oxido Reductase (FNR). Three major protein complexes are involved in running the "Z" scheme: (1) PSII; (2) Cytochrome bf complex (containing Cytb6; FeS; and Cytf) and (3) PSI. The diagram does not show where and how ATP is made (Images from Govindjee and Wilbert Veit)

transition in PSI lead to production of NADPH. In the reactions, Ferredoxin and Ferredoxin NADP oxido Reductase are intermediates. Overall, the net reaction is electron transition resulting from water oxidation in PSII to NADP+ and formation of NADPH form in PSI. These electrons are shuttled through an electron transport chain, the so-called Z-scheme shown in Fig. 1.3.

The proton gradient across the chloroplast membrane is used by ATP synthase for the concomitant synthesis of ATP. The chlorophyll molecule regains the lost electron from a water molecule and oxidizes it to dioxygen (O_2) by a complicated mechanism:

$$2H_2O + 2NADP^+ + 3ADP + 3P_i + \text{light} \rightarrow 2NADPH + 2H^+ + 3ATP + O_2$$

1.3 Structures and Mechanisms of the Water Oxidation Nanomachines

Recently, the crystal structure of the PSII WOC at atomic resolution was determined (Umena et al. 2011; Suga et al. 2015). In this structure, one calcium and four manganese ions are bridged by five oxygen atoms. The structure also shows four water molecules, two of which are

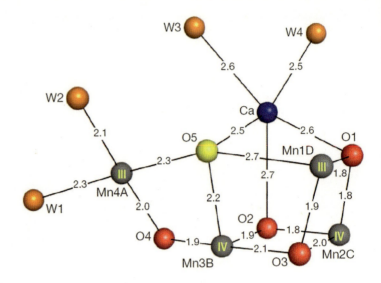

Fig. 1.4 The structure of PSII WOC (Suga et al. 2015) (Reprinted with permission from Suga et al. 2015. Copyright (2015) by McMilan publications)

suggested as the substrates for water oxidation (Fig. 1.4).

Joliot et al. (1969) showed that short flashes produced an oscillating pattern in the oxygen evolution and a maximum of water oxidation to oxygen production occurred on every fourth flash. Because water oxidation needs the removal of four electrons, these patterns were very interesting. Kok was the first scientist to propose an explanation for the observed oscillation of the oxygen evolution pattern (Kok et al. 1970). Kok's hypothesis was that in a cycle of water oxidation, a succession of oxidizing equivalents is stored on each separate and independent water oxidizing complex, and when four oxidizing equivalents have been accumulated oxygen is spontaneously evolved (Kok et al. 1970). Each oxidation state of the water-oxidizing complex is known as an "S-state" and S_0 being the most reduced state and S_4 the most oxidized state in the catalytic cycle (Fig. 1.5) (Kok et al. 1970). The S_1 state is dark-stable. The $S_4 \rightarrow S_0$ transition is light independent and in this state oxygen is evolved. Other S-state transitions are induced by the photochemical oxidation by P_{680}^+ (Blankenship 2013).

Light is also used to synthesize ATP and NADPH in photosynthesis. Cyclic and non-cyclic are two forms of the light-dependent reaction. In the non-cyclic reaction, the photons are captured in the light-harvesting antenna complexes of PSII by different pigments (Blankenship 2013). The NADPH and ATP is used by the photosynthetic organisms to drive the synthesis in the Calvin – Benson cycle in the light-independent or dark reactions (Fig. 1.6) [22].

By the Calvin – Benson, the enzyme RuBisCO captures CO_2 from the atmosphere and in a process that requires the newly formed NADPH, releases three-carbon sugars, which are later combined to form sucrose and starch. The overall equation for the light-independent reactions in green plants is (Lundqvist and Schneider 1991):

$$3CO_2 + 9ATP + 6NADPH + 6H^+$$
$$\rightarrow C_3H_6O_3\text{-phosphate} + 9ADP + 8P_i$$
$$+ 6NADP^+ + 3H_2O$$

There is a Mg(II) ion in the active site of RuBisCO, which stabilizes formation of an enediolate carbanion equivalent that is

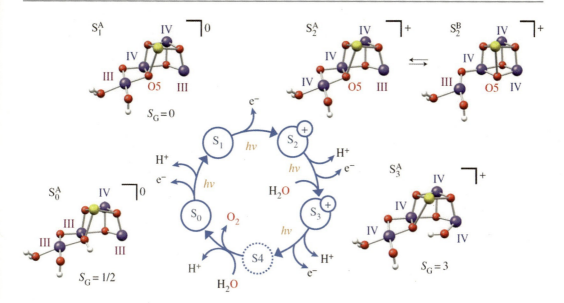

Fig. 1.5 S-state cycle of photosynthetic water oxidation (Kok cycle). Starting in the dark-stable S1 state, absorption of a photon and subsequent electron transfer leads to oxidation of tyrosine YZ, which in turn removes an electron from the WOC advancing it from state Si to Si + 1. The electron transfer steps are accompanied by charge-compensating deprotonation steps. A plausible set of oxidation-state combinations of the four Mn ions is shown for the various Si states. Models show the optimized geometry, protonation pattern and Mn oxidation states of the inorganic core. The cofactor exists in two forms in the S2 state, an open cubane (S2A) and a closed cubane (S2B) structure, which differ in their core connectivity by the reorganization of O5 (Reprinted with permission from Cox et al. 2013. Copyright (2013) by the American Chemical Society)

subsequently carboxylated by CO_2 (Lundqvist and Schneider 1991).

1.4 Applications of Artificial Photosynthetic Nanomechines

Our food, energy, environment and culture, directly or indirectly, depend on the photosynthesis. Because the solar energy absorbed by Earth is huge, approximately 3,850,000 exajoules per year, which is more energy in one hour than the world used in one year!, recently, artificial photosynthesis has been used as a model for how a sustainable supply of energy could be obtained. The term artificial photosynthesis is a general term (Najafpour et al. 2015; Tachibana et al. 2012; Nocera 2012). Recently, the phenomenon have been a blueprint to obtain a sustainable fuel for human as artificial photosynthesis, which is a general term to cover all strategies to mimic the photosynthetic process especially to provide a solar-based sustainable energy supply in the future (Fig. 1.6) (Najafpour et al. 2015; Tachibana et al. 2012; Nocera 2012). In other words, photosynthesis can provide paradigms for sustainable global energy production and efficient energy transformation (Najafpour et al. 2015; Tachibana et al. 2012; Nocera 2012; Ort et al. 2015; Lin et al. 2014; Blankenship et al. 2011). As an excellent example, artificial leaf, which is a machine that can harness sunlight to split water into molecular hydrogen and oxygen without needing any external connection, is being developed (Nocera 2012). The artificial leaf (Fig. 1.6) may be comparable to real leaves, which converts the energy of sunlight into chemical form, and represents one of the most promising directions in solar fuel and food production. In addition, several inspiring cases using different approaches in artificial photosynthesis are presented and discussed in Chaps. 15, 16, 17, 18, 19 and 20 in this book.

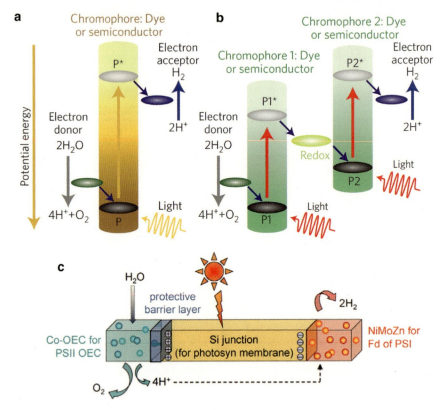

Fig. 1.6 An artificial photosynthetic system: single-step reactions (**a**), and two-step (Z-scheme) reactions (**b**). P: chromophore of a single-step reaction system; P*: excited state of P; P1: the first chromophore of a two-step reaction system; P1*: excited state of P1; P2: second chromophore of a two-step reaction system; P2*: excited state of P2 (Adapted with permission from ref. 24. Copyright (2013) by McMillan publications. Construction of an artificial leaf. The photosynthetic membrane is replaced by a Si junction, which performs the light capture and conversion to a wireless current. The oxygen-evolving complex and ferredoxin reductase of the photosynthetic membrane are replaced by Co-OEC and NiMoZn OER and HER catalysts, respectively, to perform water splitting (**c**). Adapted with permission from (Nocera 2012). Copyright (2012) by the American Chemical Society)

Acknowledgements The work is supported by the Institute for Advanced Studies in Basic Sciences (to MMN), the Alabama State University (to HJMH), and the National Elite Foundation (to SIA) for financial support. This work was supported by the Russian Science Foundation (Grant No. 14-14-00039 to SIA). We thank Drs. Gary Moore and Art Van der Est for helpful comments to improve the manuscript.

References

Allen JF, de Paula WBM, Puthiyaveetil S, Nield J (2011) A structural phylogenetic map for chloroplast photosynthesis. Trends Plant Sci 16:645–655

Blankenship RE. Molecular mechanisms of photosynthesis. John Wiley & Sons 2013

Blankenship RE, Madigan MT, Bauer CE. Anoxygenic photosynthetic bacteria. Springer, 1995

Blankenship RE, Tiede DM, Barber J, Brudvig GW, Fleming G, Ghirardi M, Gunner MR, Junge W, Kramer DM, Melis A, Moore TA, Moser CC, Nocera DG, Nozik AJ, Ort DR, Parson WW, Prince RC, Sayre RT (2011) Comparing the efficiency of photosynthesis with photovoltaic devices and recognizing opportunities for improvement. Science 332:805–809

Bryant DA, Frigaard NU (2006) Prokaryotic photosynthesis and phototrophy illuminated. Trends Microbiol 14:488–496

Canfield DE (2005) The early history of atmospheric oxygen: homage to Robert M. Garrels. Annu Rev Earth Planet Sci 33:1–36

Cox N, Pantazis DA, Neese F, Lubitz W (2013) Biological water oxidation. Acc Chem Res 46:1588–1596

Feiler U, Hauska G (1995) The reaction center from green sulfur bacteria. Anoxygenic Photosyn Bacteria 2:665–685

Fenchel T, Blackburn H, King GM. Bacterial biogeochemistry: The ecophysiology of mineral cycling. Academic Press 2012.

Ferreira KN, Iverson TM, Maghlaoui K, Barber J, Iwata S (2004) Architecture of the photosynthetic oxygen-evolving center. Science 303:1831–1838

Gould SB, Waller RF, Mcfadden GI (2008) Plastid evolution. Annu Rev Plant Biol 59:491–517

Joliot P, Barbieri G, Chabaud R (1969) Un nouveau modele des centres photochimiques du systeme II. Photochem Photobiol 10:309–329

Kazmierczak J, Altermann W (2002) Neoarchean biomineralization by benthic cyanobacteria. Science 298:2351–2351

Kok B, Forbush B, McGloin M (1970) Cooperation of charges in photosynthetic O_2 evolution: I. A linear four-step mechanism. Photochem Photobiol 11:457–475

Konhauser KO, Lalonde SV, Planavsky NJ, Pecoits E, Lyons TW, Mojzsis SJ, Rouxel OJ, Barley ME, Rosìere C, Fralick PW, Kump LR, Bekker A (2011) Aerobic bacterial pyrite oxidation and acid rock drainage during the Great Oxidadtion Event. Nature 478:369–373

Lin MT, Occhialini A, Andralojc PJ, Parry MAJ, Hanson MR (2014) A faster Rubisco with potential to increase photosynthesis in crops. Nature 513:547–550

Lundqvist T, Schneider G (1991) Crystal structure of activated ribulose-1,5-bisphosphate carboxylase complexed with its substrate, ribulose-1,5-bisphosphate. J Biol Chem 266:12604–12611

McFadden GI (2001) Chloroplast Origin and Integration. Plant Physiol 125:50–53

Najafpour MM, Zarei Ghobadi M, Larkum AW, Shen JR, Allakhverdiev SI (2015) The biological water-oxidizing complex at the nano–bio interface. Trend Plant Sci 20:559–568

Nocera DG (2012) The artificial leaf. Acc Chem Res 45:767–776

Olson JM (2006) Photosynthesis in the archean era. Photosyn Res 88:109–117

Olson JM, Blankenship RE (2004) Thinking about the evolution of photosynthesis. Photosynth Res 2004;80:373–386

Ort DR, Merchant SS, Alric J, Barkan A, Blankenship RE, Bock R, Croce R, Hanson MR, Hibberd JM, Long SP, Moore TA, Moroney J, Niyogi KK, Parry MAJ, Peralta-Yahya PP, Prince RC, Redding KE, Spalding MH, van Wijk KJ, Vermaas WFJ, von Caemmerer S, Weber APM, Yeates TO, Yuan JS, Guang Zhu XG (2015) Redesigning photosynthesis to sustainably meet global food and bioenergy demand. Proc Natl Acad Sci USA 112:8529–8536

Schopf JW. Evolutionary biology, Springer 1975;1–43

Suga M, Akita F, Hirata K, Ueno G, Murakami H, Nakajima Y, Shimizu T, Yamashita K, Yamamoto M, Ago H, Shen JR (2015) Native structure of photosystem II at 1.95 Å resolution viewed by femtosecond X-ray pulses. Nature 517:99–103

Summons RE, Jahnke LL, Hope JM, Logan GA (1999) 2-Methylhopanoids as biomarkers for cyanobacterial oxygenic photosynthesis. Nature 400:554–557

Tachibana Y, Vayssieres L, Durrant JR (2012) Artificial photosynthesis for solar water-splitting. Nature Photonics 6:511–518

Umena Y, Kawakami K, Shen JR, Kamiya N (2011) Crystal structure of oxygen-evolving photosystem II at a resolution of 1.9 Å. Nature 473:55–60

Structure and Function of the Reaction Centre – Light Harvesting 1 Core Complexes from Purple Photosynthetic Bacteria

2

Pu Qian

Summary

Reaction centre-light harvesting 1 core complex is a fundamental unit in photosynthetic bacterium. It is the place where light energy is collected and used to power photosynthetic redox reaction, leading to the synthesis of ATP ultimately. The reaction centre is surrounded by elliptical LH1 complex. The subunit of the LH1 ring is a heterodimer of α-, β-polypeptide pair, to which pigment molecules, BChl *a* or BChl *b* and carotenoid are non-covalently bonded. There are at least three different types of the RC-LH1 core complexes found in photosynthetic bacteria so far. The core complex from *Rps. palustris* is a monomer. Its LH1 ring consists of 15 pairs of α/β-polypeptide with an extra protein 'W' located between two α-polypeptides, forming an incomplete ring. The gap of the LH1 ring was proposed as a gate to facilitate quinone/quinol exchange between reaction centre and cytochrome *bc*1 complex. A dimeric core complex was found in PufX-containing species, such as *Rba. sphaerides*. Two RCs are associated by 28 α/β-apoprotein pairs and two pufX proteins, forming an S-shaped RC-LH1-PufX core complex. The pufX protein causes incomplete LH1 ring and dimerization of the core complex.

Monomeric RC-LH1 from *Tch. tepidum* has a complete elliptical LH1 ring that is composed of 16 pair α/β-apoprotein pairs without pufX-like protein. Sixteen Ca^{2+} are coordinated on C-terminal region of the α/β-polypeptide to stabilize the core complex and cause BChl *a* Qy absorption redshift to 915 nm. Carotenoid, spirilloxthanin contacts with α/β-apoproteins intimately to form an inter subunit interaction within the core complex, providing a further stability of the complex.

Keywords

Photosynthesis • Photosynthetic bacteria • Reaction center • Light harvesting 1 core complex • RCLH1 • Carotenoid • Bacteriochlorophyll

Contents

2.1	Introduction	12
2.2	The Purple Photosynthetic Bacteria	12
2.3	Mechanism of Photosynthesis in the Purple Photosynthetic Bacteria	13
2.4	Building Block of the Core Complex of Purple Photosynthetic Bacteria	14
2.4.1	Pigment Molecules	14
2.4.2	LH1 Subunit, α/β-Polypeptide Pair	17
2.4.3	Reaction Centre	19
2.4.4	PufX Protein	19
2.5	Structural Diversity of the RC-LH1 Complexes	21
2.5.1	X-Ray Crystal Structure of the Core Complex from *Rps. palustris*	22

P. Qian (✉)
Department of Molecular Biology and Biotechnology, The University of Sheffield, Sheffield, UK
e-mail: p.qian@sheffilate.ac.uk

2.5.2   X-Ray Crystal Structure of the Core
        Complex from *Rba. sphaeroides* 22
2.5.3   X-Ray Crystal Structure of Core Complex
        from *Tch. tepidum* 25

References 27

2.1   Introduction

Purple photosynthetic bacteria refer to a unique group of microorganism that use sun light as their energy source. The light energy is absorbed by pigment molecules, such as bacteriochlorophyll (BChl) and carotenoids (Car), which are non-covalently bound to so-called light harvesting complexes. In general, there are two major classes of light harvesting complex in the purple photosynthetic bacteria, light harvesting complex 1 (LH1) and light harvesting complex 2 (LH2). The LH2, sometime called peripheral antenna complex, is composed of oligomer of two short polypeptides (α and β) with associated pigments, Car and BChl *a* or BChl *b*. The α/β-polypeptide pair, therefore, is a building block of this cylinder-like complex (Koepke et al. 1996; McDermott et al. 1995). The building block of the LH1 is constructed similar as molecular architecture as that in the LH2 complexes. The LH1 complex is always intimately interacted with the reaction centre (RC) in a fixed stoichiometry. The term of core complex usually refers to the combination of the RC and the light harvesting complex 1 in the purple photosynthetic bacteria. A short name, such as RC-LH1 or RC-LH1-PufX is often used in literatures. As its name implies that the photosynthetic core complex or RC-LH1 is the central part of bacterial photosynthesis. This chapter will focus on the recent development on structural determination of the RC-LH1 core complexes from the purple photosynthetic bacteria by starting with basic background of bacterial photosynthesis and building blocks of the core complex. For the readers who are interesting in general works on the purple photosynthetic bacteria, two recent books edited by Blankenship, R. E. and Hunter, C. N. provide more detailed and comprehensive information on the bacterial photosynthesis. (Blankenship 2014; Hunter et al. 2009).

2.2   The Purple Photosynthetic Bacteria

The purple photosynthetic bacteria, much like the name suggested, are a group of dark coloured bacteria, with different morphologies such as rod, spirilla, cocci or vibrios, which can convert light energy to chemical energy to maintain their metabolism. The pigment molecules that are involved in the light absorption are usually BChl *a* or BChl *b* and carotenoids. It is these pigments that give the purple bacteria such gorgeous colours, from purple, red to green, depending on the amount and type of different carotenoids in an individual purple bacterium.

The purple photosynthetic bacteria can be divided into two groups, i.e., purple non-sulphur bacteria and purple sulphur bacteria according to their tolerance and utilization of sulphide (Imhoff et al. 1984). The purple sulphur bacteria use sulphur or sulphide, such as hydrogen sulphide H_2S as an electron donor for carbon dioxide reduction in its respiration, while the purple non-sulphur bacteria use organic electron-donor, such as succinate or malate instead. Usually, sulphide is toxic for the purple non-sulphur bacteria although the most species of the purple non-sulphur bacteria can still grow at low level of sulphide (<0.5 mM). Different from higher plant or cyanobacteria photosynthesis, photosynthesis in the purple bacteria does not give off oxygen, and it only occurs under anoxic conditions, which is called as anoxygenic photosynthesis. Therefore, an environment having abundance of oxygen hinders their photosynthetic growing. That is why they are typically found in hot sulphuric spring (for purple sulphur bacteria especially) or stagnant water.

The purple non-sulphur bacteria, such as *Rhodobacter* (*Rba.*) *sphaeroides*, *Rhodopseudomonas* (*Rps.*) *palustris*, *Blastochloris* (*Blc.*) *virids* can grow photoheterotrophically or even photoautotrophically and chemoheterotrophs in darkness as well. It is therefore relatively easy to grow them in laboratory conditions. Having such versatile metabolisms, the purple non-sulphur bacteria become the most intensively studied species for the bacterial photosynthesis.

Growing conditions for the purple sulphur bacteria, in the other hand, is relatively stricter than that in the purple non-sulphur bacteria. Firstly, they need sulphide as electron donor. That is why the large population of the purple sulphur bacteria species are found in hot springs containing sulphide. Secondly, many of them prefer high illuminated condition, implying they grow phototrophically in nature. Thirdly, some species are isolated from extreme growing conditions, such as halophilic, high or low temperature, acidic, alkaline etc. These extremophilic purple bacteria provide us chances to study photosynthesis under harsh conditions related to molecular adaptation, protein stability etc. A high thermo-stability of the core complex from a thermophilic purple sulphur bacterium, *Thermochromatium* (*Tch*.) *tepidum*, leading to a successful 3.0 Å resolution 3D structure determination of the RC-LH1 core complex, is a good example (Niwa et al. 2014; Suzuki et al. 2007).

2.3 Mechanism of Photosynthesis in the Purple Photosynthetic Bacteria

Photosynthesis is one of the most crucial reactions taking place on the Earth. By converting light energy from the Sun to chemically useful form that is used to fuel the organisms' activities, it provides almost all foods, energies we need alone with oxygen we breathe. Although the concept of photosynthesis is commonly related to oxygenic higher plant, much of milestone results revealing the mechanism of the photosynthesis in nature come from the purple photosynthetic bacteria due to their relatively simpler photosynthetic system (Cogdell et al. 2006).

In the purple bacteria, photosynthesis takes place in bacterial cell membrane, which is located near the surface of the cell. The major protein complexes involved in the reaction chain are embedded in lipid bilayer. These include peripheral light harvesting antenna complexes, such as LH2; RC-LH1; cytochrome bc1 complex and ATP synthase. Water soluble cytochrome c2 and quinone/quinol are needed to complete proton and electron cycles in the reaction chain. A schematic arrangement of all required components is shown in Fig. 2.1. Photosynthetic reaction starts from absorption of incident light photons by antenna system, e.g., LH2 (Vangrondelle et al. 1994). The energy absorbed by LH2 complexes is then rapidly transferred to LH1 in ~5 ps. Accepted excitation energy both by transferred from the LH2 and absorbed by LH1 itself is stored in LH1 ring by delocalization in ~80 fs. Finally, the energy is delivered to the RC special pair of BChl *a* (B870) in a relatively longer time constant of ~35 ps due to a longer distance between LH1 BChl *a* (B875) and RC B870. Subsequently, the RC special pair B870 is excited. When the excited B870 returns to its stale ground state, it releases an electron to bacteriophephytin (BPhe) *via* accessory BChl *a*. This electron travels continuously down to ubiquinone site (Q_B), where ubiquinone is reduced. By the second cycle of electron transferring, the ubiquinone is fully reduced to ubiquinol (QH_2). In the meantime, two protons are taken from cytoplasmic side. After fully reduction, QH_2 molecule is released from the RC Q_B site to quinone pool toward cytochrome bc1 complex, where quinol is oxidised to quinone by releasing two protons to periplasmic side and two electrons to a mobile electron carrier cytochrome c2 that brings electrons back to the RC special pair, completing the electron cycle. Successive electron cycle companied with proton translocation from cytoplasmic side to periplasmic side leads to the formation of electric potential, proton motive force (pmf), across the membrane. It is this pmf that is used to power a variety of energy-requiring biological reactions in cells, for example, the synthesis of adenosine triphosphate (ATP), which is the most commonly used as "energy currency" of cells. Companied the synthesis of ATP, protons are pumped back to cytoplasm across the membrane, completing the proton cycle in the process of photosynthesis.

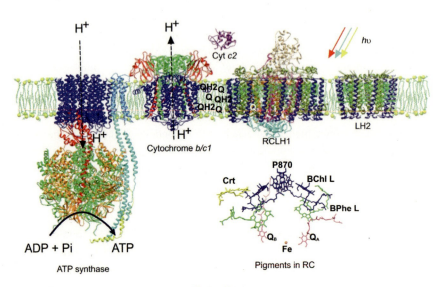

Fig. 2.1 Schematic arrangement of major protein complexes involved in the bacterial photosynthesis. On right side is LH2 of *Rps. acidophila* (McDermott et al. 1995) that consists of nine α/β subunits, forming a cylinder-like complex embedded in lipid bilayer. Next to the LH2 is RC-LH1 core complex from *Tch. Tepidum* (Niwa et al. 2014). RC, composed of subunit H in cyan, M in magenta, L in orange and C in dark khaki, is surrounded by 16 α/β LH1 subunits with α-polypeptide in olive drab and β-polypeptide in medium blue. Pigment organization in the RC is shown just below the RC-LH1 core complex. 3D crystal structure of cytochrome $bc1$ complex from *Rba. sphaeroides* was used (Esser et al. 2008). This dimeric complex comprises cytochrome b in blue, cytochrome $c1$ in green and Rieske Fe-S protein in red. On left side is an ATP synthase from *E. coli*. (Rastogi and Girvin 1999). A water-soluble protein, cytochrome $c2$ (Paddock et al. 2005) and a putative quinone pool are shown as well

2.4 Building Block of the Core Complex of Purple Photosynthetic Bacteria

2.4.1 Pigment Molecules

The RC-LH1 core complex plays an important role in the process of bacterial photosynthesis. The LH1 complex not only transfers excitation energy from LH2 complexes to RC but also absorbs light energy alone. Actually, in some of non-sulphur purple bacteria, such as *Rhodospirillum (Rsp.) rubrum* and *Blc. virids*, there are no peripheral antenna LH2 complexes at all, and the LH2 deletion mutant of *Rba. sphaeroides* can still grow photosynthetically. Strong absorbance caused by pigment molecules in the core complex in visible and near infrared regions ensure that they still have enough absorbed energy to maintain cell's biological processes.

Figure 2.2 shows the absorption spectra of four different RC-LH1 core complexes purified from (A) *Rba. sphaeroides*, (B) *Rps. palustris*, (C) *Tch. tepidum* and (D) *Blc. virids* respectively. The spectra show that the major pigment molecules in the core complexes from the purple bacteria are BChl *a* or BChl *b* and carotenoids. Absorbance between ~425 and 550 nm is caused by carotenoids. BChl *a* possesses a strong Qy absorbance band in the near infrared region, ~875 nm (B875). The Qy can red-shift depending on different molecular environments. For example, the Qy band of the BChl *a* in the RC-LH1 core complex of *Tch. tepidum* is red-shifted to 915 nm (B915). In the case of BChl *b*-containing

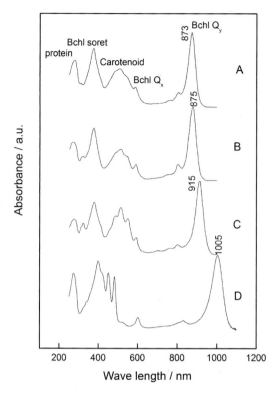

Fig. 2.2 Absorption spectra of purified RC-LH1 core complexes from (*A*) *Rba. sphaeroides*; (*B*) *Rps. palustric*; (*C*) *Tch. Tepidum* and (*D*) *Blc. virids*, which contains BChl *b*. Peaks at ~280 nm are ascribed to protein; Peaks at 376 nm are Soret band of BChl *a* in the LH1, which is overlapped with the Soret band of BChl *a* of special pair, accessory BChl *a* and Bphe in the RC; 425–550 nm absorption peaks are attributed to carotenoids; BChl *a* Qx band appears at ~590 nm; Qy bands of Bphe, accessory BChl *a* and special pair BChl *a* are located at 752, 800 and 863 nm; The strongest absorbance band in near infrared region is Qy band of BChl *a*. The Soret, Qx and Qy bands of BChl *b* in the LH1 complex of *Blc. virids* are red-shifted

is esterified to an acid side chain of the tetrapyrrole ring. The alternative arrangement of C–C and C = C bonds in the tetrapyrrole ring of BChl *a* and BChl *b* decides that they are strong and efficient photoreceptor molecules. In organic solution, such as diethyl ether, BChl *a* has a maximum absorption (Qy) at ~771 nm, and BChl *b* at ~796 nm. Once associated into the core complexes, however, their Qy absorption red-shift to ~875 nm (e.g. in *Rba. sphaeroides*) and ~1005 nm (e.g. in *Blc. virids*) (Jay et al. 1984) in purified core complexes respectively. The difference between BChl *a* and BChl *b* is very small indeed. Actually, they share the same biosynthetic pathway until a branch-point at 8-vinyl-Chlide *a*. Tsukatani and co-workers found that chlorophyllide *a* oxidoreductase (COR) from BChl *b*-producing species, such as *Blc. visids*, use 8-vinyl-Chlide *a* to synthesize 8-ethylidene group on BChlide g, a precursor to final product of BChl *b*. The COR enzyme from BChl *a*-producing species, such as *Rba. capsulatus*, in the other hand, catalyses 8-vinyl reduction, leading to BChl *a* as its final product in its biosynthetic pathway (Tsukatani et al. 2013). Therefore, it only has one bond different between BChl *a* and BChl *b* on C8 position. In the former case, an ethyl group is attached to C8. In the later case, however, an ethylidene group is connected to C8. It is this difference that provides BChl *b* a wider carbon-carbon conjugated region than BChl *a*, resulting in further red-shift, which makes BChl *b* containing species to utilise infrared light energy in longer wavelength. Figure 2.3 shows their structural difference and absorption spectra in organic solvent.

The second major photoreceptor pigment in the purple bacteria is carotenoid. Generally speaking, carotenoids in photosynthetic bacteria possess three main functions. Light harvesting, light protection and light harvesting complex structural stability (Cogdell and Frank 1987; Frank and Cogdell 1996; Polivka and Frank 2010). By strong absorbing light in the blue-green spectral region (450–550 nm), where BChls have weaker absorbance, and transferring absorbed energy fast and efficiently to BChls, they actually act as an important energy donor in the photosynthesis. By dissipation excess amount of light exposure in antenna

core complex, the Qy is red-shifted further to ~1005 nm, e.g., in *Blc. virids*. BChl *a/b* Soret band appears at ~376/398 nm and Qx band at ~590/600 nm.

Two types of bacteriochlorophyll occur in the purple bacterial core complexes, either BChl *a* or BChl *b* depending on different species. Both of them are a substituted tetrapyrrole. The tetrapyrrole ring is stabilised by a central magnesium atom that coordinates with four nitrogen atoms in the tetrapyrrole ring. A highly hydrophobic C20 alcohol tail, either phytol or geranylgerniol

Fig. 2.3 Chemical structures of BChl *a* and BChl *b*. (**a**) BChl *a*. IUPAC nomenclature is used for the numbering of the carbon atoms and the rings. C20 hydrophobic geranylgernyl tail is attached. Absorption spectra of BChl a/b in diethyl ether are shown in dashed lines. (**b**) BChl *b* with Phytyl tail. Qy transition dipole is indicated with a *red arrowed line*

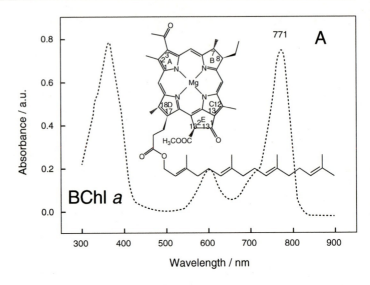

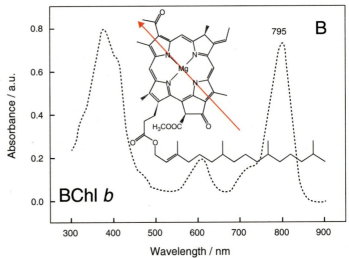

complexes, they also act as a photo-protector. Finally, by interacting with polypeptides in the antenna complexes, the carotenoid molecules stabilize the structure of the light harvesting complexes (Davis et al. 1995; Fiedor and Scheer 2005; Karrasch et al. 1995; Niwa et al. 2014).

Molecular structure of the carotenoid can be considered as a result from joining together of eight isoprene units. Four of the units connected in head-to-tail manner form a half carbon skeleton of the carotenoid molecule, and two halves of this carbon skeletons join together in reverse way to form a whole C40 prototype carotenoid, which looks central symmetry. By structural modification, such as hydrogenation, dehydrogenation, cyclization, hydration, methylation etc. on half or two halves of the prototype, a huge number of different carotenoids can be synthesised. In reality, each structural modification is controlled by an individual enzyme, which is encoded by a gene in the bacterium cell. In *Rba. sphaeroids*, for example, the carotenoid genes (*crt*) are clustered within a 9.1 kb region of the chromosome and mapped to a region with the 45.7 kb photosynthetic gene cluster by transposon mutagenesis (Coomber et al. 1990).

Similar to BChl molecules, all carotenoid molecules possess a conjugated structure. The length of conjugated double bonds and various function groups attached decide the colour of individual carotenoid molecule. Generally speaking, the greater of the number of conjugated double bond, the further red-shift of maximum absorption wavelength (λmax) (Khoo et al. 2011). As a result, the colour of carotenoids found in the purple bacteria vary from yellow, orange to deep red due to the different molecular structure they have.

There are more than 50 different carotenoids found in the purple bacteria. Most of their chemical structures are quite distinct from those found in higher plant. They are synthesised in the purple bacteria cell following two different biosynthetic pathways, i.e., spirilloxanthin pathway (normal spirilloxanthin, unusual spirilloxanthin, spheroidene and cartenal pathways) and Okenone pathway (okenone, and R. g.-keto carotenoid pathways) (Takaichi 1999). Usually, almost all carotenoids in a biosynthetic pathway, whatever final product or its precursors, can be detected in purified protein complexes, indicating that all carotenoids have ability to bound light harvesting complexes. The preference binding of carotenoid to the light harvesting complexes, however, exists in nature (Qian et al. 2001). By artificial reconstitution or mutagenesis modification of the carotenoid biosynthesis pathway, selected carotenoid can be inserted into light harvesting complexes (Akahane et al. 2004; Chi et al. 2015). The carotenoids listed in Fig. 2.4 are some commonly found in the purple photosynthetic bacteria.

2.4.2 LH1 Subunit, α/β-Polypeptide Pair

Bacterial light harvesting 1 complex is consisted of two polypeptides called α and β, which forms a heterodimer. The heterodimer, together with associated carotenoids and BChls consists a subunit of the LH1 complexes called as α/β-polypeptide pair or even shortly as α/β pair. Early researches on diverse purple bacteria revealed abundant structural information on the α and β polypeptides (Brunisholz and Zuber 1992; Zuber 1985; Zuber and Cogdell 1995). (1) They have relative small molecular weight, ~5.5 kDa, corresponding 40–70 amino acid residues. (2) All of them possess a central hydrophobic core that is long enough to cross the photosynthetic membrane once as a single transmembrane α-helix, with both polar N- and C-termini. (3) A histidine residue is reserved, which was proposed as a ligand to the central Mg atom of the BChl. Amino acid sequences of α-, β-polypeptides from seven different purple bacteria are aligned against the His residue. This is shown in Fig. 2.4.

Reversible dissociation of LH1 complex to its subunit, α/β pair, which has a maximum absorbance at ~820 nm (B820) revealed more structural details of the subunit. To dissociate LH1 ring to its subunit, the carotenoid involved in the LH1 ring need to be removed first by either extraction or blocking its biosynthetic pathway genetically because the stabilization of the LH1 from carotenoids is so strong that no any suitable detergent can dissociate it (Loach and Parkes-Loach 1995; Miller et al. 1987). By the use of detergent such as n-octyl β-D-glucopyranoside (β-OG), and careful control dissociation conditions, LH1 ring subunit, B820, can be obtained from a variety of species (Chang et al. 1990b; Heller and Loach 1990; Jirsakova and Reiss-Husson 1993; Kerfeld et al. 1994; Meckenstock et al. 1992; Miller et al. 1987; Parkes-Loach et al. 1994). The B820 has shown that it has a composition of $\alpha_1\beta_1BChl_2$ (Chang et al. 1990a; Loach and Parkes-Loach 1995; Miller et al. 1987). If carotenoid molecules were added back, the B820 subunit can be re-associated to LH1 ring (Davis et al. 1995; Fiedor and Scheer 2005; Karrasch et al. 1995). Further reversible dissociation of B820 subunit to fundamental components and association back to B820 using a series of BChl analogues revealed that C3 acetyl and $C13^2$ carbonyl groups are required to form subunit and LH1 complexes (Davis et al. 1996; Parkes-Loach et al. 1990). Site-directed mutagenesis in the LH1 complex of *Rba. sphaeroides* showed that C3 acetyl group of BChl *a* interact with Trp

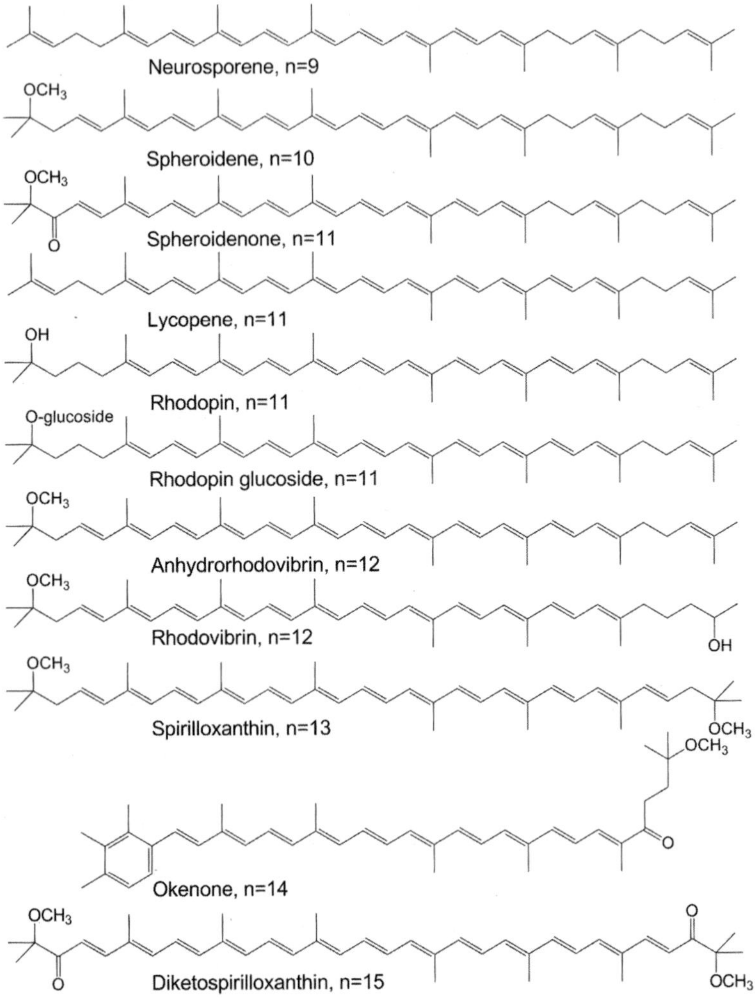
Neurosporene, n=9
OCH3
Spheroidene, n=10
OCH3
O
Spheroidenone, n=11
Lycopene, n=11
OH
Rhodopin, n=11
O-glucoside
Rhodopin glucoside, n=11
OCH3
Anhydrorhodovibrin, n=12
OCH3
Rhodovibrin, n=12
OH
OCH3
Spirilloxanthin, n=13
OCH3
OCH3
O
Okenone, n=14
OCH3
O
Diketospirilloxanthin, n=15
O
OCH3

Fig. 2.4 Selected carotenoids found in the purple photosynthetic bacteria. The number of *n* refers to the number of conjugated double band

residues, one with αTrp + 11 and another with βTrp + 9 (Davis et al. 1997; Kehoe et al. 1998; Olsen et al. 1994; Sturgis et al. 1997). The role of His0 acting as an electron donor to coordinate central Mg atom of the BChl *a* was proved by mutagenesis work. Mutation to other amino acid residues resulted in no or very low level of LH1 expression (Olsen et al. 1997).

According to mutation experimental data on *Rba. sphaeroides* LH1 complex, a cross-hydrogen bonding via His0 residue was proposed as well (Olsen et al. 1997). In addition to providing ligand to BChl *a*, His0 residue in one polypeptide may also form a hydrogen bond with the $C13^1$ keto group of BChl *a* that coordinates with the His0 in other polypeptide, to stabilize B820 and LH1 complex (Olsen et al. 1997). NMR determination of BChl *a* in the B820 subunit of *Rsp. rubrum* revealed the relative position of two BChl *a* molecules (Wang et al. 2002). They form a dimer with overlap of ring C and ring E, which is very similar to the B850 BChl *a* in the LH2 crystal structures of *Phs. molischianum* (Koepke et al. 1996) *and Rps. acidophila* (McDermott et al. 1995). All experimental data from reversible dissociation, site-direct mutagenesis, NMR, spectroscopy, electron microscopy etc. on the LH1 subunit build up a clear framework of the LH1 complex before its high resolution 3D structure available.

2.4.3 Reaction Centre

As described above, RC is the place where a light-induce charge separation across membrane takes place. The basic components in a simplest RC include three polypeptide chains associated with pigment molecules. A high resolution 3D structure of the RC from *Blc. virids* became available in 1985 as the first membrane protein structure solved by X-ray crystallography (Deisenhofer et al. 1985; Deisenhofer and Michel 1989). Except for three polypeptides called H (heavy), M (medium) and L (light) based on their apparent molecular weight as determined by electrophoresis, the RC from *Blc. virids* contains a subunit C (cytochrome). L and M subunits both have five membrane-spanning helices, forming a core of the complex by associating four BChl *b*, two Bphe *b*, one non-heam iron, two quinones and one 15-*cis*-carotenoid. The segment polypeptides connecting transmembrane helices form a flat surface parallel to membrane surface. H subunit has two distinctive parts: a single membrane spanning helix on N-terminal side, and a globular domain on C-terminal, which attaches M and L subunit on cytoplasmic side. On the opposite side, M and L subunit are attached by the cytochrome. The total height from tip of the cytochrome to tip of the H subunit is ~130 Å. Importantly, it has an elliptical cross section with long and short axis of ~70 and ~30 Å respectively. Later, we will see that it is this ellipse that decides the overall shape of the RC-LH1 complexes. Shortly after the RC of *Blc. virids*, 3D structure of the RC from *Rba. sphaeroides* were published (Allen and Holmes 1986; Chang et al. 1986). Overall, two structures have been shown to be very similar each other except that there is no cytochrome attached in the RC of *Rba. sphaeroides* and it associates BChl *a* molecules instead of BChl *b*. The third 3D high resolution RC from a purple sulphur bacteria *Thc. tepidum* showed a similar architecture compared with the RC from *Blc. virids*. It has a C subunit, but use BChl *a* and BPhe *a* as its major pigments (Nogi et al. 2000). These structures showing in Fig. 2.5 help people to understand how photosynthetic bacteria convert light energy to proton motive force by light powered charge separation, redox reactions of quinone/quinol and cytochrome movement etc.

2.4.4 PufX Protein

With basic building blocks, i.e., B820, RC and pigment molecules, the core complex RC-LH1

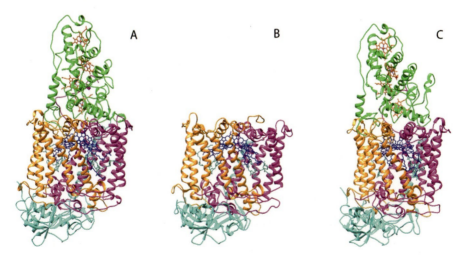

Fig. 2.5 3D structural models of RC from purple photosynthetic bacteria. (**a**) *Blc. Virids*; (**b**) *Rba. sphaeroides*; (**c**) *Tch. tepidum*. Subunits are coloured as follows: RC–H in *cyan*, RC–M in *magenta*, RC–L in *orange*, RC–C in *green*, hem in *red* and BChl in *blue*. Orientations of (**b**) and (**c**) are aligned against (**a**) with cytoplasmic side on *bottom* and periplasmic side on *top*

can be reconstituted (Bustamante and Loach 1994), although the assembly mechanism of the RC-LH1 core complexes in vivo and in vitro could be different from each other not only on the assembly procedure (Pugh et al. 1998) but also on different building blocks. In *Rba. sphaeroides*, for example, an extra polypeptide called PufX was found in its core complex, which was shown to be important for the photosynthetic growth of the cell and molecular architecture of the core complex(Holden-Dye et al. 2008).

PufX is a small transmembrane polypeptide encoded by *PufX* gene that is located at the position downstream of the *pufM* gene, and was first found in *Rba. capsulatus* and *Rba. sphaeroides* (DeHoff et al. 1988; Youvan et al. 1984a, b; Zhu et al. 1986). Later, this gene was known in other *Rhodobacter* genus, including species of *blusticus*, *veldkampi* and *azotoformans* (Tsukatani et al. 2004). The *pufX* encodes a polypeptide that consists of 78 and 82 amino acids in *Rba. sphaeroides* and *Rba. capsulatus* respectively (Lee et al. 1989). The analysis of the isolated mature PufX polypeptides has shown that they are processed at C- and N-termini. Met on N-termini encoded by *pufX* gene are not presented in both species, 12 and 8 amino acids on C-termini were removed as well respectively (Parkes-Loach et al. 2001). Intriguingly, amino acid alignment of PufX polypeptides from both species showed a very low degree of identity being between 23 and 29% depending on the alignment details (Fulcher et al. 1998; Lilburn et al. 1992). This little homology is also shown in all other *Rhodobacter* species (Tsukatani et al. 2004), implying that some polypeptides could be pufX-like proteins that have low identity compared with pufX but act as it. The W protein found in the core complex of *Rps. palustris* is a good example (Roszak et al. 2003).

The *pufX* gene was shown to be essential for photosynthetic growth in *Rba. sphaeroides* because *pufX* gene deleted mutants failed to grow photosynthetically (Barz et al. 1995a, b; Farchaus and Oesterhelt 1989; Lilburn and Beatty 1992; Lilburn et al. 1992). By deleting LH1 gene or reducing size of LH1, however, photosynthetic growth can be restored (McGlynn et al. 1994, 1996). It was suggested, therefore, that instead of directly facilitating cyclic electron transfer between RC and the cytochrome *bc*1 complex, pufX protein provides a gate to allow the qunione/quinol to pass through the LH1 helixes (Cogdell et al. 1996; Parkes-Loach et al. 2001). It was demonstrated that the pufX protein is co-purified with RC-LH1 core complex, leading to a conclusion that the pufX is a component of the core complex(Recchia et al. 1998), with a 1:1 stoichiometry of pufX/RC in the core complex(Francia et al. 1999). Two independent determined solution structures of the PufX from *Rba. sphaeroides* by NMR spectroscopy confirmed that the PufX has a single transmembrane α-helix of 34 amino acid residues, which is much longer than typical transmembrane α-helix (Tunnicliffe et al. 2006; Wang et al. 2007). However, the conformation of the α-helix they provided is different. One is approximately straight (Wang et al. 2007), resulting in protruding of the α-helix out of membrane a couple of turns at both the C- and N-terminal end. The other is bended (Tunnicliffe et al. 2006), permitting the α-helix to accommodate in membrane. This conformation difference could reflect the flexibility of this α-helix hinged by Gly residue and imply structural function of the PufX.

It is clear now that PufX affects the assembly of the core complexes. Electron microscopy image analysis on negatively stained tubular membrane isolated from a mutant *Rba. sphaeroide* lacking of LH2 antenna showed a dimeric structure of the core complex. Two RCs, each of which is associated by an incomplete C-shaped LH1 ring join together to form an S-shaped dimer (Jungas et al. 1999). Deletion of *pufX* gene results in the formation of monomer RC-LH1 core complex, in which the RC is enclosed by a complete LH1 ring (Siebert et al. 2004). In the later section, we will see more detailed structural description of this core complex.

2.5 Structural Diversity of the RC-LH1 Complexes

The light harvesting core complex from photosynthetic bacteria was first visualized in *Blc. virids* membrane by the use of electron microscopy (Miller 1979, 1982). This BChl *b*-containing core complex is shown rather regular arrangement in membrane due to the lack of LH2 complexes. Image synthesis using Furrier Transform on this well ordered membrane revealed its subunit structural details of the core complex. Central density protruded on both cytoplasmic and periplasmic sides was attributed to RC and the density surrounding RC, with a diameter of 110 Å, was interpreted as LH1. Similar characteristic feature was observed in the core complex from *Phs. molischianum* (Boonstra et al. 1994) and *Rhodobium (Rbi.) marinum* (Meckenstock et al. 1992) using electron microscopy. However, the resolution of ~20–30 Å provided by negatively stained EM was not high enough to resolve subunits in the LH1 ring. The precise number of subunit in the LH1 ring was not known unambiguously until in 1995. Karrasch and co-workers reconstituted a LH1 complex of *Rsp. rubrum* from its building block, B820. 2D crystal of the reconstituted LH1 was applied for electron crystallography using cryo-EM. Image processing resulted in an 8.5 Å resolution projection map, which is good enough to show individual electron density caused by single transmembrane helix. Sixteen B820 subunits or α/β-apoprotein pairs form a closed LH1 ring with a hole in the middle which can just hold a RC (Karrasch et al. 1995). An 8.5 Å cryo-EM projection structure of the core complex purified from wild type *Rsp. rubrum* showed a same number of α/β subunit of the LH1 ring surrounding one RC in middle (Jamieson et al. 2002). Two different shapes of the LH1 ring, circle and ellipse, reflect the fact that the LH1 ring is quite flexible (Jamieson et al. 2002; Qian et al. 2003). The model of RC-LH1 being a RC enclosed by a closed LH1 ring also was observed by atomic force microscopy in 2D crystals or photosynthetic membranes (Scheuring and Sturgis 2009). Recent published 3.0 Å resolution X-ray crystal structure of the RC-LH1 core complex from *Tcl. tepidum* confirmed the model unambiguously, i.e., an elliptical LH1 ring consisting of 16 α/β subunits encloses one RC in the middle to form a monomeric RC-LH1 core complex.

An exceptional monomeric RC-LH1 complex was found in *Rps. palustris*. Its 4.8 Å resolution X-ray crystal structure showed that the existence of an extra small polypeptide called W causes an incomplete LH1 ring which only consists of 15 α/β subunits (Roszak et al. 2003). Their structural details will be described in later sections.

As mentioned above, the PufX protein affects the assembly of the core complexes: the PufX protein hinders the completion of the LH1 ring and facilitates the dimerization of the core complex. During the past decade, there have been controversial interpretations on the structure of the dimeric core complex RC-LH1-PufX (Bullough et al. 2008; Cogdell et al. 2006; Holden-Dye et al. 2008). Initially, two halves of dimeric core complex found in tubular membrane of *Rba. sphaeroides* was proposed being associated by a cytochrome *bc1* complex (Jungas et al. 1999). However, cytochrome *bc1* complex cannot be detected in purified tubular membrane, ruling out the existence of the cytochrome *bc1* complex in the dimeric core complex(Siebert et al. 2004). By measuring the length of LH1 arc on its projection map from 2D crystal of the dimer core complex, 24 α/β subunits of the LH1 from *Rba. sphaeroides* was estimated, and a density in the centre of the dimer was attributed to a pair of pufX based on 26 Å resolution projection map (Scheuring et al. 2004a). An 8.5 Å projection map of the same complex, however, showed 28 density pairs which were assigned as α/β subunits. An extra density between RC and LH1 and near the LH1 gap was attributed to PufX (Qian et al. 2005). 28.4 ± 2.2 BChl *a* molecules per RC obtained from quantitative pigment analysis of purified dimeric core complex supported the assignment of 14 α/β subunits per RC. Furthermore, the orientation of the RCs in the dimeric core complex was determined by comparing its projection map with the simulated RC projection map at 8.5 Å derived from 3D X-ray structure of the RC from *Rba. sphaeroides*. This produced an orientation of the planes of the special pair BChl *a* porphyrin rings relative to the long axis of the dimer is ~17.5°, a good agreement with the orientation determined by LD measurements (Frese et al. 2000). Based on AFM topographs, a model of the dimeric pufX containing core complex from *Rba. blasticus* was presented (Scheuring et al. 2005). The model contains 26 α/β pairs, a dimer pufX in centre of the core, and an orientation of special pair relative to long axis is ~ − 40.0°. PufX protein was also identified in an alkaliphilic non-sulphur purple bacterium *Rhodobaca (Rca.) bogoriensis* (Milford et al. 2000), a sole example that PufX is found out of *Rhodobacter* genus. A 13 Å projection map of dimeric core was interpreted in a different way—two C-shaped LH1 ring, which compose of 13 α/β pairs and two PufXs, surround the RCs. They form a dimer through an interface of two LH1 subunits (Semchonok et al. 2012).

Obviously, 3D structures of the core complex, with resolution high enough to resolve transmembrane helix at least, are absolutely necessary for an unambiguously interoperation. By the time of this writing, only three X-ray crystal structures of the core complex from the purple bacteria are available in protein data bank (PDB). Their structural features will be described in the next section in the order of their publication time.

The inner α-apoprotein elliptical ring with a long axis of ~78 Å matches with overall shape of the RC very well, so that the LH1 subunits appears to be wrapped around the RC. These features are similar with that seen in EM projection maps and AFM topographs of the RC-LH1 core complexes (Jamieson et al. 2002; Scheuring et al. 2006, 2004b). However, the LH1 ring is interrupted by a transmenbrane helix 'W', which is located between α polypeptides, resulting in a gap in the LH1 ring. This W protein, having a molecular weight of ~11.0 kDa, has been suggested as an analogous protein to the PufX although very little is known about it, even including its amino acid sequence. Nevertheless, the unique orientation of the W protein respect to the RC provided us a possible way to understand how ubiquinone (Q_B)/ubiquinol (Q_BH_2) shuttle between RC and cytochrome *bc*1 complexes. Located on the opposite of RC-H single transmembrane helix, and near the gap of LH1 ring, the W protein is suggested as a part of gate through which Q_BH_2 can be released from its binding side to membrane lipid phase outside of the RC. It is also likely that W plays a key role for assembly of the core complex. It should be emphasized that at 4.8 Å resolution, it is impossible to resolve individual amino acid and pigment molecules in the core complex. To fully understand structure and function of this core complex, a higher resolution 3D structural data and more biochemical measurements are absolutely necessary.

2.5.1 X-Ray Crystal Structure of the Core Complex from *Rps. palustris*

The crystal structure of the core complex from the purple bacterium *Rps. palustris* was determined at 4.8 Å resolution by X-ray crystallography. The phase was obtained by molecular replacement with the RC of *Rba. sphaeroides* (McAuley et al. 1999). The model of this RC-LH1 core complex is shown in Fig. 2.6. An elliptical LH1 ring has a dimension of ~110 Å by ~95 Å measured as centre to centre distance of opposite β-helices.

2.5.2 X-Ray Crystal Structure of the Core Complex from *Rba. sphaeroides*

It is clear that the core complex in *Rba. sphaeroides* takes dimeric form, i.e., two monomeric core complexes, each of which consists of one RC surrounded by an incomplete C-shaped LH1, are associated together to form a bended S-shaped dimer (Ashby et al. 1987; Bahatyrova et al. 2004; Jungas et al. 1999; Qian et al. 2008, 2005; Scheuring et al. 2004a; Siebert et al. 2004;

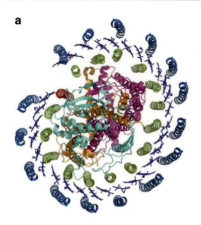 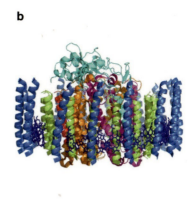

Fig. 2.6 3D structure of the RC-LH1 core complex from *Rps. palustris* at 4.8 Å resolution (**a**) *Top view* from cytoplasmic side, β-apoprotein in *blue*, α-apoprotein in *green*, RC-H in *cyan*, RC-M in *magenta*, RC-L in *orange* and W in *red*. A pair of BChl *a* (in *blue* stick) without phytal tail was modelled into the complex based on the density between α- and β-helices. For the purpose of clarity, pigment molecules in the RC are not shown (**b**) *Side view* by rotating (**a**) 90° so that cytoplasmic side is on *top*. Only transmembrane part of the polypeptides of the LH1 is modelled using ploy alanine

Westerhuis et al. 2002). However, controversies are never stopped during the last decade on the issues, such as stoichiometry of the pufX, size of LH1 ring, orientation of RC, conformation of LH1 polypeptides, and location of the pufX. An 8.0 Å X-ray crystal structure of the core complex from *Rba. sphaeroides*, which clearly shows the interaction among individual polypeptides provides us clearer insight into the structural-function relationship of the complex (Qian et al. 2013).

Overall Structure Having a molecular weight of ~521 kDa, the dimeric core complex of *Rba. sphaeroides* is composed of 64 polypeptides. Each half of the dimer consists of a PufX and a RC surrounded by 14 LH1 α/β subunits, with two BChl *a* molecules sandwiched between α- and β-transmembrane helices. Two halves of the complex are associated together through the PufX protein, which intimately interacts with RC-H subunit in one half monomer and 14th β-helix of the other half monomer, forming a S-shaped LH1 ring if viewed from periplasmic side. Two halves of monomer inclines each other ~11 degrees against periplasmic side, forming a V-shaped molecule if viewed parallel to the membrane surface. Its 3D view is showed in Fig. 2.7.

Polypeptide Interactions Involved the PufX It has long been suggested that the PufX plays a key role for dimerization of the RC-LH1-PufX core complex. The 3D X-ray crystal model of the core complex shows three important interactions that involve the PufX protein (see Fig. 2.8).

The 3D model of the dimeric complex shows that orientation of LH1 α/β subunits, among α2/β2 to α14/β14 are quite similar, being nearly straight transmembrane part, but α1/β1 is significant different from the others. Figure 2.8a shows that PufX contacts with the N-termini of LH1 α1/β1 on the cytoplasmic side of the membrane forming a α1/β1/pufX cluster. In this cluster, the bended confirmation of the PufX polypeptide chain can be superimposed onto its minimized averaged solution structure determined by NMR (Tunnicliffe et al. 2006) with an RMS deviation of 2.08 Å. Distinct bended confirmation of β1 polypeptide was observed in the cluster as well, reflecting a flexibility of the β polypeptide, which could possess different confirmations depending on different environments. The PufX polypeptide also intimately contacts with extrinsic domain of the RC-H subunit, with PufX N-terminal residues ~Asn 8 and RC-H C-terminal residues ~244–247 (see Fig. 2.8b). It is likely that this interaction forms a start point of the LH1 complex, with

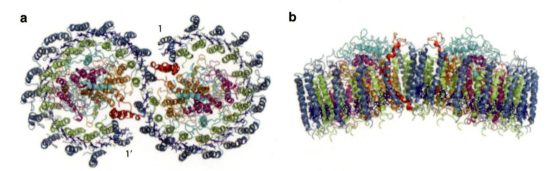

Fig. 2.7 The structure of dimeric core complex from *Rba. sphaeroides* (**a**) *Top view* from cytoplasmic side. PufX protein is coloured in *red*. For clarity, only α1/β1 and α1'/β1' pairs are labelled (**b**) *Side view* parallel to the membrane surface with cytoplasmic side on *top*

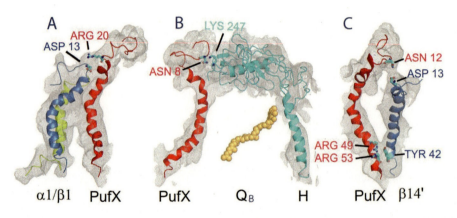

Fig. 2.8 PufX interactions that stabilize the dimer core complex (**a**) *Side view* of the interaction of the pufX with the α1/β1 pair on the cytoplasmic side. PufX in *red*, β1 in *blue* and α1 in *yellow* (**b**) Interaction of pufX with RC-H on the cytoplasmic side of the membrane. RC-H in *cyan*, ubiquinone in *brown* (**c**) Interactions of the pufX on one half of the dimer with β14' on the other half of the dimer

which the rest of LH1 subunits can follow. Given relative bigger volume of this cluster compared with normal α/β pair, this is also likely a point that stops encirclement of the LH1 ring. The third interaction comes from PufX and β14' polypeptide both on the N- and C-termini extrinsic regions (see Fig. 2.8c). It was suggested that these interactions are crucial for dimerization of the core complex. The model predicts that Asn12 of the PufX on the one half of the dimer is close to Asp13 of LH1 β14' on the other half. This explains why the truncation of 12 or more amino acids from N-terminus of the PufX stops dimerization of the core complex (Francia et al. 2002; Ratcliffe et al. 2011). C-terminal truncation of the PufX also affects dimerization, suggesting this part is involved protein-protein interactions. The model provides such evidence showing that PufX C-terminal from residues 49 contacts with β14' C-terminal residues from 38. Unfortunately, 8.0 Å resolution structural data cannot provide precise contact information. However, alternation of PufX Arg49 and Arg53 to Leu or even only PufX Gly52 to Leu abolishes the formation of the dimer, indicating that all alternation of C-terminal confirmation of the

PufX will likely change its interaction with β14′, and affect dimerization of the core complex (Ng 2008).

Quinone Channel To allow the turnover of RC photochemistry, a shuttle of quinol/quinone between RC Q_B binding side and quinone pool out of the complex through LH1 barrier is needed. A continuous 3D volume enclosed by the LH1 ring inside of the complex was found. This volume, as a putative quinone channel, has two gates. One at LH1 ring gap, allowing Q/QH_2 to be in and out of the complex directly, and the other at the interface of the dimer, allowing Q/QH_2 migrate between two halves of monomer possibly. By the use of this channel, it is possible for the migration of Q/QH_2 within spherical membrane or even through long tubular membrane that is densely packed with dimeric core complexes. A 3D model of the quinone channel can be viewed on youtube website (https://www.youtube.com/watch?v=vsaYyjfNGmI)

It should bear in mind that only with massive different experimental data support the 3D X-ray structure of the core complex of *Rba. sphaeroides* at 8.0 Å resolution can be solved. These data include electron microscopy (Qian et al. 2005, 2008), NMR (Conroy et al. 2000; Ratcliffe et al. 2011; Tunnicliffe et al. 2006), available high resolution structures of RC (Ermler et al. 1994) and LH2(Koepke et al. 1996; McDermott et al. 1995), and site-direct mutagenesis (Olsen et al. 1994, 1997; Sturgis et al. 1997). At this resolution, individual amino acid, carotenoid and phytyl tails of BChl *a* cannot be resolved from its electron density map. Therefore, we still need to wait for a higher resolution data for the precise description of the structure-function relationship of the complex.

2.5.3 X-Ray Crystal Structure of Core Complex from *Tch. tepidum*

Thermochromatium tepidum is a thermophilic purple sulphur photosynthetic bacterium found originally from a hot spring in Yellowstone National Park. Growing anaerobically at optimum temperature between 48 and 58 °C, its BChl *a*–containing RC-LH1 core complex presents a unique optical property, a maximum absorbance at 915 nm (B915). The 3D X-ray crystal structure of the core complex from *Tch. tepidum* was solved to 3.0 Å. This is the first core complex structure showing at near-atomic resolution level so far. It provides us more details to understand its molecular mechanism involved in primary photosynthetic reactions.

Overall Structure The core complex of *Tch. tepidum* is composed of a LH1, a RC, a cytochrome and 80 cofactors with a molecular weight ~ 380 kDa. Sixteen LH1 α/β subunits form a double complete elliptical LH1 ring surrounding the RC which has four subunits L, H, M and C, a cytochrome attached on periplasmic side of the RC. The major and minor axis length of outer elliptical LH1 ring is 105 Å and 96 Å, and 82 Å and 73 Å for inner LH1 elliptical ring were measured. Each α/β heterodimer subunit associates two BChl *a* molecules and one carotenoid molecule, spirilloxanthin. No PufX-like protein has been found in the core complex of *Tch. tepidum*. Figure 2.9 shows its 3D structure.

Protein-Protein Interactions in the LH1 The structure of the LH1 subunit α/β pair is very similar to those in LH2 complexes, especially to that in *Phs. molischianum* LH2(Koepke et al. 1996). Each α/β sandwiches two BChl *a* molecules and one spirilloxanthin. In C-terminal region, a Ca^{2+} ion was identified. Figure 2.10 shows their relative position in LH1 ring. Central Mg atom of the β-B915 is coordinated by β-His 36; its C3-acetyl group forms a hydrogen bond with β-Trp 45; its carbonyl oxygen of the phytyl ester group is hydrogen bonded to α-Gln 28 and β-Trp 28. Similarly, α-His 36 coordinates the central Mg atom of α-B915; its C3-acetyl group forms a H-bond with α-Trp 46. No H-bonds were found on C13-keto groups of the B915. All of these intra-subunit interactions not only fix BChl *a* molecules onto

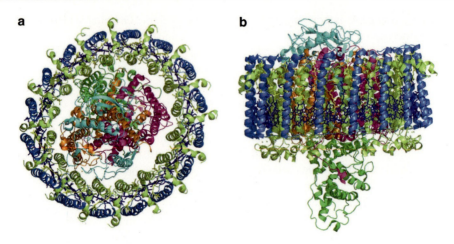

Fig. 2.9 The structure of RC-LH1 core complex of *Tch. tepidum* (**a**) *Top view* from cytoplasmic side. Colour codes are same as in Fig. 2.7, except for cytochrome in green (**b**) *Side view* parallel to the membrane surface with cytoplasmic side on *top*

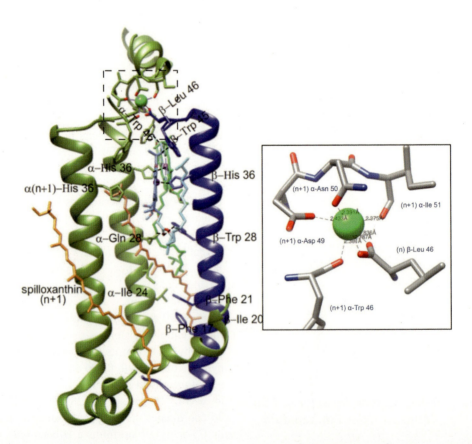

Fig. 2.10 Intra- and inter-subunit interactions in the LH1 ring. α-polypeptides in *olive green*, β-polypeptide in *blue*, spirilloxanthin in *brown*, β-B915 in *cyan*, α-B915 in *green*, Mg^{2+} in *purple* and Ca^{2+} in *bright green*. Insert is an enlargement of the area covered by a *dashed square*

the scaffold of α/β-polypeptide in a correct orientation but also keep α/β-polypeptides together to stabilizes LH1 subunits.

The structure also reveals two important inter-subunit interactions that strengthen stability of the LH1 ring further. Sixteen Ca^{2+} binding sites were identified in the C-terminal regions of α/β-apoproteins. Six coordinated bonds can assign to a Ca^{2+}, four of them from C-terminal of α-polypeptide ($α_{n+1}$-Trp46, $α_{n+1}$-Asp 49, $α_{n+1}$-Asn 50, $α_{n+1}$-Ile 51) and two others from its neighbouring β-polypeptide Leu 46's carboxyl group, forming an inter-subunit linker within the LH1 ring. It is suggested that these linkages give the RC-LH1 of *Tch. tepidum* an extra thermo-stability and cause BChl *a* Qy red-shift. It should note that under current resolution the number of coordinated bond to each Ca^{2+} ion and corresponding bond lengths vary slightly within the LH1 ring. A recent 1.9 Å resolution structure of the core complex provides more details of the coordinations (unpublished personal communication).

All-*trans* spirilloxanthin curves through the α/β pair at approximately 30° to the membrane normal. In addition to intimate interactions with α/β polypeptides and BChl *a* tails, one of the methoxy groups of the spirilloxthanin is in close proximity to the upstream neighbouring (n + 1) α-His 36, and the other methoxy group to the downstream neighbouring N-termini of (n−1) α/β-polypeptides, forming another inter-subunit linker. It has long been known that carotenoid has a function to stabilize light harvesting complexes. This structure for the first time shows such molecular mechanism clearly.

Putative channels for Q/QH_2 shuttling between RC and cytochrome *bc*1 complex are located on the cytoplasmic side of the transmembrane region between adjacent α/β pairs. Given the intrinsic flexibility of the LH1 ring the Q/QH_2 channel with an averaged dimension that approximately equals to the head of ubiquinone could let Q/QH_2 pass through the apparent LH1 barrier. The structure confirms the conclusions from previous electron microscopic study and molecular dynamic simulation that proposed a quinone diffusion mode through LH1 ring by LH1 'breathing' movement (Aird et al. 2007; Jamieson et al. 2002).

Acknowledgement The author gratefully acknowledges funding from the Biotechnology and Biological Research Council, UK. Author also thanks Dr. Seiji Akimoto for his critical comments on the manuscript. Prof. Wang-Otomo Z.Y. provided absorption spectrum of purified core complex of *Tch. tepidum*.

References

Aird A, Wrachtrup J, Schulten K, Tietz C (2007) Possible pathway for ubiquinone shuttling in *Rhodospirillum rubrum* revealed by molecular dynamics simulation. Biophys J 92: 23–33

Akahane J, Rondonuwu FS, Fiedor L, Watanabe Y, Koyama Y (2004) Dependence of singlet-energy transfer on the conjugation length of carotenoids reconstituted into the LH1 complex from *Rhodospirillum rubrum* G9. Chem Phys Lett 393: 184–191

Allen JF, Holmes NG (1986) A general model for regulation of photosynthetic unit function by protein phosphorylation. FEBS Lett 202: 175–181

Ashby MK, Coomber SA, Hunter CN (1987) Cloning, nucleotide sequence and transfer of genes for the B800-850 light harvesting complex of *Rhodobacter sphaeroides*. FEBS Lett 213: 245–248

Bahatyrova S, Frese RN, Siebert CA, Olsen JD, van der Werf KO, van Grondelle R, Niederman RA, Bullough PA, Otto C, Hunter CN (2004) The native architecture of a photosynthetic membrane. Nature 430: 1058–1062

Barz WP, Francia F, Venturoli G, Melandri BA, Vermeglio A, Oesterhelt D (1995a) Role of PufX protein in photosynthetic growth of *Rhodobacter sphaeroides*. 1. PufX is required for efficient light-driven electron transfer and photophosphorylation under anaerobic conditions. Biochemistry 34: 15235–15247

Barz WP, Vermeglio A, Francia F, Venturoli G, Melandri BA, Oesterhelt D (1995b) Role of the PufX protein in photosynthetic growth of the *Rhodobacter sphaeroides*. 2. PufX is required for efficient ubiquinone/ubiquinol exchange between the reaction center QB site and the cytochrome *bc*1 complex Biochemistry 34: 15248–15258

Blankenship RE (2014) Molecular mechanisms of photosynthesis John Wiley & Sons, Oxford, U.K.

Boonstra AF, Germeroth L, Boekema EJ (1994) Structure of the light-harvesting antenna from *Rhodospirillum molischianum* studied by electron microscopy. Biochim Biophys Acta 1184: 227–234

Brunisholz RA, Zuber H (1992) Structure, function and organization of antenna polypeptides and antenna complexes from the three families of *Rhodospirillaneae*. J Photoch Photobio B 15: 113–140

Bullough PA, Qian P, Hunter CN (2008) Reaction Center-Light-Harvesting Core Complexes of Purple Bacteria. In The Purple Phototrophic Bacteria, Hunter CN, Daldal F, Thurnauer MC, Beatty JT (eds) pp 155–179. Springer Netherlands

Bustamante PL, Loach PA (1994) Reconstitution of a functional photosynthetic receptor complex with isolated subunits of core light-harvesting complex and reaction centers. Biochemistry 33: 13329–13339

Chang CH, Tiede D, Tang J, Smith U, Norris J, Schiffer M (1986) Structure of *Rhodopseudomonas sphaeroides* R-26 reaction center. FEBS Lett 205: 82–86

Chang MC, Callahan PM, Parkes-Loach PS, Cotton TM, Loach PA (1990a) Spectroscopic characterization of the light-harvesting complex of *Rhodospirillum rubrum* and its structural subunit. Biochemistry 29: 421–429

Chang MC, Meyer L, Loach PA (1990b) Isolation and characterization of a structural subunit from the core light-harvesting complex of *Rhodobacter sphaeroides* 2.4.1 and puc 705-BA. PhotochemPhotobiol 52: 873–881

Chi SC, Mothersole DJ, Dilbeck P, Niedzwiedzki DM, Zhang H, Qian P, Vasilev C, Grayson KJ, Jackson PJ, Martin EC, Li Y, Holten D, Hunter CN (2015) Assembly of functional photosystem complexes in *Rhodobacter sphaeroides* incorporating carotenoids from the spirilloxanthin pathway. Biochim Biophys Acta 1847: 189–201

Cogdell RJ, Frank HA (1987) How carotenoids function in photosynthetic bacteria. Biochim Biophys Acta 895: 63–79

Cogdell RJ, Fyfe PK, Barrett SJ, Prince SM, Freer AA, Isaacs NW, McGlynn P, Hunter CN (1996) The purple bacterial photosynthetic unit. Photosynth Res 48: 55–63

Cogdell RJ, Gall A, Kohler J (2006) The architecture and function of the light-harvesting apparatus of purple bacteria: from single molecules to in vivo membranes. Q Rev Biophys 39: 227–324

Conroy MJ, Westerhuis WH, Parkes-Loach PS, Loach PA, Hunter CN, Williamson MP (2000) The solution structure of *Rhodobacter sphaeroides* LH1beta reveals two helical domains separated by a more flexible region: structural consequences for the LH1 complex. J Mol Biol 298: 83–94

Coomber SA, Chaudhri M, Connor A, Britton G, Hunter CN (1990) Localized transposon Tn 5 mutagenesis of the photosynthetic gene cluster of *Rhodobacter sphaeroides*. Mol Microbiol 4: 977–989

Davis CM, Bustamante PL, Loach PA (1995) Reconstitution of the bacterial core light-harvesting complexes of Rhodobacter sphaeroides and *Rhodospirillum rubrum* with isolated α- and β-polypeptides, bacteriochlorophyll a, and carotenoid. J Biol Chem 270: 5793–5804

Davis CM, Parkes-Loach PS, Cook CK, Meadows KA, Bandilla M, Scheer H, Loach PA (1996) Comparison of the structural requirements for bacteriochlorophyll binding in the core light-harvesting complexes of *Rhodospirillum rubrum* and *Rhodobacter sphaeroides* using reconstitution methodology with bacteriochlorophyll analogs. Biochemistry 35: 3072–3084

Davis CM, Bustamante PL, Todd JB, Parkes-Loach PS, McGlynn P, Olsen JD, McMaster L, Hunter CN, Loach PA (1997) Evaluation of structure-function relationships in the core light-harvesting complex of photosynthetic bacteria by reconstitution with mutant polypeptides. Biochemistry 36: 3671–3679

DeHoff BS, Lee JK, Donohue TJ, Gumport RI, Kaplan S (1988) In vivo analysis of puf operon expression in *Rhodobacter sphaeroides* after deletion of a putative intercistronic transcription terminator. J Bacteriol 170: 4681–4692

Deisenhofer J, Michel H (1989) The photosynthetic reaction centre from the purple bacterium *Rhodopseudomonas viridis*. EMBO J 8: 2149–2170

Deisenhofer J, Epp O, Miki K, Huber R, Michel H (1985) Structure of the protein subunits in the photosynthetic reaction centre of *Rhodopseudomonas viridis* at 3 Å resolution. Nature 318: 618–624

Ermler U, Fritzsch G, Buchanan SK, Michel H (1994) Structure of the photosynthetic reaction centre from Rhodobacter sphaeroides at 2.65 Å resolution: cofactors and protein-cofactor interactions. Structure 2: 925–936

Esser L, Elberry M, Zhou F, Yu CA, Yu L, Xia D (2008) Inhibitor-complexed structures of the cytochrome bc1 from the photosynthetic bacterium *Rhodobacter sphaeroides*. J Biol Chem 283: 2846–57

Farchaus JW, Oesterhelt D (1989) A *Rhodobacter sphaeroides* pufL, M and X deletion mutant and its complementation in trans with a 5.3 Kb puf operon shuttle fragment. EMBO J 8: 47–54

Fiedor L, Scheer H (2005) Trapping of an assembly intermediate of photosynthetic LH1 antenna beyond B820 subunit. Significance for the assembly of photosynthetic LH1 antenna. J Biol Chem 280: 20921–6

Francia F, Wang J, Venturoli G, Melandri BA, Barz WP, Oesterhelt D (1999) The reaction center-LH1 antenna complex of *Rhodobacter sphaeroides* contains one PufX molecule which is involved in dimerization of this complex. Biochemistry 38: 6834–6845

Francia F, Wang J, Zischka H, Venturoli G, Oesterhelt D (2002) Role of the N- and C-terminal regions of the PufX protein in the structural organization of the photosynthetic core complex of *Rhodobacter sphaeroides*. European Journal of Biochemistry 269: 1877–1885

Frank HA, Cogdell RJ (1996) Carotenoids in photosynthesis. Photochem Photobiol 63: 257–264

Frese RN, Olsen JD, Branvall R, Westerhuis WH, Hunter CN, van Grondelle R (2000) The long-range supraorganization of the bacterial photosynthetic unit: A key role for PufX. Proc Natl Acad Sci USA 97: 5197–202

Fulcher TK, Beatty JT, Jones MR (1998) Demonstration of the key role played by the PufX protein in the functional and structural organization of native and hybrid bacterial photosynthetic core complexes. J Bacteriol 180: 642–646

Heller BA, Loach PA (1990) Isolation and characterization of a subunit form of the B875 light-harvesting complex from *Rhodobacter capsulatus* Photochem Photobiol 51: 621–627

Holden-Dye K, Crouch LI, Jones MR (2008) Structure, function and interactions of the PufX protein. Biochim Biophys Acta 1777: 613–30

Hunter CN, Daldal F, Thurnauer MC, Beatty JT (2009) The Purple Phototrophic Bacteria. Springer, Dordrecht: 1013

Imhoff JF, Truper HG, Pfennig N (1984) Rearrangement of the species and genera of the phototrophic "purple nonsulfur bacteria". Int J Syst Bacteriol 34: 340–343

Jamieson SJ, Wang P, Qian P, Kirkland JY, Conroy MJ, Hunter CN, Bullough PA (2002) Projection structure of the photosynthetic reaction centre-antenna complex of *Rhodospirillum rubrum* at 8.5 Å resolution. EMBO J 21: 3927–3935

Jay F, Lambillotte M, Stark W, Muhlethaler K (1984) The Preparation and Characterization of Native Photoreceptor Units from the Thylakoids of *Rhodopseudomonas-viridis*. EMBO J 3: 773–776

Jirsakova V, Reiss-Husson F (1993) Isolation and characterization of the core light-harvesting complex B875 and its subunit form, B820, from *Rhodocyclus gelatinosus*. Biochim Biophys Acta 1183: 301–308

Jungas C, Ranck JL, Rigaud JL, Joliot P, VermÇglio A (1999) Supramolecular organization of the photosynthetic apparatus of *Rhodobacter sphaeroides*. EMBO J 18: 534–542

Karrasch S, Bullough PA, Ghosh R (1995) The 8.5 Å projection map of the light-harvesting complex I from *Rhodospirillum rubrum* reveals a ring composed of 16 subunits. EMBO J 14: 631–638

Kehoe JW, Meadows KA, Parkes-Loach PS, Loach PA (1998) Reconstitution of core light-harvesting complexes of photosynthetic bacteria using chemically synthesized polypeptides. 2. Determination of structural features that stabilize complex formation and their implications for the structure of the subunit complex. Biochemistry 37: 3418–3428

Kerfeld CA, Yeates TO, Thornber JP (1994) Biochemical and spectroscopic characterization of the reaction-center LH1 complex and the carotenoid-containing B820 subunit of *Chromatium purpuratum*. Biochim Biophys Acta 1185: 193–202

Khoo HE, Prasad KN, Kong KW, Jiang YM, Ismail A (2011) Carotenoids and Their Isomers: Color Pigments in Fruits and Vegetables. Molecules 16: 1710–1738

Koepke J, Hu XC, Muenke C, Schulten K, Michel H (1996) The crystal structure of the light-harvesting complex II (B800-B850) from *Rhodospirillum molischanum*. Structure 4: 581–597

Lee JK, DeHoff BS, Donohue TJ, Gumport RI, Kaplan S (1989) Transcriptional analysis of puf operon expression in *Rhodobacter sphaeroides* 2.4.1. and an intercistronic transcription terminator mutant. J Biol Chem 264: 19354–19365

Lilburn TG, Beatty JT (1992) Suppressor mutants of the photosynthetically incompetent pufX deletion mutant *Rhodobacter capsulatus* D RC6(pTL2). FEMS Microbiol Lett 100: 155–159

Lilburn TG, Haith CE, Prince RC, Beatty JT (1992) Pleiotropic effects of pufX gene deletion on the structure and function of the photosynthetic apparatus of *Rhodobacter capsulatus*. Biochim Biophys Acta 1100: 160–170

Loach PA, Parkes-Loach PS (1995) Structure-function relationships in core light-harvesting compelxes (LHI) as determined by characterization of the structural subunit and by reconstitution experiments. In Anoxygenic Photosynthetic Bacteria, Blankenship RE, Madigan MT, Bauer CE (eds) pp 433–471. The Netherlands: Kluwer Academic Publishers

McAuley KE, Fyfe PK, Cogdell RJ, Isaacs N, Jones MR (1999) Structural details of an interaction between cardiolipin and an integral membrane protein. Proc Natl Acad Sci USA 96: 14706–14711

McDermott G, Prince SM, Freer AA, Hawthornthwaite-Lawless AM, Papiz MZ, Cogdell RJ, Isaacs NW (1995) Crystal structure of an integral membrane light-harvesting complex from photosynthetic bacteria. Nature 374: 517–521

McGlynn P, Hunter CN, Jones MR (1994) The Rhodobacter sphaeroides PufX protein is not required for photosynthetic competence in the absence of a light harvesting system. FEBS Lett 349: 349–353

McGlynn P, Westerhuis WH, Jones MR, Hunter CN (1996) Consequences for the organisation of reaction center-light harvesting antenna 1 (LH1) core complexes of *Rhodobacter sphaeroides* arising form deletion of amino acid residues at the C terminus of the LH1 α polypeptide. J Biol Chem 271: 3285–3292

Meckenstock RU, Brunisholz RA, Zuber H (1992) The light-harvesting core-complex and the B820-subunit from *Rhodopseudomonas marina*.1. Purification and characterization. FEBS Lett 311: 128–134

Milford AD, Achenbach LA, Jung DO, Madigan MT (2000) *Rhodobaca bogoriensis* gen. nov and sp nov., an alkaliphilic purple nonsulfur bacterium from African Rift Valley soda lakes. Arch Microbiol 174: 18–27

Miller KR (1979) Structure of a bacterial photosynthetic membrane. Proc Natl Acad Sci USA 76: 6415–6419

Miller KR (1982) 3-Dimensional Structure of a Photosynthetic Membrane. Nature 300: 53–55

Miller JF, Hinchigeri SB, Parkes-Loach PS, Callahan PM, Sprinkle JR, Riccobono JR, Loach PA (1987)

Isolation and characterization of a subunit form of the light-harvesting complex of *Rhodospirillum rubrum*. Biochemistry 26: 5055–5062

Ng IW (2008) A structural and functional study of the RC-LH1-PufX core complex from *Rhodobacter sphaeroides*. In University of Sheffield

Niwa S, Yu LJ, Takeda K, Hirano Y, Kawakami T, Wang-Otomo ZY, Miki K (2014) Structure of the LH1-RC complex from *Thermochromatium tepidum* at 3.0 Å. Nature 508: 228–32

Nogi T, Fathir I, Kobayashi M, Nozawa T, Miki K (2000) Crystal structures of photosynthetic reaction center and high-potential iron-sulfur protein from *Thermochromatium tepidum*: thermostability and electron transfer. Proc Natl Acad Sci USA 97: 13561–6

Olsen JD, Sockalingum GD, Robert B, Hunter CN (1994) Modification of a hydrogen bond to a bacteriochlorophyll a molecule in the light harvesting 1 antenna of *Rhodobacter sphaeroides*. Proc Natl Acad Sci USA 91: 7124–7128

Olsen JD, Sturgis JN, Westerhuis WH, Fowler GJS, Hunter CN, Robert B (1997) Site-directed modification of the ligands to the bacteriochlorophylls of the light-harvesting LH1 and LH2 complexes of *Rhodobacter sphaeroides*. Biochemistry 36: 12625–12632

Paddock ML, Weber KH, Chang C, Okamura MY (2005) Interactions between Cytochrome c2 and the Photosynthetic Reaction Center from *Rhodobacter sphaeroides*: The Cation-Pi Interaction. Biochemistry 44: 9619–9625

Parkes-Loach PS, Michalski TJ, Bass WJ, Smith U, Loach PA (1990) Probing the bacteriochlorophyll binding site by reconstitution of the light-harvesting complex of *Rhodospirillum rubrum* with bacteriochlorophyll a analogues. Biochemistry 29: 2951–2960

Parkes-Loach PS, Jones SM, Loach PA (1994) Probing the structure of the core light-harvesting complex (LH1) of *Rhodopseudomonas viridis* by dissociation and reconstitution methodology. Photosynth Res 40: 247–261

Parkes-Loach PS, Law CJ, Recchia PA, Kehoe J, Nehrlich S, Chen J, Loach PA (2001) Role of the core region of the PufX protein in inhibition of reconstitution of the core light-harvesting complexes of *Rhodobacter sphaeroides* and *Rhodobacter capsulatus*. Biochemistry 40: 5593–5601

Polivka T, Frank HA (2010) Molecular Factors Controlling Photosynthetic Light Harvesting by Carotenoids. Accounts Chem Res 43: 1125–1134

Pugh RJ, McGlynn P, Jones MR, Hunter CN (1998) The LH1-RC core complex of Rhodobacter sphaeroides: interaction between components, time-dependent assembly, and topology of the PufX protein. Biochim Biophys Acta 1366: 301–316

Qian P, Saiki K, Mizoguchi T, Hara K, Sashima T, Fujii R, Koyama Y (2001) Time-dependent changes in the carotenoid composition and preferential binding of spirilloxanthin to the reaction center and anhydrorhodovibrin to the LH1 antenna complex in *Rhodobium marinum*. Photochem Photobiol 74: 444–452

Qian P, Addlesee HA, Ruban AV, Wang P, Bullough PA, Hunter CN (2003) A reaction center-light-harvesting 1 complex (RC-LH1) from a *Rhodospirillum rubrum* mutant with altered esterifying pigments: characterization by optical spectroscopy and cryo-electron microscopy. J Biol Chem 278: 23678–85

Qian P, Hunter CN, Bullough PA (2005) The 8.5Å projection structure of the core RC-LH1-PufX dimer of *Rhodobacter sphaeroides*. J Mol Biol 349: 948–60

Qian P, Bullough PA, Hunter CN (2008) Three-dimensional reconstruction of a membrane-bending complex: the RC-LH1-PufX core dimer of *Rhodobacter sphaeroides*. J Biol Chem 283: 14002–11

Qian P, Papiz MZ, Jackson PJ, Brindley AA, Ng IW, Olsen JD, Dickman MJ, Bullough PA, Hunter CN (2013) Three-dimensional structure of the *Rhodobacter sphaeroides* RC-LH1-PufX complex: dimerization and quinone channels promoted by PufX. Biochemistry 52: 7575–85

Rastogi VK, Girvin ME (1999) Structural changes linked to proton translocation by subunit c of the ATP synthase. Nature 402: 263–268

Ratcliffe EC, Tunnicliffe RB, Ng IW, Adams PG, Qian P, Holden-Dye K, Jones MR, Williamson MP, Hunter CN (2011) Experimental evidence that the membrane-spanning helix of PufX adopts a bent conformation that facilitates dimerisation of the *Rhodobacter sphaeroides* RC-LH1 complex through N-terminal interactions. Biochim Biophys Acta 1807: 95–107

Recchia PA, Davis CM, Lilburn TG, Beatty JT, Parkes-Loach PS, Hunter CN, Loach PA (1998) Isolation of the PufX protein from *Rhodobacter capsulatus* and *Rhodobacter sphaeroides*: Evidence for its interaction with the α-polypeptide of the core light-harvesting complex. Biochemistry 37: 11055–11063

Roszak AW, Howard TD, Southall J, Gardiner AT, Law CJ, Isaacs NW, Cogdell RJ (2003) Crystal structure of the RC-LH1 core complex from *Rhodopseudomonas palustris*. Science 302: 1969–1972

Scheuring S, Sturgis JN (2009) Atomic force microscopy of the bacterial photosynthetic apparatus: plain pictures of an elaborate machinery. Photosynth Res 102: 197–211

Scheuring S, Francia F, Busselez J, Melandri BA, Rigaud JL, Levy D (2004a) Structural role of PufX in the dimerization of the photosynthetic core complex of *Rhodobacter sphaeroides*. J Biol Chem 279: 3620–6

Scheuring S, Sturgis JN, Prima V, Bernadac A, Levy D, Rigaud JL (2004b) Watching the photosynthetic apparatus in native membranes. Proc Natl Acad Sci USA 101: 11293–7

Scheuring S, Busselez J, Levy D (2005) Structure of the dimeric PufX-containing core complex of *Rhodobacter blasticus* by in situ atomic force microscopy. J Biol Chem 280: 1426–31

Scheuring S, Goncalves RP, Prima V, Sturgis JN (2006) The photosynthetic apparatus of *Rhodopseudomonas palustris*: structures and organization. J Mol Biol 358: 83–96

Semchonok DA, Chauvin JP, Frese RN, Jungas C, Boekema EJ (2012) Structure of the dimeric RC-LH1-PufX complex from *Rhodobaca bogoriensis* investigated by electron microscopy. Philos T R Soc B 367: 3412–3419

Siebert CA, Qian P, Fotiadis D, Engel A, Hunter CN, Bullough PA (2004) Molecular architecture of photosynthetic membranes in *Rhodobacter sphaeroides*: the role of PufX. EMBO J 23 690–700

Sturgis JN, Olsen JD, Robert B, Hunter CN (1997) Functions of conserved tryptophan residues of the core light-harvesting complex of *Rhodobacter sphaeroides*. Biochemistry 36: 2772–2778

Suzuki H, Hirano Y, Kimura Y, Takaichi S, Kobayashi M, Miki K, Wang ZY (2007) Purification, characterization and crystallization of the core complex from thermophilic purple sulfur bacterium *Thermochromatium tepidum*. Biochim Biophys Acta 1767: 1057–63

Takaichi S (1999) The Photochemistry of Carotenoids. In Carotenoids and Carotogenesis in Anoxygenic Photosynthetic Bacteria, Frank HA, Young AJ, Britton G, Cogdell RJ (eds) pp 39–69.

Tsukatani Y, Matsuura K, Masuda S, Shimada K, Hiraishi A, Nagashima KVP (2004) Phylogenetic distribution of unusual triheme to tetraheme cytochrome subunit in the reaction center complex of purple photosynthetic bacteria. Photosynth Res 79: 83–91

Tsukatani Y, Yamamoto H, Harada J, Yoshitomi T, Nomata J, Kasahara M, Mizoguchi T, Fujita Y, Tamiaki H (2013) An unexpectedly branched biosynthetic pathway for bacteriochlorophyll *b* capable of absorbing near-infrared light. Sci Rep-Uk 3

Tunnicliffe RB, Ratcliffe EC, Hunter CN, Williamson MP (2006) The solution structure of the PufX polypeptide from *Rhodobacter sphaeroides*. FEBS Lett 580: 6967–71

Vangrondelle R, Dekker JP, Gillbro T, Sundstrom V (1994) Energy-transfer and trapping in photosynthesis. Biochim Biophys Acta 1187: 1–65

Wang ZY, Muraoka Y, Shimonaga M, Kobayashi M, Nozawa T (2002) Selective Detection and Assignment of the Solution NMR Signals of Bacteriochlorophyll *a* in a Reconstituted Subunit of a Light-Harvesting Complex. Journal of the American Chemical Society 124: 1072–1078

Wang ZY, Suzuki H, Kobayashi M, Nozawa T (2007) Solution Structure of the Rhodobacter sphaeroides PufX Membrane Protein: Implications for the Quinone Exchange and Protein-Protein Interactions. Biochemistry 46: 3635–3642

Westerhuis WH, Sturgis JN, Ratcliffe EC, Hunter CN, Niederman RA (2002) Isolation, size estimates, and spectral heterogeneity of an oligomeric series of light-harvesting 1 complexes from *Rhodobacter sphaeroides*. Biochemistry 41: 8698–8707

Youvan DC, Alberti M, Begusch H, Bylina EJ, Hearst JE (1984a) Reaction center and light-harvesting I genes from *Rhodopseudomonas capsulata*. Proc Natl Acad Sci USA 81: 189–192

Youvan DC, Bylina EJ, Alberti M, Begusch H, Hearst JE (1984b) Nucleotide and deduced polypeptide sequences of the photosynthetic reaction center, B870 antenna and flanking polypeptides from *Rhodopseudomonas capsulata*. Cell 37: 949–957

Zhu YS, Kiley PJ, Donohue TJ, Kaplan S (1986) Origin of the mRNA stoichiometry of the puf operon in *Rhodobacter sphaeroides*. J Biol Chem 261: 10366–10374

Zuber H (1985) Structure and function of light-harvesting complexes and their polypeptides. Photochem Photobiol 42: 821–844

Zuber H, Cogdell RJ (1995) Structure and organization of purple bacterial antenna complexes. In Anoxygenic Photosynthetic Bacteria, Blankenship RE, Madigan MT, Bauer CE (eds) pp 315–348. The Netherlands: Kluwer Academic Publishers

Recombinant Light Harvesting Complexes: Views and Perspectives

Erica Belgio and Alexander V. Ruban

Summary

This review introduces to the method of *in vitro* reconstitution of pigment-protein complexes of higher plants, a technique which allows for the assembly of functional antenna proteins starting from free pigments and bacterially-expressed apoprotein. After discussing the reconstitution method itself, the key elements required for it (xanthophylls and chlorophyll *b*) and the timescales of the process, a few examples of the achievements made by using recombinant proteins are presented. Site-directed mutagenesis of chlorophyll-binding residues of recombinant complexes provided an important contribution to the field of photosynthesis by allowing the identification of the transition energy levels of individual chromophores. Progress has also been made employing recombinant antenna complexes in photovoltaic applications (quantum dots and Ti_2O catalyst), a recent and still largely unexplored field of research. Finally, the recent use of luminal loop mutants of LHCII for the study of the non-photochemical quenching (NPQ) mechanism, one of the most studied phenomena in photosynthesis, revealed insights into how NPQ is triggered by low pH. It is proposed that reconstituting the NPQ locus *in vitro* in liposomes with a natural thylakoid membrane lipid composition, containing purified/recombinant LHCII, minor antennae and PsbS in various combinations and concentrations may clarify how the NPQ mechanism works at a molecular level

Keywords

Pigment-protein reconstitution • Site-directed mutagenesis • Chlorophyll binding site • Mixed sites • Non-photochemical quenching • Fluorescence • Aggregation quenching • Molecular switch • Proteoliposomes • Artificial photosynthesis

Contents

3.1 **Pigment-Protein Reconstitution: The Technique** 34
3.2 **Achievements Obtained by Using Recombinant Complexes** 37

E. Belgio (✉)
School of Biological and Chemical Sciences, Queen Mary University of London, London, UK

Department of Autotrophic Microorganisms – ALGATECH, Institute of Microbiology ASCR, Třeboň, Czech Republic
e-mail: belgio@alga.cz

A.V. Ruban
School of Biological and Chemical Sciences, Queen Mary University of London, London, UK

© Springer International Publishing AG 2017
H.J.M. Hou et al. (eds.), *Photosynthesis: Structures, Mechanisms, and Applications*,
DOI 10.1007/978-3-319-48873-8_3

3.3 **Recombinant LHCII Proteins for the Study of NPQ** 39

3.4 **Future of NPQ Research** 42

3.5 **Biology Serving Nanotechnology: Applications of Reconstitution Technique** ... 44

3.6 **Concluding Summary** 46

References 46

Abbreviations

CP	chlorophyll binding protein
D	aspartate
DCCD	dicyclohexylcarbodiimide
DM	dodecylmaltoside
E	glutamate
FFEM	freeze-fracture electron microscopy
G	glycine
H	histidine
HOMO	highest occupied molecular orbital
HPLC	high pressure liquid chromatography
IPTG	Isopropil-β-D-1-tiogalattopiranoside
LHCII	light harvesting complex II
LUMO	lowest unoccupied molecular orbital
Ni-NTA	Nickel- Nitrilotriacetic acid
NMR	Nuclear magnetic resonance
NPQ	non-photochemical quenching
OGP	Octyl β-D-glucopyranoside
PSII	photosystem II
Q	glutamine
QD	quantum dot
R	arginine
RT	room temperature
SDS	sodium-dodecyl sulphate
V	valine
W	tryptophan
WT	wild type
Y	tyrosine

3.1 Pigment-Protein Reconstitution: The Technique

When Plumley and Schmidt (Plumley and Schmidt 1994) published a paper entitled "Reconstitution of chlorophyll *a/b* light-harvesting complexes: xanthophyll-dependent assembly and energy transfer", they provided for the first time a method for the *in vitro* assembly of photosynthetic proteins starting from denatured thylakoid membranes and extracted pigments. The main steps of the procedure consisted of: (1) poaching purified thylakoid proteins; (2) solubilisation of extracted pigments with ethanol; (3) protein incubation with pigments (4) series of freeze (-20 °C) and thaw (RT) cycles. Since the procedure used total thylakoid protein extracts, it did not allow for the selective reconstitution of specific complexes and therefore a further step was required in order to separate the mixture of different reconstituted complexes (major and minor LHCIIs). This was done by non-denaturing LiDodSO$_4$/PAGE, which unavoidably induced a decrease in the yield and/or damage of the complexes. For this reason, the application of genetic engineering to the method was highly advantageous, because it allowed for the production of a specific recombinant Lhcb apoprotein which was readily usable in the reconstitution process (Paulsen et al. 1990). Briefly, *E. coli* bacterial strain JM101 was transformed with the pea CAB gene AB80 (Cashmore 1994) trimmed (BamHI/DraI) and ligated into a pQE52 expression vector (pDS series, Quiagen). Upon induction with IPTG, a protein of the expected molecular weight (~31 kDa) reacting positively with antibodies raised against LHCII, was produced. The apoprotein thus obtained was successfully used for pigment-protein reconstitution following the same procedure of Plumley and Schmidt, yielding a reconstituted LHCII complex similar to the native complex (see below). This method is currently broadly adopted for the reconstitution of various photosynthetic complexes. A schematic diagram of the procedure is presented in Fig. 3.1. In terms of yields, Paulsen et al. (Plumley and Schmidt 1994) report that the amount of LHCII apoprotein is usually no more than 20% of the total *E. coli* protein extract. Of this, only ~30% assembles into a folded pigment-protein complex (Remelli et al. 1999), revealing the rather low yield of the process. It should be mentioned, however, that replacement of the pQE52 expression vector by pET-3d was shown to double apoprotein production from BL21(DE3)

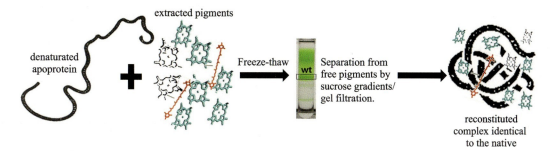

Fig. 3.1 Diagram of the pigment-protein reconstitution procedure

bacterial strain (Qinmiao et al. 2000). This group also reported an increased efficiency of reconstitution and a shortening of the procedure time when freeze ($-20\,°C$)/thaw cycles are performed using liquid nitrogen.

Regarding pigment-protein complex stability, although $LiDodSO_4$ is fundamental for keeping the apoprotein fully denatured, once reconstitution is achieved (i.e. after freeze-thaw cycles), it is important to replace it with OGP via KCl precipitation (Giuffra et al. 1996). This is to avoid chlorophyll detachment from the reconstituted complex induced by the harshness of $LiDodSO_4$. The following ratios in the reconstitution mixture have been reported to give maximal reconstitution yields for LHCII: chlorophyll/protein $= 20$, carotenoid/protein $= 7$, chlorophyll $a/b = 2.3$ (Croce et al. 1999).

Insights into the folding process have been obtained from time-resolved circular dichroism spectroscopy in the far-UV region (Horn and Paulsen 2004). Different time steps have been identified that can be divided into *fast*, *intermediate* and *slow* components. The *fast* component (timescale, 10 ms) is independent of pigment and protein concentration and involves protein rearrangements in the lipid/detergent micelles. The *intermediate* component lasts a few seconds and involves pigment binding to the protein. The *slow* component lasts up to several minutes and corresponds to the establishment of efficient energy transfer from chlorophyll *b* to chlorophyll *a* along with the formation of a folded state similar to the native complex.

The rationale behind the *in vitro* folding process still remains partially unknown. For example, it is puzzling why the *in vitro* folding capacity appears to be a prerogative of the outer antenna complexes of PSII – LHCII, CP29, CP26, CP24 whilst similar PSII antenna complexes like the inner antennae CP43 and CP47, despite many repeated attempts, could not be reconstituted (Casazza et al. 1797). It is interesting to notice in this context, that all the complexes which have not been reconstituted so far (reaction centers and inner antennae), lack chlorophyll *b*. Chlorophyll *b* is indeed thought to promote the reconstitution process (Horn and Paulsen 2004; Storf et al. 2005). What is also known is that: (1) the protein must be completely denatured when it comes into contact with the pigments – this is because the chlorophyll binding sites must be exposed to the pigments in order to be bound; (2) chlorophyll *b* and xanthophylls are indispensable to the folding (see below); (3) freeze and thaw cycles provide the small "energy kicks" required for the protein to travel along the imaginary potential energy hyper-surface of all possible protein conformations of which the global energy minimum corresponds to the folded native conformation. Finally, since the α-helix content in the protein secondary structure was found to increase with time constants similar to those observed for the establishment of efficient energy transfer, it was concluded that pigment binding is the trigger for protein folding.

From the very first report (Plumley and Schmidt 1994) it was clear that xanthophylls are required in order for chlorophyll *a* and *b* to associate with the apoprotein. When β-carotene was used instead of xanthophylls, no reassembly was observed. Rearranging the epoxide bonds of xanthophylls with HCl (Goodwin 1980) eliminated reconstitution. These evidences pointed at a specific effect induced by xanthophylls which promotes the folding process. Not all parts of the protein are important for reconstitution. Removing the N-terminus of LHCII, for example, did not affect the reconstitution result, in terms of yield and pigment complement, proving that this portion of the protein is not involved in reconstitution or pigment binding (Plumley and Schmidt 1994). Pigment binding to the protein can be demonstrated in several ways, all assaying the efficiency of excitation energy transfer among chlorophylls. Absorption, fluorescence, excitation spectroscopy and circular dichroism are all useful means for verifying that the configurations/orientations of the pigments resemble those of the native complex. As an example, the absorption spectrum of unbound chlorophyll *a* and *b* is shown in comparison to that of a native and reconstituted LHCII complex (Fig. 3.2). The similarity between the spectra of reconstituted and native complex is evident. In particular, one can observe that chlorophyll binding to the protein induces a red shift of the main absorption spectrum peak from ~663 nm to 670 nm.

This is because the spectral characteristics of protein-bound chlorophyll are modulated by interactions with their environment (Renge and Avarmaa 1985; Gudowska-Nowak et al. 1990; Giuffra et al. 1997; Rogl and Kühlbrandt 1999; van Amerongen and van Grondelle 2001; Nishigaki et al. 2001), giving rise to spectroscopically different chlorophyll forms. The origin of chlorophyll spectral forms, besides chlorophyll-chlorophyll excitonic interactions, has been also explained as due to distortions of the chlorophyll ring imposed by the protein binding, which has the effect of modifying the chlorophyll transition energies (Zucchelli et al. 2007). In the crystal structure of LHCII, all chlorophylls (*a* and *b*) display distortions of the ring. The LHCII chlorophyll macrocycle deformations were shown to shorten the HOMO-LUMO energy gaps, inducing an estimated red shift of up to 17 nm for chlorophyll *a*, and 11 nm for chlorophyll *b*, with respect to the unperturbed reference transition energies (Zucchelli et al. 2007). Other groups, based on *ab initio* calculations, considered solvatochromic effects and excitonic interactions as the main source for the observed site energies shifts (Müh et al. 2010). In both scenarios, the similarity of the spectrum of the reconstituted sample to the native one, in terms of both position of the peaks and *a/b* ratio, is therefore a good indication of correct binding of the chlorophylls to the protein scaffold.

Low reconstitution yield preparations containingloosely bound pigments usually display a 1–2 nm blue shift, and a decreased dichroic signal between 670 and 683 nm (Plumley and Schmidt 1994). Unbound pigments have in fact a weak dichroic signal. Since free pigments and aggregates of pigments cannot efficiently transfer excitation energy, another way to confirm that reconstitution has been achieved is to verify that most of the emission comes from chlorophyll *a* (~680 nm) when chlorophyll *b* is selectively excited (470 nm). These spectroscopic measurements altogether prove that the

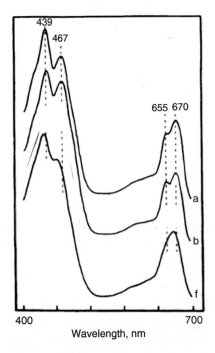

Fig. 3.2 Room temperature absorption spectra of native (*curve a*) and reconstituted (*curve b*) LHCIIs. *Curve f*, the sample presented in *b*) was heated at 100 °C for 1 min (Modified from Plumley and Schmidt 1994)

reconstitution procedure induces a pigment-binding pattern to the protein similar to the one which is typical of the native complex.

3.2 Achievements Obtained by Using Recombinant Complexes

The possibility of *in vitro* assembly of pigment-protein complexes paved the way to a whole new range of investigations impossible before. The technique allowed for the study of the theoretical foundations of the process, like how the reconstitution process works, what key elements are required for it, the timescales and, most importantly, the fundamental problem of determining the spectrum of proteinbound chlorophylls. From a purely experimental perspective, it provided a new tool for protein engineering in the context of applied research. The present paragraph will focus on the achievements made in the field of pure research in photosynthesis.

Binary competition experiments using two xanthophylls at varying ratios proved that three specific xanthophyll binding sites are present in LHCII (N1, L1 and L2) (Croce et al. 1999; Hobe et al. 2000). For example, in a lutein/neoxanthin competition experiment two binding sites, L1 and L2, showed a strong preference (> 200-fold) for lutein, whereas the third binding site, N1, was highly selective for neoxanthin and its occupancy was not essential for protein folding. These findings have been confirmed by the high resolution (2.72 Å) crystal structure of LHCII (Liu et al. 2004). Of all xanthophylls, lutein was the one strictly necessary for obtaining a minimum detectable level of reconstitution, which was then further increased by violaxantin and, lastly, by neoxanthin. Experiments of heat denaturation showed that a reconstituted LHCII sample in which the N1 site was occupied and one of the L sites was empty, denatured at lower temperature with respect to the control, in which the three xanthophyll-binding sites were occupied. Thus, both L sites seem to contribute to pigment-protein stability (Croce et al. 1999).

Insights on how the mechanism of trimerization of LHCII occurs have been obtained in (Rogl and Kühlbrandt 1999; van Amerongen and van Grondelle 2001). By analysing specific circular dichroism signals of recombinant LHCII complexes in the visible range, it is apparent that trimerization occurs spontaneously and is dependent on the presence of lipids. The same group found that the C-terminal deletions did not abolish pigment binding to LHCII nor affect trimerization (Hobe et al. 1995). By contrast, deletion of 61 amino acids from the N-terminus had no significant effects on pigment binding although 15 amino acids were found to be indispensable for formation of a trimer. This indicated that the protein motif between amino acids 16–61 is involved in the stabilization of LHCII trimers but not in that of the monomers. Closer inspection of this protein domain using a more detailed mutation analysis revealed that amino acids W16 and/or Y17 as well as R21 are essential for the formation of LHCII trimers. These amino acids are conserved in virtually all known sequences of LHCII apoproteins, but only in some of the minor chlorophyll *a/b* complexes (Hobe et al. 1995). This point could be of potential interest for projects aimed at increasing the stability of LHCII (see last session of the chapter).

Perhaps the most important contribution to the understanding of the structure and energetics of the photosynthetic antenna was the identification of chlorophyll binding sites and energies (Remelli et al. 1999; Bassi et al. 1999; Morosinotto et al. 2002; Yang et al. 1999; Belgio et al. 2010). In those works, a series of mutant apoproteins were constructed in which individual chlorophyll-binding residues, derived on the basis of the crystal structure of LHCII (Giuffra et al. 1997), were substituted by residues unable to coordinate porphyrins. This approach in most of the cases enabled the identification of the types of chlorophylls (*a* or *b*) bound by each protein site to besides deducing their spectral properties by comparison with the wild type complex. For example, the A2 chlorophyll of CP29, bound to hystidine 216, was found to be one of the lowest energy chlorophylls of CP29,

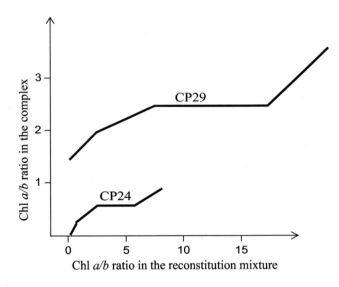

Fig. 3.3 Diagram showing the dependence of the chlorophyll *a/b* ratio in reconstituted complexes CP29 and CP24 upon the *a/b* ratio of the reconstitution mixture (Modified from Sandonà et al. 1998)

peaking at 680 nm (Bassi et al. 1999; Belgio et al. 2010). Mutagenesis of certain sites induced simultaneous changes in both chlorophyll *a* and *b* regions. This fact was consistent with the previous observation that the relative amounts of chlorophyll *a* and chlorophyll *b* bound to LHCII could be modified by varying the chlorophyll *a/b* pigment ratio in the reconstitution mixture (see the diagrams for CP29 and CP24 in Fig. 3.3). An extreme example of this was an LHCII complex with an *a/b* ratio of 0.03 obtained using a very low chlorophyll *a/b* ratio in the reconstitution mixture (Kleima et al. 1999).

It was therefore proposed that sites like the A3, B3, B5, and B6 of CP29 are mixed with respect to their capacity to bind both chlorophyll *a* and *b* (Sandonà et al. 1998). The *mixed binding sites* hypothesis, while in principle not unreasonable, has recently been criticised (Belgio et al. 2010). In the latest paper, a single gene mutation for the A3 binding site (Glutamate 230, G230) caused absorption changes in both the chlorophyll *a* and *b* spectral regions. The observed change in the chlorophyll *a* absorbing region was remarkably red-shifted, which was interpreted as due to a perturbation of the chlorin ring induced by the protein mojety around the A3 site (Zucchelli et al. 2007). However, no shift was observed in the chlorophyll *b* interval, which was unchanged with respect to the control, peaking at 640 nm. A similar result was found for the B3 site conjugated to Histidine 245, H235. These observations led to the conclusion that the A3 and B3 chlorophyll sites were not mixed, because in that case, the same distortion/electrostatic effect imposed by the local protein scaffold on chlorophyll *a* was also expected to be observed for chlorophyll *b* in terms of a red shift of its absorption peak. This conclusion has recently been proven valid on the basis of the crystal structure of CP29, where G230 and H245 were shown to be specific sites for a certain type of chlorophyll (Pan et al. 2008). Similarly, the crystallographic structure of a number of LHCII trimers showed no evidence for the existence of mixed sites. In the ten LHCII monomers examined (Liu et al. 2004) not a single mixed site was detected – though it has been suggested that this may be due to the *in vivo* chlorophyll binding conditions being different from the *in vitro* situation. It therefore seems reasonable to conclude that while chlorophyll *b* may in principle bind to chlorophyll *a* sites when the reconstitution conditions are "forced", the binding affinity of chlorophyll site ensures selection for a certain residue under physiological conditions.

Concerning the limits of the method, one is that the reconstitution yield is generally low (~ 50 μg of protein per litre of *E. coli*), – far below that for extraction of native proteins (usually of the order of milligrams). Another problem is determining whether the reconstituted complex is absolutely identical to the purified one. This is mainly due to the lack of a reliable method to quantify the absolute amount of pigment and protein in a preparation. Pigments can easily be extracted with 80% acetone, which denatures and precipitates the protein, whereas the non-covalently bound pigments end up in the supernatant (Paulsen et al. 2010) and can subsequently be separated, analyzed and quantified either spectro-photometrically, using, for example, Porra's method (Porra et al. 1989), or by high performance liquid chromatography (HPLC). HPLC analysis is normally integrated by matching the absorption spectrum of 96% ethanol pigment extracts with the composite spectra made up of individual purified pigments (Connelly et al. 1997). SDS-page and the ninhydrin reaction are classic methods for measuring protein concentration (Hirs 1967). Quantification of proteins in detergent solution however is often difficult, which in turn renders chlorophyll/protein ratios unreliable. Moreover, it must be borne in mind that the unfolded protein is often a contaminant of the preparation and to some extent (3–4%) free pigments can also be present. The problem has recently been exacerbated by structural data revealing that the CP29 complex contains 13 chlorophylls (Pan et al. 2008), i.e., five chlorophylls more than those previously estimated for reconstituted complexes (Giuffra et al. 1996, 1997; Belgio et al. 2010; Pieper et al. 2000). Accordingly, the group of Jankowiak concluded that the previously studied CP29 complex from spinach (Pieper et al. 2000) had lost at least two chlorophylls during the preparation procedure, and it is unlikely that the spinach CP29 protein contains only eight chlorophylls as previously thought (Feng et al. 2013). It is therefore compelling at this stage to develop new methods for absolute pigment/protein ratios.

3.3 Recombinant LHCII Proteins for the Study of NPQ

In higher plants, high light conditions trigger the activation of non-photochemical quenching (NPQ), a process of light energy dissipation involving structural rearrangements within the peripheral PSII antenna complex LHCII (Holzwarth et al. 2009; Betterle et al. 2009; Johnson et al. 2011). The protein rearrangements observed via freeze fracture electron microscopy (FFEM) consisted in the formation of LHCII clusters during the establishment of the NPQ state, leading to the formation of quenching aggregates (see Fig. 3.4). About 20 years of research into this mechanism proved that: (1) it is triggered by a delta pH across the photosynthetic membrane (Shikanai et al. 1999), (2) it is reversible (Krieger et al. 1992); (3) it can occur in the dark (Gilmore and Yamamoto 1992); (4) its rate depends upon certain xanthophylls and upon the concentration of the PsbS protein (Gilmore 1997).

On the basis of the spectroscopic similarities between *in vivo* NPQ and quenching induced by the aggregation of isolated major LHCII complexes under low detergent/low pH conditions (Ruban 2012; Ruban et al. 1992, 1997; Phillip et al. 1996), the "LHCII aggregation model" of NPQ was proposed. However, the conditions used to achieve a highly-quenched state of LHCII *in vitro* inevitably caused pronounced aggregation of the protein, making it difficult to disentangle whether aggregation is the cause or consequence of quenching. Therefore, determining whether control of the quenching mode occurs at the level of interactions between neighbouring proteins rather than at the level of single LHCIIs, has been a matter of debate for some time. The second case implies the existence of a conformational switch within the LHCII monomer upon induction of the NPQ state, as shown in Fig. 3.5.

To investigate this point, a system to prevent protein-protein aggregation has recently been implemented (Belgio et al. 2013) by exploiting

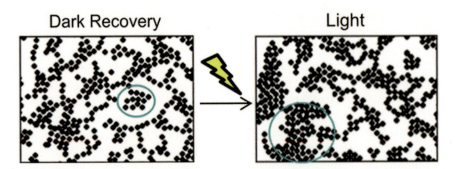

Fig. 3.4 Representative coordinate maps (obtained via FFEM analysis) showing the spatial distribution of LHCII trimers within the grana membrane following light exposure and subsequent dark recovery. The *blue circles* highlight typical aggregated domains of LHCII, indicating the significant increase in *in vivo* LHCII aggregation during the switch to the photoprotective state (Figure adapted from Ruban 2012)

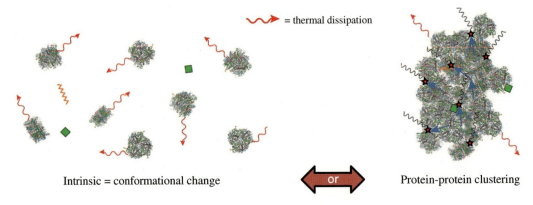

Fig. 3.5 Schematic diagram representing the NPQ hypothesis: in one case NPQ is due to a conformational change within a single LHCII, in the other it is induced by protein-protein aggregation

a recombinant LHCII protein His-tag affinity to the Nickel of a Ni-NTA resin (see Fig. 3.6).

After incubation of the protein with the resin at a concentration which excluded protein-protein aggregation (the smallest complex-to-complex distance being >100 nm), the mix was loaded into a column and fluorescence quenching induced by washing with the appropriate buffer (Fig. 3.7). In the same way, washing the column with a high detergent/high pH buffer restored the "harvesting" state of the complex. The reversibility of the process (see Fig. 3.7) indicated that the assumption of negligible aggregation was realistic. The experiment confirmed the intrinsic capacity of the protein to adopt a quenched conformation, giving credit to the idea that aggregation is a consequence rather than a cause of quenching. This is in agreement with experiments performed on LHCII immobilised in a gel matrix (Ilioaia et al. 2008) and with a number of observation indicating that conformational changes do occur within LHCII (van Oort et al. 2007; Krüger et al. 2012). In addition to its compatibility with these previous experiments the "in column quenching set up" went further, allowing for the study of the "single complex" pH-sensitivity in the absence of aggregation (Belgio et al. 2013). Moreover, the technique offered the possibility to immobilise LHCII to a substrate, which has possible future applications connected to the field of artificial photosynthesis (see last section).

Understanding the mechanism by which pH triggers NPQ at the molecular level is still one of the main goals of the field. In particular, it is still unclear whether certain lumen-exposed residues of LHCII can effectively be protonated upon illumination, when the lumen pH drops from

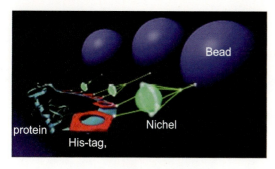 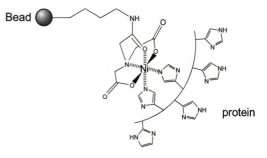

Fig. 3.6 Diagram showing how the recombinant LHCII complex was immobilised to a substrate (Ni-NTA conjugated agarose bead) to avoid aggregation, by exploiting the affinity of the protein His-tag to the Nickel atoms of the substrate. The atoms involved in the interactions are also shown on the *right* (Modified from Ni-NTA Magnetic Agarose Beads Handbook from Quiagen)

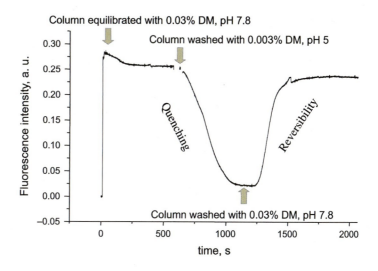

Fig. 3.7 Diagram representing the "*in column* quenching experiment" of LHCII. Quenching was induced by washing the LHCII bound to the column with 0.003% DM and 10 mM Hepes adjusted to the desired pH (Modified from Belgio et al. 2013)

pH 7 to 5. Evidence for LHCII antenna complexes as protonation sites has been obtained by using the carboxyl-modifying agent dicyclohexylcarbodiimide (DCCD), which binds to aspartic acid and glutamate residues in relatively hydrophobic domains of membrane proteins (Walters et al. 1994). Due to the build-up and decay of a pH gradient during photosynthesis the lumenal loop of LHCII is exposed to a broadly variable pH environment. Hence, the negatively charged amino acids in the lumenal loop might play important roles in adjusting the structure and functions of LHCII. In the luminal loop of LHCII, between transmembrane helices B/C there are three negatively charged amino acids (E94, E107 and D111, see Fig. 3.8), two of which (E94, D111) form ion pairs (E94–Q103, D111–H120). They are thought to be important for stabilising the secondary structure loop of lumen-exposed regions (Liu et al. 2004; Yang et al. 1777).

Recently, two recombinant LHCII complexes mutated at the level of the lumenal loop showed an altered pH-induced quenching capacity with respect to the control (Belgio et al. 2013; Yang et al. 1777). In one case, replacement of the acidic lumenal-facing residue aspartate 111 (D111) by neutral valine (V111) yielded a recombinant complex with increased quenching capacity due to a shift of the quenching pK by 1 pH unit from 5 to 6. The increase in total quenching was consistent with a 40% reduction

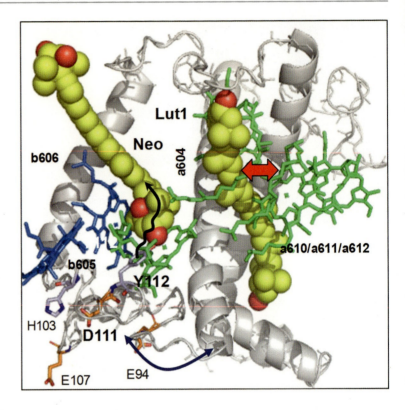

Fig. 3.8 Putative structural domains of LHCII involved in photoprotective energy dissipation. It has been suggested that the conformational change induced during NPQ (Ruban et al. 2007) involves a twist in the neoxanthin molecule and is suggested to be sensed also by the central a-helices (*blue arrow*) bringing about a quenching interaction between Lut1 and Chl a610–612 (*red arrow*) (Modified from Belgio et al. 2013)

in the relative chlorophyll fluorescence lifetime and was accompanied by the appearance of a lower-energy emitting state, as demonstrated by a red shift of the low temperature emission spectrum in the mutant. On the other hand, replacement of acidic glutamate 94 (E94) with glycine (G94) resulted in a decreased quenching capacity attained at low pH, despite the virtually unchanged pigment composition of the mutant (Fig. 3.9).

The fact that a subtle change in the apoprotein structure at the level of the lumenal loop affected the protein sensitivity to pH, was interpreted as a strong indication of involvement of the luminal loop in LHCII protonation. The potential of such studies is evident. Finding LHCII reconstituted complexes with a higher or less efficient quenching capability than the wild type or even complexes that are totally incapable of quenching, may in the future allow to find which protein domains/pigments are responsible for the photoprotective state, something like the Holy Grail on NPQ research. Interesting mutants, like E94G or D111V mentioned above, that reduce or increase the efficiency of NPQ, can in future be objects of NMR studies. This technique proved to be valid for determining local structural and electronic perturbations of the protein backbone. Recently, the first solid-state NMR analysis of uniformly ^{13}C–enriched major light harvesting complexes from *Chlamydomonas reinhardtii* and identification of protein and cofactor spin clusters have been performed by Pandit and co-workers (Pandit et al. 1827). An NMR comparison of the wild type and mutant proteins will be a much desirable future work that may allow an understanding of the structural basis for the modified quenching behaviour of the recombinant proteins.

3.4 Future of NPQ Research

Despite the many *in vivo* and *in vitro* studies aimed at investigating the mechanism of NPQ, a full agreement on how it works at the molecular

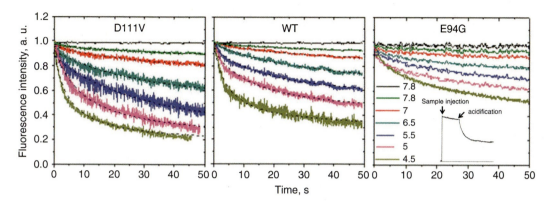

Fig. 3.9 Representative fluorescence time course traces of WT, D111V and E94G recombinant complexes in function of pH. Samples were re-suspended in 0.03% DM (*top trace*) or 0.003% DM (all other traces), 10 mM Hepes, 10 mM sodium citrate HCl-buffered to the pH indicated in figure legend on the right side. Data were normalised to 1 just prior to the acid addition. A sketch of the experimental procedure is reported also on the *right* (Modified from Belgio et al. 2013)

level has not yet been achieved. The general consensus is that LHCII, PsbS and xanthophylls are involved. However, whilst for some groups these factors participate at different stages of the process and with different kinetics, generating two main independent NPQ mechanisms (Holzwarth et al. 2009), for others the only site of quenching is LHCII, with PsbS and xanthophylls being only modulators/enhancers of the process (Ruban et al. 1817, 2007; Horton et al. 1991; Pascal et al. 2005). The latter "LHCII aggregation model" was mainly based on the spectroscopic similarities between *in vivo* NPQ and quenching induced by aggregation of isolated major LHCII complexes under low detergent/ low buffer conditions (Ruban 2012; Ruban et al. 1992, 1997; Phillip et al. 1996). Indeed, both NPQ and aggregation of all LHC complexes, including those of the minor antenna, are accompanied by formation of red-shifted fluorescence emission bands indicative of chlorophyll–chlorophyll interactions (Ruban et al. 1991, 1993, 1996; Miloslavina et al. 2008; Johnson and Ruban 2009; Ballottari et al. 2009; Belgio et al. 2012). It must be mentioned however that the experiments conducted *in vitro* on solubilised LHCII cannot be considered as truly representative of the *in vivo* situation. This is because the fluorescent (light-harvesting) state of LHCII *in vivo* (lifetime ~2 ns) is different from the detergent-solubilised one (4 ns). Such a long emitting state has never been observed *in vivo* and is therefore considered to be an artefact induced by both the unnatural detergent-aqueous environment LHCII is exposed to and a pre-aggregated state of the complex in the photosynthetic membrane (Belgio et al. 2012; Petrou et al. 2014). Moreover, it is not feasible to induce a ΔpH across the detergent micelle system. To reproduce the NPQ locus, it would be more desirable to set up a system in which LHCII is solubilised by a lipid membrane and in which the luminal loop can be exposed to an acid environment, mimicking what happens in chloroplasts under light. Reconstituting the process in liposomes with a natural thylakoid membrane lipid composition, containing purified/recombinant LHCII, minor antennae and PsbS in various combinations and concentrations in the presence and absence of zeaxanthin and ΔpH could in principle allow for the reproduction of NPQ and perhaps the determination of which conditions and factors are indispensable for attaining a ΔpH-dependent, reversible NPQ mechanism.

Recent pioneers in this context are Kühlbrandt and co-workers who created a proteoliposome system for the study of NPQ (Wilk et al. 2013). This system showed that the chlorophyll fluorescence of reconstituted LHCII can be quenched by the presence of both PsbS and the xantophyll zeaxanthin. Two-photon excitation measurements revealed new electronic interactions between the carotenoid S_1 state and chlorophyll states that correlated directly with chlorophyll fluorescence quenching. The authors therefore proposed a carotenoid-dependent model of NPQ based upon the direct interactions of LHCs with PsbS monomers. The reported approach gives way to new potential experiments that could test how the NPQ mechanism works. It will be worth assaying, for example, the effect of ΔpH on the fluorescence state of LHCII solubilised in such a proteo-liposome system and combining it with the use of the reconstitution and mutagenesis approaches mentioned above.

3.5 Biology Serving Nanotechnology: Applications of Reconstitution Technique

The benefits of using recombinantion and reconstitution techniques go beyond pure research. Recently, these approaches also delivered some progress in the applied field. Three main examples of this will be reported here: (1) a hybrid structure made up of a recombinant LHCII protein and inorganic type II quantum dots (QDs) (see Fig. 3.10); (2) a LHCII-TiO_2 hybrid for methane/hydrogen production; (3) silica encapsulation and stabilization of a recombinant LHCII.

Semiconductor nanocrystal quantum dots (QDs) are a class of nanomaterial broadly used in photovoltaic applications due to their stability and efficient excitation energy transfer both as electron acceptors and donors (Nabiev et al. 2010). QDs of type II are capable of light-driven electron-hole separation. Light harvesting complexes offer a valid antenna system to be attached to QDs because: (i) they are remarkably efficient in transferring the absorbed energy (> 80%), a process which occurs with virtually no energy loss apart from the Stokes shift (Jennings et al. 1709); (ii) they can be engineered to contain anchors such as cysteine sulfhydryls or hexahistidyl (His6x) tags for labelling or attachment to surfaces. Recently, a paper has been published concerning a synthesised hybrid structure made of recombinant LHCII protein and inorganic type II QDs (Werwie et al. 2012). The absorption and fluorescence spectroscopy demonstrated effective energy transfer (~50%) from LHCII to the QDs (Fig. 3.11).

Binding of LHCII to quantum dots demonstrated an increase in the light-energy utilization of the semiconductor nanocrystals, particularly in the red spectral domain where QD absorption is relatively low. The increase was further enhanced by attaching "green absorbing" dyes to LHCII which filled the chlorophyll absorption gap. Since these dyes were bound to the protein His-tag, the use of recombinant LHCIIs was instrumental to this project.

A similar example, still a work in progress, is the construction of a (TiO_2)-LHCII hybrid. TiO_2

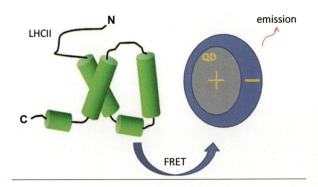

Fig. 3.10 Sketch showing LHCII recombinant complex acting as an antenna system for the attached quantum dot (QD). The quantum dot in turns performs electron-hole separation (Modified from Werwie et al. 2012)

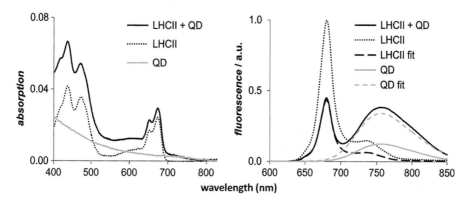

Fig. 3.11 LHCII transfers energy to QDs. Absorption (*left*) and emission (*right*) spectra of LHCII-QDs hybrid (*solid black line*) compared to QDs only (*gray line*) and pure LHCII (*dotted black line*). For didactic porpoise the LHCII and QDs components contributing to the spectrum of the hybrid complex emission spectrum are also shown (*dashed lines*) (Modified from Werwie et al. 2012)

is commonly used to oxidise organic molecules in water treatment as well as in selective oxidation reactions. TiO_2 can reduce CO_2 to CO, formic acid, formaldehyde, methanol and methane (Inoue et al. 1979; Yamashita et al. 1994; Anpo et al. 1995; Kaneco et al. 1998; Wu 2009). As any catalyst, it can perform both the back and forward reactions, meaning that it can reduce as well as oxidise organic compounds. Its involvement in CO_2 photo-reduction is very interesting in this context. Though TiO_2 is efficient relative to other photocatalysts, it is still very inefficient relative to the amounts of energy required for the process; in fact it absorbs only UV light, which is a small portion of the solar radiation profile. For this reason it is often bound to organic dyes, to shift the absorption spectrum to the visible but these are used up by the process because they pass electrons to the catalyst, instead of excitation energy, and regeneration via the extraction of dye from water is not possible. A recent project from a group in Nottingham (Lee et al. 2013) attempted the binding of LHCII to the catalyst to provide a visible range absorbing antenna for the TiO_2. The results demonstrated effective energy transfer from LHCII to the catalyst. The hybrid catalyst produced, under the visible light regime, 150% times more methane than that produced by the catalyst alone. This proved that the interaction between the LHC and the catalyst was functional and that LHC passed the visible light energy to the catalyst.

The effective implementation of the LHCII-TiO_2 hybrid requires an increased stability of LHCII to sustain days of prolonged illumination and/or temperatures above 10–15 °C. One method could be site-directed mutagenesis/chemical treatment of key-stabilising residues. In this context a number of studies provided useful insights (Mick et al. 2004). This mutagenesis work showed that the stromal loop domain has a significant impact on LHCII formation and/or stability *in vitro*. In particular, mutagenesis of W97 in the N-proximal section of the loop gave low reconstitution yields, even under very mild conditions. W97 may therefore be a target of future research aimed at increasing the stability of LHCII by covalent binding, perhaps by means of cross-likers like glutaraldehyde. In a similar way, cross-linking of residues involved in the trimerisation of LHCII (W16, Y17 as well as R21, see second paragraph of this chapter) could in principle yield an enhanced stability of the trimer.

A different, novel approach aimed at increased LHCII stability (Hobe and Röder 2013) consisted of the encapsulation and stabilization of the protein by means of spermine promoted condensation. The encapsulation dramatically increased stability: from a few minutes of resistance at 50 °C for the non-encapsulated complex, up to 24 hours at the same temperature for the encapsulated one. LHCII within silica showed efficient inter-complex energy transfer, supporting the idea of an intact light-harvesting and energy delivering organic-inorganic hybrid material. As it is evident from this range of examples, recombinant proteins brought about wide progress, especially in pure photosynthesis research. If, in the future, researchers will manage to overcome the two limits of the method, i.e. a relatively low yield and lack of stability under prolonged light regimes, this will render a robust antenna system available to the scientific community, a system to be broadly employed in fields of applied research such as building new photovoltaic systems.

3.6 Concluding Summary

A range of examples has been presented in order to show that recombinant complexes provided important contributions to various fields of photosynthetic research. The review has been divided into six sections, here summarised:

1. "Pigment-protein reconstitution: the technique", where the reconstitution method is presented with technical details;
2. "Achievements obtained by using recombinant complexes", describing how site-directed mutagenesis of recombinant complexes for the first time allowed the spectral properties of protein bound pigments, like the A2 chlorophyll of CP29, to be obtained.
3. "Recombinant LHCII proteins for the study of NPQ". The case of two recombinant LHCII mutant complexes with altered quenching capacity (D111V, E94G) is here reported. This study provided insights into how pH could trigger NPQ *in vivo*, a hot topic in photosynthesis.
4. "Future scenarios for NPQ research". An ambitious project, aimed at reproducing the NPQ *locus*, is presented. The new system, consisting of proteoliposomes containing recombinant LHCIIs, by attaining a ΔpH-dependent, reversible *in vitro* NPQ, could in future clarify the molecular details of the mechanism.
5. "Biology serving nanotechnology: applications of the reconstitution technique". This conclusive paragraph presents two examples in which recombinant LHCII have been successfully used in applications related to the photovoltaic research.

Acknowledgements This work was supported by The Leverhulme Trust Research Grant RPG-2012-478 awarded to AVR.

Conflict of Interest None Declared.

References

Anpo M, Yamashita H, Ichihashi Y, Ehara S. Photocatalytic reduction of CO_2 with H_2O on various titanium oxide catalysts. Journal of Electroanalytical Chemistry 1995; 396: 21–26.

Ballottari M, Girardon J, Betterle N, Morosinotto T, Bassi R. Identification of the chromophores involved in aggregation-dependent energy quenching of the monomeric photosystem II antenna protein Lhcb5. J Biol Chem 2009; 285: 28309–28321.

Bassi R, Croce R, Cugini D, Sandonà D. Mutational analysis of a higher plant antenna protein provides identification of chromophores bound into multiple sites. Proc Natl Acad Sci 1999; 18: 10056–10061.

Belgio E, Casazza AP, Zucchelli G, Garlaschi FM, Jennings RC. Band shape heterogeneity of the low-energy chlorophylls of CP29: absence of mixed binding sites and excitonic interactions, Biochemistry 2010; 5: 882–92.

Belgio E, Johnson MP, Juric S, Ruban AV. Higher plant photosystem II light-harvesting antenna, not the reaction center, determines the excited state lifetime -both the maximum and the nonphotochemically quenched. Biophys J 2012; 102: 2761–2771.

Belgio E, Duffy CDP, Ruban AV. Switching light harvesting complex II into photoprotective state involves the lumen-facing apoprotein loop. PCCP 2013; 15: 12253–12261.

Betterle N, Ballottari M, Zorzan S, de Bianchi S, Cazzaniga S, Dall'Osto L, Morosinotto T, Bassi R. Light-induced dissociation of an antenna heterooligomer is needed for non-photochemical quenching induction. J Biol Chem 2009; 284:15255–15266.

Casazza AP, Szczepaniak M, Müller MG, Zucchelli G, Holzwarth AR. Energy transfer processes in the isolated core antenna complexes CP43 and CP47 of photosystem II. BBA 2010; 1797: 1606–16.

Cashmore AR. Structure and expression of a pea nuclear gene encoding a chlorophyll *a/b* binding polypeptide. Proc Natl Acad Sci 1994; 81: 2960–2964.

Connelly JP, Müller MG, Bassi R, Croce R, Holzwarth A.R. Femtosecond transient absorption study of carotenoid to chlorophyll energy transfer in the light-harvesting complex II of photosystem II. Biochemistry 1997; 36: 281–287.

Croce R, Weiss S, Bassi R. Carotenoid-binding sites in the major light-harvesting complex (LHCII) of higher plants. J Biol Chem 1999; 274: 29613–29623.

Feng X, Pan X, Li M, Pieper J, Chang W, Jankowiak R. Spectroscopic study of the light-harvesting CP29 antenna complex of Photosystem II – Part I. Phys Chem B 2013; 117: 6585–92.

Gilmore AM. Mechanistic aspects of xanthophyll cycle-dependent photoprotection in higher plant chloroplasts and leaves. Physiol Plant 1997; 99: 197–209.

Gilmore AM, Yamamoto HY. Dark induction of zeaxanthin-dependent non-photochemical fluorescence quenching mediated by ATP. PNAS 1992; 89: 1899–1903.

Giuffra E, Cugini D, Croce R, Bassi R. Reconstitution and Pigment Binding Protein of Recombinant CP29. Eur J Biochem 1996; 238: 112–120.

Giuffra E, Zucchelli G, Sandonà D, Croce R, Garlaschi FM, Bassi R, Jennings RC. Analysis of some optical properties of native and reconstituted photosystem II antenna complex, CP29: pigment binding sites can be occupied by chlorophyll a or chlorophyll b and determine the spectral forms. Biochemistry 1997. 36: 12984–12993.

Goodwin TW. The Biochemistry of the Carotenoids, Chapman & Hall, New York 1980.

Gudowska-Nowak E, Newton MD, Fajer J. Conformational and environmental effects on bacteriochlorophyll optical spectra: correlations of calculated spectra with structural results. J Phys Chem 1990; 94: 5795–5801.

Hirs CHW. Reduction and S-carboxymethylation of proteins. In: Methods in Enzymology. Academic Press, New York, 1967; 11: 325–329.

Hobe S, Förster R, Klingler J, Paulsen H. N-proximal sequence motif in light-harvesting chlorophyll a/b-binding protein is essential for the trimerization of light-harvesting chlorophyll a/b complex. Biochemistry 1995; 34: 10224–10228.

Hobe S, Niemeier H, Bender A, Paulsen H. Carotenoid binding sites in LHCII – Relative affinities towards major xanthophylls of higher plants. Eur J Biochem 2000; 267: 616–624.

Hobe S, Prytulla S, Kühlbrandt W, Paulsen H. Trimerization and crystallization of reconstituted light-harvesting chlorophyll a/b complex. EMBO J 2004; 13: 3423–3429.

Hobe S, Röder S, Paulsen H. Biomimetic encapsulation and stabilization of a recombinant light-harvesting membrane protein complex. 12th International Symposium on Biomineralization 27–30 August 2013, Freiberg, Saxony/Germany.

Holzwarth AR, Miloslavina Y, Nilkens M, Jahns P. Identification of two quenching sites active in the regulation of photosynthetic light-harvesting studied by time-resolved fluorescence. Chem Phys Lett 2009; 489: 262–267.

Horn R, Paulsen H. Early steps in the assembly of light-harvesting chlorophyll a/b complex: time-resolved fluorescence measurements. J Biol Chem 2004; 279: 44400–6.

Horton P, Ruban AV, Rees D, Pascal A, Noctor GD, Young A. Control of the light-harvesting function of chloroplast membranes by the proton concentration in the thylakoid lumen: aggregation states of the LHCII complex and the role of zeaxanthin. FEBS Lett 1991; 292: 1–4.

Ilioaia C, Johnson MP, Horton P, Ruban AV. Induction of efficient energy dissipation in the isolated light harvesting complex of photosystem II in the absence of protein aggregation. J Biol Chem 2008; 283: 29505–29512.

Inoue T, Fujishima A, Konishi S, Honda K. Photoelectrocatalytic reduction of carbon dioxide in aqueous suspensions of semiconductor powders. Nature 1979; 277: 637–638.

Jennings RC, Engelmann E, Garlaschi F, Casazza AP, Zucchelli G. Photosynthesis and negative entropy production. Biochim Biophys Acta 2005; 1709: 251–5.

Johnson MP, Goral TK, Duffy CDP, Brain APR, Mullineaux CW, Ruban AV. Photoprotective energy dissipation involves the reorganization of photosystem II light harvesting complexes in the grana membranes of spinach chloroplasts. Plant Cell 2011; 23: 1468–1479.

Johnson, MP, Ruban AV. Photoprotective energy dissipation in higher plants involves alteration of the excited state energy of the emitting chlorophyll(s) in LHCII. J Biol Chem 2009; 284: 23592–23601.

Kaneco S, Shimizu Y, Ohta K, Mizuno T. Photocatalytic reduction of high pressure carbon dioxide using TiO_2 powders with a positive hole scavenger. J Photochem Photobiol A: Chemistry 1998; 115: 223–226.

Kleima FJ, Hobe S, Calkoen F, Urbanus ML, Peterman EJG, van Grondelle R, Paulsen H, van Amerongen H. Decreasing the chlorophyll a/b ratio in reconstituted LHCII: structural and functional consequences. Biochemistry 1999; 38: 6587–6596.

Krieger A, Moya I, Weis E. Energy-dependent quenching of chlorophyll a fluorescence: effect of pH on stationary fluorescence and picosecond-relaxation kinetics in thylakoid membranes and photosystem II preparations. Biochim Biophys Acta 1992; 1102: 167–176.

Krüger TPJ, Ilioaia C, Johnson MP, Ruban AV, Papagiannakis E, Horton P, van Grondelle R. Controlled disorder in plant light-harvesting complex II explains its photoprotective role. Biophys J. 2012, ,102, 2669–2676.

Lee CW, Antoniou-Kourounioti R, Chi-Sheng Wu J, Murchie E, Maroto-Valer M, Jensen OE, Huang CW, Ruban AV. Photocatalytic conversion of CO_2 to hydrocarbons by light-harvesting complex assisted

Rh-doped TiO$_2$ photocatalyst. Journal of CO$_2$ utilisation 2013; submitted.

Liu Z, Yan H, Wang K, Kuang T, Zhang J, Gui L, An X, Chang W. Crystal structure of spinach major light-harvesting complex at 2.72 Å resolution. Nature 2004; 6980: 287–92.

Mick V, Geister S, Paulsen H. Folding state of the lumenal loop determines the thermal stability of light-harvesting chlorophyll a/b protein (LHCIIb). Biochemistry 2004; 43: 14704–14711.

Miloslavina Y, Wehner A, Lambrev PH, Wientjes E, Reus M, Garab G, Croce R, Holzwarth AR. Far-red fluorescence: a direct spectroscopic marker for LHC II oligomer formation in non-photochemical quenching. FEBS Lett 2008; 582: 3625–3631.

Morosinotto T, Castelletti S, Breton J, Bassi R, Croce R. Mutation analysis of Lhca1 antenna complex. J Biol Chem 2002; 277: 36253–36261.

Müh F, Madjet ME, Renger T. Structure-Based Identification of Energy Sinks in Plant Light-Harvesting Complex II. J. Phys. Chem. B 2010; 114: 13517–13535.

Nabiev I, Rakovich A, Sukhanova A, Lukashev E, Zagidullin V, Pachenko V, Rakovich YP, Donegan JF, Rubin AB, Govorov AO. Fluorescent quantum dots as artificial antennas for enhanced light harvesting and energy transfer to photosynthetic reaction centers. Angew Chem 2010; 49: 7217–7221.

Nishigaki A, Ohshima S, Nakayama K, Okada M, Nagashima U. Application of molecular orbital calculations to interpret the chlorophyll spectral forms in pea photosystem II. Photochem Photobiol 2001; 73: 245–248.

Pan X, Li M, Wan T, Wang L, Jia C, Hou Z, Zhao X, Zhang J, Chang W. Structural insights into energy regulation of light-harvesting complex CP29 from spinach. Nat Struct Mol Biol 2008; 18: 309–15.

Pandit A, Reus M, Morosinotto T, Bassi R, Holzwarth AR, de Groot HJM. An NMR comparison of the light-harvesting complex II (LHCII) in active and photoprotective states reveals subtle changes in the chlorophyll a ground-state electronic structures. Biochem Biophys Acta 2013; 1827: 738–744.

Pascal AA, Liu Z, Broess K, van Oort B, van Amerongen H, Wang C, Horton P, Robert B, Chang W, Ruban AV. Molecular basis of photoprotection and control of photosynthetic light-harvesting. Nature 2005; 436: 134–137.

Paulsen H, Rümler U, Rüdiger W. Reconstitution of pigment-containing complexes from light-harvesting chlorophyll a/b binding protein over-expressed in Escherichia coli. Planta 1990; 181: 204–211.

Paulsen, H, Dockter C, Volkov A, Jeschke G. Folding and pigment binding of light-harvesting chlorophyll a/b protein. In: The Chloroplast. Advances in photosynthesis and respiration, Springer, Dordrecht, The Netherlands 2010; 31: 231–244.

Petrou K, Belgio E, Ruban AV. pH sensitivity of chlorophyll fluorescence quenching is determined by the detergent/protein ratio and the state of LHCII aggregation. Biochim Biophys Acta 2014, recently accepted.

Phillip D, Ruban AV, Horton P, Asato A, Young AJ. Quenching of chlorophyll fluorescence in the major light-harvesting complex of photosystem II: a systematic study of the effect of carotenoid structure. Proc Natl Acad Sci 1996; 93: 1492–1497.

Pieper J, Irrgang KD, Rätsep M, Voigt J, Renger G, Small GJ. Assignment of the lowest Q_y state and spectral dynamics of the CP29 chlorophyll a/b antenna complex of green plants: A hole burning study. Photochem Photobiol 2000; 71: 574–581.

Plumley FG, Schmidt GW. Reconstitution of chlorophyll a/b light-harvesting complexes: xanthophyll-dependent assembly and energy transfer. Proc Natl Acad Sci 1994; 84: 146–150.

Porra RJ, Thompson WA, Kriedmann PA. Determination of accurate extinction coefficients and simultaneous equations for assaying chlorophylls a and b extracted with four different solvents: verification of the concentration of chlorophyll standards by atomic absorption spectroscopy. Biochem. Biophys. Acta 1989; 975: 384–394.

Qinmiao S, Liangbi L, Dazhang M, Tingyun K. High efficient expression of Lhcb2 gene from pea in E. coli and reconstitution of its expressed product with pigment in vitro. Science China 2000; 43: 464–471.

Remelli R, Varotto C, Sandonà D, Croce R, Bassi R. Chlorophyll binding sites of monomeric light harvesting complex (LHCII) reconstituted in vitro: a mutation analysis of chromophore binding residues. J Biol Chem 1999; 274: 33510–33521.

Renge I, R Avarmaa. Specific solvation of chlorophyll a: solvent nucleophility, hydrogen bonding and steric effects on absorption spectra. Photochem Photobiol 1985; 42: 253–260.

Rogl H, Kühlbrandt W. Mutant trimers of light-harvesting complex II exhibit altered pigment content and spectroscopic features. Biochemistry 1999; 42: 16214–16222.

Ruban AV. The photosynthetic membrane: molecular mechanisms and biophysics of light harvesting, Wiley-Blackwell, Chichester 2012.

Ruban AV, Rees D, Noctor GD, Young A, Horton P. Long wavelength chlorophyll species are accociated with amplification of high-energy-state excitation quenching in higher plants. Biochim Biophys Acta 1991; 1059: 355–360.

Ruban AV, Rees D, Pascal AA, Horton P. Mechanism of ΔpH-dependent dissipation of absorbed excitation energy by photosynthetic membranes II: the relationships between LHCII aggregation in vitro and qE in isolated thylakoids. BBA 1992; 1102: 39–44.

Ruban AV, Young A, Horton P. Induction of nonphotochemical energy dissipation and absorbance changes in leaves; evidence for changes in the state of the light harvesting system of photosystem II in vivo. Plant Physiology 1993; 102: 741–750.

Ruban AV, Young, AJ and Horton P. Dynamic properties of the minor chlorophyll a/b binding proteins of photosystem II – an in vitro model for photoprotective energy dissipation in the photosynthetic membrane of green plants. Biochemistry 1996; 35: 674–678.

Ruban AV, Calkoen F, Kwa SLS, van Grondelle R, Horton P, Dekker JP. Characterisation of the aggregated state of the light harvesting complex of photosystem II by linear and circular dichroism spectroscopy. BBA 1997; 1321: 61–70.

Ruban AV, Berera R, Ilioaia C, van Stokkum IHM, Kennis JTM, Pascal AA, van Amerongen H, Robert B, Horton P, van Grondelle R. Identification of a mechanism of photoprotective energy dissipation in higher plants. Nature 2007a; 450: 575–579.

Ruban AV, Berera R, Ilioaia C, van Stokkum IHM, Kennis JTM, Pascal AA, van Amerongen H, Robert B, Horton P, van Grondelle R. Identification of a mechanism of photoprotective energy dissipation in higher plants. Nature 2007b; 450, 575–578.

Ruban AV, Johnson MP, Duffy CPD. Photoprotective molecular switch in photosystem II. Biochim Biophys Acta 2012; 1817: 167–181.

Sandonà D, Croce R, Pagano A, Crimi M, Bassi R. Higher plants light harvesting proteins. Structure and function as revealed by mutation analysis of either protein or chromophore moieties. Biochim Biophys Acta 1998; 1365: 207–214.

Shikanai T, Munekage Y, Shimizu K, Endo T, Hashimoto T. Identification and characterization of Arabidopsis mutants with reduced quenching of chlorophyll fluorescence. Plant Cell Physiol 1999; 40: 1134–1142.

Storf S, Jansson S, Schmid VHR. Pigment binding, fluorescence properties, and oligomerization behavior of Lhca5, a novel light-harvesting protein. J Biol Chem 2005; 280: 5163–5168.

van Amerongen H, van Grondelle R. Understanding the energy transfer function of LHCII, the major light-harvesting complex of green plants. J Phys Chem B 2001; 105: 604–617.

van Oort B, van Hoek A, Ruban AV, van Amerongen H. Equilibrium between quenched and nonquenched conformations of the major plant light-harvesting complex studied with high-pressure time-resolved fluorescence. J Phys Chem B 2007; 111: 7631–7637.

Walters RG, Ruban AV, Horton P. Light-harvesting complexes bound by dicyclohexylcarbodiimide during inhibition of protective energy dissipation. Eur J Biochem 1994; 226: 1063–69.

Werwie M, Xiangxing X, Haase M, Basche T, Paulsen H. Bio Serves Nano. Biological Light-Harvesting Complex as Energy Donor for Semiconductor Quantum Dots. Langmuir 2012; 28: 5810–5818.

Wilk L, Grunwald M, Liao PN, Walla PJ, Kühlbrandt W. Direct interaction of the major light-harvesting complex II and PsbS in nonphotochemical quenching. Proc Natl Acad Sci 2013; 14: 5452–6.

Wu JCS. Photocatalytic reduction of greenhouse gas CO_2 to fuel. Catal Surv Asia 2009; 13: 30–40.

Yamashita H., Nishiguchi H, Kamada N, Anpo M., Teraoka Y, Hatano H, Ehara S, Kikui K, Palmisano K, Sclafani A, Schiavello A, Fox MA. Photocatalytic reduction of CO_2 with H_2O on TiO_2 and cu/TiO_2 catalysts. Res Chem Intermed 1994; 20: 815–823.

Yang C, Kosemund K, Cornet C, Paulsen H. Exchange of pigment-binding amino acids in light-harvesting chlorophyll a/b protein. Biochemistry 1999; 38: 16205–16213.

Yang C, Lambrev P, Chen Z, Jávorfi T, Kiss AZ, Paulsen H and Garab G. The negatively charged amino acids in the lumenal loop influence the pigment binding and conformation of the major light-harvesting chlorophyll a/b complex. Biochim Biophys Acta 2008; 1777: 1463–1470.

Zucchelli G, Brogioli D, Casazza AP, Garlaschi FM, Jennings RC. Chlorophyll ring deformation modulates Q_y electronic energy in chlorophyll-protein complexes and generates spectral forms. Biophys J 2007; 93: 2240–2254.

Alternative Electron Acceptors for Photosystem II

Jessica Wiwczar and Gary W. Brudvig

Summary

Photosystem II (PSII) is conserved in all oxygenic photosynthetic organisms and is important for its unique ability to use energy from light to split water, generate molecular oxygen in the Earth's atmosphere and drive electrons into the photosynthetic electron transport chain by reducing the plastoquinone (PQ) pool in the thylakoid membrane. The focus of this chapter is on alternative electron-transfer pathways on the acceptor side of PSII. Upon close examination of the literature there is evidence of exogenous electron acceptors that are reduced directly by the primary PQ electron acceptor (Q_A), bypassing the canonical terminal PQ-reduction (Q_B) site. These herbicide-insensitive electron-acceptor molecules include but are not limited to ferricyanide, synthetic cobalt coordination complexes, and cytochrome c. We also discuss experimental treatments to PSII such as cation exchange and herbicide treatment that have been shown to alter the redox midpoint potential (E_m) of Q_A and impact electron transfer from Q_A to Q_B. The results described in this chapter provide a platform for understanding how electrons generated in PSII by photochemical water oxidation can be extracted from the electron-acceptor side of PSII for energy applications.

Keywords

Photosystem II • Structure • Function • Quinone • Alternative electron acceptor • Ferricyanide • Silicomolybdate • Herbicide • Oxygen evolution • Lipid • Bicarbonate • Cytochrome b_{559}

Contents

4.1	**Introduction**	52
4.1.1	The Structure and Function of Photosystem II	52
4.2	**Investigating the Reductase Activity of Photosystem II**	54
4.2.1	The Hill Reaction	54
4.2.2	Native Electron Acceptors in the Plastoquinone B Pocket	54
4.2.2.1	Quinone Analogs: Artificial Electron Acceptors from the Q_B Site	54
4.2.2.2	Herbicides that Inhibit at the Q_B Binding Site	55
4.3	**DCMU-Insensitive Electron Acceptors**	55
4.3.1	Ferricyanide	55
4.3.2	Hg^{2+}	56
4.3.3	Silicomolybdate	57
4.3.4	PSII Mutagenesis to Redirect Electron Flow	57

J. Wiwczar
Department of Molecular Biophysics and Biochemistry, Yale University, New Haven, CT 06520-8114, USA

G.W. Brudvig (✉)
Department of Chemistry, Yale University, New Haven, CT 06520-8107, USA
e-mail: gary.brudvig@yale.edu

© Springer International Publishing AG 2017
H.J.M. Hou et al. (eds.), *Photosynthesis: Structures, Mechanisms, and Applications*,
DOI 10.1007/978-3-319-48873-8_4

4.3.5	Designer Electron Acceptors – Co[(terpy)$_2$]$^{3+}$	57
4.3.6	Electrodes	57
4.4	**Factors Influencing PSII Reductase Activity and Redox Potential of Q_A**	58
4.4.1	Herbicides	58
4.4.2	Quinone Pocket Hydrogen-Bonding Interactions	58
4.4.3	D1 Isoforms	59
4.4.4	OEC Perturbations	59
4.4.4.1	Challenging the "Long-Range Effect" of OEC Perturbations	60
4.5	**Factors Influencing Electron Transfer from Q_A to Q_B**	60
4.5.1	Non-Heme Iron and Bicarbonate	60
4.5.2	Additional Quinones and Cytochrome b_{559}	61
4.5.3	Stromal Side Lipids	61
4.5.3.1	Phosphatidylglycerol	61
4.5.4	Small Transmembrane Polypeptides	62
4.6	**Conclusion**	62
References		63

4.1 Introduction

4.1.1 The Structure and Function of Photosystem II

The protein structure and cofactor arrangement of Photosystem II (PSII) is strongly conserved in all oxygenic photosynthetic organisms. In recent years, PSII has been purified and crystallized from thermophilic cyanobacteria, providing X-ray crystal structures at 1.95 Å – 1.85 Å resolution (Umena et al. 2012; Suga et al. 2015; Tanaka et al. 2017). PSII is a large homo-dimeric pigment-protein complex. The membrane bound dimer is roughly 700 kDa in size with surface area dimensions of 90 Å × 200 Å and 110 Å wide (Ferreira et al. 2004; Zouni et al. 2001). In cyanobacteria, each of the two monomers contains 17 unique alpha helical polypeptides that span the thylakoid membrane and three polypeptides bound facing the thylakoid lumen. Redox-active cofactors are embedded in the protein complex and facilitate the light-harvesting, charge-transfer and subsequent water-splitting reactions that are central to photosynthesis. According to the 1.9 Å crystal structure, the redox-active cofactors include 35 chlorophyll a molecules, 2 pheophytins, 2 plastoquinones, a non-heme iron with a bicarbonate ligand, 11 β-carotenes, 2 hemes (cytochromes b_{559} and c_{550}) and the Mn_4CaO_5 metal cluster in the oxygen-evolving complex (OEC) (Umena et al. 2012).

Crystal structures have provided the orientation of the electron-transport chain in PSII with respect to the thylakoid membrane (Kamiya and Shen 2003). A cartoon diagram representing PSII and its cofactors is represented in Fig. 4.1. Each cofactor in PSII is positioned so that its redox midpoint potential is finely tuned to accept and transfer electrons along the internal electron-transport chain, avoiding charge recombination and free-radical damage to the surrounding protein. If the primary chlorophyll electron donor called P_{680} is considered the center of the reaction, the cofactors can be divided into two groups based on their orientation towards the stromal side (acceptor side) or lumen side (donor side) of the protein.

When PSII is activated by light, the chlorophyll a molecules populating the CP43 and CP47 polypeptides harvest the photon energy and transfer it to P_{680}. P_{680} uses the energy to separate charge by transfer of an electron towards the stromal side of PSII, as indicated by arrows in Fig. 4.1. The electron is transferred via a nearby Pheophytin (Pheo) to the primary quinone electron acceptor, plastoquinone A (Q_A), forming Q_A^- and stabilizing the charge separation across the protein complex. Q_A^- is tightly bound in the D1 peptide and must be oxidized before it can be reduced again. In normal photosynthesis, the second plastoquinone, Q_B is the terminal electron acceptor for PSII and will accept two electrons from Q_A, become protonated to form plastoquinol (PQH_2), and exchange with an oxidized plastoquinone from the thylakoid membrane. Q_B exchange is the often the rate-limiting step of PSII turnover and directly governs the rate of oxidation events in the OEC.

The OEC is a Mn_4CaO_5 cluster near the lumenal side of PSII. Its structure was not clear until the aforementioned 1.9 Å structure was solved (Umena et al. 2012). The structure of the OEC is made of four Mn ions with oxo bridges and one Ca ion, forming a cuboid with a dangling Mn, resembling a chair and held in place by

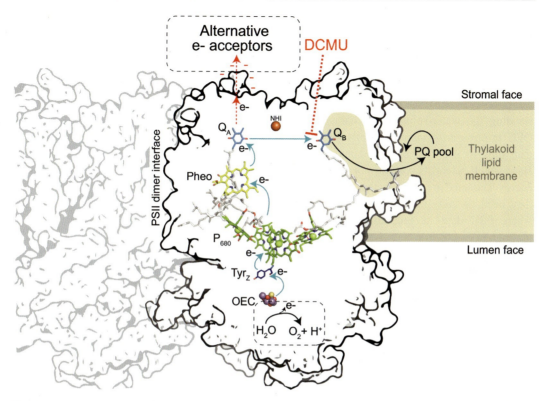

Fig. 4.1 Representation of the PSII electron transport chain and pathway of alternative electron transfer to exogenous acceptors in the presence of DCMU as described in text. Light-activated charge separation occurs on P_{680} (special chlorophyll electron donor) and reduces the nearby pheophytin (Pheo), reducing the primary plastoquinone A (PQ_A), which transfers electrons past the non-heme iron (NHI) to the terminal plastoquinone B (Q_B) that is exposed to the lipid bilayer and exchanges with an oxidized plastoquinone (PQ) from the PQ pool. Meanwhile the hole formed by charge separation on P_{680} is replaced by oxidizing the redox active tyrosine D1-Y161 (Tyr_Z), which oxidizes the oxygen-evolving complex (OEC), catalyzing the water-splitting reaction

ligand contacts to the D1 and CP43 proteins. The OEC is connected to the nearby thylakoid lumen by a network of channels that enable diffusion of water and protons; for a recent review of these channels see (Vogt et al. 2015).

The OEC catalyzes the water-splitting chemical reaction that PSII is responsible for in the scheme of photosynthesis (Eqs. 4.1–4.2).

$$\text{Photosynthesis}: 6CO_2 + 6H_2O \xrightarrow{h\nu} C_6H_{12}O_6 + 6O_2 \quad (4.1)$$

$$\text{Photosystem II}: 2H_2O \xrightarrow{h\nu} O_2 + 4e^- + 4H^+ \quad (4.2)$$

The details of the water-splitting OEC reaction have been studied extensively and are discussed in recent relevant reviews (McEvoy and Brudvig 2006; Vinyard et al. 2013a). In short, when PSII uses light to create a charge separation at P_{680}, the hole oxidizes a redox-active tyrosine (Tyr_Z), which oxidizes the OEC, which in turn oxidizes water. As the OEC functions to oxidize water, it incrementally cycles through intermediate oxidation states known as S_n-states, (n = 0 through 4, with S_1 being the dark-adapted S state). After one complete cycle, the OEC has split two water molecules to liberate 4 electrons, 4 protons, and one molecule of dioxygen.

In order to advance the field of solar-energy conversion by studying, exploiting, and mimicking photosynthesis, we must fully understand both the electron-accepting and electron-donating sides of the PSII enzyme described above. PSII is enzymatically a water-plastoquinone/oxidoreductase, using solar energy to move electrons from water to plastoquinones and into the photosynthetic metabolic network. Understanding the water-splitting mechanism in the OEC informs artificial photosynthesis mimicry that aims to use synthetic dyes and catalysts. However, if scientists want to *harness* the solar energy collected by PSII in the form of fuels, they need to study the electron-accepting side, where PSII deposits reducing equivalents on plastoquinone electron acceptors. The following sections focus on alternative terminal electron acceptors and factors that influence the direction of electron transfer on the stromal side of PSII. First, we discuss results that demonstrate that electrons in PSII can deviate from their natural path and reduce non-plastoquinone electron acceptors. Then, we discuss factors that have been shown to influence Q_A–Q_B electron transfer on the stromal side of PSII. These studies provide important contributions towards understanding of the redox activity at the stromal surface of PSII.

4.2 Investigating the Reductase Activity of Photosystem II

4.2.1 The Hill Reaction

Long before scientists had crystal structures of PSII or methods to purify PSII from chloroplast membranes, they knew that PSII reduced plastoquinone and that artificial electron acceptors were necessary for the activity of PSII under conditions where the oxidizing plastoquinone pool had been compromised. In 1937, biochemist Robin Hill reported that isolated chloroplasts could produce oxygen in the presence of an electron acceptor, ferric oxalate and soon introduced the more common acceptor ferricyanide, FeCy (Hill and Scarisbrick 1940; Hill 1937). This pioneering discovery showed that the activity of PSII with light served to move an electron from a donor to acceptor and is classically named the Hill Reaction. In the Hill reaction, the donor is naturally water and the acceptor is naturally plastoquinone. When PSII is active and H_2O is the electron donor, activity can be monitored by detecting light-driven oxygen evolution using a Clark-type electrode. If the OEC had been removed, an artificial electron donor can be used with an artificial electron acceptor and their redox states can be monitored spectroscopically. In that way, PSII activity is measured as the electron acceptor is reduced. For a review of electron donors and acceptors used in early photosynthesis research, see (Hauska 1977). The Hill reaction demonstrates one part of the electron energetics of photosynthesis, which is graphically described by the Z-scheme that was first assembled in 1968 and described by Bendall and Hill (1968).

4.2.2 Native Electron Acceptors in the Plastoquinone B Pocket

4.2.2.1 Quinone Analogs: Artificial Electron Acceptors from the Q_B Site

The most common electron acceptors used in cell-free assays of PSII are those that mimic the native plastoquinone (PQ) binding properties in the Q_B pocket to act as the terminal electron acceptor. The most analogous to the native PQ is 2,3-dimethyl-*p*-benzoquinone (DMBQ), E_m = 174 mV (Petrouleas and Diner 1987; Izawa 1980), which has the same head group as PQ but lacks the isoprene tail. DMBQ has been shown to receive electrons from the Q_B pocket and exchange rapidly when reduced. The rapid exchange is due to the lack of the isoprene chain that normally partitions the PQ into the lipid environment and promotes tighter binding to the Q_B pocket. DMBQ can also be reduced indirectly by the reduced PQH_2 (or $DBMQH_2$) pool in the thylakoid membrane (Koike et al. 1996). Each benzoquinone analog has slightly different head group substitutions that affect their binding affinities and redox potentials. Other examples of benzoquinone analogs used as electron acceptors are 2,5-dichloro-*p*-benzoquinone

(DCBQ), $E_m = 309$ mV (Petrouleas and Diner 1987) and phenyl-p-benzoquinone (PPBQ), $E_m = 279–290$ mV (Petrouleas and Diner 1987; Izawa 1980). PPBQ has been shown to support the highest turnover frequency of PSII, which is attributed to the lack of bulky methyl groups on its head group that allows for faster exchange in the Q_B site (Shevela and Messinger 2012).

4.2.2.2 Herbicides that Inhibit at the Q_B Binding Site

In the field of biochemistry, it is beneficial to have access to inhibitors in order to probe the mechanism of an enzyme. In the field of photosynthesis, inhibitors are commonly commercial herbicides that have been heavily studied and are readily available (Oettmeier 1999; Trebst 2007; Büchel 1972). The most common herbicide for probing PSII activity is the urea-type herbicide DCMU (3-(3,4-dichlorophenyl)-1,1-dimethylurea), commercially knows as Diuron. DCMU inhibits Q_A–Q_B electron transfer by competing for the Q_B hydrogen-bonding partner, D1-Ser264 (Lavergne 1982; Takahashi et al. 2010). Another class of herbicides that interact at the Q_B site is the phenolic-type herbicides such as bromoxynil (3,5-dibromo-4-hydroxybenzonitrile). Bromoxynil competes for the hydrogen-bonding partner D1-His215 and is stabilized in the pocket by a second hydrogen bond to the backbone near D1-Ser265 (Takahashi et al. 2010). Many types of herbicides and their mechanisms can be found in comprehensive reviews (Oettmeier 1999; Trebst 2007; Büchel 1972).

4.3 DCMU-Insensitive Electron Acceptors

Although DCMU inhibits Q_A–Q_B electron transfer, there is some evidence that electron acceptors can receive electrons even in the presence of DCMU. DCMU-insensitive electron acceptors are controversial because many believe there is no binding site for electron acceptors on the stromal side of PSII other than the Q_B pocket. Arguments against the DCMU-insensitive electron acceptors are that they are: (1) Degrading DCMU in solution or (2) Competing with DCMU or modifying the Q_B binding niche.

Examples of DCMU-insensitive electron acceptors that were observed in early years are ferricyanide (Sugiura and Inoue 1999; Kirilovsky et al. 1994), mercuric chloride (Hg^{2+}) (Miles et al. 1973; Mohanty et al. 1989), and silicomolybdate (Izawa 1980). Also, in recent years, researchers have been interested in *promoting* electron transfer directly from Q_A^- to an exogenous electron acceptor, thereby bypassing the Q_B site. Electron transfer from Q_A^- has been observed with acceptors such as cytochrome c (Larom et al. 2010, 2015), a synthetic Co [(terpy)$_2$]$^{3+}$complex (terpy = 2,2′;6′,2″-terpyridine) (Ulas and Brudvig 2011; Khan et al. 2015), and electrodes (Rao et al. 1990; Kato et al. 2012a). The following sections will discuss those examples in detail. Figure 4.2 shows an energy diagram of the electron-transfer path in the presence of alternative electron acceptors.

4.3.1 Ferricyanide

Ferricyanide (FeCy), $E_m = 358$ mV (Ulas and Brudvig 2011), is a classic electron acceptor commonly used in combination with a benzoquinone analog to study the activity of PSII. FeCy is added to the reaction as a secondary non-competitive electron acceptor to oxidize the soluble quinones, thereby keeping a constant concentration of oxidized quinone electron acceptors so the rate of quinone reduction from the Q_B site remains constant for oxygen-evolution measurements (Takasaka et al. 2010; Kern et al. 2005). However some studies show that FeCy can be reduced independently from the Q_B site on the PSII stromal side.

A study with PSII core complexes isolated from *Thermosynechococcus elongatus* with a CP43-His tag demonstrated that FeCy could be used alone as an electron acceptor (Sugiura and Inoue 1999). In fact, not only did FeCy support better oxygen evolution activities than with

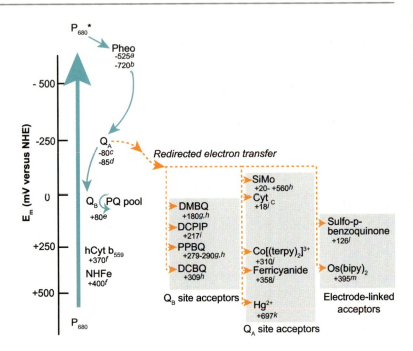

Fig. 4.2 Energetic diagram of PSII acceptor side electron transfer to electron acceptors. Values listed under each electron acceptor are reduction potentials (mV vs. NHE). Notes: a, Allakhverdiev et al. 2011; b, Alcantara et al. 2006; c, Krieger et al. 1995; d, Zhang et al. 2013; e, Ashizawa and Noguchi 2014; f, Buser et al. 1990; g, Petrouleas and Diner 1987; h, Izawa 1980; i, De Wael et al. 2010; j, Ulas and Brudvig 2011; k, Zhu et al. 2009; l, Maly et al. 2005; m, Badura et al. 2008

quinones, it was also insensitive to the inhibition treatment of DCMU. Surprisingly, DCMU-treated PSII activity was inhibited only by 64% when using FeCy as an acceptor versus >92% with quinones as electron acceptors. This implied that FeCy must be able to accept electrons directly from Q_A in the presence of DCMU when using PSII isolated with a CP43-His tag from *Thermosynechococcus elongatus*. The DCMU-insensitive effect of FeCy has not been seen with untreated PSII isolated from spinach but it has been shown in trypsin-treated spinach PSII complexes (Van Rensen et al. 1977).

Another example of the DCMU-insensitive behavior of FeCy is its ability to quench the Q_A^-/Fe^{2+} EPR signal. When the Q_A^-/Fe^{2+} EPR signal is quenched by FeCy, it indicates either: (1) The Q_A^- semiquinone has been oxidized, most likely by electron transfer to FeCy, or (2) The Q_A^- semiquinone has been reduced by a second electron, which is unlikely in the presence of FeCy. If the semiquinone has been reduced by a second electron, forming plastoquinol, it will be accompanied by the formation of a triplet chlorophyll signal due to electrons getting backed up on the PSII redox chain. In the presence of FeCy, the Q_A^-/Fe^{2+} signal disappears without forming the chlorophyll triplet signal, which normally occurs when DCMU is present, inhibiting forward electron transfer (Kirilovsky et al. 1994). This experiment suggests that FeCy is quenching the Q_A^-/Fe^{2+} signal by oxidizing Q_A^- directly.

4.3.2 Hg^{2+}

Some heavy metals have also been shown to act as electron acceptors on the stromal side of PSII. A very early study in 1973 demonstrated that Hg^{2+}, $E_m = 697$ mV (Zhu et al. 2009), acted as an electron acceptor near Q_A. In this study, Hg^{2+} was able to inhibit FeCy reduction and sustain oxygen evolution, acting as the sole electron acceptor for PSII (Miles et al. 1973).

Studies regarding heavy metals are important from an environmental point of view and specifically in the PSII community. Although the effect of Hg^{2+} has been demonstrated, the more general role of divalent cations on the acceptor side is a controversial topic. Researchers are still investigating the exact roles of various divalent cationic metals on the acceptor side (Mohanty et al. 1989; Khan et al. 2015; Yruela et al. 1991; Sigfridsson et al. 2004). There is a mixture of explanations about how cations are interfering with electron transfer. They may be accepting electrons directly from Q_A^-, interfering with Q_B protonation, or affecting the PQ redox potentials from long-range conformational effects transmitted from the OEC to Q_A. This will be discussed further below.

4.3.3 Silicomolybdate

Another classic electron acceptor is silicomolybdate (SiMo), $E_m = 560, 430, 190,$ and 20 mV for each reduction site (Izawa 1980). SiMo is a large, branching anionic molecule. Early studies suggest that it is a DCMU-insensitive electron acceptor. (Barr et al. 1975; Giaquinta and Dilley 1975). In fact, for many years SiMo was used as a standard electron acceptor, assumed to interact with PSII along the Pheo-Q_A-Fe^{2+} region. The precise manner and location that SiMo receives an electron are unknown. It has been suggested that it non-competitively displaces the bicarbonate that is bound to the non-heme Fe^{2+} through non-specific interactions with the stromal surface (Izawa 1980; Giaquinta and Dilley 1975).

4.3.4 PSII Mutagenesis to Redirect Electron Flow

Recently, there have been studies of larger complexes that accept electrons from the stromal side of PSII in a DCMU-insensitive manner. With information from the crystal structures and using site-directed mutagenesis on cyanobacterial PSII, scientists can speculate on the sites of reduction with precision. The crystal structures have exposed a group of negatively charged glutamate residues on the stromal surface of the D1 peptide that is located only ~15 Å above Q_A. This negatively charged group of glutamate residues has been the target for modifications and designing electron acceptors to oxidize Q_A^- directly.

A study done in on *Synechocystis* showed that PSII could directly transfer electrons from Q_A^- to cytochrome c if the aforementioned negative patch near Q_A was made larger by mutating a nearby positive lysine K238 to a glutamate (Larom et al. 2010). The K238E mutation enabled DCMU *dependent* reduction of cytochrome c, $E_m = 18$ mV (De Wael et al. 2010), which could be monitored spectroscopically (Larom et al. 2010). The mechanism for turnover of the mutant is thought to be due to the creation of a favorable binding site for cytochrome c.

4.3.5 Designer Electron Acceptors – $Co[(terpy)_2]^{3+}$

Another recent study made use of the negative patch on the stromal surface of PSII to direct electrons to synthetic soluble electron acceptors (Ulas and Brudvig 2011). This study examined the electrostatic potential map of the stromal surface of PSII and found a very negative patch, in an area similar to the negative patch identified by Larom et al. (2010). It was hypothesized that this negative patch could be a site for electron transfer to synthetic positively charged coordination complexes. A series of cobalt complexes designed to target the observed site were screened for DCMU-insensitive turnover of PSII. The terpyridine ligands on the cobalt complexes were selected so that the redox midpoint potential would be in the range to favorably accept electrons from Q_A^-, and also to bind via electrostatic interactions to the stromal surface of the PSII complex. $Co[(terpy)_2]^{3+}$, with the proper midpoint potential and a positive charge, was the most successfully complex. The cobalt-terpy complex rescued the DCMU-inhibited PSII activity by 30% (Ulas and Brudvig 2011; Khan et al. 2015).

4.3.6 Electrodes

There has been a growing interest to study the behavior or redox active enzymes, including

PSII, on electrodes. For recent reviews of experiments with PSII on electrodes, see references (Kato et al. 2012a; Yehezkeli et al. 2013). Several of these studies demonstrate DCMU-insensitive photocurrent from PSII to electrodes using soluble mediating complexes.

In 2012, Kato et al. layered PSII onto a mesoporous indium-tin oxide (meso-ITO) surface and measured the photocurrent produced under red light. When irradiated with red light and an applied bias potential of -0.5 V, PSII produced a photocurrent of 1.6 μA/cm². In the presence of DCMU, the photocurrent only dropped to 0.5 μA/cm², showing residual DCMU-independent current. An experiment in their supporting information also clearly demonstrates that the PSII/meso-ITO electrode setup could still retain 30% of the photocurrent when treated with DCMU (Kato et al. 2012a). The authors propose that the Q_A site is competing with the Q_B site to inject electrons towards the meso-ITO electrode, as seen by the DCMU-insensitive Q_A-driven photocurrent (Kato et al. 2012a).

In 1990, there was an early study with PSII on a titanium dioxide (TiO_2)-coated electrode that indicated DCMU-insensitive photocurrent (Rao et al. 1990). The total photocurrent obtained by the setup was about 35 μA and ~10% remained when DCMU was added to the cell. These results support a DCMU-insensitive photocurrent from PSII.

As suggested in the Kato et al. (2012a) paper, there is a potential electron-transfer path to the stromal surface from Q_A. If electron transfer to an exogenous acceptor is competing between the Q_A and Q_B sites, it would be beneficial for the electron to be extracted from the Q_A site because the electron could have a lower reduction potential. Also, the turnover rate of PSII could potentially be increased by bypassing the rate-limiting Q_A–Q_B/PQ-pool electron-transfer and quinone-exchange steps. Therefore, electron extraction directly from Q_A could yield a better PSII biohybrid-photovoltaic cell.

4.4 Factors Influencing PSII Reductase Activity and Redox Potential of Q_A

Many factors maintain the natural Q_A–Q_B electron-transfer reactions and redox properties of the cofactors on the stromal side of PSII. The energetics for reduction of Q_A and electron exchange between Q_A and Q_B are highly conserved and tightly regulated. Factors that influence the electron-transfer behavior on the stromal side of PSII include but are not limited to: the redox potentials of Q_A and Q_B, the properties of the non-heme iron and its bicarbonate ligand that lie between the quinones, the influence of stromal side lipids, and the influence of small polypeptides. Figure 4.3 illustrates the variations of the E_m of Q_A caused by different treatments.

4.4.1 Herbicides

Not only do the Q_B-site herbicides mentioned above directly affect the binding of Q_B, they also affect the reduction potential of Q_A, $E_m = -140$ mV for cyanobacteria (Shibamoto et al. 2009) and ~ -80 mV in spinach (Krieger et al. 1995), through the hydrogen-bonding network shared along Q_A-Fe-Q_B (Takahashi et al. 2010). In spinach, the reduction potential of Q_A is shifted from -80 mV to $+52$ mV or -45 mV in the presence of DCMU or bromoxynil, respectively (Krieger-Liszkay and Rutherford 1998). The effect of DCMU on the Q_A reduction potential has been used to modify the driving force for electron transfer from Q_A^- to alternative electron acceptors in a DCMU-insensitive manner (Khan et al. 2015).

4.4.2 Quinone Pocket Hydrogen-Bonding Interactions

A large impact on the Q_A and Q_B redox potentials comes from the hydrogen-bonding network associated with the plastoquinone head

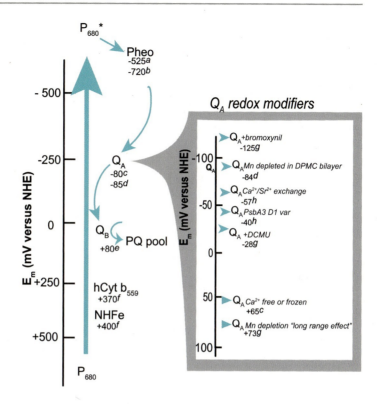

Fig. 4.3 Energetic diagram of experimental modification of Q_A redox potential. Values listed under each electron acceptor are reduction potentials (mV vs. NHE). Notes: a, Allakhverdiev et al. 2011; b, Alcantara et al. 2006; c, Krieger et al. 1995; d, Zhang et al. 2013; e, Ashizawa and Noguchi 2014; f, Buser et al. 1990; g, Krieger-Liszkay and Rutherford 1998; h, Müh and Zouni 2013

groups. It has been shown that the hydrogen bonding interactions to the C = O double bonds contribute less to the binding stability and more to shifting the redox potentials of the quinones (Hasegawa and Noguchi 2012). Increasing the number of hydrogen bonds on the quinone head group shifts the redox midpoint potential for the quinone up by 100–200 mV (Ishikita and Knapp 2005; Ashizawa and Noguchi 2014). The pi stacking interactions of aromatic trp and phe residues with the head groups of Q_A and Q_B and van der Waals interactions with the phytyl tails and nearby lipids contribute the most energetically to the quinone-PSII binding affinity (Hasegawa and Noguchi 2012; Lambreva et al. 2014).

4.4.3 D1 Isoforms

Another contributing factor that tunes the redox potential of Q_A is the presence of different D1 isoforms. Cyanobacteria have three different genes that code for the D1 protein. The three genes, psbA1, psbA2 and psbA3, are differentially expressed based on varying conditions such as light (Kós et al. 2008). It has been shown that the various isoforms can modulate the midpoint potential of Q_A. The D1:PsbA3 isoform is considered to be the more dominant form under normal light conditions; for recent reviews, see (Vinyard et al. 2013b; Mulo et al. 2012). The midpoint potentials of Q_A in the D1:PsbA1 and D1:PsbA3 isoforms are −124 and −83 mV, respectively (Kato et al. 2012b).

4.4.4 OEC Perturbations

Some studies have indicated that perturbations in the structure of the Mn_4CaO_5 composition of the OEC of PSII cause a dramatic change in the redox midpoint potential of Q_A (Krieger et al. 1995; Krieger-Liszkay and Rutherford 1998;

Kato and Noguchi 2014; Fufezan et al. 2007; Allakhverdiev et al. 2011). These studies suggest that depletion of the cationic Mn and Ca^{2+} ions in the OEC and structural perturbations cause the midpoint potential of Q_A to increase by +150 mV. Furthermore, substituting Ca^{2+} with Sr^{2+} in the OEC induced a shift in the Q_A midpoint potential by +27 mV (Kato et al. 2012b). Based on these results, it has been proposed that the composition of the OEC modulates the reduction potential of Q_A, which is over 20 Å away, via long-range conformational changes. Although this effect has been observed many times with careful measurements, the precise molecular mechanism of this shift has yet to be determined.

4.4.4.1 Challenging the "Long-Range Effect" of OEC Perturbations

Several lines of evidence challenge the model of a long-range influence of the OEC on the redox properties of Q_A. The same conditions that shift the Q_A redox midpoint potential by +150 mV caused a shift of only +18 mV in the potential of the non-heme Fe (Kato and Noguchi 2014). Also, Sr^{2+}-substitution in the OEC shows no structural changes near Q_A (Koua et al. 2013), as would be expected for a long-range conformational change when Ca^{2+} in the OEC is substituted with Sr^{2+}. The same large shift of the Q_A potential could be reproduced by simply freezing the sample in the presence of Mn and Ca^{2+} ions, indicating the effect might not result from a change in the ion occupancy of the OEC (Krieger et al. 1995). Furthermore, when the PSII complexes were embedded in a synthetic membrane made of zwitterionic 1,2-dimyristoyl-sn-glycero-3-phosphocholine (DMPC) lipids, there was no detected shift in the redox midpoint potential when Mn was depleted from the sample (Zhang et al. 2013). DMPC is a zwitterionic lipid with a head group similar to phosphatidylcholine and it may have a protective role on Q_A during cation depletion. Another observation is that Cl^- depletion from the OEC has no effect on the Q_A midpoint potential (Kato et al. 2012b). which argues against a long-range effect mediated by electrostatic interactions. Taken together, these observations warrant a closer examination of the molecular mechanism by which Mn and Ca^{2+} depletion from PSII, both from the OEC and other potential cation binding sites, affects the redox midpoint potential of Q_A.

4.5 Factors Influencing Electron Transfer from Q_A to Q_B

4.5.1 Non-Heme Iron and Bicarbonate

The non-heme iron (NHI) is bound symmetrically between Q_A and Q_B, and its location would seem to be ideal to "intercept" an electron from Q_A^-. However, the NHI is not an electron-transfer intermediate between Q_A and Q_B because its redox midpoint potential, E_m = +400 mV, is too high (Petrouleas and Diner 1987). The NHI works with its bicarbonate ligand to modulate electron transfer from Q_A^- to Q_B (Chernev et al. 2011). When Q_A is reduced, the NHI is magnetically coupled to Q_A^-, forming an EPR-active Q_A^-/Fe^{2+} species (Miller and Brudvig 1991).

In its Q_A^-/Fe^{2+} reduced form, the NHI can participate in the reduction of high-potential exogenous quinone electron acceptors, such as PPBQ (Petrouleas and Diner 1987). Studies have looked at the redox state of the NHI, Fe^{II} and Fe^{III} during flashes in the presence of PPBQ. After one flash, the Q_A/NHI center is reduced to Q_A^-/Fe^{II}. This is followed by electron transfer to PPBQ, forming a strongly oxidizing semiquinone that oxidizes the NHI (reduction-induced oxidation) (Petrouleas and Diner 1987). In the presence of DCMU, the NHI cannot get oxidized by any exogenous quinones, and FeCy oxidizes it only very slowly, suggesting that electron transfer from the Q_A/NHI center works through the Q_B pocket (Petrouleas and Diner 1987).

4.5.2 Additional Quinones and Cytochrome b_{559}

Additional quinones have been identified to reside near the Q_B pocket in a small internal lipid bilayer. There is a PQ binding site called Q_C that is located next to the lipid filled Q_B diffusion pocket, present in the 2.9 Å structure. The Q_C head is ~15 Å from Q_B and its tail forms many van der Waals contacts with lipids and cofactors occupying the Q_B pocket (Guskov et al. 2009). Q_C is very labile in its binding pocket, which is indicated by the finding that it was not present in all crystal structures (Umena et al. 2012). The loose binding affinity can be attributed to the lack of a head group pi-pi stacking partner that both Q_A and Q_B have (Kós et al. 2008). It has been postulated that Q_C is involved with modulation of the redox potential of nearby cytochrome b_{559} (Kaminskaya et al. 2007). Cytochrome b_{559} is involved in cyclic electron transfer and has been discussed in a recent review (Shinopoulos and Brudvig 2012).

4.5.3 Stromal Side Lipids

Cyanobacterial thylakoid membranes consist of four major lipids. Those lipids can be separated into uncharged lipids including monogalactosyldiacylglyerol (MGDG) and digalactosyldiacylglycerol (DGDG) and anionic lipids including sulfoquinovosyldiacylglycerol (SQDG) and phosphatiylglycerol (PG). A recent review has focused on the roles of lipids in photosynthetic thylakoid membranes (Mizusawa and Wada 2012). One generalization that has been made is that the neutral MDGD and DGDG serve as bulk lipids that support dimerization and assembly of the small transmembrane polypeptides of PSII; the anionic lipids SQDG and PG have more specialized roles regarding electron transfer in PSII. It was shown that enriching the PSII lipid environment with either of the two anionic lipids caused major structural perturbations that impact electron transfer, such as loss of CP43 and reduced electron transfer along the Pheo-Q_A–Q_B path (Kansy et al. 2014).

4.5.3.1 Phosphatidylglycerol

To point out the importance of PG, a study done in 1999 investigated changes of relative lipid ratios when the thermophilic cyanobacterium *Thermosynechococcus vulcanus* was exposed to various growth temperatures. Of the four lipids mentioned above, PG was the only lipid that was conserved during the temperature changes and was also the smallest amount at 5% of the thylakoid lipid content (Kiseleva et al. 1999). The resistance of PG content to growth conditions indicates that PG has an important and protective role in photosynthesis.

To investigate the large role of PG in photosynthesis and PSII stromal side electron transfer, there have been many studies of PSII targeting PG. For a review of phosphatidylglycerol in photosynthesis, see (Wada and Murata 2010). For example, when the synthesis pathway of PG is depleted in *Synechocystis*, the oxygen-evolution rate measured in whole cells is decreased and can only be rescued by adding exogenous PG to the medium. This study also demonstrated that PG depletion affected the binding properties of the Q_B pocket, indicating that PG impacts the stromal-side electron transfer (Itoh et al. 2012). In accordance with the crystal structures, the location of PG molecules in PSII was shown to be on the stromal side because depletion of PG affected the binding properties of common Q_B-site inhibitors and acceptors such as the plastoquinone analog PBQ (p-benzoquinone). Interestingly, under PG-depleted conditions, the PBQ electron acceptor *inhibits* electron transfer (Itoh et al. 2012). Similarly, in another study where PG was depleted enzymatically with phospholipase A2, causing a drop in oxygen evolution, addition of DCMU *increased* electron transfer (Leng et al. 2008). Together, these two studies support the conclusion that PG is acting as a gate for electron transfer to acceptors bound in the Q_B pocket.

Other roles of PG are still left to be investigated. The studies of phospholipase treatment on PSII left 2–3 PG molecules untouched, indicating more roles for PG other than regulation of electron transfer at the Q_B site. For example, there are three PG molecules close to the Q_A site that are modeled in the 1.9 Å crystal structure, PG664, PG694 and PG702. A recent study used site-directed mutagenesis to perturb two of the three PG head groups, PG664 and 694 (Endo et al. 2015). To date, no studies directly target PG702, which is the closest to Q_A, only 8 Å away, as mentioned above. The proximity of the lipids to the Q_A site would suggest a role on the modulation of the Q_A redox potential.

4.5.4 Small Transmembrane Polypeptides

The smaller, alpha-helical polypeptides of PSII also contribute to the electron transfer behavior of the stromal side. For a recent review of the functional roles of PSII polypeptides, see (Pagliano et al. 2013). PsbM has been shown to be important for dimerization of PSII due to its location on the dimer interface, as evidenced by a PsbM-less crystal structure (Kawakami et al. 2011). The same study found that PsbI was necessary for PSII assembly but dispensable for oxygen evolution activity (Kawakami et al. 2011). PsbT appears to be involved with the repair of Q_A (Ohnishi et al. 2006) as well as dimerization and binding of PsbM (Iwai et al. 2004). PsbJ is located near the Q_B pocket and its deletion in plants showed that it affects Q_A–Q_B electron transfer, causing a longer Q_A^- lifetime and slower oxygen evolution (Regel et al. 2001).

PsbL appears to have the strongest influence on the redox properties of Q_A and the most interesting effect on electron transfer. PsbL is positioned between Q_A and the dimer interface, holding three PG, PG664, PG694 and PG702, molecules close to Q_A. Plant mutagenesis studies showed a new phenomenon where PsbL deletion mutants exhibited a back electron transfer from Q_B to Q_A during dark incubations (Ohad et al. 2004). The authors hypothesized that the Q_B site must be perturbed by the hydrophilic region of PsbL extending over the stromal surface. The Q_B perturbation is allowing for a reverse of the plastoquinone/plastoquinol reaction where reduced PQH_2 is binding, getting deprotonated, and reducing Q_A. In this way, they explain their observation that PsbL has a very strong influence on the behavior of forward electron transfer from Q_A to Q_B in PSII (Ohad et al. 2004).

An alternative explanation to the PsbL/Q_A redox influence could be that the PsbL polypeptide, located on the PSII interface and protecting the three PG molecules mentioned above from the lipid membrane, is also protecting Q_A from exposure to the lipid domain containing the reduced PQH_2 pool. Loss of PsbL could expose Q_A and allow electron exchange directly from Q_A to the PQ pool, explaining why the observed reduction of Q_A by PQH_2 is temperature and diffusion-dependent (Ohad et al. 2004).

4.6 Conclusion

In this chapter, we have reviewed electron transport on the acceptor side of PSII with a focus on alternative electron acceptors and factors that modulate the redox properties of Q_A and Q_B. There is a growing number of examples of artificial electron acceptors that function at sites away from the Q_B-binding pocket, in a DCMU-insensitive manner. These DCMU-insensitive electron acceptors may be functioning by binding at or near a cluster of glutamate residues on the stromal surface of PSII near Q_A. Interestingly, this group of DCMU-insensitive electron acceptors includes the "classic" DCMU-insensitive electron acceptor silicomolybdate, whose function may be understood in terms of an interaction at the stromal surface of PSII rather than in the Q_B site. We also describe factors that influence electron transfer between the natural primary electron acceptors, Q_A and Q_B, and artificial electron acceptors. When considered together, the experiments and results described in this chapter provide a platform for understanding how the electron transfer behavior on the acceptor side of PSII is controlled both naturally and artificially.

References

Alcantara K, Munge B, Pendon Z, et al. (2006) Thin film voltammetry of spinach photosystem II. Proton-gated electron transfer involving the Mn_4 cluster. J Am Chem Soc 128:14930–14937. doi: 10.1021/ja0645537

Allakhverdiev SI, Tsuchiya T, Watabe K, et al. (2011) Redox potentials of primary electron acceptor quinone molecule QA- and conserved energetics of photosystem II in cyanobacteria with chlorophyll a and chlorophyll d. P Natl Acad Sci USA 108:8054–8058. doi: 10.1073/pnas.1100173108

Ashizawa R, Noguchi T (2014) Effects of hydrogen bonding interactions on the redox potential and molecular vibrations of plastoquinone as studied using density functional theory calculations. Phys Chem Chem Phys 16:11864–11876. doi: 10.1039/c3cp54742f

Badura A, Guschin D, Esper B, et al. (2008) Photo-Induced Electron Transfer Between Photosystem 2 via Cross-linked Redox Hydrogels. Electroanal 20:1043–1047. doi: 10.1002/elan.200804191

Barr R, Crane FL, Giaquinta RT (1975) Dichlorophenylurea-insensitive reduction of silicomolybdic acid by chloroplast Photosystem II. Plant Physiol 55:460–462. doi:10.1104/pp.55.3.460

Bendall DS, Hill R (1968) Haem-proteins in photosynthesis. Ann Rev Plant Phys 167–186.

Büchel KH (1972) Mechanisms of action and structure activity relations of herbicides that inhibit photosynthesis. Pestic Sci 3:89–110. doi: 10.1002/ps.2780030113

Buser CA, Thompson LK, Diner BA, Brudvig GW (1990) Electron-transfer reactions in manganese-depleted photosystem II. Biochemistry-US 29:8977–8985. doi: 10.1021/bi00490a014

Chernev P, Zaharieva I, Dau H, Haumann M (2011) Carboxylate shifts steer interquinone electron transfer in photosynthesis. J Biol Chem 286:5368–5374. doi: 10.1074/jbc.M110.202879

De Wael K, De Belder S, Van Vlierberghe S, et al. (2010) Electrochemical study of gelatin as a matrix for the immobilization of horse heart cytochrome c. Talanta 82:1980–1985. doi: 10.1016/j.talanta.2010.08.019

Endo K, Mizusawa N, Shen J-R, et al. (2015) Site-directed mutagenesis of amino acid residues of D1 protein interacting with phosphatidylglycerol affects the function of plastoquinone Q_B in photosystem II. Photosynth Res 126:385–397. doi: 10.1007/s11120-015-0150-9

Ferreira KN, Iverson TM, Maghlaoui K, et al. (2004) Architecture of the photosynthetic oxygen-evolving center. Science 303:1831–1838. doi: 10.1126/science.1093087

Fufezan C, Gross CM, Sjodin M, et al. (2007) Influence of the Redox Potential of the Primary Quinone Electron Acceptor on Photoinhibition in Photosystem II. J of Biol Chem 282:12492–12502. doi: 10.1074/jbc.M610951200

Giaquinta RT, Dilley RA (1975) A partial reaction in photosystem II: reduction of silicomolybdate prior to the site of dichlorophenyldimethylurea inhibition. Biochim Biophys Acta 387:288–305. doi:10.1016/0005-2728(75)90111-5

Guskov A, Kern J, Gabdulkhakov A, et al. (2009) Cyanobacterial photosystem II at 2.9-Å resolution and the role of quinones, lipids, channels and chloride. Nat Struct Mol Biol 16:334–342. doi: 10.1038/nsmb.1559

Hasegawa K, Noguchi T (2012) Molecular interactions of the quinone electron acceptors Q_A, Q_B, and Q_C in photosystem II as studied by the fragment molecular orbital method. Photosynth Res 120:113–123. doi: 10.1007/s11120-012-9787-9

Hauska G (1977) Artificial Acceptors and Donors. In: Photosynthesis I. Springer Berlin/Heidelberg, pp 253–265

Hill R (1937) Oxygen Evolved by Isolated Chloroplasts. Nature 139:881–882. doi: 10.1038/139881a0

Hill R, Scarisbrick R (1940) Production of Oxygen by Illuminated Chloroplasts. Nature 146:61–62. doi: 10.1038/146061a0

Ishikita H, Knapp E-W (2005) Control of quinone redox potentials in photosystem II: Electron transfer and photoprotection. J Am Chem Soc 127:14714–14720. doi: 10.1021/ja052567r

Itoh S, Kozuki T, Nishida K, et al. (2012) Two functional sites of phosphatidylglycerol for regulation of reaction of plastoquinone Q_B in photosystem II. Biochim Biophys Acta 1817:287–297. doi: 10.1016/j.bbabio.2011.10.002

Iwai M, Katoh H, Katayama M, Ikeuchi M (2004) PSII-Tc protein plays an important role in dimerization of photosystem II. Plant Cell Physiol 45:1809–1816. doi: 10.1093/pcp/pch207

Izawa S (1980) Acceptors and donors for chloroplast electron transport. Method Enzymol 69:413–434.

Kaminskaya O, Shuvalov VA, Renger G (2007) Evidence for a novel quinone-binding site in the photosystem II (PS II) complex that regulates the redox potential of cytochrome b_{559}. Biochemistry-US 46:1091–1105. doi: 10.1021/bi0613022

Kamiya N, Shen J-R (2003) Crystal structure of oxygen-evolving Photosystem II from Thermosynechococcus vulcanus at 3.7Å resolution. Proc Natl Acad Sci U S A 100:98–103. doi: 10.1073/pnas.0135651100

Kansy M, Wilhelm C, Goss R (2014) Influence of thylakoid membrane lipids on the structure and function of the plant photosystem II core complex. Planta 240:781–796. doi: 10.1007/s00425-014-2130-2

Kato Y, Noguchi T (2014) Long-range interaction between the Mn_4CaO_5 cluster and the non-heme iron center in photosystem II as revealed by FTIR spectroelectrochemistry. Biochemistry-US 53:4914–4923. doi: 10.1021/bi500549b

Kato M, Cardona T, Rutherford AW, Reisner E (2012a) Photoelectrochemical Water Oxidation with Photosystem II Integrated in a Mesoporous Indium–Tin Oxide Electrode. J Am Chem Soc 134:8332–8335. doi: 10.1021/ja301488d

Kato Y, Shibamoto T, Yamamoto S, et al. (2012b) Influence of the PsbA1/PsbA3, Ca^{2+}/Sr^{2+} and Cl^-/Br^- exchanges on the redox potential of the primary quinone Q_A in Photosystem II from Thermo*synechococcus elongatus* as revealed by spectroelectrochemistry. Biochim Biophys Acta 1817:1998–2004. doi: 10.1016/j.bbabio.2012.06.006

Kawakami K, Umena Y, Iwai M, et al. (2011) Roles of PsbI and PsbM in photosystem II dimer formation and stability studied by deletion mutagenesis and X-ray crystallography. Biochim Biophys Acta 1807:319–325. doi: 10.1016/j.bbabio.2010.12.013

Kern J, LOLL B, Lüneberg C, et al. (2005) Purification, characterisation and crystallisation of photosystem II from *Thermosynechococcus elongatus* cultivated in a new type of photobioreactor. Biochim Biophys Acta 1706:147–157. doi: 10.1016/j.bbabio.2004.10.007

Khan S, Sun JS, Brudvig GW (2015) Cation effects on the electron-acceptor side of Photosystem II. J Phys Chem B 119:7722–7728. doi: 10.1021/jp513035u

Kirilovsky D, Rutherford AW, Etienne AL (1994) Influence of DCMU and ferricyanide on photodamage in photosystem II. Biochemistry 33:3087–3095. doi: 10.1021/bi00176a043

Kiseleva LL, HorvÁth I, Vigh LS, Los DA (1999) Temperature-induced specific lipid desaturation in the thermophilic cyanobacterium *Synechococcus vulcanus*. FEMS Microbiology Letters 175:179–183. doi: 10.1111/j.1574-6968.1999.tb13617.x

Koike H, Yoneyama K, Kashino Y, Satoh K (1996) Mechanism of Electron Flow through the Q_B Site in Photosystem II. 4. Reaction Mechanism of Plastoquinone Derivatives at the Q_B Site in Spinach Photosystem II Membrane Fragments. Plant Cell Physiol 37:983–988. doi: 10.1093/oxfordjournals.pcp.a029048

Kós PB, Deák Z, Cheregi O, Vass I (2008) Differential regulation of psbA and psbD gene expression, and the role of the different D1 protein copies in the cyanobacterium Thermo*synechococcus elongatus* BP-1. Biochim Biophys Acta 1777:74–83. doi:10.1016/j.bbabio.2007.10.015

Koua FHM, Umena Y, Kawakami K, Shen J-R (2013) Structure of Sr-substituted photosystem II at 2.1 Å resolution and its implications in the mechanism of water oxidation. P Natl Acad Sci USA 110:3889–3894. doi: 10.1073/pnas.1219922110

Krieger A, Rutherford AW, Johnson GN (1995) On the determination of redox midpoint potential of the primary quinone electron acceptor, Q_A, in Photosystem II. BBA – Bioenergetics 1229:193–201. doi: 10.1016/0005-2728(95)00002-Z

Krieger-Liszkay A, Rutherford AW (1998) Influence of herbicide binding on the redox potential of the quinone acceptor in photosystem II: relevance to photodamage and phytotoxicity. Biochemistry 37:17339–17344. doi: 10.1021/bi9822628

Lambreva MD, Russo D, Polticelli F, et al. (2014) Structure/Function/Dynamics of Photosystem II Plastoquinone Binding Sites. Curr Protein Pept Sci 15:285–295.

Larom S, Salama F, Schuster G, Adir N (2010) Engineering of an alternative electron transfer path in photosystem II. P Natl Acad Sci USA 107:9650–9655. doi: 10.1073/pnas.1000187107

Larom S, Kallmann D, Saper G, et al. (2015) The Photosystem II D1-K238E mutation enhances electrical current production using cyanobacterial thylakoid membranes in a bio-photoelectrochemical cell. Photosynth Res 126:161–169. doi: 10.1007/s11120-015-0075-3

Lavergne J (1982) Mode of action of 3-(3,4-dichlorophenyl)-1,1-dimethylurea. Evidence that the inhibitor competes with plastoquinone for binding to a common site on the acceptor side of Photosystem II. BBA – Bioenergetics 682:345–353. doi:10.1016/0005-2728(82)90048-2

Leng J, Sakurai I, Wada H, Shen J-R (2008) Effects of phospholipase and lipase treatments on photosystem II core dimer from a thermophilic cyanobacterium. Photosynth Res 98:469–478. doi: 10.1007/s11120-008-9335-9

Maly J, Masojídek J, Masci A, et al. (2005) Direct mediatorless electron transport between the monolayer of photosystem II and poly(mercapto-p-benzoquinone) modified gold electrode—new design of biosensor for herbicide detection. Biosens Bioelectron 21:923–932. doi: 10.1016/j.bios.2005.02.013

McEvoy JP, Brudvig GW (2006) Water-splitting chemistry of photosystem II. Chem Rev 106:4455–4483. doi: 10.1021/cr0204294

Miles D, Bolen P, Farag S, et al. (1973) Hg^{++} – A DCMU independent electron acceptor of Photosystem II. Biochem Biophys Res Co 50:1113–1119.

Miller AF, Brudvig GW (1991) A guide to electron paramagnetic resonance spectroscopy of Photosystem II membranes. BBA – Bioenergetics 1056:1–18. doi:10.1016/S0005-2728(05)80067-2

Mizusawa N, Wada H (2012) The role of lipids in photosystem II. BBA – Bioenergetics 1817:194–208. doi: 10.1016/j.bbabio.2011.04.008

Mohanty N, Vass I, Demeter S (1989) Impairment of photosystem 2 activity at the level of secondary quinone electron acceptor in chloroplasts treated with cobalt, nickel and zinc ions. Physiol Plant 76:386–390. doi: 10.1111/j.1399-3054.1989.tb06208.x

Müh F, Zouni A (2013) The nonheme iron in photosystem II. Photosynth Res 116:295–314. doi: 10.1007/s11120-013-9926-y

Mulo P, Sakurai I, Aro E-M (2012) Strategies for psbA gene expression in cyanobacteria, green algae and higher plants: From transcription to PSII repair.

BBA – Bioenergetics 1817:247–257. doi: 10.1016/j.bbabio.2011.04.011

Oettmeier W (1999) Herbicide resistance and supersensitivity in photosystem II. Cell Mol Life Sci 55:1255–1277. doi: 10.1007/s000180050370

Ohad I, Dal Bosco C, Herrmann RG, Meurer J (2004) Photosystem II proteins PsbL and PsbJ regulate electron flow to the plastoquinone pool. Biochemistry-US 43:2297–2308. doi: 10.1021/bi0348260

Ohnishi N, Kashino Y, Satoh K, et al. (2006) Chloroplast-encoded polypeptide PsbT Is Involved in the repair of primary electron acceptor Q_A of Photosystem II during photoinhibition in *Chlamydomonas reinhardtii*. J Biol Chem 282:7107–7115. doi: 10.1074/jbc.M606763200

Pagliano C, Saracco G, Barber J (2013) Structural, functional and auxiliary proteins of photosystem II. Photosynth Res 116:167–188. doi: 10.1007/s11120-013-9803-8

Petrouleas V, Diner BA (1987) Light-induced oxidation of the acceptor-side Fe(II) of Photosystem II by exogenous quinones acting through the Q_B binding site. I. Quinones, kinetics and pH-dependence. BBA – Bioenergetics 893:126–137. doi: 10.1016/0005-2728(87)90032-6

Rao KK, Hall DO, Vlachopoulos N, et al. (1990) Photoelectrochemical responses of photosystem II particles immobilized on dye-derivatized TiO_2 films. J Photoch Photobio B 5:379–389. doi: 10.1016/1011-1344(90)85052-X

Regel RE, Ivleva NB, Zer H, et al. (2001) Deregulation of electron flow within photosystem II in the absence of the PsbJ protein. J Biol Chem 276:41473–41478. doi: 10.1074/jbc.M102007200

Shevela D, Messinger J (2012) Probing the turnover efficiency of photosystem II membrane fragments with different electron acceptors. Biochim Biophys Acta 1817:1208–1212. doi: 10.1016/j.bbabio.2012.03.038

Shibamoto T, Kato Y, Sugiura M, Watanabe T (2009) Redox Potential of the Primary Plastoquinone Electron Acceptor Q_A in Photosystem II from *Thermosynechococcus elongatus* Determined by Spectroelectrochemistry. Biochemistry-US 48:10682–10684. doi: 10.1021/bi901691j

Shinopoulos KE, Brudvig GW (2012) Cytochrome b_{559} and cyclic electron transfer within Photosystem II. BBA – Bioenergetics 1817:66–75. doi:10.1016/j.bbabio.2011.08.002

Sigfridsson KGV, Bernat G, Mamedov F, Styring S (2004) Molecular interference of Cd^{2+} with Photosystem II. BBA – Bioenergetics 1659:19–31. doi: 10.1016/j.bbabio.2004.07.003

Suga M, Akita F, Hirata K, et al. (2015) Native structure of photosystem II at 1.95 Å resolution viewed by femtosecond X-ray pulses. Nature 517:99–103. doi: 10.1038/nature13991

Sugiura M, Inoue Y (1999) Highly purified thermo-stable oxygen-evolving photosystem II core complex from the thermophilic cyanobacterium *Synechococcus elongatus* having His-tagged CP43. Plant Cell Physiol 40:1219–1231.

Takahashi R, Hasegawa K, Takano A, Noguchi T (2010) Structures and binding sites of phenolic herbicides in the Q_B pocket of photosystem II. Biochemistry-US 49:5445–5454. doi: 10.1021/bi100639q

Takasaka K, Iwai M, Umena Y, et al. (2010) Structural and functional studies on Ycf12 (Psb30) and PsbZ-deletion mutants from a thermophilic cyanobacterium. BBA – Bioenergetics 1797:278–284. doi: 10.1016/j.bbabio.2009.11.001

Tanaka A, Fukushima Y, Kamiya N (2017) Two different structures of the oxygen-evolving complex in the same polypeptide frameworks of photosystem II. J Am Chem Soc 139:1718–1721. doi: 10.1021/jacs.6b09666

Trebst A (2007) Inhibitors in the functional dissection of the photosynthetic electron transport system. Photosynth Res 92:217–224. doi: 10.1007/s11120-007-9213-x

Ulas G, Brudvig GW (2011) Redirecting Electron Transfer in Photosystem II from Water to Redox-Active Metal Complexes. J Am Chem Soc 133:13260–13263. doi: 10.1021/ja2049226

Umena Y, Kawakami K, Shen J-R, Kamiya N (2012) Crystal structure of oxygen-evolving Photosystem II at a resolution of 1.9 Å. Nature 473:55–60. doi: 10.1038/nature09913

Van Rensen J, van der Vet W, van Vliet W (1977) Inhibition and uncoupling of electron transport in isolated chloroplasts by the herbicide 4,6-dinitro-o-cresol. Photochem Photobiol 25:579–583.

Vinyard DJ, Ananyev GM, Dismukes GC (2013a) Photosystem II: the reaction center of oxygenic photosynthesis. Annu Rev Biochem 82:577–606. doi: 10.1146/annurev-biochem-070511-100425

Vinyard DJ, Gimpel J, Ananyev GM, et al. (2013b) Natural Variants of Photosystem II Subunit D1 Tune Photochemical Fitness to Solar Intensity. J Biol Chem 288:5451–5462. doi: 10.1074/jbc.M112.394668

Vogt L, Vinyard DJ, Khan S, Brudvig GW (2015) Oxygen-evolving complex of Photosystem II: an analysis of second-shell residues and hydrogen-bonding networks. Curr Opin Chem Bio 25:152–158. doi: 10.1016/j.cbpa.2014.12.040

Wada H, Murata N (2010) Lipids in thylakoid Membranes and Photosynthetic Cells. In: Lipids in Photosynthesis. Springer Netherlands, Dordrecht, pp 1–9

Yehezkeli O, Tel-Vered R, Michaeli D, et al. (2013) Photosynthetic reaction center-functionalized electrodes for photo-bioelectrochemical cells. Photosynth Res 120:71–85. doi: 10.1007/s11120-013-9796-3

Yruela I, Montoya G, Alonso PJ, Picorel R (1991) Identification of the pheophytin-Q_A-Fe domain of the reducing side of the photosystem II as the Cu(II)-inhibitory binding site. J Biol Chem 266:22847–22850.

Zhang Y, Magdaong N, Frank HA, Rusling JF (2013) Protein film voltammetry and co-factor electron transfer dynamics in spinach photosystem II core complex. Photosynth Res 120:153–167. doi: 10.1007/s11120-013-9831-4

Zhu Z, Su Y, Li J, et al. (2009) Highly sensitive electrochemical sensor for mercury(II) ions by using a mercury-specific oligonucleotide probe and gold nanoparticle-based amplification. Anal Chem 81:7660–7666. doi: 10.1021/ac9010809

Zouni A, Witt HT, Kern J, et al. (2001) Crystal structure of Photosystem II from *Synechococcus elongatus* at 3.8 Å resolution. Nature 409:739–743. doi: 10.1038/35055589

The Cl⁻ Requirement for Oxygen Evolution by Photosystem II Explored Using Enzyme Kinetics and EPR Spectroscopy

Alice Haddy, Vonda Sheppard, Rachelle Johnson, and Eugene Chen

Summary

Chloride is a well-known activator of oxygen evolution activity in photosystem II. Its effects have been characterized over several decades of research, as methods have developed and improved. By replacing chloride with other small anions with a range of chemical properties, a picture of the requirements of a successful anion activator can be formulated. In this review, the results of experiments on the chloride effect using enzyme kinetics methods and electron paramagnetic resonance spectroscopy are described, with summaries for the major anion activators and inhibitors that have been studied.

Keywords

Chloride • Photosystem II • Oxygen evolution • Electron paramagnetic resonance spectroscopy • Water oxidation • Water splitting • Inhibitor • Activator • Kinetics

Contents

5.1 Introduction 67
5.2 Chloride in PSII 68
    5.2.1 Early Studies of the Cl⁻ Dependence in O₂ Evolution 68
    5.2.2 Cl⁻ Depletion of PSII Preparations from Higher Plants 69
    5.2.3 S₂ State EPR Signals and Chloride 70
    5.2.4 XRD and EXAFS Structural Characterizations of Cl⁻ Sites 72
5.3 Activators 74
    5.3.1 Studies of Cl⁻ Activation in Cyanobacteria 74
    5.3.2 Quantifying the Affinity of Cl⁻ Binding Sites in Higher Plant PSII 76
    5.3.3 Bromide and Nitrate Activators 79
5.4 Inhibitors 80
    5.4.1 Ammonia 80
    5.4.2 Fluoride 82
    5.4.3 Acetate 83
    5.4.4 Azide 85
    5.4.5 Iodide 86
    5.4.6 Nitrite 88
5.5 Concluding Remarks 89
References 89

5.1 Introduction

Chloride is an important inorganic ion cofactor required for oxygen evolution by photosystem II (PSII) (for earlier reviews, see: van Gorkom and Yocum 2005; Popelkova and Yocum 2007; Yocum 2008; Pokhrel et al. 2011). With recent advances in X-ray diffraction studies of PSII from thermophilic cyanobacteria, two chloride binding sites (Murray et al. 2008; Guskov et al. 2009; Kawakami et al. 2009; Umena et al. 2011) have

A. Haddy (✉) • V. Sheppard • R. Johnson • E. Chen
Department of Chemistry and Biochemistry, University of North Carolina at Greensboro, Greensboro, NC 27402, USA
e-mail: aehaddy@uncg.edu

been identified at approximately 6–7 Å from the catalytic Mn_4CaO_5 cluster, where O_2 is produced from two H_2O molecules. Although the efforts of many researchers have been devoted to unraveling the details of its function, the mode of how it activates the catalysis of oxygen evolution is still not well understood. This is not surprising, given the complexity of the oxygen production reaction, a high energy demanding reaction that carries out a four-electron oxidation of water molecules. To accomplish this reaction, the Mn_4CaO_5 cluster cycles through five major oxidation states or S-states, designated S_0 through S_4, releasing molecular oxygen from the transient S_4 state. Electrons are withdrawn from the Mn cluster to a nearby tyrosine radical, Tyr Z, promoted by the absorption of light at the reaction center of PSII, P680. Oxygen evolution also requires the well-controlled intake of water molecules and coordinated release of O_2 and protons.

In this review will be described the various studies to explore the Cl^- requirement for the activation of oxygen evolution using the kinetics of oxygen evolution and electron paramagnetic resonance (EPR) spectroscopy. Since EPR spectroscopy is based on the spin states of paramagnetic species, i.e., those with one or more unpaired electrons, it is able to give details about the state of the Mn_4CaO_5 cluster and other electron transfer centers at various points in catalysis. Researchers have employed a variety of anion activators and inhibitors to probe the Cl^- site. Given the different properties the anions possess, this approach has helped characterize the features of Cl^- that contribute to activation. These studies suggest that Cl^- contributes to oxygen evolution in multiple related ways, such as promotion of proton movement, H-bond coordination, and stabilization of the ligation of the Mn_4CaO_5 cluster.

5.2 Chloride in PSII

5.2.1 Early Studies of the Cl^- Dependence in O_2 Evolution

The importance of chloride in the function of water oxidation by photosystem II was first observed by Warburg and Lüttgens and described in a brief note (Warburg and Lüttgens 1944). In addition to Cl^-, they found that Cl^-, Br^-, I^-, and NO_3^- could support oxygen evolution in chloroplast fragments, but that thiocyanate (SCN^-), SO_4^{2-} and PO_4^{3-} could not. A later more extensive study published in Russian showed a concentration-dependent stimulating effect of KCl (described by Homann (2002)). Similar observations were made by Arnon and Whatley (1949), however they concluded that Cl^- was not directly required (probably because of incomplete removal of Cl^- from chloroplast samples), but rather it probably protected the photosynthetic apparatus from damage by excess light. However Gorham and Clendenning (1952) found that chloride did indeed have a direct stimulating effect on oxygen evolution. This marked the first time that an inorganic ion was shown to be a cofactor in a biochemical reaction, a point of controversy at the time. In the course of their studies, all of these researchers tested and ranked several other anions for effectiveness in substituting for chloride; Gorham and Clendenning also examined the pH dependence with and without chloride.

With the increased use of various artificial electron donors and acceptors that absorb in the visible region, the discovery of an array of inhibitors of photosynthesis, and the development of the Clark-type oxygen electrode (Fork 1972), the steps of electron transfer during photosynthesis could be separated into discrete reactions. Studies to narrow down the role of chloride were carried out by Izawa and coworkers using chloroplasts and successfully demonstrated that the site of Cl^- activation was closely associated with water oxidation (Izawa et al. 1969; Kelley and Izawa 1978). Among other observations, electron transfer from donors that bypass the site of O_2 evolution was found to show no dependence on Cl^- (Izawa et al. 1969) and Cl^- was found to slow inactivation of O_2 evolution by hydroxylamine, which was known to reduce functional Mn (Kelley and Izawa 1978). The concentration dependence of the Cl^- effect indicated that Cl^- could be viewed as an enzyme activator that followed simple steady state kinetics (i.e., the Michaelis-Menten equation), for which an activation constant (K_M) of 0.9 mM at pH 7.2 was found (Kelley and Izawa 1978). Although standard for the study of enzyme activators, the approach had seldom if ever been applied to the treatment of small ionic activators. Full activation was found when 10 mM Cl^- was present, as had been observed in earlier studies.

Another important step was taken when it was found that Cl^- was required for advancement beyond the S_2 state, but that it does not inhibit the transitions leading up to that oxidation state. This was first discovered in studies of chlorophyll fluorescence, which occurs during normal electron transfer primarily in photosystem II due to the build-up of excited-state chlorophyll as the quinone acceptors Q_A and Q_B become reduced. In normal electron transfer, each S-state transition is preceded by the reduction of $P680^+$ by Tyr Z within 10 μs of light absorption and formation of $P680^+$ by electron transfer from excited state P680* to the quinone acceptor. Reduction of $P680^+$ by Tyr Z can be seen as a transient increase in chlorophyll fluorescence upon reduction of $P680^+$ in response to a flash of light. If normal electron transfer from Tyr Z is prevented, for example by preventing its re-reduction by the Mn_4CaO_5 complex, then the increase in chlorophyll fluorescence will be slowed. Thus in studies of Cl^--depleted chloroplasts (by treatment at pH 7.8 with 100 mM Na_2SO_4), study of chlorophyll fluorescence transients showed that the S_0-to-S_1 and S_1-to-S_2 transitions occurred normally, but that further transitions could not take place (Itoh et al. 1984; Theg et al. 1984). Later studies confirmed and expanded this finding using additional methods, as will be described below.

Advances in preparative methods during the early 1980s led to more detailed studies of all aspects of photosystem II, including its Cl^- requirement. Of major importance was the introduction of the Triton X-100 method for extraction of oxygen-evolving PSII membrane fragments from spinach and other higher plants ((Berthold et al. 1981), with modifications such as those in Ford and Evans (1983) or Kuwabara and Murata (1982)). Soon the characterization of higher plant PSII preparations led to various methods for removal of extrinsic subunits that were found to control the access of ions to the oxygen evolving complex (OEC). (For reviews of the extrinsic subunits of PSII, see: Roose et al. 2007 and Bricker et al. 2012). Particularly significant for the study of the Cl^- requirement was the removal of the extrinsic subunits PsbP and PsbQ (also known as the 23 kDa and 17 kDa subunits, respectively) by NaCl-washing in 1–2 M NaCl (Kuwabara and Murata 1983; Miyao and Murata 1983; Ghanotakis et al. 1984a). This treatment led to a preparation that was found to have a previously unsuspected dependence on Ca^{2+} (which we now recognize to be that associated with the Mn_4CaO_5 cluster) as well as an increased requirement for Cl^-, both of which were closely associated with the PsbP subunit (Kuwabara and Murata 1983; Andersson et al. 1984; Ghanotakis et al. 1984a, 1984b, 1985; Miyao and Murata 1984, 1985). In the case of the Cl^- requirement, it was found that the binding of PsbP to NaCl-washed PSII decreased the optimum Cl^- concentration for oxygen evolution activity from 30 to 10 mM, whereas PsbQ, which requires PsbP to bind, had a relatively minor effect on the Cl^- requirement (Miyao and Murata 1985). It was also found that when the extrinsic subunit PsbO (also known as the 33 kDa subunit or manganese stabilizing protein) was removed in addition to PsbP and PsbQ, 150 mM Cl^- or more was required to promote activity (Miyao and Murata 1985; Ono and Inoue 1986). Observation of the effect of removal of the extrinsic subunits led to the descriptions of their regulatory role in terms of creating a barrier to Cl^- (and Ca^{2+}) movement or serving as Cl^- (and Ca^{2+}) concentrators. This description has been reinforced by numerous subsequent observations by other researchers.

5.2.2 Cl^- Depletion of PSII Preparations from Higher Plants

To study enzyme activation, it is necessary to remove the activator from the enzyme and solution so that the effects of adding it back can be monitored. To accomplish this for the study of Cl^-

activation in PSII, a number of Cl^- depletion methods have been developed for higher plant PSII, with or without removal of the extrinsic subunits PsbP and PsbQ. Early methods for Cl^- depletion of both chloroplast thylakoids and PSII preparations often involved an increase in pH. One such method was to briefly expose PSII to pH ~ 10 conditions followed by reduction of the pH within seconds back to a more neutral range (van Vliet and Rutherford 1996; Homann 1985). Another commonly used method involved treatment of PSII at pH 7.5 to 8.0 in the presence of Na_2SO_4 at a concentration of 50 mM or so (Sandusky and Yocum 1983; Homann 1988a; Wincencjusz et al. 1997; Kuwabara and Murata 1982). This method, it was soon realized, also removes the extrinsic PsbP and PsbQ subunits while leaving some Ca^{2+} bound (Wincencjusz et al. 1997). Indeed removal of the PsbP and PsbQ subunits by either pH 7.5/50 mM Na_2SO_4 or by treatment with 1–2 M NaCl greatly facilitates Cl^- depletion because, as noted above, the ion diffusion barrier presented by these subunits is thereby removed allowing essentially complete Cl^- removal. Moreover the activation constant (or dissociation constant) for Cl^- is found to be in the mM range without the complication of higher affinity Cl^- binding (described further below). Under these conditions, the exchange of Cl^- or other ions at the site of activation is essentially unrestricted so that equilibrium with the surrounding solution occurs easily. While valid concerns can be raised that loss of the extrinsic subunits alters the properties of the OEC, the degree to which their removal simplifies the study of Cl^- activation makes this concern of secondary importance for many researchers.

On the other hand, depletion of Cl^- from intact PSII without increasing the pH or removing the extrinsic subunits appears to be quite difficult. Intact PSII assayed directly in Cl^- free buffer shows essentially full activity and even wash by centrifugation to remove Cl^- succeeds in lowering the activity by little more than 10–20%. However, more extensive dialysis for up to 24 h using purified reagents can remove enough Cl^- to decrease the activity significantly, although at least 30% activity still remains in the best of cases (Lindberg and Andréasson 1996; Olesen and Andréasson 2003, our observations). An important factor for intact PSII, in addition to the presence of the subunit "ion diffusion barrier", is that in intact PSII a fraction of the preparation exhibits a high affinity Cl^- dissociation constant in the micromolar range (Lindberg and Andréasson 1996; Olesen and Andréasson 2003). A dissociation constant in this range would be difficult to quantify even under unrestricted exchange conditions because the trace concentration of Cl^- in "Cl^- free buffers" is generally also in the micromolar range. Researchers who test the background Cl^- concentration in "Cl^- free buffer" have found values ranging from 10 to 100 μM depending on the reagents used (our observations; also see, for example: Lindberg and Andréasson 1993, Miyao and Murata 1985, Wincencjusz et al. 1998, Gorham and Clendenning 1952). Furthermore, there is no way at this time to remove Cl^- from an aqueous solution that will not leave the solution contaminated with an unwanted cation (such as Ag^+); for Cl^-, there is no equivalent to the Ca^{2+} chelators EDTA and EGTA. Therefore it stands to reason that the activity remaining after extensive dialysis of intact PSII against Cl^- free buffer may represent the equilibrium fraction of PSII with Cl^- bound to a high affinity site.

In some cases, Cl^--depleted PSII with the PsbP and PsbQ subunits bound has been prepared by first depleting of subunits and Cl^- by a standard method, then allowing the PsbP and PsbQ subunits to rebind (Boussac et al. 1989; Rachid and Homann 1992). This not only addresses concerns that the removal of extrinsic subunits alters the characteristics of the Cl^- site, but this method also facilitates the preparation of PSII with Cl^- or other anion that is specific bound.

5.2.3 S_2 State EPR Signals and Chloride

The S_2 state of higher plant PSII shows two well-known electron paramagnetic resonance signals that arise from the same overall oxidation state of the Mn_4CaO_5 cluster, which is a combination of $(Mn^{4+})_3(Mn^{3+})$. (For reviews see: Peloquin and

Britt 2001, Haddy 2007, and Brynda and Britt 2010). Discovered in the early 1980s (Dismukes and Siderer 1980, 1981; Casey and Sauer 1984; Zimmermann and Rutherford 1984), the signals both confirmed the presence of a coupled manganese cluster at the core of the OEC and opened avenues to study of its mechanism. One EPR signal is a multiline signal centered at $g = 2$ consisting of at least 19 main lines and arising from a spin $S = 1/2$ state of the cluster. The other is a broad featureless signal of line width 350–400 G centered at $g = 4.1$ and arising from a spin $S = 5/2$ state of the cluster. The identity of both signals as originating from the S_2 state was confirmed in studies of signal intensity in response to light flashes or by limiting the transfer of electrons using inhibitors such as DCMU (Dismukes and Siderer 1981; Brudvig et al. 1983; Zimmermann and Rutherford 1986). The relative amounts of the two signals are affected by many factors, including the cryoprotectant used, the temperature of illumination, the presence of alcohols, and the presence of Cl^- or other anions. Both signals can appear in the spectrum of active higher plant PSII, although the cryoprotectant and temperature of illumination may influence the representation of each signal.

It was found in early studies of the S_2 state EPR signals that Cl^- is required for formation of the multiline signal, but that the $g = 4.1$ signal can form both in the presence or absence of Cl^- or when Cl^- is substituted with a variety of inhibitory anions. Many of the first studies of the effects of anion substitution on the S_2 state signal employed Cl^- depletion of PSII using pH 7.5/50 mM Na_2SO_4 treatment and related methods, followed by addition of a series of anions at a concentration of 10–25 mM. Thus it was found that Br^- supported formation of the multiline signal approximately as well as Cl^-, whereas other anions tested, including NO_3^-, F^-, SO_4^-, I^-, and CH_3COO^-, did not (Damoder et al. 1986; Yachandra et al. 1986; Ono et al. 1987). In general, the effects of anions on the S_2 state multiline signal tended to follow the same trend as their effects on oxygen evolution activity. While the $g = 4.1$ signal in the presence of most anions was similar to that in Cl^--depleted PSII, F^- caused a distinct enhancement and line width narrowing of the signal (Casey and Sauer 1984; Yachandra et al. 1986). Details of the effects of various anions will be described further for individual anions below. At this point it is of interest to recall that it was known from other studies that the absence of Cl^- did not prevent advancement to the S_2 state, but did prevent advancement beyond the S_2 state. One study of the EPR signals showed that if the OEC was advanced using light flashes to the S_2 state in Cl^--depleted PSII (pH 7.5/50 mM Na_2SO_4), then subsequent rapid addition of Cl^- led to formation of the multiline signal (Ono et al. 1986). A similar result was obtained when Cl^--depleted PSII was advanced to the S_2 state using 200 K illumination, then rapidly reconstituted with Cl^- (Boussac and Rutherford 1994). Evidently, the S_2 oxidation state achieved in the absence of Cl^- does not include the coupled state associated with the multiline signal ($S = 1/2$). Since these results did not correspond to noteworthy changes in the $g = 4.1$ signal, they led to the suggestion that an EPR-silent S_2 state is formed.

In the absence of the extrinsic PsbP and PsbQ subunits, the multiline signal can form in high yield as long as the concentrations of Ca^{2+} and Cl^- are sufficiently high, i.e. 10–25 mM range (de Paula et al. 1986; Imaoka et al. 1986), as expected from oxygen evolution measurements. The multiline signal is essentially the same as that observed in intact PSII. The $g = 4.1$ signal can also be observed, although formation may be impeded when using 130 K illumination in the presence of ethylene glycol (de Paula et al. 1986). The multiline signal is able to form in the absence of all three extrinsic subunits, including PsbO, however in this case the Cl^- concentration must be increased to about 200 mM (Styring et al. 1987). In another study, it was found that in $CaCl_2$-washed PSII lacking all three extrinsic subunits, the removal of Cl^- resulted in the uncoupling of a portion of Mn, which was observable as the appearance of protein-bound Mn^{2+} by Q-band EPR spectroscopy at room temperature (Mavankal et al. 1986); this Mn^{2+} could be reincorporated into a coupled state by the addition of very high Cl^- concentrations (> 400 mM).

Another EPR signal from the OEC to receive attention because of its relationship to anion activation is the $S_2Y_Z^\bullet$ signal from inhibited PSII. This signal appears as a broad signal consisting of two major lines, i.e. a "split" signal, at $g = 2$. Production of the signal is generally done by illuminating the inhibited sample at a temperature of 0 °C or above in the presence of an electron acceptor, since it requires the transfer of two electrons. This type of signal was first observed in Ca^{2+}-depleted PSII (Boussac et al. 1989, 1990b; Sivaraja et al. 1989; Ono and Inoue 1990) and was recognized as arising from the interaction between the S_2 state Mn_4CaO_5 cluster and an organic radical, which was later identified as Tyr Z. The treatment of PSII with anion inhibitors such as F^- (Baumgarten et al. 1990; DeRose et al. 1995), ammonia (Andréasson and Lindberg 1992; Hallahan et al. 1992), or acetate (MacLachlan and Nugent 1993, Szalai and Brudvig 1996a, b) also leads to similar $S_2Y_Z^\bullet$ signals, with line width depending on the treatment. The presence of the signal indicates that electron transfer from the Mn_4CaO_5 cluster to the Tyr Z radical has been blocked, a result that correlates well with the known effect of Cl^- depletion in preventing advancement beyond the S_2 state.

5.2.4 XRD and EXAFS Structural Characterizations of Cl^- Sites

Before the first X-ray diffraction (XRD) structure of PSII was determined, Cl^- bound near the Mn_4CaO_5 cluster eluded detection as a scatterer of Mn K-edge electrons by EXAFS methods (Sauer et al. 2008; Yano and Yachandra 2009; Grundmeier and Dau 2012). In samples in which Cl^- was replaced with F^-, using conditions in which a strong $g = 4.1$ EPR signal was observed, the EXAFS revealed subtle changes in the Mn cluster distances in the S_1 and S_2 states due to F^- binding (DeRose et al. 1995). Those changes for the S_2 state were similar to changes seen for the samples in which the $g = 4.1$ signal was induced at low temperature. Comparison of the Mn-EXAFS of PSII from cyanobacteria grown on Br^- and Cl^- suggested differences that could be due to ligation of the anion (Klein et al. 1993). These ambiguous results using Mn-EXAFS suggested that Cl^- was outside the first coordination sphere of Mn, although several researchers pointed out that a single halide scatterer would show only a faint signal in the first coordination sphere of a cluster of 4 Mn atoms. Later a study using Br EXAFS was carried out in intact higher plant PSII, in which Br^- replaced Cl^- after Cl^- depletion by the dialysis method (Haumann et al. 2006). This shifted the focus to the halide as the central ion with Mn in the role of scatterer and took advantage of the relatively high X-ray absorption edge energy of Br to enhance sensitivity. It was shown that Br was not directly bound to Mn or Ca in the S_1 state, however a distance to a single Mn/Ca atom of ~5 Å or more was revealed (Haumann et al. 2006).

Since 2008, two binding sites for Cl^- near the Mn_4CaO_5 cluster have been characterized in XRD studies of PSII core complexes from thermophilic cyanobacteria (*Thermosynechococcus elongatus* or *T. vulcanus*), which have proven amenable to crystallization. Two early studies at relatively low resolution (~4 Å) using Br^- substituted PSII preparations revealed the two Cl^- binding sites at 6–7 Å from the Mn_4CaO_5 cluster and identified several ligands to the anion (Murray et al. 2008; Kawakami et al. 2009). In the study by Murray and coworkers, the sites were located and characterized using X-ray anomalous diffraction, which is carried out using the X-ray energy at the absorption edge of the atom of interest, in this case Br. In the study by Kawakami and coworkers, the two sites were revealed in PSII crystals prepared using a crystallization buffer in which Cl^- was replaced with Br^- or I^- (Kawakami et al. 2009). The two anion sites were found using difference Fourier maps comparing the anion substituted crystals with Cl^- containing PSII crystals. In the I^- substituted crystals, three additional I^-

sites were revealed at more distant locations from the Mn_4CaO_5 cluster, including one at the redox active Tyr D residue (D2-Tyr160); these additional sites were thought not to be associated with functional Cl^- sites. A higher resolution study (2.9 Å) showed a single bound Cl^-, referred to as Cl-1, located 6.5 Å from Mn-4 of the Mn_4CaO_5 cluster (Guskov et al. 2009). The second Cl^- site was not observed in this study, perhaps because of the sites' differing binding affinities. The site of bound Cl^- was confirmed in crystals in which Cl^- was replaced with Br^- by growing the cyanobacteria on Br^- containing media.

Finally an XRD study carried out at 1.9 Å resolution, revealing the placement of water molecules for the first time, showed the two Cl^- binding sites in their best detail so far with Cl-1 located 6.7 Å from Mn-4 and Cl-2 located 7.4 Å from Mn-2 (Kawakami et al. 2011; Umena et al. 2011). In this study 1.75 Å wavelength X-rays were used to assist in the location of Cl^- sites with data processed to 2.5 Å resolution, while the main data set was taken using 0.9 Å wavelength X-rays. A more recent structure has provided data at 1.95 Å resolution using X-ray free electron laser (XFEL) (Suga et al. 2015), which by using pulsed X-rays avoids damage to crystals that occurs during sustained X-ray radiation. Based on these structures and subsequent analyses by other researchers, the coordination environment of the two Cl^- ions has become well characterized. As is typical for biological Cl^-, each site shows coordination of Cl^- by ligands from main-chain amide nitrogens, side-chain nitrogens, and water molecules. The coordination environment of Cl-1 is made up of D2-Lys317 (side-chain), D1-Glu333 (main-chain), D1-Asn181 (side-chain), and two water molecules, while the coordination environment of Cl-2 is made up of CP43-Glu354 (main-chain), D1-Asn338 (main-chain), and either D1-His337 (main-chain) or D1-Phe339 (main-chain), and two water molecules (Rivalta et al. 2011; Shoji et al. 2015). The two sites show an interesting similarity in that they are each coordinated to the main-chain nitrogen of a residue that also coordinates two Mn ions through a side-chain carboxylate (Mn-3 and Mn-4 in the case of Cl-1 and Mn-2 and Mn-3 in the case of Cl-2). This characteristic is suggestive of a role in stabilizing the structure of the Mn_4CaO_5 cluster (Kawakami et al. 2011; Umena et al. 2011). A distinct difference between the two binding environments is that Cl-1 is coordinated by an ionic group, D2-Lys317, whereas the coordination environment of Cl-2 includes all neutral groups. This is expected to lead to a major difference in the affinity of binding of Cl^- between the two sites. The binding affinity at the Cl-1 site has been predicted to be about 10 kcal mol^{-1} higher than at the Cl-2 site (Rivalta et al. 2011).

A third Cl^- binding site (Cl-3) was identified in the 1.9 Å resolution XRD study, located near the extrinsic subunits of cyanobacteria PSII, PsbU and PsbV, about 25 Å away from the Mn_4CaO_5 cluster (Kawakami et al. 2011). This chloride was found to be coordinated by six water molecules, indicative of weak binding affinity. Its location in a water channel suggests a route for exchange of the activating anion, as well as substrate water.

In a 3.2 Å resolution study of the binding of the herbicide terbutryn to PSII, two separate positions were found for Cl^- binding in the region of Cl-1 (Broser et al. 2011), suggesting some flexibility in the binding of Cl^-. One Cl^- position (Cl-1A), located 7.1 Å from Mn-4, was essentially the same as that reported in previous studies and represented 30% occupancy in the sample. A second position (Cl-1B), located 8.7 Å from Mn-4 and 6.7 Å from the Cl-1A site, was reported for the first time and represented 70% occupancy. This site also involved coordination by D2-Lys317 (side-chain), but the other residues were different and included D1-Arg334 (side-chain) and D1-Asn335 (side-chain). The authors suggested that the side chain of D2-Lys317 may shift position depending on the Cl^- position, although this was not resolved in the structure. The finding of two possible Cl-1 positions suggests that the Cl^- position may be influenced by the presence of herbicide on the acceptor side of PSII;

alternatively, Cl^- may move between sites during its function, in a manner that corresponds to a change in affinity.

It was noted from the first studies that the Cl-1 binding site is positioned near a likely proton channel, sometimes called the broad channel (Murray et al. 2008; Guskov et al. 2009; Kawakami et al. 2009; Umena et al. 2011). In addition, the observation of long bonding distances to protonated species from higher resolution studies is suggestive of hydrogen bonding and weak binding (Kawakami et al. 2011; Shoji et al. 2015; Vogt et al. 2015). The position of Cl-1 places it near D1-Asp 61, which is a residue often favored for the role of proton transfer based on mutagenesis studies (Dilbeck et al. 2012; Debus 2014). The shift in position of Cl-1 related to the binding of terbutryn also supports a role in proton transfer, since the alternative position for Cl-1B is more clearly associated with the carboxylates from D1-Asp61 and D1-Glu65 within the proton channel (Broser et al. 2011).

Structural studies have contributed much to the development of recent hypotheses for how Cl^- promotes activation of oxygen evolution. There is general agreement that Cl^- (particularly Cl-1) is involved in facilitating proton and/or water movement by influencing the hydrogen bonding of waters and nearby amino acid residues. Visualization of its placement with respect to channels allows examination of this in detail. In one hypothesis for the mechanism of action of Cl^-, it is proposed that Cl^- prevents the formation of a salt bridge between D2-Lys317 and D1-Asp61; in the absence of Cl^-, formation of the salt bridge causes movement of residues that influence the Mn_4CaO_5 cluster and interferes with the ability of D1-Asp61 to participate in proton transfer (Pokhrel et al. 2011; Rivalta et al. 2011). Other researchers have proposed that Cl^- depletion is likely to involve replacement of the missing anion with hydroxyl ion, which would be readily available from nearby water channels. In this hypothesis OH^-, like other Lewis base inhibitors, is unable to function in the place of Cl^- and thus the role of Cl^- may actually be to exclude OH^- from the site (van Gorkom and Yocum 2005; Yocum 2008). Recent modeling studies have optimized the hydrogen bond network, including Cl^- in the channels near the OEC. Results suggest that Cl^-, along with other factors, affects the relative stability of two conformations of the Mn_4CaO_5 complex, which differ by inter-Mn bond distances and are associated with the two S_2 state EPR signals (Shoji et al. 2015).

5.3 Activators

5.3.1 Studies of Cl^- Activation in Cyanobacteria

While the extrinsic PsbP and PsbQ subunits of higher plant PSII were recognized early in the investigations of the PSII subunit structure, homologous subunits of cyanobacterial PSII, called CyanoP and CyanoQ, were found much later because of the difficulty in their detection. Their functions with respect to inorganic ion requirements are less clear than in higher plant PSII. Inactivation of the *psbQ* or *psbP* genes alone was found to either have little effect or a slight slowing effect on the growth of *Synechocystis* sp. PCC 6803 in the absence of Cl^- or Ca^{2+} (Thornton et al. 2004; Summerfield et al. 2005a, b). However O_2 evolution activity in the absence of $CaCl_2$ was significantly reduced for both the $\Delta psbQ$ and $\Delta psbP$ mutants (Thornton et al. 2004). In addition, combinations of $\Delta psbQ$ or $\Delta psbP$ mutants with other subunit deletions led to additional effects related to the inorganic ion requirements (Summerfield et al. 2005a, b). Overall, the results lead to the conclusion that these subunits have roles in the assembly and maintenance of PSII complexes related to Ca^{2+} and Cl^- requirement.

Of importance in controlling the Ca^{2+} and Cl^- requirements in cyanobacteria are two other extrinsic subunits PsbU (12 kDa) and PsbV (15 kDa), also known as cytochrome c_{550}. These subunits, along with the PsbO subunit, are released from PSII by treatment with 1 M $CaCl_2$ or 1 M Tris at pH 8.5, whereas treatment with high NaCl concentration alone does not release any of the extrinsic subunits (Shen et al.

1992; Shen and Inoue 1993). As for higher plant PSII, when all extrinsic subunits are bound, there is little observable dependence on Ca^{2+} or Cl^-. It appears to be especially difficult to Cl^- deplete PSII core complexes from wild type cyanobacteria; for example, in one study even after extensive dialysis in the absence of Cl^- about 75–80% activity remained relative to that observed in the presence of sufficient Cl^- (Pokhrel et al. 2013), whereas the activity of intact PSII from plants can be reduced to 30–35% using the same treatment (Lindberg and Andréasson 1996, our observations). Thus few studies of Cl^- activation have been carried out using wild type cyanobacterial PSII without subunit removal, although values of K_M for Cl^- activation in the mM range have been found for the activatable portion in core complexes from *Synechocystis* sp. PCC 6803 (Pokhrel et al. 2013; Suzuki et al. 2013). Studies of Cl^- activation using $CaCl_2$-washed cyanobacterial PSII are also difficult, since removal of all three subunits by this treatment results in a preparation that, while showing no activity in the absence of Cl^-, is also poorly activated by Cl^- (or Ca^{2+}) addition. However some studies have explored the characteristics of the Cl^- (and Ca^{2+}) requirement in $CaCl_2$-washed cyanobacterial PSII to which selected subunits have rebound. For example, in $CaCl_2$-washed *Thermosynechococcus* PSII with various combinations of subunits rebound, $CaCl_2$ was required for the highest activity in each case tested, while $MgCl_2$ could partially restore activity (Shen and Inoue 1993). A study of red algae *Cyanidium caldarium*, which releases four extrinsic proteins upon treatment with high $CaCl_2$ (where the fourth is a 20 kDa homologue of PsbQ), found a similar pattern of Ca^{2+} and Cl^- requirements with the rebinding of various subunits; in the absence of PsbU and PsbV, the Ca^{2+} and Cl^- requirements were the highest (Enami et al. 1998).

Many studies of the Cl^- and Ca^{2+} requirements in cyanobacterial PSII have involved manipulation of one or more of the genes for extrinsic subunits, including deletion, truncation, or point mutation, and observation of the effects on the Ca^{2+} or Cl^- requirements for growth (among other things). For example, using this approach it was found that PsbO is especially important for Ca^{2+} function, since in deletion mutants of *Synechocystis* there is no growth in the absence of Ca^{2+} (Philbrick et al. 1991; Summerfield et al. 2005b). However, in Cl^- limiting media, some mutants showed growth that was close to that of wild-type (Philbrick et al. 1991; Summerfield et al. 2005b), while others showed essentially no growth (Morgan et al. 1998; Inoue-Kashino et al. 2005), thus there appear to be unidentified factors related to cyanobacterial PsbO that influence Cl^- dependence.

PsbV, or cyt c_{550}, has an important role in maintaining a high affinity for Ca^{2+} and Cl^-, probably by controlling the ion environment or access to the Mn_4CaO_5 cluster. In this way its role is analogous to that of the PsbP subunit in higher plants. Studies of cyanobacteria in which PsbV is deleted show an inability to grow photoautotrophically in the absence of Ca^{2+} or Cl^-, indicating a marked decrease in affinity for these ions. In particular, deletion mutants of PsbV in *Synechocystic* sp. PCC 6803 were found to be unable to grow at all in the absence of Ca^{2+} or Cl^-, although wild-type could (Morgan et al. 1998; Shen et al. 1998). In addition, the loss of PsbV led to unusual S-state patterns, indicating that it is important for function of the Mn_4CaO_5 cluster (Shen et al. 1998). Similar results were found in *T. elongatus*, where disruption of the *psbV* gene (leading to absence of PsbV) resulted in no growth in the absence of Cl^-, while growth of wild-type was only slightly decreased in the absence of Cl^- (Katoh et al. 2001; Kirilovsky et al. 2004). Study of the PsbV homologue PsbV2 revealed that it could not replace PsbV in this function, emphasizing the specialization of PsbV in Cl^- affinity. It was also found that mutation of a His residue coordinating the cytochrome heme center of PsbV, thereby modifying its redox properties, did not result in a phenotype that was growth-impaired in the absence of Cl^- and did not impair activity (Kirilovsky et al. 2004).

It has been found that deletion of the PsbU subunit affects the Cl^- requirement for growth, although not as profoundly as deletion of PsbV. A deletion mutant of *psbU* in *Synechocystis* sp. PCC 6803 could grow as well as wild-type in complete media, but when Ca^{2+} or Cl^- was omitted growth was slower (Shen et al. 1997). A similar deletion mutant was found to be unable to grow in the absence of both Ca^{2+} and Cl^- and to show a longer S_2 state lifetime (Inoue-Kashino et al. 2005), indicating a role for PsbU in optimizing O_2 evolution catalysis. This study also found that Cl^- was required for O_2 evolution in the $\Delta psbU$ mutant and that NO_3^- could promote about 60% of the activity that Cl^- could.

The S_2 state in PSII from cyanobacteria shows a multiline signal that is very similar to that in higher plant PSII, but the $g = 4.1$ EPR signal is generally not observed in wild-type cyanobacteria (McDermott et al. 1988; Yachandra et al. 1996). However, the $g = 4.1$ signal is observed after Sr^{2+} substitution of Ca^{2+} (Strickler et al. 2005; Boussac et al. 2012), along with the altered multiline signal typical of Sr^{2+} treatment in higher plant PSII. While the latter treatment targets the essential Ca^{2+} site, other treatments associated with the site of Cl^- activation also lead to formation of the $g = 4.1$ signal. Ammonia treatment of *T. elongatus* PSII core complexes at pH 7.5 resulted in a $g = 4.1$-type signal that was shifted slightly to $g = 4.25$, along with a multiline signal, after illumination at 190 K (Boussac et al. 2000); the $g = 4.25$ signal disappeared upon warming the sample to 250 K, with concurrent appearance of the NH_3-altered multiline signal, reminiscent of higher plant PSII. Replacement of Cl^- with I^- by an exchange treatment of PSII core complexes was also found to produce the $g = 4.1$ signal (Boussac et al. 2012). Modification of the Cl-1 binding site by mutagenesis to convert the coordinating lysine residue, D2-K317, to arginine in *Synechocystis* sp. PCC 6803 resulted in a PSII core preparation that showed a $g = 4.1$ signal (Pokhrel et al. 2013); because this core preparation could be activated by Cl^- from a level of essentially zero O_2 evolution activity in the absence of Cl^-, whereas the wild-type preparation showed about 75–80% of maximum activity after extensive dialysis to remove Cl^-, the authors proposed that the appearance of the $g = 4.1$ signal was related to lower binding affinity for Cl^- in the D2-K317R mutant. On the other hand, a study of core preparations from two species of cyanobacteria (*Synechocystis* sp. PCC 6803 and *T. lividus*) concluded that the appearance of the $g = 4.1$ signal was correlated with the binding of the extrinsic subunit PsbV (Lakshmi et al. 2002), which based on photoautotrophic growth studies is believed to increase the affinity of PSII for Cl^-. This result is reminiscent of results from higher plant PSII suggesting that the absence of the PsbP and PsbQ subunits suppressed the appearance of the $g = 4.1$ signal. In summary, while studies suggest that the appearance of the $g = 4.1$ signal is significantly influenced by the site of Cl^- activation, as in high plant PSII, characteristics of this influence do not exactly parallel those of higher plants.

5.3.2 Quantifying the Affinity of Cl^- Binding Sites in Higher Plant PSII

In an important set of studies to determine the presence of high affinity or intrinsically bound Cl^-, Lindberg and coworkers used ^{36}Cl to trace its presence in higher plant PSII. Spinach was grown hydroponically on medium supplemented with $Na^{36}Cl$ and used to prepare intact PSII (i.e., retaining the PsbP and PsbQ subunits). Direct measurement of bound ^{36}Cl indicated that PSII contained about one high affinity Cl^- per PSII, with a release half-time of a few hours or less (Lindberg et al. 1990). The estimate of the total bound $^{36}Cl^-$ took into account the PSII preparation time, assuming that Cl^- dissociation was initiated by exposure to the detergent Triton X-100. This study also demonstrated that there was no non-exchangeable Cl^- bound to PSII. The slow-exchanging Cl^- was also studied by monitoring the binding and release of $^{36}Cl^-$ in intact PSII at 0 °C in the dark, revealing half-times of about 1 h in each case (Lindberg et al. 1993). This result was thought to indicate that binding and release took place through the same rate-limiting intramolecular process, leading to the proposal of open and closed conformations for the Cl^- binding site. Scatchard analysis of equilibrium $^{36}Cl^-$ binding showed that under these conditions the number of Cl^- binding sites was about 0.6 per PSII and the dissociation constant K_d was 20 μM. However it was also found that the high affinity Cl^- binding site was lost when PsbP and PsbQ subunits were removed.

Subsequent studies led to the hypothesis that the binding affinity of the activating Cl^- site in intact PSII could be either high or low depending on the exposure to Cl^- in solution (Lindberg and Andréasson 1996; Olesen and Andréasson 2003). In this one-site/two-state model, in the absence of Cl^- the site goes into a low affinity/fast exchange state while in the presence of Cl^- the site goes into a high affinity/slow exchange state. This was supported by data in which intact PSII was dialyzed against Cl^- free buffer to remove Cl^-, then exposed to various concentrations of Cl^- for either long or short times, followed by measurement of O_2 evolution activity. In this way, the high affinity state was found to have a K_d (or K_M) of 20 μM, while the low affinity state had a K_d (or K_M) of 0.5 mM (Lindberg and Andréasson 1996). A similar study that included 1.4 M glycerol instead of sucrose in the buffer system found values for K_d (or K_M) of 13 μM for the high affinity state and 0.8 mM for the low affinity state (Olesen and Andréasson 2003). Induction of the closed or high affinity/slow exchange site required incubation with anion for 2 or more hours (Lindberg and Andréasson 1996; Olesen and Andréasson 2003). As noted previously, the dialysis of intact PSII against Cl^- free buffer at pH 6.3 lowered the activity in the absence of Cl^- to 30–40%. Lindberg and Andréasson argued that this remaining activity was not due to a fraction of the sample retaining Cl^-, but to an inherent activity of intact PSII without bound Cl^-, a suggestion that was supported by light dependence measurements of PSII activity (Lindberg and Andréasson 1996). However outside of the light-dependence measurements, the residual activity and virtually all other data related to Cl^- activation in intact PSII can be explained by a portion of the sample retaining a high affinity for Cl^-, coupled with trace Cl^- concentrations in the micromolar range that always persists in buffers.

Despite the difficulty in determining the affinity of the activating chloride binding site in intact PSII, numerous reports in the literature can be found with values usually in the mM range. Because of the difficulty in measuring the direct binding of Cl^-, affinity is often reported as the Michaelis constant K_M found for the Cl^- concentration dependence of O_2 evolution. This has the advantage of corresponding to Cl^- that is associated with activation only. In addition, if activator or substrate binding is the step that is rate limiting, then K_M is approximately equal to the dissociation constant K_d. In the case of PSII, it appears that the two are very close based on the studies by Lindberg and coworkers, who measured both direct binding of $^{36}Cl^-$ and activation by Cl^- (Lindberg et al. 1993; Lindberg and Andréasson 1996). For intact PSII, the K_M of activation reported usually corresponds only to the activatable portion of PSII, which varies depending on the Cl^- depletion method used. As already noted, one of the earliest determinations was made in Cl^--depleted thylakoids, for which a K_M of 0.9 mM at pH 7.2 was found (Kelley and Izawa 1978). In their early studies of subunit removal, Miyao and Murata showed Cl^- dependence data at pH 6.5 for O_2 evolution of intact PSII and PSII to which extrinsic subunits had rebound; for the activatable portion (25–30% of total activity), the K_M can be estimated to be about 1 mM (Miyao and Murata 1985). Similar results were found in another study, but with somewhat higher K_M values (Imaoka et al. 1986). Based on our own studies of intact PSII after removal of Cl^- by the dialysis method, the K_M of Cl^- activation is found to be between 0.5 and 2 mM at pH 6.3 for the activatable portion (40–70% of total activity) depending on the sample (Fig. 5.1a). These various results probably correspond to the low affinity/fast exchange site described by Lindberg and Andréasson with K_d of 0.5 mM (Lindberg and Andréasson 1996).

Removal of the extrinsic PsbP and PsbQ subunits leads to a PSII preparation that no longer shows high affinity Cl^- binding in the

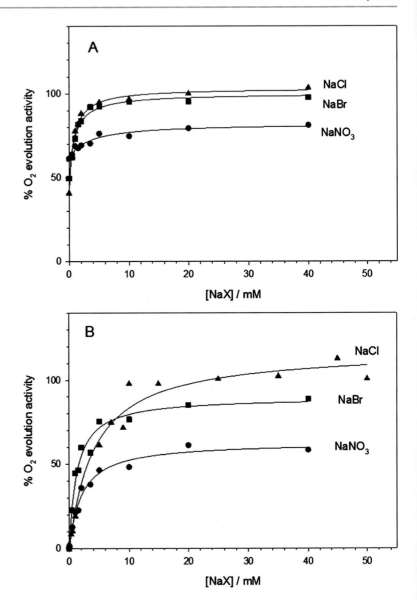

Fig. 5.1 Activation of oxygen evolution by anions chloride (▲), bromide (■), and nitrate (●) in Cl^--depleted intact PSII (**a**) and PSII lacking the PsbP and PsbQ subunits (**b**). For Cl^- depletion, PSII was dialyzed against buffer at pH 6.3 in the absence of added anion for 24–26 h at 4 °C. For removal of PsbP and PsbQ subunits, intact PSII was treated with 1.5 M NaCl in buffer at pH 6.3, followed by successive washes to remove Cl^-. O_2 evolution activity was assayed at pH 6.3 and 25 °C, in the presence of 6 mM Ca^{2+} in the case of NaCl-washed PSII, and the activity was normalized to 100% using the activity in 20 NaCl. *Curves* represent fits using the Michaelis-Menten equation, with the addition of a constant for nonzero initial activity for Cl^--depleted intact PSII. For intact PSII, fits gave $K_M = 0.75$, 0.95, and 3.4 mM for Cl^-, Br^-, and NO_3^-, respectively; for NaCl-washed PSII, $K_M = 4.1$, 1.3, and 2.0 mM for Cl^-, Br^-, and NO_3^-, respectively

10–20 μM range and is therefore completely activatable, for practical purposes. Under these conditions, the activation of PSII by Cl^- is very well modeled by simple Michaelis-Menten kinetics. In one of the first studies of the Cl^- dependence in NaCl-washed PSII, the data presented allows an estimate of K_M of about 3 mM (Miyao and Murata 1985). In a study of PSII depleted of Cl^- using the pH 7.5/50 mM Na_2SO_4 method (and therefore probably lacking PsbP and PsbQ), the K_M of Cl^- activation was found to be 2 mM at pH 6.5 (Itoh and Uwano 1986); this study also examined the pH dependence, showing that K_M increased with pH. Using a similar preparation in which subunit removal was confirmed, the half maximal activation value (equivalent to K_M) was found at be 6.5 mM at pH 7.5 (Wincencjusz et al. 1998). In a study to look at the FTIR effects of Cl^- and other anions, a K_M of 2 mM for Cl^- activation was found in NaCl-washed PSII at pH 6.5 (Hasegawa et al. 2002). Similarly, our own determinations of K_M in NaCl-washed PSII have revealed values in the range of 1–4 mM at pH 6.3, with variations appearing to depend on the preparation (Fig. 5.1b).

As noted previously, the importance of the PsbO subunit in controlling the affinity of higher plant PSII for Cl^- can be seen in the need for Cl^- concentrations in the range 100–200 mM when PsbO is removed (Miyao and Murata 1985; Ono and Inoue 1986). The role of the PsbO subunit in Cl^- activation was further explored by site directed mutagenesis studies within a loop region around residues 151–162. In one study, arginine residues R151 and R161 were mutated with the result that the mutant PsbO did not bind effectively to PSII and O_2 evolution activity was impaired (Popelkova et al. 2006). These mutants all showed increased values of K_M for Cl^- activation, in the range 1.5–2.5 mM, compared with 0.4 mM for NaCl-washed PSII. A second set of mutants of aspartate residue D157 all showed binding of PsbO to PSII that was the same as in wild-type PsbO, but O_2 evolution was still impaired (Popelkova et al. 2009). These mutants showed values of K_M for Cl^- activation of 1.5–1.6 mM compared with 0.9 mM for NaCl-washed PSII. These studies suggest participation of this loop region of the PsbO subunit in controlling the access of Cl^- to the OEC, probably via a proton channel.

A classic study of the pH dependence of Cl^- binding was carried out by Homann using PSII depleted of chloride by a 5 s treatment at pH 9.6–9.7, followed by rapid decrease to the desired pH (Homann 1985, 1988b). The data were analyzed using a model in which Cl^- binding was preceded by a protonation step. The pH-independent K_M was found to be 70 μM, with an associated protonation constant, pK_H, of 6 (Homann 1985). In a later study, the possible loss of the extrinsic PsbP and PsbQ subunits under those conditions was noted and a pH dependence study including data at lower pH was undertaken using both intact and NaCl-washed PSII (Homann 1988b). It was found that the affinity for chloride was about tenfold lower in the absence of the PsbP and PsbQ subunits. In addition, the value of the pK_H was reconsidered, leading to the conclusion that pK_H was less than 5 for the protonation step (Homann 1988b).

Although several studies have confirmed that Cl^- is required for the water oxidation cycle to advance beyond the S_2 state, few studies have examined the affinity of PSII for Cl^- in the individual S-states because of the challenging nature of the measurements. The exception is a set of studies carried out by Wincencjusz and coworkers (Wincencjusz et al. 1997, 1998, 1999), using measurements of the flash dependence of UV-Vis absorbance to monitor Mn oxidation state changes during S-state transitions. This study employed PSII depleted of Cl^- by a limited treatment with 50 mM Na_2SO_4 at pH 7.5, thereby removing the PsbP and PsbQ subunits with minimal loss of Ca^{2+}. It was found that Cl^- was required for the S_2-to-S_3 and the S_3-to-S_0 transitions, but not for the earlier two transitions (Wincencjusz et al. 1997). Although the S_2 state formed in the absence of Cl^-, it was longer lived than normal. This was thought to be explained by the hypothesis that Cl^- is required for electron transfer within the Mn_4CaO_5 complex. While the exchange of Cl^- was found to be slowed by the presence of the extrinsic subunits, the affinity for Cl^- at its site of activation was not affected, rather the pH and S-state were the major factors in determining affinity (Wincencjusz et al. 1998). The K_d for Cl^- was found to increase with Mn oxidation state, with estimated values of 0.08 mM in S_1, 1.0 mM in S_2, and 130 mM in S_3 at pH 6.0; the affinity for Cl^- in the S_1 and S_2 states was found to be about tenfold lower at pH 7.5.

5.3.3 Bromide and Nitrate Activators

Bromide and nitrate are generally found to activate oxygen evolution with kinetics that are very close to those of chloride. Numerous studies have compared the O_2 evolution activity in the presence of these and other anions using a "sufficient" anion concentration, i.e. 20–25 mM (see for example: Damoder et al. 1986; Yachandra et al. 1986; Ono et al. 1987; Hasegawa et al. 2002; Olesen and Andréasson 2003). Most find that Br^- can activate in the range of 90–100% as well as Cl^-, whereas NO_3^- lags relatively far behind in the 40–70% range compare with Cl^-. In a study of NaCl-washed PSII, Hasegawa found K_M values of 3 and 7 mM for activation by Br^- and NO_3^-, respectively (Hasegawa et al. 2002); this study also found that the level of light saturation had an effect on the relative maximum activity observed. In studies that compared the activation kinetics, we have found that the values of K_M are very similar for the three ions particularly when comparing measurements from the same preparation (Fig. 5.1). As for Cl^- activation, Br^- and NO_3^- generally show K_M values of 0.5–2 mM for intact PSII (Cl^- depleted by dialysis) and 1–4 mM for PSII in the absence of PsbP and PsbQ (NaCl-washed). Thus NO_3^- seems to differ more in the V_{max} values obtained than in K_M values when compared with Cl^- and Br^-.

In intact PSII depleted of Cl^- by dialysis at pH 6.3, the presence of both the open (or low affinity/fast exchange) and closed (or high affinity/slow exchange) states were characterized for Br^- and NO_3^- binding in the presence of 1.4 M glycerol (Olesen and Andréasson 2003). The open conformation was found to have K_M values of 2.8 mM and 0.8 mM for Br^- and NO_3^-, respectively, compared with 0.8 mM for Cl^-; the closed conformation was found to have K_M values of 31 µM and 2.8 mM for Br^- and NO_3^-, respectively, compare with 13 µM for Cl^-. Thus while Br^- bound to the anion site in both open and closed states, for NO_3^- the closed and open states had about the same affinity, perhaps because incubation with NO_3^- was not able to induce the closed state. In another study that employed a similar preparation, the K_d for Br^- in the closed state was found to be 53 µM (Haumann et al. 2006).

In their studies of UV-Vis absorbance changes during S-state changes, Wincencjusz and coworkers studied the effects of Br^- and NO_3^-, as well as I^- and NO_2^-, in PSII lacking the PsbP and PsbQ subunits (Wincencjusz et al. 1999). All were found to replace Cl^- functionally to varying extents, with Br^- the most effective substitute. NO_3^- and the other anions slowed the S_4-to-S_0 transition by several fold, while Br^- had very little effect on the rate of this transition. The anions I^- and NO_2^- were also found to accelerate the decay of the S_2 and S_3 states from a site different from the Cl^- activation site and even NO_3^- and Br^- had this effect to a lesser degree.

When Br^- is substituted for Cl^- in intact PSII, the S_2 state EPR signals are essentially indistinguishable from those found in the presence of Cl^- (Damoder et al. 1986; Yachandra et al. 1986; Ono et al. 1987; Olesen and Andréasson 2003). In NaCl-washed PSII lacking PsbP and PsbQ, treatment with either Br^- or Cl^- in the presence of sufficient Ca^{2+} also leads to similar S_2 state signals (Boussac 1995) (Fig. 5.2). In a study using electron spin echo envelope modulation (ESEEM) to detect nearby ligands, differences in the presence of Cl^- or Br^- suggested magnetic interaction between the anion and Mn (Boussac 1995). In both intact and NaCl-washed PSII, NO_3^- does not support formation of the multiline signal, although the $g = 4.1$ signal is observed (Ono et al. 1987) (Fig. 5.2).

5.4 Inhibitors

5.4.1 Ammonia

While not an anion inhibitor, ammonia (NH_3) has been recognized as a Lewis base inhibitor of oxygen evolution that competes with Cl^-, with numerous studies of its inhibition characteristics. Since the early investigations into its effects, it has provided a model for studies of anion inhibitors. Ammonia was of much interest because it is considered to be a water analogue, since it is has a similar electronic structure and size to that of H_2O. In early studies, it was found using flash-induced fluorescence that NH_3 binds in the higher S-states, with rapid binding in S_2 and slower binding in S_3 (Velthuys 1975). In addition there were indications from thylakoid studies that it was also associated with the Cl^- site (Izawa et al. 1969).

Although the optimal pH range is 6.0–6.5 for oxygen evolution by PSII, studies of the effects of NH_3 are usually carried out at a pH of 7.5 or so to ensure a significant presence of the NH_3

Fig. 5.2 EPR spectra of the S_2 state signals in PSII lacking the PsbP and PsbQ subunits (NaCl-washed) in the presence of 25 mM NaCl, 25 mM NaBr, 25 mM NaNO$_3$, or no added anion. Samples were prepared in buffer at pH 6.3 with 7 mM Ca^{2+}. The S_2 state was produced by illumination at 195 K. Spectra were taken at 10 K using 20 mW microwave power and 18 G modulation amplitude. Difference spectra of the illuminated sample minus the dark-adapted sample are shown

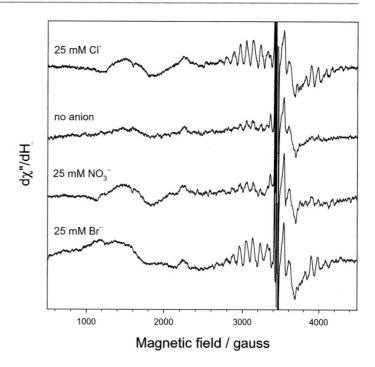

species. Even at pH 7.5, neutral NH$_3$ represents only about 2% of the total NH$_3$/NH$_4^+$ species, since NH$_4^+$ has pK$_a$ of 9.25. Studies by Sandusky and Yocum (1984, 1986) were the first to apply inhibitor enzyme kinetics with respect to the activator Cl$^-$ in the water oxidation reaction. They were able to show that ammonia inhibits through both competitive (K_i) and uncompetitive (K_i') modes (Sandusky and Yocum 1986), with inhibition constants that were both around 0.4–0.5 mM (see Table 5.1). However rather than representing noncompetitive inhibition (i.e., inhibition from a single site that is unaffected by binding of the substrate/activator), this was interpreted as two separate sites, one competitive with Cl$^-$ activation and the other associated with a water binding site. This was partly based on the observation that other larger amines showed inhibition that was more clearly Cl$^-$ competitive; the various amines tested also showed that effectiveness as an inhibitor correlated with Lewis base character, i.e., the pK$_a$ of the acid form (Sandusky and Yocum 1986). Interestingly, the size of the amine, even those as large as Tris, did not seem to be a significant factor in its effectiveness as a Cl$^-$ competitive inhibitor.

Numerous subsequent studies of ammonia were carried out to clarify its effect and solidify the characterization of two separate inhibitory sites. This was one of the first effectors found to alter the spacing of the hyperfine lines of the S_2 state multiline EPR signal (Beck and Brudvig 1986a, b; Andréasson et al. 1988; Ono and Inoue 1988; Britt et al. 1989; Boussac et al. 1990a), reducing the spacing to 67–68 G compared with 88–92 G for the normal multiline signal. This observation indicated that although the type of coupling remained the same, leading to an $S = 1/2$ spin state, NH$_3$ had an effect on the degree of coupling between Mn ions. Most investigators found that the binding of NH$_3$ leading to the altered multiline signal took place in the S_2 state but not the S_1 state. Thus 200 K illumination of samples containing NH$_3$ in the S_1 state did not lead to the altered multiline signal in the S_2 state, but subsequent warming (annealing) of the sample to 0 °C or above did; on the other hand illumination at 0 °C produced the altered multiline signal directly. Andréasson and coworkers (1988) found

Table 5.1 Inhibition constants for anion inhibitors based on steady state enzyme kinetics studies

Inhibitor, PSII type	K_i/mM	K_i'/mM	pH	Inhibitor type	References
NH_3, intact PSII	0.39	0.54	7.5	Mixed	a
F^-, intact PSII	~4	>40	7.6	Competitive	a
F^-, intact PSII	1.8 ± 1.0	79 ± 7	6.3	Competitive	b
F^-, intact PSII	<0.2	10–40	5.45	Competitive	b
N_3^-, intact	0.6	11	6.3	Competitive	c
N_3^-, NaCl-washed PSII	0.26	2.3	6.3	Competitive	c
Acetate, intact PSII	16	130	6.0	Competitive	d
I^-, intact PSII	>200	37	6.3	uncompetitive	e
I^-, intact PSII	na	8.8–17	6.3	Substrate inhibitor	e
I^-, NaCl-washed PSII	na	1.1–3.9	6.3	Substrate inhibitor	e
NO_2^-, intact PSII	nr	1.3, 7.5	6.0	Uncompetitive	f
NO_2^-, intact PSII	na	14.5 ± 0.6	6.3	Substrate inhibitor	g
NO_2^-, NaCl-washed PSII	na	4.5 ± 3.1	6.3	Substrate inhibitor	g

Most results were found from inhibitor enzyme kinetics studies with respect to competition with Cl^-. Some of the results for I^- and NO_2^- were found from activation studies in the absence of Cl^-, using the substrate inhibition model; in this case the competitive constant K_i is not modeled

References: (a) Sandusky and Yocum (1986), (b) Kuntzleman and Haddy (2009), (c) Haddy et al. (1999), (d) Kühne et al. (1999), (e) Bryson et al. (2005), (f) Pokhrel and Brudvig (2013), (g) Chen (2008)

na not applicable, *nr* not reported

that the binding of NH_3 associated with the altered multiline signal did take place in the S_1 state, but with a relatively weak affinity. Using ESEEM to detect ^{14}N and ^{15}N couplings in the NH_3-altered multiline signal, Britt and coworkers (1989) showed that NH_3 directly coordinates to the Mn_4CaO_5 cluster in the S_2 state, an important first demonstration of direct ligand binding. The altered multiline signal was not affected by increasing concentrations of Cl^-, indicating that NH_3 did not compete with Cl^- at this site (Beck and Brudvig 1986b; Andréasson et al. 1988). The binding site for NH_3 represented by the altered multiline signal was therefore associated with an H_2O binding site at the Mn_4CaO_5 cluster.

If the concentration of Cl^- was low, the $g = 4.1$ EPR signal was also found to be affected by NH_3 that bound in the S_1 state. This was observed as an increase in signal intensity and/or a decrease in the peak-to-trough linewidth of the signal to 280–300 G in the presence of ammonia compared with 360–400 G in its absence (Beck and Brudvig 1986b; Andréasson et al. 1988; Ono and Inoue 1988; Boussac et al. 1990a). This effect on the signal is similar to that observed for other weak base inhibitors that compete with Cl^-, such as F^-. The binding site represented by alterations in the $g = 4.1$ signal was therefore associated with Cl^- competition.

5.4.2 Fluoride

Fluoride is an inhibitor of oxygen evolution that is primarily competitive with Cl^- activation. Like ammonia it is a weak base or Lewis base, although the pK_a of hydrogen fluoride (HF) is much lower than that of NH_4^+, with a value of 3.17; at pH 6.3, it is virtually all in the anion form with less than 0.1% in the form of HF. The first steady state kinetics study to show F^- is competitive with Cl^- was carried out at pH 7.6 by Sandusky and Yocum (1986), as a part of their enzyme kinetics analyses of ammonia and other amine inhibitors. In this study, the competitive inhibition constant K_i was found to be around 4 mM (Table 5.1). The Cl^- competitive nature of F^- was also shown at pH 6.3, which is within the optimal pH range for oxygen evolution activity, revealing a competitive constant K_i of 1.8 mM and an uncompetitive constant K_i' of about 79 mM (Kuntzleman and Haddy 2009). The latter study also included an examination of PSII preparations in which PsbQ and PsbP had been partially removed using combinations of NH$_4$Cl and CaCl$_2$ at pH 7.5–7.6; this showed that with progressive removal of the extrinsic subunits, the value of K$_i$ increased, but the value of K$_i'$ was relatively insensitive. (Studies of the effects of F$^-$ in PSII lacking PsbP and PsbQ are complicated by the Ca^{2+} requirement found in the absence of PsbP, because of the formation of insoluble CaF$_2$). Whether the uncompetitive site was located on the donor or acceptor side of PSII was not clear, but F$^-$ has been found to associate with the acceptor side FeQ$_A$ site (Stemler and Murphy 1985; Sanakis et al. 1999). Examination of the kinetics at pH 5.45 indicated that they were more complex at this pH and that the value of K$_i$ was at least tenfold lower than at pH 6.3 (Kuntzleman and Haddy 2009).

While results of one study suggested that flash oscillations were supported by F$^-$ in PSII that had been Cl$^-$ depleted by brief pH 10 treatment (van Vliet and Rutherford 1996), another study found that F$^-$ blocked the S$_2$-to-S$_3$ transition in PSII Cl$^-$ depleted by the dialysis method (Olesen and Andréasson 2003). In addition, Wincencjusz and coworkers (Wincencjusz et al. 1999) found that F$^-$ was unable to support any of the S-state transitions in PSII lacking PsbP and PsbQ (pH 7.5/50 mM Na$_2$SO$_4$ treatment). We have also been unable to find any conditions under which F$^-$ can support steady state oxygen evolution in PSII, either intact or NaCl-washed lacking PsbP and PsbQ. We conclude that fluoride is unable to activate oxygen evolution in place of Cl$^-$.

If Cl$^-$ is replaced with F$^-$ in PSII, the S$_2$ state multiline EPR signal is not able to form and the $g = 4.1$ signal intensity is enhanced (Casey and Sauer 1984; Yachandra et al. 1986; Baumgarten et al. 1990; Haddy et al. 1992; DeRose et al. 1995; van Vliet and Rutherford 1996; Haddy et al. 2000; Olesen and Andréasson 2003), as shown in Fig. 5.3a. The $g = 4.1$ signal also becomes narrower by 10–20%, similar to that observed for ammonia treatment. Using a wash procedure at pH 6.0 to replace Cl$^-$ with F$^-$ in intact PSII, DeRose and coworkers tracked the concentration dependence of the changes in the two signals (DeRose et al. 1995). It was found that the multiline signal decreased in height as the $g = 4.1$ signal increased, with the changes in both half complete at 5–10 mM F$^-$ and the loss in O$_2$ evolution activity lagging somewhat behind. Thus the changes promoted by F$^-$ can be seen as an exchange of population between the two coupled states associated with the S$_2$ state signals. The S$_2$Y$_Z^\bullet$ signal can be observed in F$^-$-treated PSII (Baumgarten et al. 1990; DeRose et al. 1995; van Vliet and Rutherford 1996), as shown in Fig. 5.3b, indicating that the transfer of electrons from the S$_2$ state Mn$_4$CaO$_5$ cluster to Tyr Z is blocked by F$^-$. In the same F$^-$ concentration dependence study of the signal heights, it was found that the S$_2$Y$_Z^\bullet$ signal continued to increase at concentrations much above that at which the effects on the multiline and $g = 4.1$ signals leveled off, showing half maximal concentration of 50 mM or more (DeRose et al. 1995). However it was also observed that the signal formed when high F$^-$ concentrations were added to dark adapted samples in the S$_1$ state under very low intensity light. This suggests a secondary effect of F$^-$ at very high concentrations.

5.4.3 Acetate

In studies using both higher plant thylakoids and cyanobacteria PSII particles, it was found that acetate (H$_3$CCOO$^-$) inhibited oxygen evolution from a Cl$^-$ sensitive site (Sinclair 1984; Saygin et al. 1986). Later, a complete enzyme kinetics study showed that its mode of inhibition is primarily competitive with Cl$^-$ activation, with K$_i$ = 16 mM at pH 6.0 in intact PSII (Kühne et al. 1999) (Table 5.1). This high inhibitor dissociation constant necessitates the preparation of samples with the acetate concentration in the 100 s of mM range in order to observe its effects. A secondary uncompetitive site of inhibition has been associated with the acceptor side of PSII (Kühne et al. 1999).

Like F$^-$ treatment and Cl$^-$ depletion, acetate induces the S$_2$Y$_Z^\bullet$ signal due to interference with the transition from the S$_2$ to S$_3$ states, when illuminated at 0–25 °C in the presence of an electron acceptor (MacLachlan and Nugent 1993; Szalai and Brudvig 1996a; Force et al.

Fig. 5.3 EPR spectra of the S_2 state signals in the presence of 100 mM NaCl or NaF (**a**) and the $S_2Y_Z^\bullet$ signal in the presence of 100 mM NaF (**b**) in intact PSII at pH 6.3. The S_2 state (**a**) was produced by illumination at 195 K; the $S_2Y_Z^\bullet$ state (**b**) was produced by illumination at 0 °C in the presence of 2 mM phenyl-*p*-benzoquinone. Other EPR conditions are as described in the legend of Fig. 5.2

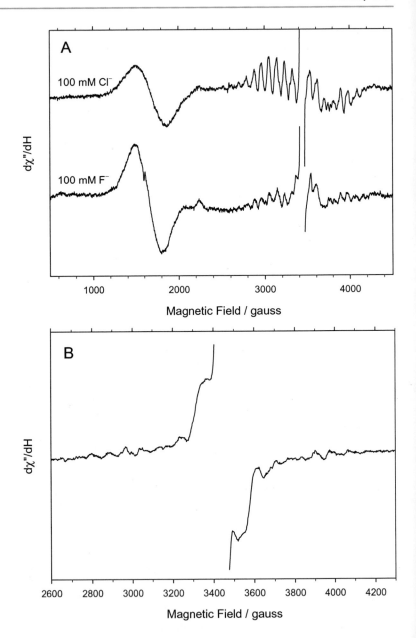

1997). The signal is among the wider $S_2Y_Z^\bullet$ signals reported, with a width of about 230–240 G (MacLachlan and Nugent 1993; Szalai and Brudvig 1996a; Lakshmi et al. 1998). Studies of the signal have generally been carried out at a pH of 5.5, at which the acid form represents about 15% of the sample ($pK_a = 4.76$ for acetic acid). An interesting feature of the acetate-inhibited $S_2Y_Z^\bullet$ state is that the multiline signal representing the S_2 state can be directly observed in addition to the broad tyrosine signal (Szalai and Brudvig 1996b); this multiline signal is similar to the normal multiline signal, but with peaks shifted either by 40 G or by 125 G depending on how the shift is viewed (Lakshmi et al. 1998; Szalai et al. 1998). In a creative approach to isolating one of the interacting spins, NO was used to quench the Tyr Z radical, revealing the multiline signal more clearly (Szalai and Brudvig 1996b). Oddly, in experiments in which only a single electron was transferred to form the S_2 state, acetate-treated PSII showed the $g = 4.1$ signal but not the multiline signal (MacLachlan and Nugent 1993) and the $g = 4.1$ signal was also found to form upon decay of the $S_2Y_Z\bullet$ signal (Szalai and Brudvig 1996a).

The acetate-induced $S_2Y_Z\bullet$ EPR signal drew much interest because of its potential to reveal details about the interaction between the Mn_4CaO_5 cluster and the nearby radical residue, which was not identified at first. These studies eventually revealed conclusively that Tyr Z was the radical interacting with the S_2 state. Approaches to studying the acetate-induced $S_2Y_Z\bullet$ signal included ESEEM of PSII prepared from *Synechocystis* grown with deuterated tyrosine (Tang et al. 1996), and theoretical modeling of the EPR, ESEEM, and ENDOR signals (Dorlet et al. 1998; Lakshmi et al. 1998; Peloquin et al. 1998). Estimates of the distance between the Mn_4CaO_5 cluster and Tyr Z were made based on the shape of the $S_2Y_Z\bullet$ signals, leading to an interspin distance of 7.7 Å (Lakshmi et al. 1998) or 8–9 Å (Dorlet et al. 1998). In addition, ESEEM was used to study the interaction between the methyl-deuterated acetate and the Tyr Z radical in spinach PSII, leading to a distance determination of about 3.1 Å between the two (Force et al. 1997). Since acetate is competitive with Cl^-, this study suggested a close interaction between Cl^- and Tyr Z. Using NO to quench the Tyr Z radical, the interaction between deuterated acetate and the modified S_2 state multiline signal was also studied using ESEEM, with the finding that acetate was within 6 Å of the Mn_4CaO_5 cluster (Clemens et al. 2002).

5.4.4 Azide

We carried out studies of azide (N_3^-) as an inhibitor of Cl^- activated O_2 evolution using steady state kinetics measurements (Haddy et al. 1999). These studies showed that azide, like F^-, is primarily competitive with Cl^- with an inhibition constant, K_i, of 0.6 mM and an uncompetitive constant, K_i', of 11 mM (Table 5.1). Values were also found for NaCl-washed PSII; these were lower than for intact PSII, suggesting that the uncompetitive site was also on the donor side. Like F^- and other competitive inhibitors, azide is a weak base, with $pK_a = 4.6$ for hydrazoic acid (N_3H); at pH 6.3 about 2% is in the protonated form.

Early observations of the inhibitory effect of azide on electron transport in photosystem II led to the characterization that it was promoted by illumination during pre-incubation (Katoh 1972). Studies by Kawamoto and coworkers explored this observation further in PSII preparations, with the finding that continuous illumination in the presence of azide resulted in irreversible inhibition of both oxygen evolution and electron transfer (Kawamoto et al. 1995). When Tris-washed PSII lacking Mn was used, the presence of the azidyl radical could be demonstrated by spin trapping. This led to the proposal that the azidyl radical was produced as a result of oxidation of azide by the Y_Z radical, followed by irreversible damage by azidyl radical at a site between Y_Z and the quinone acceptor Q_A. This interesting effect of azide is probably associated with the uncompetitive site of inhibition found in enzyme kinetics studies.

EPR studies of the S_2 state signals in PSII in which Cl^- was substituted with N_3^- showed very similar effects to those observed in F^--treated PSII, namely an inability to support the formation of the multiline signal and a narrowing and slight enhancement of the $g = 4.1$ signal (Haddy et al. 2000). These effects are associated with the competitive binding of azide at the site of Cl^- activation. In the presence of Cl^-, the addition of N_3^- led to a decrease in both S_2 state EPR signals, implying that it prevented formation of the S_2 state (Haddy et al. 2000). This effect was correlated with the binding of azide to the uncompetitive inhibition site. Finally, in a study by Yu and coworkers (Yu et al. 2005) using ESEEM of the multiline signal, a weak coupling was observed between the Mn_4CaO_5 cluster and ^{15}N–labelled N_3^- in the absence of Cl^-,

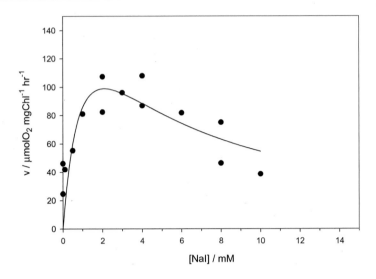

Fig. 5.4 Activation of oxygen evolution by iodide in PSII lacking PsbP and PsbQ subunits (NaCl-washed). Activity assays were carried out as described in the legend of Fig. 5.1. The curve was fitted using a substrate inhibition model, with the fitted constants of $K_M = 1.3$ mM, $K_I = 3.5$ mM, and $V_{max} = 217$ μM

conditions that favor the binding of N_3^- at the Cl^- competitive site. The results indicated that the azide anion was within about 5 Å of the Mn cluster.

5.4.5 Iodide

Iodide is an anion that shows both activating and inhibitory effects on oxygen evolution. It activates oxygen evolution at low concentrations when chloride is removed from the medium, but this is overcome by its inhibitory properties as the concentration is increased above the low mM range (Papageorgiou and Lagoyanni 1991; Hasegawa et al. 2002; Bryson et al. 2005). This enzyme kinetics behavior can be described by the substrate inhibition model, or activator inhibition in this case, in which the substrate or activator functions to promote activity from the substrate/activator site, but inhibits activity from a second site. The inhibitory site is generally an uncompetitive binding site, so it is observed only after activity has been promoted by binding of the substrate/activator. For this reason, the study of the inhibitor by steady state kinetics in the presence of the normal high-affinity substrate/activator (chloride in this case) will lead to the conclusion that the effector is an uncompetitive inhibitor. This is indeed the case for iodide, where study of the inhibition kinetics in the presence of chloride in intact PSII revealed an uncompetitive inhibition constant, K_i', of 37 mM and a competitive constant that was undetermined ($K_i > 200$ mM) (Bryson et al. 2005) (Table 5.1). Study of iodide activation in the absence of chloride in intact PSII showed a Michaelis constant, K_M, of 0.6–1.5 mM and an inhibition constant, K_I, of 8.8–17 mM (Bryson et al. 2005); this site of inhibition is presumably the same as the uncompetitive site characterized by kinetics studies in the presence of Cl^-, although the value of K_I was somewhat lower than K_i'. Study of iodide activation in the absence of chloride in PSII lacking Psb/PsbQ (NaCl-washed) showed Michaelis constant $K_M = 1.5$–5.3 mM and inhibition constant $K_I = 1.1$–3.9 mM (Bryson et al. 2005), suggesting that the affinity of the activating site decreased in the absence of the extrinsic subunits, while the affinity of the inhibitor site increased. A similar study of I^- activation of O_2 evolution using NaCl-washed PSII is shown in Fig. 5.4, where fitted values of K_M and K_I are 1.3 and 3.5 mM, respectively. The K_M values found for intact and NaCl-washed PSII are comparable to those found for other activators, such as Cl^- and Br^-.

Both inhibitory and activating effects were also found for iodide under conditions that produced the open (low affinity/fast exchange) and closed (high affinity/slow exchange) states of the Cl^- site (Olesen and Andréasson 2003). In the closed state the K_d for activation by I^- was 13 µM, while in the open state I^- was inhibitory with K_i of 10 mM for the major phase (90%) and 7 µM for the minor phase (10%).

The results described above using conventional enzyme kinetics methods are complementary to those obtained by Wincencjusz and coworkers, who studied iodide among other anions using UV-Vis absorption changes (Wincencjusz et al. 1999). Their methods had the ability to distinguish the changes in Mn oxidation state upon S-state changes in response to light flashes. As previously noted, iodide was found to activate oxygen evolution in place of Cl^-, but also to deactivate the higher S-states, S_2 and S_3, by reduction of the Mn cluster from a second site. While this suggests possible direct binding of I^- at the Mn cluster, the authors pointed out that this is not necessarily the case, since the single electron reduction may be promoted through more remote effects on the electron transfer chain. Using a PSII preparation that reduced the effects of excess I^-, Rachid and Homann (Rachid and Homann 1992) examined the charge recombination for the S_2 and S_3 states in PSII from which the extrinsic PsbP and PsbQ subunits had been removed by pH 7.2/100 mM Na_2SO_4 treatment, then allowed to rebind in the presence of NaCl or NaI, followed by removal of excess anion. Based on the temperature shift of thermoluminescence emission, the S_2 and S_3 states of the I^--reconstituted PSII were found to be more stable, with lower oxidation potentials. While this may seem at first glance to be contradictory to the results of Wincencjusz and coworkers, who found reduction of the S_2 and S_3 states by I^-, the findings of the two groups are most likely characterizations of different I^- binding sites. A different method of substituting I^- for Cl^- without excess I^- was used for PSII core complexes from the cyanobacterium *T. elongatus*, where the conditions for anion exchange were optimized based on minimization of secondary effects, including destabilization of the S_2 and S_3 states (Boussac et al. 2012). In this study, I^- was found to slow the S_3-to-S_0 transition, probably due to a longer lifetime of the intermediate $(S_3 Tyr_Z \bullet)'$ state.

Evidence for the direct inhibitory effect of iodide on the electron transfer chain came from early studies in which ^{125}I was found to label the D1 and D2 subunits of PSII (Takahashi et al. 1986; Ikeuchi and Inoue 1987; Takahashi and Styring 1987; Ikeuchi et al. 1988). Labeling of the D2 subunit took place in the dark in a PSII core complex preparation, with concurrent loss of the Tyr D radical EPR signal (Takahashi et al. 1986; Takahashi and Styring 1987). On the other hand, labeling of the D1 subunit in preparations depleted of Mn (by various means) required illumination to promote electron transfer and was insensitive to the presence of Cl^- (Takahashi et al. 1986; Ikeuchi and Inoue 1987, 1988). These results were interpreted to mean that I^- reduced the Tyr D radical or Tyr Z radical in the absence of the Mn cluster. In addition it was found that the D1 subunit was labelled in NaCl-washed Mn-containing PSII and that this labelling could be prevented by Cl^-, F^- or acetate, suggesting that under these conditions I^- reduced the Mn cluster from the Cl^- site (Ikeuchi et al. 1988). In this case, a K_M of 0.2–0.5 mM was determined for DCIP reduction by I^- at pH 6.0 in NaCl-washed PSII, which is similar to values found for the activation of oxygen evolution; higher K_M values were determined for preparations in which Mn or the PsbO subunit had been removed.

Iodide is usually found to support the formation of both S_2 state EPR signals, with possible suppression of the multiline signal at high concentrations. Using PSII lacking the PsbP and PsbQ subunits (NaCl-washed), we found that I^- in place of Cl^- supported the formation of both the multiline and $g = 4.1$ signals when the concentration of I^- was either low (3 mM) or higher (25 mM) (Bryson 2005); thus even when the concentration of I^- was high enough to inhibit oxygen evolution, the S_2 state appeared to be normal. In a similar experiment, but using intact PSII depleted of Cl^- by dialysis, we observed a partial suppression of the multiline signal in the presence of 25 mM NaI (Fig. 5.5). Using intact PSII, Olesen and Andréasson (Olesen and

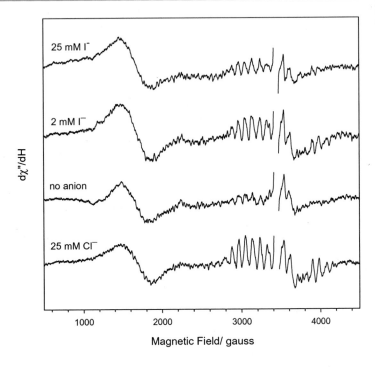

Fig. 5.5 EPR spectra of the S_2 state signals in the presence of 25 mM NaI, 2 mM NaI, no anion, or 25 mM NaCl in Cl^--depleted intact PSII. Other conditions are as described in the legend of Fig. 5.2

Andréasson 2003) observed formation of both S_2 state EPR signals in the presence of 10 mM NaI, when the anion site was prepared in either the open or the closed state. In an earlier study, it was found that only the $g = 4.1$ signal formed in PSII after treatment with 50 mM NaI at pH 7.5 (Ono et al. 1987), thus the absence of the multiline signal may be related to the high concentration of I^- in this case. In the study of cyanobacterial PSII in which I^- was exchanged for activating Cl^-, the S_2 state EPR spectrum showed a normal multiline signal and a prominent $g = 4.1$ signal, although the latter signal is not usually observed in cyanobacteria (Boussac et al. 2012).

5.4.6 Nitrite

Nitrite (NO_2^-) is an inhibitor of oxygen evolution with many properties that are similar to those of iodide. Unlike iodide, nitrite is a weak base with pK_a of 3.25 for nitrous acid (NO_2H). It is able to activate oxygen evolution at low concentrations, but its inhibitory effects dominate at higher concentrations, indicating that it is also a substrate inhibitor (or activator inhibitor). In our lab, a study of the nitrite concentration dependence of O_2 evolution in the absence of Cl^- showed that the K_M of activation was 0.33 ± 0.09 mM for intact PSII and 2.1 ± 1.4 mM for NaCl-washed PSII at pH 6.3 (Chen 2008) (Table 5.1). At the same time, the inhibition constant K_I was found to be 14.5 ± 0.6 mM for intact PSII and 4.5 ± 3.1 mM for NaCl-washed PSII. A study of the inhibition kinetics in the presence of Cl^- carried out by Pokhrel and coworkers using intact PSII at pH 6.0 showed that the mode of inhibition was uncompetitive, consistent with its behavior as a substrate inhibitor (Pokhrel and Brudvig 2013). With an overall uncompetitive inhibition constant, K_i', of about 3 mM, the curvature observed in the kinetics plots was thought to be better interpreted using two values for K_i', 1.3 and 7.5 mM. As for iodide, studies of the UV/Vis absorption changes in response to light flashes showed that nitrite reduced the higher S-states, S_2 and S_3, from a binding site not associated with Cl^- activation (Wincencjusz et al. 1999). This conclusion was supported by flash studies of O_2 evolution which revealed reduction of the OEC by NO_2^-, probably in the higher S-states (S_2 and/or S_3) (Pokhrel and Brudvig 2013).

EPR investigation of PSII treated with nitrite revealed inhibitory effects on both the donor and acceptor sides of the electron transfer path (Pokhrel and Brudvig 2013). Both S_2 state signals could form in the presence of nitrite, but as the concentration of NO_2^- increased, the intensity of the multiline signal decreased relative to the $g = 4.1$ signal, particularly at concentrations of 30 mM NO_2^- and above. This effect on the S_2 state is similar to that of iodide described above. EPR evidence was also found for the binding of NO_2^- to the $Fe^{2+}Q_A^-$ site of PSII. Electron transfer measurements indicated that the inhibitory effect of NO_2^- was mainly on the donor side, but both effects probably contribute to the uncompetitive inhibition observed in steady state kinetics experiments.

5.5 Concluding Remarks

Study of the effects of a variety of anions can reveal the properties that are necessary to activate oxygen evolution in photosystem II. Small anion effectors seem to fall into three main groups (Table 5.1) including activators, inhibitors that are competitive with respect to Cl^- activation, and inhibitors that are uncompetitive with respect to Cl^- activation (or alternatively are substrate inhibitors). While chloride is the natural activator of O_2 evolution, bromide is nearly as good but nitrate is only partially effective as an activator. All three anions are completely unprotonated in aqueous solution, since they are conjugate bases of strong acids (HCl, HBr, HNO_3).

The competitive inhibitors are represented by F^-, N_3^-, and acetate, which all share the property that they are weak bases with a tendency to partially protonate. Although not an anion, ammonia is also a member of this group since it competes effectively with Cl^-, even though it has a second mode of inhibition that is equally effective. The weak base character of this group of inhibitors is probably key to their inability to function as activators from the chloride site.

They are probably unable to properly function in proton coordination and transfer, a role frequently proposed for Cl^-. These competitive inhibitors also share an inability to support the formation of the S_2-state multiline signal, although the $g = 4.1$ signal is able to form in a slightly modified form, with narrower line width. The apparent exception to this is acetate, for which an altered multiline signal is able to exist as a part of the $S_2Y_Z^\bullet$ state.

The uncompetitive inhibitors, I^- and nitrite, may actually be viewed as activators with an unfortunate secondary inhibitory effect. Like Cl^-, I^- has no weak base properties, but NO_2^- is a weak base and does not fit in with the other activators in this respect. It is difficult to assess how much this interferes with its ability as an activator, because the inhibitory effects complicate the evaluation of nitrite as an activator. For both I^- and NO_2^-, inhibition takes place from a site that is not identified with the Cl^- activation site. The inhibiting reaction occurs during a later part of the S-state cycle, when Mn is in a higher oxidation state, and evidently involves reduction of Mn. Both of the S_2 state EPR signals can form in the presence of low concentrations of each, but current evidence indicates that the multiline signal is more poorly supported when the anion concentration is high. Thus while inhibition by these anions does not reveal properties about the Cl^- activation site, it does reflect the properties of a site that is able to influence the Mn_4CaO_5 cluster.

Acknowledgements This work was supported by grants from the National Science Foundation, the Camille and Henry Dreyfus Foundation, and the UNCG Office of Research. Many thanks to Sergei Baranov, David Bryson, and Hong Qian for technical assistance.

References

Andersson, B., C. Critchley, I. J. Ryrie, C. Jansson, C. Larsson and J. M. Anderson (1984). "Modification of the chloride requirement for photosynthetic O_2 evolution. The role of the 23 kDa polypeptide." FEBS Letters 168: 113–117.

Andréasson, L.-E. and K. Lindberg (1992). "The inhibition of photosynthetic oxygen evolution by ammonia probed by EPR." Biochimica et Biophysica Acta 1100: 177–183.

Andréasson, L.-E., Ö. Hansson and K. von Schenck (1988). "The interaction of ammonia with the photosynthetic oxygen-evolving system." Biochimica et Biophysica Acta 936: 351–360.

Arnon, D. I. and F. R. Whatley (1949). "Is chloride a coenzyme of photosynthesis?" Science 110: 554–556.

Baumgarten, M., J. S. Philo and G. C. Dismukes (1990). "Mechanism of photoinhibition of photosynthetic water oxidation by Cl^- depletion and F^- substitution: Oxidation of a protein residue." Biochemistry 29: 10814–10822.

Beck, W. F. and G. W. Brudvig (1986a). "Ammonia binds to the manganese site of the O_2-evolving complex of photosystem II in the S_2 state." Journal of the American Chemical Society 108: 4018–4022.

Beck, W. F. and G. W. Brudvig (1986b). "Binding of amines to the O_2-evolving center of Photosytem II." Biochemistry 25: 6479–6486.

Berthold, D. A., G. T. Babcock and C. F. Yocum (1981). "A highly resolved, oxygen-evolving photosystem II preparation from spinach thylakoid membranes." FEBS Letters 134: 231–234.

Boussac, A. (1995). "Exchange of chloride by bromide in the manganese Photosystem-II complex studied by cw- and pulsed-EPR." Chemical Physics 194: 409–418.

Boussac, A. and A. W. Rutherford (1994). "Electron transfer events in chloride-depleted photosystem II." Journal of Biological Chemistry 269: 12462–12467.

Boussac, A., J.-L. Zimmermann and A. W. Rutherford (1989). "EPR signals from modified charge accumulation states of the oxygen evolving enzyme in Ca^{2+}-deficient Photosystem II." Biochemistry 28: 8984–8989.

Boussac, A., A. W. Rutherford and S. Styring (1990a). "Interaction of ammonia with the water splitting enzyme of photosystem II." Biochemistry 29: 24–32.

Boussac, A., J.-L. Zimmermann and A. W. Rutherford (1990b). "Factors influencing the formation of modified S_2 EPR signal and the S_3 EPR signal in Ca^{2+}-depleted photosystem II." FEBS Letters 277: 69–74.

Boussac, A., M. Sugiura, Y. Inoue and A. W. Rutherford (2000). "EPR study of the oxygen evolving complex in His-tagged photosystem II from the cyanobacterium *Synechococcus elongatus*." Biochemistry 39: 13788–13799.

Boussac, A., N. Ishida, M. Sugiura and F. Rappaport (2012). "Probing the role of chloride in photosystem II from *Thermosynechococcus elongatus* by exchanging chloride for iodide." Biochimica et Biophysica Acta 1817: 802–810.

Bricker, T. M., J. L. Roose, R. D. Fagerlund, L. K. Frankel and J. J. Eaton-Rye (2012). "The extrinsic proteins of photosystem II." Biochimica et Biophysica Acta 1817: 121–142.

Britt, R. D., J.-L. Zimmermann, K. Sauer and M. P. Klein (1989). "Ammonia binds to the catalytic Mn of the oxygen evolving complex of photosystem II: Evidence by electron spin echo envelope modulation spectroscopy." Journal of the American Chemical Society 111: 3522–3532.

Broser, M., C. Glöckner, A. Gabdulkhakov, A. Guskov, J. Buchta, J. Kern, F. Müh, H. Dau, W. Saenger and A. Zouni (2011). "Structural basis of cyanobacterial photosystem II inhibition by the herbicide terbutryn." Journal of Biological Chemistry 286: 15964–15972.

Brudvig, G. W., J. L. Casey and K. Sauer (1983). "The effect of temperature on the formation and decay of the multiline EPR signal species associated with photosynthetic oxygen evolution." Biochimica et Biophysica Acta 723: 366–371.

Brynda, M. and R. D. Britt (2010). The manganese-calcium cluster of the oxygen-evolving system: Synthetic models, EPR studies, and electronic structure calculations. Metals in Biology: Applications of High-Resolution EPR. G. Hanson and L. Berliner, Springer Science: 203–271.

Bryson, D. I., N. Doctor, R. Johnson, S. Baranov and A. Haddy (2005). "Characteristics of iodide activation and inhibition of oxygen evolution by photosystem II." Biochemistry 44: 7354–7360.

Casey, J. L. and K. Sauer (1984). "EPR detection of a cryogenically photogenerated intermediate in photosynthetic oxygen evolution." Biochimica et Biophysica Acta 767: 21–28.

Chen, X. (2008). Mathematical models for the pH dependence of oxygen evolution under fluoride inhibition and effects of nitrite on oxygen evolution in photosystem II, Master's Thesis, UNC-Greensboro.

Clemens, K. L., D. A. Force and R. D. Britt (2002). "Acetate binding at the Photosystem II oxygen evolving complex: An S_2-state multiline signal ESEEM study." Journal of the American Chemical Society 124: 10921–10933.

Damoder, R., V. V. Klimov and G. C. Dismukes (1986). "The effect of Cl^- depletion and X^- reconstitution on the oxygen-evolution rate, the yield of the multiline manganese EPR signal and EPR Signal II in the isolated photosystem-II complex." Biochimica et Biophysica Acta 848: 378–391.

de Paula, J. C., P. M. Li, A.-F. Miller, B. W. Wu and G. W. Brudvig (1986). "Effect of the 17- and 23-kilodalton polypeptides, calcium, and chloride on electron transfer in photosystem II." Biochemistry 25: 6487–6494.

Debus, R. J. (2014). "Evidence from FTIR difference spectroscopy that D1-Asp61 influences the water reactions of the oxygen-evolving Mn_4CaO_5 cluster of photosystem II." Biochemistry 53: 2941–2955.

DeRose, V. J., M. J. Latimer, J.-L. Zimmermann, I Mukerji, V. K. Yachandra, K. Sauer and M. P. Klein (1995). "Fluoride substitution in the Mn cluster from photosystem II: EPR and X-ray absorption spectroscopy studies." Chemical Physics 194: 443–459.

Dilbeck, P. L., H. J. Hwang, I. Zaharieva, L. Gerencser, H. Dau and R. L. Burnap (2012). "The D1-D61N mutation in *Synechocystis* sp. PCC 6803 allows the observation of pH-sensitive intermediates in the formation and release of O_2 from photosystem II." Biochemistry 51: 1079–1091.

Dismukes, G. C. and Y. Siderer (1980). "EPR spectroscopic observations of a manganese center associated with water oxidation in spinach chloroplasts." FEBS Letters 121: 78–80.

Dismukes, G. C. and Y. Siderer (1981). "Intermediates of a polynuclear manganese center involved in photosynthetic oxidation of water." Proceedings of the National Academy of Science 78: 274–278.

Dorlet, P., M. Di Valentin, G. T. Babcock and J. L. McCracken (1998). "Interaction of $Y_Z\bullet$ with its environment in acetate-treated photosystem II membranes and reaction center cores." Journal of Physical Chemistry B 102: 8239–8247.

Enami, I., S. Kikuchi, T. Fukuda, H. Ohta and J.-R. Shen (1998). "Binding and functional properties of four extrinsic proteins of photosystem II from a red alga, *Cyanidium caldarium*, as studied by release-reconstitution experiments." Biochemistry 37: 2787–2793.

Force, D. A., D. W. Randall and R. D. Britt (1997). "Proximity of acetate, manganese, and exchangeable deuterons to Tyrosine Y_Z-dot in acetate-inhibited Photosystem II membranes: Implications for the direct involvement of Y_Z-dot in water-splitting." Biochemistry 36: 12062–12070.

Ford, R. C. and M. C. W. Evans (1983). "Isolation of a photosystem 2 preparation from higher plants with highly enriched oxygen evolution activity." FEBS Letters 160: 159–164.

Fork, D. C. (1972). "Oxygen electrode." Methods in Enzymology 24: 113–122.

Ghanotakis, D. F., G. T. Babcock and C. F. Yocum (1984a). "Calcium reconstitutes high rates of oxygen evolution in polypeptide depleted Photosystem II preparations." FEBS Letters 167: 127–130.

Ghanotakis, D. F., J. N. Topper, G. T. Babcock and C. F. Yocum (1984b). "Water-soluble 17 and 23 kDa polypeptides restore oxygen evolution activity by creating a high-affinity binding site for Ca^{2+} on the oxidizing side of Photosystem II." FEBS Letters 170: 169–173.

Ghanotakis, D. F., G. T. Babcock and C. F. Yocum (1985). "On the role of water-soluble polypeptides (17, 23 kDa), Fcalcium and chloride in photosynthetic oxygen evolution." FEBS Letters 192: 1–3.

Gorham, P. R. and K. A. Clendenning (1952). "Anionic stimulation of the Hill reaction in isolated chloroplasts." Archives of Biochemistry and Biophysics 37: 199–223.

Grundmeier, A. and H. Dau (2012). "Structural models of the manganese complex of photosystem II and mechanistic implications." Biochimica et Biophysica Acta 1817: 88–105.

Guskov, A., J. Kern, A. Gabdulkhakov, M. Broser, A. Zouni and W. Saenger (2009). "Cyanobacterial photosystem II at 2.9-Å resolution and the role of quinones, lipids, channels and chloride." Nature Structural and Molecular Biology 16: 334–342.

Haddy, A. (2007). "EPR spectroscopy of the manganese cluster of photosystem II." Photosynthesis Research 92: 357–368.

Haddy, A., W. R. Dunham, R. H. Sands and Roland Aasa (1992). "Multifrequency EPR investigations into the origin of the S_2-state signal at $g = 4$ of the O_2-evolving complex." Biochimica et Biophysica Acta 1099: 25–34.

Haddy, A., J. A. Hatchell, R. A. Kimel and R. Thomas (1999). "Azide as a competitor of chloride in oxygen evolution by photosystem II." Biochemistry 38: 6104–6110.

Haddy, A., R. A. Kimel and R. Thomas (2000). "Effects of azide on the S_2 state EPR signals from photosystem II." Photosynthesis Research 63: 35–45.

Hallahan, B. J., J. H. A. Nugent, J. T. Warden and M. C. W. Evans (1992). "Investigation of the origin of the "S3" EPR signal from the oxygen-evolving complex of Photosystem 2: The role of Tyrosine Z." Biochemistry 31: 4562–4573.

Hasegawa, K., Y. Kimura and T.-a. Ono (2002). "Chloride cofactor in the photosynthetic oxygen-evolving complex studied by Fourier transform infrared spectroscopy." Biochemistry 41: 13839–13850.

Haumann, M., M. Barra, P. Loja, S. Löscher, R. Krivanek, A. Grundmeier, L.-E. Andreasson and H. Dau (2006). "Bromide does not bind to the Mn_4Ca complex in its S_1 state in Cl^--depleted and Br^--reconstituted oxygen-evolving photosystem II: Evidence from X-ray absorption spectroscopy at the Br K-edge." Biochemistry 45: 13101–13107.

Homann, P. H. (1985). "The association of functional anions with the oxygen-evolving center of chloroplasts." Biochimica et Biophysica Acta 809: 311–319.

Homann, P. H. (1988a) "Structural effects of Cl^- and other anions on the water oxidizing complex of chloroplast photosystem II". Plant Physiology 88, 194–199.

Homann, P. H. (1988b). "The chloride and calcium requirement of photosynthetic water oxidation: effects of pH." Biochimica et Biophysica Acta 934: 1–13.

Homann, P. H. (2002). "Chloride and calcium in Photosystem II: from effects to enigma." Photosynthesis Research 73: 169–175.

Ikeuchi, M. and Y. Inoue (1987). "Specific ^{125}I labeling of D1 (herbicide-binding protein)." FEBS Letters 210: 71–76.

Ikeuchi, M., H. Koike and Y. Inoue (1988). "Iodination of D1 (herbicide-binding protein) is coupled with photooxidation of $^{125}I^-$ associated with Cl^--binding site in Photosystem-II water-oxidation system." Biochimica et Biophysica Acta 932: 160–169.

Imaoka, A., K. Akabori, M. Yanagi, K. Izumi, Y. Toyoshima, A. Kawamori, H. Nakayama and J. Sato (1986). "Roles of three lumen-surface proteins in the formation of S_2 state and O_2 evolution in

Photosystem II particles from spinach thylakoid membranes." Biochimica et Biophysica Acta 848: 201–211.

Inoue-Kashino, N., Y. Kashino, K. Satoh, I. Terashima and H. B. Pakrasi (2005). "PsbU provides a stable architecture for the oxygen-evolving system in cyanobacterial photosystem II." Biochemistry 44: 12214–12228.

Itoh, S. and S. Uwano (1986). "Characteristics of the Cl^- action site in the O_2 evolving reaction in PS II particles: Electrostatic interaction with ions." Plant and Cell Physiology 27: 25–36.

Itoh, S., C. T. Yerkes, Y. Koike, H. H. Robinson and A. R. Crofts (1984). "Effects of chloride depletion on electron donation from the water-oxidizing complex to the photosystem II reaction center as measured by the microsecond rise of chlorophyll fluorescence in isolated pea chloroplasts." Biochimica et Biophysica Acta 766: 612–622.

Izawa, S., R. L. Heath and G. Hind (1969). "The role of chloride ion in photosynthesis. III. The effect of artificial electron donors upon electron transport." Biochimica et Biophysica Acta 180: 388–398.

Katoh, H., S. Itoh, J.-R. Shen and M. Ikeuchi (2001). "Functional analysis of $psbV$ and a novel c-type cytochrome gene $psbV2$ of the thermophilic cyanobacterium $Thermosynechococcus\ elongatus$ strain BP-1." Plant and Cell Physiology 42: 599–607.

Katoh, S. (1972). "Inhibitors of electron transport associated with photosystem II in chloroplasts." Plant and Cell Physiology 13: 273–286.

Kawakami, K., Y. Umena, N. Kamiya and J.-R. Shen (2009). "Location of chloride and its possible functions in oxygen-evolving photosystem II revealed by X-ray crystallography." Proceedings of the National Academy of Sciences of the United States of America 106: 8567–8572.

Kawakami, K., Y. Umena, N. Kamiya and J.-R. Shen (2011). "Structure of the catalytic, inorganic core of oxygen-evolving photosystem II at 1.9 A resolution." Journal of Photochemistry and Photobiology, B: Biology 104: 9–18.

Kawamoto, K., J. Mano and K. Asada (1995). "Photoproduction of the azidyl radical from the azide anion on the oxidizing side of Photosytem II and suppression of photooxidation of Tyrosine Z by the azidyl radical." Plant and Cell Physiology 36: 1121–1129.

Kelley, P. and S. Izawa (1978). "The role of chloride ion in photosystem II: I. Effects of chloride ion on photosystem II electron transport and on hydroxylamine inhibition." Biochimica et Biophysica Acta 502: 198–210.

Kirilovsky, D., M. Roncel, A. Boussac, A. Wilson, J. L. Zurita, J.-M. Ducruet, H. Bottin, M. Sugiura, J. M. Ortega and A. W. Rutherford (2004). "Cytochrome c_{550} in the cyanobacterium $Thermosynechococcus\ elongatus$." Journal of Biological Chemistry 279: 52869–52880.

Klein, M. P., K. Sauer and V. K. Yachandra (1993). "Perspectives on the structure of the photosynthetic oxygen evolving manganese complex and its relation to the Kok cycle." Photosynthesis Research 38: 265–277.

Kühne, H., V. A. Szalai and G. W. Brudvig (1999). "Competitive binding of acetate and chloride in Photosystem II." Biochemistry 38: 6604–6613.

Kuntzleman, T. S. and A. Haddy (2009). "Fluoride inhibition of photosystem II and the effect of removal of the PsbQ subunit." Photosynthesis Research 102: 7–19.

Kuwabara, T. and N. Murata (1982). "Inactivation of photosynthetic oxygen evolution and concomitant release of three polypeptides in the photosystem II particles of spinach chloroplasts." Plant and Cell Physiology 23: 533–539.

Kuwabara, T. and N. Murata (1983). "Quantitative analysis of the inactivation of photosynthetic oxygen evolution and the release of polypeptides and manganese in the photosystem II particles of spinach chloroplast." Plant and Cell Physiology 24: 741–747.

Lakshmi, K. V., S. S. Eaton, G. R. Eaton, H. A. Frank and G. W. Brudvig (1998). "Analysis of dipolar and exchange interactions between manganese and Tyrosine Z in the S_2Y_Z-dot state of acetate-inhibited Photosystem II via EPR spectral simulations at X- and Q-bands." Journal of Physical Chemistry B 102: 8327–8335.

Lakshmi, K. V., M. J. Reifler, D. A. Chisholm, J. Y. Wang, B. A. Diner and G. W. Brudvig (2002). "Correlation of the cytochrome c_{550} content of cyanobacterial Photosystem II with the EPR properties of the oxygen-evolving complex." Photosynthesis Research 72: 175–189.

Lindberg, K. and L.-E. Andréasson (1996). "A one-site, two-state model for the binding of anions in photosystem II." Biochemistry 35: 14259–14267.

Lindberg, K., T. Wydrzynski, T. Vänngård and L.-E. Andréasson (1990). "Slow release of chloride from ^{36}Cl-labeled photosystem II membranes." FEBS Letters 264: 153–155.

Lindberg, K., T. Vänngård and L.-E. Andréasson (1993). "Studies of the slowly exchanging chloride in photosystem II of higher plants." Photosynthesis Research 38: 401–408.

MacLachlan, D. J. and J. H. A. Nugent (1993). "Investigation of the S3 electron paramagnetic resonance signal from the oxygen-evolving complex of Photosystem 2: Effect of inhibition of oxygen evolution by acetate." Biochemistry 32: 9772–9780.

Mavankal, G., D. C. McCain and T. M. Bricker (1986). "Effects of chloride on paramagnetic coupling of manganese in calcium chloride-washed photosystem II preparations." FEBS Letters 202: 235–239.

McDermott, A. E., V. K. Yachandra, R. D. Guiles, J. L. Cole, S. L. Dexheimer, R. D. Britt, K. Sauer and M. P. Klein (1988). "Characterization of the manganese O_2-evolving complex and the iron-quinone acceptor complex in photosystem II from a thermophilic cyanobacterium by electron paramagnetic resonance and X-ray absorption spectroscopy." Biochemistry 27: 4021–4031

Miyao, M. and N. Murata (1983). "Partial disintegration and reconstitution of the photosynthetic oxygen evolution system." Biochimica et Biophysica Acta 725: 87–93.

Miyao, M. and N. Murata (1984). "Calcium ions can be substituted for the 24-kDa polypeptide in photosynthetic oxygen evolution." FEBS Letters 168: 118–120.

Miyao, M. and N. Murata (1985). "The Cl^- effect on photosynthetic oxygen evolution: interaction of Cl^- with 18-kDa, 24-kDa and 33-kDa proteins." FEBS Letters 180: 303–308.

Morgan, T. R., J. A. Shand, S. M. Clarke and J. J. Eaton-Rye (1998). "Specific requirements for cytochrome c-550 and the manganese-stabilizing protein in photoautotrophic strains of *Synechocystis* sp. PCC 6803 with mutations in the domain Gly-351 to Thr-436 of the chlorophyll-binding protein CP47." Biochemistry 37: 14437–14449.

Murray, J. W., K. Maghlaoui, J. Kargul, N. Ishida, T.-L. Lai, A. W. Rutherford, M. Sugiura, A. Boussac and J. Barber (2008). "X-ray crystallography identifies two chloride binding sites in the oxygen evolving centre of photosystem II." Energy and Environmental Science 1: 161–166.

Olesen, K. and L.-E. Andréasson (2003). "The function of the chloride ion in photosynthetic oxygen evolution." Biochemistry 42: 2025–2035.

Ono, T.-A. and Y. Inoue (1986). "Effects of removal and reconstitution of the extrinsic 33, 24 and 16 kDa proteins on flash oxygen yield in Photosystem II particles." Biochimica et Biophysica Acta 850: 380–389.

Ono, T.-A. and Y. Inoue (1988). "Abnormal S-state turnovers in NH_3-binding Mn centers of photosynthetic O_2 evolving system." Archives of Biochemistry and Biophysics 264: 82–92.

Ono, T.-A. and Y. Inoue (1990). "Abnormal redox reactions in photosynthetic O_2-evolving centers in NaCl/EDTA-washed PS II. A dark-stable EPR multiline signal and an unknown positive charge accumulator." Biochimica et Biophysica Acta 1020: 269–277.

Ono, T.-A., J. L. Zimmermann, Y. Inoue and A. W. Rutherford (1986). "EPR evidence for a modified S-state transition in chloride-depleted photosystem II." Biochimica et Biophysica Acta 851: 193–201.

Ono, T.-A., H. Nakayama, H. Gleiter, Y. Inoue and A. Kawamori (1987). "Modification of the properties of S_2 state in photosynthetic O_2-evolving center by replacement of chloride with other anions." Archives of Biochemistry and Biophysics 256: 618–624.

Papageorgiou, G. C. and T. Lagoyanni (1991). "Interactions of iodide ions with isolate photosystem 2 particles." Archives of Biochemistry and Biophysics 285: 339–343.

Peloquin, J. M. and R. D. Britt (2001). "EPR/ENDOR characterization of the physical and electronic structure of the OEC Mn cluster." Biochimica et Biophysica Acta 1503: 96–111.

Peloquin, J. M., K. A. Campbell and R. D. Britt (1998). "^{55}Mn pulsed ENDOR demonstrates that the Photosystem II "split" EPR signal arises from a magnetically-coupled mangano-tyrosyl complex." Journal of the American Chemical Society 120: 6840–6841.

Philbrick, J. B., B. A. Diner and B. A. Zilinskas (1991). "Construction and characterization of cyanobacterial mutants lacking the manganese-stabilizing polypeptide of photosystem II." Journal of Biological Chemistry 266:13370–13376

Pokhrel, R. and G. W. Brudvig (2013). "Investigation of the inhibitory effect of nitrite on photosystem II." Biochemistry 52: 3781–3789.

Pokhrel, R., I. L. McConnell and G. W. Brudvig (2011). "Chloride regulation of enzyme turnover: Application to the role of chloride in photosystem II." Biochemistry 50: 2725–2734.

Pokhrel, R., R. J. Service, R. J. Debus and G. W. Brudvig (2013). "Mutation of Lysine 317 in the D2 subunit of photosystem II alters chloride binding and proton transport." Biochemistry 52: 4758–4773.

Popelkova, H. and C. F. Yocum (2007). "Current status of the role of Cl^- ion in the oxygen-evolving complex." Photosynthesis Research 93: 111–121.

Popelkova, H., S. D. Betts, N. Lydakis-Symantiris, M. M. Im, E. Swenson and C. F. Yocum (2006). "Mutagenesis of basic residues R151 and R161 in manganese-stabilizing protein of photosystem II causes inefficient binding of chloride to the oxygen-evolving complex." Biochemistry 45: 3107–3115.

Popelkova, H., A. Commet and C. F. Yocum (2009). "Asp157 is required for th function of PsbO, the photosystem II manganese stabilizing protein." Biochemistry 48, 11920–11928.

Rachid, A. and P. H. Homann (1992). "Properties of iodide-activated photosynthetic water-oxidizing complexes." Biochimica et Biophysica Acta 1101: 303–310.

Rivalta, I., M. Amin, S. Luber, S. Vassiliev, R. Pokhrel, Y. Umena, K. Kawakami, J.-R. Shen, N. Kamiya, D. Bruce, G. W. Brudvig, M. R. Gunner and V S. Batista (2011). "Structural-functional role of chloride in photosystem II." Biochemistry 50: 6312–6315.

Roose, J. L., K. M. Wegener and H. B. Pakrasi (2007). "The extrinsic proteins of photosystem II." Photosynthesis Research 92: 369–387.

Sanakis, Y., D. Petasis, V. Petrouleas and M Hendrich (1999). "Simultaneous binding of fluoride and NO to the nonheme iron of photosystem II: Quantitative EPR evidence for a weak exchange interaction between the semiquinone Q_A^- and the iron-nitrosyl complex." Journal of the American Chemical Society 121: 9155–9164.

Sandusky, P. O. and C. F. Yocum (1983). "The mechanism of amine inhibition of the photosynthetic oxygen evolving complex." FEBS Letters 162: 339–343.

Sandusky, P. O. and C. F. Yocum (1984). "The chloride requirement for photosynthetic oxygen evolution:

analysis of the effects of chloride and other anions on amine inhibition of the oxygen-evolving complex." Biochimica et Biophysica Acta 766: 603–611.

Sandusky, P. O. and C. F. Yocum (1986). "The chloride requirement for photosynthetic oxygen evolution: factors affecting nucleophilic displacement of chloride from the oxygen-evolving complex." Biochimica et Biophysica Acta 849: 85–93.

Sauer, K., J. Yano and V. K. Yachandra (2008). "X-ray spectroscopy of the photosynthetic oxygen-evolving complex." Coordination Chemistry Reviews 252: 318–335.

Saygin, Ö., S. Gerken, B. Meyer and H. T. Witt (1986). "Total recovery of O_2 evolution and nanosecond reduction kinetics of chlorophyll-a_{II}^+ (P-680$^+$) after inhibition of water cleavage with acetate." Photosynthesis Research 9: 71–78.

Shen, J.-R. and Y. Inoue (1993). "Binding and functional properties of two new extrinsic components, cytochrome c-550 and a 12-kDa protein, in cyanobacterial photosystem II." Biochemistry 32: 1825–1832.

Shen, J.-R., M. Ikeuchi and Y. Inoue (1992). "Stoichiometric association of extrinsic cytochrome c_{550} and 12 kDa protein with a highly purified oxygen-evolving photosystem II core complex from Synechococcus vulcanus." FEBS Letters 301: 145–149.

Shen, J.-R., M. Ikeuchi and Y. Inoue (1997). "Analysis of the psbU gene encoding the 12-kDa extrinsic protein of photosystem II and studies on its role by deletion mutagenesis in Synechocystis sp. PCC 6803." Journal of Biological Chemistry 272: 17821–17826.

Shen, J.-R., M. Qian, Y. Inoue and R. L. Burnap (1998). "Functional characterization of Synechocystis sp. PCC 6803 del-psbU and del-psbV mutants reveals important roles of cytochrome c-550 in cyanobacterial oxygen evolution." Biochemistry 37: 1551–1558.

Shoji, M., H. Isobe, S. Yamanaka, Y. Umena, K. Kawakami, N. Kamiya, J.-R. Shen, T. Nakajima and K. Yamaguchi (2015). "Theoretical modelling of biomolecular systems I. Large-scale QM/MM calculations of hydrogen bonding networks of the oxygen evolving complex of photosystem II." Molecular Physics 113: 359–384.

Sinclair, J. (1984). "The influence of anions on oxygen evolution by isolated spinach chloroplasts." Biochimica et Biophysica Acta 764: 247–252.

Sivaraja, M., J. Tso and G. C. Dismukes (1989). "A calcium-specific site influences the structure and activity of the manganese cluster responsible for photosynthetic water oxidation." Biochemistry 28: 9459–9464.

Stemler, A. and J. B. Murphy (1985). "Bicarbonate-reversible and irreversible inhibition of photosystem II by monovalent anions." Plant Physiology 77: 974–977.

Strickler, M. A., L. M. Walker, W. Hillier and R. J. Debus (2005). "Evidence from biosynthetically incorporated strontium and FTIR difference spectroscopy that the C-terminus of the D1 polypeptide of photosystem II does not ligate calcium." Biochemistry 44: 8571–8577.

Styring, S., M. Miyao and A. W. Rutherford (1987). "Formation and flash-dependent oscillation of the S_2-state multiline EPR signal in an oxygen-evolving photosystem-II preparation lacking the three extrinsic proteins in the oxygen-evolving system." Biochimica et Biophysica Acta 890: 32–38.

Suga, M., F. Akita, K. Hirata, G. Ueno, H. Murakami, Y. Nakajima, T. Shimizu, K. Yamashita, M. Yamamoto, H. Ago and J.-R. Shen (2015). "Native structure of photosystem II at 1.95 Å resolution viewed by femtosecond X-ray pulses." Nature (London) 517: 99–103.

Summerfield, T. C., J. A. Shand, F. K. Bentley and J. J. Eaton-Rye (2005a). "PsbQ (Sll1638) in Synechocystis sp. PCC 6803 is required for photosystem II activity in specific mutants and in nutrient-limiting conditions." Biochemistry 44: 805–815.

Summerfield, T. C., R. T. Winter and J. J. Eaton-Rye (2005b). "Investigation of a requirement for the PsbP-like protein in Synechocystis sp. PCC 6803." Photosynthesis Research 84: 263–268.

Suzuki, H., J. Yu, T. Kobayashi, H. Nakanishi, P. J. Nixon and T. Noguchi (2013). "Functional roles of D2-Lys317 and the interacting chloride ion in the water oxidation reaction of photosystem II as revealed by Fourier transform infrared analysis." Biochemistry 52: 4748–4757.

Szalai, V. A. and G. W. Brudvig (1996a). "Formation and decay of the S3 EPR signal species in acetate-inhibited photosystem II." Biochemistry 35: 1946–1953.

Szalai, V. A. and G. W. Brudvig (1996b). "Reversible binding of nitric oxide to tyrosyl radicals in photosystem II. Nitric oxide quenches formation of the S3 EPR signal species in acetate-inhibited photosystem II." Biochemistry 35: 15080–15087.

Szalai, V. A., H. Kühne, K. V. Lakshmi and G. W. Brudvig (1998). "Characterization of the interaction between manganese and tyrosine Z in acetate-inhibited photosystem II." Biochemistry 37: 13594–13603.

Takahashi, Y. and S. Styring (1987). "A comparative study of the reduction of EPR signal II_{slow} by iodide and the iodo-labeling of the D2-protein in photosystem II." FEBS Letters 223: 371–375.

Takahashi, Y., M.-a. Takahashi and K. Satoh (1986). "Identification of the site of iodide photooxidation in the photosystem II reaction center complex." FEBS Letters 208: 347–351.

Tang, X.-S., D. W. Randall, D. A. Force, B. A. Diner and R. D. Britt (1996). "Manganese-tyrosine interaction in the Photosystem II oxygen-evolving complex." Journal of the American Chemical Society 118: 7638–7639.

Theg, S. M., P. A. Jursinic and P. H. Homann (1984). "Studies on the mechanism of chloride action on photosynthetic water oxidation." Biochimica et Biophysica Acta 766: 636–646.

Thornton, L. E., H. Ohkawa, J. L. Roose, Y. Kashino, N. Keren and H. B. Pakrasi (2004). "Homologs of plant PsbP and PsbQ proteins are necessary for regulation of photosystem II activity in the cyanobacterium *Synechocystis* 6803." Plant Cell 16: 2164–2175.

Umena, Y., K. Kawakami, J.-R. Shen and N. Kamiya (2011). "Crystal structure of oxygen-evolving photosystem II at a resolution of 1.9 Å." Nature (London) 473: 55–59.

van Gorkom, H. J. and C. F. Yocum (2005). The calcium and chloride cofactors. Photosystem II: The light-driven water: plastoquinone oxidoreductase. T. Wydrzynski and K. Satoh. The Netherlands, Springer: 307–327.

van Vliet, P. and A. W. Rutherford (1996). "Properties of the chloride-depleted oxygen-evolving complex of photosystem II studied by electron paramagnetic resonance." Biochemistry 35: 1829–1839.

Velthuys, B. R. (1975). "Binding of the inhibitor NH_3 to the oxygen-evolving apparatus of spinach chloroplasts." Biochimica et Biophysica Acta 396: 392–401.

Vogt, L., D. J. Vinyard, S. Khan and G. W. Brudvig (2015). "Oxygen-evolving complex of Photosystem II: An analysis of second-shell residues and hydrogen-bonding networks." Current Opinion in Chemical Biology 25: 152–158.

Warburg, O. and W. Lüttgens (1944). "Weitere experimente zur kohlensaureassimilation." Naturwissenschaften 40: 301.

Wincencjusz, H., H. J. van Gorkom and C. F. Yocum (1997). "The photosynthetic oxygen evolving complex requires chloride for its redox state S_2-to-S_3 and S_3-to-S_0 transitions but not for S_0-to-S_1 or S_1-to-S_2 transitions." Biochemistry 36: 3663–3670.

Wincencjusz, H., C. F. Yocum and H. J. van Gorkom (1998). "S-state dependence of chloride binding affinities and exchange dynamics in the intact and polypeptide-depleted O_2 evolving complex of photosystem II." Biochemistry 37: 8595–8604.

Wincencjusz, H., C. F. Yocum and H. J. van Gorkom (1999). "Activating anions that replace Cl^- in the O_2-evolving complex of photosystem II slow the kinetics of the terminal step in water oxidation and destabilize the S_2 and S_3 states." Biochemistry 38: 3719–3725.

Yachandra, V. K., R. D. Guiles, K. Sauer and M. P. Klein (1986). "The state of manganese in the photosynthetic apparatus. 5. The chloride effect in photosynthetic oxygen evolution." Biochimica et Biophysica Acta 850: 333–342.

Yachandra, V. K., K. Sauer and M. P. Klein (1996). "Manganese cluster in photosynthesis: Where plants oxidize water to dioxygen." Chemical Reviews 96: 2927–2950.

Yano, J. and V. K. Yachandra (2009). "X-ray absorption spectroscopy." Photosynthesis Research 102: 241–254.

Yocum, C. F. (2008). "The calcium and chloride requirements of the O_2 evolving complex." Coordination Chemistry Reviews 252: 296–305.

Yu, H., C. P. Aznar, X. Xu and R. D. Britt (2005). "Evidence that azide occupies the chloride binding site near the manganese cluster in photosystem II." Biochemistry 44: 12022–12029.

Zimmermann, J.-L. and A. W. Rutherford (1984). "EPR studies of the oxygen-evolving enzyme of Photosystem II." Biochimica et Biophysica Acta 767: 160–167.

Zimmermann, J.-L. and A. W. Rutherford (1986). "Electron paramagnetic resonance properties of the S_2 state of the oxygen-evolving complex of photosystem II." Biochemistry 25: 4609–4615.

Vectorial Charge Transfer Reactions in the Protein-Pigment Complex of Photosystem II

6

Mahir D. Mamedov and Alexey Yu Semenov

Summary

The pigment-protein complex of photosystem II localized in the thylakoid membranes of higher plants, algae, and cyanobacteria is the main source of oxygen on Earth. The light-induced functioning of photosystem II is directly linked to electron and proton transfer across the membrane, which results in the formation of transmembrane electric potential difference ($\Delta\Psi$). In this review, we describe the electrogenic reactions related to charge transfer on the donor side of photosystem II complexes which expand current understanding of the nature and mechanisms of vectorial processes and provide a necessary step in development of efficient systems of transformation of solar energy.

Keywords

Photosystem II • S-state transitions • Vectorial charge transfer • Transmembrane electric potential difference • Electron donors • Synthetic complex • Proteoliposomes • Direct electrometry

Contents

6.1	Structural Peculiarities of Photosystem II	98
6.2	Functioning of Photosystem II	98
6.2.1	Photochemical Processes	99
6.2.2	The Catalytic Cycle of Water Oxidation ...	100
6.2.3	Reduction of Tyrosine Cation Radical Y_Z ..	103
6.2.4	Lipophilic Electron Donors	103
6.2.5	Synthetic Mn-Containing Complex	104
6.2.6	Hydrophilic Electron Donors	104
6.2.7	Mechanism of Electron Transfer	104
6.2.8	Vectorial Charge Transfer Reactions in Photoactivated Apo-OEC-PS II Core Complexes	106
6.3	Concluding Remarks	107
References		107

Abbreviations

Chl	chlorophyll
DAD	2,3,5,6-Tetramethyl-p-phenylenediamine
DCPIP	2,6-Dichlorophenolindophenol
DPC	1,5-Diphenylcarbazide
OEC	oxygen evolving complex
P680	Primary electron donor
PMS	Phenazine methosulfate
Q_A	Primary quinone acceptor of PS II
Q_B	Secondary quinone acceptor of PS II
RC	Reaction center
S_i	redox states of OEC

M.D. Mamedov (✉) • A.Y. Semenov
A.N. Belozersky Institute of Physical-Chemical Biology, Lomonosov Moscow State University, 119991 Moscow, Leninskie gory, Russia
e-mail: mahirmamedov@yandex.ru

© Springer International Publishing AG 2017
H.J.M. Hou et al. (eds.), *Photosynthesis: Structures, Mechanisms, and Applications*,
DOI 10.1007/978-3-319-48873-8_6

TMPD	N,N,N′,N′-Tetramethyl-*p*-phenylenediamine
Y_Z	Redox-active tyrosine of D1 polypeptide
$\Delta\Psi$	Transmembrane electric potential difference.

6.1 Structural Peculiarities of Photosystem II

Photosystem II (PS II) is one of the two photosystems of oxygenic photosynthetic organisms that is responsible for oxygen evolution. This enzyme captures light energy by small core antenna pigments, which is consequently transferred to the reaction center (RC) pigments, where it triggers the charge separation process and secondary electron transfer steps.

The core complex of PS II is considered as a minimal unit capable of oxidizing water molecules. The structure of the PS II core complexes isolated from thermophilic cyanobacteria, was determined by X-ray structure analysis at a resolution of 3.8–1.9 Å (Ferreira et al. 2004; Guskov et al. 2009; Umena et al. 2011). Each monomer of the enzyme contains 20 individual protein subunits (total molecular weight of ~350 kDa), with 17 of them being integral membrane proteins. In addition, one monomer comprises ~100 cofactors, 25 lipid molecules and >1300 water molecules. Note that all organic and inorganic cofactors involved in the charge transfer reactions are located on subunits D1 and D2 of RC (Fig. 6.1).

PS II can be functionally subdivided into three structural domains: central photochemical, plastoquinone- reducing, and water-oxidizing. Four chlorophyll and two pheophytin molecules are arranged into two branches associated with proteins D1 and D2 and located in the central part of the enzyme (Ferreira et al. 2004; Guskov et al. 2009; Umena et al. 2011). The quinone acceptor complex of PS II consists of two plastoquinone (PQ) molecules with significantly different properties. The primary (Q_A) and secondary (Q_B) quinone acceptors function as one and two electron carriers, respectively (see Shinkarev 2004 for review). The water-oxidizing site in thylakoids is located on the lumenal side of PS II. Based on the latest X-ray data with resolution of 1.9 Å, all components of the oxygen-evolving complex (OEC) described by the formula Mn_4CaO_5 have been detected (Umena et al. 2011). In addition, all amino acid ligands of the OEC have been identified, and the bound water molecules have been found for the first time. The PS II core complex, due to availability of its 3D structure and of site-specific mutants, is a convenient object for the study of structure–function interactions.

6.2 Functioning of Photosystem II

Light energy absorbed by the pigments of integral antenna proteins CP43 and CP47 is transferred in sub-nanosecond time range to the primary electron donor P680, which is a chlorophyll *a* dimer. The primary charge separation in RC, including a series of highly optimized electron transfer processes (see Fig. 6.1 and (Shelaev et al. 2011)), has been studied by various methods such as absorption and fluorescent spectroscopy (including femtosecond time-resolved spectroscopy), electrochromic shift of absorption bands and experiments with hole burning, etc. The data obtained by differential absorption spectroscopies at 20-fs resolution under physiological conditions (278 K) in the PS II core complex show that the primary electron transfer between P680* and monomeric chlorophyll Chl_{D1} occurs with a lifetime of ~0.9 ps (Shelaev et al. 2011) and is determined by charge separation within P680. The subsequent electron transfer from Chl_{D1}, which occurs during 13–14 ps, corresponds to formation of the secondary ion-radical pair $P680^+Phe^-$ (Shelaev et al. 2011) and is stabilized as a result of rapid electron transfer to the tightly bound quinone Q_A on the stromal/cytoplasmic side of the membrane in ~200 ps. Thus, excitation energy (P680)* is used for electron transfer to a distance of ~23 Å. It should be noted that only the D1 branch is functionally active in PS II (Novoderezhkin et al.

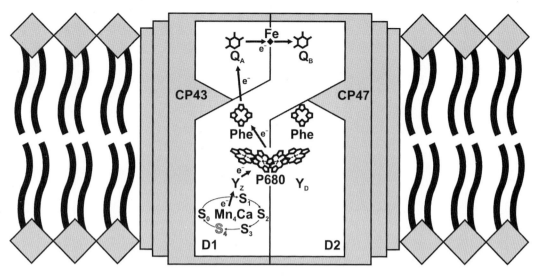

Fig. 6.1 Redox cofactors of PS II core complex

2007; Shelaev et al. 2011). The hole on P680 is filled by electron transfer from redox-active tyrosine-161 (Y_Z) of the PS II subunit D1 (Tommos and Babcock 1991). This electron transfer is coupled with the proton escape from Y_Z to the closest histidine (His190), which results in formation of the neutral tyrosine radical ($Y_Z^\bullet$) (Rappaport and Lavergne 1997; Renger 2004; Sproviero et al. 2008). The reduction of P680$^+$ in PS II complex with intact OEC occurs during 25 ns–50 µs (Renger and Renger 2008).

The electron from Q_A is further transferred to a distance of ~17 Å to the secondary quinone acceptor Q_B within 0.1–0.3 ms without detected oxidation–reduction involvement of an iron ion (Shinkarev 2004).

The subsequent turnovers of the enzyme include the same reactions but with different kinetics at some stages due to charge accumulation on the manganese cluster and on Q_B. The transfer of the second electron to Q_B^- causes the uptake of two protons from the water phase, the release of formed plastoquinol Q_BH_2 from the binding site in the protein, and its substitution by a molecule of oxidized plastoquinone from the membrane pool (See Shinkarev 2004 for review).

During each catalytic cycle, two water molecules are transformed into an oxygen molecule and four protons in the cycle of five intermediates called S-states ($S_0 \to S_4$) (Fig. 6.1). Since the S_1 state (basic state) is most stable in the dark, the first four flashes cause transitions $S_1 \to S_2$, $S_2 \to S_3$, $S_3 \to S_0$, and $S_0 \to S_1$. Molecular oxygen is released during the $S_3 \to S_0$ transition via intermediate state S_4 (Haumann et al. 2005; Dau and Haumann 2007). The kinetics of S_1 to S_3 transitions vary from 50 to 300 µs, while the terminal reaction $Y_Z^\bullet S_3 \to Y_Z S_0 + O_2 + 2H^+$ occurs in the millisecond time scale. The overall catalytic cycle can be presented as a scheme including eight stages of alternating reactions of electron transfer and proton release followed by the final stage – release of an oxygen molecule (Haumann et al. 2005; Dau and Haumann 2007).

Thus, single-electron transfer processes (P680* $\to Q_A$, $Y_Z \to$ P680$^+$) in PS II are coupled with four-electron oxidation of water ($2H_2O \to 4e^- + 4H^+ + O_2$) and two-electron plastoquinone reduction (PQ + 2e$^-$ + 2H$^+$ $\to$ PQH$_2$), which makes this enzyme one of the most complex energy converters.

6.2.1 Photochemical Processes

Light induced excitation of chlorophyll P680 in the PS II RC causes electron transfer across the dielectric layer of the thylakoid membrane to a

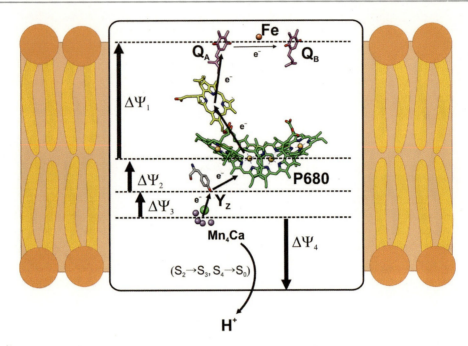

Fig. 6.2 Electrogenic stages of charge transfer in PS II. The scheme does not include electrogenesis due to protonation of doubly reduced Q_B

distance of ~35 Å. As a result, the release and uptake of protons occur at the donor and acceptor regions of the enzyme, respectively. Since these vectorial processes have components perpendicular to the plane of the membrane, they are coupled with generation of transmembrane electric potential difference ($\Delta\Psi$) (Fig. 6.2). Only under certain conditions (the pathway of the charge transfer reaction must have a component perpendicular to the plane of the membrane; transfer must occur in a region of the protein with dielectric permittivity of less than 40 (Semenov et al. 2006) and the transported charge must not be locally compensated)), the electron and proton transfers are accompanied by $\Delta\Psi$ generation. The generation of $\Delta\Psi$ coupled to the vectorial reactions of charge transfer in PS II has been registered in different photosynthetic preparations (chloroplasts, thylakoids, PS II-enriched membranes, PS II core complexes) by different methods (microelectrodes, modified patch clamp, electroluminescence, electrochemical shift of pigment absorption bands in light-harvesting complexes, and direct electrometry) (Semenov et al. 2006).

The kinetics of $\Delta\Psi$ generation determined in the early stages of electron transfer in RC was measured by the light gradient method (based on $\Delta\Psi$ measurement by silver chloride electrodes located in the lower and upper parts of the measuring cell) on chloroplasts and PS II-enriched membrane fragments (Trissl and Leibl 1989; Pokorny et al. 1994). The authors arrived at a conclusion that the primary charge separation in PS II includes two electrogenic stages with approximately the same contributions to the total amplitude of transmembrane electric potential – electron transfer between P680 and Phe and further electron transfer from Phe to Q_A (Trissl and Leibl 1989) (Fig. 6.2).

6.2.2 The Catalytic Cycle of Water Oxidation

The typical rate of water oxidation by the PS II complex under stationary illumination is 100–200 turnovers of the enzyme per second. During this process, each turnover of the enzyme is characterized by a set of distinguishable intermediates formed during time periods from picoseconds to several milliseconds from the starting moment of the reaction. Registration of the kinetics of charge transfer between the intermediates during a single turnover of the enzyme allows the study of molecular mechanisms of separate charge transfer reactions, while measurements under equilibrium conditions gives only the overall view of the light-dependent processes in PS II. Significant advantages for studying the mechanism of $\Delta\Psi$ generation associated with charge transfer within the enzyme are provided by isolated PS II core complexes incorporated into phospholipid liposomes studied by the direct electrometric method (Semenov et al. 2006, 2008). The principle of direct electrometry developed in our laboratory consists in the fusion of closed vesicles or liposomes containing protein complexes with a lipid-impregnated thin collodion film and $\Delta\Psi$ measurement with macroelectrodes immersed in electrolyte buffer solution on the two sides of the artificial membrane. This method is exceptionally sensitive and allows registration of intraprotein charge transfer to a distance of >0.5 Å in the direction perpendicular to the plane of the membrane. Note that the kinetics of electron transfer can be recorded by different methods of spectroscopy, while the kinetics of vectorial proton transfer can be measured with high time resolution only by electrometry. The results obtained by this method have shown that the relative contribution to the total $\Delta\Psi$ (~17%) attributed to the electron transfer from tyrosine Y_Z to photooxidized P680 in the PS II core complexes incorporated into liposomes (Haumann et al. 1997; Mamedov et al. 1999) is close to the $\Delta\Psi$ value measured previously by the light gradient method in PS II-enriched membrane fragments (~16%) oriented in a microcoaxial cell (Pokorny et al. 1994). In the present work we have reviewed and summarized the results of recording the kinetics of separate stages of $\Delta\Psi$ generation determined by charge transfer on the donor side of PS II core complexes incorporated into proteoliposomes under conditions of single actuation of the enzyme (Fig. 6.2).

It is well known that OEC in dark-adapted PS II samples are in the state S_1 until the first light flash. Under these conditions, in the kinetics of the photoelectric response induced by the first laser flash, in addition to rapid $\Delta\Psi$ generation due to charge separation between P680 and quinone Q_A and re-reduction of P680$^+$ by an electron transfer from Y_Z, an additional electrogenicity with a characteristic time (τ) of 30–65 µs (pH 6.5) and relative contribution of ~2.5–3.5% of the amplitude of kinetically-unresolved $Y_Z^{\bullet}Q_A^-$ major phase was observed (Haumann et al. 1997; Mamedov et al. 1999). Such characteristic times are close to the rate of manganese ion oxidation during the $S_1 \rightarrow S_2$ transition measured by X-ray absorption spectroscopy with high time resolution (Haumann et al. 2005).

Comparison of the kinetics of photoelectric responses of PS II core complex depleted of the Mn_4Ca cluster and the preparation treated with iron (II) ions (in the latter case, the high-affinity manganese-binding site is inhibited) suggested that manganese oxidation at the low-affinity site was non-electrogenic (Kurashov et al. 2009a, b). This can be considered as an evidence that the electrogenic reduction of tyrosine $Y_Z^{\bullet}$ occurs by vectorial electron transfer from manganese bound at the high-affinity site.

Note that one of the methods for distinguishing between the protolytic reactions and the other processes involved in generation of $\Delta\Psi$ is the study of pH-dependence of the charge transfer reaction or comparison of the reaction rate constants in the solutions of H_2O and D_2O. The absence of the isotope effect on the kinetics of the electrogenic phase during the $S_1 \rightarrow S_2$ transition of the OEC demonstrates that proton release does not take place on the donor side of the enzyme.

In response to the second laser flash (transition $S_2 \rightarrow S_3$), the kinetics of the photoelectric response shows an additional electrogenic phase with a characteristic time of 240–300 µs (pH 6.5) and relative contribution ~5–7% of the kinetically-unresolved fast-phase amplitude $Y_Z^{\bullet}Q_A^-$ (Haumann et al. 1997; Mamedov et al. 1999). The latest electrometric experiments show that the reaction rate constant of the $S_2 \rightarrow S_3$ transition decreases in the presence of D_2O, which allows to ascribe this transition to the electrogenic proton transfer to the bulk water phase. The dependence of the kinetics of the $S_2 \rightarrow S_3$ transition on D_2O has been shown previously for PS II-enriched membrane fragments by pulse absorption spectroscopy (Renger 2007).

The photoelectric response induced by the third laser flash (transition $S_3 \rightarrow S_4 \rightarrow S_0$) contains an additional electrogenic component with $\tau \sim$ 4.5–6 ms and relative contribution ~4–6% of the amplitude of the $Y_Z^{\bullet}Q_A^-$ phase (Haumann et al. 1997; Mamedov et al. 1999; Semenov et al. 2008). It is assumed that in the intact PS II complexes two protons are cleaved from water molecules during the terminal transition, and these data are in agreement with the role of amino acid groups close to the OEC in proton release (Haumann et al. 2005; Ishikita et al. 2006; Dau and Haumann 2007; Shimada et al. 2011). However, the formation of state S_4 does not include electron transfer from the manganese complex to radical $Y_Z^{\bullet}$, and it seems that the central event is deprotonation of the manganese complex (or its environment) with a characteristic time of ~200 μs (Haumann et al. 2005; Dau and Haumann 2007). The absence of an electrogenic phase with the characteristic time of ~200 μs in the kinetics of the photoelectric response induced by the third laser flash demonstrates that the proton transfer during the $S_3 \rightarrow S_4$ transition is not vectorial, i.e. electrically neutral. Also, the reduction of manganese through electron transfer from the water molecule during the $S_4 \rightarrow S_0$ transition is most likely non-electrogenic. Therefore, the electrogenesis observed in response to the third laser flash is probably determined by proton transfer from the manganese complex or surrounding amino acids into the water phase during the $S_4 \rightarrow S_0$ transition. The similar amplitudes of electrogenic reactions during OEC transitions $S_2 \rightarrow S_3$ and $S_4 \rightarrow S_0$ proves that protons cover equal distances. In contrast, the kinetics of the $S_2 \rightarrow S_3$ transition shows a more significant (nearly twofold) isotopic effect compared to the $S_4 \rightarrow S_0$ transition (Dau and Haumann 2007).

It should be noted that the contributions of the $S_2 \rightarrow S_3$ and $S_4 \rightarrow S_0$ transitions to the total electrogenesis in thylakoids measured by the electrochromic shift of the absorption bands of antenna carotenoids is approximately twofold higher than those revealed by direct electrometry in the PS II core complexes (Haumann et al. 1997). In addition, as distinct from the intact PS II core complexes with the terminal transition $S_4 \rightarrow S_0$ characterized by the time of 4.6–6 ms, the value of this parameter in thylakoids and PS II-enriched membrane fragments is 1–1.2 ms (Haumann et al. 1997). This difference may be due to the increase in dielectric permittivity around the manganese cluster as a result of removal of peripheral proteins.

During the catalytic cycle of oxygen formation from water in PS II, protons must be released into the water phase from the manganese complex immersed deep in the protein matrix. Therefore, there must be special pathways for proton transfer in the hydrophobic part of the enzyme (Muh and Zouni 2011; Ho 2012). Such a pathway for proton release must be effective enough not to retard the S-cycle. In addition to effectiveness, there is also a question of direction. It should be noted that the pathways of proton transfer in PS II have been studied much less than the pathways of electron transfer (Belevich and Verkhovsky 2008). In addition to the identification of amino acid residues around the manganese cluster, similar but not identical predictions for proton channels have been obtained in three works on intraprotein channel modeling (Murray and Barber 2007; Ho and Styring 2008; Gabdulkhakov et al. 2009). Recent studies of molecular dynamics of protein has revealed a series of residues and water molecules linked by hydrogen bonds that leads from the manganese cluster to the lumen (Swanson and Simons 2009). Three redox-active manganese atoms, an oxygen atom O(5) (probably OH^-), two water molecules bound to Mn(4), and the side chains of charged residues $D61^-/R357^+$ are included in the proton release pathway (McEvoy and Brudvig 2004). Although the authors of work (Hoganson and

Babcock 1997) supposed that Y_Z was a part of the pathway of proton release to the lumen and that there was a potential pathway of proton release from Y_Z in the latter 3D structure of PS II core complexes (Umena et al. 2011). It should be noted that the experimental data clearly support the "fluctuating" model, according to which the Y_Z proton never leaves the Y_Z-His190 site (Rappaport and Lavergne 1997).

6.2.3 Reduction of Tyrosine Cation Radical Y_Z

Photooxidized P680 is reduced by electron transfer from tyrosine Y_Z that in turn, accepts an electron from the manganese complex. Y_Z is usually considered as a part of the single-electron "wiring" of the OEC but not as a part of the pentametal Mn_4Ca cluster (Hoganson and Babcock 1997). The important question is how Y_Z couples the single-electron photochemical reaction and the four-electron catalytic oxidation of water, effectively controlling the water oxidation process. In this context, it is important to know in detail the protein environment of Y_Z and its interaction with water molecules and the manganese cluster in PS II. It is supposed that Y_Z is located in a hydrophobic environment and does not directly interact with the substrate water in PS II (Hillier et al. 1998). It is also known that the properties of Y_Z in PS II lacking the manganese cluster may be crucially different from those in the intact PS II. It has been supposed that in such preparations Y_Z is located in a hydrophilic environment and contacts the water phase (Babcock and Sauer 1975; Conjeaud and Mathis 1980; Blubaugh and Cheniae 1992; Hillier et al. 1998; Chroni and Ghanotakis 2001; Semin et al. 2002; Dasgupta et al. 2008; Gopta et al. 2008). Some substances, such as manganese, ascorbate, N,N,N′,N′-tetramethyl-*p*-phenylenediamine (TMPD), 2,3,5,6-tetramethyl-*p*-phenylenediamine (DAD), 2,6-dichlorophenolindophenol (DCPIP), phenazine methosulfate (PMS), 1,5-diphenylcarbazide (DPC), benzidine, hydroxylamine, and hydrazine are able to act as electron donors in the absence of the manganese cluster (Blubaugh and Cheniae 1992; Chroni and Ghanotakis 2001; Semin et al. 2002; Dasgupta et al. 2008; Gopta et al. 2008). Ascorbate seems to be the only alternative electron donor capable of supplying electrons to the PS II RC in sufficient amounts in the absence of the manganese cluster *in vivo*.

6.2.4 Lipophilic Electron Donors

The extremely asymmetric orientation of PS II core complexes in liposomes (the donor side outside the membrane) makes it possible to use the direct electrometric method for studying the mechanism of interaction between the PS II RC depleted of Mn_4Ca cluster and artificial electron donors. Increase in relative contributions of the slow components of membrane potential decrease associated with recombination of charges between Q_A and $Y_Z^{\bullet}$ (20–200 ms) in the presence of both Mn^{2+} (4 Mn per P680) and the reduced forms of lipophilic redox mediators (TMPD, DCPIP, DAD, PMS) and DPC can be considered as evidence of their ability to effectively interact with Y_Z radical. At certain concentrations of these substances, an additional slow electrogenic phase appears in the kinetics of the photoelectric response to a light flash, contributing 15–25% to the total electrogenesis (Gopta et al. 2008).

Experimental results lead to the conclusion that artificial electron donors are arranged in the following order in accordance with the degree of efficiency in reducing the radical Y_Z: PMS > TMPD > DAD > DPC > DCPIP. Under anaerobic conditions, PMS proved to be even more effective. However, it is still unclear whether the reaction rate is saturated at increased PMS concentration. Saturation is observed in the experiments, e.g. with TMPD, where the maximum K_v value of 400–500 s^{-1} is reached at 4 mM of the mediator.

6.2.5 Synthetic Mn-Containing Complex

In recent years, various Mn-containing complexes have been synthesized as models of the OEC manganese cluster. The direct electrometric method applied for proteoliposomes containing PS II core complexes depleted of manganese ions and three peripheral proteins showed that the addition of a synthetic trinuclear manganese complex (complex 173–1) to the measuring medium resulted in extra generation of photoelectric response with an amplitude of ~25% of the fast $(Y_Z^{\bullet}Q_A^-)$ phase and was characterized by $\tau \sim 160$ ms (Kurashov et al. 2009a, b). This phase was attributed to electron transfer from the protein–water interface to the oxidized cofactor immersed deep in the protein. Previously, it was demonstrated that the rate of oxygen release in PS II membrane fragments depleted of manganese ions in the presence of 173–1 synthetic complex is higher than in the presence of $MnCl_2$ (Nagata et al. 2008).

6.2.6 Hydrophilic Electron Donors

There are substantial differences in the mechanism of reduction of tyrosine Y_Z between small hydrophilic NH_2OH/NH_2NH_2 and large lipophilic compounds DPC/TMPD/DCPIP/DAD. The increase in relative contributions of the slow components of membrane potential decay in the presence of NH_2OH/NH_2NH_2 implies the prevention of charge recombination between Q_A^- and $Y_Z^{\bullet}$ by way of effective electron transfer to tyrosine $Y_Z^{\bullet}$. However, the absence of additional electrogenesis in the kinetics of photoelectric response is the evidence of the non-electrogenic character of this reaction.

6.2.7 Mechanism of Electron Transfer

Photosynthetic RCs are generally an ideal object for studying long-distance electron transfer (Savitsky et al. 2010; Novoderezhkin et al. 2011). Before discussing the mechanism of electron transfer in PS II RC in the presence of artificial electron donors, it is important to note that the direct electrometric method used in our previous experiments showed the presence of an extra electrogenic phase in the millisecond time domain in the kinetics of photoelectric response during the reduction of the photooxidized primary electron donors P870 in bacterial RC (Drachev et al. 1986a, b) and P700 in cyanobacterial complexes of PS I (Gourovskaya et al. 1997), by artificial redox-active compounds such as TMPD, DCPIP, and PMS. Since the contribution of this phase (~20%) to the total photoelectric response was approximately equal to the contribution of the phase observed in the presence of the native electron donor – cytochrome c_2 in the case of bacterial RC (Drachev et al. 1986b) and cytochrome c_6 (Mamedov et al. 1996) or plastocyanin (Mamedov et al. 2001) in PS I, it was concluded that the electrogenic reduction of $P870^+/P700^+$ by redox-active compounds results from the vectorial transfer of electrons from the protein–water interface to $P870^+/P700^+$ immersed in the protein matrix.

We believe that additional electrogenesis observed in the presence of TMPD, DAD, DCPIP, DPC, and synthetic trinuclear manganese complex is also determined by the vectorial electron transfer from the protein–water interface (Fig. 6.3). This assumption is supported by the results of modeling the structure of the donor region of PS II core complexes depleted of Mn_4Ca cluster and three peripheral proteins (Mamedov et al. 2010a, b). Removal of these subunits from PS II structure creates a cavity on the donor side of the enzyme. It has been shown that the TMPD molecule may rather tightly adjoins the cavity bottom, with a distance of ~17 Å between the edges of molecular π-orbitals of Y_Z and TMPD (the oxygen atom of Y_Z and the nearest nitrogen atom of TMPD) (Mamedov et al. 2010a, b).

It should be noted that electron transfer inside the protein is a complicated process determined by many factors. According to theory (Moser

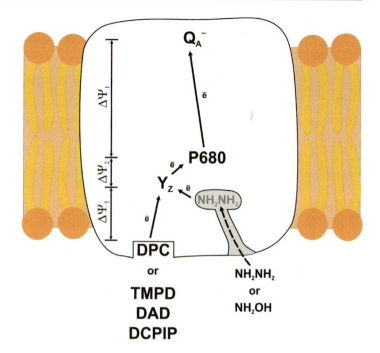

Fig. 6.3 Scheme of electron transport in the donor region of PS II depleted of Mn$_4$Ca cluster in the presence of artificial donors. *Solid arrows* show the electrogenic stages of electron transfer; the *dashed arrow* shows the hypothetical diffusion of NH$_2$NH$_2$ (NH$_2$OH) through the hydrophilic channel leading from the protein-water boundary to the binding site of the Mn$_4$Ca cluster (The scheme was adapted from work Mamedov et al. 2010a, b)

et al. 2003), the rate of electron transfer depends, in particular, on the distance between the donor and the acceptor, the difference of their redox potentials, and the reorganization energy. Electron transfer inside the protein is supposed to occur through specific tunneling and may be modulated by conformational changes in the secondary protein structure on the protein–water interface (Gray and Winkler 1996). As mentioned in work (Moser et al. 1992), effective tunneling is not limited by any specifically formed pathway inside the protein but rather occurs via several trajectories inside the protein matrix.

As regards the mechanism of electron transfer in the presence of low molecular weight donors in PS II preparations depleted of Mn$_4$Ca cluster and peripheral proteins, let us note that the modeling of structure of the donor side of such preparations (Mamedov et al. 2010a, b) has shown three channels with a minimal diameter of about 2.0–3.0 Å that link the binding site of the manganese cluster to the water–protein interface. It is obvious that the sizes of TMPD, DAD, DCPIP, and DPC molecules (4 × 14 Å) noticeably exceed the diameter of these channels, while the hydrophilic, low molecular weight electron donors NH$_2$NH$_2$ and NH$_2$OH (2.0 × 2.4 Å) can diffuse through these channels to the binding site of the Mn4Ca cluster (Fig. 6.3).

The results obtained by the direct electrometric method with a single turnover of the enzyme make it possible to follow the transfer of charges (electron and proton) inside the protein in real time. It has been shown that electrogenesis observed during the catalytic cycle of the OEC in dark-adapted PS II samples is determined by the electron transfer between manganese bound in the high-affinity site of subunit D1 and tyrosine radical Y$_Z$ (transition S$_1$ → S$_2$), as well as with the transfer of protons in the opposite direction from the manganese complex to the aqueous phase (transitions S$_2$ → S$_3$ and S$_4$ → S$_0$) (Fig. 6.2).

The data obtained with the preparations of PS II core complexes depleted of Mn$_4$Ca cluster and peripheral proteins show that the effective reduction of oxidized Y$_Z$ radical from artificial electron donors can be both electrogenic and non-electrogenic. Hydrophobic artificial electron donors (PMS, TMPD, DAD, DCPIP, and DPC) and synthetic trinuclear manganese complex designated as complex 173–1 reduce the tyrosine radical Y_Z electrogenically due to vectorial electron transfer from the binding site on the protein–water interface, while more hydrophilic and low molecular weight donors (NH_2OH, NH_2NH_2) can diffuse through the channels with minimum diameter of 2.0–3.0 Å passing from the water (lumen) surface of the protein to the Mn_4Ca cluster, followed by non-electrogenic reduction of tyrosine radical Y_Z (Fig. 6.3).

All these results are important for understanding the mechanism of interaction between the artificial electron donors and PS II RC.

6.2.8 Vectorial Charge Transfer Reactions in Photoactivated Apo-OEC-PS II Core Complexes

Water oxidation and O_2 evolution are completely inhibited upon depletion of Mn ions from PS II. The Mn-depleted complexes are usually designated as apo-OEC-PS II. The OEC function can be assembled in the presence of the free inorganic ions (Mn^{2+}, Ca^{2+}, Cl^-) and PS II, which are capable of performing this complex process (Ananyev and Dismukes 1996; Rova et al. 1998; Dasgupta et al. 2008; Petrova et al. 2013 and references therein). This light-dependent process, which also occurs upon assembling of a manganese cluster *in vivo* (Rova et al. 1998; Dasgupta et al. 2008), is called photoactivation.

Recently, we applied for the first time a sensitive electrometric technique to study electrogenic reactions due to intraprotein vectorial electron and proton transfer during the catalytic cycle of water oxidation in manganese-depleted and reconstituted PS II core complexes (Petrova et al. 2013). In dark-adapted samples, the electrogenic reactions observed in response to the 1st, 2nd, and 3rd laser flashes are presumably associated with the $S_1 \rightarrow S_2$ (electron transfer from Mn to $Y_Z^\cdot$), $S_2 \rightarrow S_3$, and $S_4 \rightarrow S_0$ (proton transfer in the opposite direction from Mn complex or its immediate environment into the aqueous bulk phase) transitions (Fig. 6.4, see also Fig. 6.1). The lack of additional voltage in the kinetics of the photoelectric response on the 4th laser flash indicates that proton release during the $S_0 \rightarrow S_1$ transition is not electrogenic. Thus, only

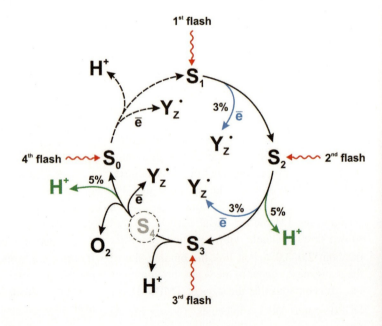

Fig. 6.4 Scheme of transitions between the S-states of PS II OEC. Percentages denote the relative contribution of electrogenic reactions. In dark-adapted PS II sample, OEC are mainly in state S_1

two of the four proton transfer reactions during the catalytic cycle of water oxidation are electrogenic ($S_2 \rightarrow S_3$ and $S_4 \rightarrow S_0$ transitions) (Fig. 6.4).

The obtained results showed that the kinetics and relative amplitudes of the electrogenic phases during the S-state transitions in Mn-reconstituted apo-OEC-PS II complexes are comparable to those obtained in intact PS II core complexes (Petrova et al. 2013). Almost the full reconstruction of electrogenic $S_1 \rightarrow S_2$, $S_2 \rightarrow S_3$, and $S_4 \rightarrow S_0$ transitions occurs even in the absence of the extrinsic proteins. However, a highly conserved extrinsic manganese-stabilizing protein (PsbO) has been suggested to be essential for maximum yield of recovering of the oxygen evolution rate (Dasgupta et al. 2008). Indeed, the relative rate of oxygen evolution upon reassembling OEC in our experiments did not exceed ~55% compared to the PS II with active OEC. Thus, one can conclude that the full reconstruction of electrogenic reactions due to S-state transitions is insufficient for achieving the entire recovery of OEC function.

6.3 Concluding Remarks

The results obtained by the direct electrometric method with a single turnover of the PS II incorporated into liposomes make it possible to follow the transfer of electron and proton inside the protein in real time. It has been shown that electrogenesis derived from the PS II donor side in dark-adapted PS II samples are due to charge transfer during S-state transitions of the OEC.

The data obtained with the preparations of PS II core complex depleted of Mn_4Ca cluster and peripheral proteins show that the effective reduction of oxidized tyrosine $Y_Z^\bullet$ from artificial electron donors can be both electrogenic and non-electrogenic.

The results described above expand current understanding of mechanisms of (i) electrogenic reactions during the S-state transitions of the OEC and (ii) interaction between the artificial electron donors and PS II RC and therefore provide a necessary step in development of efficient systems of transformation of solar energy.

Acknowledgments We are grateful to Drs. Dmitry Cherepanov, Vasily Kurashov and Irina Petrova for stimulating and fruitful discussions. This work has the support by the Russian Science Foundation (Grant 14-14-00789). Experimental results concerning the electrogenic reactions derived from Mn-depleted samples were obtained with support from the Russian Foundation for Basic Research (Grants 14-04-00519 and 15-04-04252).

References

Ananyev GM, Dismukes GC (1996) High-resolution kinetic studies of the reassembly of the tetramanganese cluster of photosynthetic water oxidation: proton equilibrium, cations, and electrostatics. Biochemistry 35:4102–4109

Babcock GT, Sauer K (1975) The electron donation sites for exogenous reductants in chloroplast photosystem II. Biochim Biophys Acta 396:48–62

Belevich I, Verkhovsky MI (2008) Molecular mechanism of proton translocation by cytochrome c oxidase. Antioxid Redox Signal 10:1–29

Blubaugh DJ, Cheniae GM (1992) Photoassembly of the Photosystem II Manganese Cluster. In Murata N (ed) Kluwer Academic Publishers: The Netherlands, pp 361–364

Chroni S, Ghanotakis DF (2001) Accessibility of tyrosine Y(.)(Z) to exogenous reductants and Mn(2+) in various Photosystem II preparations. Biochim Biophys Acta 1504:432–437

Conjeaud H, Mathis P (1980) The effects of pH on the reductions kinetics of P-680 in Tris-treated chloroplasts. Biochim Biophys Acta 590:353–359

Dau H, Haumann M (2007) Eight steps preceding O–O bond formation in oxygenic photosynthesis—A basic reaction cycle of the Photosystem II manganese complex. Biochim Biophys Acta 1767:472–483

Dasgupta J, Ananyev GM, Dismukes GC (2008) Photoassembly of the water-oxidizing complex in photosystem II. Coord Chem Rev 252:347–360

Drachev LA, Kaminskaya OP, Konstantinov AA, Mamedov MD, Samuilov VD, Semenov A. Yu, Skulachev VP (1986a) Effects of electron donors and acceptors on the kinetics of the photoelectric responses in *Rhodospirillum rubrum* and *Rhodopseudomonas sphaeroides* chromatophores. Biochim Biophys Acta 850:1–9

Drachev LA, Kaminskaya OP, Konstantinov AA, Kotova EA, Mamedov MD, Samuilov VD, Semenov AYu, Skulachev VP (1986b) The effect of cytochrome c, hexammineruthenium and ubiquinone-10 on the kinetics of photoelectric responses of *Rhodospirillum rubrum* reaction centres. Biochim Biophys Acta 848:137–146

Ferreira KN, Iverson TM, Maghlaoui K, Barber J, Iwata S (2004) Architecture of the photosynthetic oxygen-evolving center. Science 303:1831–1838

Gabdulkhakov A, Guskov A, Broser M, Kern J, Muh F, Saenger W, Zouni A (2009) Probing the accessibility of the Mn(4)Ca cluster in photosystem II: channels calculation, noble gas derivatization, and cocrystallization with DMSO. Structure 17:1223–1234

Gopta OA, Tyunyatkina AA, Kurashov VN, Semenov AYu, Mamedov MD (2008) Effect of redox-mediators on the flash-induced membrane potential in Mn-depleted photosystem II core particles. Eur Biophys J 37:1045–1050

Gourovskaya KN, Mamedov MD, Vassiliev IR, Golbeck JH, Semenov AY (1997) Electrogenic reduction of the primary electron donor P700+ in photosystem I by redox dyes. FEBS Lett 414:193–196

Gray H.B. Winkler JR (1996) Electron transfer in proteins. Annu Rev Biochem 65:537–561

Guskov A, Kern J, Gabdulkhakov A, Broser M, Zouni A, Saenger W (2009) Cyanobacterial photosystem II at 2.9-Å resolution and the role of quinones, lipids, channels and chloride. Nat Struct Mol Biol 16:334–342

Haumann M, Mulkidjanian A, Junge W (1997) Electrogenicity of electron and proton transfer at the oxidizing side of photosystem II. Biochemistry 36:9304–9315

Haumann M, Liebisch P, Muller C, Barra M, Grabolle M, Dau H (2005) Photosynthetic O_2 formation tracked by time-resolved X-ray experiments. Science 310:1019–1021

Hillier W, Messinger J, Wydrzynski T (1998) Kinetic determination of the fast exchanging substrate water molecule in the S3 state of photosystem II. Biochemistry 37:16908–16914

Ho FM (2012) Structural and mechanistic investigations of photosystem II through computational methods. Biochim Biophys Acta 1817:106–120

Ho FM, Styring S (2008) Access channels and methanol binding site to the CaMn4 cluster in Photosystem II based on solvent accessibility simulations, with implications for substrate water access. Biochim Biophys Acta 1777:140–153

Hoganson CW, Babcock GT (1997) A metalloradical mechanism for the generation of oxygen from water in photosynthesis. Science 277:1953–1956

Ishikita H, Saenger W, Loll B, Biesiadka J, Knapp EW (2006) Energetics of a possible proton exit pathway for water oxidation in photosystem II. Biochemistry 45:2063–2071

Kurashov VN, Allakhverdiev SI, Zharmukhamedov SK, Nagata T, Klimov VV, Semenov AYu, Mamedov MD (2009a) Electrogenic reactions on the donor side of Mn-depleted photosystem II core particles in the presence of $MnCl_2$ and synthetic trinuclear Mn-complexes. Photochem Photobiol Sci 8:162–166

Kurashov VN, Lovyagina ER, Shkolnikov DY, Solntsev MK, Mamedov MD, Semin BK (2009b) Investigation of the low-affinity oxidation site for exogenous electron donors in the Mn-depleted photosystem II complexes. Biochim Biophys Acta 1787:1492–1498

Mamedov MD, Gadzhieva RM, Gourovskaya KN, Drachev LA, Semenov AYu (1996) Electrogenicity at the donor/acceptor sides of cyanobacterial photosystem I. J Bioenerg Biomembr 28:517–522

Mamedov MD, Beshta OP, Gurovskaya KN, Mamedova AA, Neverov KA, Samuilov VD, Semenov AYu (1999) Photoelectric responses of oxygen-evolving complexes of photosystem II. Biochemistry (Moscow) 64:504–509

Mamedov MD, Mamedova AA, Chamorovsky SK, Semenov AYu (2001) Electrogenic reduction of the primary electron donor P700 by plastocyanin in photosystem I. FEBS Lett 500:172–176

Mamedov MD, Kurashov VN, Petrova IO, Zaspa AA, Semenov AYu (2010a) Electron transfer between exogenous electron donors and reaction center of photosystem II. Biochemistry (Moscow) 75:579–584

Mamedov MD, Kurashov VD, Cherepanov DA, Semenov AYu (2010b) Photosysem II: where does the light-induced voltage come from? Front Biosci 15:1007–1017

McEvoy JP, Brudvig GW (2004) Structure-based mechanism of photosynthetic water oxidation. Phys Chem Chem Phys 6:4754–4763

Moser CC, Keske JM, Warncke K, Farid RS, Dutton PL (1992) Nature of biological electron transfer. Nature 355:796–802

Moser CC, Page CC, Cogdell RJ, Barber J, Wraight CA, Dutton PL (2003) Length, time, and energy scales of photosystems. Adv Protein Chem 63:71–109

Murray JW, Barber J (2007) Structural characteristics of channels and pathways in photosystem II including the identification of an oxygen channel. J Struct Biol 159:228–237

Muh F, Zouni A (2011) Light-induced water oxidation in photosystem II. Front Biosci 17:3072–3132

Nagata T, Zharmukhamedov SK, Khorobrykh AA, Klimov VV, Allakhverdiev SI (2008) Reconstitution of the water-oxidizing complex in manganese-depleted photosystem II preparations using synthetic Mn complexes: a fluorine-19 NMR study of the reconstitution process. Photosyn Res 98:277–284

Novoderezhkin VI, Dekker JP, van Grondelle R (2007) Mixing of exciton and charge-transfer states in Photosystem II reaction centers: modeling of Stark spectra with modified Redfield theory. Biophys J 93:1293–1311

Novoderezhkin VI, Romero E, Dekker JP, Grondelle R (2011) Multiple charge-separation pathways in photosystem II: modeling of transient absorption kinetics. ChemPhysChem 12:681–688

Petrova IO, Kurashov VN, Zaspa AA, Semenov AYu, Mamedov MD (2013) Vectorial charge transfer reactions on the donor side of manganese-depleted and reconstituted photosystem II core complexes. Biochemistry (Moscow) 78:395–402

Pokorny A, Wulf K, Trissl H-W (1994) An electrogenic reaction associated with the re-reduction of P680 by tyr Z in photosystem II. Biochim Biophys Acta 1184:65–70

Rappaport F, Lavergne J (1997) Charge recombination and proton transfer in manganese-depleted photosystem II. Biochemistry 36:15294–15302

Renger G (2004) Coupling of electron and proton transfer in oxidative water cleavage in photosynthesis. Biochim Biophys Acta 1655:195–204

Renger G (2007) Oxidative photosynthetic water splitting: energetics, kinetics and mechanism. Photosyn Res 92:407–425

Renger G, Renger T (2008) Photosystem II: The machinery of photosynthetic water splitting. Photosynth Res 98:53–80

Rova M, Mamedov F, Magnuson A, Fredriksson P-O, Styring S (1998) Coupled activation of the donor and the acceptor side of photosystem II during photoactivation of the oxygen evolving cluster. Biochemistry 37:11039–11045

Savitsky A, Malferrari M, Francia F, Venturoli G, Möbius K (2010) Bacterial photosynthetic reaction centers in trehalose glasses: coupling between protein conformational dynamics and electron-transfer kinetics as studied by laser-flash and high-field EPR spectroscopies. J Phys Chem B 114:12729–12743

Semenov AYu, Mamedov MD, Chamorovsky SK (2006) Electrogenic reactions associated with electron transfer in photosystem I. In: Golbeck JH (ed) Photosystem I: the Light-driven, plastocyanine:ferredoxin oxidoreductase. Springer, Dordrecht, pp 319–424

Semenov A, Cherepanov D, Mamedov M (2008) Electrogenic reactions and dielectric properties of photosystem II. Photosynth Res 98:121–130

Semin BK, Ghirardi ML, Seibert M (2002) Blocking of electron donation by Mn(II) to YZ* following incubation of Mn-depleted photosystem II membranes with Fe(II) in the light. Biochemistry 41:5854–5864

Shelaev IV, Gostev FE, Vishnev MI, Shkuropatov AY, Ptushenko VV, Mamedov MD, Sarkisov OM, Nadtochenko VA, Semenov AYu, Shuvalov VA (2011) P680 (P(D1)P(D2)) and Chl(D1) as alternative electron donors in photosystem II core complexes and isolated reaction centers. Photochem Photobiol 104:44–50

Shimada Y, Suzuki H, Tsuchiya T, Mimuro M, Noguchi T (2011) Structural coupling of an arginine side chain with the oxygen-evolving Mn4Ca cluster in photosystem II as revealed by isotope-edited Fourier transform infrared spectroscopy. J Am Chem Soc 133:3808–3811

Shinkarev VP (2004) Photosystem II: Oxygen evolution and chlorophyll a fluorescence induced by multiple flashes. In: Papageorgiou G, Govindjee (eds) Chlorophyll fluorescence: a signature of photosynthesis. Kluwer Academic Publishers, Netherlands, pp 197–229

Sproviero EM, Gascon JA, McEvoy JP, Brudvig GW, Batista VS (2008) A model of the oxygen-evolving center of photosystem II predicted by structural refinement based on EXAFS simulations. J Am Chem Soc 130: 3428–3442

Swanson JMJ, Simons J (2009) Role of charge transfer in the structure and dynamics of the hydrated proton. J Phys Chem B 113:5149–5161

Tommos C, Babcock GT (1991) Proton and hydrogen currents in photosynthetic water oxidation. Biochim Biophys Acta 1458:199–219

Trissl HW, Leibl W (1989) Primary charge separation in photosystem II involves two electrogenic steps. FEBS Lett 244:85–88

Umena Y, Kawakami K, Shen J-R, Kamiya N (2011) Crystal structure of oxygen-evolving photosystem II at a resolution of 1.9 Å. Nature 473:55–60

Function and Structure of Cyanobacterial Photosystem I

Wu Xu and Yingchun Wang

Summary

Photosystem I, an essential membrane complex in photosynthesis, is a light–driven reducing power generator that accepts electrons from plastocyanin on the luminal side, and donates electrons to ferredoxin on the stromal side of plants and cyanobacteria. The 2.5 Å crystal structure of the photosystem I of cyanobacterium *Thermosynechccocus elongates*, containing twelve subunits, 96 chlorophyll *a* molecules, 22 β–carotenes, two phylloquinones, three [4Fe–4S] clusters and four lipids, demonstrated one of the most fascinating membrane protein–cofactor complexes in biology. Twelve subunits interact with each other and bind cofactors and provide unique local environment for each cofactor to form a stable and dynamical complex for efficient energy and electron transfer. Cyanobacterial photosystem I is remarkably similar to its counterpart in the chloroplast of plants and algae. Structural information combined with physiological, biochemical and spectroscopic characterization of subunit–deficient and site–directed cyanobacterial mutants have revealed functions of individual subunits and cofactors. However, there are still many unresolved questions, such as initial charge separation, assembly, degradation and regulation of photosystem I. In the past 15 years, multidisciplinary research efforts have advanced our understanding of structure and function of photosystem I that is discussed in this chapter.

Keywords

Photosystem I • Cyanobacteria • Structure • Function • Assembly • Degradation • Regulation • Chlorophyll • Carotenoid • Lipid • Reaction center • Electron transfer

Contents

7.1	Introduction	112
7.2	**Peripheral Subunits of PS I**	114
7.2.1	PsaC	115
7.2.2	PsaD	116
7.2.3	PsaE	119

W. Xu (✉)
Department of Chemistry, University of Louisiana at Lafayette, Lafayette, LA 70504, USA
e-mail: wxx6941@louisiana.edu

Y. Wang
State Key Laboratory of Molecular Developmental Biology, Institute of Genetics and Developmental Biology, Chinese Academy of Sciences, Beijing 100101, China

7.3	**Integral Membrane Subunits**	121
7.3.1	PsaA and PsaB	121
7.3.1.1	Introduction	121
7.3.1.2	Antenna Systems	125
7.3.1.3	Lipids	130
7.3.1.4	Electron Transfer Chain	131
7.3.1.5	Docking of Plastocyanin/ Cytochrome c_6	144
7.3.2	PsaF	145
7.3.3	PsaI	148
7.3.4	PsaJ	150
7.3.5	PsaK	151
7.3.6	PsaL	153
7.3.7	PsaM	155
7.3.8	PsaU or PsaX	156
7.4	**Concluding Remarks**	158
References		159

7.1 Introduction

All chemical forms of energy and oxygen on the earth are generated via photosynthesis. Photosynthesis is a process in which higher plants, eukaryotic algae, and cyanobacteria convert inorganic CO_2 to chemical forms of energy and produce O_2 using sunlight. Although it only captures somewhere between 0.05 and 3% of the light energy received by the earth's surface, photosynthesis is sufficient to power the entire biological world (De Marais 2000; Rye and Holland 1998). Virtually all oxygen in the atmosphere is thought to be generated through photosynthetic process (Allen 2005; Dismukes et al. 2001). Cellular respiration, a chemical process opposite to photosynthesis, requires oxygen to generate ATP and produce CO_2 using chemical forms of energy. It occurs in all living organisms and plays a fundamental role in sustaining life (Kalckar 1991; Saraste 1999). Photosynthesis and respiration are interlinked processes. Each depends on the products of the other. In higher plants, photosynthesis occurs in chloroplasts whereas respiration occurs in mitochondria.

Cyanobacteria, which have been dated back about 3.5 billion years, presumably are the earliest known group of organisms that contributed greatly to the formation of the atmospheric oxygen that we breathe today. They can be found in diverse habitats from Antarctica to hot springs, and account for approximately 40% of planetary oxygen production (Paumann et al. 2005; Kolber et al. 2000). More broadly, cyanobacteria are also considered as a perfect candidate for counteracting the greenhouse effect, one major source of global warming, since they consume CO_2 and evolve pure O_2 by photosynthesis. The concentration of CO_2 in the atmosphere has increased by approximately 30% since the middle of the nineteenth century and probably will continue to increase unless societies choose to change their ways. Besides oxygen and carbohydrate production, cyanobacteria are considered to have both beneficial and harmful effects on human life. They are regarded as harmful primarily due to their association with environmental problems, for example, fouling water bodies and producing toxins (Mwaura et al. 2004). In contrast, they have also been used for beneficial causes: such as food additives (Lem and Glick 1985), biofertilizers (Mekonnen et al. 2002; Irisarri et al. 2001; Dutta et al. 2005; Tsygankov et al. 2002), renewable fuels (e.g., hydrogen (Ghirardi et al. 2007; Rupprecht et al. 2006; Ducat et al. 2012), ethanol (Deng and Coleman 1999), butanol (Lan and Liao 2011), ethylene (Ungerer et al. 2012; Guerrero et al. 2012; Eckert et al. 2014), isobutyraldehyde (Atsumi et al. 2009), fatty acids (Liu et al. 2011), or isoprene (Lindberg et al. 2009)), and bioactive natural compounds (Corbett et al. 1996; Tan 2007; Singh et al. 2005; Burja et al. 2001). Interestingly, it was reported recently that many cyanobacteria are naturally able to produce alkanes, which are the major constituents of gasoline, diesel, and jet fuels (Schirmer et al. 2010; KlÃ¤hn et al. 2014). To our knowledge, biochemical function and biodegradation of alkane in cyanobacteria remains completely unknown. The research efforts will advance our fundamental understanding of photosynthesis in cyanobacteria. This in turn will enable scientists to accurately model this organism for improving our living environments and life quality.

The photosynthetic process can be divided into two types of coordinated reactions: light reactions and dark reactions. In the light reactions, light energy is harnessed to synthesize

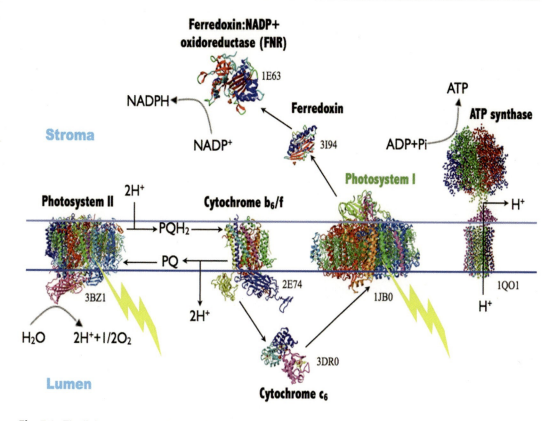

Fig. 7.1 The light reactions of photosynthesis

ATP and NADPH, which are used in the dark (light–independent) reactions to drive the synthesis of carbohydrates from CO_2 and H_2O. During the light reactions, four membrane–protein complexes, photosystem II (PS II), cytochrome b_6/f, photosystem I (PS I) and ATP synthase function in a coordinated way to drive photosynthetic electron transport (Fig. 7.1). Photosynthetic reaction centers of light reactions are classified according to their terminal electron acceptor as either type I, with an iron sulfur cluster acceptor, or type II, with a quinone terminal acceptor (Mazor et al. 2014). Striking similarities between type I and II photosystems recently became apparent in the 3–D structures of PS I, PS II and reaction center type II (Baymann et al. 2001). We currently know only two versions of the type I reaction centers: the relatively simple bacterial homodimer found in green sulfur bacteria (Buttner et al. 1992), *heliobacteria* and *Chloracidobacterium* or the much more complex PS I with its 11–15 subunits found in cyanobacteria and all photosynthetic eukaryotes (Buttner et al. 1992; Nelson and Ben-Shem 2005; Nelson and Yocum 2006). PS I is a protein–pigment complex that mediates the light–driven electron transfer from plastocyanin on the luminal side to ferredoxin on the stromal side of cyanobacteria. Cyanobacterial PS I is remarkably similar to its counterpart in the chloroplast of plants and algae. Therefore, it has served as a prototype for the type I reaction centers of photosynthesis.

The PS I complex of cyanobacterium *Thermosynechcocus elongatus* contains twelve subunits (PsaA, B, C, D, E, F, I, J, K, L, M and X), 96 chlorophyll *a* molecules, 22 β–carotenes, two phylloquinones, three [4Fe–4S] clusters and four lipids (Jordan et al. 2001). PsaA and PsaB are the core subunits that harbor the most antenna chlorophyll *a* molecules and the primary electron donor P700, a dimer of chlorophyll *a* and *a'* molecules (eC1A/eC1B), and a chain of electron acceptors A_0 (a chlorophyll *a* molecule, eC3A or 3C3B), A_1 (a phylloquinone) and F_X (a [4Fe–4S] cluster). The peripheral subunit PsaC binds the terminal electron acceptors F_A and F_B, two [4Fe–4S] clusters. The electron transfer pathway in PS I begins with the reaction center P700, which receives the excitation energy from a photon of light and forms the excited state, $P700^*$ ($eC1^*$), leading to a charge separation. Subsequently, an electron from $P700^*$ is transferred to A_0 (eC3), and then to A_1, and from there the electron is transferred to a series of [4Fe–4S] clusters. Ultimately, the electron is used to reduce ferredoxin. The electron lost by P700 is gained by plastocyanin from the luminal side.

Despite the remarkable conservation, there are some differences between the cyanobacterial and chloroplast PS I complexes. First, the higher order organization of the PS I complexes is different in cyanobacteria and chloroplasts. The plant and algal PS I complexes associate with membrane–bound light–harvesting complexes (LHC I) and can be isolated as core complexes or holocomplexes (the PS I complex along with LHC I). Cyanobacteria do not contain membrane–bound LHC I complexes. Second, chloroplast and cyanobacterial PS I complexes have differences in their subunit composition regarding some accessory proteins. Eukaryotic PS I contains three additional proteins. PsaG is an integral membrane protein with two transmembrane regions (Kalckar 1991). Its primary sequence shows homology to that of PsaK (Kalckar 1991), which is found in both cyanobacteria and chloroplasts. PsaH and PsaN are peripheral proteins on the stromal and luminal side of PS I, respectively (Saraste 1999). In contrast, the presence of PsaM, a 3kDa hydrophobic protein, has been demonstrated only in cyanobacterial PS I complexes. Lastly, the plant PS I complexes are segregated from PS II complexes in the thylakoid membranes of chloroplasts. PS I is located mainly in the unstacked stroma lamellae, where contact with the stroma provides easy access to ferredoxin and $NADP^+$. PS II is located almost exclusively in the closely stacked grana, whereas cytochrome b_6f complex is distributed uniformly throughout the membrane (Paumann et al. 2005). In contrast, PS I complexes in some cyanobacteria (e.g. *Synechococcus* sp. PCC 7942) is distributed in a radial asymmetric fashion, with higher concentration in the outermost thylakoids of a cyanobacterial cell (Kolber et al. 2000). However, this is not universally true for cyanobacteria, for example *Synechocystis* sp. PCC 6803, does not show radial asymmetry (Kolber et al. 2000).

Cyanobacteria provide many advantages in PS I studies. The absence of membrane–bound light–harvesting complexes and the presence of trimeric quaternary organization of PS I allow rapid purification of PS I complexes that can be used in sophisticated spectroscopic investigations. The genome of *Synechocystis* sp. PCC 6803 is the first completely sequenced genome of a photosynthetic organism. Since then, genomes of additional cyanobacterial species are being sequenced. This information and the availability of reverse genetic system have allowed manipulation of cyanobacterial genes for PS I proteins and for biosynthesis of PS I cofactors. Cyanobacteria are easily transformed via a homologous recombination system. Many genes have been selectively inactivated and some of the knockout strains serve as valuable reagents for functional studies. *Synechocystis* sp. PCC 6803 can grow under a number of different conditions ranging from photoautotrophic to fully heterotrophic modes, permitting research into how the genetic modifications may alter fundamental processes such as photosynthesis and/or respiration (Ungerer et al. 2012; Nakamura et al. 2000; Vermaas 1996; Wang et al. 2012).

7.2 Peripheral Subunits of PS I

Three peripheral subunits (PsaC, PsaD and PsaE) are located at the stromal side of PS I and are involved in the docking of ferredoxin. The structure of the three subunits showed that they are in the stromal hump and provide the potential docking site of ferredoxin.

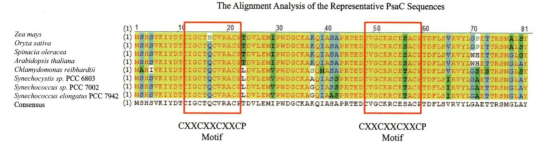

Fig. 7.2 The sequence alignment of PsaC

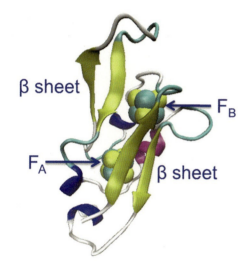

Fig. 7.3 The PsaC structure (PDB ID: 1JB0)

7.2.1 PsaC

PsaC, a 86-aa acidic polypeptide (also designated subunit VII) with a predicted isoelectric point of 5.68, is encoded by *ssl0563* in *Synechocystis* sp. PCC 6803. Cyanobacterial, green algal and plant PsaC are highly conserved. Our alignment analysis of the selected representative PsaC sequences showed that they share nearly 100% similarity and about 80% identity (Fig. 7.2), demonstrating their similar function in plants, green algae and cyanobacteria. *psaC* gene in plants locates in chloroplast genome (Scheller et al. 2001). The main functions of PsaC are to bind to a core of PsaA and PsaB to accept electrons from F_X, to efficiently outcompete the backreaction from F_X to P700 (Naver et al. 1996)

and to donate electron to ferredoxin. PsaC harbors the two [4Fe-4S] clusters: F_A and F_B, at the terminal of iron-sulfur clusters in type I reaction center. PsaC contains two conserved sequence motifs CXXCXXCXXXCP (Fig. 7.2) in which the cysteines provide the ligands to the Fe atoms. The redox potentials of F_A and F_B were estimated as $-520mV$ and $-580mV$ respectively (Nelson and Yocum 2006). They are defined by distinct lines in the EPR spectra of PS I (Evans et al. 1974). It exhibits pseudo-twofold symmetry similar to bacterial two [4Fe-4S] ferredoxins. The long C-terminus of PsaC interacts with PsaA, PsaB and PsaD and appears to be important for the proper assembly of PsaC into the PS I complex (Jordan et al. 2001).

The gene of a homology of subunit PsaC to bacterial ferredoxins also containing two [4Fe-4S] clusters was suggested from strong sequence similarity and homology models (Dunn et al. 1988; Golbeck 1993) and was confirmed by the similarity of both structures (Jordan et al. 2001; Adman et al. 1973). The structure of PsaC is shown in Fig. 7.3. The central part of PsaC consists of two short α helices connecting the two iron-sulfur clusters: F_A and F_B. This part is very similar in PsaC and ferredoxin. PsaC contains two β sheets, each with two antiparallel β stands that surround F_A and F_B (Fig. 7.3). The preeminent deviations between the two structures are the N- and C-termini, elongated by 2 and 14 residues in PsaC, respectively, and an extension of 10 residues pointing towards the putative ferredoxin/flavodoxin docking site in the internal loop region exposed to the stromal surface of the PS I complex. The

C−terminus of PsaC is very important for the correct docking of PsaC to the PS I core.

Biochemical analysis of the PsaC−less strain of *Chlamydomonas reinhardtii* indicates that neither PS I reaction center subunits nor the seven small subunits belonging to PS I accumulate stably in the thylakoid membranes. Pulse−chase labeling of cell proteins shows that the PS I reaction center subunits are synthesized normally but turn over rapidly in the PsaC−less strain. It was concluded that the iron−sulfur binding protein encoded by the *psaC* gene is an essential component, both for photochemical activity and for stable assembly of PS I (Takahashi et al. 1991). To elucidate the exact roles of F_A and F_B, two site−directed mutant strains of the cyanobacterium *Anabaena variabilis* ATCC 29413 were created. In one mutant, cysteine 13, a ligand for F_B was replaced by an aspartic acid (C13D); in the other mutant, cysteine 50, a ligand for F_A was modified similarly (C50D). Low temperature electron paramagnetic resonance studies demonstrated that the C50D mutant has a normal F_B center and a modified F_A center. In contrast, the C13D strain has normal F_A, but failed to reveal any signal from F_B. The room−temperature optical studies showed that C13D has only one functional electron acceptor in PsaC, whereas two such acceptors are functional in the C50D and the wild−type strains. Although both mutants grow under photoautotrophic conditions, the rate of PS I−mediated electron transfer in C13D under low light levels is about half that of C50D or the wild type. These data showed that (i) F_B is not essential for the assembly of the PsaC protein in PS I and (ii) F_B is not absolutely required for electron transfer from the PS I reaction center to ferredoxin (Mannan et al. 1996). In a separate report a year later, cysteine ligands in positions 14 or 51 to F_B and F_A in *Synechocystis* sp. PCC 6803, respectively, were replaced with aspartate, serine, or alanine, and the effect on the genetic, physiological, and biochemical characteristics of PS I complexes from the mutant strains were studied. Contrary to the phenotypes of the similar mutations in *Anabaena variabilis*, all mutant strains were unable to grow photoautotrophically, and compared with the wild type, mixotrophic growth was inhibited under normal light intensity. The thylakoids isolated from the aspartate and serine mutants have lower levels of PS I subunits PsaC, PsaD, and PsaE and lower rates of PS I−mediated substrate photoreduction compared with the wild type. The alanine and double aspartate mutants have no detectable levels PsaC, PsaD, and PsaE. Electron transfer rates, measured by cytochrome c_6−mediated $NADP^+$ photoreduction, were lower in purified PS I complexes from the aspartate and serine mutants. By measuring the $P700^+$ kinetics after a single turnover flash, a large percentage of the backreaction in the aspartate and serine mutants was found to be derived from A_1 and F_X, indicating an inefficiency at the $F_X \rightarrow F_A/F_B$ electron transfer step. The alanine and double aspartate mutants failed to show any backreaction from $[F_A/F_B]^-$. These results indicate that the various mutations of the cysteine 14 and 51 ligands to F_B and F_A affect biogenesis and electron transfer differently depending on the type of substitution (Yu et al. 1997). Species−specific phenotype differences between *Anabaena variabilis* ATCC 29413 and *Synechocystis* sp. PCC 6803 may be due to their minor structural differences of PsaC and/or physiological differences. For example, *Anabaena variabilis* 29413 is a filamentous organism and a natural heterotroph and does not require special light−pulse treatment as does *Synechocystis* sp. PCC 6803 to grow heterotrophically (Anderson and McIntosh 1991).

7.2.2 PsaD

PsaD, a 141−aa basic polypeptide (also designated subunit II) with a predicted isoelectric point of 8.95, is encoded by *slr0737* in *Synechocystis* sp. PCC 6803. In plants, PsaD is encoded in the nuclear genome and is therefore synthesized as a preprotein with a transit peptide (Scheller et al. 2001). The PsaD subunit is a conserved peripheral protein on the reducing side of PS I. Our alignment analysis of the selected representative sequences showed that plant, green algal and cyanobacterial PsaD

The Alignment Analysis of the Representative PsaD Sequences

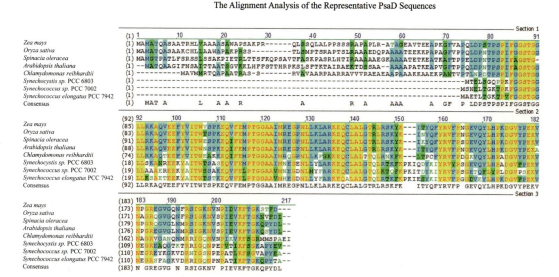

Fig. 7.4 The sequence alignment of PsaD

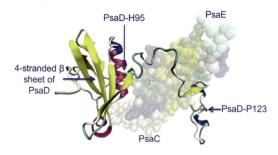

Fig. 7.5 The PsaD structure (PDB ID: 1JB0). PsaD: New cartoon presentation; PsaC: Glass1 and VDW presentation; PsaE: Glass2 and VDW presentation

share ~80% amino acid similarity and ~28% identity (Fig. 7.4). The main function of PsaD is essential for the electron transfer from PS I to ferredoxin (Chitnis et al. 1996; Chitnis et al. 1997; Barth et al. 1998; Setif 2001; Setif et al. 2002) through maintaining stability of PS I complex and facilitating the dock of ferredoxin. PsaD forms an antiparallel, four-stranded β sheet, in which the loop connecting the third and fourth strands contains an α helix, followed by a two-stranded β sheet (Fig. 7.5). The loop segment reaching from residue PsaD-His95 to PsaD-Pro123 is attached by numerous hydrogen bonds to the stromally exposed sides of PsaC and PsaE, and seems to help in the positioning of these subunits, as suggested by the importance of PsaD for electron transfer from F_X to F_A/F_B (Jordan et al. 2001). This loop segment confirms the important function of PsaD as a critical stabilization factor of the electron acceptor sites in PS I and its important role in holding PsaC in its correct orientation (Li et al. 1991; Lagoutte et al. 2001). The crystal structure showed that PsaD interacts with PsaB and PsaL (Jordan et al. 2001).

As stated earlier, PsaD has two major functions. First, it enables the stability and proper assembly of PS I. The gene encoding the PsaD subunit of PS I from *Synechocystis* sp. PCC 6803 was the first cyanobacterial PS I gene that was inactivated by targeted mutagenesis (Chitnis et al. 1989). The cells of *Synechocystis* sp. PCC 6803 without PsaD can grow photoautotrophically and are sensitive to high light. However, in the PsaD-deficient mutant, PsaC and F_A/F_B clusters are lost more readily after treatment with Triton X-100 or with a chaotropic agent (Chitnis et al. 1996). PsaD interacts with the PS I core, PsaC, PsaE, and PsaL. The site-directed mutagenesis study indicated that the basic residues in the basic domain of PsaD are crucial in the assembly of PsaD. The mutations in this domain disturb the interaction between PsaD and PsaL, thereby causing abnormal assembly of

PsaL (Xu et al. 1994a). The PsaD–deficient mutant has reduced trimers due to the loss of the interaction between PsaD and PsaL. These interactions are critical in the protective role of PsaD. The availability of overproduced PsaD has allowed studies of the structure of the free, unassembled protein and of the assembly of PsaD into the PS I complex. When the assembly of overproduced PsaD was studied using microcalorimetry, thermodynamic parameters associated with the assembly of this protein into PS I could be determined (Jin et al. 1999). Circular dichroism spectroscopy revealed that PsaD of *Synechocystis* sp. PCC 6803 contains a small proportion of α helical conformation in its soluble form. Size–exclusion chromatography, dynamic light scattering and measurement of ^{15}N transverse relaxation times showed that the unassembled PsaD protein of *Nostoc* sp. is a stable dimer in solution, whereas there is only one copy of PsaD per reaction center (Xia et al. 1998). The NMR experiments showed that the dimer is symmetrical and that each PsaD monomer contains a central structured region and unstructured C and N–termini (Xia et al. 1998). Therefore, assembly of PsaD into the PS I complex confers structural rigidity to the protein through extensive interprotein interactions with the PsaA, PsaB, PsaL, PsaM, and PsaC proteins.

Second, PsaD provides a ferredoxin docking site (Lelong et al. 1994; Xu et al. 1994b). Modeling of the interaction of ferredoxin from *Spirulina platensis* (Tsukihira et al. 1981) to the 6 Å electron density map of PS I from *Synechococcus elongatus* (Krauss et al. 1993) led to the suggestion of a binding site of ferredoxin, which is located close to the terminal [4Fe–4S] cluster of PS I (Fromme et al. 1994). The distance between the [4Fe–4S] cluster of PS I to the [2Fe–2S] cluster of ferredoxin was estimated to be ~14 Å (center–to–center distance). This would lead to an edge–to–edge distance of 11–12 Å. This distance is in reasonable agreement with the fastest kinetics of the electron transport from PS I to ferredoxin, which was determined to exhibit a halftime of 500 ns (Setif and Bottin 1994, 1995). The ferredoxin–mediated $NADP^+$ photoreduction is severely inhibited in the membrane of PsaD deficient mutant (Xu et al. 1994b). The first order reduction of ferredoxin cannot be observed in the PsaD–deficient mutant (Hanley et al. 1996). In general, PsaD is a basic protein and thus able to interact with acidic ferredoxin (Lelong et al. 1994). More specifically, the negatively charged ferredoxin may be guided towards by the positively charged patch provided mainly by PsaD and PsaC. Cross–linking study showed that K106 of PsaD from *Synechocystis* sp. PCC 6803 can be cross–linked to E93 in ferredoxin (Lelong et al. 1994). The site–directed mutagenesis study revealed that K106 residue of PsaD from *Synechocystis* sp. PCC 6803 is a dispensable site for ferredoxin docking (Chitnis et al. 1996). Various mutations of H97 showed that the histidyl residue is involved in the increased affinity of PS I for ferredoxin when pH is lowered. This histidyl residue could be central in regulating *in vivo* the rate of ferredoxin reduction as a precise sensor of local proton concentration (Hanley et al. 1996). Since the single site–directed mutations in the basic residues of PsaD do not alter electron transfer drastically (Chitnis et al. 1996; Hanley et al. 1996), it is possible that the interaction between PsaD and ferredoxin contains several alternative components or an overall electrostatic field may be more important than specific interactions between charged residues.

The dissociation constant for the complex between the PsaD–less PS I and ferredoxin at pH 8 is increased 25 times as compared to the wild type. However, the presence of fast kinetic components in the electron transfer from the mutant PS I to ferredoxin indicated that the relative positions of ferredoxin and of the terminal PS I acceptor are not significantly disturbed by the absence of PsaD. The second–order rate constant of ferredoxin reduction is lowered tenfold for PsaD–less PS I. Assuming a simple binding equilibrium between PS I and ferredoxin, PsaD appears to be important for the guiding of ferredoxin to its binding site (main effect on the association rate) (Barth et al. 1998). The same binding site was found by electron microscopy of cross-linked complexes of PS I with either ferredoxin (Lelong et al. 1996) or flavodoxin (Muhlenhoff et al. 1996). At this binding pocket, ferredoxin would get in contact with all three stromal subunits PsaC, PsaD, and PsaE. This is in good agreement with the functional studies (Setif 2001). Co-crystals between PS I and ferredoxin have been reported which may serve as a basis for a structure of the PS I-ferredoxin complex (Fromme et al. 2002).

7.2.3 PsaE

Cyanobacterial PsaE is a slightly basic (the predicted isoelectric point: 8.27), water-soluble protein (subunit IV) that contains 69-75 residues, 74 residues encoded by *ssr2831* in *Synechocystis* sp. PCC 6803. Our alignment analysis of the selected representative sequences showed that plant, green algal and cyanobacterial PsaE share ~46% amino acid similarity and ~14% identity (Fig. 7.6). The NMR structures of PsaE in solutions from three cyanobacterial species: *Synechococcus* sp. PCC 7002 (Falzone et al. 1994), *Nostoc* sp. PCC 8009 (Mayer et al. 1999) and *Synechocystis* sp. PCC 6803 (Barth et al. 2002) are known. PsaE structure of *Synechococcus* sp. PCC 7002 was determined in solution at pH 5.8 and room temperature using over 900 experimental restraints derived from two- and three-dimensional NMR experiments. The structure is comprised of a well-defined five-stranded β sheet with (+1, +1, +1, −4 alpha) topology and four loops (designated the A-B, B-C, C-D, and D-E loops) that connect these β strands (Fig. 7.7). There is no helical region except for a single turn of 3_{10} helix between the βD and βE strands. PsaE also exhibits a large unrestrained loop spanning residues 42-56. A comparison to the known protein structures revealed similarity with the Src homology 3 (SH3) domain, a membrane-associated protein involved in signal transduction in eukaryotes. The match is remarkable as 47 of the alpha-carbons of PsaE can be superimposed onto those of the SH3 domain from chicken brain alpha-spectrin with a root-mean-square deviation of 2.3 Å. Although the amino acid sequences have low identity and the loops are different in both proteins, the topology of the β sheet and the 3_{10} turn is conserved. SH3 domains from other sources show a similar structural homology. The structure of PsaE was used to suggest approaches for elucidating its roles within PS I (Falzone et al. 1994). The second structure of PsaE of *Nostoc* sp. PCC 8009 (Mayer et al. 1999) in solution is similar to the one of *Synechococcus* sp. PCC 7002. However, variability in loop lengths, as well as N- or C-terminal extensions, suggests that the structure of a second representative PsaE subunit would be useful to characterize the interactions among PS I polypeptides. This PsaE has a seven-residue deletion in the loop connecting strands βC and βD, and an eight-residue

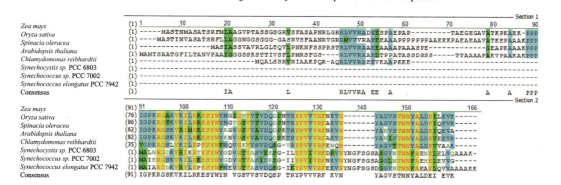

Fig. 7.6 The sequence alignment of PsaE

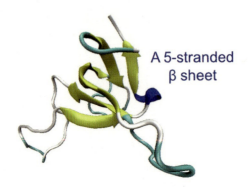

Fig. 7.7 The PsaE structure (PDB ID: 1JB0)

C−terminal extension. Differences between the two cyanobacterial proteins are mostly confined to the CD loop region; the C−terminal extension is disordered. The ΔG degrees of unfolding at room temperature is 12.4 +/− 0.3 kJ mol(−1) (pH 5), and the thermal transition midpoint is 59 +/− 1 degrees C (pH 7), suggesting that interactions with other proteins in the PS I complex may aid in maintaining PsaE in its native state under physiological conditions. A third solution structure of PsaE from *Synechocystis* sp. PCC 6803 was investigated by NMR with a special emphasis on its protein dynamic properties. As compared to previously determined PsaE structures, they all have similar overall structures, suggesting a strongly conserved across all oxygen−evolving photosynthetic organisms. Conformational differences are observed in the first three loops. The flexibility of the loops was investigated using ^{15}N relaxation experiments. This flexibility is small in amplitude for the A−B and B−C loops, but is large for the C−D loop, particularly in the region corresponding to the missing sequence of *Nostoc* sp. PCC 8009. The plasticity of the connecting loops in the free subunit is compared to that when bound to the PS I and discussed in relation to the insertion process and the function(s) of PsaE. The core structure of PsaE attached to the PS I complex consists of five β strands (Jordan et al. 2001). Compared three PsaE structures (Falzone et al. 1994; Mayer et al. 1999; Barth et al. 2002) in solutions with the crystal structure of PsaE in PS I complex (Jordan et al. 2001), the cores of the structures are essentially the same except conformations of the loops and N− and C−termini.

Three different roles of PsaE have been reported in the literatures. First, The turnover of PS I was increased in the PsaE−deficient mutant from *Synechocystis* sp. PCC 6803 (Chitnis 2001; Xu et al. 2001). Therefore, it is suggested that PsaE is important for the stable assembly of PS I supported by the analysis of the structures. The main difference between the free and bound PsaE are the conformations of the loops and the C and N−termini. The flexible loop connecting the β−sheets β3 (C) and β4 (D) were not well resolved in the NMR structure and therefore seems to be flexible and involved in the interaction with the PsaA/PsaB core. This loop has a different conformation and is twisted when PsaE binds to the core complex. The twist of this loop already was reported at 4 Å (Klukas et al. 1999) and is fully confirmed in the structural model at 2.5 Å resolution (Zouni et al. 2000). This loop is involved in interactions with PsaA, PsaB and PsaC, suggesting a change of the loop conformation during assembly of the PS I complex. In addition, the site−directed mutagenesis in PsaE in *Synechocystis* sp. PCC 6803 showed that the C−terminal 8 amino acids are necessary for precise anchorage of PsaC into PS I. The recent studies showed that PsaE can assemble into the PS I complex without help of assembly factors and that it is driven by electrostatic interactions (Lushy et al. 2002). Interactions between PsaE (loop β1/β2) (AB) and the C−terminal region of the partially membrane integral subunit PsaF also exist, in good agreement with mutagenesis and cross−linking studies (Xu et al. 1994c). The interactions of PsaE with PsaC and PsaD are relatively weak, and this explains the finding that the geometry of the stromal structure formed by PsaC and PsaD is not dramatically changed in the absence of PsaE (Barth et al. 1998). However, the C−terminal region of PsaD, which forms a clamp surrounding PsaC, is in direct contact with PsaE (loop β2/β3), confirming previous cross−linking studies (Muhlenhoff et al. 1996).

Second, PsaE facilitates the interaction between ferredoxin and PS I. The first evidence for the location of subunit PsaE at the periphery of the stromal hump came from electron microscopy of a mutant lacking the gene of PsaE in cyanobacteria (Kruip et al. 1997). This was confirmed by the 4 Å and 2.5 Å structural models of PS I (Jordan et al. 2001; Klukas et al. 1999). The loop β2β3 (BC–loop) connecting strands β2 (B) and β3 (C) of PsaE, which points towards the putative docking site of ferredoxin, is close to the loop insertion of PsaC. The PsaE–deficient mutant had much reduced activity of ferredoxin reduction (Xu et al. 1994b). The site–directed mutagenesis study of arginine9 to glutamine in PsaE from *Synechocystis* sp. PCC 6803 showed severely affected the activity of ferredoxin reduction, indicating that this residue is involved in the interaction with ferredoxin (Chitnis 2001; Xu et al. 2001). It has been proposed that R39 of PsaE controlled the electrostatic interaction with ferredoxin (Barth et al. 2000). The fact that PsaE is involved in docking of ferredoxin and flavodoxin was questioned by the finding that the PsaE deletion mutants are still able to grow photoautotrophically. This contradiction was solved by the discovery that the PsaE deletion mutants increased the level of ferredoxin in the cells by orders of magnitude to compensate defects caused by the lack of PsaE (van Thor et al. 1999). The expression of the genes encoding catalase (*katG*) and iron superoxide dismutase (*sodB*) was upregulated in the PsaE–deficient *Synechocystis* sp. PCC 6803 cells, and the increase in *katG* expression was correlated with an increase in catalase activity of the cells. The double mutant of *katG* and *psaE* genes was more photosensitive than the single mutants, showing cell bleaching and lipid peroxidation in high light. These results showed that the presence of the PsaE polypeptide at the reducing side of PS I has a function in avoidance of electron leakage to oxygen in the light and the resulting formation of toxic oxygen species. The PsaE–deficient cells can counteract the chronic photoreduction of oxygen by increasing their capacity to detoxify reactive oxygen species (Jeanjean et al. 2008).

Third, PsaE may be required for the electron transfer around PS I (Yu et al. 1993). The PsaE–deficient mutant from *Synechococcus* sp. PCC 7002 grew much more slowly than the wild–type strain at low light intensities (Zhao et al. 1993). This mutant was also unable to grow under stringent photoheterotrophic conditions (with glucose and DCMU in light). This phenotype suggested that PsaE might affect the cyclic electron transfer around PS I. When methyl viologen was added as an inhibitor of cyclic electron flow, the $P700^+$ reduction rates for the wild type and the PsaE–deficient mutant were similar (Yu et al. 1993). Slower $P700^+$ reduction in the mutant is found in the presence of DCMU and adding CN^- in the presence of DCMU increased the rate of $P700^+$ reduction in the mutant, but remained slower than in the wild type. These observations show that the cyclic electron transfer pathway depends on the presence of PsaE in PS I. The PsaE–deficient mutant growth of *Synechocystis* sp. PCC 6803 also showed similar, but less severe phenotype.

7.3 Integral Membrane Subunits

7.3.1 PsaA and PsaB

7.3.1.1 Introduction

PsaA and PsaB, also called subunit Ia and Ib respectively, show a large homology to each other. Our analysis of the representative protein sequences showed that PsaA and PsaB of plants, green algae and cyanobacteria share ~91% amino acid similarity and ~31% amino acid identity (Fig. 7.8). If we compare only PsaA amino acid sequences, they share ~97% similarity and 67% identity. Similarly, PsaB subunits have 99.6% similarity and ~70% identity. Both contain 11 transmembrane α helices, and form the central core of PS I. The reaction center core shows some similarity to the arrangement of the L and M proteins in bacterial reaction centers (Deisenhofer et al. 1984, 1995; Lancaster and Michel 1999) and the D1 and D2 proteins in PS II (Umena et al. 2011; Grotjohann et al. 2004; Ferreira et al. 2004; Kamiya and Shen 2003;

Zouni et al. 2001). Our alignment of the selected representative PsaA, PsaB, D1, and D2 sequences showed ~65% amino acid similarity. D1/D2 proteins are smaller than PsaA/B, and they were aligned against the N-termini of PsaA/B in our sequence alignment analysis. However, the arrangement of type II reaction centers resembles a more open "S-shaped"

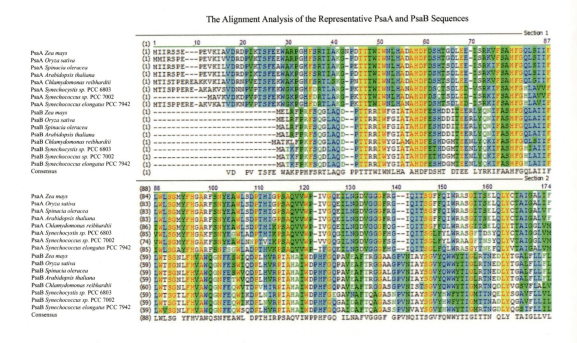

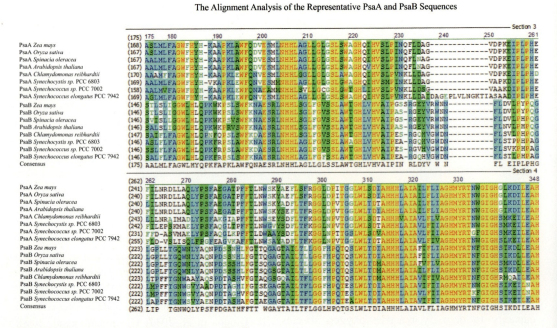

Fig. 7.8 The sequence alignment of PsaA and PsaB

7 Function and Structure of Cyanobacterial Photosystem I

The Alignment Analysis of the Representative PsaA and PsaB Sequences

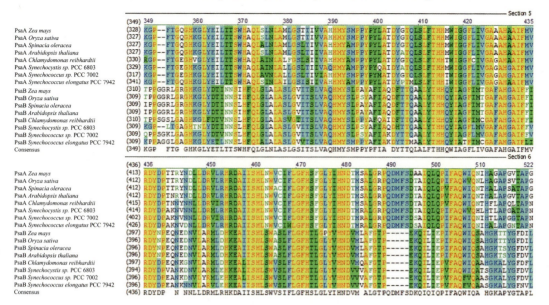

The Alignment Analysis of the Representative PsaA and PsaB Sequences

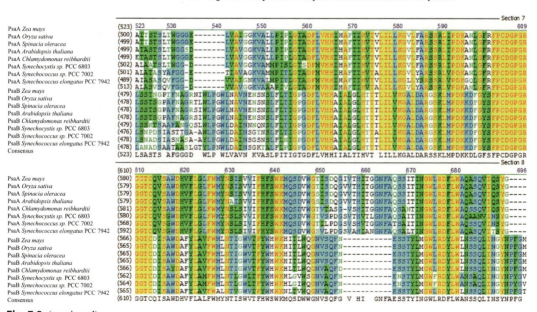

Fig. 7.8 (continued)

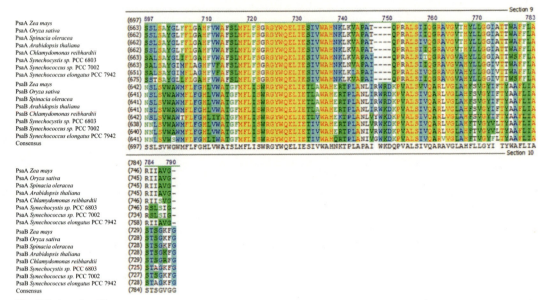

Fig. 7.8 (continued)

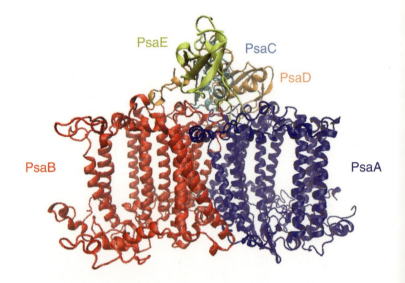

Fig. 7.9 The structures of PsaA and PsaB with PsaC, PsaD and PsaE (PDB ID: 1JB0). PsaA: *blue*; PsaB: *red*; PsaC: *cyan*; PsaD: *orange*; PsaE: *yellow*

conformation whereas the 5 C−terminal helices surround the electron transfer chain in PS I like a fence (Grotjohann et al. 2004). Interestingly, our sequence analysis showed that cyanobacterial D1 and D2 share ~71% amino acid sequence similarity with bacterial subunits L and M. PsaA and PsaB are the largest two subunits in PS I complex, and PsaA is slightly large than PsaB. PsaA (the predicted isoelectric point: 7.39) encoded by *slr1834 Synechocystis sp.* PCC 6803 with 751 amino acids is slightly longer than PsaB (the predicted isoelectric point: 6.44), which has 731 amino acids and is encoded by *slr1835*. PsaA and PsaB of other cyanobacteria, algae and plants have similar sizes. PsaA and PsaB form the core of the PS I complex. Deletion of either

PsaA or PsaB destroys the whole PS I complex in the thylakoid membranes (Shen et al. 1993; Sun et al. 1999) and the cells cannot grow photoautotrophically.

PsaA and PsaB have similar transmembrane topology, each containing eleven transmembrane α helices and twelve extramembrane loops (Fig. 7.9). The eleven transmembrane α helices can be divided into two domains: the N−terminal domain containing six α helices and the C−terminal domain with five α helices. The major function of the N−terminal six transmembrane α helices is to bind antenna chlorophyll a molecules whereas the last five transmembrane α helices coordinate the electron transfer cofactors from P700 to F_X (Jordan et al. 2001). Most extramembrane loops are not well conserved between PsaA and PsaB. Therefore, the loops make the system more asymmetric than transmembrane α helices, with striking differences in sequence, length and secondary structural elements. The introduction of the asymmetry allows PS I to attach the stromal subunits PsaC, PsaE and PsaD and the small membrane intrinsic subunits at an appropriate position relative to the core of PS I (Fig. 7.9). The peripheral membrane intrinsic subunits are attached to the core of PsaA and PsaB by the specific interactions (salt bridges and H−bonds) within the asymmetric loops of PsaA and PsaB. In addition, the loops contact with the cofactors (chlorophylls and carotenoids) through hydrophobic interactions (Grotjohann and Fromme 2005). Furthermore, the loops provide binding sites for plastocyanin or cytochrome c_6 (Sun et al. 1999).

7.3.1.2 Antenna Systems

In contrast to the purple bacterial reaction center (PbRC), which collects light energy through separate membrane−intrinsic light−harvesting complexes, PS I features a core antenna system formed by 90 chlorophyll a molecules and 22 carotenoids (Jordan et al. 2001). PsaA and PsaB bind the majority of antenna chlorophylls, 79 of 90 in PS I through coordination interactions. In addition, PsaA or PsaB binds most of the carotenoids through hydrophobic interactions. Most of the chlorophylls coordinated by PsaA and PsaB follow the twofold symmetry. All of these chlorophylls are coordinated by histidines and were conserved over millions of years during evolution between plants and cyanobacteria (Jordan et al. 2001; Ben-Shem et al. 2003). The small subunits PsaJ, PsaK, PsaL, PsaM, PsaX and a phosphatidylglycerol, coordinate to Mg^{2+} of the rest 11 antenna chlorophyll a molecules either directly or indirectly through water molecules (Jordan et al. 2001).

Antenna Chlorophylls

The major pigments of the antenna system are 90 antenna chlorophylls. The arrangement of the antenna chlorophylls in PS I is a clustered network (Jordan et al. 2001) in contrast to the symmetric arrangement of light harvesting pigments in the external antenna system of purple bacteria (McDermott et al. 1995). In the network, each of the chlorophylls has several neighbors at a distance of less than 15 Å, which is preferable for fast Foerster energy transfer rates between the chlorophylls (Grotjohann and Fromme 2005). The antenna system in PS I is nearly perfect in respect to efficiency and robustness, with some pigments being highly optimized for fast excitation energy transfer from the antenna system to the electron transfer chain (Grotjohann and Fromme 2005). The core antenna system of PS I can be divided into three domains: a central domain and two peripheral domains (Fig. 7.10). The central domain surrounds the electron transfer chain. One peripheral domain is on the one side of the core of the electron transfer chain and the central domain and another peripheral domain locates roughly symmetrically on another side. Each peripheral domain is arranged into two layers: the stromal surface layer and the lumenal surface layer. Most strongly excitonically coupled chlorophylls in PS I are oriented parallel to the membrane and are coordinated by amino acids of the loop regions. The central domain shows that many chlorophylls are located in the middle of the membrane, thereby structurally and functionally connecting the stromal and lumenal layers (Jordan et al. 2001;

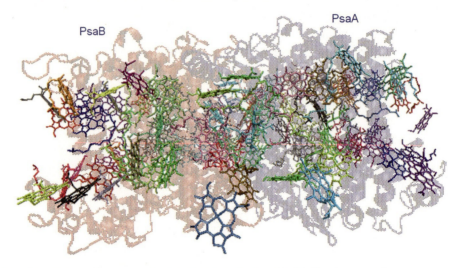

Fig. 7.10 Chlorophylls in photosystem I (PDB ID: 1JB0). PsaA and PsaB: Glass1 presentation

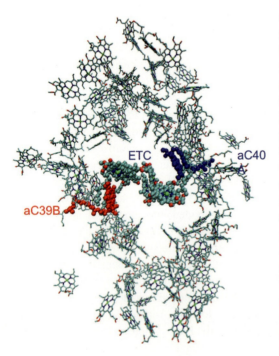

Fig. 7.11 The Electron Transfer Chain (ETC) and antenna of photosystem I (PDB ID: 1JB0)

Grotjohann and Fromme 2005). Under conditions when peripheral antenna chlorophyll is excited, the energy will be first transferred from this 'two dimensional' layer to the central domain. In the central domain, chlorophylls are distributed over the full depth of the membrane, i.e. the excitation energy can be exchanged between the two layers (Grotjohann and Fromme 2005). The excitation energy is then transferred from the chlorophylls of the central domain to the electron transfer chain (Grotjohann and Fromme 2005). Remarkably, the positions of plant PS I antenna chlorophylls of cyanobacterium are virtually the same as those in plants (Ben-Shem et al. 2003; Mazor et al. 2015; Qin et al. 2015). Out of the 96 chlorophyll molecules reported in the model of the cyanobacterial PS I reaction center, only three are missing in the plant PS I reaction center: two bound to PsaM and PsaX, and one bound to PsaJ. From the remaining 93 chlorophylls, 92 are identified at the same position in the plant reaction center, including 15 chlorophylls with their Mg^{2+} coordinated by water (Ben-Shem et al. 2003). A new chlorophyll bound by PsaH was identified in a recent paper (Mazor et al. 2015).

Two of the chlorophyll *a* molecules (aC40A and aC39B) are special as they structurally and perhaps functionally connect eC2A/eC3B and eC2B/eC3A of the electron transfer chain to the antenna (Fig. 7.11). They are called connecting chlorophylls in this chapter. The center–to–center distances are 12.77 Å between aC40A and eC3A, and 10.85 Å between aC39B

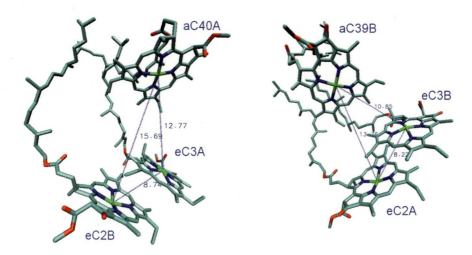

Fig. 7.12 aC40A is close to eC3A and eC2B (*left panel*) and aC39B is close to eC3B and eC2A (*right panel*) (PDB ID: 1JB0)

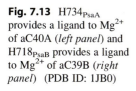
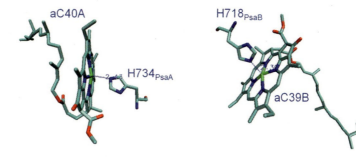

Fig. 7.13 $H734_{PsaA}$ provides a ligand to Mg^{2+} of aC40A (*left panel*) and $H718_{PsaB}$ provides a ligand to Mg^{2+} of aC39B (*right panel*) (PDB ID: 1JB0)

and eC3B, and 15.69 Å between aC40A and eC2B and 13.76 Å between aC39B and eC2A (Fig. 7.12). Otherwise, the electron transfer chain and antenna are well separated and isolated from each other. It is not yet clear whether transfer of energy proceeds through these two chlorophyll *a* molecules from the antenna to eC2/eC3 or P700. If this were the case, eC2A, eC3A, eC2B and eC3B could be engaged both in excitation energy and in electron transfer. It will be interesting to generate aC40A and aC39B mutants to understand how antenna and electron transfer chain are functionally connected.

PsaA–His734 provides a ligand to Mg^{2+} of aC40A, the connecting chlorophyll *a*, in the PsaA side and PsaB–His718 provides a ligand to Mg^{2+} of aC39B, the connecting chlorophyll *a*, in the PsaB side (Fig. 7.13). PsaA–His734 of *Synechocystis* sp. PCC 6803 can be mutated to PsaA–Gln734, PsaA–Met734, PsaA–Asn734 and PsaA–Leu734. Similarly, PsaB–His718 of *Synechocystis* sp. PCC 6803 can be mutated to PsaB–Gln718, PsaB–Met718, PsaB–Asn718 and PsaB–Leu718. Gln, Met or Asn can directly provide a ligand to the Mg^{2+} of the connecting chlorophylls if PS I complex can adjust its entire structure to fit into these local structural changes. In contrast, Leu cannot provide a ligand to the Mg^{2+} of the connecting chlorophylls. Beside the ligand interaction between the connecting chlorophylls and PsaA/PsaB, one unique hydrogen bond exists between PsaA–Gln726 and aC40A (Fig. 7.14). Interestingly, PsaB–Gln710 is little far to provide a hydrogen bond to aC39B (Fig. 7.14). Instead, the amino acid closest to aC39B is PsaB–Val714 although the distance

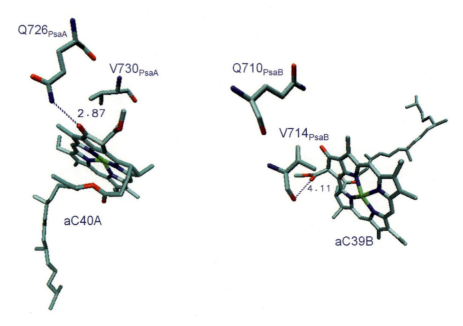

Fig. 7.14 A hydrogen bond interaction between Q726$_{PsaA}$ and aC40A (*left panel*) and no hydrogen bond interaction between Q710$_{PsaB}$ and aC39B (*right panel*) (PDB ID: 1JB0)

(4.11 Å) is longer than a hydrogen bond. PsaA–Gln726 of *Synechocystis* sp. PCC 6803 can be mutated to PsaA–His726, PsaA–Asn726, and PsaA–Leu726 and PsaB–Gln710 of *Synechocystis* sp. PCC 6803 can be considered to mutate to PsaB–His710, PsaB–Asn710, and PsaB–Leu710 as the controls. The same mutagenesis strategy can be used as reported previously (Xu et al. 2003a; Dashdorj et al. 2005; Cohen et al. 2004).

The absorption spectra of PS I from green plants, algae, and cyanobacteria showed the existence of chlorophyll *a* molecules absorbing at lower energy than the usual Qy absorption at 680 nm of chlorophyll *a* in solution and, in particular, at lower excitation energy than the primary donor P700. These redshifted chlorophyll *a* are often called the "redmost" chlorophylls (Brecht et al. 2008). Cyanobacteria have more likely adapted to low light conditions by the introduction of pigments absorbing at longer wavelength. The major function of the long–wavelength chlorophylls may lie in increasing the spectral width of the light absorbed by PS I (Grotjohann and Fromme 2005). However, if the excitation energy is localized within the red chlorophylls, the stored energy is no longer sufficient to directly excite P700 to P700*. Additional activation energy, e.g., thermal energy of the photon bath, is necessary to excite P700 and start the charge separation process (Brecht et al. 2008). The question concerning the origin of the red shift and the physiological role of the red chlorophylls is puzzling (Gobets et al. 2003; Byrdin et al. 2002). The red shift of the chlorophyll absorption can be caused by protein–chlorophyll or chlorophyll–cofactor interaction. Strong excitonical coupling with neighboring chlorophylls may provide the largest contribution to the red shift, but the protein environment may also play an important role, as the red shift can also be caused by a variation of the 5th ligand of the Mg^{2+} or electrostatic fields provided by the protein. The studies showed that the excitation energy transfer in PS I is trap limited and is highly optimized for robustness and efficiency (Byrdin et al. 2002; Sener et al. 2004). Most of the studies agreed that the chlorophyll trimer B31, B32, B33 as well as the linker dimers B37, B38 and A38, A39 that are located close to the electron transfer chain may be red shifted.

Fig. 7.15 Six clusters of β carotenoids in photosystem I (PDB ID: 1JB0)

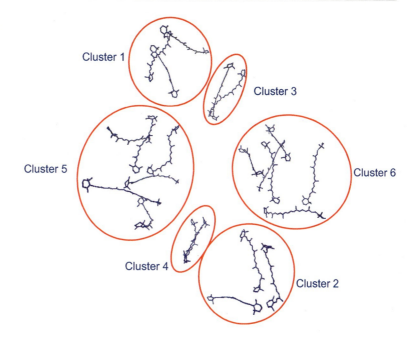

The latter ones may have a special function in funneling of the excitation energy to the reaction center (Grotjohann and Fromme 2005).

Antenna Carotenoids

Carotenoids are a group of pigments that perform several important physiological functions in all kingdoms of living organisms (Domonkos et al. 2013). The X–ray crystallographic localization of carotenoids revealed that they are present at functionally and structurally important sites of both photosystems: PS I (Jordan et al. 2001) and PS II (Guskov et al. 2009). Twenty–two carotenoids have been identified in the structure of PS I from *Thermosynechccocus elongatus*. The carotenoids are deeply inserted in the membrane, with only few closer to the lumenal or stromal surface. Their arrangement may be subdivided into six clusters (Fig. 7.15). Clusters 1, 2, 3 and 4 contain 10 carotenoids and are bound to PsaA and PsaB by hydrophobic interactions, with 1 forming a few additional contacts to PsaK. Clusters 5 and 6, each consisting of six carotenoids, are in contact with PsaA and PsaB; 5 is also bound to PsaF and PsaJ and 6 to PsaI, PsaL and PsaM. The carotenoids in 1 and 3 are roughly related by the pseudo–C2 axis to those in 2 and 4, respectively, but the pseudosymmetry is less obvious for 5 and 6 (Jordan et al. 2001). They fulfill three functions: (i) they are important for maintaining the structure of PS I; (ii) they function as additional antenna pigments and (iii) they prevent the system from damage by over–excitation caused by excess light (photoinhibition). The photo–protective function is most critical for the stability of PS I and also for the whole system of the electron transport chain. The carotenoids prevent the system from over–excitation via quenching of highly damaging triplet states of chlorophylls. Chlorophylls in the triplet state are very reactive molecules that react with oxygen to form the highly toxic singlet oxygen. In order to quench the chlorophyll triplet states, the carotenoids are distributed over the whole antenna system forming multiple interactions with the chlorophylls. Even under high light conditions, the system works very efficiently as the chlorophyll triplet state cannot be detected in the intact PS I (Grotjohann and Fromme 2005). Another unsolved question in PS I is the function of the cis–carotenoids in PS I. Whereas seventeen of the 22 carotenoids are in the energetically favored all–trans configuration, five carotenoids contain one or two cis−double bonds. The 'fit−into−space' consideration would be the easiest explanation for the incorporation of the cis carotenoids. Other options may include the possibilities that there may be a different efficiency for quenching for the cis− and trans−carotenoids, due to the higher energy level for the cis− compared to the trans−carotenoids (Grotjohann and Fromme 2005).

In a recent study, the light harvesting role of β carotenes in the cyanobacterium *Synechococcus* sp. PCC 7942 was investigated using selective β−carotene excitation and selective photosystem detection of photo−induced electron transport to and from the intersystem plastoquinones (the plastoquinone pool). Interestingly, β carotenes enable only the oxidation of the plastoquinone pool by PS I but not its reduction by PS II although selectively excited β carotenes transfer electronic excitation to the chlorophyll *a* of both photosystems. This study may suggest a light harvesting role for the β carotenes of the PS I core complex but not for those of the PS II core complex (Stamatakis et al. 2014). Another study investigated the importance of carotenoids on the accumulation and function of the photosynthetic apparatus using a mutant of the green alga *Chlamydomonas reinhardtii* lacking carotenoids. The mutant is deficient in phytoene synthase, the first enzyme of the carotenoid biosynthesis pathway, and therefore is unable to synthesize any carotenes and xanthophylls. This mutant is unable to accumulate the light−harvesting complexes associated with the cores of both photosystems as well as the subunits of PS II. The accumulation of the cytochrome b_6f complex is also strongly reduced to a level approximately 10% that of the wild type. However, the residual fraction of assembled cytochrome b_6f complexes exhibits single−turnover electron transfer kinetics comparable to those observed in the wild−type strain. Very surprisingly, PS I is assembled to significant levels in the absence of carotenoids in the mutant and possesses functional properties that are very similar to those of the wild−type complex (Santabarbara et al. 2013). The result is in agreement with the previous study in the *Synechocystis* sp. PCC 6803 mutants that contained only short−chain carotenes (Bautista et al. 2005).

7.3.1.3 Lipids

The recent X−ray crystallographic analyses of PS I and PS II complexes from *Thermosynechococcus elongatus* revealed the presence of 4 (Jordan et al. 2001) and 25 (Guskov et al. 2009) lipid molecules per PS I and PS II monomer, respectively, indicating the enrichment of lipids in PS II. Four lipids in PS I of *Thermosynechococcus elongatus* are three molecules of phosphatidylglycerole and one molecule of monoglalactosyldiacylglycerole (Jordan et al. 2001) and they all are closer to the stromal side of PS I (Fig. 7.16). One of the phosphatidylglyceroles and the monoglalactosyldiacylglycerole are located close to the electron

Fig. 7.16 Four lipids (VDW presentation) in photosystem I (PDB ID: 1JB0)

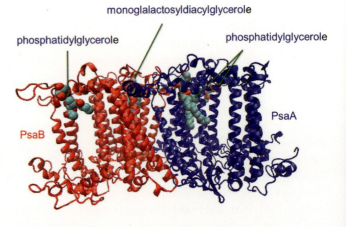

transfer chain and may have a role in branching of the electron transfer chain (Grotjohann and Fromme 2005). Their head groups are completely covered by the loops of PsaA and PsaB and the three stromal subunits PsaC, PsaD, and PsaE (Grotjohann and Fromme 2005). They are, therefore, not solvent accessible and must be incorporated into the PS I complex at a very early stage of the assembly process. Two further phosphatidylglycerole molecules are located at the periphery of PS I, one at the monomer–monomer interface, and the other at the membrane exposed surface of PS I in tight interaction with PsaX. The interaction between PsaX and phosphatidylglycerole PL2 are mainly hydrophobic and both molecules interact so tightly that the role of this phosphatidylglycerole might be the stabilization of PsaX within the complex (Grotjohann and Fromme 2005).

The availability of the complete genomic sequence of *Synechocystis* sp. PCC 6803 (Kaneko and Tabata 1997) opened the way for studying the structural and functional roles of phosphatidylglycerole via molecular genetic approaches. The *pgsA* gene encoding phosphatidylglycerole phosphate synthase was inactivated in *Synechocystis* sp. PCC 6803 cells by inserting a kanamycin resistance gene cassette (Hagio et al. 2000). Maintenance of the recently generated *pgsA* mutant strain requires exogenously supplied phosphatidylglycerole. A 40% decrease in photosynthetic oxygen–evolving activity could be detected following a 3–day depletion of phosphatidylglycerole, which resulted in an approximately 50% decrease in the amount of phosphatidylglycerole molecules in the cellular membranes (Domonkos et al. 2004). In summary, the *pgsA* mutant demonstrated the important role of phosphatidylglycerol in PS II dimer formation and in electron transport between the primary and secondary electron–accepting plastoquinones of PS II. Using a long–term depletion of phosphatidylglycerol from *pgsA* mutant cells, we could induce an activity decrease not only in PS II but also in PS I. Simultaneously with the decrease in PS I activity, dramatic structural changes of the PS I complex were detected. A 21–day phosphatidylglycerol depletion resulted in the degradation of PS I trimers and concomitant accumulation of monomer PS I. The analyses of PS I particles isolated by MonoQ chromatography showed that, following the 21–day depletion, PS I trimers were no longer detectable in the thylakoid membranes. The immunoblot analyses revealed that the PS I monomers accumulating in the phosphatidylglycerol–depleted mutant cells do not contain PsaL, the protein subunit thought to be responsible for the trimer formation. Nevertheless, the trimeric structure of PS I reaction center could be restored by readdition of phosphatidylglycerol, even in the presence of the protein synthesis inhibitor lincomycin, indicating that free PsaL was present in thylakoid membranes following the 21–day phosphatidylglycerol depletion. The data suggest an indispensable role for phosphatidylglycerol in the PsaL–mediated assembly of the PS I reaction center (Domonkos et al. 2004). Another way to investigate lipid specific functions by mutating the amino acids that interact with lipids in *Synechocystis* sp. PCC 6803, in which extensive information about various mutants has already been accumulated and the methods for gene manipulation have been well established.

7.3.1.4 Electron Transfer Chain

The functions of PS I cofactors have been addressed and the aspects of electron transfer modes have been established in the model systems of *Chlamydomonas reinhardtii* and *Synechocystis* sp. PCC 6803 by the investigators in the field. The electron transfer chain of PS I consists of six chlorophylls, two phylloquinones and three [4Fe–4S] clusters. The chlorophylls and phylloquinones are arranged in two branches, named A– and B–branch, based on the coordination of the majority of the cofactors by either PsaA or PsaB (Jordan et al. 2001). Although the mechanism of primary charge separation is still under debate, it is uncontested that within several picoseconds after excitation, a radical pair forms between P700(eC1) and $A_{0A}(eC3A)/A_{0B}(eC3B)$ (Savikhin et al. 1999; Muller et al. 2003; Holzwarth et al. 2006a; Savikhin et al. 2000). Within 30 ps, the electron is passed forward to A_{1A}/A_{1B}. It should be pointed out that the individual cofactors are

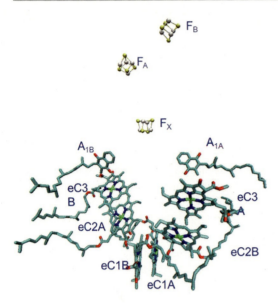

Fig. 7.17 Electron transfer chain in photosystem I (PDB ID: 1JB0)

labeled with their respective structural and spectroscopic names since electron transfer cofactors are located on both sides of a pseudo–C2 axis of symmetry (Fig. 7.17). The use of 'A' or 'B' in the structural name specifies the protein (PsaA or PsaB) that ligates the cofactors. The spectroscopic names, such as A_0 and A_1, refer to the electron transfer pathway in which each cofactor participates (Srinivasan and Golbeck 2009a). For instance, the cofactor Q_KA (A_{1A}) is bound by the PsaA subunit and is therefore on the A–branch.

Electron transfer from A_{1A}^-/A_{1B}^- to F_X has been studied extensively in both cyanobacteria and eukaryotic algae. Two exponential phases with lifetime of ~20 ns (the fast phase) and ~200 ns (the slow phase) are present, with relative ratio of amplitudes differing from 1:1 to 1:4 in various species and preparations (reviewed in (Srinivasan and Golbeck 2009b)). The fast and slow kinetic phases are attributed to electron transfer through the B– and A–branches of cofactors, respectively. The electron is transferred serially from through F_X, F_A and F_B after which it donated to soluble ferredoxin or flavodoxin. $P700^+$, in turn, is reduced by plastocyanin or soluble cytochrome c_6. When no electron acceptors or donors are present, the charge separated state between $P700^+$ and $[F_A/F_B]^-$ recombines within ~50 ms. Protein structural changes could be accompanied and/or required for efficient electron transfer in PS I. A study showed that nanosecond absorption dynamics at 685 nm after excitation of PS I from Synechocystis sp. PCC 6803 is consistent with electrochromic shift of absorption bands of the chlorophyll a pigments in the vicinity of the secondary electron acceptor A_1. Based on the experimental optical data and structure–based simulations, the effective local dielectric constant has been estimated to be between 3 and 20, suggesting that electron transfer in PS I is accompanied by considerable protein relaxation. Similar effective dielectric constant values have been previously observed for the bacterial photosynthetic reaction center and indicate that protein reorganization leading to effective charge screening may be a necessary structural property of proteins that facilitate the charge transfer function (Dashdorj et al. 2004).

P700, The Primary Electron Donor

In all photosynthetic organisms, conversion of sunlight into chemical energy occurs in transmembrane pigment–protein complexes called reaction centers. The reaction centers of both oxygenic and nonoxygenic organisms exhibit a common general architecture and share the same basic functional principles. Following absorption of a photon, a separation of electric charges takes place between a primary electron donor, a dimer of chlorophyll or bacteriochlorophyll molecules, and a series of electron acceptors situated at increasing distances away from the primary donor (Breton et al. 2005). Owing to their prominent role in the process of transmembrane separation and stabilization of the electric charges, the electronic structure of the primary donors has been investigated in great detail using various forms of optical, vibrational, and magnetic resonance spectroscopy (Breton et al. 2005).

A pair of two chlorophylls is located close to the lumenal surface of PS I. This pair of chlorophylls assigned to P700, the primary electron donor. PS I uses light energy to convert P700 to an excited state, a powerful reductant, P700*, that can transfer an electron to the primary electron acceptor A_0. As a result, P700* becomes $P700^+$ that is reduced to P700 by plastocyanin or cytochrome c_6. The redox potentials of P700 and P700* are approximately +400 mV (Nelson and Yocum 2006) and −1,200 mV respectively. Thus, P700* probably is the most reducing compound found in natural systems (Webber and Lubitz 2001). The X−ray structure of *Thermosynechococcus elongatus* PS I at 2.5 Å resolution shows interesting features of the asymmetry of the interactions between the primary electron donor (called P700) and the two homologous polypeptides, PsaA and PsaB, that form the core of the reaction center. First, the two chlorophylls are chemically different: the chlorophyll that PsaB binds is the 'common' chlorophyll *a* molecule, chemically identical to all the other 95 chlorophylls in PS I, whereas the chlorophyll that PsaA binds is chlorophyll a', the epimer at the C13 position of the chlorin ring system. The chlorophyll a' is a constituent in cyanobacterial, algal and plant PS I. Even algae that contain chlorophyll *d* instead of chlorophyll *a* have a chlorophyll *d*/chlorophyll d' heterodimer as the primary electron donor (Akiyama et al. 2002). Chlorophyll a' is specific for PS I and is found only in the primary electron donor. An interesting question is how chlorophyll a' is synthesized and then assembled into PS I. The substrate specificity of several enzymes participating in chlorophyll biosynthesis has been examined. Helfrich et al. have shown that there is no naturally occurring chlorophyll a' (Helfrich et al. 1994, 1996). To date, no enzyme or activity has been found that would catalyze epimer formation. This raises the intriguing possibility that chlorophyll a' is formed during assembly of the complex and perhaps mediated by the reaction center itself (Webber and Lubitz 2001). Second, the asymmetry of the chemical nature is extended to the asymmetry of the chemical environment. Even if both chlorophylls are axially coordinated to histidine residues (PsaA-His680 and PsaB-His660 of *Thermosynechococcus elongatus*) (Fig. 7.18), their hydrogen bonding pattern is asymmetric: whereas the chlorophyll a' molecule at the A−branch of P700 (eC−A1) forms three hydrogen bonds to the side chains of transmembrane α helices A−i

and A−k and a water molecule, and no hydrogen bonds are formed between the surrounding protein and the chlorin head group of P700 (eC−B1) (Grotjohann and Fromme 2005). More specifically, the residue PsaA−Thr743 of *Thermosynechococcus elongatus* is proposed to interact both with the 9−keto C=O group of chlorophyll a' and with a bound water molecule. This water (H_2O−19) appears at the center of a network of hydrogen bonds that involves, in addition to PsaA−Thr743, the 10a−carbomethoxy group of chlorophyll a' and, at least, PsaA−Tyr603 and PsaA−Ser607 (Fig. 7.19). The homologous

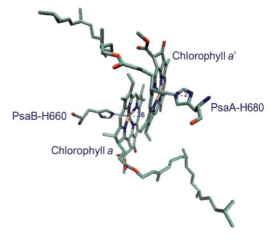

Fig. 7.18 The axial ligands of P700 in photosystem I (PDB ID: 1JB0)

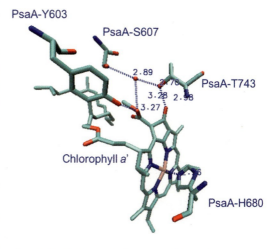

Fig. 7.19 The hydrogen bonds of P700 in photosystem I (PDB ID: 1JB0)

residues on the PsaB side are Tyr727, Leu590, and Gly594, respectively. None of the latter three residues appears to form hydrogen bonds with the carbonyls of chlorophyll a (Jordan et al. 2001). The structural asymmetry of P700 reflects import and functional differences of both chlorophylls. The ENDOR studies in solution (Käβ et al. 1995) and on single crystals of PS I (Käss et al. 2001) showed that the spin density in $P700^{+\cdot}$ is asymmetric, with more than 85% of the spin density located on the B−branch chlorophyll a of P700. The distribution of the spin density over both halves of P700 was also shown by the FTIR studies on $P700/P700^+$ (Breton et al. 2002; Pantelidou et al. 2004). This asymmetry of the spin density is caused by the interplay between the asymmetric hydrogen bonding, differences in the protein environment and the chemically different nature of the two chlorophyll molecules (Grotjohann and Fromme 2005).

To investigate the influence of three hydrogen bonds on P700, the site−directed mutagenesis has been employed in *Chlamydomonas reinhardtii* (Witt et al. 2002; Wang et al. 2003; Li et al. 2004) and *Synechocystis* sp. PCC 6803 (Breton et al. 2005; Pantelidou et al. 2004). In *Chlamydomonas reinhardtii*, the threonine at position 739 of PsaA, which donates a putative hydrogen bond to the 13(1)−keto group of chlorophyll a', was replaced with valine, histidine, and tyrosine using a site−directed mutagenesis approach. Growth of the mutants was not impaired (Witt et al. 2002). The PsaA−T739A mutant was capable of assembling active PS I. Furthermore, PsaA−T739A mutant PS I contained approximately one chlorophyll a' molecule per reaction center, indicating that P700 was still a chlorophyll a/a' heterodimer in the mutant. However, this mutation induced several band shifts in the visible $P700^+ − P700$ absorbance difference spectrum. Redox titration of P700 revealed a 60 mV decrease in the $P700/P700^+$ midpoint potential of the PsaA−T739A mutant, consistent with loss of a H−bond. The ENDOR spectroscopy showed that the hyperfine coupling of the methyl protons at position 12 of the spin−carrying chlorophyll a is decreased due to the removal of the hydrogen bond to chlorophyll a' (Wang et al. 2003). Comparison of the FTIR difference band shifts upon $P700^+$ formation in the wild type and mutant PS I suggests that the mutation modifies the charge distribution over the pigments in the $P700^+$ state, with approximately 14−18% of the positive charge on chlorophyll a in the wild type being relocated onto chlorophyll a' in the mutant. Interestingly, the mutation−induced change in asymmetry of P700 did not cause an observable change in the directionality of electron transfer within PS I (Li et al. 2004), indicating that the function of the asymmetry of P700 may be not the gating between the electron transfer branches (Grotjohann and Fromme 2005).

In *Synechocystis* sp. PCC 6803, the effect on the $P700^+/P700$ FTIR difference spectra of replacing the three PsaA residues involved in the network of hydrogen bonds (PsaA−Thr739, PsaA−Ser603, and PsaA−Tyr599) by their PsaB homologues (Tyr718, Gly585, and Leu581, respectively) was consistent with the rupture, or at least a significant loosening, of the hydrogen bond to the 9−keto C=O of chlorophyll a' and with an increased freedom of a bound 10a−ester of chlorophyll a' (Breton et al. 2002). When only residue PsaA−Thr739 was changed to Phe, the perturbation of the hydrogen bond to the 9−keto C=O group of chlorophyll a' was still observed with almost no changes at the level of the 10a−ester C=O (Pantelidou et al. 2004). Very comparable experimental results have been reported in *Chlamydomonas reinhardtii* PS I when PsaA−Thr739 was changed to Tyr, His, or Val (Witt et al. 2002) or to Ala (Wang et al. 2003; Li et al. 2004). Another site−directed mutagenesis attempts to introduce hydrogen bonds to the carbonyl groups of chlorophyll a on PsaB side in *Synechocystis* sp. PCC 6803. The present FTIR study showed that the PsaB−Y718T mutation leads to hydrogen bonding of the 9−keto C=O group of chlorophyll a and chlorophyll a^+ in a significant fraction of the centers. On the other hand, the mutations PsaB−L581Y and PsaB−G585S have only a small impact on the $P700^+/P700$ FTIR difference spectra. None of these three mutations perturb the hydrogen−bonding interactions assumed by the 9−keto and 10a−ester carbonyl groups of chlorophyll a' and chlorophyll a'^+ with the protein. These mutations have only a limited effect on the relative charge distribution between chlorophyll a'^+ and chlorophyll a^+ (Breton et al. 2005).

The distance between the Mg^{2+} ions of the two chlorophylls of P700 is with only 6.3 Å (Jordan et al. 2001; Grotjohann and Fromme 2005), much shorter than the corresponding distance between the bacteriochlorophylls in the special pair of PbRC (Deisenhofer et al. 1995). The orientation of the ring system and their overlap differs between P700 in PS I (Jordan et al. 2001) and the special pair of PbRC (Deisenhofer et al. 1995). The rings I overlap perfectly in the PbRC (Deisenhofer et al. 1995), whereas both rings I and II of the chlorophylls only partially overlap in P700 (Jordan et al. 2001). The molecular orbital studies of the electronic structure of P700, based on semi−empirical density functional calculations on the 2.5 Å structure showed, that the two chlorophylls are tightly coupled and P700 is a 'super−molecule' (Plato et al. 2003), i.e. both chlorophylls behave spectroscopically as a single molecule.

Understanding primary charge separation and electron transfer processes in the reaction centers has long been a key objective in photosynthesis research. In bacterial reaction center, the primary donor of electrons is the first singlet excited state of a pair of excitonically coupled bacteriochlorophyll a (BChl), termed P*. The first clearly resolved electron acceptor is a bacteriopheophytin a (BPhe, a demetallated bacteriochlorin) termed H_A (Zinth and Wachtveitl 2005). An intervening monomeric BChl, B_A, is between P and H_A and makes direct atomic contact with both P and H_A. It is generally accepted that P* decays with a lifetime of ~3 ps to form the radical pair $P^+B_A^-$, followed by a more rapid electron transfer to H_A in ~1 ps to form $P^+H_A^-$ (Arlt et al. 1993). A third electron transfer to a ubiquinone (Q_A) with a lifetime of ~200 ps produces a membrane−spanning radical pair, $P^+Q_A^-$, that is stable on a millisecond timescale (Zhu et al. 2013). For intact PS II, a component of ~1.5 ps reflects the dominant energy−trapping kinetics from the antenna by the reaction center. A 5.5−ps component reflects the apparent lifetime of primary charge separation. The 35−ps component represents the apparent lifetime of formation of a secondary radical pair, and the ~200−ps component represents the electron transfer to the Q_A acceptor (Holzwarth et al. 2006b). The mechanism and kinetics of the ultrafast events of energy transfer from the antenna to P700 and for the electron transfer within the reaction center are still a matter of intensive debate and controversy (Holzwarth et al. 2006a). Since the antenna chlorophylls and the reaction center chlorophylls are both bound to the same polypeptides, an intact well−defined PS I reaction center devoid of antenna chlorophylls cannot be isolated. In view of this situation, the electron transfer processes cannot be studied separately from the energy transfer processes (Holzwarth et al. 2006a). The kinetic sequence of electron transfer events in PS I after oxidation of P700 is not well established. A review by Brettel (1997) suggested that electron transfer from P700* to the chlorophylloid primary acceptor (A_0) occurs with 1−3 ps kinetics, followed by 20−50 ps electron transfer from A_0 to the phylloquinone secondary acceptor, A_1 (Brettel 1997). However, these electron transfer timescales are necessarily based on indirect or cumulative measurements since the energy transfer and the electron transfer dynamics are tightly linked and intertwined, and this situation presents the key difficulty in the analysis and interpretation of ultrafast optical data for PS I. The problem is aggravated further by the fact that energy and electron transfer processes occur on comparable picosecond timescales, which make the assignment of the different lifetime components even more difficult (Holzwarth et al. 2006a). The excitation transport and trapping kinetics of core antenna−reaction center complexes from PS I of the wild−type *Synechocystis* sp. PCC 6803 were investigated under annihilation−free conditions in complexes with open and closed reaction centers. The results support a scenario in which trapping time from antenna to P700 is ~24 ps (Savikhin et al. 2000) and is the intrinsic time constants for $P700^+A_0^-$ formation and subsequent $A_0 \rightarrow A_1$ electron transfer of 1.3 ps and 13 ps, respectively (Savikhin et al. 2001). Similar results were obtained earlier by White et al. (1996) in PS I complexes from spinach. A more recent study showed that the main events in the reaction center of PS I include very fast (<100 fs) charge separation with the formation of the $P700^+A_0^-A_1$ state in approximately one half of the reaction centers, the approximately 5—ps energy transfer from antenna excited chlorophylls to $P700A_0A_1$ in the remaining reaction centers, and approximately 25—ps formation of the secondary radical pair $P700^+A_0A_1^-$ (Shelaev et al. 2010).

The PsaA and PsaB axial His residues in *Chlamydomonas reinhardtii* have been replaced by a range of different amino acids, and the characterizations of those mutants have led to a general understanding of the types of amino acids that may substitute for His as a ligand to P700. Replacement of His with smaller uncharged or polar amino acids was found to have a moderate effect (50—60% decrease) on the accumulation of PS I. However, replacement with charged amino acids, or large hydrophobic residues, always resulted in a strong decrease or complete loss of PS I. Interestingly, when similar substitutions are compared between the PsaA and PsaB proteins, mutation of the PsaA subunit always had a more profound effect on PS I accumulation. This probably reflects the more rigid structure of the chlorophyll a' due to H—bonding, which may make this region less flexible to changes in local protein structure (Webber and Lubitz 2001).

The ultrafast transient absorption study of the site—directed mutations of *Chlamydomonas reinhardtii* near the P700 reaction center chlorophylls provided new insight into the nature of the primary electron donor. In the mutant PsaB—H656C, which lacks the ligand to the central metal of chlorophyll a on the B—side of P700, the rate constant of the secondary electron transfer process is slowed down by a factor of ~2. For the mutant PsaA—T739V, which breaks the hydrogen bond to the keto carbonyl of chlorophyll a', only a slight slowing down of the secondary electron transfer is observed. For mutant PsaA—W679A, which has the Trp near chlorophyll a' of P700 replaced, either no pronounced effect or, at best, a slight increase on the secondary electron transfer rate constants is observed. The effective charge recombination rate constant is modified in all mutants to some extent, with the strongest effect observed in mutant PsaB—H656C. The data strongly suggested that the chlorophylls a and a' pair, constituting what is traditionally called the "primary electron donor P700", are not oxidized in the first electron transfer process, but rather only in the secondary electron transfer step. The authors thus proposed a new electron transfer mechanism for PS I where the accessory chlorophyll (s) function as the primary electron donor(s) and the A_0 chlorophyll(s) are the primary electron acceptor(s). This new mechanism also resolves in a straightforward manner the difficulty with the previous mechanism, where an electron would have to overcome a distance of ~14 Å in < 1 ps in a single step (Holzwarth et al. 2006a). In addition, the data suggested that the B—branch is the active branch, although parallel A—branch activity cannot be excluded (Holzwarth et al. 2006a). It is interesting to note that the new mechanism proposed is in fact analogous to the electron transfer mechanism in PS II, where the accessory chlorophyll also plays the role of the primary electron donor, rather than the special chlorophyll pair P680 (Prokhorenko and Holzwarth 2000).

Accessory Chlorophylls

The two accessory chlorophylls, A_{-1A} and A_{-1B}, are located between P700 and the primary acceptor(s), A_{0A} and A_{0B}. These two chlorophylls are the only cofactors in the electron transfer chain, in which PsaA subunit binds to the accessory chlorophyll in the B—branch and *vice versa* (Jordan et al. 2001). In both branches, a water molecule provides the fifth ligand to the central Mg^{2+} ions of the two accessory chlorophylls. The water molecule serving as the axial ligand to the

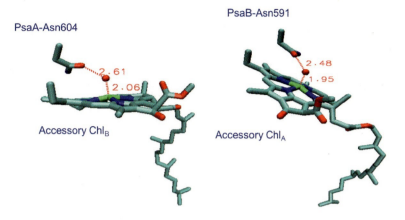

Fig. 7.20 The axial ligands of accessory chlorophyll (Chl) in photosystem I (PDB ID: 1JB0)

accessory chlorophyll *a* in the A−branch, is hydrogen bonded to PsaB−Asn591 in *Synechococcus elongates*, thus this accessory chlorophyll is named as A_{-1B}. Similarly, the B−branch accessory chlorophyll *a* is liganded to a water, which is hydrogen bonded to PsaA−Asn604, thus this accessory chlorophyll *a* is named as A_{-1A} (Fig. 7.20). Coordination of the accessory chlorophylls is unique for the PS I complex because His provides the fifth ligand to the Mg^{2+} ion of bacterioaccessory chlorophyll. It is also interesting that the "accessory chlorophylls" in PS II have the same ligand as that in PS I (Grotjohann and Fromme 2005; Guskov et al. 2009). However, the orientation of the "accessory chlorophylls" in PS II is similar to bacterioaccessory chlorophylls (Grotjohann and Fromme 2005; Guskov et al. 2009). Characterization of PS I accessory chlorophylls has been done by the investigators in the field using experimental and theoretical approaches although the understanding on these two chlorophylls is still limited. Exciton calculations, based on the recent PS I structure (Jordan et al. 2001), indicated that the accessory chlorophylls are strongly excitonically coupled to A_0 (Gibasiewicz et al. 2003). In addition, the kinetic modeling of ultrafast spectroscopic study of the PS I complexes from *Chlamydomonas reinhardtii* suggested that the accessory chlorophyll(s) might be even involved in the primary charge separation (Muller et al. 2003; Gibasiewicz et al. 2003; Muller et al. 2010).

The data provided support for the previously published model in which the initial charge separation event occurs within an A_{-1}/A_0 pair, generating a primary A_{-1}^+/A_0^- radical pair, followed by rapid reduction by P700 in the second electron transfer step. In contrast, the accessory chlorophylls are suggested not to directly involve in this primary charge separation based on their close proximity and the extent of excitonic coupling. At this time, it is unclear whether the accessory chlorophylls are involved in the charge separation similar to the mechanisms in PS II and the purple bacterial reaction center. It is also unclear whether the accessory chlorophylls are involved in electron and energy transfer in PS I of plants, algae and cyanobacteria since they have not been detected by spectroscopy to date.

As discussed earlier, in PS I, it was recently proposed that the true primary donor is in fact the accessory chlorophyll(s) positioned in between P700 and A_0 (Holzwarth et al. 2006a). According to this model, P700 is a secondary electron donor and gives the electron to accessory chlorophyll$^+$ only in the secondary electron transfer step, forming the state $P700^+A_0^-$. A similar sequence of primary electron transfer events was also proposed for PS II (Prokhorenko and Holzwarth 2000; van Brederode and van Grondelle 1999; Groot et al. 2005). It will be very interesting and important to generate site−directed mutants around the accessory chlorophylls using the same mutagenesis strategy as reported previously (Xu et al. 2003a; Dashdorj et al. 2005;

Cohen et al. 2004; Xu et al. 2011). PsaA−Asn604 of *Synechocystis* sp. PCC 6803 can be mutated to PsaA−Gln604, Leu604, Met604 and His604, and similarly, PsaB−Asn591 can be mutated to Gln591, Lue591, Met591 and His591. It should be pointed out that the nomenclature of *Synechococcus elongates* will be used for *Synechocystis* sp. PCC 6803 in this chapter to avoid confusion from the description of the PS I structure. Gln, Met and His can directly provide ligand to the Mg^{2+} of the accessory chlorophylls if the PS I complex can adjust its entire structure to fit into these local structural changes. If yes, the Asn → His mutations will mimic bacterioaccessory chlorophylls. Gln has similar side chains to Asn. However, Gln is longer than Asn, thus Asn → Gln mutations might remove the water molecule from the accessory chlorophyll binding pockets. The side chains of Met and Leu are also larger than Asn. The detailed functions of the two accessory chlorophylls can be studied using these mutations as a research tool through a combination of biochemical and spectroscopic approaches in the future.

A_0, The Primary Electron Acceptor

The second pair of chlorophylls named eC−A3 and eC−B3, are located in the middle of the membrane at a position (Grotjohann and Fromme 2005) which roughly shows similarities to the position of pheophytin both in the PbRC (Deisenhofer et al. 1995) and in PS II (Guskov et al. 2009). One or both the chlorophylls eC−A3 and eC−B3 are assumed to be A_0. The second pair of chlorophylls are in close vicinity to the chlorophylls of the second pair of chlorophylls, with the edge−to−edge distance between eC−B2 <=> eC−A3 and eC−A2 <=> eC−B3 being as short as 3.8 Å (Jordan et al. 2001; Grotjohann and Fromme 2005). Even if eC−A3 and eC−B3 are supposed to correspond to the spectroscopically identified electron acceptor A_0, it is very likely that the spectroscopic and redox properties of eC−A3 and eC−B3 may be influenced by eC−B2 and eC−A2, respectively (Grotjohann and Fromme 2005), and this argument is supported by the spectroscopic result that the difference spectrum A_0/A_0^- of the PS I particles isolated from cyanobacteria, green algae, and higher plants contains contributions from more than one chlorophyll (Hastings et al. 1995).

The axial ligands to A_{0A} and A_{0B} are PsaA−Met688 and PsaB−Met668 respectively. The interaction between a hard acid like Mg^{2+} and the soft base sulfur is very surprising and could account for the part of unusual redox potential of A_0. The A−branch A_0 has two hydrogen bonds, one is between the keto oxygen of ring and PsaA−Tyr696 and a second between the phytyl ester carbonyl and backbone oxygen of PsaB−Ser429. The B−branch A_0 has only one hydrogen bond between the keto oxygen of ring and PsaB−Tyr676 (Jordan et al. 2001) (Fig. 7.21). The specific interactions between A_0 and PsaA, and PsaB probably lead to a low redox potential of A_0 that was estimated as −1,000mV (Nelson and Yocum 2006) or −1,100 mV (Jordan et al. 2001; Fromme et al.

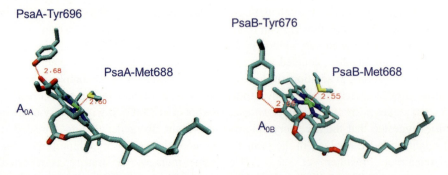

Fig. 7.21 The axial ligands and hydrogen bonds of A_0 in photosystem I (PDB ID: 1JB0)

2001). Both A_0 ligand hydrogen bond mutants (Tyr → Phe) have been generated in *Chlamydomonas reinhardtii*. The transient absorption and EPR spectra showed that the A−branch A_0 ligand and hydrogen bond mutants resulted in a greater proportion of fast decay from A_1 to F_X while the B−branch A_0 ligand and hydrogen-bond mutants resulted in a smaller proportion of fast decay. However, the relative amplitudes of the two fractions predicted from EPR differed somewhat from those obtained from the optical data (Ramesh et al. 2004, 2007; Li et al. 2006). It was found the rate of the primary charge separation was lowered in both mutants, providing evidence that the primary electron transfer event can be initiated independently in each branch (Muller et al. 2010). The A_0 hydrogen-bond mutants of *Synechocystis* sp. PCC 6803, in which PsaB−Tyr676 was mutated to Ala and Phe, have been generated (unpublished data). Replacement of Tyr667 with Phe did not affect assembly and accumulation of PS I, but its replacement with Ala resulted in dramatically reduced cellular levels of PS I. PS I complex cannot be isolated from PsaB−Y667A mutant while high quality of PS I trimers can be purified from the PsaA−Y667F mutant. The corresponding A_0 hydrogen bond mutant Tyr → Phe in PsaA has been generated recently to understand functions of both primary electron acceptors in *Synechocystis* sp. PCC 6803.

In *Chlamydomonas reinhardtii*, PsaA−M688 and PsaB−M668 have been replaced with Leu, Ser and His (Ramesh et al. 2004; Fairclough et al. 2003; Santabarbara et al. 2005; Byrdin et al. 2006; Giera et al. 2009; Berthold et al. 2012) and in *Synechocystis* sp. PCC 6803 with Leu, Asn and His (Dashdorj et al. 2005; Cohen et al. 2004; van der Est et al. 2010; Savitsky et al. 2010; Santabarbara et al. 2010; Sun et al. 2014). In *Chlamydomonas reinhardtii*, the Met → His mutants have been studied the most extensively. In an ultrafast optical study in the red, long−lived difference spectra observed in the PsaA−M688H and PsaB−M668H mutants were assigned to $(A_{0A}^- - A_{0A})$ and $(A_{0B}^- - A_{0B})$, respectively and were interpreted to indicate that forward electron transfer beyond A_0 was either blocked or slowed in the branch carrying the mutation (Ramesh et al. 2004). The amplitudes of the $(A_{0A}^- - A_{0A})$ and $(A_{0B}^- - A_{0B})$ difference spectra were nearly identical in the two mutants, suggesting roughly equal use of both branches. A subsequent study at 390 nm showed that formation of phyllosemiquinone in both mutants decreased to about one−half of that in the wild type, a result implying that electron transfer is blocked between A_0 and A_1 in the affected branch (Giera et al. 2009). An EPR study of the PsaA−M688H mutant at 265 K showed the absence of an electron spin polarized (ESP) signal, suggesting that the $P700^+A_{1A}^-$ radical pair cannot be formed and that B−branch transfer, if present, does not produce an ESP signal (Fairclough et al. 2003). In a more recent study in deuterated whole cells of the PsaA−M688H mutant, a spin−polarized spectrum was detected at 100 K and assigned to the radical pair $P700^+A_{1B}^-$ (Berthold et al. 2012). The pulse EPR studies revealed that in the presence of reduced F_X, the decay of the out−of−phase spin polarized signal in the wild type was biphasic but that it was monophasic in the PsaA−M688H and PsaB−M668H mutants, with lifetimes of ~3 μs and ~17 μs, respectively (Fairclough et al. 2003; Santabarbara et al. 2005). The echo modulation frequencies were different in the two mutants and were explained as a result of the difference in the spin−spin coupling in $P700^+A_{1A}^-$ and $P700^+A_{1B}^-$. The wild−type echo decays and modulation curves could be reconstructed as a linear combination of the signals of the two radical pairs. These data are consistent with a blockage of electron transfer from A_0^- to A_1 in Met to His mutants of *Chlamydomonas reinhardtii*. The A_0 ligand mutants from *Chlamydomonas reinhardtii* were also investigated using femtosecond laser flash, in which both branch mutants showed an additional bleaching with a maximum at ~681 nm (Ramesh et al. 2004). The time−resolved fluorescence studies with a 3−ps temporal resolution using PS I core samples isolated from the wild type and Met → His or Ser mutants of *Chlamydomonas reinhardtii* supported the model in which P700 is not the primary electron donor, but rather a secondary electron donor, with the primary charge separation event occurring between the accessory chlorophyll, A, and A_0 (Giera et al. 2010).

In a recent study, the A_0 ligand mutants of Met → His and Cys have been generated in *Synechocystis* sp. PCC 6803 by another group in the field (Sun et al. 2014). It is predicted that His could provide a ligand to A_0 in the Met → His mutants. Cys could provide a ligand to A_0 as well, although the PS I complex needs to adjust itself to fit the changes caused by this mutation since the side chain of Cys is smaller than Met. The interesting and unexpected results have been obtained from A_0 ligand mutants of Met → His (Sun et al. 2014). The X−band spin−polarized transient EPR spectra of PS I trimers isolated from A_0 Met → His mutants at 80 K showed that $P700^+A_{1A}^-$ radical pair of PsaA−M688H mutant differed from the wild type and PsaB−M668H mutant, suggesting that PsaA−M688H provides an additional H bond to A_{1A}. Interestingly, the X−band spin−polarized transient EPR spectra of the wild type and PsaB−M668H mutant were very similar. The room−temperature transient EPR spectra demonstrated that the electron transfer was blocked at A_{1A} and electron cannot transfer to F_X in A branch in the PS I complex of PsaA−M688H. PsaA−M688H mutant can still exhibit photoautotrophic growth although in a slightly slower rate under normal light intensity but is highly sensitive to high light intensity. It is still unclear and also interesting why PsaB−M668H does not show any obvious difference in EPR spectra and growth rate (Sun et al. 2014). There are inconsistencies of function of the two−branch A_0 between *Chlamydomonas reinhardtii* and *Synechocystis* sp. PCC 6803. It could be due to minor local structural differences of species−specificity.

A_1, The Second Electron Acceptor

The two phylloquinones, named Q_KA (A_{1A}) and Q_KB (A_{1B}), (Fig. 7.22) represent the spectroscopically identified electron acceptor A_1. They are located at the stromal side of the membrane, in close vicinity to the membrane surface (Grotjohann and Fromme 2005). The three cofactors, A_0, A_1, and F_X are linked by an intricate network of contacts and hydrogen bonds that bind them to the protein and promote electron transfer (Srinivasan and Golbeck 2009a). Specifically, PsaA−M688 (PsaB−M668) is the axial ligand to A_{0A} (A_{0B}) and is H−bonded via its backbone oxygen to the side chain oxygen of PsaA−S692 (PsaB−S672). The side chain oxygen of PsaA−S692 (PsaB−S672) is also H−bonded to the indole ring nitrogen of PsaA−W697 (PsaB−W677) (Srinivasan and Golbeck 2009a). This Trp is π−stacked with the phylloquinone A_{1A} (A_{1B}), which, in turn, is H−bonded to PsaA−L722 (PsaB−L706) (Jordan et al. 2001). The backbone oxygen of PsaA−L722 (PsaB−L706) is H−bonded to the F_X binding loop through PsaA−R694 (PsaB−R674) (Srinivasan and Golbeck 2009a). The two most striking features of the A_{1A} (A_{1B}) quinone binding pocket are the π−stacked arrangement with PsaA−W697 (PsaB−W677) and the presence of only one hydrogen bond to the protein backbone with PsaA−L722 (PsaB−L706). The other oxygen atom of each of two phylloquinone is not H−bonded at all. The interaction with the Trp residue is assumed to destabilize the negative charge on the semiquinone anion radical, thereby lowering its

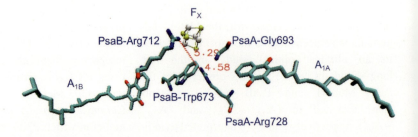

Fig. 7.22 The A_1 sites in photosystem I (PDB ID: 1JB0)

redox potential. In contrast, the H−bond withdraws electron density and stabilizes the negative charge on the semiquinone anion radical, thereby raising its redox potential. The question is how these and other factors conspire to confer an appropriate redox potential to the A_{1A} and A_{1B} quinones (Srinivasan and Golbeck 2009a). The redox potential of −810 mV was thereby estimated for A_1 (Iwaki and Itoh 1994).

Despite high degree of overall symmetry, there are subtle differences between the two phylloquinone binding sites (Grotjohann and Fromme 2005; Srinivasan and Golbeck 2009a). First, the most apparent difference is the orientation of the phytyl tails of the phylloquinones in the A− and B−branches. Second, two lipid molecules are located close to the pathway from A_{1A} and A_{1B} to F_X that could be another main factor in the establishment of the asymmetry. At the slower A−branch, a negatively charged phosphatidylglycerol is located in vicinity to F_X; thereby the electron has to be transferred against this negative charge, whereas on the faster B−branch a neutral monoglalactosyldiacylglycerol replaces the phospholipid. Third, one of the less obvious differences includes the presence of carotenoids with different orientations and configurations in the vicinity of each of the two phylloquinone molecules. Forth, there are differences in the water clusters that are located between the quinone−binding site and F_X. Five water molecules located in a pocket at the A−branch, which show a non−specific arrangement. In contrast, five of the total of six water molecules located between the quinone−binding pocket and F_X at the B−branch phylloquinone form the structure of a hexagon, which is a well defined low energy arrangement of a water cluster. This hexagon structure of the water cluster at the B−branch may also contribute to the lowering of the reorganization energy and the activation energy barrier at the B−branch (Grotjohann and Fromme 2005). Fifth, the presence of a unique Trp between A_1 and F_X is from PsaB. The corresponding residue in PsaA is Gly (Srinivasan and Golbeck 2009a). The redox potential of F_X was calculated to be −680 mV, which is within 10 mV of the consensus midpoint potential of −688 mV determined experimentally. The redox potentials of A_{1A} and A_{1B} were calculated to be −671 mV and −696 mV, respectively. In this formulation, electron transfer from A_{1A} to F_X would be endothermic by 9 mV and electron transfer from A_{1B} to F_X would be exothermic by 16 mV (Srinivasan and Golbeck 2009a).

To understand effect of local protein structure on phylloquinone properties, and rate and directionality of electron transfer in PS I, a number of site−directed mutants were generated in *Chlamydomonas reinhardtii* and *Synechocystis* sp. PCC 6803. As discussed earlier, the forward electron transfer from A_1 to F_X has been studied extensively in both prokaryotes (cyanobacteria) and eukaryotes (spinach or algae). Two exponential phases with lifetime of ~20 ns and ~200 ns were revealed, with relative ratio of amplitudes differing from 1:1 to 1:4 in various species. The phylloquinones in the A_{1A} and A_{1B} sites are difficult to distinguish spectroscopically because they exist in near−equivalent environments.

The studies involve changing the π−stacked Trp in *Chlamydomonas reinhardtii*, showed that the Trp → Phe mutation on the PsaB side slowed the 18−ns fast kinetic phase to 97 ns, and the Trp → Phe mutation on the PsaA side slowed the 218−ns slow kinetic phase to 609 ns (Guergova-Kuras et al. 2001). Only the relative kinetics of electron transfer appeared to be changed, not the amplitudes of the two kinetic phases. The results indicated that both branches participate in the electron transfer with the fast phase assigned to the A_{1B} re−oxidation, and the slow phase to the A_{1A} re−oxidation (Guergova-Kuras et al. 2001). Similarly, the Trp involving the π−stacked with phylloquinone were changed to Phe in PsaA or PsaB in *Synechocystis* sp. PCC 6803 (Xu et al. 2003a, b). Both Trp → Phe mutants can grow photoautotrophically. The ENDOR studies showed that the PsaA−W697F mutation leads to a 5% increase in the hyperfine coupling of the methyl group on the phylloquinone ring (Xu et al. 2003a), demonstrating the effect of minor local structure on cofactor properties. The change correlates with spectral alterations observed by CW EPR spectroscopy of photoaccumulated PS I complexes as well as those studied by transient

EPR spectroscopy. Thus, electron transfer detected by EPR spectroscopy in this cyanobacterium occurs on cofactors associated with the PsaA side. The transient EPR studies performed at low temperature correlate with those at room temperature, indicating that conclusions reached at low temperatures are valid at physiological temperatures. In contrast, the PsaB−W677F mutant yielded the same spectra as the wild type (Xu et al. 2003a). This agrees with an assessment that those electron transfer steps detected by EPR spectroscopy in the eukaryotic organism *Chlamydomonas reinhardtii* also involve cofactors associated with the PsaA side (Purton et al. 2001; Boudreaux et al. 2001). The further studies by time−resolved optical difference spectroscopy and transient EPR spectroscopy demonstrated that the slow kinetic phase resulted from electron transfer from A_{1A} to F_X and that this accounted for at least 70% of the electrons, indicating asymmetrical electron transfer along two branches (Xu et al. 2003b). Another study of the A_1 binding pocket was carried out by Srinivasan et al. in 2009 in *Synechocystis sp.* PCC 6803. In that study, a site−specific mutant (PsaA−L722W) was generated, and was proven to weaken the hydrogen bond between the protein backbone and the O_4 position of A_{1A} (Srinivasan et al. 2009). Such a change had a large impact on the spin distribution and redox potential of the phylloquinone. It also demonstrated that in the wild−type PS I, there was only one hydrogen bond on the phylloquinone. The altered reduction behavior in the PsaA−L722W mutant suggested that the primary purpose of the H−bond is to tie up the C (4) carbonyl group of phylloquinone in a H−bond so as to prevent protonation and hence lower the probability of double reduction during periods of high light intensity (Srinivasan et al. 2009).

F_X, The First FeS Cluster

F_X, a [4Fe−4S] iron−sulfur cluster, plays an important structural and functional role in PS I. Its main function is to accept electrons from A_1 to the terminal iron−sulfur cluster F_A (Fig. 7.23). More specifically, the function of F_X along with F_A and F_B is to serve as a molecular wire, lengthening the time of charge separation at the expense of a fraction of the transiently stored Gibbs free energy, and vectoring the electron out of the membrane into the soluble phase (Srinivasan and Golbeck 2009a). The electron transfer to F_A is faster than the electron transfer from the phylloquinones to F_X, therefore the F_X intermediate is difficult to detect spectroscopically in intact PS I complexes. The twofold axis between PsaA and PsaB runs through F_X. It is coordinated by both subunits PsaA and PsaB, being thereby a rare example of an inter−protein iron−sulfur cluster. The sequence analysis showed that PsaA and PsaB have only three and two conserved cysteine residues respectively.

Fig. 7.23 F_X in photosystem I (PDB ID: 1JB0)

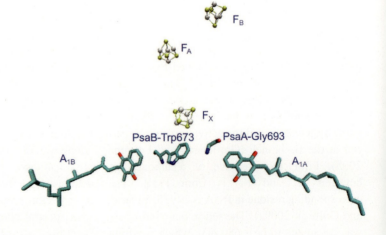

Cys578 and Cys587 of PsaA and Cys565 and Cys574 of PsaB provide the ligands to the iron atoms of the cluster. The ligands are located in the loop connecting the transmembrane helices 7h and 8i (Jordan et al. 2001; Fromme et al. 2001). The redox potential of F_X was estimated as -705 mV (Nelson and Yocum 2006), and is one of the most reducing iron–sulfur clusters known in biology (Parrett et al. 1989). This loop contains the most conserved sequence between all PS I species (FPCDGPGRGGT CXXSAWDH). This sequence was first suggested to be the coordination site for F_X by (Fish et al. 1985) and further supported by mutagenesis studies (Webber et al. 1993; Hallahan et al. 1995) in *Chlamydomonas reinhardtii* and in *Synechocystis* sp. PCC 6803 (Vassiliev et al. 1995) and structural studies (Jordan et al. 2001). The photoreduction of F_X was largely inhibited as seen from direct measurement of the extent of electron transfer from A_1 to F_X in the PsaC–deficient mutant of *Synechocystis* sp. PCC 6803 (Gong et al. 2003), suggesting that the physical chemical parameters of F_X influenced by the presence of PsaC is required for efficient electron transfer in PS I (Grotjohann and Fromme 2005). The experiment demonstrated that Rubredoxin encoded by *rubA* gene is required for assembly of F_X. The *rubA* gene was insertionally inactivated in *Synechococcus* sp. PCC 7002, and the properties of PS I complexes were characterized spectroscopically. The X–band EPR spectroscopy at low temperature showed that the three terminal iron–sulfur clusters, F_X, F_A, and F_B, were missing in whole cells, thylakoids, and PS I complexes of the *rubA*–deficient mutant. It is proposed that rubredoxin is specifically required for the assembly of the F_X iron–sulfur cluster since the PsaC protein requires the presence of F_X for binding, and the absence of F_A and F_B may be an indirect result of the absence of F_X. Interestingly, F_X is not required for the assembly of trimeric P700–A_1 cores since electron transfer from P700 to A_1 can be detected in the *rubA*–deficient mutant using the flash–induced decay kinetics of both P700$^+$ in the visible and A_1^- in the near–UV (Shen et al. 2002).

The residues PsaA–Gly693 and PsaB–Trp673 of PS I point towards each other and are located between two phylloquinones and F_X (Fig. 7.23) (Jordan et al. 2001). This asymmetric structural arrangement between the A– and B–branches is well conserved among species. The aromatic Trp673 has been suggested to play a direct role in the electron transfer acting as an electron acceptor between A_{1B} and F_X (Ivashin and Larsson 2003). The PsaB–Trp673 $\rightarrow$ Gly673 mutant has been generated to mimic the phylloquinone binding pocket around A_{1A} in *Chlamydomonas reinhardtii*, which showed slower electron transfer from A_{1B} to F_X (Ali et al. 2006). The PsaB–Trp673 $\rightarrow$ Ala673 mutation of *Synechocystis* sp. PCC 6803 has been generated. However, this mutant severely affects PS I accumulation in the thylakoid membranes (unpublished data). Due to this reason, no spectroscopic data has obtained from this mutant because no PS I trimers can be isolated with high purity. A more conserved mutation of PsaB–Trp673 $\rightarrow$ Phe673 in *Synechocystis* sp. PCC 6803, which is predicted not to affect PS I assembly and accumulation but could affect the rate of electron transfer from A_1 to F_X, has been generated and it will be interesting to characterize this mutant in the future.

After carefully reviewing the literature, the interaction between PsaB–Trp673 and PsaA–Arg728 has been proposed based on the calculated electron coupling (Ivashin and Larsson 2003) (Fig. 7.22). If this is the case and PsaB–Trp673 is indeed involved in the electron transfer from A_1 to F_X, it is reasonable to hypothesize that the disruption of the PsaB–Trp673 π system with the extension to PsaA–Arg728 will affect the rate of electron transfer from A_1 to F_X. The PsaA–Arg728 to Lys, Glu and Met mutations can be generated. Lys has the same charge as Arg while Glu has an opposite charge as Arg. Met has a neutral charge. The corresponding Arg to Lys, Glu and Met in PsaB can also be generated to serve as the controls. All these mutants can be investigated by biochemical and spectroscopic approaches to test the hypothesis.

7.3.1.5 Docking of Plastocyanin/ Cytochrome c_6

PsaA and PsaB also play an essential role in the docking of the soluble electron donors to PS I that re-reduce the primary electron donor, $P700^+$. In cyanobacteria and algae, cytochrome c_6 can functionally replace plastocyanin-deficient strains (Clarke and Campbell 1996) as well as under copper-deficient conditions (Zhang et al. 1992). However, *Arabidopsis* plants mutated in both plastocyanin-coding genes and with a functional cytochrome c_6 cannot grow photoautotrophically because of a complete blockade in light-driven electron transport, demonstrating that in *Arabidopsis* only plastocyanin can donate electrons to PS I *in vivo* (Weigel et al. 2003). Plastocyanin or cytochrome c_6 interacts directly with the PsaA and PsaB core subunits (Sun et al. 1999; Sommer et al. 2002, 2004) and donates electrons to $P700^+$ (Sun et al. 1999; Mamedov et al. 2001; Finazzi et al. 2005; Duran et al. 2004, 2006). The kinetic data of PS I reduction by plastocyanin corresponds to a monophasic process while the PS I reduction by cytochrome c_6 follows biphasic kinetics with the first fast component in the microsecond range (Duran et al. 2004; Hervas et al. 1995). This fast phase of PS I reduction has been typically described by a kinetic model involving transient complex formation before the electron-transfer step (Duran et al. 2004; Hervas et al. 1995), suggesting that cytochrome c_6 interacts with PS I and donates an electron to $P700^+$ *in vivo* following a mechanism more complex and more efficient than that of plastocyanin although there are discrepancies between *in vivo* and *in vitro* results (Duran et al. 2005). The docking site for plastocyanin and cytochrome c_6 is located at an indentation on the lumenal side of PS I (Grotjohann and Fromme 2005).

Docking of plastocyanin or cytochrome c_6 to the PS I in plants or green algae is mainly promoted by two highly conserved structural interaction patterns, which are (i) long range electrostatic attractions between basic patches of PsaF and acidic regions of plastocyanin or cytochrome c_6 (Hippler et al. 1996, 1997, 1998; Nordling et al. 1991; Haehnel et al. 1994) and (ii) a hydrophobic region around the electron transfer site of the donors interacting with a hydrophobic region site on PS I including PsaA-Trp651 and PsaB-Trp627 in *Chlamydomonas reinhardti* (Sommer et al. 2002, 2004; Haehnel et al. 1994). The function of the positively charged residues in the eukaryotic N-terminus of PsaF in binding of both donors has been studied extensively by cross-linking, knock-out, and reverse genetics experiments (Hippler et al. 1996, 1998, 1999; Farah et al. 1995; Haldrup et al. 2000). The details will be discussed later. The recent plant PS I structures showed that plastocyanin binding site in plant is buried deeper in the complex, this is achieved by the extension of the PsaF N-terminal and also by the new position of the N-terminus of PsaH, which forms a loop mirroring the conformation of the conserved luminal PsaA loop, suggesting a direct role for PsaH in plastocyanin binding (Mazor et al. 2015; Qin et al. 2015). In conclusion, these studies showed that the basic patch present in the N-terminal domain of PsaF is crucial for proper binding, complex formation between donor and PS I, and fast electron transfer. In contrast to eukaryotic organisms, efficient binding and electron transfer between PS I and plastocyanin or cytochrome c_6 in *Synechocystis* sp. PCC 6803 does not depend on the PsaF subunit since the basic N-terminal region of plant and green algal PsaF is absent in cyanobacteria. Therefore, the cyanobacterial subunit PsaF is not involved in the docking of plastocyanin/ cytochrome c_6 (see the more detailed discussion later).

The docking site for plastocyanin and cytochrome c_6 is mainly hydrophobic and forms two surface helices (Sommer et al. 2004) in the loop between the transmembrane helices *i* and *j*. Both plastocyanin and cytochrome c_6 have hydrophobic faces that match the hydrophobic docking site of PS I (Frazao et al. 1995). On the lumenal side of thylakoid membranes, P700 is separated from the lumenal space by two α helices, *l'* and *l*, formed by loops *j'* and *j* in PsaA and PsaB, respectively, which are arranged in parallel to the membrane plane. One tryptophan residue in loop j' of PsaA and the corresponding tryptophan in loop j of PsaB that are partially exposed to the aqueous phase are also prominent features of the docking site (Fromme et al. 2003). The hydrophobic interaction site of the PS I core formed by PsaB has been studied by site−directed mutagenesis. Sun et al. (1999) introduced short stretches of mutations in the luminal loop j of the PsaB protein from *Synechocystis* sp. PCC 6803 and could isolate a double mutant (W622C/A623R), which was highly photosensitive and showed a severe defect in the interaction with plastocyanin or cytochrome c_6. A more conservative mutation of Trp627 (corresponding to Trp622 in *Synechocystis* sp. PCC 6803 and Trp631 in *Synechoccocus elongatus*) to Phe in PsaB of *Chlamydomonas reinhardtii* also displayed a strong effect on cell growth (Sommer et al. 2002). The cells became strongly photosensitive, and the *in vitro* analysis of the electron transfer reactions revealed a differential effect on the binding constants of plastocyanin and cytochrome c_6. No complex formation was observed for the interaction of plastocyanin with the altered PS I, whereas it was still present with cytochrome c_6, displaying a tenfold decreased electron transfer rate. Interestingly, as seen from the crystal structure, Trp631 in loop j of PsaB (corresponding to Trp627 in *Chlamydomonas reinhardtii*) forms a sandwich complex with the corresponding Trp655 of PsaA (corresponding to Trp651 in *Chlamydomonas reinhardtii*). This stacked π−electron system is located in close distance to P700. As predicted, the mutation PsaA−W651F of *Chlamydomonas reinhardtii* completely abolished the formation of a first order electron transfer complex between plastocyanin and the altered PS I and increased the dissociation constant for binding of cytochrome c_6 by more than a factor of 10 as compared with the wild type (Sommer et al. 2004). These results demonstrated that the highly conserved structural recognition motif that is formed by PsaA−Trp651 and PsaB−Trp627 confers a differential selectivity in binding of both donors to PS I (Sommer et al. 2004).

In the 2.5 Å resolution crystal structure of PS I, two chlorophyll molecules of P700 are confined by symmetrically positioned four α helices, A−j, A−k and B−j, B−k (Jordan et al. 2001). Since P700 is positioned 10–15 Å away from lumenal surface (Jordan et al. 2001), the protein components filling the space between the electron donor and P700 may provide a pathway for directly or indirectly migrating electrons that reduce P700$^+$. It is also possible that these aromatic residues are integral components of the electron transfer path between the redox centers of plastocyanin or cytochrome c_6 and P700$^+$ since it was reported that side chains of aromatic residues can serve as an electron tunneling bridge (Shih et al. 2008; Nishioka et al. 2005). A report indicated that photo−oxidation of the chlorophyll a/a' heterodimer, P700, causes shifts in the vibrational frequencies of two or more tryptophan residues in PS I, demonstrating that role of aromatic residues in electron transfer (Chen et al. 2009). To determine the role of conserved aromatic residues adjacent to the histidyl molecule in the helix of B−j, six site−directed mutants of the *psaB* gene in *Synechocystis* sp. PCC 6803 were generated. Three mutant strains with W645C, W643C/A644I and S641C/V642I substitutions could grow photoautotrophically and showed no obvious reduction in the PS I activity. Kinetics of P700 re−reduction by plastocyanin remained unaltered in these mutants. In contrast, the strains with H651C/L652M, F649C/G650I and F647C substitutions could not grow under photoautotrophic conditions because those mutants had low PS I activity. The molecular analysis of the spontaneous revertants suggested that an aromatic residue at F647 may be necessary for maintaining the structural integrity of PS I (Xu et al. 2011).

7.3.2 PsaF

The *psaF* gene, *sll0819*, in *Synechocystis* sp. PCC 6803 encodes a mature protein (subunit III) of 15,705 Da that is synthesized with a

23 amino acid extension (Chitnis et al. 1991). The predicted isoelectric point of PsaF from *Synechocystis* sp. PCC 6803 is 7.02. The structure of *Thermosynechococcus elongatus* PS I has been determined at 2.5 Å resolution (Jordan et al. 2001) and it was shown that PsaA and PsaB form the core reaction center located in the center of the monomer, surrounded by seven small integral membrane proteins. PsaF is situated in the periphery of the PS I complex according to the crystal structure. The structure of subunit PsaF (see Fig. 7.24) consists of three domains. The N−terminal domain is located in the lumen, followed by a transmembrane domain with one transmembrane helix and two short helical pieces in a V−shaped arrangement. The C−terminus is located in the stroma and is sandwiched between PsaA and PsaE (Jordan et al. 2001). Besides the lumenal loops of PsaA and B, subunit PsaF contributes prominent structural features to this surface of PS I with two hydrophilic α helices F−c and F−d at the N−terminus of transmembrane helix F−f (Fig. 7.24). PsaF function has been well studied in cyanobacteria and green algae, as well as in plants. As the shortest distance between their helix axes and the pseudo−C2 axis is 27 Å (Jordan et al. 2001), direct interaction with cytochrome c_6 or plastocyanin is unlikely, consistent with the observation that deletion of PsaF does not influence the kinetics of electron transfer in cyanobacteria (Xu et al. 1994d). The cyanobacterial PsaF−null mutant showed normal photoautotrophic growth when compared with the wild type, suggesting that PsaF has dispensable accessory roles in the function and organization of PS I complex (Chitnis et al. 1991; Xu et al. 1994d). However, the detailed study of the PsaF−deficient and PsaF mutant strains of cyanobacteria and plants yielded interesting observations.

In a mutant strain of the cyanobacterium *Synechocystis* sp. PCC 6803 that contains a deletion of the *psaF* gene, a *psaJ* gene is also transcriptionally inactive since *psaF* and *psaJ* belong to the same cistron. PS I complexes were assembled in the cells lack PsaF and PsaJ. The cells of the mutant and wild−type strains have similar rates of photosynthetic electron transfer and $P700^+$ re−reduction under linear and cyclic electron transfer conditions. Analysis of flash−induced absorption transients at 700 nm demonstrated that absence of PsaF in the purified mutant PS I did not affect the rate of $P700^+$ re−reduction by cytochrome c_{553}. Therefore, PsaF is not essential for docking of cytochrome c_{553} and possible plastocyanin neither. The PsaA−PsaB subunits were more easily degraded by thermolysin in the mutant PS I. Thermolysin cleavage of PsaB yielded two major fragments that were immunoreactive with an antibody raised against the C−terminus of PsaB. The N−termini of these PsaB peptides mapped at Ile482 and Ile498 residues, thus identifying a surface−exposed domain of the core of PS I. The PsaE subunit could be removed by 1 M NaI and was rapidly digested by thermolysin in the mutant but not in the wild−type PS I. Also, the removal of PsaF from PS I complex of *Synechococcus elongatus* had no effect on electron transfer from cytochrome c_6 to P700 (Baymann et al. 2001). Therefore, PsaF and PsaJ subunits of PS I have dispensable accessory roles in the function and organization of the complex (Xu et al. 1994d). Deletion of *psaJ* of *Synechocystis* sp. PCC 6803 led to a reduction in the steady state RNA level from *psaF* which is located upstream from *psaJ*. Immunoquantification using an anti−PsaF antibody revealed a significant decrease in the amount of PsaF in membranes of the mutant strain. Trimeric PS I

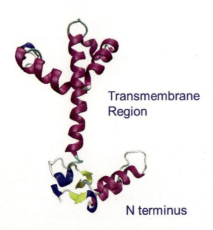

Fig. 7.24 The PsaF structure (PDB ID: 1JB0)

complexes isolated from the mutant strain using n−dodecyl β−D−maltoside lacked PsaJ, contained 80% less PsaF, but maintained the wild−type levels of other PS I subunits. In contrast, the PS I purified using Triton X−100 contained less than 2% PsaF when compared to the wild type, showing the more extractable nature of PsaF in PsaJ−less PS I in the presence of Triton X−100. PsaE was more accessible to removal by NaI in a mutant strain lacking PsaF and PsaJ than in the wild type. The presence of PsaF in PS I from the PsaJ−less strain did not alter the increased susceptibility of PsaE to removal by NaI. These results demonstrated an interaction between PsaJ and PsaF in the organization of the complex (Xu et al. 1994c).

A separate study showed that in the *psaFJ*−null mutant of *Synechocystis* sp. strain PCC 6803, electron transfer from plastocyanin to PS I was not affected. Instead, a restraint in full chain photosynthetic electron transfer was correlated to malfunction of PS I at its stromal side (Jeanjean et al. 2003). It is hypothesized that absence of PsaF causes oxidative stress, which triggers the induction of the "iron stress inducible" operon *isiAB*. Products are the IsiA chlorophyll−binding protein (CP43′), part of the external antenna system, and the *isiB* gene product flavodoxin. Supporting evidence was obtained by similar *isiAB* induction in the wild−type cells artificially exposed to oxidative stress (Jeanjean et al. 2003). In addition, the *psaFJ*−null mutant formed large aggregates of IsiA, however some PS I−IsiA complexes can still be isolated that contain 17 instead of 19 IsiA proteins attached to the PS I complex. These results showed that PsaF and PsaJ were important but not absolutely essential for the interaction of PS I with the IsiA ring (Kouril et al. 2003).

The membrane intrinsic domain of PsaF contains only one transmembrane α helix F−f, followed by a short hydrophilic helix F−g and two shorter hydrophobic α helices, F−h and F−i. This region of PsaF is very unusual. A helix F−h enters the membrane from the stromal side and ends in the first third of the membrane. It is followed after a crease by α helix F−i, running back to the stromal side, where the C−terminus is located and forms contacts with PsaE, PsaA and also PsaB. PsaF does not axially coordinate chlorophylls, but forms hydrophobic interactions with chlorophylls and several carotenoids. A possible role of the transmembrane part of PsaF could be a shielding of the carotenoids and of chlorophylls from the lipid phase. The electron spectrum study suggested that the subunits PsaF/J/I interacted with about five chlorophyll *a* molecules of PS I antenna (Soukoulis et al. 1999).

Unlike the cyanobacterial PsaF, chemical cross−linking and site−directed mutagenesis experiments (Hippler et al. 1989, 1996; Wynn and Malkin 1988; Wynn et al. 1989) demonstrated that the eukaryotic PsaF is essential for docking of plastocyanin or cytochrome c_6 to the oxidizing side of PS I. It is of note that the existence and the discussed function of the eukaryotic N−terminal domain of PsaF is supported by the new crystal structure data on plant PS I (Ben-Shem et al. 2003; Mazor et al. 2015; Qin et al. 2015). Inactivation of the *psaF* gene from *Chlamydomonas reinhardtii* resulted a mutant that still assembled functional PS I complex and was capable of photoautotrophic growth. However, the electron transfer from plastocyanin to P700$^+$ was dramatically reduced in the mutant, thereby providing the evidence that PsaF plays an important role in docking plastocyanin to PS I in chloroplasts (Hippler et al. 1997; Farah et al. 1995). This controversial function of PsaF between plants and cyanobacteria was addressed by the peculiarities of molecular recognition between plastocyanin and PsaF (Hippler et al. 1996). Compared to cyanobacterial PsaF sequences, a basic N−terminal region only present in plant and green algae (Fig. 7.25), was demonstrated important for plastocyanin or cytochrome c_6 binding. Our alignment analysis of the selected representative sequences showed that plant, green algal and cyanobacterial PsaF share ~73% amino acid similarity and ~16% identity (Fig. 7.25). Two conserved negative patches in plant plastocyanin were cross−linked with lysine residues in the N−terminal domain of PsaF in plants. The lysine rich region is absent in cyanobacteria. Adding

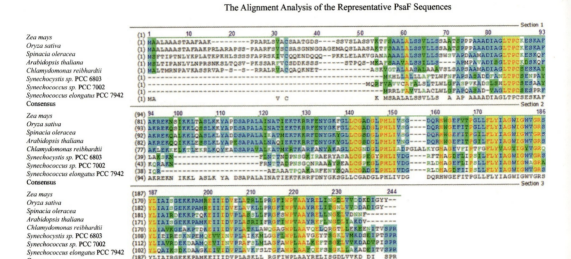

Fig. 7.25 The sequence alignment of PsaF

this lysine rich region, first 83 amino acids from *Chlamydomonas reinhardtii* to PsaF of *Synechocystis elongatus* resulted in an increased rate of $P700^+$ reduction due to the interaction between this lysine rich region and plastocyanin (Hippler et al. 1999). The study indicated that the lysine–rich region is sufficient for the binding of the donor proteins, but not for the electron transfer within the intermolecular complex. A recent study showed that deletion of LepB1(Sll0716), leader peptidases, in *Synechocystis* sp. PCC 6803 resulted in an inability to grow photoautotrophically and an extreme light sensitivity. In this mutant, PsaF was found incorporated into PS I complex in its unprocessed form, which could influence the assembly and/or stability of PS I. The presence of the signal peptide introduces an extra N–terminal domain on the lumenal side of PS I which could result in a slower assembly rate or a lower stability of PS I (Zhang et al. 2013). Similarly observed in the chimeric PsaF, containing the N–terminal domain of PsaF from *Chlamydomonas reinhardtii* and the C–terminal sequence of the cyanobacterial PsaF was found to assemble into PS I complexes at a reduced level in *Synechococcus* (Hippler et al. 1999). In addition, a direct contact of PsaF with the light harvesting systems in plants has been suggested by experiments, in which plant subunit PsaF was isolated as a chlorophyll–protein complex with LHC I proteins (Anandan et al. 1989). Interaction sites between PsaF and the LHC I proteins were also identified in the structure of PS I from pea (Ben-Shem et al. 2004). The different functions of plant PsaF and cyanobacterial PsaF indicates that a faster electron transfer from plastocyanin to PsaF achieves in plants.

7.3.3 PsaI

This is one of the 3–4 very small proteins in the cyanobacterial PS I complexes that contain only one transmembrane helix and very small extramembraneous region. The deduced amino acid sequences of PsaI from plants and cyanobacteria share high degree of conservation (Xu et al. 1995). Our alignment analysis of the selected representative sequences showed that plant, green algal and cyanobacterial PsaI share ~77.5% amino acid similarity and ~22.5% identity (Fig. 7.26). The *psa*I gene, *smr0004*, in *Synechocystis* sp. PCC 6803 encodes a mature protein (subunit VIII) with a predicted isoelectric point of 4.00. The mutagenesis study showed that PsaI had a crucial role in aiding normal structural

The Alignment Analysis of the Representative PsaI Sequences

```
                              (1)  1         10        20        30        40
Zea mays                      (1) ----MTDFNLPSIFVPLVGLVFPAIAMTSLFLYVQKNKIV
Oryza sativa                  (1) ----MMDFNLPSIFVPLVGLVFPAIAMASLFLYVQKNKIV
Spinacia oleracea             (1) -------MNFPSIFVPLVGLVFPAIAMASLFLYVQKNKIV
Arabidopsis thaliana          (1) ---MTTFNNLPSIFVPLVGLVFPAIAMASLFLHIQKNKIF
Synechocystis sp. PCC 6803    (1) MDGSYAASYLPWILIPMVGWLFPAVTMGLLFIHIESEGEG
Synechococcus sp. PCC 7002    (1) MNGAYAASFLPVILVPLAGVVFPALAMGLLFNYIESDA--
Synechococcus elongatus PCC 7942 (1) MSGDFAAAFLPTIFVPLVGLGLPAVLMSLLFTYIESEA--
Consensus                     (1)     NLPSIFVPLVGLVFPAIAMASLFLYIQKNKI
```

Fig. 7.26 The sequence alignment of PsaI

organization of PsaL within the PS I complex and the absence of PsaI altered PsaL organization, leading to a small, but physiologically significant, defect in PS I function. Inactivation of *psaI* in *Synechocystis* sp. PCC 6803 led to an 80% decrease in the PsaL level in the photosynthetic membranes and to a complete loss of PsaL in the purified PS I preparations, but had little effect on the accumulation of other PS I subunits (Xu et al. 1995). The PsaI−less membranes from the cells grown at either 25°C or 40°C showed a small decrease in NADP$^+$ photoreduction rate when compared to the wild−type membranes. Therefore, a structural interaction between PsaL and PsaI may stabilize association of PsaL with the PS I core. One year later, a separate study of a PsaI−deficient mutant of *Synechococcus* sp. PCC 7002 showed that the growth rate of the mutant was identical to that of the wild type under low or high intensity of white light but slower than the wild type in green light (Schluchter et al. 1996). The compositional analyses of the mutant PS I complexes confirmed that PsaI plays a role in stabilizing the binding of PsaL in PS I complex (Schluchter et al. 1996). In addition, PsaI could interact with PsaM (Schluchter et al. 1996), and it is agreed with the crystal structure of PS I complex (Jordan et al. 2001). PsaI is located between PsaL and PsaM (Fig. 7.27), and forms direct contact with both subunits (Jordan et al. 2001).

Besides interacting with PsaL and PsaM, PsaI forms hydrophobic interactions with carotenoid molecules but not chlorophyll *a* molecules. The structure of higher plant PS I has been

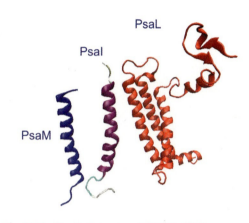

Fig. 7.27 The PsaI structure (PDB ID: 1JB0)

determined at 4.4 Å resolution (Ben-Shem et al. 2003), 3.4 Å resolution (Amunts et al. 2007) and 2.8 Å resolution (Mazor et al. 2015; Qin et al. 2015). The existence of close interactions of PsaI and PsaL in higher plants suggested that the arrangement of these small subunits is a motif that is conserved during evolution (Grotjohann and Fromme 2005; Ben-Shem et al. 2004). Even if the sequence was not assigned due to the limited resolution, the structure shows that PsaI and PsaL also form close contacts in plant PS I. This is remarkable, taking into account the fact that plant PS I is a monomer and the region of PsaI and PsaL (and PsaH) may function in forming interactions with LHC II (Scheller et al. 2001; Grotjohann and Fromme 2005). In plants, an additional subunit (PsaH) is located in close vicinity to PsaI and PsaL (Ben-Shem et al. 2003; Mazor et al. 2015; Qin et al. 2015). This region may be the contact site between PS I and the LHC II complex (Zhang and Scheller 2004).

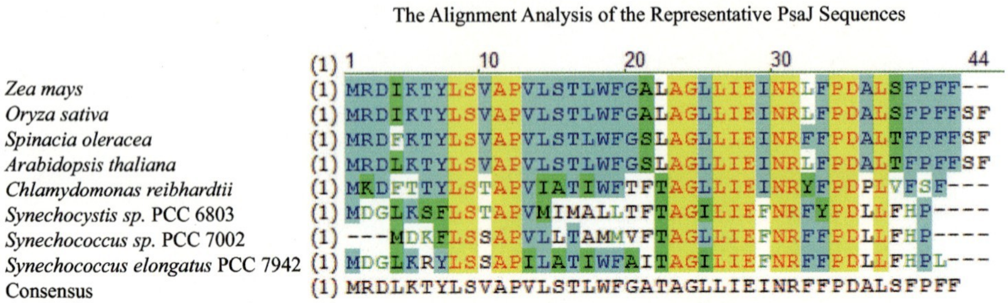

The Alignment Analysis of the Representative PsaJ Sequences

```
                              (1) 1        10        20        30        44
Zea mays                      (1) MRDIKTYLSVAPVLSTLWFGALAGLLIEINRLFPDALSFPFF--
Oryza sativa                  (1) MRDIKTYLSVAPVLSTLWFGALAGLLIEINRLFPDALSFPFFSF
Spinacia oleracea             (1) MRDFKTYLSVAPVLSTLWFGSLAGLLIEINRFFPDALTFPFFSF
Arabidopsis thaliana          (1) MRDLKTYLSVAPVLSTLWFGSLAGLLIEINRLFPDALTFPFFSF
Chlamydomonas reibhardtii     (1) MKDFTTYLSTAPVIATIWFTFTAGLLIEINRYFPDPLVFSF---
Synechocystis sp. PCC 6803    (1) MDGLKSFLSTAPVMIMALLTFTAGILIEFNRFYPDLLFHP----
Synechococcus sp. PCC 7002    (1) ---MDKFLSSAPVLLTAMMVFTAGLLIEFNRFFPDLLFHP----
Synechococcus elongatus PCC 7942 (1) MDGLKRYLSSAPILATIWFAITAGILIEFNRFFPDLLFHPL---
Consensus                     (1) MRDLKTYLSVAPVLSTLWFGATAGLLIEINRFFPDALSFPFF
```

Fig. 7.28 The sequence alignment of PsaJ

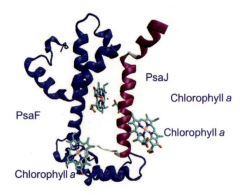

Fig. 7.29 The PsaJ structure (PDB ID: 1JB0)

The idea can be put forward that PsaL stabilized by PsaI might form an entrance gate for the excitation energy from the external antenna complexes in plants (Grotjohann and Fromme 2005; Busch and Hippler 2011). This would correspond to excitation energy transfer between the monomers in the trimeric PS I. Significant excitation energy transfer between monomers have been spectroscopically determined (Karapetyan et al. 1999).

7.3.4 PsaJ

PsaJ is a 4.4 kDa hydrophobic subunit, which has been identified in PS I preparations from cyanobacteria and higher plants (Xu et al. 1994c; Ikeuchi et al. 1991). The protein is chloroplast encoded in plants as is the case also for PsaI, which has a similar size and hydrophobicity (Scheller et al. 2001). Our alignment analysis of the selected representative sequences showed that plant, green algal and cyanobacterial PsaJ share 95.5% amino acid similarity and 31.8% identity (Fig. 7.28). PsaJ contains one transmembrane α helix (Jordan et al. 2001). The structure is shown in Fig. 7.29. A location of PsaJ close to PsaF was predicted by the mutagenesis and cross−linking experiments on the cyanobacterial (Xu et al. 1994c, 1994d) and plant PS I (Fischer et al. 1999; Jansson et al. 1996). The N−terminus of PsaJ is located in the stroma, and the C−terminus is located in the lumen (Jordan et al. 2001). The *psaJ* gene, *sml0008*, in *Synechocystis* sp. PCC 6803 encodes a mature protein (subunit IX) with a predicted isoelectric point of 5.38. The analysis of the PsaJ−deficient mutant in *Synechocystis* sp. PCC 6803 showed that PsaJ was not required for the electron transfer in PS I, but may interact with PsaF since the deletion of the *psaJ* gene resulted in PS I particles containing only 20% of the normal level of PsaF (Xu et al. 1994c), suggesting that PsaJ may maintain the stable PS I structure. In *Chlamydomonas reinhardtii*, the lack of PsaJ, although no decrease in the content of PsaF, resulted in a functional heterogeneity where only 30% of PS I exhibited the typical fast kinetics of plastocyanin and cytochrome c_6 oxidation. In the remaining 70% of the PS I complexes, the oxidation of plastocyanin was as slow as in PS I devoid of PsaF (Fischer et al. 1999). The double mutant lacking both PsaJ and PsaF was similar to the mutant lacking only PsaF (Fischer et al. 1999). Thus, PsaJ in eukaryotes has a function in maintaining PsaF in a conformation that enables efficient electron transfer from plastocyanin.

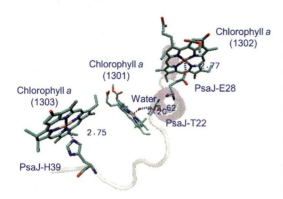

Fig. 7.30 PsaJ provides ligands to chlorophylls (PDB ID: 1JB0)

In addition to interact with PsaF, PsaJ provides coordination of three antenna chlorophylls (Fig. 7.29) (Jordan et al. 2001). Specifically, PsaJ-Thr22 directly interacts with a water molecule that is coordinated with a chlorophyll *a*. PsaJ-Glu28 and PsaJ-His39 provide ligands to two chlorophyll *a* molecules (Fig. 7.30). PsaJ may play an important role in the stabilization of the pigment clusters located at the interface between PsaJ/PsaF and the PsaA/PsaB core. The three chlorophylls, that are coordinated by this subunit are supposed to play an important role in the excitation energy transfer from the IsiA ring to the PS I core (Jordan et al. 2001; Grotjohann and Fromme 2005). The room temperature absorption difference spectra (the wild−type − PsaF/J less mutant) of PS I trimer isolated from the mutants lacking the PsaF/J subunits suggested that the mutant was deficient in core antenna chlorophylls absorbing near 685 nm and also 665 nm (Soukoulis et al. 1999).

7.3.5 PsaK

PsaK is an integral membrane protein in PS I complex. The complete sequence of *Synechocystis* sp. PCC 6803 genome revealed the presence of two unlinked *psaK* genes (Kaneko and Tabata 1997). The orf *ssr0390* of *Synechocystis* sp. PCC 6803 encodes an 86−amino acid subunit which has been identified PsaK1 (subunit X) with a predicted isoelectric point of 9.79 in the PS I complex, whereas the orf *sll0629* encodes a protein with 126 amino acids (an alternative subunit X) which was recently shown to be present in substoichiometric amounts in the PS I complexes (Naithani et al. 2000). Only one PsaK subunit is present in the crystal structure of *Synechococcus elongates* (Jordan et al. 2001). Our alignment analysis of the selected representative sequences showed that plant, green algal and cyanobacterial PsaK1 and PsaK2 share ~60% amino acid similarity and low identity (~8%) (Fig. 7.31). Based on the sequence analysis, the PsaK subunit is PsaK1. This subunit seems to be the least ordered subunit in the PS I complex, as indicated by high temperature factors, so that an unambiguous sequence assignment was not possible and the structure was modeled with polyalanine. It only forms protein contacts with PsaA (Jordan et al. 2001; Grotjohann and Fromme 2005). PsaK1 has two helices (Fig. 7.32) connected in the stroma (Grotjohann and Fromme 2005), and this arrangement is in agree with the prediction from the sequence analysis (Muhlenhoff et al. 1993), so that both the C− and N−termini are located in the lumen. It is located peripherally, close to the interface between the monomers of a PS I trimer. It is predicted that two residues of PsaK, although not definitely determined, provide ligands to two antenna chlorophyll *a* molecules (Fig. 7.32).

The role of PsaK in PS I has been studied by using the PsaK−deficient mutants. Inactivation of *psaK1*, or *psaK2* did not affect photoautotrophic growth and accumulation of other subunits of the PS I complex. The *psaK1*−deficient, *psaK2*−deficient and *psaK1*/*psaK2*−double deficient strains showed normal levels of PS I trimers. A 6.2 kDa polypeptide with a predicted isoelectric point of 9.57 was observed in the PS I preparations from the wild type, but not from the *psaK2*−deficient strain, suggesting the presence of PsaK2 in the PS I complexes under normal light conditions. Another study showed that PsaK2 subunit was absent in the purified PS I complexes under low light condition, but was present in the purified PS I complexes during acclimation to high light (Fujimori et al. 2005).

The Alignment Analysis of the Representative PsaK Sequences

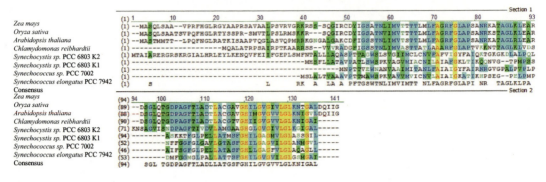

Fig. 7.31 The sequence alignment of PsaK

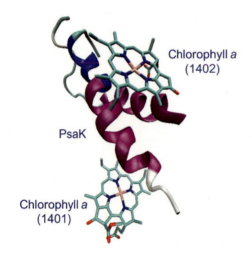

Fig. 7.32 The PsaK structure (PDB ID: 1JB0)

In conclusion, both *psaK1* and *psaK2* are expressed in *Synechocystis* sp. PCC 6803 and the absence of both proteins results in only a small reduction in PS I electron transport (Naithani et al. 2000). The crystal structure of PS I complex showed that PsaK1 coordinates two chlorophylls and forms contacts with carotenoids (Jordan et al. 2001). It may also play an important role in the interaction with the IsiA antenna ring under iron deficiency (Grotjohann and Fromme 2005). The room temperature absorption difference spectra (the wild−type − PsaK1/K2 less mutant) of the PS I trimer isolated from the mutants lacking the PsaK1/K2 subunits suggested that the mutant was deficient in core antenna chlorophylls absorbing near 665 nm and 680 nm (Soukoulis et al. 1999).

Why does *Synechocystis* sp. PCC 6803 contain two *psaK* genes, *psaK1* and *psaK2*? Fujimori et al. (2005) searched for the two types of PsaK proteins in other cyanobacteria. They found that cyanobacteria such as *Synechococcus elongatus* PCC 7942 and *Trichodesmium erythraeum* IMS 101 contain two types of the *psaK* gene, whereas the marine cyanobacterial strains *Prochlorococcus* and *Synechococcus* have only one *psaK* gene that forms a distinct clade in a phylogenetic tree. The *Anabaena* sp. PCC 7120 genome encodes even three *psaK* genes, one of the *psaK1* type, whereas the other two are quite divergent from *psaK2* and the marine type genes. The DNA microarray study showed that the *psaK2* mRNA was the only transcript of a PS I gene that accumulates under high light conditions (Hihara et al. 2001), suggesting that it is involved in responses to high light. To avoid the photodamage, cyanobacteria regulate the distribution of light energy absorbed by phycobilisome antenna either to PS II or to PS I upon high light acclimation by the process so−called state transition. In cyanobacteria, PsaK2 was shown to be involved in excitation energy transfer from phycobilisomes to PS I under high light conditions (Fujimori et al. 2005). These data imply that the assembly mechanism of different PsaK subunits into the PS I monomer is not specific to *Synechocystis* sp. PCC 6803 but is rather common in cyanobacterial strains that have to adapt to changing light conditions and that perform state transitions.

7 Function and Structure of Cyanobacterial Photosystem I

Interestingly, PsaK2 shares around 30% amino acid identity with PsaG, a plant PS I subunit that is lacking in cyanobacteria (Duhring et al. 2007). The studies of *psaG* and *psaK* *Arabidopsis* mutants (Scheller et al. 2001; Zygadlo et al. 2005; Jensen et al. 2000; Varotto et al. 2002) suggested a stabilizing role of these subunits for the PS I core and the peripheral antenna in specific light conditions in plants. PsaG as well as PsaK are located on the outer edge of the plant PS I complex (Ben-Shem et al. 2003; Mazor et al. 2015; Qin et al. 2015). Since the space that is occupied by PsaG in plant PS I is empty in cyanobacterial PS I as judged from the crystal structure, Fujimori et al. (2005) suggested that PsaK2 could occupy this site under high light conditions.

7.3.6 PsaL

Comparison of the deduced primary sequences indicates that the PsaL subunits contain a greater diversity than seen in other subunits (Allen 2005). Our alignment analysis of the selected representative sequences showed that plant, green algal and cyanobacterial PsaL share 75.3% amino acid similarity and 16.9% identity (Fig. 7.33). The structure of PsaL is shown in Fig. 7.34. This subunit contains three transmembrane helices, named L–d, L–e and L–g with L–d and L–e forming hydrophobic contact sites between the monomers within the trimer (Jordan et al. 2001). Most of the further contact sites between the monomers in the trimerization domain are provided by hydrogen bonds and electrostatic interactions within the loop regions. The N–terminal loop is located on the stromal side, harboring three small β strands and one α helix (Fig. 7.34). This loop forms various contacts with loop regions of PsaA and is also in contact with PsaD, thereby attaching PsaL to the core of PS I. A short lumenal loop connects the first and second transmembrane α helices. Correspondingly the second and third transmembrane α helices are connected by a short stromal loop. The C–terminus is folded into a short α helix located in the lumen. The electron density map suggests that a metal ion, possibly a Ca^{2+}, is coordinated by two residues of PsaL (Fig. 7.34) in two adjacent PS I monomers and by PsaA (Jordan et al. 2001). Based on the crystal structure, Ca^{2+} is coordinated by side chain of PsaL–Asp70, main chain of PsaL–Pro167 and two water molecules (Fig. 7.34). Possibly, it could be required for stabilization of the PS I trimer, in agreement with the observations in *Synechocystis sp.* PCC 6803 that the addition of

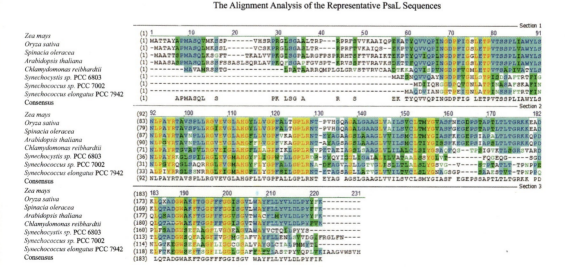

Fig. 7.33 The sequence alignment of PsaL

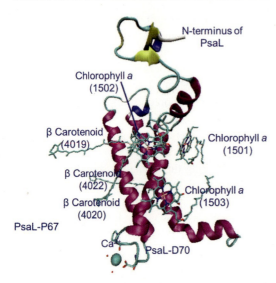

Fig. 7.34 The PsaL structure (PDB ID: 1JB0)

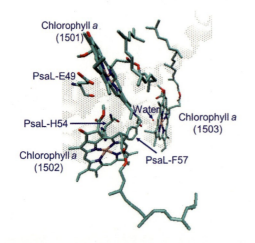

Fig. 7.35 PsaL provides ligands to chlorophylls (PDB ID: 1JB0)

Ca^{2+} stimulated formation of PS I trimers. The *psaL* gene, *slr1655*, in *Synechocystis* sp. PCC 6803 encodes a mature protein (subunit XI) with a predicted isoelectric point of 4.68.

The main function of PsaL is essential for formation of PS I trimers revealed by the inactivation of the *psaL* gene in *Synechocystis* sp. PCC 6803 (Chitnis and Chitnis 1993). No trimers can be detected in the *psaL* deletion mutant. This essential role of PsaL in trimer formation was later confirmed in a *psaL*−deficient mutant from *Synechococcus* sp. PCC 7002 (Schluchter et al.

1996). The crystal structure of cyanobacterial PS I complex showed that PsaL is located close to the C3 axis in the "trimerization domain" (Jordan et al. 2001). As discussed earlier, it forms most of the contacts: hydrophobic interactions between helices of monomer PsaL subunits and hydrogen bonds and electrostatic interactions between loops of monomer PsaL subunits (Jordan et al. 2001). In contrast to cyanobacterial PS I trimer, plants have only PS I monomer. This raises the question of the function of the trimer in cyanobacteria. The mutants of *Synechococcus elongatus*, which lack PsaL, showed normal growth at high light intensity, whereas growth under low light was decreased by a factor of 10 compared to the wild type (Muhlenhoff and Chauvat 1996). These results suggested that the trimer was essential for optimal light capturing in cyanobacteria (Jordan et al. 2001; Grotjohann and Fromme 2005; Fromme et al. 2001).

Another role of PsaL is in binding to antenna chlorophyll *a* (Fromme et al. 2001; Soukoulis et al. 1999) and carotenoid molecules (Jordan et al. 2001). The crystal structure of PS I complex showed that PsaL coordinates three antenna chlorophyll *a* molecules (Fig. 7.34). PsaL−Glu49 and PsaL−His54 coordinate with two chlorophylls. The third chlorophyll is coordinated with a water molecule that interacts with main chain of PsaL−Phe57 (Fig. 7.35). In addition, PsaL forms hydrophobic contacts with three carotenoids (Fig. 7.34), two are located at the interface between PsaL and PsaA and PsaI, the third one is located at the monomer−monomer interface in the trimerization domain. The latter carotenoid may also play an important role in the stabilization of the trimeric PS I complex. In conclusion, interactions with chlorophyll *a* and carotenoids of PsaL suggest that may be therefore important for the excitation energy transfer between the monomers.

Though structure and function of the PS I has been elucidated in detail (Jordan et al. 2001; Chitnis 2001; Grotjohann and Fromme 2005), its assembly, and degradation are poorly understood. It is known that the early steps of photosystem assembly occur in the plasma membrane in cyanobacteria (Zak et al. 2001). Several factors that are involved in assembly of PS I have been identified in cyanobacteria. The known PS I assembly factors like Ycf3 and Ycf4 were found mainly in the plasma membrane, probably for PS I assembly at the early stage while Ycf37 and also Pyg7 are located in the thylakoid membrane (Stockel et al. 2006). Therefore Ycf37 should rather act on later steps of PS I assembly. Most probably Ycf37 stabilizes late PS I assembly intermediates in the thylakoid membrane and protects them from premature degradation through proteolysis (Duhring et al. 2007). Labeling of *Synechocystis* sp. PCC 6803 cells followed by a BN−PAGE analysis to examine the dynamics of assembly and disassembly intermediates of PS I complexes suggested that first assembly intermediate detected following the synthesis of individual subunits and formation of the PS I core was a PsaL/PsaK−less PS I monomer. The next step is the integration of the PsaL subunit into the complex resulting in the recently discovered PS I intermediate. Finally, addition of PsaK leads to a complete PS I monomer consisting of all 11 PS I subunits (Duhring et al. 2007). PsaF should consequently be inserted in an earlier step of the monomer assembly at least before the PsaL subunit is inserted, although the dynamics leading to early PS I complex formation are still unknown (Duhring et al. 2007; Ozawa et al. 2009).

Hippler et al. suggested that in *Chlamydomonas* PsaK is also the last subunit assembled into PS I complexes (Hippler et al. 2002). The pulse−chase experiments suggest a homeostasis of trimeric and the two "late" monomeric PS I complexes. The question remains why PsaK, not PsaL, is the last subunit that is incorporated into the complex. PsaL is essential (but not sufficient) for trimerization of PS I in cyanobacteria (Duhring et al. 2007). Thus, association of PsaL with a PS I monomer as a final step in the assembly of trimers would agree with this role. However, in plants, PS I is a monomeric complex, although it contains PsaL. The association of PsaL with PS I subassemblies does not always directly lead to trimerization but may also depend on the attachment of PsaK (Duhring et al. 2007).

7.3.7 PsaM

PsaM, the smallest subunit (subunit XII with a predicted isoelectric point of 6.07), has 29 amino acids (3.4 kDa) in *Synechocystis* sp. PCC 6803 (Chitnis 2001). *smr0005* encodes PsaM in *Synechocystis* sp. PCC 6803. Our alignment analysis of the selected representative PsaM sequences showed that cyanobacteria share nearly 100% amino acid similarity and ~52% identity (Fig. 7.36). It contains a hydrophobic domain flanked by hydrophilic termini (Chitnis 2001). This subunit contains only one transmembrane α−helix (see Fig. 7.37) as predicted by (Muhlenhoff et al. 1993). PsaM is located close to the monomer/monomer interface, in the neighborhood of PsaI and PsaB. The N−terminus is located in the lumen, and the C−terminus in the stroma (Jordan et al. 2001). The PsaM−deficient mutant of *Synechocystis sp.* PCC 6803 has similar phenotype to the wild type. The other subunits PsaA/B, PsaC, PsaD, PsaE, PsaF, PsaL, PsaI, PsaJ and PsaK are present in the membrane and PS I preparations from the PsaM−deficient mutant. This mutant has less PS I trimers than the wild type, it suggests that PsaM could function in stabilizing the trimers (Naithani et al. 2000).

The Alignment Analysis of the Representative PsaM Sequences

Fig. 7.36 The sequence alignment of PsaM

PsaM is detected only in cyanobacterial PS I (Anderson and McIntosh 1991). Although an open reading frame for this subunit was also found in the liverwort chloroplast genome (Ohyama et al. 1986), this subunit was not identified thus far in any preparations of plant PS I and is also not present in the plant PS I structures (Ben-Shem et al. 2003; Mazor et al. 2015; Qin et al. 2015; Amunts et al. 2007). PsaM forms hydrophobic contacts with one carotenoid molecule and is involved in the coordination of one chlorophyll *a* (Jordan et al. 2001). One chlorophyll *a* is coordinated with a water molecule that directly interacts with PsaM-Arg24 (Fig. 7.37). This chlorophyll (M−1) may play an important role in excitation energy transfer between monomers (Grotjohann and Fromme 2005). The room temperature absorption difference spectra (the wild−type − PsaM−less mutant) of PS I trimer isolated from the mutants lacking the PsaM subunits suggested that the mutant is deficient in core antenna chlorophylls absorbing near 675nm (major) and 700 nm (minor) (Soukoulis et al. 1999). It is strongly functionally coupled to chlorophylls of the neighboring monomer, i.e. functionally may be considered to be part of the clustered network of this adjacent monomeric unit (Sener et al. 2004). In this respect, it is remarkable that the protein side chains of PsaM do not form direct contacts between monomers. However it may play a role in the stabilization of the trimeric structure by forming hydrophobic interaction with the carotenoid that is involved in trimerization (Jordan et al. 2001; Grotjohann and Fromme 2005), and it agrees with the experimental data (Naithani et al. 2000) discussed earlier.

7.3.8 PsaU or PsaX

This is a small peptide in PS I with one transmembrane helix in the neighborhood of PsaF. This protein has been detected in the crystal structure of PS I complexes from *Synechococcus elongatus* and is referred to as PsaU or PsaX (Jordan et al. 2001). The structural model of PsaX contains 35 residues with a predicted isoelectric point of 9.82. The six stromally located N−terminal amino acids were not identified in the structure, possibly because this part of the structure is flexible (Jordan et al. 2001; Grotjohann and Fromme 2005). It is present at the membrane exposed surface of PS I. It coordinates one chlorophyll and forms hydrophobic contacts with several carotenoid molecules and one of the lipids (Jordan et al. 2001). The crystal structure *Thermosynechococcus elongates* (Jordan et al. 2001) showed that PsaX has an α helix and PsaX-Asn23 coordinates with one antenna chlorophyll *a* (Fig. 7.38). It has been reported that PsaX was identified so far only in PS I from the

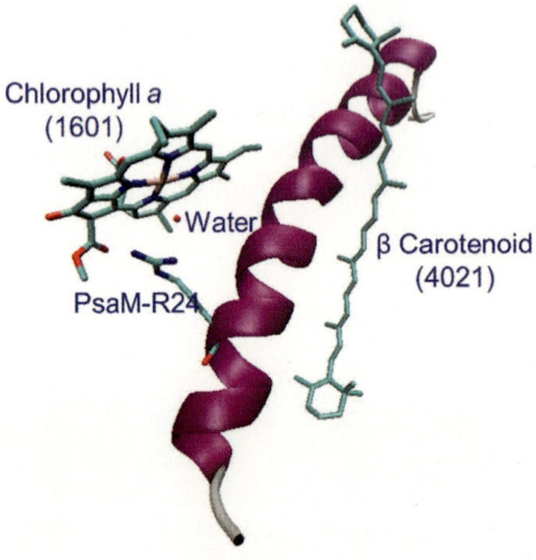

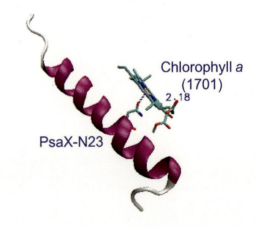

Fig. 7.37 The PsaM structure (PDB ID: 1JB0)

Fig. 7.38 The PsaX structure (PDB ID: 1JB0)

7 Function and Structure of Cyanobacterial Photosystem I

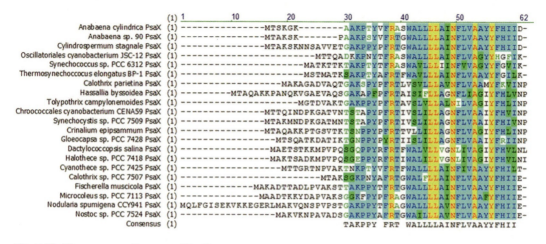

Fig. 7.39 The sequence alignment of PsaX

thermophilic cyanobacteria *Thermosynechococcus vulcanus* and *Anabaena variabilis* (Ikeuchi et al. 1991; Koike et al. 1989) including the crystal structure of PS I complex from *Synechococcus elongates* (Jordan et al. 2001). We performed blast search and identified twenty-one putative PsaX that share ~47% amino acid similarity (Fig. 7.39). This protein is not present in plant PS I, as recently shown by the structures of the PS I from the pea (Ben-Shem et al. 2003; Mazor et al. 2015; Qin et al. 2015; Amunts et al. 2007). Also, no gene sequence for PsaX has yet been assigned in the mesophilic cyanobacterium *Synechocystis* sp. PCC 6803 (Xu et al. 2001).

A common theme in PS I is the occurrence of red chlorophylls, which absorb (and fluoresce) light at longer wavelengths than P700 (Gobets et al. 2001; Palsson et al. 1998). Red chlorophylls slow the trapping kinetics of P700, however, the quantum yield of the complex is not affected by this at room temperature; at higher light intensities red traps may have a photoprotective role (Mazor et al. 2014). In addition, red chlorophylls increase the absorbance cross section of the entire complex, especially in shaded environments (Mazor et al. 2014). The recent 3–D crystal structure of *Synechocystis* sp. PCC 6803 identified a new chlorophyll trimer. Chlorophyll B40 forms a stromal chlorophyll trimer with B19 and B18 in *Synechocystis* sp. PCC 6803 (Mazor et al. 2014), but a chlorophyll dimer in *Thermosynechococcus elongates* (Jordan et al. 2001) since the PsaB loop on the stromal side (PsaB 301–319) extends to coordinate chlorophyll B40 and a phospholipid. Interestingly, the oxygen of the head group of the phospholipid coordinates with the magnesium atom of B40 (Mazor et al. 2014). In contrast, the position of this chlorophyll (B40) is occupied by subunit PsaX in *Thermosynechococcus elongates* (Jordan et al. 2001). Surprisingly, and in spite of the large structural changes observed in this region, this lipid is still clearly visible, this high degree of conservation demonstrates the important role of the PS I lipids. The question as to whether PsaX is a unique subunit of thermophilic cyanobacteria, necessary for stability of the PS I complex at higher temperatures, has been raised. The fact that the newly discovered chlorophyll trimer in *Synechocystis* sp. PCC 6803 relates to its counterpart by pseudo–C2 symmetry suggested that this configuration preceded the association of PsaX, which may have been an adaptation to high temperature. Possibly the most important role of PsaX is to protect this lipid at higher temperatures. The occurrence of this chlorophyll trimer suggested that these trimers have a structural role and the appearance of the red absorption results from fine–tuning of their position. It is likely that the pigment environment contributes significantly to its spectral properties (Mazor et al. 2014).

7.4 Concluding Remarks

The crystal structure of PS I complex from cyanobacterium *Thermosynechcoccus elongatus* was solved at 2.5 Å resolution (PDB ID: 1JB0) 14 years ago, along with plant PS I structures at 4.4 Å resolution (PDB ID: 1QZV), 3.4 Å resolution (PDB ID: 2O01) and 2.8 Å resolution (PDB IDs: 4XK8 and 4Y28), and the recent PS I structure from *Synechocystis* sp. PCC 6803 at 2.8 Å resolution (PDB ID: 4L6V), allow the architecture of pigments, cofactors and proteins to be accurately modeled at atomic level. In *Synechocystis* sp. PCC 6803, the mutants for all proteins of PS I have been reported. The studies of these mutants have revealed the roles of PS I proteins in the organization and function of PS I. However, roles of some proteins of PS I are not evidenced from the phenotypes of the cyanobacterial mutants. The cyanobacterial mutants that lack PsaK1, PsaK2, PsaM, PsaL, PsaI, PsaJ, and PsaF do not affect photoautotrophic growth or photosynthetic electron transfer rates. Therefore, in these mutants the deficiency is not rate limiting for photosynthesis. In most of these mutants, the electron transfer activity of PS I and rates of electron transfer within PS I are also not affected. Algae and higher plants contain homologous PsaF, PsaK, PsaI, PsaJ, and PsaL proteins. The structural conservation of these implies that their role provides critical advantage to the organism. The structural information has shown that these integral membrane proteins bind chlorophyll *a* and carotenoid molecules. However, the absence of these proteins and consequent lack of associated cofactors do not significantly decrease the efficiency of PS I complexes by a detectable margin. Only PsaA, PsaB and PsaC are directly involved in the electron transfer from plastocyanin on the luminal side to ferredoxin on the stromal side. However, the recent studies showed that electron transfer and energy transfer are made efficient only in the presence of the proper assembly of the other subunits.

The PsaA and PsaB polypeptides assemble as a heterodimer and provide ligands for the majority of the electron transfer cofactors, which are located on both sides of a pseudo—C2 axis of symmetry. Thus, a common overall binding frame exists for the two branches of electron transfer cofactors. The current data from *Chlamydomonas reinhardtii* and *Synechocystis* sp. PCC 6803 support both branches may be asymmetrically involved in electron transfer. There exist two models of initial charge separation. Initiating charge separation of PS I has historically been considered as the primary asymmetrical chlorophyll pair P700 although the details of the process remain obscure. It is generally assumed that the accessory chlorophyll A_A (or A_B) plays a role as a transient electron transfer intermediate. Recently, a new model was proposed in which the initial charge separation occurs between A_A and A_{0A} (or A_B and A_{0B}), resulting in the primary radical pair $A_A^+ A_{0A}^-$ (or $A_B^+ A_{0B}^-$). P700 quickly donates an electron to A_A^+ (or A_B^+), thereby initiating the first step in the stabilization of the charge—separated states and resulting in the first readily observable state, $P700^+ A_{0A}^-$ (or $P700^+ A_{0B}^-$). The amino acids of PsaA and PsaB coordinating accessory chlorophylls have been changed (unpublished data), and characterization of these mutants will be of important to understand exact process of the initial charge separation. Two of the chlorophyll *a* molecules (aC40A and aC39B) are special as they structurally and perhaps functionally connect eC2A/eC3B and eC2B/eC3A of the electron transfer chain to the antenna. It will be interesting to investigate the exact functions of these two chlorophylls as well.

The most studies of electron transfer and spectroscopic properties of cofactors were carried out using isolated PS I particles under room or low temperature conditions. Exact energy and electron transfer processes under physiological conditions remains large unknown, especially in nature, where the cyanobacteria experience much greater fluctuations in the environmental and nutritional conditions than what could be tested in the laboratory. In addition, mechanisms of assembly, degradation and regulation of PS I complex under different conditions are not clear.

Acknowledgements The authors thank the support (NSF(2010)-PFUND-217 and LEQSF(2013-16)-RD-A-15) from US National Science Foundation's EPSCoR Program and Louisiana RCS Program.

References

Adman, E.T., Sieker, L.C. and Jensen, L.H. (1973). Structure of a bacterial ferredoxin. J Biol Chem 248, 3987–96.

Akiyama, M., Miyashita, H., Kise, H., Watanabe, T., Mimuro, M., Miyachi, S. and Kobayashi, M. (2002). Quest for minor but key chlorophyll molecules in photosynthetic reaction centers – unusual pigment composition in the reaction centers of the chlorophyll d-dominated cyanobacterium Acaryochloris marina. Photosynth Res 74, 97–107.

Ali, K., Santabarbara, S., Heathcote, P., Evans, M.C. and Purton, S. (2006). Bidirectional electron transfer in photosystem I: replacement of the symmetry-breaking tryptophan close to the PsaB-bound phylloquinone A1B with a glycine residue alters the redox properties of A1B and blocks forward electron transfer at cryogenic temperatures. Biochim Biophys Acta 1757, 1623–33.

Allen, J.F. (2005). A redox switch hypothesis for the origin of two light reactions in photosynthesis. FEBS Lett 579, 963–8.

Amunts, A., Drory, O. and Nelson, N. (2007). The structure of a plant photosystem I supercomplex at 3.4 A resolution. Nature 447, 58–63.

Anandan, S., Vainstein, A. and Thornber, J.P. (1989). Correlation of some published amino acid sequences for photosystem I polypeptides to a 17 kDa LHCI pigment-protein and to subunits III and IV of the core complex. FEBS Lett 256, 150–4.

Anderson, S.L. and McIntosh, L. (1991). Light-activated heterotrophic growth of the cyanobacterium Synechocystis sp. strain PCC 6803: a blue-light-requiring process. J Bacteriol 173, 2761–7.

Arlt, T., Schmidt, S., Kaiser, W., Lauterwasser, C., Meyer, M., Scheer, H. and Zinth, W. (1993). The accessory bacteriochlorophyll: a real electron carrier in primary photosynthesis. Proc Natl Acad Sci U S A 90, 11757–61.

Atsumi, S., Higashide, W. and Liao, J.C. (2009). Direct photosynthetic recycling of carbon dioxide to isobutyraldehyde. Nat Biotechnol 27, 1177–80.

Barth, P., Lagoutte, B. and Setif, P. (1998). Ferredoxin reduction by photosystem I from Synechocystis sp. PCC 6803: toward an understanding of the respective roles of subunits PsaD and PsaE in ferredoxin binding. Biochemistry 37, 16233–41.

Barth, P., Guillouard, I., Setif, P. and Lagoutte, B. (2000). Essential role of a single arginine of photosystem I in stabilizing the electron transfer complex with ferredoxin. J Biol Chem 275, 7030–6.

Barth, P., Savarin, P., Gilquin, B., Lagoutte, B. and Ochsenbein, F. (2002). Solution NMR structure and backbone dynamics of the PsaE subunit of photosystem I from Synechocystis sp. PCC 6803. Biochemistry 41, 13902–14.

Bautista, J.A., Rappaport, F., Guergova-Kuras, M., Cohen, R.O., Golbeck, J.H., Wang, J.Y., Beal, D. and Diner, B.A. (2005). Biochemical and biophysical characterization of photosystem I from phytoene desaturase and zeta-carotene desaturase deletion mutants of Synechocystis Sp. PCC 6803: evidence for PsaA- and PsaB-side electron transport in cyanobacteria. J Biol Chem 280, 20030–41.

Baymann, F., Brugna, M., Muhlenhoff, U. and Nitschke, W. (2001). Daddy, where did (PS)I come from? Biochim Biophys Acta 1507, 291–310.

Ben-Shem, A., Frolow, F. and Nelson, N. (2003). Crystal structure of plant photosystem I. Nature 426, 630–5.

Ben-Shem, A., Frolow, F. and Nelson, N. (2004). Evolution of photosystem I – from symmetry through pseudo-symmetry to asymmetry. FEBS Lett 564, 274–80.

Berthold, T. et al. (2012). Exploring the electron transfer pathways in photosystem I by high-time-resolution electron paramagnetic resonance: observation of the B-side radical pair P700(+)A1B(-) in whole cells of the deuterated green alga Chlamydomonas reinhardtii at cryogenic temperatures. J Am Chem Soc 134, 5563–76.

Boudreaux, B. et al. (2001). Mutations in both sides of the photosystem I reaction center identify the phylloquinone observed by electron paramagnetic resonance spectroscopy. J Biol Chem 276, 37299–306.

Brecht, M., Radics, V., Nieder, J.B., Studier, H. and Bittl, R. (2008). Red antenna states of photosystem I from Synechocystis PCC 6803. Biochemistry 47, 5536–43.

Breton, J., Xu, W., Diner, B.A. and Chitnis, P.R. (2002). The two histidine axial ligands of the primary electron donor chlorophylls (P700) in photosystem I are similarly perturbed upon P700+ formation. Biochemistry 41, 11200–10.

Breton, J., Chitnis, P.R. and Pantelidou, M. (2005). Evidence for hydrogen bond formation to the PsaB chlorophyll of P700 in photosystem I mutants of Synechocystis sp. PCC 6803. Biochemistry 44, 5402–8.

Brettel, K. (1997). Electron transfer and arrangement of the redox cofactors in photosystem I. Biochimica et Biophysica Acta (BBA) – Bioenergetics 1318, 322-373.

Burja, A.M., Banaigs, B., Abou-Mansour, E., Grant Burgess, J. and Wright, P.C. (2001). Marine cyanobacteria—a prolific source of natural products. Tetrahedron 57, 9347–9377.

Busch, A. and Hippler, M. (2011). The structure and function of eukaryotic photosystem I. Biochim Biophys Acta 1807, 864-77.

Buttner, M., Xie, D.L., Nelson, H., Pinther, W., Hauska, G. and Nelson, N. (1992). Photosynthetic reaction center genes in green sulfur bacteria and in photosystem 1 are related. Proc Natl Acad Sci U S A 89, 8135–9.

Byrdin, M., Jordan, P., Krauss, N., Fromme, P., Stehlik, D. and Schlodder, E. (2002). Light harvesting in photosystem I: modeling based on the 2.5-A structure of photosystem I from Synechococcus elongatus. Biophys J 83, 433–57.

Byrdin, M., Santabarbara, S., Gu, F., Fairclough, W.V., Heathcote, P., Redding, K. and Rappaport, F. (2006). Assignment of a kinetic component to electron transfer between iron-sulfur clusters F(X) and F(A/B) of Photosystem I. Biochim Biophys Acta 1757, 1529-38.

Chen, J., Bender, S.L., Keough, J.M. and Barry, B.A. (2009). Tryptophan as a probe of photosystem I electron transfer reactions: a UV resonance Raman study. J Phys Chem B 113, 11367-70.

Chitnis, P.R. (2001). PHOTOSYSTEM I: Function and Physiology. Annu Rev Plant Physiol Plant Mol Biol 52, 593–626.

Chitnis, V.P. and Chitnis, P.R. (1993). PsaL subunit is required for the formation of photosystem I trimers in the cyanobacterium Synechocystis sp. PCC 6803. FEBS Lett 336, 330-4.

Chitnis, P.R., Reilly, P.A. and Nelson, N. (1989). Insertional inactivation of the gene encoding subunit II of photosystem I from the cyanobacterium Synechocystis sp. PCC 6803. J Biol Chem 264, 18381–5.

Chitnis, P.R., Purvis, D. and Nelson, N. (1991). Molecular cloning and targeted mutagenesis of the gene psaF encoding subunit III of photosystem I from the cyanobacterium Synechocystis sp. PCC 6803. J Biol Chem 266, 20146-51.

Chitnis, V.P., Jungs, Y.S., Albee, L., Golbeck, J.H. and Chitnis, P.R. (1996). Mutational analysis of photosystem I polypeptides. Role of PsaD and the lysyl 106 residue in the reductase activity of the photosystem I. J Biol Chem 271, 11772–80.

Chitnis, V.P., Ke, A. and Chitnis, P.R. (1997). The PsaD subunit of photosystem I. Mutations in the basic domain reduce the level of PsaD in the membranes. Plant Physiol 115, 1699–705.

Clarke, A.K. and Campbell, D. (1996). Inactivation of the petE gene for plastocyanin lowers photosynthetic capacity and exacerbates chilling-induced photoinhibition in the cyanobacterium Synechococcus. Plant Physiol 112, 1551-61.

Cohen, R.O. et al. (2004). Evidence for asymmetric electron transfer in cyanobacterial photosystem I: analysis of a methionine-to-leucine mutation of the ligand to the primary electron acceptor A0. Biochemistry 43, 4741–54.

Corbett, T.H. et al. (1996). Preclinical anticancer activity of cryptophycin-8. J Exp Ther Oncol 1, 95–108.

Dashdorj, N., Xu, W., Martinsson, P., Chitnis, P.R. and Savikhin, S. (2004). Electrochromic shift of chlorophyll absorption in photosystem I from Synechocystis sp. PCC 6803: a probe of optical and dielectric properties around the secondary electron acceptor. Biophys J 86, 3121-30.

Dashdorj, N., Xu, W., Cohen, R.O., Golbeck, J.H. and Savikhin, S. (2005). Asymmetric electron transfer in cyanobacterial Photosystem I: charge separation and secondary electron transfer dynamics of mutations near the primary electron acceptor A0. Biophys J 88, 1238–49.

De Marais, D.J. (2000). Evolution. When did photosynthesis emerge on Earth? Science 289, 1703–5.

Deisenhofer, J., Epp, O., Miki, K., Huber, R. and Michel, H. (1984). X-ray structure analysis of a membrane protein complex. Electron density map at 3 A resolution and a model of the chromophores of the photosynthetic reaction center from Rhodopseudomonas viridis. J Mol Biol 180, 385–98.

Deisenhofer, J., Epp, O., Sinning, I. and Michel, H. (1995). Crystallographic refinement at 2.3 A resolution and refined model of the photosynthetic reaction centre from Rhodopseudomonas viridis. J Mol Biol 246, 429–57.

Deng, M.D. and Coleman, J.R. (1999). Ethanol synthesis by genetic engineering in cyanobacteria. Appl Environ Microbiol 65, 523–8.

Dismukes, G.C., Klimov, V.V., Baranov, S.V., Kozlov, Y.N., DasGupta, J. and Tyryshkin, A. (2001). The origin of atmospheric oxygen on Earth: the innovation of oxygenic photosynthesis. Proc Natl Acad Sci U S A 98, 2170–5.

Domonkos, I. et al. (2004). Phosphatidylglycerol is essential for oligomerization of photosystem I reaction center. Plant Physiol 134, 1471–8.

Domonkos, I., Kis, M., Gombos, Z. and Ughy, B. (2013). Carotenoids, versatile components of oxygenic photosynthesis. Prog Lipid Res 52, 539–61.

Ducat, D.C., Sachdeva, G. and Silver, P.A. (2012). Rewiring hydrogenase-dependent redox circuits in cyanobacteria. Proc Natl Acad Sci U S A 108, 3941–6.

Duhring, U., Ossenbuhl, F. and Wilde, A. (2007). Late assembly steps and dynamics of the cyanobacterial photosystem I. J Biol Chem 282, 10915–21.

Dunn, P.P., Packman, L.C., Pappin, D. and Gray, J.C. (1988). N-terminal amino acid sequence analysis of the subunits of pea photosystem I. FEBS Lett 228, 157–61.

Duran, R.V., Hervas, M., De La Rosa, M.A. and Navarro, J.A. (2004). The efficient functioning of photosynthesis and respiration in Synechocystis sp. PCC 6803 strictly requires the presence of either cytochrome c6 or plastocyanin. J Biol Chem 279, 7229–33.

Duran, R.V., Hervas, M., De la Rosa, M.A. and Navarro, J.A. (2005). In vivo photosystem I reduction in thermophilic and mesophilic cyanobacteria: the thermal resistance of the process is limited by factors other than the unfolding of the partners. Biochem Biophys Res Commun 334, 170–5.

Duran, R.V., Hervas, M., De la Cerda, B., De la Rosa, M.A. and Navarro, J.A. (2006). A laser flash-induced kinetic analysis of in vivo photosystem I reduction by site-directed mutants of plastocyanin and cytochrome c6 in Synechocystis sp. PCC 6803. Biochemistry 45, 1054–60.

Dutta, D., De, D., Chaudhuri, S. and Bhattacharya, S.K. (2005). Hydrogen production by Cyanobacteria. Microb Cell Fact 4, 36.

Eckert, C. et al. (2014). Ethylene-forming enzyme and bioethylene production. Biotechnology for Biofuels 7, 11.

Evans, M.C., Reeves, S.G. and Cammack, R. (1974). Determination of the oxidation-reduction potential of the bound iron-sulphur proteins of the primary electron acceptor complex of photosystem I in spinach chloroplasts. FEBS Lett 49, 111–4.

Fairclough, W.V., Forsyth, A., Evans, M.C., Rigby, S.E., Purton, S. and Heathcote, P. (2003). Bidirectional electron transfer in photosystem I: electron transfer on the PsaA side is not essential for phototrophic growth in Chlamydomonas. Biochim Biophys Acta 1606, 43–55.

Falzone, C.J., Kao, Y.H., Zhao, J., MacLaughlin, K.L., Bryant, D.A. and Lecomte, J.T. (1994). 1H and 15N NMR assignments of PsaE, a photosystem I subunit from the cyanobacterium Synechococcus sp. strain PCC 7002. Biochemistry 33, 6043–51.

Farah, J., Rappaport, F., Choquet, Y., Joliot, P. and Rochaix, J.D. (1995). Isolation of a psaF-deficient mutant of Chlamydomonas reinhardtii: efficient interaction of plastocyanin with the photosystem I reaction center is mediated by the PsaF subunit. EMBO J 14, 4976–84.

Ferreira, K.N., Iverson, T.M., Maghlaoui, K., Barber, J. and Iwata, S. (2004). Architecture of the photosynthetic oxygen-evolving center. Science 303, 1831–8.

Finazzi, G., Sommer, F. and Hippler, M. (2005). Release of oxidized plastocyanin from photosystem I limits electron transfer between photosystem I and cytochrome b6f complex in vivo. Proc Natl Acad Sci U S A 102, 7031–6.

Fischer, N., Boudreau, E., Hippler, M., Drepper, F., Haehnel, W. and Rochaix, J.D. (1999). A large fraction of PsaF is nonfunctional in photosystem I complexes lacking the PsaJ subunit. Biochemistry 38, 5546–52.

Fish, L.E., Kuck, U. and Bogorad, L. (1985). Two partially homologous adjacent light-inducible maize chloroplast genes encoding polypeptides of the P700 chlorophyll a-protein complex of photosystem I. J Biol Chem 260, 1413–21.

Frazao, C. et al. (1995). Ab initio determination of the crystal structure of cytochrome c6 and comparison with plastocyanin. Structure 3, 1159–69.

Fromme, P., Schubert, W.-D. and Krauß, N. (1994). Structure of Photosystem I: Suggestions on the docking sites for plastocyanin, ferredoxin and the coordination of P700. Biochimica et Biophysica Acta (BBA) – Bioenergetics 1187, 99–105.

Fromme, P., Jordan, P. and Krauss, N. (2001). Structure of photosystem I. Biochim Biophys Acta 1507, 5–31.

Fromme, P., Bottin, H., Krauss, N. and Setif, P. (2002). Crystallization and electron paramagnetic resonance characterization of the complex of photosystem I with its natural electron acceptor ferredoxin. Biophys J 83, 1760–73.

Fromme, P., Melkozernov, A., Jordan, P. and Krauss, N. (2003). Structure and function of photosystem I: interaction with its soluble electron carriers and external antenna systems. FEBS Lett 555, 40–4.

Fujimori, T., Hihara, Y. and Sonoike, K. (2005). PsaK2 subunit in photosystem I is involved in state transition under high light condition in the cyanobacterium Synechocystis sp. PCC 6803. J Biol Chem 280, 22191–7.

Ghirardi, M.L., Posewitz, M.C., Maness, P.C., Dubini, A., Yu, J. and Seibert, M. (2007). Hydrogenases and Hydrogen Photoproduction in Oxygenic Photosynthetic Organisms (*). Annu Rev Plant Biol 58, 71–91.

Gibasiewicz, K., Ramesh, V.M., Lin, S., Redding, K., Woodbury, N.W. and Webber, A.N. (2003). Excitonic interactions in wild-type and mutant PSI reaction centers. Biophys J 85, 2547–59.

Giera, W., Gibasiewicz, K., Ramesh, V.M., Lin, S. and Webber, A. (2009). Electron transfer from A to A (1) in Photosystem I from Chlamydomonas reinhardtii occurs in both the A and B branch with 25-30-ps lifetime. Phys Chem Chem Phys 11, 5186–91.

Giera, W., Ramesh, V.M., Webber, A.N., van Stokkum, I., van Grondelle, R. and Gibasiewicz, K. (2010). Effect of the P700 pre-oxidation and point mutations near A(0) on the reversibility of the primary charge separation in Photosystem I from Chlamydomonas reinhardtii. Biochim Biophys Acta 1797, 106–12.

Gobets, B., van Stokkum, I.H., Rogner, M., Kruip, J., Schlodder, E., Karapetyan, N.V., Dekker, J.P. and van Grondelle, R. (2001). Time-resolved fluorescence emission measurements of photosystem I particles of various cyanobacteria: a unified compartmental model. Biophys J 81, 407–24.

Gobets, B., van Stokkum, I.H., van Mourik, F., Dekker, J.P. and van Grondelle, R. (2003). Excitation wavelength dependence of the fluorescence kinetics in Photosystem I particles from Synechocystis PCC 6803 and Synechococcus elongatus. Biophys J 85, 3883–98.

Golbeck, J.H. (1993). The structure of photosystem I: Current Opinion in Structural Biology 1993, 3:508–514. Current Opinion in Structural Biology 3, 508–514.

Gong, X.M., Agalarov, R., Brettel, K. and Carmeli, C. (2003). Control of electron transport in photosystem I by the iron-sulfur cluster FX in response to intra- and intersubunit interactions. J Biol Chem 278, 19141–50.

Groot, M.L., Pawlowicz, N.P., van Wilderen, L.J., Breton, J., van Stokkum, I.H. and van Grondelle, R. (2005). Initial electron donor and acceptor in isolated Photosystem II reaction centers identified with femtosecond mid-IR spectroscopy. Proc Natl Acad Sci U S A 102, 13087–92.

Grotjohann, I. and Fromme, P. (2005). Structure of cyanobacterial photosystem I. Photosynth Res 85, 51–72.

Grotjohann, I., Jolley, C. and Fromme, P. (2004). Evolution of photosynthesis and oxygen evolution: Implications from the structural comparison of Photosystems I and II. Physical Chemistry Chemical Physics 6, 4743–4753.

Guergova-Kuras, M., Boudreaux, B., Joliot, A., Joliot, P. and Redding, K. (2001). Evidence for two active branches for electron transfer in photosystem I. Proc Natl Acad Sci U S A 98, 4437-42.

Guerrero, F., Carbonell, V., Cossu, M., Correddu, D. and Jones, P.R. (2012). Ethylene Synthesis and Regulated Expression of Recombinant Protein in Synechocystis sp. PCC 6803. PLoS One 7, e50470.

Guskov, A., Kern, J., Gabdulkhakov, A., Broser, M., Zouni, A. and Saenger, W. (2009). Cyanobacterial photosystem II at 2.9-A resolution and the role of quinones, lipids, channels and chloride. Nat Struct Mol Biol 16, 334–42.

Haehnel, W. et al. (1994). Electron transfer from plastocyanin to photosystem I. EMBO J 13, 1028–38.

Hagio, M., Gombos, Z., Varkonyi, Z., Masamoto, K., Sato, N., Tsuzuki, M. and Wada, H. (2000). Direct evidence for requirement of phosphatidylglycerol in photosystem II of photosynthesis. Plant Physiol 124, 795–804.

Haldrup, A., Simpson, D.J. and Scheller, H.V. (2000). Down-regulation of the PSI-F subunit of photosystem I (PSI) in Arabidopsis thaliana. The PSI-F subunit is essential for photoautotrophic growth and contributes to antenna function. J Biol Chem 275, 31211–8.

Hallahan, B.J., Purton, S., Ivison, A., Wright, D. and Evans, M.C. (1995). Analysis of the proposed Fe-SX binding region of Photosystem 1 by site directed mutation of PsaA in Chlamydomonas reinhardtii. Photosynth Res 46, 257–64.

Hanley, J., Setif, P., Bottin, H. and Lagoutte, B. (1996). Mutagenesis of photosystem I in the region of the ferredoxin cross-linking site: modifications of positively charged amino acids. Biochemistry 35, 8563–71.

Hastings, G., Hoshina, S., Webber, A.N. and Blankenship, R.E. (1995). Universality of energy and electron transfer processes in photosystem I. Biochemistry 34, 15512–22.

Helfrich, M., Schoch, S., Lempert, U., Cmiel, E. and RÜDiger, W. (1994). Chlorophyll synthetase cannot synthesize chlorophyll a'. European Journal of Biochemistry 219, 267–275.

Helfrich, M., Schoch, S., Schäfer, W., Ryberg, M. and Rüdiger, W. (1996). Absolute Configuration of Protochlorophyllide a and Substrate Specificity of NADPH−Protochlorophyllide Oxidoreductase. Journal of the American Chemical Society 118, 2606–2611.

Hervas, M., Navarro, J.A., Diaz, A., Bottin, H. and De la Rosa, M.A. (1995). Laser-flash kinetic analysis of the fast electron transfer from plastocyanin and cytochrome c6 to photosystem I. Experimental evidence on the evolution of the reaction mechanism. Biochemistry 34, 11321–6.

Hihara, Y., Kamei, A., Kanehisa, M., Kaplan, A. and Ikeuchi, M. (2001). DNA microarray analysis of cyanobacterial gene expression during acclimation to high light. Plant Cell 13, 793–806.

Hippler, M., Ratajczak, R. and Haehnel, W. (1989). Identification of the plastocyanin binding subunit of photosystem I. FEBS Letters 250, 280–284.

Hippler, M., Reichert, J., Sutter, M., Zak, E., Altschmied, L., Schroer, U., Herrmann, R.G. and Haehnel, W. (1996). The plastocyanin binding domain of photosystem I. EMBO J 15, 6374–84.

Hippler, M., Drepper, F., Farah, J. and Rochaix, J.D. (1997). Fast electron transfer from cytochrome c6 and plastocyanin to photosystem I of Chlamydomonas reinhardtii requires PsaF. Biochemistry 36, 6343–9.

Hippler, M., Drepper, F., Haehnel, W. and Rochaix, J.D. (1998). The N-terminal domain of PsaF: precise recognition site for binding and fast electron transfer from cytochrome c6 and plastocyanin to photosystem I of Chlamydomonas reinhardtii. Proc Natl Acad Sci U S A 95, 7339–44.

Hippler, M., Drepper, F., Rochaix, J.D. and Muhlenhoff, U. (1999). Insertion of the N-terminal part of PsaF from Chlamydomonas reinhardtii into photosystem I from Synechococcus elongatus enables efficient binding of algal plastocyanin and cytochrome c6. J Biol Chem 274, 4180–8.

Hippler, M., Rimbault, B. and Takahashi, Y. (2002). Photosynthetic complex assembly in Chlamydomonas reinhardtii. Protist 153, 197–220.

Holzwarth, A.R., Muller, M.G., Niklas, J. and Lubitz, W. (2006a). Ultrafast transient absorption studies on photosystem I reaction centers from Chlamydomonas reinhardtii. 2: mutations near the P700 reaction center chlorophylls provide new insight into the nature of the primary electron donor. Biophys J 90, 552–65.

Holzwarth, A.R., Muller, M.G., Reus, M., Nowaczyk, M., Sander, J. and Rogner, M. (2006b). Kinetics and mechanism of electron transfer in intact photosystem II and in the isolated reaction center: pheophytin is the primary electron acceptor. Proc Natl Acad Sci U S A 103, 6895–900.

Ikeuchi, M., Nyhus, K.J., Inoue, Y. and Pakrasi, H.B. (1991). Identities of four low-molecular-mass subunits of the photosystem I complex from Anabaena variabilis ATCC 29413. Evidence for the presence of the psaI gene product in a cyanobacterial complex. FEBS Lett 287, 5–9.

Irisarri, P., Gonnet, S. and Monza, J. (2001). Cyanobacteria in Uruguayan rice fields: diversity, nitrogen fixing ability and tolerance to herbicides and combined nitrogen. J Biotechnol 91, 95–103.

Ivashin, N. and Larsson, S. (2003). Electron transfer pathways in photosystem I reaction centers. Chemical Physics Letters 375, 383–387.

Iwaki, M. and Itoh, S. (1994). Reaction of Reconstituted Acceptor Quinone and Dynamic Equilibration of Electron Transfer in the Photosystem I Reaction Center. Plant and Cell Physiology 35, 983–993.

Jansson, S., Andersen, B. and Scheller, H.V. (1996). Nearest-neighbor analysis of higher-plant photosystem I holocomplex. Plant Physiol 112, 409–20.

Jeanjean, R., Zuther, E., Yeremenko, N., Havaux, M., Matthijs, H.C. and Hagemann, M. (2003). A photosystem 1 psaFJ-null mutant of the cyanobacterium Synechocystis PCC 6803 expresses the isiAB operon under iron replete conditions. FEBS Lett 549, 52–6.

Jeanjean, R., Latifi, A., Matthijs, H.C. and Havaux, M. (2008). The PsaE subunit of photosystem I prevents light-induced formation of reduced oxygen species in the cyanobacterium Synechocystis sp. PCC 6803. Biochim Biophys Acta 1777, 308–16.

Jensen, P.E., Gilpin, M., Knoetzel, J. and Scheller, H.V. (2000). The PSI-K subunit of photosystem I is involved in the interaction between light-harvesting complex I and the photosystem I reaction center core. J Biol Chem 275, 24701–8.

Jin, P., Sun, J. and Chitnis, P.R. (1999). Structural features and assembly of the soluble overexpressed PsaD subunit of photosystem I. Biochim Biophys Acta 1410, 7–18.

Jordan, P., Fromme, P., Witt, H.T., Klukas, O., Saenger, W. and Krauss, N. (2001). Three-dimensional structure of cyanobacterial photosystem I at 2.5 A resolution. Nature 411, 909–17.

Kalckar, H.M. (1991). 50 years of biological research--from oxidative phosphorylation to energy requiring transport regulation. Annu Rev Biochem 60, 1–37.

Kamiya, N. and Shen, J.R. (2003). Crystal structure of oxygen-evolving photosystem II from Thermosynechococcus vulcanus at 3.7-A resolution. Proc Natl Acad Sci U S A 100, 98–103.

Kaneko, T. and Tabata, S. (1997). Complete genome structure of the unicellular cyanobacterium Synechocystis sp. PCC6803. Plant Cell Physiol 38, 1171–6.

Karapetyan, N., Shubin, V. and Strasser, R. (1999). Energy exchange between the chlorophyll antennae of monomeric subunits within the Photosystem I trimeric complex of the cyanobacterium Spirulina. Photosynthesis Research 61, 291–301.

Käss, H., Fromme, P., Witt, H.T. and Lubitz, W. (2001). Orientation and Electronic Structure of the Primary Donor Radical Cation in Photosystem I: A Single Crystals EPR and ENDOR Study. The Journal of Physical Chemistry B 105, 1225–1239.

Käβ, H., Bittersmann-Weidlich, E., Andréasson, L.E., Bönigk, B. and Lubitz, W. (1995). ENDOR and ESEEM of the 15N labelled radical cations of chlorophyll a and the primary donor P700 in photosystem I. Chemical Physics 194, 419–432.

KlÃhn, S., Baumgartner, D., Pfreundt, U., Voigt, K., Schoen, V., Steglich, C. and Hess, W.R. (2014). Alkane biosynthesis genes in cyanobacteria and their transcriptional organization. Frontiers in BIOENGINEERING AND BIOTECHNOLOGY 2

Klukas, O., Schubert, W.D., Jordan, P., Krauss, N., Fromme, P., Witt, H.T. and Saenger, W. (1999). Photosystem I, an improved model of the stromal subunits PsaC, PsaD, and PsaE. J Biol Chem 274, 7351–60.

Koike, H., Ikeuchi, M., Hiyama, T. and Inoue, Y. (1989). Identification of photosystem I components from the cyanobacterium, Synechococcus vulcanus by N-terminal sequencing. FEBS Lett 253, 257–63.

Kolber, Z.S., Van Dover, C.L., Niederman, R.A. and Falkowski, P.G. (2000). Bacterial photosynthesis in surface waters of the open ocean. Nature 407, 177–9.

Kouril, R., Yeremenko, N., D'Haene, S., Yakushevska, A. E., Keegstra, W., Matthijs, H.C., Dekker, J.P. and Boekema, E.J. (2003). Photosystem I trimers from Synechocystis PCC 6803 lacking the PsaF and PsaJ subunits bind an IsiA ring of 17 units. Biochim Biophys Acta 1607, 1–4.

Krauss, N. et al. (1993). Three-dimensional structure of system I of photosynthesis at 6 Å resolution. Nature 361, 326–331.

Kruip, J., Chitnis, P.R., Lagoutte, B., Rogner, M. and Boekema, E.J. (1997). Structural organization of the major subunits in cyanobacterial photosystem 1. Localization of subunits PsaC, -D, -E, -F, and -J. J Biol Chem 272, 17061–9.

Lagoutte, B., Hanley, J. and Bottin, H. (2001). Multiple functions for the C terminus of the PsaD subunit in the cyanobacterial photosystem I complex. Plant Physiol 126, 307–16.

Lan, E.I. and Liao, J.C. (2011). Metabolic engineering of cyanobacteria for 1-butanol production from carbon dioxide. Metab Eng 13, 353–63.

Lancaster, C.R. and Michel, H. (1999). Refined crystal structures of reaction centres from Rhodopseudomonas viridis in complexes with the herbicide atrazine and two chiral atrazine derivatives also lead to a new model of the bound carotenoid. J Mol Biol 286, 883–98.

Lelong, C., Setif, P., Lagoutte, B. and Bottin, H. (1994). Identification of the amino acids involved in the functional interaction between photosystem I and ferredoxin from Synechocystis sp. PCC 6803 by chemical cross-linking. J Biol Chem 269, 10034–9.

Lelong, C., Boekema, E.J., Kruip, J., Bottin, H., Rogner, M. and Setif, P. (1996). Characterization of a redox active cross-linked complex between cyanobacterial photosystem I and soluble ferredoxin. EMBO J 15, 2160–8.

Lem, N.W. and Glick, B.R. (1985). Biotechnological uses of cyanobacteria. Biotechnol Adv 3, 195–208.

Li, N., Warren, P.V., Golbeck, J.H., Frank, G., Zuber, H. and Bryant, D.A. (1991). Polypeptide composition of the Photosystem I complex and the Photosystem I core protein from Synechococcus sp. PCC 6301. Biochim Biophys Acta 1059, 215–25.

Li, Y. et al. (2004). Mutation of the putative hydrogen-bond donor to P700 of photosystem I. Biochemistry 43, 12634–47.

Li, Y. et al. (2006). Directing electron transfer within Photosystem I by breaking H-bonds in the cofactor branches. Proc Natl Acad Sci U S A 103, 2144–9.

Lindberg, P., Park, S. and Melis, A. (2009). Engineering a platform for photosynthetic isoprene production in cyanobacteria, using Synechocystis as the model organism. Metab Eng 12, 70–9.

Liu, X., Sheng, J. and Curtiss, R., 3rd. (2011). Fatty acid production in genetically modified cyanobacteria. Proc Natl Acad Sci U S A 108, 6899–904.

Lushy, A., Verchovsky, L. and Nechushtai, R. (2002). The stable assembly of newly synthesized PsaE into the photosystem I complex occurring via the exchange mechanism is facilitated by electrostatic interactions. Biochemistry 41, 11192–9.

Mamedov, M.D., Mamedova, A.A., Chamorovsky, S.K. and Semenov, A.Y. (2001). Electrogenic reduction of the primary electron donor P700 by plastocyanin in photosystem I complexes. FEBS Lett 500, 172–6.

Mannan, R.M., He, W.Z., Metzger, S.U., Whitmarsh, J., Malkin, R. and Pakrasi, H.B. (1996). Active photosynthesis in cyanobacterial mutants with directed modifications in the ligands for two iron-sulfur clusters on the PsaC protein of photosystem I. EMBO J 15, 1826–33.

Mayer, K.L., Shen, G., Bryant, D.A., Lecomte, J.T. and Falzone, C.J. (1999). The solution structure of photosystem I accessory protein E from the cyanobacterium Nostoc sp. strain PCC 8009. Biochemistry 38, 13736–46.

Mazor, Y., Nataf, D., Toporik, H. and Nelson, N. (2014). Crystal structures of virus-like photosystem I complexes from the mesophilic cyanobacterium Synechocystis PCC 6803. eLife 3, e01496.

Mazor, Y., Borovikova, A. and Nelson, N. (2015). The structure of plant photosystem I super-complex at 2.8 A resolution. eLife 4, e07433.

McDermott, G., Prince, S.M., Freer, A.A., Hawthornthwaite-Lawless, A.M., Papiz, M.Z., Cogdell, R.J. and Isaacs, N.W. (1995). Crystal structure of an integral membrane light-harvesting complex from photosynthetic bacteria. Nature 374, 517–521.

Mekonnen, A.E., Prasanna, R. and Kaushik, B.D. (2002). Cyanobacterial N2 fixation in presence of nitrogen fertilizers. Indian J Exp Biol 40, 854–7.

Muhlenhoff, U. and Chauvat, F. (1996). Gene transfer and manipulation in the thermophilic cyanobacterium Synechococcus elongatus. Mol Gen Genet 252, 93–100.

Muhlenhoff, U., Haehnel, W., Witt, H. and Herrmann, R.G. (1993). Genes encoding eleven subunits of photosystem I from the thermophilic cyanobacterium Synechococcus sp. Gene 127, 71–8.

Muhlenhoff, U., Kruip, J., Bryant, D.A., Rogner, M., Setif, P. and Boekema, E. (1996). Characterization of a redox-active cross-linked complex between cyanobacterial photosystem I and its physiological acceptor flavodoxin. EMBO J 15, 488–97.

Muller, M.G., Niklas, J., Lubitz, W. and Holzwarth, A.R. (2003). Ultrafast transient absorption studies on Photosystem I reaction centers from Chlamydomonas reinhardtii. 1. A new interpretation of the energy trapping and early electron transfer steps in Photosystem I. Biophys J 85, 3899–922.

Muller, M.G., Slavov, C., Luthra, R., Redding, K.E. and Holzwarth, A.R. (2010). Independent initiation of primary electron transfer in the two branches of the photosystem I reaction center. Proc Natl Acad Sci U S A 107, 4123–8.

Mwaura, F., Koyo, A.O. and Zech, B. (2004). Cyanobacterial blooms and the presence of cyanotoxins in small high altitude tropical headwater reservoirs in Kenya. J Water Health 2, 49–57.

Naithani, S., Hou, J.-M. and Chitnis, P. (2000). Targeted inactivation of the psaK1, psaK2 and psaM genes encoding subunits of Photosystem I in the cyanobacterium Synechocystis sp. PCC 6803. Photosynthesis Research 63, 225–236.

Nakamura, Y., Kaneko, T. and Tabata, S. (2000). CyanoBase, the genome database for Synechocystis sp. strain PCC6803: status for the year 2000. Nucleic Acids Res 28, 72.

Naver, H., Scott, M.P., Golbeck, J.H., Moller, B.L. and Scheller, H.V. (1996). Reconstitution of barley photosystem I with modified PSI-C allows identification of domains interacting with PSI-D and PSI-A/B. J Biol Chem 271, 8996–9001.

Nelson, N. and Ben-Shem, A. (2005). The structure of photosystem I and evolution of photosynthesis. Bioessays 27, 914–22.

Nelson, N. and Yocum, C.F. (2006). Structure and function of photosystems I and II. Annu Rev Plant Biol 57, 521–65.

Nishioka, H., Kimura, A., Yamato, T., Kawatsu, T. and Kakitani, T. (2005). Interference, fluctuation, and alternation of electron tunneling in protein media. 1. Two tunneling routes in photosynthetic reaction center alternate due to thermal fluctuation of protein conformation. J Phys Chem B 109, 1978–87.

Nordling, M., Sigfridsson, K., Young, S., Lundberg, L.G. and Hansson, O. (1991). Flash-photolysis studies of the electron transfer from genetically modified spinach plastocyanin to photosystem I. FEBS Lett 291, 327–30.

Ohyama, K. et al. (1986). Chloroplast gene organization deduced from complete sequence of liverwort Marchantia polymorpha chloroplast DNA. Nature 322, 572–574.

Ozawa, S., Nield, J., Terao, A., Stauber, E.J., Hippler, M., Koike, H., Rochaix, J.D. and Takahashi, Y. (2009).

Biochemical and structural studies of the large Ycf4-photosystem I assembly complex of the green alga Chlamydomonas reinhardtii. Plant Cell 21, 2424–42.

Palsson, L.O., Flemming, C., Gobets, B., van Grondelle, R., Dekker, J.P. and Schlodder, E. (1998). Energy transfer and charge separation in photosystem I: P700 oxidation upon selective excitation of the long-wavelength antenna chlorophylls of Synechococcus elongatus. Biophys J 74, 2611–22.

Pantelidou, M., Chitnis, P.R. and Breton, J. (2004). FTIR spectroscopy of synechocystis 6803 mutants affected on the hydrogen bonds to the carbonyl groups of the PsaA chlorophyll of P700 supports an extensive delocalization of the charge in P700+. Biochemistry 43, 8380–90.

Parrett, K.G., Mehari, T., Warren, P.G. and Golbeck, J.H. (1989). Purification and properties of the intact P-700 and Fx-containing Photosystem I core protein. Biochim Biophys Acta 973, 324–32.

Paumann, M., Regelsberger, G., Obinger, C. and Peschek, G.A. (2005). The bioenergetic role of dioxygen and the terminal oxidase(s) in cyanobacteria. Biochim Biophys Acta 1707, 231–53.

Plato, M., Krauß, N., Fromme, P. and Lubitz, W. (2003). Molecular orbital study of the primary electron donor P700 of photosystem I based on a recent X-ray single crystal structure analysis. Chemical Physics 294, 483–499.

Prokhorenko, V.I. and Holzwarth, A.R. (2000). Primary Processes and Structure of the Photosystem II Reaction Center: A Photon Echo Study†,‡. The Journal of Physical Chemistry B 104, 11563–11578.

Purton, S., Stevens, D.R., Muhiuddin, I.P., Evans, M.C., Carter, S., Rigby, S.E. and Heathcote, P. (2001). Site-directed mutagenesis of PsaA residue W693 affects phylloquinone binding and function in the photosystem I reaction center of Chlamydomonas reinhardtii. Biochemistry 40, 2167–75.

Qin, X., Suga, M., Kuang, T. and Shen, J.R. (2015). Photosynthesis. Structural basis for energy transfer pathways in the plant PSI-LHCI supercomplex. Science 348, 989–95.

Ramesh, V.M., Gibasiewicz, K., Lin, S., Bingham, S.E. and Webber, A.N. (2004). Bidirectional electron transfer in photosystem I: accumulation of A0- in A-side or B-side mutants of the axial ligand to chlorophyll A0. Biochemistry 43, 1369–75.

Ramesh, V.M., Gibasiewicz, K., Lin, S., Bingham, S.E. and Webber, A.N. (2007). Replacement of the methionine axial ligand to the primary electron acceptor A0 slows the A0- reoxidation dynamics in photosystem I. Biochim Biophys Acta 1767, 151–60.

Rupprecht, J., Hankamer, B., Mussgnug, J.H., Ananyev, G., Dismukes, C. and Kruse, O. (2006). Perspectives and advances of biological H2 production in microorganisms. Appl Microbiol Biotechnol 72, 442–9.

Rye, R. and Holland, H.D. (1998). Paleosols and the evolution of atmospheric oxygen: a critical review. Am J Sci 298, 621–72.

Santabarbara, S., Kuprov, I., Fairclough, W.V., Purton, S., Hore, P.J., Heathcote, P. and Evans, M.C. (2005). Bidirectional electron transfer in photosystem I: determination of two distances between P700+ and A1- in spin-correlated radical pairs. Biochemistry 44, 2119–28.

Santabarbara, S., Kuprov, I., Poluektov, O., Casal, A., Russell, C.A., Purton, S. and Evans, M.C.W. (2010). Directionality of Electron-Transfer Reactions in Photosystem I of Prokaryotes: Universality of the Bidirectional Electron-Transfer Model. The Journal of Physical Chemistry B 114, 15158–15171.

Santabarbara, S. et al. (2013). The requirement for carotenoids in the assembly and function of the photosynthetic complexes in Chlamydomonas reinhardtii. Plant Physiol 161, 535–46.

Saraste, M. (1999). Oxidative phosphorylation at the fin de siecle. Science 283, 1488–93.

Savikhin, S., Xu, W., Soukoulis, V., Chitnis, P.R. and Struve, W.S. (1999). Ultrafast primary processes in photosystem I of the cyanobacterium Synechocystis sp. PCC 6803. Biophysical journal 76, 3278–88.

Savikhin, S., Xu, W., Chitnis, P.R. and Struve, W.S. (2000). Ultrafast primary processes in PS I from Synechocystis sp. PCC 6803: roles of P700 and A(0). Biophys J 79, 1573–86.

Savikhin, S., Xu, W., Martinsson, P., Chitnis, P.R. and Struve, W.S. (2001). Kinetics of charge separation and A0- --> A1 electron transfer in photosystem I reaction centers. Biochemistry 40, 9282–90.

Savitsky, A., Gopta, O., Mamedov, M., Golbeck, J., Tikhonov, A., Möbius, K. and Semenov, A. (2010). Alteration of the Axial Met Ligand to Electron Acceptor A0 in Photosystem I: Effect on the Generation of P 700 ·+ A 1 ·− Radical Pairs as Studied by W-band Transient EPR. Applied Magnetic Resonance 37, 85–102.

Scheller, H.V., Jensen, P.E., Haldrup, A., Lunde, C. and Knoetzel, J. (2001). Role of subunits in eukaryotic Photosystem I. Biochimica et Biophysica Acta (BBA) – Bioenergetics 1507, 41–60.

Schirmer, A., Rude, M.A., Li, X., Popova, E. and del Cardayre, S.B. (2010). Microbial biosynthesis of alkanes. Science 329, 559–62.

Schluchter, W.M., Shen, G., Zhao, J. and Bryant, D.A. (1996). Characterization of psaI and psaL mutants of Synechococcus sp. strain PCC 7002: a new model for state transitions in cyanobacteria. Photochem Photobiol 64, 53–66.

Sener, M.K., Park, S., Lu, D., Damjanovic, A., Ritz, T., Fromme, P. and Schulten, K. (2004). Excitation migration in trimeric cyanobacterial photosystem I. J Chem Phys 120, 11183–95.

Setif, P. (2001). Ferredoxin and flavodoxin reduction by photosystem I. Biochim Biophys Acta 1507, 161–79.

Setif, P.Q. and Bottin, H. (1994). Laser flash absorption spectroscopy study of ferredoxin reduction by photosystem I in Synechocystis sp. PCC 6803: evidence for submicrosecond and microsecond kinetics. Biochemistry 33, 8495–504.

Setif, P.Q. and Bottin, H. (1995). Laser flash absorption spectroscopy study of ferredoxin reduction by photosystem I: spectral and kinetic evidence for the existence of several photosystem I-ferredoxin complexes. Biochemistry 34, 9059–70.

Setif, P., Fischer, N., Lagoutte, B., Bottin, H. and Rochaix, J.D. (2002). The ferredoxin docking site of photosystem I. Biochim Biophys Acta 1555, 204–9.

Shelaev, I.V., Gostev, F.E., Mamedov, M.D., Sarkisov, O.M., Nadtochenko, V.A., Shuvalov, V.A. and Semenov, A.Y. (2010). Femtosecond primary charge separation in Synechocystis sp. PCC 6803 photosystem I. Biochimica et Biophysica Acta (BBA) – Bioenergetics 1797, 1410–1420.

Shen, G., Boussiba, S. and Vermaas, W.F. (1993). Synechocystis sp PCC 6803 strains lacking photosystem I and phycobilisome function. Plant Cell 5, 1853–63.

Shen, G. et al. (2002). Assembly of photosystem I. II. Rubredoxin is required for the in vivo assembly of F(X) in Synechococcus sp. PCC 7002 as shown by optical and EPR spectroscopy. J Biol Chem 277, 20355–66.

Shih, C. et al. (2008). Tryptophan-accelerated electron flow through proteins. Science 320, 1760–2.

Singh, S., Kate, B.N. and Banerjee, U.C. (2005). Bioactive compounds from cyanobacteria and microalgae: an overview. Crit Rev Biotechnol 25, 73–95.

Sommer, F., Drepper, F. and Hippler, M. (2002). The luminal helix l of PsaB is essential for recognition of plastocyanin or cytochrome c6 and fast electron transfer to photosystem I in Chlamydomonas reinhardtii. J Biol Chem 277, 6573–81.

Sommer, F., Drepper, F., Haehnel, W. and Hippler, M. (2004). The hydrophobic recognition site formed by residues PsaA-Trp651 and PsaB-Trp627 of photosystem I in Chlamydomonas reinhardtii confers distinct selectivity for binding of plastocyanin and cytochrome c6. J Biol Chem 279, 20009–17.

Soukoulis, V., Savikhin, S., Xu, W., Chitnis, P.R. and Struve, W.S. (1999). Electronic spectra of PS I mutants: the peripheral subunits do not bind red chlorophylls in Synechocystis sp. PCC 6803. Biophys J 76, 2711–5.

Srinivasan, N. and Golbeck, J.H. (2009a). Protein–cofactor interactions in bioenergetic complexes: The role of the A1A and A1B phylloquinones in Photosystem I. Biochimica et Biophysica Acta (BBA) – Bioenergetics 1787, 1057–1088.

Srinivasan, N. and Golbeck, J.H. (2009b). Protein-cofactor interactions in bioenergetic complexes: The role of the A(1A) and A(1B) phylloquinones in Photosystem I. Biochim et Biophys Acta-Bioenergetics 1787, 1057–1088.

Srinivasan, N., Karyagina, I., Bittl, R., van der Est, A. and Golbeck, J.H. (2009). Role of the Hydrogen Bond from Leu722 to the A1A Phylloquinone in Photosystem I†‡. Biochemistry 48, 3315–3324.

Stamatakis, K., Tsimilli-Michael, M. and Papageorgiou, G.C. (2014). On the question of the light-harvesting role of beta-carotene in photosystem II and photosystem I core complexes. Plant Physiol Biochem 81, 121–7.

Stockel, J., Bennewitz, S., Hein, P. and Oelmuller, R. (2006). The evolutionarily conserved tetratrico peptide repeat protein pale yellow green7 is required for photosystem I accumulation in Arabidopsis and copurifies with the complex. Plant Physiol 141, 870–8.

Sun, J., Xu, W., Hervas, M., Navarro, J.A., Rosa, M.A. and Chitnis, P.R. (1999). Oxidizing side of the cyanobacterial photosystem I. Evidence for interaction between the electron donor proteins and a luminal surface helix of the PsaB subunit. J Biol Chem 274, 19048–54.

Sun, J. et al. (2014). Evidence that histidine forms a coordination bond to the A(0A) and A(0B) chlorophylls and a second H-bond to the A(1A) and A(1B) phylloquinones in M688H(PsaA) and M668H(PsaB) variants of Synechocystis sp. PCC 6803. Biochim Biophys Acta 1837, 1362–75.

Takahashi, Y., Goldschmidt-Clermont, M., Soen, S.Y., Franzen, L.G. and Rochaix, J.D. (1991). Directed chloroplast transformation in Chlamydomonas reinhardtii: insertional inactivation of the psaC gene encoding the iron sulfur protein destabilizes photosystem I. EMBO J 10, 2033–40.

Tan, L.T. (2007). Bioactive natural products from marine cyanobacteria for drug discovery. Phytochemistry 68, 954–79.

van Brederode, M.E. and van Grondelle, R. (1999). New and unexpected routes for ultrafast electron transfer in photosynthetic reaction centers. FEBS Lett 455, 1–7.

van der Est, A., Chirico, S., Karyagina, I., Cohen, R., Shen, G. and Golbeck, J. (2010). Alteration of the Axial Met Ligand to Electron Acceptor A0 in Photosystem I: An Investigation of Electron Transfer at Different Temperatures by Multifrequency Time-Resolved and CW EPR. Applied Magnetic Resonance 37, 103–121.

van Thor, J.J., Geerlings, T.H., Matthijs, H.C. and Hellingwerf, K.J. (1999). Kinetic evidence for the PsaE-dependent transient ternary complex photosystem I/Ferredoxin/Ferredoxin:NADP(+) reductase in a cyanobacterium. Biochemistry 38, 12735–46.

Tsukihira, T. et al. (1981). X-ray analysis of a [2Fe-2S] ferrodoxin from Spirulina platensis. Main chain fold and location of side chains at 2.5 A resolution. J Biochem 90, 1763–73.

Tsygankov, A.A., Fedorov, A.S., Kosourov, S.N. and Rao, K.K. (2002). Hydrogen production by cyanobacteria in an automated outdoor photobioreactor under aerobic conditions. Biotechnol Bioeng 80, 777–83.

Umena, Y., Kawakami, K., Shen, J.R. and Kamiya, N. (2011). Crystal structure of oxygen-evolving photosystem II at a resolution of 1.9 A. Nature 473, 55–60.

Ungerer, J., Tao, L., Davis, M., Ghirardi, M., Maness, P.-C. and Yu, J. (2012). Sustained photosynthetic conversion of CO_2 to ethylene in recombinant cyanobacterium Synechocystis 6803. Energy & Environmental Science 5, 8998–9006.

Varotto, C., Pesaresi, P., Jahns, P., Lessnick, A., Tizzano, M., Schiavon, F., Salamini, F. and Leister, D. (2002). Single and double knockouts of the genes for photosystem I subunits G, K, and H of Arabidopsis. Effects on photosystem I composition, photosynthetic electron flow, and state transitions. Plant Physiol 129, 616–24.

Vassiliev, I.R., Jung, Y.S., Smart, L.B., Schulz, R., McIntosh, L. and Golbeck, J.H. (1995). A mixed-ligand iron-sulfur cluster (C556SPaB or C565SPsaB) in the Fx-binding site leads to a decreased quantum efficiency of electron transfer in photosystem I. Biophys J 69, 1544–53.

Vermaas, W. (1996). Molecular genetics of the cyanobacteriumSynechocystis sp. PCC 6803: Principles and possible biotechnology applications. Journal of Applied Phycology 8, 263–273.

Wang, R., Sivakumar, V., Li, Y., Redding, K. and Hastings, G. (2003). Mutation induced modulation of hydrogen bonding to P700 studied using FTIR difference spectroscopy. Biochemistry 42, 9889–97.

Wang, B., Wang, J., Zhang, W. and Meldrum, D.R. (2012). Application of synthetic biology in cyanobacteria and algae. Front Microbiol 3, 344.

Webber, A.N. and Lubitz, W. (2001). P700: the primary electron donor of photosystem I. Biochim Biophys Acta 1507, 61–79.

Webber, A.N., Gibbs, P.B., Ward, J.B. and Bingham, S.E. (1993). Site-directed mutagenesis of the photosystem I reaction center in chloroplasts. The proline-cysteine motif. J Biol Chem 268, 12990–5.

Weigel, M., Varotto, C., Pesaresi, P., Finazzi, G., Rappaport, F., Salamini, F. and Leister, D. (2003). Plastocyanin is indispensable for photosynthetic electron flow in Arabidopsis thaliana. J Biol Chem 278, 31286–9.

White, N.T.H., Beddard, G.S., Thorne, J.R.G., Feehan, T.M., Keyes, T.E. and Heathcote, P. (1996). Primary Charge Separation and Energy Transfer in the Photosystem I Reaction Center of Higher Plants. The Journal of Physical Chemistry 100, 12086–12099.

Witt, H. et al. (2002). Hydrogen bonding to P700: site-directed mutagenesis of threonine A739 of photosystem I in Chlamydomonas reinhardtii. Biochemistry 41, 8557–69.

Wynn, R.M. and Malkin, R. (1988). Interaction of plastocyanin with photosystem I: a chemical cross-linking study of the polypeptide that binds plastocyanin. Biochemistry 27, 5863–9.

Wynn, R.M., Luong, C. and Malkin, R. (1989). Maize Photosystem I : Identification of the Subunit which Binds Plastocyanin. Plant Physiol 91, 445–9.

Xia, Z., Broadhurst, R.W., Laue, E.D., Bryant, D.A., Golbeck, J.H. and Bendall, D.S. (1998). Structure and properties in solution of PsaD, an extrinsic polypeptide of photosystem I. Eur J Biochem 255, 309–16.

Xu, Q., Armbrust, T.S., Guikema, J.A. and Chitnis, P.R. (1994a). Organization of Photosystem I Polypeptides (A Structural Interaction between the PsaD and PsaL Subunits). Plant Physiol 106, 1057–1063.

Xu, Q., Jung, Y.S., Chitnis, V.P., Guikema, J.A., Golbeck, J.H. and Chitnis, P.R. (1994b). Mutational analysis of photosystem I polypeptides in Synechocystis sp. PCC 6803. Subunit requirements for reduction of NADP+ mediated by ferredoxin and flavodoxin. J Biol Chem 269, 21512–8.

Xu, Q., Odom, W.R., Guikema, J.A., Chitnis, V.P. and Chitnis, P.R. (1994c). Targeted deletion of psaJ from the cyanobacterium Synechocystis sp. PCC 6803 indicates structural interactions between the PsaJ and PsaF subunits of photosystem I. Plant Mol Biol 26, 291–302.

Xu, Q., Yu, L., Chitnis, V.P. and Chitnis, P.R. (1994d). Function and organization of photosystem I in a cyanobacterial mutant strain that lacks PsaF and PsaJ subunits. J Biol Chem 269, 3205–11.

Xu, Q., Hoppe, D., Chitnis, V.P., Odom, W.R., Guikema, J.A. and Chitnis, P.R. (1995). Mutational analysis of photosystem I polypeptides in the cyanobacterium Synechocystis sp. PCC 6803. Targeted inactivation of psaI reveals the function of psaI in the structural organization of psaL. J Biol Chem 270, 16243–50.

Xu, W., Tang, H., Wang, Y. and Chitnis, P.R. (2001). Proteins of the cyanobacterial photosystem I. Biochim Biophys Acta 1507, 32–40.

Xu, W. et al. (2003a). Electron transfer in cyanobacterial photosystem I: I. Physiological and spectroscopic characterization of site-directed mutants in a putative electron transfer pathway from A0 through A1 to FX. J Biol Chem 278, 27864–75.

Xu, W. et al. (2003b). Electron transfer in cyanobacterial photosystem I: II. Determination of forward electron transfer rates of site-directed mutants in a putative electron transfer pathway from A0 through A1 to FX. J Biol Chem 278, 27876–87.

Xu, W., Wang, Y., Taylor, E., Laujac, A., Gao, L., Savikhin, S. and Chitnis, P.R. (2011). Mutational Analysis of Photosystem I of Synechocystis sp. PCC 6803: The Role of Four Conserved Aromatic Residues in the j-helix of PsaB. PLoS ONE 6, e24625.

Yu, L., Zhao, J., Muhlenhoff, U., Bryant, D.A. and Golbeck, J.H. (1993). PsaE Is Required for in Vivo Cyclic Electron Flow around Photosystem I in the Cyanobacterium Synechococcus sp. PCC 7002. Plant Physiol 103, 171–180.

Yu, J., Vassiliev, I.R., Jung, Y.S., Golbeck, J.H. and McIntosh, L. (1997). Strains of synechocystis sp. PCC 6803 with altered PsaC. I. Mutations incorporated in the cysteine ligands of the two [4Fe-4S] clusters FA and FB of photosystem I. J Biol Chem 272, 8032–9.

Zak, E., Norling, B., Maitra, R., Huang, F., Andersson, B. and Pakrasi, H.B. (2001). The initial steps of biogenesis of cyanobacterial photosystems occur in plasma membranes. Proc Natl Acad Sci U S A 98, 13443–8.

Zhang, S. and Scheller, H.V. (2004). Light-harvesting complex II binds to several small subunits of photosystem I. J Biol Chem 279, 3180–7.

Zhang, L., McSpadden, B., Pakrasi, H.B. and Whitmarsh, J. (1992). Copper-mediated regulation of cytochrome c553 and plastocyanin in the cyanobacterium Synechocystis 6803. J Biol Chem 267, 19054–9.

Zhang, L., Selao, T.T., Pisareva, T., Qian, J., Sze, S.K., Carlberg, I. and Norling, B. (2013). Deletion of Synechocystis sp. PCC 6803 leader peptidase LepB1 affects photosynthetic complexes and respiration. Mol Cell Proteomics 12, 1192–203.

Zhao, J., Snyder, W.B., Muhlenhoff, U., Rhiel, E., Warren, P.V., Golbeck, J.H. and Bryant, D.A. (1993). Cloning and characterization of the psaE gene of the cyanobacterium Synechococcus sp. PCC 7002: characterization of a psaE mutant and overproduction of the protein in Escherichia coli. Mol Microbiol 9, 183–94.

Zhu, J., van Stokkum, I.H., Paparelli, L., Jones, M.R. and Groot, M.L. (2013). Early bacteriopheophytin reduction in charge separation in reaction centers of Rhodobacter sphaeroides. Biophys J 104, 2493–502.

Zinth, W. and Wachtveitl, J. (2005). The first picoseconds in bacterial photosynthesis--ultrafast electron transfer for the efficient conversion of light energy. Chemphyschem 6, 871–80.

Zouni, A., Jordan, R., Schlodder, E., Fromme, P. and Witt, H.T. (2000). First photosystem II crystals capable of water oxidation. Biochim Biophys Acta 1457, 103–5.

Zouni, A., Witt, H.T., Kern, J., Fromme, P., Krauss, N., Saenger, W. and Orth, P. (2001). Crystal structure of photosystem II from Synechococcus elongatus at 3.8 A resolution. Nature 409, 739–43.

Zygadlo, A., Jensen, P.E., Leister, D. and Scheller, H.V. (2005). Photosystem I lacking the PSI-G subunit has a higher affinity for plastocyanin and is sensitive to photodamage. Biochim Biophys Acta 1708, 154–63.

How Light-Harvesting and Energy-Transfer Processes Are Modified Under Different Light Conditions: STUDIES by Time-Resolved Fluorescence Spectroscopy

Seiji Akimoto and Makio Yokono

Summary

Photosynthetic organisms contain specific pigment-protein complexes that absorb light energy and subsequently transfer excitation energy to the photosynthetic reaction centers. Changing the quality and/or quantity of the complexes is how light-harvesting and energy-transfer processes adapt to environments. Cyanobacteria and red algae form a unique peripheral membrane complex, phycobilisome, whereas integral membrane complexes containing specific carotenoids are found in green algae and higher plants. We examine light-harvesting and energy-transfer processes in different types of complexes by time-resolved fluorescence spectroscopy. Changes in these processes in response to different environments are also discussed.

Keywords

Antenna • Carotenoid • Chlorophyll • Energy transfer • Fluorescence • Light harvesting • Photosystem • Phycobilisome • Time-resolved spectroscopy

Contents

8.1	Introduction	169
8.2	Method	172
8.2.1	Time-Resolved Fluorescence Spectroscopy	172
8.2.2	Global Fitting Analysis	172
8.3	Modification of Energy Transfer	172
8.3.1	Excitation Energy Regulation in Phycobilisomes	172
8.3.2	Modification of Energy Transfer from Phycobilisomes in the Cyanobacterium Arthrospira platensis	175
8.3.3	Excitation Balance in Photochemical Reaction Centers in Red Algae	176
8.3.4	Light Harvesting by Carotenoids in the Higher Plant Arabidopsis thaliana and Green Algae Codium fragile	178
8.4	Conclusions	180
References		181

8.1 Introduction

Photosynthetic organisms contain specific pigment-protein complexes as peripheral or integral membrane complexes that absorb light energy and subsequently transfer excitation energy to the photosynthetic reaction centers in photosystems (PSs) (Rabinowitch and Govindjee 1969; Blankenship 2002). Besides

S. Akimoto (✉)
Graduate School of Science, Kobe University, Kobe, Japan
e-mail: akimoto@hawk.kobe-u.ac.jp

M. Yokono
Institute of Low Temperature Science, Hokkaido University, Sapporo, Japan

chlorophylls (Chls) or bacteriochlorophylls as major pigments in photosynthesis, phycobilins and/or carotenoids (Cars) are found in pigment-protein complexes. To adapt light-harvesting and energy-transfer processes to environments, organisms can modify the quality of pigment-protein complexes, the quantity of pigment-protein complexes, the interaction between pigments and protein, and/or the interaction between pigment-protein complexes.

In oxygenic photosynthetic organisms, two PSs (PSII and PSI) function in parallel, and their excitation levels must be balanced to maintain optimal photosynthesis (Minagawa 2011). Equal excitation of two PSs is needed to achieve linear electron flow from PSII to PSI, which generates both ATP and NADPH (Evans 1987). Cyclic electron flow around PSI is also essential for photosynthesis (Munekage et al. 2004). Frequent excitation of PSI is achieved during cyclic electron flow, causing ATP, which is required for photorespiration, to be preferentially synthesized (Kuvykin et al. 2011; Kozaki and Takeba 1996). Excitation balance can be achieved in several ways, which itself indicates the importance of the balance of excitation between PSI and PSII in photosynthesis. The simplest way to achieve this balance is regulation of amount of each reaction center (Cunningham et al. 1990). Another well-known mechanism is "Mobile antenna", where a light-harvesting antenna moves and preferentially binds to a reaction center that needs more excitation energy (Minagawa 2011). In addition, various quenchers including Cars and Car-Chl heterodimers can extort excitation energy from PSII (Holt et al. 2005; Yokono et al. 2008a; Cheng et al. 2008; Ruban and Duffy 2012; Sutter et al. 2013a; Tian et al. 2012), and then PSI is preferentially excited (Kuvykin et al. 2011). Direct PSII→PSI energy transfer in a process known as "spillover" has also been proposed (Murata 1969; Satoh et al. 1976). Each mechanism has its corresponding advantages and disadvantages. For example, the regulation of accumulation levels could be a fundamental solution, but requires time to synthesize the reaction centers. In addition, the metals Fe and Mn are needed to synthesize PSI and PSII, respectively (Ben-Shem et al. 2003; Umena et al. 2011). Fe and Mn supply is the factor limiting of growth rate in green and red plastid lineages, respectively (Falkowski et al. 2004; Raven 1990). The mobile antenna mechanism does not require protein synthesis, but the mobility of an antenna depends on its size; a larger antenna moves more slowly than a smaller one (Sarcina et al. 2001). Various quenchers can effectively dissipate excitation energy as heat. Under weak light conditions, excitation energy should be efficiently transferred to reaction centers instead of dissipating. The spillover mechanism can be very effective, but it is regulated by the redox state of the plastoquinone pool in photosynthetic membranes. In cyanobacteria, the redox state is affected not only by photosynthesis activity but also respiration; therefore, cyanobacteria cannot maintain excitation balance by spillover alone.

Phycobilisome (PBS) is a peripheral antenna found in cyanobacteria and red algae that contains phycoerythrin (PE), phycocyanin (PC), and allophycocyanin (APC), and absorbs light energy in the visible region instead of Chl (Gantt 1981; Mimuro and Kikuchi 2003). The light energy captured by PBS is transferred to Chl as excitation energy. In cyanobacteria, the quality and/or quantity of PBSs can be modified in response to light conditions (Ghosh and Govindjee 1966; Boardman et al. 1966; Stowe-Evans and Kehoe 2004; Yokono et al. 2008b). For example, the cyanobacterium *Anacystis nidulans* decreases its PC/Chl *a* ratio under strong orange light, and increases it under strong red light (Ghosh and Govindjee 1966). It has also been reported that *A. nidulans* can alter the extent of PBS→PSII energy transfer without affecting the lifetime of PSII fluorescence (Boardman et al. 1966). The cyanobacterium *Fremyella diplosiphon* changes its phycobiliprotein composition in response to light quality; PE, PC, and APC are synthesized during cultivation under green light, whereas only PC and APC were synthesized under red light (Stowe-Evans and Kehoe 2004). In the cyanobacterium *Gloeobacter violaceus* PCC 7421, the short-wavelength form of PC (PC_{615}) was produced during cultivation under red light without changing the energy transfer processes of PC and APC (Stowe-Evans and Kehoe 2004). Energy transfer in the PBS in the cyanobacterium *Arthrospira platensis* (*Spirulina platensis*) is modified depending on the light quality; the PBS transferred excitation energy without loss of energy in cells grown under green and far-red light, whereas rapid energy transfer was established in the PBS in cells grown under yellow light (Akimoto et al. 2012).

Cars are found in all organisms, where they play important roles in photoprotective activities including radical scavenging and singlet oxygen trapping (Mimuro and Katoh 1991; Frank and Cogdell 1993). In photosynthesis, Cars have an additional function; they absorb light energy and transfer it to nearby Chl molecules. The functions of Cars are closely related with the properties of their electronically excited states, energy level and lifetime; therefore, conjugation structures of Cars can be examined to determine the function(s) of individual Cars. With respect to conjugation structure, two kinds of Cars are found in photosynthetic organisms. One possesses conjugated polyenes ($-(C=C)_n-$) and the other contains a keto carbonyl group ($>C=O$) in the conjugated double-bond system. Among the polyene-type Cars, those containing 9–10 conjugated double bonds are expected to function as efficient antenna because of the relatively long lifetimes of their second excited state (Akimoto et al. 2000). Lutein (Lut) and spheroidene have conjugation of $n=10$ and act as an antenna pigment in higher plants (Holt et al. 2003; Akimoto et al. 2005) and photosynthetic bacterium *Rhodobacter sphaeroides* (Ricci et al. 1996), respectively. By containing longer conjugation, the polyene-type Cars exhibit a photoprotective function. In the xanthophyll cycle, violaxanthin (the short-conjugation form, $n=9$) works as an antenna pigment, whereas zeaxanthin (the long-conjugation form, $n=9+2$) behaves as a quencher of Chl (Horton et al. 1996). Keto-Cars, which have the conjugation of eight $C=C$ bonds and one $C=O$ bond, are present in specific classes of algae and function as efficient antenna pigments (Mimuro 2003; Song et al. 1976; Akimoto et al. 2004, 2007), including fucoxanthin in brown algae and diatoms, and peridinin in photosynthetic dinoflagellates. A marine green alga, *Codium fragile*, contains a specific keto-Car, siphonaxanthin (Siph), which shows a characteristic *in vivo* absorption band at 535 nm (Akimoto et al. 2004, 2007). This band is advantageous for light harvesting in green light-rich underwater conditions (Akimoto et al. 2004, 2007).

Recent developments in laser technology enable us to precisely study ultrafast processes in photosynthetic systems. Time-resolved fluorescence spectroscopy is a most useful technique because it detects signals only from excited states. By observing fluorescence signals as a function of time after short-pulse laser excitation, excited state dynamics are resolved with a time resolution of approximately 3 ps (picosecond) by the time-correlated single photon counting method (O'Connor and Phillips 1984), and 20 fs (femtosecond) by the up-conversion method (Shah 1988; Kahlow et al. 1988) after convolution calculation. One application of time-resolved fluorescence spectroscopy is observation of pigment-to-pigment energy-transfer processes. By monitoring fluorescence signals from pigments as a function of time, energy transfer is observed as a time-dependent decrease in fluorescence intensity from an energy donor, and a concomitant increase in the intensity of emission from an acceptor (Mimuro et al. 1989). Another application is observation of energy transfer between two PSs. By detecting delayed fluorescence signals after charge recombination in PSII, the PSII→PSI energy transfer process may be examined (Yokono et al. 2011, 2015). The delayed fluorescence observed for this process in the 10-ns (nanosecond) region at 77 K proceeds by the following mechanism. When the reaction center of PSII is excited, charge separation takes place. It is well known that charge recombination between the primary electron donor and primary electron acceptor occurs with a given probability. After charge recombination, an excited state is regenerated at the reaction center, and delayed fluorescence is emitted. Reported lifetimes of delayed fluorescence are in the order of 10 ns, so they are easily distinguished from prompt fluorescence with a lifetime in the order of 1 ns or less (Mimuro et al. 2007). Studies of isolated PSI revealed that its longest lifetime is less than 5 ns; therefore, PSI alone does not emit delayed fluorescence (Mimuro et al. 2010). In some cases, excitation energy regenerated by charge recombination in the PSII reaction center transfers to the PSII core antenna, which is observed as long-lived fluorescence at 685 and/or 695 nm. Furthermore, long-lived fluorescence is observed in the region of PSI fluorescence when PSII is present, indicating that a certain amount of PSII attached to PSI undergo a PSII→PSI energy transfer (Yokono et al. 2011, 2015). At 77 K, energy transfer occurs only from pigments with higher transition energy to those with lower (downhill energy transfer), *i.e.*, PSII→PSI energy transfer. In contrast, at room temperature, uphill energy transfer including PSI→PSII energy transfer may become possible. Therefore, by detecting delayed fluorescence at 77 K, the mechanisms of how two PSs share excitation energy are examined (Yokono et al. 2015).

8.2 Method

8.2.1 Time-Resolved Fluorescence Spectroscopy

Fluorescence kinetics in the picosecond to nanosecond region were measured using a time-correlated single-photon counting system at 77 K (Akimoto et al. 2012). The excitation wavelength was 400 or 425 nm; Chl *a*, Chl *b*, and Car were simultaneously excited at 425 nm, while all pigments including PBS were simultaneously excited at 400 nm. Polyethylene glycol 3350 (final concentration 15% (w/v)) was added to sample solutions to obtain homogeneous ice samples at 77 K. Fluorescence kinetics in the femtosecond to picosecond region were measured using a fluorescence up-conversion system at room temperature (Akimoto and Mimuro 2007). The excitation wavelength was 425 nm. To avoid polarization effects, the angle between the polarizations of the excitation laser and fluorescence was set to the magic angle (54.7°).

8.2.2 Global Fitting Analysis

Fluorescence rise and decay curves at different wavelengths ($F(\lambda, t)$) were fitted by sums of exponentials with common time constants as follows:

$$F(\lambda, t) = \sum_{i=1}^{n} A_i \, \exp\left(-\frac{t}{\tau_i}\right) \quad (8.1)$$

The fluorescence decay-associated spectrum (FDAS) for each time constant (τ_i) is given by plotting the amplitudes ($A_i(\lambda)$) as a function of wavelength (λ). Positive and negative values of amplitude indicate fluorescence decay and rise, respectively. For discussion on energy transfer, a pair of positive and negative amplitudes indicates energy transfer from a pigment with positive amplitude to one with negative amplitude. The FDAS with the longest τ_i (>10 ns) gives the delayed fluorescence spectrum. The PSI/PSII ratio in the delayed fluorescence spectrum is an index of how two PSs share excitation energy (Yokono et al. 2011, 2015).

8.3 Modification of Energy Transfer

8.3.1 Excitation Energy Regulation in Phycobilisomes

Cyanobacteria and red algae possess PBSs as light harvesting antennae that are located outside of the photosynthetic membranes (Gantt 1981). Phycobiliprotein discs (trimers and hexamers) are assembled with the aid of linker polypeptides into PBSs (Apt et al. 1995). The molecular architecture of a PBS depends on the linker polypeptides (Guan et al. 1996).

The simplest architecture is found in cyanobacteria. *Acaryochloris marina* has been isolated from inside ascidian (Miyashita et al.

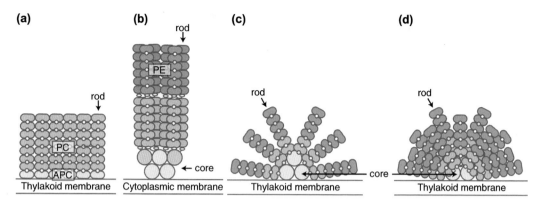

Fig. 8.1 Architectures of a (**a**) rod-shaped PBS, (**b**) bundle-like PBS, (**c**) hemidiscoidal PBS, and (**d**) hemiellipsoidal PBS. *PE* phycoerythrin, *PC* phycocyanin, *APC* allophycocyanin

1996), and possesses a simple rod-shaped PBS (Fig. 8.1a) (Chen et al. 2009) that can achieve efficient energy transfer (Theiss et al. 2011). The rod-shaped structure enables a dense distribution of light-harvesting pigments on the thylakoid membranes of *A. marina*, which might be an advantage under weak steady light conditions.

Many cyanobacteria possess hemidiscoidal PBSs (Fig. 8.1c) (Arteni et al. 2009). A hemidiscoidal PBS contains two distinct structural domains, the core and rods. The core accepts excitation energy from the rods, and then efficiently transfers this energy to Chl (Gillbro et al. 1985; Mullineaux and Holzwarth 1991). Hemidiscoidal PBSs possess lower pigment density per unit area compared with simple rod-shaped PBSs; however, the core structure is a large advantage. Many cyanobacteria accumulate the soluble Car binding protein called orange carotenoid protein (OCP), which binds to the core structure of PBS so that excess excitation energy is thermally dissipated under high light conditions (Kirilovsky 2007; Sutter et al. 2013b; Tian et al. 2013). This regulation mechanism is important for survival of these bacteria under high light conditions in shallow water (Wilson et al. 2006).

The most primordial cyanobacterium, *G. violaceus*, has a bundle-like PBS (Fig. 8.1b) (Rippka et al. 1974; Koyama et al. 2006). The bundle-like PBS also contains a core and rods. The existence of the core structure allows OCP to be used (Bernát et al. 2012). In the rod domain, unlike hemidiscoidal PBSs, three rods are bundled together by special linker polypeptides (Koyama et al. 2006). This bundle-like structure enables a dense distribution of light-harvesting pigments on cytoplasmic membranes, which overcomes the shortcomings of area constraint caused by the lack of a thylakoid membrane.

The bundle structure accelerates excitation energy migration between rods (Yokono et al. 2008b). Figure 8.2 shows representative fluorescence decay curves of some PBSs. The bundle-like PBS (solid lines) shows slower kinetics than the hemidiscoidal PBS (dotted lines), which may be caused by energy migration in the rod assemblies.

The slower energy transfer does not seem appropriate. However, the lifetime of the lowest excited state (S_1 state) of isolated phycobiliprotein is known to be much longer than that of the bundle-like PBS (Yamazaki et al. 1994), so the overall efficiency of energy transfer remains reasonable.

Many red algae contain hemiellipsoidal PBSs with a large amount of PE at their periphery (Fig. 8.1d) (Gantt 1981; Arteni et al. 2008). The hemiellipsoidal structure may enable these PBSs to form a dense distribution of light-harvesting pigments, and accelerate excitation energy migration between rods, leading to slow energy transfer in such PBSs (Yokono et al. 2012). In the core domain, unlike a hemidiscoidal PBS, the core structure is covered with a large amount of PE, so quenchers including OCP may not be able

Fig. 8.2 Architectures and fluorescence decay curves of PBSs from *Gloeobacter violaceus* and *Fremyella diplosiphon*

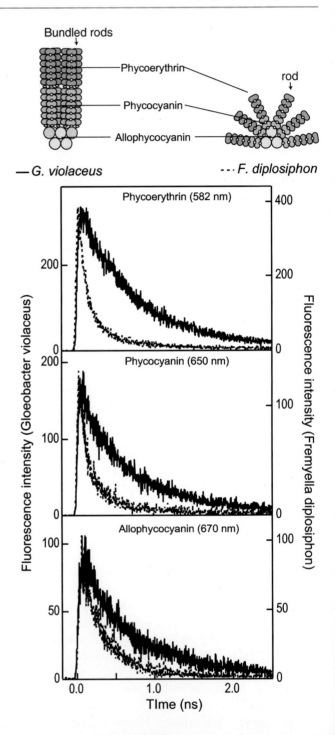

to access it. Actually, OCP has not been found in any red algae. Instead, red algae can adjust energy transfer using PE (Yokono et al. 2012). Under high stress conditions, some PE detaches from the PBSs (Yokono et al. 1817a; Liu et al. 2008a), and an unknown quencher dissipates the excess energy in PE (Fig. 8.3) (Yokono et al. 2012). In cyanobacteria, the PBS itself detaches

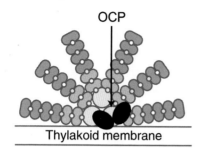

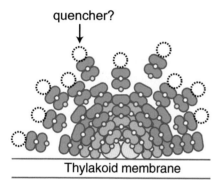

Fig. 8.3 Quenching of PBSs in cyanobacteria and red algae. OCP: *orange* carotenoid protein

from the thylakoid membrane under high light conditions (Tamary et al. 2012); therefore, cyanobacteria and red algae might employ different mechanisms to regulate their excitation energy distributions.

8.3.2 Modification of Energy Transfer from Phycobilisomes in the Cyanobacterium Arthrospira platensis

Not only the quality and/or quantity of pigment–protein complexes, but also the interactions among these complexes are essential to adapt light-harvesting and energy-transfer processes to light conditions. The cyanobacterium *A. platensis* possesses an extremely low energy Chl *a* in PSI (F760), but does not contain PE in its PBSs (Akimoto et al. 2012; Boussiba and Richmond 1979; Shubin et al. 1991, 1992). Excitation energy transfers in the PBSs and those from PSII to PSI are modified depending on the light quality (Akimoto et al. 2012). Light energy captured by the PBSs is transferred to both PSI and PSII. PBS→PSII energy transfer occurs directly, whereas two pathways are conceivable for PBS→PSI energy transfer: direct energy transfer from PBS to PSI (direct PBS→PSI energy transfer) (Mullineaux 1992), and energy transfer from PBS to PSI via PSII (PBS → PSII→PSI energy transfer) (Butler and Kitajima 1975; Bruce et al. 1985). Energy transfer process from PBS and PSII to PSI were examined for *A. platensis* cells grown under lights with different spectral profiles and different light intensities (Akimoto et al. 2013). Before cultivation under different light qualities, *A. platensis* cells were grown under a fluorescent lamp with a light intensity of 50 µmol photons $m^{-2}\ s^{-1}$ for 7 days (whole-day illumination) as controls. To examine the effects of light quantity/quality on energy-transfer processes, *A. platensis* cells were also grown under the following light conditions: high light (270 µmol photons $m^{-2}\ s^{-1}$) for 2 days (whole-day illumination), low light (20 µmol photons $m^{-2}\ s^{-1}$) for 2 days (whole-day illumination), and low light for 14 days (whole-day illumination). Under high light conditions, the growth rate of *A. platensis* cells decreased after 2.5 days (Aikawa et al. 2012). Eight different light sources were used to cultivate cells (Fig. 8.4b, c, and d): a white fluorescent lamp, white light-emitting diode (LED), and six different single-color LEDs with the spectral profiles shown in Fig. 8.4d. The single-color LEDs emit a single peak at 461 nm (blue), 527 nm (green), 590 nm (yellow), 637 nm (red light with shorter wavelength emission), 666 nm (red light with longer wavelength emission), and 740 nm (far-red). This set of single-color LEDs covers the visible to the far-red region of the electromagnetic spectrum. In the absorption spectrum of *A. platensis* cells (Fig. 8.4a), four peaks are recognized, which are assigned to the Chl Soret (~435 nm), Car (~500 nm), PBS (~620 nm), and Chl Qy (~676 nm) bands. The light from the fluorescent lamp and white LED is

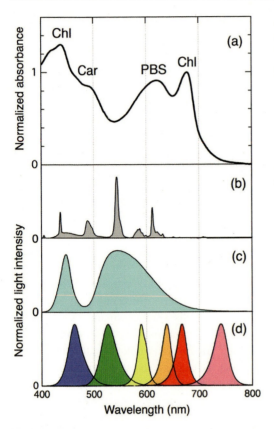

Fig. 8.4 Absorption spectrum of *Arthrospira platensis* and spectral profiles of lights used for cultivation: (**a**) absorption spectrum of control cells, and spectra of (**b**) white fluorescent light, (**c**) white LED, and (**d**) six kinds of single-color LEDs: blue LED (*blue*), green LED (*green*), yellow LED (*yellow*), red LED with shorter wavelength emission (*orange*), red LED with longer wavelength emission (*red*), and far-red LED (*pink*). In (a), absorption bands caused by chlorophyll *a* (Chl), phycobilisome (PBS), and carotenoid (Car) are indicated

absorbed by all photosynthetic pigments in *A. platensis* cells. In contrast, the pigments in these cells do not necessarily absorb the light from single-color LEDs; blue light can be absorbed by Chl and Car, green light by Car, yellow light mainly by PBS and slightly by Car, red light with shorter wavelength emission mainly by PBS and slightly by Chl, red light with longer wavelength emission by PBS and Chl, and far-red by low-energy Chl.

The fluorescence intensity ratio of PSI to PSII changes depending on the PSI-PSII interaction. By examining the PSI/PSII ratios in the delayed and steady-state fluorescence spectra as a function of light conditions, it is possible to elucidate how energy transfer to PSI changes with the light conditions. In the delayed fluorescence, the PSI/PSII ratio increases when PSII transfers energy to PSI (PSII→PSI). Connection of PBSs to the PSs was not apparent. Conversely, in the steady-state fluorescence signal formed by excitation at the PBS absorption band, the PSI/PSII ratio becomes larger when PSI can receive energy from PBS. To achieve this, one of the following energy-transfer pathways is required: PBS→PSI or PBS → PSII→PSI. The latter case can also be detected from the delayed fluorescence of these cells. Therefore, the difference in the PSI/PSII ratio between the steady-state fluorescence and delayed fluorescence (ΔPSI/PSII), which is calculated by subtracting the PSI/PSII ratio of the delayed fluorescence from that of the steady-state fluorescence, gives us information about the energy donors to PSI. To increase ΔPSI/PSII, there should be a greater contribution of PBS → PSI energy transfer; to decrease ΔPSI/PSII, there should be a greater contribution of PSII → PSI energy transfer without PBS. Figure 8.5 shows ΔPSI/PSII as a function of the PSI/PSII ratio of delayed fluorescence. There is a linear relationship between ΔPSI/PSII and PSI/PSII in the delayed fluorescence spectra; ΔPSI/PSII decreases linearly with increasing PSI/PSII of delayed fluorescence with a slope of −1. This suggests that the total energy transfer to PSI remains constant independent of light conditions and cultivation period. The contribution of PBS → PSI energy transfer becomes greater as that of PSII → PSI energy transfer decreases, and *vice versa*. Therefore, it is likely that PSI can control energy flow to itself by using PSII as an antenna complex even when PBS is not energetically connected to PSI.

8.3.3 Excitation Balance in Photochemical Reaction Centers in Red Algae

Red algae are thought to have arisen from the whole-cell engulfment of cyanobacteria by a

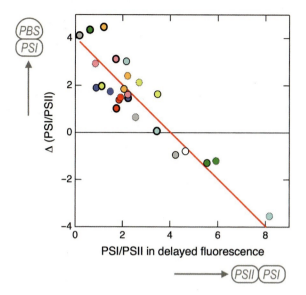

Fig. 8.5 Difference of PSI/PSII ratio between steady-state fluorescence and delayed fluorescence (ΔPSI/PSII) plotted against PSI/PSII ratio of delayed fluorescence in cells grown under low light conditions for 2 days (circle with *thin solid line*) and 14 days (circle with *thin dotted line*), and under high light conditions for 2 days (circle with *thick solid line*). Light sources used were as follows: white fluorescent light (gray), white LED (*light blue*), blue LED (*blue*), green LED (*green*), yellow LED (*yellow*), *red* LED with shorter wavelength emission (*orange*), red LED with longer wavelength emission (*red*), and far-red LED (*pink*). Control: white circle with *thin solid line*. The *red* line is the best-fit function: $Y = -X + 4$. See details in (Akimoto et al. 2013)

eukaryotic host cell (Falkowski et al. 2004), so photosynthesis and respiration occur in separate places. Red algae proliferated in coastal benthic habitats, where consistently oxic conditions would have first been established (Falkowski et al. 2004), indicating low light and low Fe conditions. In addition, as mentioned above (see the section *Excitation Energy Regulation in Phycobilisomes*), many red algae use a large light-harvesting antenna, hemiellipsoidal PBS, which covers the surface of the thylakoid membrane (Liu et al. 2008b). This feature is favorable under low light conditions, but the mobility of the antenna might be low. It has been reported that red algae predominantly use the spillover mechanism to achieve excitation balance between PSII and PSI (Yokono et al. 2011; Kowalczyk et al. 2013).

Red algae showed a wealth of colors such as green, purple, and red. The color reflects the antenna size of PSII, where red or purple indicates numerous PBSs. The spillover mechanism was found to dominate when the amount of PBS was high (Fig. 8.6) (Yokono et al. 2011). In the most prominent example, about 75% of PSII was physically connected to PSI, and excitation energy transfer occurred to ensure the balance between them (Yokono et al. 2011).

There are two possible roles of the spillover mechanism: maintenance of the excitation balance between two PSs, and protection of PSII from photo-damage (Yokono et al. 1807). The former role is important in red algae. In red algae, PSII-binding antennae contain not only Chl a, but also PBSs as light-harvesting pigments, whereas PSI-binding antennae do not contain phycobilins (Gardian et al. 2007). Here, PBSs absorb light in the wavelength region of 500–650 nm, which cannot be absorbed by Chl a. Therefore, PSII-binding antennae can harvest a broader range of light, and then share the excitation energy with two reaction centers by the spillover mechanism. Conversely, in diatoms, green algae, moss, and higher plants, the light-harvesting antennae of both PSs contain similar types of pigments. In these organisms, protection

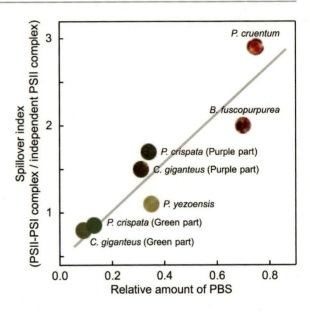

Fig. 8.6 Estimated relationship between the relative amount of PBS and spillover. The color of each point represents the color of each red alga. The relative amount of PBS is represented by the absorbance ratio (A550/A680). The spillover index was estimated from the delayed fluorescence intensity and mean lifetime in the PSI and PSII wavelength region. See details in (Yokono et al. 2011)

of PSII might be more important than maintaining energy balance. In higher plants, the thylakoid membranes are arranged in a continuous network of non-appressed regions interconnected with appressed regions of the grana stacks, and the appressed regions contain only the PSII reaction center (Anderson and Andersson 1988). Therefore, spillover could occur only in non-appressed regions, where the repair cycle of PSII also takes place (Yokono et al. 2015; Tikkanen et al. 2008; Nixon et al. 2010).

8.3.4 Light Harvesting by Carotenoids in the Higher Plant Arabidopsis thaliana and Green Algae Codium fragile

In the light-harvesting Chl-protein complexes of PSII (LHCII), green algae and higher plants possess Chl *a*, Chl *b*, and specific Cars. As light-harvesting Cars, Lut and Siph are found in the LHCIIs of *Arabidopsis thaliana* and *Codium fragile*, respectively (Fig. 8.7). Differences in the Car contents and pigment ratios of Chl *a*/Chl *b* affect the ability of LHCII to absorb light. Chl *a* exhibits peaks around 438 nm (Soret band) and 673 nm (Qy band), whereas Chl *b* absorbs around 475 nm (Soret band) and 653 nm (Qy band). Relative absorbances at 650 and 476 nm are larger in the LHCII of *C. fragile* than in *A. thaliana*, reflecting a larger relative amount of Chl *b* in green algae (Fig. 8.8). In addition, a green absorption band appears in the absorption spectrum of the LHCII of *C. fragile* (green band in Fig. 8.8).

Lut is a typical polyene-type Car that contains nine conjugated double bonds with an additional conjugation by the β-end group, so the total conjugation number is ten (Fig. 8.7). In contrast, Siph is a keto-Car containing eight C = C bonds and one C = O bond, so in total, $n = 9$ (Fig. 8.7). In the non-polar solvent *n*-hexane Lut and Siph exhibit absorption spectra with clear vibrational bands, and the spectrum of Siph is shifted to longer wavelength by 7 nm compared with that of Lut (Fig. 8.9). In methanol, Lut shows almost the same absorption spectrum as that in *n*-hexane (not shown). In contrast, the absorption spectrum of Siph shows an increase in intensity in the longer wavelength region and completely loses its vibrational structure, which is not observed for the polyene-type Cars (LeRosen and Reid 1952). Although the dielectric constants (ε) of acetonitrile ($\varepsilon_{\text{Acetonitrile}} = 37$) and methanol ($\varepsilon_{\text{Methanol}} = 33$) are almost the

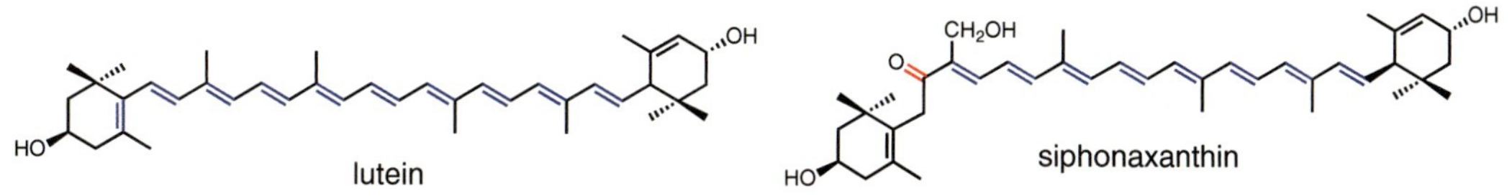

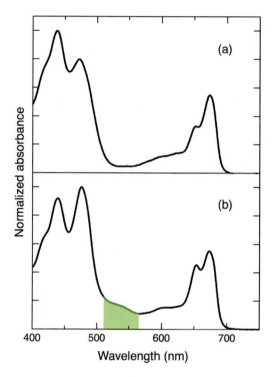

Fig. 8.7 Molecular structures of the carotenoids lutein and siphonaxanthin

Fig. 8.8 Steady-state absorption spectra of (**a**) *Arabidopsis thaliana* LHCII and (**b**) *Codium fragile* LHCII. The specific absorption band of siphonaxanthin is highlighted in *green*

same, this solvent-induced change was not observed in acetonitrile (Akimoto et al. 2008). Therefore, it is suggested that the increase in intensity in the longer wavelength region of Siph absorption is not related to the polarity of the surrounding molecules, but to a specific interaction between Siph and the surrounding molecule(s), such as hydrogen bonding between the C = O group of Siph and OH group of the solvent.

Time-resolved fluorescence spectra of Lut in *n*-hexane, Siph in *n*-hexane, and Siph in methanol are presented in Fig. 8.9. In *n*-hexane, time-dependent changes are relatively small, although spectral shapes and peak wavelengths change over time. Conversely, in methanol, fluorescence intensity in the longer wavelength region increases over time. The fluorescence lifetimes of Siph were analyzed by triple exponential functions with decay lifetimes of 30–35 fs, 180–200 fs, and >10 ps (Akimoto et al. 2008). It was found that the difference in time-dependent change in the spectra of Siph in *n*-hexane and methanol is caused by a difference in the contribution from the longest-lived component (>10 ps); the contribution from this component is larger in methanol than in *n*-hexane. The most probable origin of this long-lived component is fluorescence from the S_1 state including vibrationally excited states in the S_1 state. In general, Cars are classified into polyenes belonging to the point group C_{2h} in which the S_1 state is dipole forbidden from the ground state (S_0). Therefore, in *n*-hexane, the contribution from S_1 fluorescence is negligible even in Siph, whose conjugation structure is not symmetric because of the presence of the C = O group. In methanol, the hydrogen bond formed between the solvent and C = O group of Siph might increase the intensity of S_1 fluorescence by increasing the deviation of Siph from C_{2h} symmetry. The increase in the intensity of S_1 fluorescence in methanol suggests that the optically forbidden S_1–S_0 transition of Cars is partially allowed for Siph and its strength increases in methanol. According to the Förster theory for energy transfer, energy transfer efficiency depends on the strength of an electronic transition of an energy donor (Förster 1959). In this sense, the efficiency of energy transfer from Siph to Chl *a* via the S_1 state of Siph seems to depend on the interaction between Siph and the surrounding environment.

Based on sequences of genes coding *Codium* LHC II apoprotein, specific substitutions of

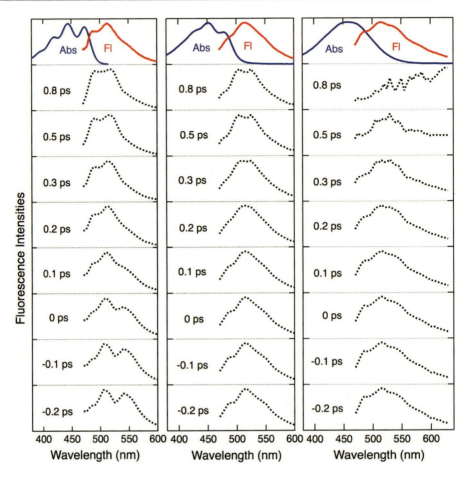

Fig. 8.9 Steady-state absorption spectra (Abs, *blue line*), steady-state fluorescence spectra (Fl, *red line*), and time-resolved fluorescence spectra (*dotted line*) of lutein in *n*-hexane (*left panel*), siphonaxanthin in *n*-hexane (*middle panel*), and siphonaxanthin in methanol (*right panel*). Each spectrum is normalized at its maximum intensity

amino acids close to the C = O group of Siph, from Ser to Ala in helix A and from Ala to Val in helix B, were indicated, although the overall similarity of the whole sequences of all LHCII is maintained (Tsuchiya et al. 2008). Therefore, it is expected that differences in protein environment around Cars play an important role in determining the properties of Cars, besides the difference in the conjugation structure of Car. To produce the characteristic light-harvesting function of keto-Cars, asymmetric Car-protein interactions including hydrogen bonding between the amino acid residues and C = O group of Car might be important. For the siphonous green alga *Bryopsis corticulans*, it has also been suggested that Siph and siphonein in a weakly hydrophobic protein environment enhance light harvesting in the green light region (Wang et al. 2013).

8.4 Conclusions

It has been found that the cyanobacterium *A. platensis* maintains energy balance between PSI and PSII by controlling the PBS–PSI and PSII–PSI interactions (Akimoto et al. 2013). Although both PSII→PSI and PBS→PSI energy transfers were affected by changes in light conditions, the total energy transferred to PSI did not depend on the light conditions. Conversely, in red algae, it appears that energy distribution is mainly controlled by PSII→PSI energy transfer (Yokono et al. 1807). A larger amount of PBS increases the contribution of PSII→PSI energy transfer. The difference in light harvesting by PBSs between cyanobacteria and red algae might be related to respiration complexes located in close proximity to photosynthetic complexes in cyanobacteria (Mullineaux and Allen 1990; Liu et al. 2012) and/or differences in the architecture of PBSs (see Fig. 8.1). Further study is required to reveal the mechanism of energy distribution control and the role(s) of spillover in different organisms.

For absorption of green light by a Car, the green alga *C. fragile* possesses Siph instead of Lut, and substitutes some amino acids close to Siph. This behavior was also found for Chl. When monovinyl Chls are replaced with divinyl Chls, two amino acid alternations are needed to improve the tolerance to strong light conditions by changing energy-transfer pathways (Ito and Tanaka 2011; Yokono et al. 2011). In addition to alternation of pigments, substitution of amino acids can be used to optimize the function of pigments.

Acknowledgements The authors would like to thank Prof. Mimuro and Prof. Tsuchiya (Kyoto University), Prof. Yamazaki and Prof. Tanaka (Hokkaido University), Prof. Takaichi (Nippon Medical School), Prof. Tomo (Tokyo University of Science), and Prof. Murakami and Prof. Kondo (Kobe University), for helpful discussion and important contributions to the research surveyed in this article.

References

Aikawa, S.; Izumi, Y.; Matsuda, F.; Hasunuma, T.; Chang, J. S.; Kondo, A. Synergistic enhancement of glycogen production in *Arthrospira platensis* by optimization of light intensity and nitrate supply. Bioresour. Technol., 2012, 108, 211–215.

Akimoto, S.; Mimuro, M. Application of time-resolved polarization fluorescence spectroscopy in the femtosecond range to photosynthetic systems. Photochem. Photobiol., 2007, 83, 163–170.

Akimoto, S.; Yokono, M.; Hamada, F.; Teshigahara, A.; Aikawa, S; Kondo, A. Adaptation of light-harvesting systems of *Arthrospira platensis* to light conditions, probed by time-resolved fluorescence spectroscopy. Biochim. Biophys. Acta Bioenerg., 2012, 1817, 1483–1489.

Akimoto, S.; Yamazaki, I.; Takaichi, S.; Mimuro, M. Excitation relaxation dynamics of linear carotenoids. J. Lumin., 2000, 87–89, 797–799.

Akimoto, S.; Yamazaki, I.; Murakami, A.; Takaichi, S.; Mimuro, M. Ultrafast excitation relaxation dynamics and energy transfer in the siphonaxanthin-containing green alga *Codium fragile*. Chem. Phys. Lett., 2004, 390, 45–49.

Akimoto, S.; Yokono, M.; Ohmae, M.; Yamazaki, I.; Tanaka, A.; Higuchi, M.; Tsuchiya, T.; Miyashita, H.; Mimuro, M. Ultrafast excitation relaxation dynamics of lutein in solution and in the light-harvesting complexes II isolated from *Arabidopsis thaliana*. J. Phys. Chem. B, 2005, 109, 12612–12619.

Akimoto, S.; Tomo, T.; Naitoh, Y.; Otomo, A.; Murakami, A.; Mimuro, M. Identification of a new excited state responsible for the in vivo absorption band of siphonaxanthin in the green alga *Codium fragile*. J. Phys. Chem. B, 2007, 111, 9179–9181.

Akimoto, S.; Yokono, M.; Higuchi, M.; Tomo, T.; Takaichi, S.; Murakami, A.; Mimuro, M. Solvent effects on excitation relaxation dynamics of a ketocarotenoid, siphonaxanthin. Photochem. Photobiol. Sci., 2008, 7, 1206–1209.

Akimoto, S.; Yokono, M.; Aikawa, S.; Kondo, A. Modification of energy transfer processes in the cyanobacterium *Arthrospira platensis* to adapt to light conditions, probed by time-resolved fluorescence spectroscopy. Photosynth. Res., 2013, 117, 235–243.

Anderson, J. M.; Andersson, B. The dynamic photosynthetic membrane and regulation of solar energy conversion. Trends Biochem. Sci., 1988, 13, 351–355.

Apt, K. E.; Collier, J. L.; Grossman, A. R. Evolution of the phycobiliproteins. J. Mol. Biol., 1995, 248, 79–96.

Arteni, A. A.; Liu, L.-N.; Aartsma, T. J.; Zhang, Y.-Z.; Zhou, B.-C.; Boekema, E. J. Structure and organization of phycobilisomes on membranes of the red alga *Porphyridium cruentum*. Photosynth. Res., 2008, 95, 169–174.

Arteni, A. A.; Ajlani, G.; Boekema, E. J. Structural organisation of phycobilisomes from *Synechocystis* sp. strain PCC6803 and their interaction with the membrane. Biochim. Biophys. Acta Bioenerg., 2009, 1787, 272–279.

Ben-Shem, A.; Frolow, F.; Nelson, N. Crystal structure of plant photosystem I. Nature, 2003, 426, 630–635.

Bernát, G.; Schreiber, U.; Sendtko, E.; Stadnichuk, I. N.; Rexroth, S.; Rögner, M.; Koenig F. Unique properties vs. common themes: the atypical cyanobacterium *Gloeobacter violaceus* PCC 7421 is capable of state transitions and blue-light-induced fluorescence quenching. Plant Cell Physiol., 2012, 53, 528–542.

Blankenship, R. E. Molecular mechanisms of photosynthesis, Blackwell Publishing: Oxford, 2002.

Boardman, N. K.; Thome S. W.; Anderson J. M. Fluorescence properties of particles obtained by digitonin fragmentation of spinach chloroplasts. Proc. Natl. Acad. Sci. USA, 1966, 56, 586–593.

Boussiba S.; Richmond, A. E. Isolation and characterization of phycocyanins from the blue-green alga *Spirulina platensis*. Arch. Microbiol., 1979, 120, 155–159.

Bruce, D.; Biggins, J.; Steiner, T.; Thewalt, M. Mechanism of the light state transition in photosynthesis. IV. Picosecond fluorescence spectroscopy of *Anacystis nidulans* and *Porphyridium crentum* in state 1 and state 2 at 77 K. Biochim. Biophys. Acta Bioenerg., 1985, 806, 237–246.

Butler, W. L.; Kitajima, M. Energy transfer between photosystem I and photosystem II in chloroplasts. Biochim. Biophys. Acta Bioenerg., 1975, 396, 72–85.

Chen, M.; Floetenmeyer, M.; Bibby, T. Supramolecular organization of phycobiliproteins in the chlorophyll *d*-containing cyanobacterium *Acaryochloris marina*. FEBS Lett., 2009, 583, 2535–2539.

Cheng, Y. C.; Ahn, T. K.; Avenson, T. J.; Zigmantas, D.; Niyogi, K. K.; Ballottari, M.; Bassi, R.; Fleming, G. R. Kinetic modeling of charge-transfer quenching in the CP29 minor complex. J. Chem. Phys. B, 2008, 112, 13418–13423.

Cunningham Jr, F.; Dennenberg, R.; Jursinic, P.; Gantt, E. Growth under red light enhances photosystem II relative to photosystem I and phycobilisomes in the red alga *Porphyridium cruentum*. Plant Physiol., 1990, 93, 888–985.

Evans, J. The dependence of quantum yield on wavelength and growth irradiance. Funct. Plant Biol., 1987, 14, 69–79.

Falkowski, P. G.; Katz, M. E.; Knoll, A. H.; Quigg, A.; Raven, J. A.; Schofield, O.; Taylor, F. J. R. The evolution of modern eukaryotic phytoplankton. Science, 2004, 305, 354–360.

Förster, T. 10th Spiers memorial lecture. Transfer mechanisms of electronic excitation. Disc. Farad. Soc., 1959, 27, 7–17.

Frank, H. A.; Cogdell, R. J. Photochemistry and function of carotenoids in photosynthesis, In: Carotenoids in photosynthesis; Young A.; Britton, G. Eds; Chapman & Fall: London, 1993, Chap. 8, pp. 252–326.

Gantt, E. Phycobilisomes. Ann. Rev. Plant Physiol., 1981, 32, 327–347.

Gardian, Z.; Bumba, L.; Schrofel, A.; Herbstova, M.; Nebesarova, J.; Vacha, F. Organisation of photosystem I and photosystem II in red alga *Cyanidium caldarium*: encounter of cyanobacterial and higher plant concepts. Biochim. Biophys. Acta Bioenerg., 2007, 1767, 725–731.

Ghosh A. K.; Govindjee Transfer of the excitation energy in *Anacystis nidulans* grown to obtain different pigment ratios. Biophys. J., 1966, 6, 611–619.

Gillbro, T.; Sandström, Å.; Sundström, V.; Wendler, J.; Holzwarth, A. Picosecond study of energy-transfer kinetics in phycobilisomes of *Synechococcus* 6301 and the mutant AN 112. Biochim. Biophys. Acta Bioenerg., 1985, 808, 52–65.

Guan, X.; Qin, S.; Zhao, F.; Zhang, X.; Tang, X. Phycobilisomes linker family in cyanobacterial genomes: divergence and evolution. Int. J. Biol. Sci., 1996, 3, 434–445.

Holt, N. E.; Kennis, J. T. M.; Osto, L. D.; Bassi, R.; Fleming, G. R. Carotenoid to chlorophyll energy transfer in light harvesting complex II from *Arabidopsis thalaina* probed by femtosecond fluorescence upconversion. Chem. Phys. Lett., 2003, 379, 305–313.

Holt, N. E.; Zigmantas, D.; Valkunas, L.; Li, X.; Niyogi, K. K.; Fleming, G. R. Carotenoid cation formation and the regulation of photosynthetic light harvesting. Science, 2005, 307, 433–436.

Horton, P.; Ruban, A. V.; Walters, R. G. Regulation of light harvesting in green plants. Ann. Rev. Plant Physiol. Plant Mol. Biol., 1996, 47, 655–684.

Ito, H.; Tanaka, A. Evolution of a divinyl chlorophyll-based photosystem in *Prochlorococcus*. Proc. Natl. Acad. Sci. USA, 2011, 108, 18014–18019.

Kahlow, M. A.; Jarzeba, W.; DuBruil, T. P.; Barbara, P. F. Ultrafast emission spectroscopy in the ultraviolet by time-gated upconversion. Rev. Sci. Instrum., 1988, 59, 1098–1109.

Kirilovsky, D. Photoprotection in cyanobacteria: the orange carotenoid protein (OCP)-related non-photochemical-quenching mechanism. Photosynth. Res., 2007, 93, 7–16.

Kowalczyk, N.; Rappaport, F.; Boyen, C.; Wollman, F-A.; Collén, J.; Joliot, P. Photosynthesis in *Chondrus crispus*: The contribution of energy spill-over in the regulation of excitonic flux. Biochim. Acta Bioenerg., 2013, 1827, 834–842.

Koyama, K.; Tsuchiya, T.; Akimoto, S.; Yokono, M.; Miyashita, H.; Mimuro, M. New linker proteins in phycobilisomes isolated from the cyanobacterium *Gloeobacter violaceus* PCC 7421. FEBS Lett., 2006, 580, 3457–3461.

Kozaki, A.; Takeba, G. Photorespiration protects C3 plants from photooxidation. Nature, 1996, 384, 557–560.

Kuvykin, I. V.; Ptushenko, V. V.; Vershubskii, A. V; Tikhonov, A. N. Regulation of electron transport in C3 plant chloroplasts in situ and in silico: Short-term effects of atmospheric CO_2 and O_2. Biochim. Biophys. Acta Bioenerg., 2011, 1807, 336–347.

LeRosen, A. L.; Reid, C. E. An investigation of certain solvent effect in absorption spectra. J. Chem. Phys., 1952, 20, 233–236.

Liu, L.-N.; Zhou, B.-C.; Zhang, Y.-Z. Light-induced energetic decoupling as a mechanism for phycobilisome-related energy dissipation in red algae: a single molecule study. Plos One, 2008a, 3, e3134.

Liu, L.; Aartsma, T. J.; Thomas, J.; Lamers, G. E.; Zhou, B.; Zhang, Y. Watching the native supramolecular architecture of photosynthetic membrane in red algae: topography of phycobilisomes and their crowding, diverse distribution patterns. J. Biol. Chem., 2008b, 283, 34946–34953.

Liu, L.-N.; Bryan, S. J.; Huang, F; Yu J.; Nixon, P. J.; Rich, P. R.; Mullineaux, C. W. Control of electron transport routes through redox- regulated redistribution of respiratory complexes. Proc. Natl. Acad. Sci. USA, 2012, 109, 11431–11436.

Mimuro, M.; Akimoto, S. Energy transfer processes from fucoxanthin and peridinin to chlorophyll, In: Advances in Photosynthesis and Respiration, vol. 14, Photosynthesis in Algae; Larkum, A. W. D.; Douglas, S. E.; Raven, J. A., Eds.; Kluwer Academic Publishers: The Netherlands, 2003, pp. 335–349.

Mimuro, M.; T. Katoh. Carotenoids in photosynthesis: absorption, transfer and dissipation of light energy. Pure Appl. Chem., 1991, 63, 123–130.

Mimuro M.; Kikuchi, H. Antenna systems and energy transfer in Cyanophyta and Rhodophyta, In: Light-harvesting antennas in photosynthesis; Green, B. R.; Parson, W. W. Eds; Kluwer Academic Publishers: Dordrecht, The Netherlands, 2003, pp. 281–306.

Mimuro, M.; Akimoto, S.; Tomo, T.; Yokono, M.; Miyashita, H.; Tsuchiya, T. Delayed fluorescence observed in the nanosecond time region at 77 K originates directly from the photosystem II reaction center. Biochim. Biophys. Acta Bioenerg, 2007, 1767, 327–334.

Mimuro, M.; Yamazaki, I.; Tamai, N.; Katoh, T. Excitation energy transfer in phycobilisomes at −196 °C isolated from the cyanobacterium *Anabaena variabilis* (M-3): evidence for the plural transfer pathways to the terminal emitters. Biochim. Biophys. Acta Bioenerg., 1989, 973, 153–162.

Mimuro, M.; Yokono, M.; Akimoto, S. Variations in photosystem I properties in the primordial cyanobacterium *Gloeobacter violaceus* PCC 7421. Photochem. Photobiol., 2010, 86, 62–69.

Minagawa, J. State Transitions-the molecular remodeling of photosynthetic supercomplexes that controls energy flow in the chloroplast. Biochim. Biophys. Acta Bioenerg., 2011, 1807, 897–905.

Miyashita, H.; Ikemoto, H.; Kurano, N. Chlorophyll *d* as a major pigment. Nature, 1996, 383, 402.

Mullineaux, C. W. Excitation energy transfer from phycobilisomes to photosystem I in a cyanobacterium. Biochim. Biophys. Acta Bioenerg., 1992, 110, 285–292.

Mullineaux, C. W.; Allen, J. F. State 1-State 2 transitions in the cyanobacterium *Synechococcus* 6301 are controlled by the redox state of electron carriers between Photosystems I and II. Photosynth. Res., 1990, 23, 297–311.

Mullineaux, C. W.; Holzwarth, A. R. Kinetics of excitation energy transfer in the cyanobacterial phycobilisome-Photosystem II complex. Biochim. Biophys. Acta Bioenerg., 1991, 1098, 68–78.

Munekage, Y.; Hashimoto, M.; Miyake, C.; Tomizawa, K.; Endo, T.; Tasaka, M.; Shikanai, T. Cyclic electron flow around photosystem I is essential for photosynthesis. Nature, 2004, 429, 579–582.

Murata, N. Control of excitation transfer in photosynthesis I. Light-induced change of chlorophyll *a* fluorescence in *Porphyridium cruentum*. Biochim. Biophys. Acta Bioenerg., 1969, 172, 242–251.

Nixon, P. J.; Michoux, F.; Yu, J., Boehm, M.; Komenda, J. Recent advances in understanding the assembly and repair of photosystem II. Ann. Bot., 2010, 106, 1–16.

O'Connor, D. V.; Phillips, D. Time-correlated single photon counting; Academic Press: London, 1984.

Rabinowitch, E.; Govindjee. Photosynthesis, Wiley: New York, 1969.

Raven, J. A. Predictions of Mn and Fe use efficiencies of phototrophic growth as a function of light availability for growth and of C assimilation pathway. New Phytol., 1990, 116, 1–18.

Ricci, M.; Bradforth, S. E.; Jimenez, R.; Fleming, G. R. Internal conversion and energy transfer dynamics of spheroidene in solution and in the LH-1 and LH-2 light harvesting complexes. Chem. Phys. Lett., 1996, 259, 381–390.

Rippka, R.; Waterbury, J.; Cohen-Bazire, G. A cyanobacterium which lacks thylakoids. Arch. Microbiol., 1974, 100, 419–436.

Ruban, A. V.; Johnson, M. P; Duffy, C. D. The photoprotective molecular switch in the photosystem II antenna. Biochim. Biophys. Acta Bioenerg., 2012, 1817, 167–181.

Sarcina, M.; Tobin, M. J.; Mullineaux, C. W. Diffusion of phycobilisomes on the thylakoid membranes of the cyanobacterium *Synechococcus* 7942: Effects of phycobilisome size, temperature, and membrane lipid composition. J. Biol. Chem., 2001, 276, 46830–46834.

Satoh, K.; Strasser, R.; Butler, W. L. A demonstration of energy transfer from photosystem II to photosystem I in chloroplasts. Biochim. Biophys. Acta, 1976, 440, 337–345.

Shah, J. Ultrafast luminescence spectroscopy using sum frequency generation. IEEE J Quantum Electroc., 1988, 24, 276–288.

Shubin, V. V.; Murthy, S. D. S.; Karapetyan, N. V.; Mohanty, P. Origin of the 77 K variable fluorescence at 758 nm in the cyanobacterium *Spirulina platensis*. Biochim Biophys Acta Bioenerg., 1991, 1060, 28–36.

Shubin, V. V.; Bezsmertnaya, I. N.; Karapetyan, N. V. Isolation from *Spirulina* membranes of two photosystem I-type complexes one of which contains chlorophyll responsible for the 77K fluorescence band at 760 nm. FEBS Lett., 1992, 309, 340–342.

Song, P.-S.; Koka, P.; Prezelin, B. B.; Haxo, F. T. Molecular topology of photosynthetic light-harvesting pigment complex, peridinin–chlorophyll-*a*–protein, from marine dinoflagellates. Biochemistry, 1976, 15, 4422–4427.

Stowe-Evans, E. L.; Kehoe, D. M. Signal transduction during light-quality acclimation in cyanobacteria: a model system for understanding phytochrome-response pathways in prokaryotes. Photochem. Photobiol. Sci., 2004, 3, 495–502.

Sutter, M.; Wilson, A.; Leverenz, R. L.; Lopez-Igual, R.; Thurotte, A.; Salmeen, A. E.; Kirilovsky, D.; Kerfeld, C. A. Crystal structure of the FRP and identification of the active site for modulation of OCP-mediated photoprotection in cyanobacteria. Proc. Natl. Acad. Sci. USA, 2013a, 110, 10022–10027.

Sutter, M.; Wilson, A.; Leverenzd, R. L.; Lopez-Igualb, R.; Thurotte, A.; Salmeena, A. E.; Kirilovsky, D.; Kerfeld, C. A. Crystal structure of the FRP and identification of the active site for modulation of OCP-mediated photoprotection in cyanobacteria. Proc. Natl. Acad. Sci. USA, 2013b, 110, 10022–10027.

Tamary, E.; Kiss, V.; Nevo, R.; Adam, Z.; Bernát, G.; Rexroth, S.; Rögner, M.; Reich, Z. Structural and functional alterations of cyanobacterial phycobilisomes induced by stress. Biochim. Biophys. Acta Bioenerg., 2012, 1817, 319–327.

Theiss, C.; Schmitta, F.-J.; Pieperb, J.; Nganoua, C.; Grehn, M.; Vitali, M.; Olliges, R.; Eichler, H. J.; Eckert, H.-J. Excitation energy transfer in intact cells and in the phycobiliprotein antennae of the chlorophyll d containing cyanobacterium *Acaryochloris marina*. J. Plant Physiol., 2011, 168, 1473–1487.

Tian, L.; van Stokkum, I. H.; Koehorst, R. B.; van Amerongen, H. Light harvesting and blue-green light induced non-photochemical quenching in two different C-phycocyanin mutants of *Synechocystis* PCC 6803. J. Phys. Chem. B, 2012, 117, 11000–11006.

Tian, L.; van Stokkum, I. H.; Koehorst, R. B.; van Amerongen, H. Light harvesting and blue-green light induced non-photochemical quenching in two different C-phycocyanin mutants of *Synechocystis* PCC 6803. J. Phys. Chem. B, 2013, 117, 11000–11006.

Tikkanen, M.; Nurmi, M.; Kangasjärvi, S.; Aro, E.-M. Core protein phosphorylation facilitates the repair of photodamaged photosystem II at high light. Biochim. Biophys. Acta Bioenerg., 2008, 1777, 1432–1437.

Tsuchiya, T.; Tomo, T.; Akimoto, S.; Murakami, A.; Mimuro, M. Unique optical properties of LHC II isolated from *Codium fragile* – its correlation to protein environment, In: Photosynthesis. Energy from the Sun; Allen, J.; Gantt, E.; Golbeck, J.; Osmond, B. Eds; Springer: Dordrecht, 2008, pp. 343–346.

Umena, Y.; Kawakami, K.; Shen, J. R.; Kamiya, N. Crystal structure of oxygen-evolving photosystem II at a resolution of 1.9 Å. Nature, 2011, 473, 55–60.

Wang, W.; Qin, X.; Sang, M.; Chen, D.; Wang, K.; Lin, R.; Lu, C.; Shen, J.-R., Kuang, T. Spectral and functional studies on siphonaxanthin-type light-harvesting complex of photosystem II from *Bryopsis corticulans*. Photosynth Res., 2013, 117, 267–279.

Wilson, A.; Ajlani, G.; Verbavatz, J.-M.; Vass, I.; Kerfeld, C. A.; Kirilovsky, D. A soluble carotenoid protein involved in phycobilisome-related energy dissipation in cyanobacteria. Plant Cell Online, 2006, 18, 992–1007.

Yamazaki, T.; Nishimura, Y.; Yamazaki, I,; Hirano, M.; Matsuura, K.; Shimada, K.; Mimuro, M. Energy migration in allophycocyanin-B trimer with a linker polypeptide: analysis by the principal multi-component spectral estimation (PMSE) method. FEBS Lett., 1994, 353, 43–47.

Yokono, M.; Murakami, A.; Akimoto, S. Excitation energy transfer between photosystem II and photosystem I in red algae: Larger amounts of phycobilisome enhance spillover. Biochim. Biophys. Acta Bioenerg., 2011, 1807, 847–853.

Yokono, M.; Akimoto, S; Tanaka, A. Seasonal changes of excitation energy transfer and thylakoid stacking in the evergreen tree *Taxus cuspidata*: How does it divert excess energy from photosynthetic reaction center? Biochim. Biophys. Acta Bioenerg., 2008a, 1777, 379–387.

Yokono, M.; Akimoto, S.; Koyama, K.; Tsuchiya, T.; Mimuro, M. Energy transfer processes in *Gloeobacter violaceus* PCC 7421 that possesses phycobilisomes with a unique morphology. Biochim. Biophys. Acta Bioenerg., 2008b, 1777, 55–65.

Yokono, M.; Uchida, H.; Suzawa, Y.; Akiomoto, S.; Murakami, A. Stabilization and modulation of the phycobilisome by calcium in the calciphilic freshwater red alga *Bangia atropurpurea*. Biochim. Biophys. Acta Bioenerg., 2012a, 1817, 306–311.

Yokono, M.; Tomo, T.; Nagao, R.; Ito, H.; Tanaka, A.; Akimoto, S. Alterations in photosynthetic pigments and amino acid composition of D1 protein change energy distribution in photosystem II. Biochim. Biophys. Acta Bioenerg., 2012b, 1817, 754–759.

Yokono, M.; Takabayashi, A.; Akimoto, S.; Tanaka, A. A megacomplex composed of both photosystem reaction centres in higher plants. Nat. Commun., 2015, 6, 6675.

9 Interaction of Glycine Betaine and Plant Hormones: Protection of the Photosynthetic Apparatus During Abiotic Stress

Leonid V. Kurepin, Alexander G. Ivanov, Mohammad Zaman, Richard P. Pharis, Vaughan Hurry, and Norman P.A. Hüner

Summary

The growth and development of higher plants depends not only upon an active photosynthetic apparatus and adequate water and mineral nutrient supply, but also tight regulation of growth by plant hormones and specific secondary metabolites. Environmental (abiotic) stresses influence plant growth via changes in the metabolism and action of plant hormones, their interactions with secondary metabolites, as well as a reduction in photosynthetic activity. An initial response to abiotic stress often includes an increasing accumulation of two "stress" hormones, abscisic acid (ABA) and salicylic acid (SA). This is followed by an activation of multiple physiological pathways which yield an increase in tolerance to the stress. Also associated with these increases in ABA and SA levels are a reduction in the biosynthesis and/or action of plant "growth" hormones, such as the gibberellins, auxin and cytokinins.

L.V. Kurepin (✉)
Department of Biology and The Biotron Center for Experimental Climate Change Research, University of Western Ontario (Western University), 1151 Richmond Street N, London, ON N6A 5B7, Canada

Department of Forest Genetics and Plant Physiology, Swedish University of Agricultural Sciences, Skogsmarksgränd 17, Umeå, Sweden
e-mail: lkurepin@uwo.ca

A.G. Ivanov (✉)
Department of Biology and The Biotron Center for Experimental Climate Change Research, University of Western Ontario (Western University), 1151 Richmond Street N, London, ON N6A 5B7, Canada

Institute of Biophysics and Biomedical Engineering, Bulgarian Academy of Sciences, Acad. G. Bonchev Street, Bl. 21, 1113 Sofia, Bulgaria
e-mail: aivanov@uwo.ca

M. Zaman
Soil and Water Management and Crop Nutrition Section, Joint FAO/IAEA Division of Nuclear Techniques in Food and Agriculture, Department of Nuclear Sciences and Applications, International Atomic Energy Agency, Vienna International Centre, PO Box 100,1400 Vienna, Austria
e-mail: M.Zaman@iaea.org

R.P. Pharis
Department of Biological Sciences, University of Calgary, Calgary, AB T2N 1N4, Canada
e-mail: rpharis@ucalgary.ca

V. Hurry
Department of Forest Genetics and Plant Physiology, Swedish University of Agricultural Sciences, Skogsmarksgränd 17, Umeå, Sweden
e-mail: vaughan.hurry@slu.se

N.P.A. Hüner
Department of Biology and The Biotron Center for Experimental Climate Change Research, University of Western Ontario (Western University), 1151 Richmond Street N, London, ON N6A 5B7, Canada
e-mail: nhuner@uwo.ca

The plant's internal resources are then diverted toward enhancing stress tolerance which is usually associated with diminished photosynthetic productivity. However, some plant varieties (genotypes) are capable of biosynthesising a unique secondary metabolite, glycine betaine (GB). These genotypes exhibit a greater tolerance to abiotic stress, and often have an enhanced growth and yield, relative to varieties which do not accumulate GB. The increased GB accumulation occurs mainly in the chloroplast and is responsible for initiating a network of interactions between the plant's photosynthetic apparatus, its "stress" and "growth" hormones, and reactive oxygen species. The end result of these GB-induced interactions is the alleviation of abiotic stress effects.

Keywords

Glycine betaine • Abiotic stress • Photosynthetic apparatus • Plant hormones

Contents

9.1 Introduction 186
9.2 Biosynthesis and Accumulation of Glycine Betaine in Response to Abiotic Stress 187
9.3 Glycine Betaine Alleviates Abiotic Stress Effects on Crop Growth and Yield via an Interaction with Photosynthesis 188
9.4 Glycine Betaine Protects the Photosynthetic Apparatus from Abiotic Stress 191
9.5 Glycine Betaine Interaction with Reactive Oxygen Species Aids in the Protection of the Photosynthetic Apparatus 192
9.6 Glycine Betaine Interaction with Plant "Stress" Hormones and Photosynthesis 192
9.7 Glycine Betaine Interactions with Plant "Growth" Hormones 195
9.8 Modelling the Interactions Between Glycine Betaine, Plant Hormones, Photosynthesis and Reactive Oxygen Species 195
References 196

9.1 Introduction

Plants cope with a range of abiotic stresses, including changes in light and water availability, or fluctuations in temperature and mineral levels, via multiple pathways which are regulated by changes in plant hormones biosynthesis and perception (Kiba et al. 2011; Dodd and Davies 2010; Kurepin and Pharis 2014; Hüner et al. 2014). One of these hormonal pathways leads to the accumulation of an osmolyte, glycine betaine (Kurepin et al. 2015a). Glycine betaine (GB) accumulation by plant tissues has been shown to occur in a range of abiotic stresses and it has often been directly correlated with a plant's increasing tolerance to the abiotic stress (Rhodes and Hanson 1993; Chen and Murata 2011). The increased abiotic stress tolerance of plants able to accumulate GB appears, in large part, to be due to the ability of chloroplast-produced GB to protect the photosynthetic apparatus (Kurepin et al. 2015a). In particular, accumulation of GB in the chloroplast in response to a stress signal protects the enzymes and lipids which are required to maintain both the flow of electrons through thylakoid membranes and the continued assimilation of CO_2 (Park et al. 2007; Chen and Murata 2011). Additionally, GB's functions in chloroplasts appear to be especially important for stability of photosystem (PS) II, the most vulnerable part of the photosynthetic apparatus and a component which is believed to play a key role in a plant's photosynthetic responses to abiotic stress (Baker 1991; Adams et al. 2013). This GB-mediated maintenance of both photosynthetic performance and PSII stability in abiotically-stressed plants is also associated with increases plant growth and reproductive yield, i.e. traits regulated by plant hormones (Kurepin et al. 2013a). Thus, a key role of GB in alleviating or mitigating abiotic stress responses by the plant can be evaluated in the context of the integration of the regulation of GB's biosynthesis by plant "stress" hormones, its function in protecting the photosynthetic apparatus, and its subsequent direct and indirect interactions with plant "growth" hormones.

9.2 Biosynthesis and Accumulation of Glycine Betaine in Response to Abiotic Stress

Glycine betaine (N,N′,N″-trimethylglycine; Fig. 9.1) is found in many organisms, including bacteria, marine invertebrates, plants and mammals. GB is a "compatible osmolyte", i.e. one of the members of a group of small molecules, such as proline, prolinebetaine, β-alaninebetaine, choline-O-sulfate and 3-dimethylsulfoniopropionate (Rhodes and Hanson 1993; Chen and Murata 2011). Compatible osmolytes are uncharged at a neutral pH, are highly soluble in water and are excluded from the hydration sphere of proteins. This allows them to function in the protection of cells against osmotic inactivation by increasing water retention and stabilizing folded protein structures (Low 1985; Ballantyne and Chamberlin 1994; Liu and Bolen 1995; Sakamoto and Murata 2002; Ashraf and Foolad 2007). The main function of GB and other compatible osmolytes (which act as non-toxic cytoplasmic osmolytes) thus appears to be increasing a plant's tolerance to abiotic stresses (Dawson et al. 1969; Wyn Jones et al. 1977).

Not all higher plants exhibit GB accumulation following the initiation of abiotic stress and this is discussed in Kurepin et al. (2015a). For example, *Arabidopsis thaliana* (Hibino et al. 2002), eggplant (*Solanum melongena* L.; de Zwart et al. 2003), potato (*Solanum tuberosum* L.; de Zwart et al. 2003), tobacco (*Nicotiana tabacum* L.; Nuccio et al. 1998), tomato (*Solanum lycopersicum* L.; Park et al. 2004) and rice (*Oryza sativa* L.; Sakamoto and Murata 1998) were all reported to have no detectable accumulation of GB in response to abiotic stresses. That said, there are conflicting results regarding *A. thaliana* as Xing and Rajashekar (2001), using the same cultivar, have shown GB accumulation in shoot tissues following the Arabidopsis plant's exposure to a cold stress. Cultivar specificity was also shown for sorghum (*Sorghum bicolor* L. Moench.) and corn (*Zea mays* L.), two species which also have cultivars capable of accumulating GB (Grote et al. 1994; Saneoka et al. 1995). Furthermore, crossing GB-accumulating with GB non-accumulating genotypes resulted in the production of greater proportion of genotypes with high GB accumulation, than genotypes with low GB accumulation (Mickelbart et al. 2003). Thus, the ability to produce higher amounts of GB in response to an abiotic stress seems likely to be genotype specific within most plant species.

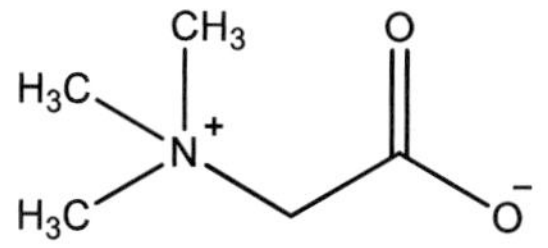

There are transgenic plants which can over-express GB biosynthetic genes of plant or bacterial origin (Chen and Murata 2011). For example, *A. thaliana*, rice and tobacco plants, which typically do not accumulate GB in response to abiotic stress, were genetically engineered to express plant-origin choline monooxygenase (CMO). These transgenic plants accumulated relatively low levels of GB unless exogenous choline was supplied and their stress tolerance was largely not increased (Nuccio et al. 1998; McNeil et al. 2001; Hibino et al. 2002; Shirasawa et al. 2006). However, when the same genotypes of these plant species were genetically engineered to express bacterial choline oxidase (COD), the plants with bacterial COD accumulated relatively high amounts of GB, even without administration of exogenous choline, and they exhibited increased stress tolerance (Hayashi et al. 1997; Nuccio et al. 1998; Sakamoto and Murata 1998; Huang et al. 2000; Mohanty et al. 2002; Hibino et al. 2002). Genotypes which do not normally accumulate GB may thus utilize GB biosynthetic precursors for other biosynthetic pathways. If so, this would explain why introducing the late stage plant-origin GB biosynthesis genes only results in significant GB accumulation when the transgenic plants are supplemented with choline. The GB biosynthetic genes in bacteria are different, as discussed below. Thus, the introduction of bacterial genes to plant genotypes lacking the ability to accumulate GB can result in significant GB

Fig. 9.1 Chemical structure of glycine betaine (N,N′,N″-trimethylglycine)

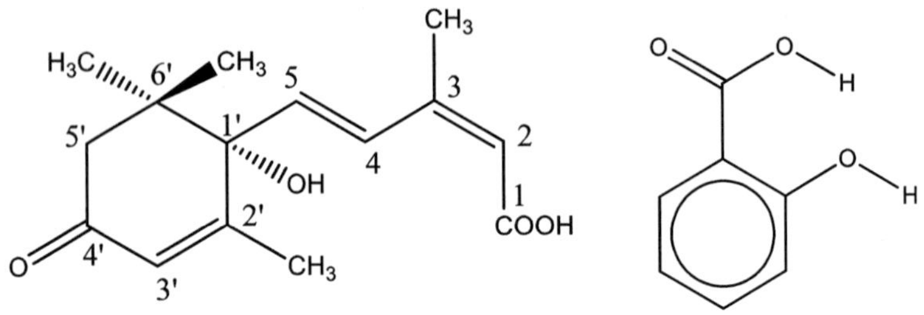

Fig. 9.2 Chemical structures of (S)-*cis*-abscisic acid (ABA, *left*) and salicylic acid (SA, *right*)

accumulation in the transgenic plants, even without supplemental choline.

The biosynthesis of GB in higher plants begins in the cytosol, when a methyl group from S-adenosylmethionine (SAM) is added to ethanolamine (EA) to produce monomethyl EA (Fig. 9.2). SAM is also involved in the biosynthesis of ethylene and polyamines (Abeles et al. 1992). The monomethyl EA is converted to choline via several biosynthetic steps (Sahu and Shaw 2009). Then, choline has to be transported to the chloroplast where it is converted to GB. The conversion of choline to GB in chloroplast is a two-step process in plants, but can be a one-, two- or three-step process in bacteria (Nyyssola et al. 2000; Sahu and Shaw 2009). In plants and some bacteria, CMO converts choline to betaine aldehyde (BA) (Rhodes and Hanson 1993). Then, betaine aldehyde dehydrogenase (BADH) converts BA to GB (Rhodes and Hanson 1993). In other bacteria species, such as *Escherichia coli*, the choline dehydrogenase (CDH) and BADH enzymes, encoded by *betA* and *betB* genes respectively, catalyse the conversion of choline to GB via BA (Landfald and Strom 1986). In contrast to higher plants and *E. coli*, the bacterium *Arthrobacter globiformis* uses the COD enzyme, encoded for by the gene *codA*, to catalyse the conversion of choline to GB in a single step (Ikuta et al. 1977). The use of bacterial *betA*, *betB* and *codA* genes to engineer transgenic plants over-producing GB has proven to be a very successful approach in creating varieties with increased abiotic stress tolerance (Chen and Murata 2011).

The main site of endogenous GB accumulation in plants following an abiotic stress event is in the chloroplasts of younger leaf tissues (Park et al. 2007; Chen and Murata 2011). Leaves are the main site of GB biosynthesis, but it can also occur in other plant tissues. For example, in barley (*Hordeum vulgare* L.), stress-induced GB biosynthesis occurs mainly in vascular tissues of leaves and the pericycle of roots, with major accumulations of GB being detected in younger leaves (Hattori et al. 2009). The transport of GB from the main biosynthetic sites occurs mainly via the phloem (Makela et al. 1996a; Hattori et al. 2009). Translocation studies with [^{14}C]-labelled GB in tomato, pea (*Pisum sativum* L.), soybean (*Glycine max* [L.] Merr.) and turnip (*Brassica rapa* ssp. *Oleifera*) plants demonstrated that leaf to root translocation via the phloem occurred within 2 h, with labelled GB being found throughout the plant by 24 h (Makela et al. 1996a).

9.3 Glycine Betaine Alleviates Abiotic Stress Effects on Crop Growth and Yield via an Interaction with Photosynthesis

Abiotic stress leads, among other responses, to reduced rates of plant growth. The reduction in growth rate may allow the plant to cope with the effects of the stress, but this often comes at the expense of vegetative and/or reproductive productivity. Application of exogenous GB in foliar form, via seed imbibition or root drench has been shown to improve plant stress tolerance and thus maintain or increase the productivity of commercially important crops grown under both field and controlled environment conditions - see Table 9.1.

These GB-mediated increases in crop productivity are attributed mainly to the role GB plays

Table 9.1 The effects of exogenously applied GB on growth and other physiological parameters on abiotically stressed crop and forage plant species grown in a controlled environment, or under field conditions

Abiotic stress	GB dose	Mode of application	Physiological effect(s)
Barley (*Hordeum vulgare* L.)			
Drought	17.5 kg ha^{-1}	Foliar (field)	Increase in leaf area index (Makela et al. 1996b)
Heat	20 mM	Imbibition (seeds)	Increase in shoot biomass (Wahid and Shabbir 2005)
Canola (*Brassica napus* L.)			
Salinity	5 mM	Imbibition (seeds)	Increase in shoot biomass (Athar et al. 2009)
Corn (*Zea mays* L.)			
Drought	4 kg ha^{-1}	Foliar (field)	Increase in grain yield (Agboma et al. 1997a)
Drought	100 mM	Foliar (pots)	Increase in shoot biomass and grain yield (Anjum et al. 2011)
Drought	30 mM	Foliar (pots)	Increase in grain quality (Ali and Ashraf 2011)
Salinity	100 mM	Foliar (pots)	Increase in shoot biomass (Navaz and Ashraf 2007)
Cold	100 mg L^{-1}	Imbibition (seeds)	Increase in plant biomass (Farooq et al. 2008a)
Cold	2.5 mM	Roots (soil drenching, pots)	Increase in shoot biomass (Chen et al. 2000)
Eggplant (*Solanum melongena* L.)			
Salinity	50 mM	Foliar (pots)	Increase in shoot biomass and fruit yield (Abbas et al. 2010)
Pea (*Pisum sativum* L.)			
Drought	200 mM	Foliar (pots and field)	Increase in shoot biomass (Makela et al. 1997)
Pepper (*Capsicum annuum* L.)			
Salinity	10 mM	Imbibition (seeds)	Increase in germination (Korkmaz and Sirikci 2011)
Perennial ryegrass (*Lolium perenne* L.)			
Salinity	20 mM	Foliar (pots)	Increase in plant biomass (Hu et al. 2012)
Rice (*Oryza sativa* L.)			
Drought	50 mg L^{-1}	Foliar (pots)	Increase in shoot biomass and grain yield (Zhang et al. 2009)
Drought	150 mg L^{-1}	Foliar (field)	Increase in shoot biomass (Farooq et al. 2010)
Drought	100 mg L^{-1}	Foliar (field)	Increase in shoot biomass (Farooq et al. 2008b)
Salinity	5 mM	Roots (hydroponically grown)	Increase in shoot biomass (Rahman et al. 2002)
Salinity	15 mM	Roots (hydroponically grown)	Increase in shoot biomass (Demiral and Turkan 2004)
Salinity	50 mM	Foliar (pots)	Increase in shoot biomass (Cha-Um et al. 2007)
Sorghum (*Sorghum bicolor* [L.] Moench)			
Drought	4 kg ha^{-1}	Foliar (field)	Increase in grain yield (Agboma et al. 1997a)
Soybean (*Glycine max* L.)			
Drought	100 mM (3 kg ha^{-1})	Foliar (field)	Increase in seed yield (Agboma et al. 1997b)
Sunflower (*Helianthus annuus* L.)			
Drought	100 mg L^{-1}	Foliar (field)	Increase in achene yield (Hussain et al. 2008)
Drought	100 mg L^{-1}	Foliar (field)	Increase in achene yield (Iqbal et al. 2008)
Salinity	50 mM	Foliar (pots)	Increase in shoot biomass (Ibrahim et al. 2006)
Tobacco (*Nicotiana tabacum* L.)			
Drought	100 mM	Foliar (pots)	Increase in shoot biomass (Agboma et al. 1997c)
Drought	80 mM	Foliar (pots)	Increase in shoot biomass (Ma et al. 2007)
Tomato (*Solanum lycopersicum* L.)			
Cold	1 mM	Foliar (pots)	Increase in shoot height (Park et al. 2006)
Turnip rape (*Brassica rapa* ssp. *Oleifera*)			
Drought	50 mM	Foliar (pots and field)	Increase in shoot growth rate (Makela et al. 1997)

(continued)

Table 9.1 (continued)

Abiotic stress	GB dose	Mode of application	Physiological effect(s)
Wheat (*Triticum aestivum* L.)			
Drought	50 mM	Seed imbibition (for field study)	Increase in shoot biomass (Mahmood et al. 2009)
Drought	50 mM	Foliar (field)	Increase in shoot biomass (Shahbaz and Zia 2011)
Cold	250 mM	Foliar (pots)	Increased freezing tolerance (Allard et al. 1998)

in the protecting the plant's photosynthetic apparatus. Following from that thesis, an abiotic stress leads to GB accumulation which, in turn, is associated with maintaining near-optimal levels of photosynthesis performance. The apparent involvement of GB in the protection of photosynthetic processes of higher plants which are subjected to an abiotic stress can also be found in other photosynthetic organisms. For example, the cyanobacterium, *Synechococcus* sp. PCC7942, when subjected to a high salinity stress, exhibited an enhanced GB accumulation which was correlated with increased stabilization of the activity of ribulose 1,5-bisphosphate carboxylase/oxygenase (RuBisCO) (Nomura et al. 1998). Listed below are several examples of the GB-mediated protection of photosynthetic performance for a range of important crops subjected to one or more abiotic stresses:

Barley – Seeds pre-treated with GB produced seedlings with a greater shoot biomass and also an increased net photosynthetic rate when germinated under heat stress conditions, relative to control, heat stressed barley seeds (Wahid and Shabbir 2005).

Bean – Plants treated with foliar GB exhibited increases in stomatal conductance, transpiration and photosynthetic rates, as well as leaf relative water content, when grown under high salinity stress (Lopez et al. 2002).

Corn – Transformation of plants with the *betA* gene resulted in increased photosynthetic efficiency and higher total soluble sugar accumulation, as well as significantly higher tolerance to low temperature stress (Quan et al. 2004a, b).

Rice – Plants treated with foliar GB exhibited no ultrastructural leaf damage, such as swelling of thylakoids, disintegration of grana stacking and intergranal lamellae and destruction of mitochondria which occurred in untreated rice (Rahman et al. 2002). GB-treated rice plants also had enhanced CO_2 assimilation and better photosynthetic performance (Cha-Um et al. 2007), when subjected to a high salinity stress.

Ryegrass – Italian ryegrass (*Lolium multiflorum* Lam.) plants expressing the BADH gene (*ZBD1*) of zoysiagrass (*Zoysia tenuifolia* Willd. ex Trin.) exhibited improved PSII photochemistry, relative to wild type plants under salt stress conditions (Takahashi et al. 2010).

Soybean – Drought stress reduced photosynthetic rate and N_2 fixation in soybean plants (Weisz et al. 1985; Frederick et al. 1989) which caused an inhibition of leaf and stem growth, decrease in biomass accumulation and a reduction in reproductive yield (Muchow et al. 1986; Sinclair et al. 1987). However, foliar spray of field-grown soybean plants with GB increased photosynthetic rate and N_2 fixation, thus increasing growth and yield parameters (Agboma et al. 1997b).

Tobacco – Plants sprayed with GB exhibited higher drought stress tolerance, including improved shoot biomass. They also showed enhanced photosynthesis, greater stomatal conductance, higher carboxylation efficiency of assimilated CO_2 and an increased efficiency of PSII (Ma et al. 2007). Similar results were demonstrated for tobacco plants transformed with the spinach BADH gene (Yang et al. 2005, 2008).

Tomato – Plants treated with foliar GB and subjected to a cold stress maintained high photosynthetic rates under the cold stress conditions, as well as during the subsequent growth period at 25°C (Park et al. 2006).

Wheat – Plants treated with GB exhibited increases in photosynthetic capacity, stomatal conductance, transpiration rate and activities of enzymes associated with antioxidant effects, when subjected to a high salinity stress (Rajasekaran et al. 1997; Ashraf et al. 2008).

9.4 Glycine Betaine Protects the Photosynthetic Apparatus from Abiotic Stress

The exposure of photosynthetic organisms to environmental stresses (low and high temperatures, excess light, water stress, etc.) may cause an imbalance between the capacity for harvesting light energy and the capacity to dissipate this energy through metabolic activity. Such an imbalance can result in excess photosystem II (PSII) excitation pressure, which is a measure of the relative redox state of Q_A, the first stable quinine electron acceptor of PSII reaction centers (Hüner et al. 1996, 2012, 2014). The imbalance between the excess reducing equivalents and the capacity of the metabolic sinks to utilize the electrons generated from the absorbed light energy, may be caused either by exposure to an irradiance that exceeds the light harvesting capacity of the photosynthetic apparatus, or by environmental constraints that may decrease the capacity of metabolic pathways downstream of the photochemistry (e.g., pathways associated with C, N, and S assimilation) to utilize photosynthetically generated reductants (Hüner et al. 1996, 1998, 2012). Such an imbalance, which can be induced by a range of stresses, has the potential to generate reactive oxygen species (ROS), such as $^1O_2$ and O^-_2, thereby leading to photoinhibition and photooxidative damage of PSII (Aro et al. 1993; Long et al. 1994; Nishiyama et al. 2006; Murata et al. 2012). Apart from damage to PSII, environmental stresses can also result in photoinhibitory damage of photosystem I (PSI) (Sonoike 1996; Ivanov et al. 1998, 2012a, b; Scheller and Haldrup 2005).

The most vulnerable part of the photosynthetic apparatus is PSII, which is believed to play a key role in a plant's photosynthetic responses to abiotic stresses (Baker 1991; Murata et al. 2012; Adams et al. 2013). Glycine betaine, synthesized in chloroplasts, has been associated with an increased protection of functional proteins, including enzymes, and also of lipids of the photosynthetic apparatus which are necessary for maintaining light-dependent electron flow (Chen and Murata 2011). Exogenously applied GB has been shown to protect thylakoid membranes against freezing stress (Coughlan and Heber 1982) and to stabilize the oxygen-evolving complex of PSII (Murata et al. 1992; Papageorgiou and Murata 1995; Busheva and Apostolova 1997). In doing this GB counteracts the stress-induced inactivation of the PSII complex (Papageorgiou et al. 1991; Mamedov et al. 1993; Allakhverdiev et al. 1996, 2003).

Genetic manipulations have also confirmed that GB functions by increasing the plant's tolerance to photoinhibition of PSII. Insertion of the *codA* gene into chloroplasts of Indian mustard (*Brassica juncea* [L.] Czern.) plants led to an increased tolerance of PSII to high light intensity under either of salt or cold stresses (Prasad and Saradhi 2004). Moreover, studies with transgenic rice (Sakamoto and Murata 1998), Arabidopsis (Alia et al. 1998) and tobacco (Holmstrom et al. 2000) plants which were transgenic for either plant or bacterial genes that regulate GB biosynthesis, demonstrated that the transgenic plants were more tolerant of the photoinhibition caused by a range of abiotic stresses.

Further, application of GB minimized photodamage to the PSI submembrane particles in cold-stressed spinach leaves (Rajagopal and Carpentier 2003). Foliar-applied GB also prevented photoinhibition in wheat during freezing (Allard et al. 1998) and drought stresses (Ma et al. 2006). Also, GB-treated wheat plants maintained a higher net photosynthesis rate during drought stress than untreated control plants and the maximal photochemistry efficiency of their PSII (Fv/Fm) was also increased (Ma et al. 2006). Furthermore, GB-treated wheat plants recovered more rapidly from a drought stress-induced photoinhibition of PSII (Ma et al.

2006). The response of the GB-treated wheat plants to freezing stress was accompanied by improved tolerance to photoinhibition of PSII and the steady-state yield of electron transport (Allard et al. 1998). In another study, a 100 mM foliar application of GB increased chlorophyll content, produced a modified lipid composition within the thylakoid membranes, and increased both gas exchange and photosynthesis in drought-stressed wheat (Zhao et al. 2007).

9.5 Glycine Betaine Interaction with Reactive Oxygen Species Aids in the Protection of the Photosynthetic Apparatus

Exposure of plants to abiotic stresses, which limit photosynthetic CO_2 fixation, increases the excitation pressure of the chloroplast since excess electrons accumulate on the acceptor-side of PSI (Hüner et al. 1998; Wilson et al. 2006). This increase in excitation pressure leads to enhanced production of ROS (Hüner et al. 1998). In turn, an increase in ROS can directly damage membranes of both the cell and its organelles, including chloroplasts (Hüner et al. 1998).

The increase in GB accumulation following the exposure of a plant to an abiotic stress may counteract ROS activity in a number of ways. First, since some of the excess electrons generated by photosynthesis in response to an abiotic stress are utilized in GB biosynthesis, excitation pressure is reduced and less ROS is being generated. Second, the actions of GB in the maintenance of osmoregulation, as well as in stabilization and protection of the biological membrane integrity can directly counteract the negative effects of ROS activity on a range of physiological parameters. Third, increased GB accumulation may decrease ROS production by maintaining higher rates of CO_2 assimilation in plants exposed to an abiotic stress. This maintenance of CO_2 assimilation may occur via GB's role as chemical chaperone activating a number of molecular chaperones (Diamant et al. 2001), helping to maintain the structure and/or functional integrity of the major CO_2-fixing enzymes, such as Rubisco, Rubisco activase, FBPase, FBP aldolase, and PRKase. Fourth, increased GB accumulation can counteract the ROS-induced inhibition of the reaction center protein of PSII (psbA, D1) at the translation step. This action of GB would be accomplished via its effect on protecting the repair cycle of photo-damaged PSII (Ohnishi and Murata 2006). Fifth, the increase in GB accumulation may activate the expression of genes for ROS-scavenging enzymes and thus enhance the protection of the photosynthetic machinery from an abiotic stress. Sixth and finally, increased GB accumulation may act directly or indirectly on ROS transport by limiting ROS-induced efflux of K^+ ions, thereby protecting membrane integrity, or by a blocking of K^+ channels.

9.6 Glycine Betaine Interaction with Plant "Stress" Hormones and Photosynthesis

To cope with abiotic stresses plants can modify biosynthesis and action of "stress" hormones, such as ABA, ethylene and possibly SA (Fig. 9.3; Abeles et al. 1992; Morgan and Drew 1997; Dodd and Davies 2010; Kurepin et al. 2013b). These stress hormones may interact with GB biosynthesis directly or indirectly (Kurepin et al. 2015a). Abscisic acid has been shown to directly upregulate GB biosynthesis (Ishitani et al. 1995; Jagendorf and Takabe 2001) and the interaction of ABA and SA is well documented (see Kurepin et al. 2013b), and may indirectly contribute to enhanced GB biosynthesis (Kurepin et al. 2015a). The interaction of GB with ethylene is potentially more complicated, as the pathway for the biosynthesis of choline is integrated with the biosynthesis of ethylene (Sahu and Shaw 2009). These defined and potential interactions of GB with plant stress hormones are discussed in detail below.

Plant tissues accumulate ABA rapidly in response to several abiotic stresses (Kurepin et al. 2008a; Dodd 2003; Dodd and Davies

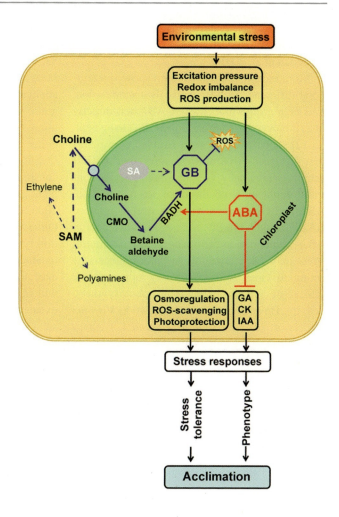

Fig. 9.3 A simplified schematic model for GB action and the interactions of GB with photosynthesis, plant hormones and ROS in alleviating abiotic stress effects in plants

2010; Qaderi et al. 2012). This rapid *de novo* synthesis of ABA, which occurs primary in leaves of stressed plants, quickly leads to stomatal closure and thus helps to protect the plant from rapid desiccation following abiotic (osmotic) stress (Schwartz and Zeevaart 2010). Treatment of barley plants with exogenous ABA caused significant increases in GB accumulation in salt-, drought-, or cold-stressed plants (Jagendorf and Takabe 2001). These increases in GB accumulation following ABA application were shown to occur due to the substantially increased BADH mRNA levels in leaves and roots of barley plants (Ishitani et al. 1995). Furthermore, application of GB increased the stress tolerance of barley a range of stresses, including high salt levels, drought and freezing (Hitz et al. 1982; Arakawa et al. 1990, 1992; Kishitani et al. 1994). It is likely that ABA acts upstream of GB and that both act in tandem when increasing plant stress tolerance. This conclusion is supported by studies with *A. thaliana*, where freezing tolerance was increased by application of either GB or ABA, and application of ABA was shown to increase endogenous GB accumulation (Xing and Rajashekar 2001). Furthermore, application of fluridone (1-methyl-3-phenyl-5-[3-[trifluoromethyl]phenyl]-4[1H]-pyridinone), an inhibitor of carotenoid, and ABA biosynthesis (Gamble and Mullet 1986; Goggin et al. 2015), not only lowered endogenous ABA levels, but also reduced BADH mRNA transcript levels (Saneoka et al. 2001).

More recent studies with corn showed a temporal difference in ABA and GB accumulation, thus supporting a model where ABA acts directly on GB biosynthesis upstream of GB. Zhang et al. (2012) showed that drought-stressed corn plants first increased ABA accumulation, then BADH activity, and finally GB accumulation. In contrast, application of fluridone decreased both ABA and GB accumulation, as well as reduced BADH activity (Zhang et al. 2012). Increases in ABA accumulation following abiotic (osmotic) stress may cause a decline in photosynthesis by acting to close stomata, thereby opposing the effects of GB in maintaining photosynthetic activity. However, it has been shown in Virginia spiderwort (*Tradescantia virginiana* L.) leaves that increasing endogenous ABA content via continuous application of exogenous ABA decreases stomatal conductance without having negative effect on photosynthetic capacity (Franks and Farquhar 2001). Thus, there may be a direct interaction between ABA and GB at the biosynthesis level of GB, with both metabolites acting, in effect, to protect the plant from abiotic stress via protection of the photosynthetic capacity.

Abiotic stresses can also increase endogenous SA levels in plants, as was shown for plants subjected to drought, extreme temperature, reductions in light irradiance or quality, or treatment with UV light (Yalpani et al. 1994; Scott et al. 2004; Wang et al. 2005; Kurepin et al. 2010a, 2012a, 2013b). The abiotic stress-induced increases in endogenous SA levels are often paralleled by increases in endogenous ABA levels (Kurepin et al. 2013b). For example, in abiotically-stressed barley plants the increase in GB accumulation was associated not only with increased ABA, but also SA content (Jagendorf and Takabe 2001). It is thus likely that SA can interact directly with GB by influencing the biosynthesis of GB, or indirectly, via an ABA-SA interaction. Based on a proteomic analysis of corn leaves, both ABA and SA induce the production of several proteins that are primarily involved in photosynthesis, stress and defense responses (Wu et al. 2013). The concept of a SA-GB interaction is supported by a recent study with drought-stressed wheat plants, where application of each of SA or GB gave increased grain yield, but their co-application was even more effective in increasing grain yield, than either applied alone (Aldesuquy et al. 2012).

Ethylene is a gaseous plant hormone and its constitutive presence is required for optimal plant growth (Lee and Reid 1997; Kurepin et al. 2006, 2016; Walton et al. 2006, 2010, 2012; Oinam et al. 2011). However, ethylene production typically shows large increases in response to abiotic stresses (Abeles et al. 1992). These exceptional increases in ethylene production will often inhibit stem elongation and leaf area expansion, coincidentally causing promotion of stem radial growth and increased leaf thickness (Abeles et al. 1992; Kurepin et al. 2007, 2010b, c). The extent to which an abiotic stress increases ethylene production is variable. For example, drought-stressed alfalfa leaves produced more ethylene in response to moderate stress, but showed no change or even reductions in ethylene production in response to severe drought stress (Irigoyen et al. 1992). Similarly, increasing the growth temperature to 30 °C from 20 °C enhanced plant ethylene production (Hansen 1945; Burg and Thimann 1959; Yu et al. 1980; Kurepin et al. 2011a), but a further increase in growth temperature decreased ethylene production (Yu et al. 1980). Finally, cold acclimation increased ethylene production in some plant species, but decreased it in others, yet all of the species showed increased freezing tolerance (Kurepin et al. 2013c).

These variable changes in plant ethylene production in response to abiotic stresses could perhaps be best explained by an interaction of ethylene with GB at the biosynthesis level, i.e. the fact that during ethylene biosynthesis SAM donates a methyl group to choline. Thus, plants subjected to a moderate abiotic stress increase ethylene production without significantly increasing GB accumulation. This could be thought of as a short-term strategy to cope with the moderate abiotic stress. In contrast, when plants are subjected to a severe abiotic stress, one which may cause irreversible damage to the photosynthetic apparatus - and thus plant growth and survival – increases in GB biosynthesis may occur at the expense of ethylene synthesis.

The above conclusions are supported by multiple examples of correlations between GB and ethylene production in response to abiotic stress. Species that increase ethylene production during cold acclimation, such as tobacco and tomato (Ciardi et al. 1997; Zhang and Huang 2010), do not accumulate GB in response to abiotic stress (Nuccio et al. 1998; Park et al. 2004). In contrast, species that show a decreased ethylene production during cold acclimation, such as bean and wheat (Field 1984; Machacckova et al. 1989; Collins et al. 1995), do accumulate GB in response to abiotic stress (Takhtajan 1980; Gadallah 1999; McDonnell and Wyn Jones 1988; Allard et al. 1998; Wang et al. 2010).

9.7 Glycine Betaine Interactions with Plant "Growth" Hormones

Growth and development events in plants, such as stem elongation and leaf growth, stem and shoot biomass accumulation, root growth, flowering and reproductive development, are regulated mainly by plant "growth" hormones (Davies 2010; Kurepin et al. 2011b; 2014; Turnbull et al. 2012; Park et al. 2015; Zaman et al. 2014, 2015, 2016). In turn, the biosynthesis and action of plant growth-promoting hormones, including gibberellins, cytokinins and auxin, are influenced by abiotic signals (Peleg and Blumwald 2011; Kurepin et al. 2008b, 2011c, 2012b, 2015b). The effect of abiotic stress on the biosynthesis and/or action of plant "growth" hormones is typically negative, resulting in a reduced growth, thus allowing the plant to better cope with the stress. This negative effect of various stresses on plant growth also appears to be regulated by plant stress-related hormones. For example, a constitutive overproduction of ABA can negatively influence the biosynthesis or action of GAs and auxin (Schwartz and Zeevaart 2010), thereby leading to reduced plant growth and delayed reproductive development (Thompson et al. 2000; Qin and Zeevaart 2002). Since increases in ABA enhance GB accumulation, it is plausible to expect an antagonistic interaction between GB and plant growth-promoting hormones.

One example is the over-expression of the spinach chloroplast CMO and BADH genes in perennial ryegrass (*Lolium perenne* L.) plants, which increased GB accumulation and enhanced abiotic stress tolerance (Bao et al. 2011). These transgenic ryegrass plants had a dwarf phenotype and significantly lower levels of the growth-active GA, GA_1 and the normal (tall) phenotype could be restored by the exogenous application of a growth-active GA (Bao et al. 2011). Another example is shown for transgenic rice plants expressing the *codA* gene. Here, abiotic stress causes GB accumulation while also increasing the expression of the cytokinin dehydrogenase 1 (*CKX1*) gene, which encodes for the enzyme that deactivates growth-active CKs (Kathuria et al. 2009). Moreover, application of a synthetic auxin (dicamba) to abiotically-stressed *Bassia scoparia* (L.) A.J. Scott plants decreased CMO expression, thereby reducing both GB accumulation and stress tolerance (Kern and Dyer 2004).

9.8 Modelling the Interactions Between Glycine Betaine, Plant Hormones, Photosynthesis and Reactive Oxygen Species

A schematic model featuring interdependent interactions of GB with photosynthesis, plant hormones and ROS in abiotic stress responses by GB-synthesizing higher plants is presented in Fig. 9.3. In this model exposure to stress increases the PSII excitation pressure PSII (Hüner et al. 1998) and results in a redox imbalance of the photosynthetic electron transport carriers and/or the redox status of the chloroplast's stroma, thus leading to production of ROS. These events are then accompanied by an increase in GB biosynthesis. The GB biosynthesis pathway shares the same precursor (SAM) with the ethylene and polyamine biosynthesis pathways. Excess photosynthetic electrons produced as a result of the increased excitation pressure can be utilized in the oxidation of choline to BA (Brouquisse et al. 1989; Rathinasabapathi et al. 1997), a process regulated by CMO. It has thus been suggested that the increased synthesis of GB may contribute to stress tolerance by decreasing the excitation pressure and redox imbalance, thus serving as an effective, albeit indirect, mechanism for photoprotection of the photosynthetic apparatus (Kurepin et al. 2015a). The biosynthesis of GB from BA, which is regulated by betaine aldehyde dehydrogenase, can be enhanced by stress-induced increases in endogenous ABA levels. Increase in excitation pressure and a redox imbalance may also contribute to this increase in endogenous ABA (Kacperska 1999; Rapacz et al. 2003). The biosynthesis of GB can also be enhanced by SA, and SA typically accumulates in response to a wide range of environmental stresses.

Enhanced GB accumulation increases the plant's tolerance to stress by maintaining cell and tissue osmoregulation and activating the systems in chloroplasts and the cytosol which scavenge ROS. The effects of an increase in ABA accumulation likely occur in an indirect manner, either by affecting the biosynthesis or action of growth-active GAs, CKs and IAA. The increased accumulation of GB is also associated with a reduction in the levels of growth-active GAs, CKs and IAA. These reductions in growth-active hormones lead to changes in plant phenotype – typically a dwarfing of the plant shoot. Taken together, interactions between GB, the photosynthetic apparatus and plant hormones, appear to be the major factors leading to an effective acclimation of higher plants to a wide range of environmental stresses.

Acknowledgements We would like to thank Dr. Shazia Zaman for drawing the chemical structures of glycine betaine, abscisic acid and salicylic acid. We would also like to acknowledge the financial support from the Ballance Agri-Nutrients, New Zealand (MZ, LVK, RPP), KEMPE Foundation, Sweden (VMH, LVK, NPAH), Natural Sciences and Engineering Research Council of Canada (NPAH), the Canada Research Chairs Program (NPAH) and the Canada Foundation for Innovation (NPAH).

References

Abbas W, Ashraf M, Akram NA. Alleviation of salt-induced adverse effects in eggplant (*Solanum melongena* L.) by glycinebetaine and sugarbeet extracts. Sci Horticul 2010; 125: 188–95.

Abeles FB, Morgan PW, Saltveit ME. Ethylene in Plant Biology, second edition. Academic Press, New York 1992.

Adams W, III, Muller O, Cohu C, Demmig-Adams B. May photoinhibition be a consequence, rather than a cause, of limited plant productivity? Photosyn Res 2013; 117: 31–44.

Agboma PC, Jones MGK, Peltonen-Sainio P, Rita H, Pehu E. Exogenous glycinebetaine enhances grain yield of maize, sorghum and wheat grown under two supplementary watering regimes. J Agron Crop Sci 1997a; 178: 29–37.

Agboma PC, Sinclair TR, Jokinen K, Peltonen-Sainio P, Pehu E. An evaluation of the effect of exogenous glycinebetaine on the growth and yield of soybean: timing of application, watering regimes and cultivars. Field Crop Res 1997b; 54: 51–64.

Agboma PC, Peltonen-Sainio P, Hinkkanen R, Pehu E. Effect of foliar application of glycinebetaine on yield components of drought-stressed tobacco plants. Exp Agric 1997c; 33: 345–52.

Aldesuquy HS, Abbas MA, Abo-Hamed SA, Elhakem AH, Alsokari SS. Glycine betaine and salicylic acid induced modification in productivity of two different cultivars of wheat grown under water stress. J Stress Physiol Biochem 2012; 8: 72–89.

Ali Q, Ashraf M. Exogenously applied glycinebetaine enhances seed and seed oil quality of maize (*Zea mays* L.) under water deficit conditions. Environ Exp Bot 2011; 71: 249–59.

Alia, Hayashi H, Sakamoto A, Murata N. Enhancement of the tolerance of Arabidopsis to high temperatures by genetic engineering of the synthesis of glycinebetaine. Plant J 1998; 16: 155–61.

Allakhverdiev SI, Feyziev YM, Ahmed A, Hayashi H, Aliev JA, Kimlov VV, Murata N, Carpentier R. Stabilization of oxygen evolution and primary electron transport reactions in photosystem II against heat stress with glycinebetaine and sucrose. J Photochem Photobiol 1996; 34: 149–57.

Allakhverdiev SI, Hayashi H, Nishiyama Y, Ivanov AG, Aliev JA, Klimov VV, Murata N, Carpentier R. Glycinebetaine protects the D1/D2/Cytb559 complex of photosystem II against photo-induced and heat-induced inactivation. J Plant Physiol 2003; 160: 41–49.

Allard F, Houde M, Krol M, Ivanov A, Huner NPA, Sarhan F. Betaine improves freezing tolerance in wheat. Plant Cell Physiol 1998; 39: 1194–202.

Anjum SA, Farooq M, Wang LC, Xue LL, Wang SG, Wang L, Zhang S, Chen M. Gas exchange and chlorophyll synthesis of maize cultivars are enhanced by exogenously-applied glycinebetaine under drought conditions. Plant Soil Environ 2011; 57: 326–31.

Arakawa K, Katayama M, Takabe T. Levels of glycinebetaine and glycinebetaine aldehyde dehydrogenase activity in the green leaves, and etiolated leaves and roots of barley. Plant Cell Physiol 1990; 31: 797–803.

Arakawa K, Mizuno K, Kishitani S, Takabe T. Immunological studies of betaine aldehyde dehydrogenase of barley. Plant Cell Physiol 1992; 33: 833–40.

Aro E-M, Virgin I, Anderson B. Photoinhibition of photosystem II: inactivation, protein damage and turnover. Biochim Biophys Acta 1993; 1143: 113–34.

Ashraf M, Foolad MR. Roles of glycinebetaine and proline in improving plant abiotic stress resistance. Environ Exp Bot 2007; 59: 206–16.

Ashraf M, Nawaz K, Athar HUR, Raza SH. Growth enhancement in two potential cereal crops, maize and wheat, by exogenous application of glycinebetaine. In: Abdelly C, Ozturk M, Ashraf M, Grignon C, Eds. Biosaline Agriculture and High Salinity Tolerance. Birkhauser Verlag, Switzerland, 2008; pp. 21–35.

Athar HUR, Ashraf M, Wahid A, Jamil A. Inducing salt tolerance in canola (Brassica napus L.) by exogenous application of glycinebetaine and proline: responses at the initial growth stages. Pakistan J Bot 2009; 41: 1311–9.

Baker NR. Possible role of photosystem II in environmental perturbations of photosynthesis. Physiol Plant 1991; 81: 563–70.

Ballantyne JS, Chamberlin ME. Regulation of cellular amino acid levels. In: Strange K, Ed. Cellular and Molecular Physiology of Cell Volume Regulation. CRC Press, Boca Raton, 1994; pp. 111–22.

Bao Y, Zhao R, Li F, Tang W, Han L. Simultaneous expression of Spinacia oleracea chloroplast choline monooxygenase (CMO) and betaine aldehyde dehydrogenase (BADH) genes contribute to dwarfism in transgenic Lolium perenne. Plant Mol Biol Rep 2011; 29: 379–388.

Brouquisse R, Weigel P, Rhodes D, Yocum CF, Hanson AD. Evidence for a ferredoxin-dependent choline monooxygenase from spinach chloroplast stroma. Plant Physiol 1989; 90: 322–329.

Burg SP, Thimann KV. The physiology of ethylene formation in apples. Proc Nat Acad Sci USA 1959; 45: 335–44.

Busheva M, Apostolova E. Influence of saccharides and glycine betaine on freezing of photosystem 2-enriched particles: a chlorophyll fluorescence study. Photosynthetica 1997, 34: 591–4.

Cha-Um S, Supaibulwatana K, Kirdmanee C. Glycinebetaine accumulation, physiological characterizations and growth efficiency in salt-tolerant and salt-sensitive lines of indica rice (Oryza sativa L. ssp. indica) in response to salt stress. J Agron Crop Sci 2007; 193: 157–66.

Chen THH, Murata N. Glycinebetaine protects plants against abiotic stress: mechanisms and biotechnological applications. Plant Cell Environ 2011; 34: 1–20.

Chen WP, Li PH, Chen THH. Glycinebetaine increases chilling tolerance and reduces chilling-induced lipid peroxidation in Zea mays L. Plant Cell Environ 2000; 23: 609–18.

Ciardi JA, Deikman J, Orzolek MD. Increased ethylene synthesis enhances chilling tolerance in tomato. Physiol Plant 1997; 101: 333–40.

Collins GG, Nie X, Saltveit ME. Heat shock increases chilling tolerance of mung bean hypocotyls tissues. Physiol Plant 1995; 89: 117–24.

Coughlan SJ, Heber U. The role of glycinebetaine in the protection of spinach thylakoids against freezing stress. Planta 1982; 156: 62–69.

Davies PJ. The Plant Hormones: Their Nature, Occurrence, and Functions. In: Davies PJ, Ed. Plant Hormones: Biosynthesis, Signal Transduction and Action!, 3rd ed. Springer Science+Business Media B.V.: Dordrecht, The Netherlands, 2010; pp. 1–15.

Dawson RWC, Elliot WH, Jones KM. Data for Biochemical research. II. Clarendon Press, Oxford, 1969; p. 12.

de Zwart FJ, Slow S, Payne RJ, Lever M, George PM, Gerrard JA, Chambers ST. Glycine betaine and glycine betaine analogues in common foods. Food Chem 2003; 83: 197–204.

Demiral T, Turkan I. Does exogenous glycinebetaine affect antioxidative system of rice seedlings under NaCl treatment? J Plant Physiol 2004; 161: 1089–100.

Diamant S, Eliahu N, Rosenthal D, Goloubinoff P. Chemical chaperones regulate molecular chaperones in vitro and in cells under combined salt and heat stresses. J Biol Chem 2001; 276: 39586–39591.

Dodd IC. Hormonal interactions and stomatal responses. J Plant Growth Regul 2003; 22: 32–46.

Dodd IC, Davies WJ. Hormones and the regulation of water balance. In: Davies PJ, Ed. Plant Hormones: Biosynthesis, Signal Transduction and Action! (revised third edition). Springer Dordrecht Heidelberg London New York, 2010; pp. 241–61.

Farooq M, Aziz T, Hussain M, Rehman H, Jabran K, Khan MB. Glycinebetaine improves chilling tolerance in hybrid maize. J Agron Crop Sci 2008a; 194: 152–60.

Farooq M, Basra SMA, Wahid A, Cheema ZA, Khalid A. Physiological role of exogenously applied glycinebetaine to improve drought tolerance in fine grain aromatic rice (Oryza sativa L.). J Agron Crop Sci 2008b; 194: 325–33.

Farooq M, Wahid A, Lee D-J, Cheema SA, Aziz T. Comparative time course action of the foliar applied glycinebetaine, salicylic acid, nitrous oxide, brassinosteroids and spermine in improving drought resistance of rice. J Agron Crop Sci 2010; 196: 336–45.

Field RJ. The role of 1-aminocyclopropane-1-carboxylic acid in the control of low temperature induced ethylene production in leaf tissue of *Phaseolus vulgaris* L. Ann Bot 1984; 54: 61–7.

Franks PJ, Farquhar GD. The effect of exogenous abscisic acid on stomatal development, stomatal mechanics, and leaf gas exchange in *Tradescantia virginiana*. Plant Physiol 2001; 125: 935–42.

Frederick JR, Alm DM, Hesketh JD. Leaf photosynthetic rates, stomatal resistances, and internal CO_2 concentrations of soybean cultivars under drought stress. Photosynthetica 1989; 23: 575–84.

Gadallah MAA. Effects of proline and glycinebetaine on *Vicia faba* responses to salt stress. Biol Plant 1999; 42: 249–57.

Gamble PE, Mullet JE. Inhibition of carotenoid accumulation and abscisic acid biosynthesis in fluridone-treated dark-grown barley. Eur J Biochem 1986; 160: 117–121.

Goggin DE, Emery RJN, Kurepin LV, Powles SB. A potential role for endogenous microflora in dormancy release, cytokinin metabolism and the response to fluridone in *Lolium rigidum* seeds. Ann Bot 2015; 115: 293–301.

Grote EM, Ejeta G, Rhodes D. Inheritance of glycinebetaine deficiency in sorghum. Crop Sci 1994; 34: 1217–20.

Hansen E. Quantitative study of ethylene production in apple varieties. Plant Physiol 1945; 20: 631–5.

Hattori T, Mitsuya S, Fujiwara T, Jagendorf AT, Takabe T. Tissue specificity of glycinebetaine synthesis in barley. Plant Sci 2009; 176: 112–8.

Hayashi H, Alia, Mustardy L, Deshnium P, Ida M, Murata N. Transformation of *Arabidopsis thaliana* with the *codA* gene for choline oxidase: accumulation of glycinebetaine and enhanced tolerance to salt and cold stress. Plant J 1997; 12: 133–42.

Hibino T, Waditee R, Araki E, Ishikawa H, Aoki K, Tanaka Y, Takabe T. Functional characterization of choline monooxygenase, an enzyme for betaine synthesis in plants. J Biol Chem 2002; 277: 41352–60.

Hitz WD, Ladyman JAR, Hanson AD. Betaine synthesis and accumulation in barley during field water stress. Crop Sci 1982; 22: 47–54.

Holmstrom K-O, Somersalo S, Mandal A, Palva TE, Welin B. Improved tolerance to salinity and low temperature in transgenic tobacco producing glycinebetaine. J Exp Bot 2000; 51: 177–85.

Hu LX, Hu T, Zhang XZ, Pang HC, Fu JM. Exogenous glycine betaine ameliorates the adverse effect of salt stress on perennial ryegrass. J Am Soc Hort Sci 2012; 137: 38–46.

Huang J, Hirji R, Adam L, Rozwadowski KL, Hammerlindl JK, Keller WA, Selvaraj G. Genetic engineering of glycinebetaine production toward enhancing stress tolerance in plants: metabolic limitations. Plant Physiol 2000; 122: 747–56.

Hüner NPA, Maxwell DP, Gray GR, Savitch LV, Krol M, Ivanov AG, Falk S. Sensing environmental change: PSII excitation pressure and redox signalling. Physiol Plant 1996; 98: 358–364.

Hüner NPA, Oquist G, Sarhan F. Energy balance and acclimation to light and cold. Trends Plant Sci 1998; 3: 224–230.

Hüner NPA, Bode R, Dahal K, Hollis L, Rosso D, Krol M, Ivanov AG. Chloroplast redox imbalance governs phenotypic plasticity: the "grand design of photosynthesis" revisited. Front Plant Sci 2012; 3: 255.

Hüner NPA, Dahal K, Kurepin LV, Savitch L, Singh J, Ivanov AG, Kane K, Sarhan F. Potential for increased photosynthetic performance and crop productivity in response to climate change: role of CBFs and gibberellic acid. Front Chem 2014; 2: 18.

Hussain M, Malik MA, Farooq M, Ashraf MY, Cheema MA. Improving drought tolerance by exogenous application of glycinebetaine and salicylic acid in sunflower. J Agron Crop Sci 2008; 194: 193–9.

Ibrahim M, Anjum A, Khaliq N, Iqbal M, Athar HUR. Four foliar applications of glycinebetaine did not alleviate adverse effects of slat stress on growth of sunflower. Pakistan J Bot 2006; 38: 1561–70.

Ikuta S, Imamura S, Misaki H, Horiuti Y. Purification and characterization of choline oxidase from *Arthrobacter globiformis*. J Biochem 1977; 82: 1741–9.

Iqbal N, Ashraf M, Ashraf MY. Glycinebetaine, an osmolyte of interest to improve water stress tolerance in sunflower (*Helianthus annuus* L.): water relations and yield. South Afr J Bot 2008; 74: 274–81.

Irigoyen JJ, Emerich DW, Sánchez-Díaz M. Alfalfa leaf senescence induced by drought stress: photosynthesis, hydrogen peroxide metabolism, lipid peroxidation and ethylene evolution. Physiol Plant 1992; 84: 67–72.

Ishitani M, Nakamura T, Han SY, Takabe T. Expression of the betaine aldehyde dehydrogenase gene in barley in response to osmotic stress and abscisic acid. Plant Mol Biol 1995; 27: 307–15.

Ivanov AG, Morgan R, Gray GR, Velitchkova MY, Hüner NPA. Temperature/light dependent development of selective resistance to photoinhibition of photosystem I. FEBS Lett 1998; 430: 288–292

Ivanov AG, Rosso D, Savitch LV, Stachula P, Rosembert M, Oquist G, Hurry V, Hüner NPA. Implications of alternative electron sinks in increased resistance of PSII and PSI photochemistry to high light stress in cold-acclimated *Arabidopsis thaliana*. Photosynth Res 2012a; 113:191–206.

Ivanov AG, Allakhverdiev SI, Hüner NPA, Murata N. Genetic decrease in fatty acid unsaturation of phosphatidylglycerol increased photoinhibition of photosystem I at low temperature in tobacco leaves. Biochim Biophys Acta 2012b; 1817: 1374–1379.

Jagendorf AT, Takabe T. Inducers of glycinebetaine synthesis in barley. Plant Physiol 2001; 127: 1827–35.

Kacperska A. Plant responses to low temperature: signaling pathways involved in plant acclimation. In: Margesin R, Schinner F, Eds. Salicylic Acid: Plant Growth and Development. Springer-Verlag Berlin Heilderberg, Germany, 1999; pp. 79–104.

Kathuria H, Giri J, Nataraja KN, Murata N, Udayakumar M, Tyagi AK. Glycinebetaine-induced water-stress tolerance in *codA*-expressing transgenic *indica* rice is associated with up-regulation of several stress responsive genes. Plant Biotech J 2009; 7: 512–526.

Kern AJ, Dyer WE. Glycine betaine biosynthesis is induced by salt stress but repressed by auxinic herbicides in *Kochia scoparia*. J Plant Growth Regul 2004; 23: 9–19.

Kiba T, Kudo T, Kojima M and Sakakibara H. Hormonal control of nitrogen acquisition: roles of auxin, abscisic acid, and cytokinin. J Exp Bot 2011; 62: 1399–1409.

Kishitani S, Watanabe K, Yasuda S, Arakawa K, Takabe T. Accumulation of glycinebetaine during cold acclimation and freezing tolerance in leaves of winter and spring barley plants. Plant Cell Environ 1994; 17: 89–95.

Korkmaz A, Sirikci R. Improving salinity tolerance of germinating seeds by exogenous application of glycinebetaine in pepper. Seed Sci Tech 2011; 39: 377–88.

Kurepin LV, Pharis RP. Light signaling and the phytohormonal regulation of shoot growth. Plant Sci 2014; 229: 280–289.

Kurepin LV, Mancell L, Reid DM, Pharis RP, Chinnappa CC. Possible roles for ethylene and gibberellin in the phenotypic plasticity of an alpine population of *Stellaria longipes*. Botany 2006; 84: 1101–1109.

Kurepin LV, Walton LJ, Reid DM. Interaction of red to far red light ratio and ethylene in regulating stem elongation of *Helianthus annuus*. Plant Growth Regul 2007; 51: 53–61.

Kurepin LV, Qaderi MM, Back TG, Pharis RP, Reid DM. A rapid effect of applied brassinolide on abscisic acid concentrations in *Brassica napus* leaf tissue subjected to short-term heat stress. Plant Growth Regul 2008a; 55: 165–7.

Kurepin LV, Emery RJN, Chinnappa CC, Reid DM. Light irradiance differentially regulates endogenous levels of cytokinins and auxin in alpine and prairie genotypes of *Stellaria longipes*. Physiol Plant 2008b; 134: 624–635.

Kurepin LV, Walton LJ, Reid DM, Chinnappa CC. Light regulation of endogenous salicylic acid levels in hypocotyls of *Helianthus annuus* seedlings. Botany 2010a; 88, 668–74.

Kurepin LV, Walton LJ, Yeung EC, Chinnappa CC, Reid DM. The interaction of light irradiance with ethylene in regulating growth of *Helianthus annuus* shoot tissues. Plant Growth Regul 2010b; 62: 43–50.

Kurepin LV, Yip WK, Fan R, Yeung EC, Reid DM. The roles and interactions of ethylene with gibberellins in the far-red enriched light-mediated growth of *Solanum lycopersicum* seedlings. Plant Growth Regul 2010c; 61: 215–22.

Kurepin LV, Walton LJ, Pharis RP, Emery RJN, Reid DM. Interactions of temperature and light quality on phytohormone-mediated elongation of *Helianthus annuus* hypocotyls. Plant Growth Regul 2011a; 64: 147–54.

Kurepin L, Haslam T, Lopez-Villalobos A, Oinam G, Yeung E. Adventitious root formation in ornamental plants: II. The role of plant growth regulators. Prop Ornam Plants 2011b; 11: 161–171.

Kurepin LV, Walton LJ, Yeung EC, Reid DM. The interaction of light irradiance with auxin in regulating growth of *Helianthus annuus* shoots. Plant Growth Regul 2011c; 65: 255–262.

Kurepin LV, Walton LJ, Hayward A, Emery RJN, Reid DM, Chinnappa CC. Shade light interaction with salicylic acid in regulating growth of sun (alpine) and shade (prairie) ecotypes of *Stellaria longipes*. Plant Growth Regul 2012a; 68: 1–8.

Kurepin LV, Farrow S, Walton LJ, Emery RJN, Pharis RP, Chinnappa CC. Phenotypic plasticity of sun and shade ecotypes of *Stellaria longipes* in response to light quality signaling: Cytokinins. Environ Exp Bot 2012b; 84: 25–32.

Kurepin LV, Ozga JA, Zaman M, Pharis RP. The physiology of plant hormones in cereal, oilseed and pulse crops. Prairie Soils Crops 2013a; 6: 7–23.

Kurepin LV, Dahal KP, Zaman M, Pharis RP. Interplay between environmental signals and endogenous salicylic acid concentration. In: Hayat S, Ahmad A, Alyemini MN, Eds. Salicylic Acid: Plant Growth and Development. Springer Science+Business Media B. V., Dordrecht, The Netherlands, 2013b; pp. 61–82.

Kurepin LV, Dahal KP, Savitch LV, Singh J, Bode R, Ivanov AG, Hurry V and Hüner NPA. Role of CBFs as integrators of chloroplast redox, phytochrome and plant hormone signaling during cold acclimation. Int J Mol Sci, 2013c; 14: 12729–63.

Kurepin LV, Zaman M, Pharis RP. Phytohormonal basis for the plant growth promoting action of naturally occurring biostimulators. J Sci Food Agric 2014; 94: 1715–1722.

Kurepin LV, Ivanov AG, Zaman M, Pharis RP, Allakhverdiev SI, Hurry V and Hüner NPA. Stress-related hormones and glycinebetaine interplay in protection of photosynthesis under abiotic stress conditions. Photosynth Res 2015a; 126: 221–235.

Kurepin LV, Pharis RP, Emery RJN, Reid DM, Chinnappa CC. Phenotypic plasticity of sun and shade ecotypes of *Stellaria longipes* in response to light quality signaling, gibberellins and auxin. Plant Physiol Biochem 2015b; 94: 174–180.

Kurepin LV, Yeung EC, Reid DM, Pharis RP. Light signaling regulates tulip organ growth and ethylene production in a tissue-specific manner. Inter J Plant Sci 2016; DOI: 10.1086/684947.

Landfald B, Strom AR. Choline-glycine betaine pathway confers a high level of osmotic tolerance in *Escherichia coli*. J Bacteriol 1986; 165: 849–55.

Lee SH, Reid DM. The role of endogenous ethylene in the expansion of *Helianthus annuus* leaves. Can J Bot 1997; 78: 501–508.

Liu Y, Bolen DW. The peptide backbone plays a dominant role in protein stabilization by naturally occurring osmolytes. Biochemistry 1995; 34: 12884–91.

Long SP, Humphries S, Falkowski PG. Photoinhibition of photosystem in nature. Ann Rev Plant Physiol Plant Mol Biol 1994; 45: 633–62.

Lopez CML, Takahashi H, Yamazaki S. Plant-water relations of kidney bean plants treated with NaCl and foliarly applied glycinebetaine. J Agron Crop Sci 2002; 188: 73–80.

Low PS. Molecular basis of the biological compatibility of nature's osmolytes. In: Gilles R, Gilles-Baillien M, Eds. Transport Processes, Iono- and Osmoregulation. Springer-Verlag, Berlin, 1985; pp. 469–77.

Ma Q-Q, Wang W, Lib Y-H, Lib D-Q, Zou Q. Alleviation of photoinhibition in drought-stressed wheat (*Triticum aestivum*) by foliar-applied glycinebetaine. J Plant Physiol 2006; 163: 165–75.

Ma XL, Wang YJ, Xie SL, Wang C, Wang W. Glycinebetaine application ameliorates negative effects of drought stress in tobacco. Rus J Plant Physiol 2007; 54: 472–9.

Machacckova I, Hanisova A, Krekule J. Levels of ethylene, ACC, MACC, ABA and proline as indicators of cold hardening and frost resistance in winter wheat. Physiol Plant 1989; 76: 603–7.

Mahmood T, Ashraf M, Shahbaz M. Does exogenous application of glycinebetaine as a pre-sowing seed treatment improve growth and regulate some key physiological attributes in wheat plants grown under water deficit conditions? Pakistan J Bot 2009; 41: 1291–302.

Makela P, Peltonen-Sainio P, Jokinen K, Pehu E, Setala H, Hinkkanen R, Somersalo S. Uptake and translocation of foliar-applied glycinebetaine in crop plants. Plant Sci 1996a; 121: 221–30.

Makela P, Mantila J, Hinkkanen R, Pehu E, Peltonen-Sainio P. Effect of foliar applications of glycinebetaine on stress tolerance, growth, and yield of spring cereals and summer turnip rape in Finland. J Agric Crop Sci 1996b; 176: 223–34.

Makela P, Kleemola J, Jokinen K, Mantila J, Pehu E, Peltonen-Sainio P. Growth response of pea and summer turnip rape to foliar application of glycinebetaine. Acta Agric Scan 1997; 47: 168–75.

Mamedov M, Hayashi H, Murata N. Effects of glycinebetaine and unsaturation of membrane lipids on heat stability of photosynthetic electron-transport and phosphorylation reactions in *Synechocystis* PCC6803. Biochim Biophys Acta 1993; 1142: 1–5.

McDonnell E, Wyn Jones RG. Glycinebetaine biosynthesis and accumulation in unstressed and salt-stressed wheat. J Exp Bot 1988; 39: 421–30.

McNeil SD, Nuccio ML, Ziemak MJ, Hanson AD. Enhanced synthesis of choline and glycine betaine in transgenic tobacco plants that overexpress phosphoethanolamine N-methyltransferase. Proc Nat Acad Sci USA 2001; 98: 10001–5.

Mickelbart MV, Peel G, Joly RJ, Rhodes D, Ejeta G, Goldsbrough PB. Development and characterization of near-isogenic lines of sorghum segregating for glycinebetaine accumulation. Physiol Plant 2003; 118: 253–61.

Mohanty A, Kathuria H, Ferjani A, Sakamoto A, Mohanty P, Murata N, Tyagi AK. Transgenics of an elite indica rice variety Pusa Basmati 1 harbouring the codA gene are highly tolerant to salt stress. Theor Appl Gen 2002; 106: 51–7.

Morgan PW, Drew MC. Ethylene and plant responses to stress. Physiol Plant 1997; 100: 620–30.

Muchow RC, Sinclair TR, Bennett JM, Hammond LC. Response of leaf growth, leaf nitrogen, and stomatal conductance to water deficits during vegetative growth of field-grown soybean. Crop Sci 1986; 26: 1190–5.

Murata N, Mohanty PS, Hayashi H, Papageorgiou GC. Glycinebetaine stabilizes the association of extrinsic proteins with the photosynthetic oxygen-evolving complex. FEBS Lett 1992; 296: 187–9.

Murata N, Allakhverdiev SI, Nishiyama Y. The mechanism of photoinhibition *in vivo*: Re-evaluation of the roles of catalase, α-tocopherol, non-photochemical quenching, and electron transport. Biochim Biophys Acta 2012; 1817: 1127–1133.

Nawaz K, Ashraf M. Improvement in salt tolerance of maize by exogenous application of glycinebetaine: growth and water relations. Pakistan J Bot 2007; 39: 1647–53.

Nishiyama Y, Allakhverdiev SI, Murata N. A new paradigm for the action of reactive oxygen species in the photoinhibition of photosystem II. Biochim. Biophys. Acta 2006; 1757: 742–749.

Nomura M, Hibino T, Takabe T, Sugiura T, Yokota A, Miyake H, Takabe T. Transgenically produced glycinebetaine protects ribulose 1,5-bisphosphate carboxylase/oxygenase from inactivation in *Synechococcus* sp. PCC7942 under salt stress. Plant Cell Physiol 1998; 39: 425–32.

Nuccio ML, Russell BL, Nolte KD, Rathinasabapathi B, Gage DA, Hanson AD. The endogenous choline supply limits glycine betaine synthesis in transgenic tobacco expressing choline monooxygenase. Plant J 1998; 16: 487–96.

Nyyssola A, Kerovuo J, Kaukinen P, von Weymarn N, Reinikainen T. Extreme halophiles synthesize betaine from glycine by methylation. J Biol Chem 2000; 275: 22196–201.

Ohnishi N, Murata N. Glycinebetaine counteracts the inhibitory effects of salt stress on the degradation and synthesis of D1 protein during photoinhibition in *Synechococcus* sp. PCC 7942. Plant Physiol 2006; 141: 758–765.

Oinam G, Yeung E, Kurepin L, Haslam T, Villalobos AL. Adventitious root formation in ornamental plants: I. General overview and recent successes. Prop Ornam Plants 2011; 11: 78–90.

Papageorgiou GC, Murata N. The unusually strong stabilizing effects of glycine betaine on the structure and function of the oxygen-evolving photosystem II complex. Photosyn Res 1995; 44: 243–52.

Papageorgiou GC, Fujimura Y, Murata N. protection of the oxygen evolving Photosystem II complex by glycinebetaine. Biochim Biophys Acta 1991; 1057: 361–366.

Park EJ, Jeknic Z, Sakamoto A, DeNoma J, Yuwansiri R, Murata N, Chen THH. Genetic engineering of glycinebetaine synthesis in tomato protects seeds, plants, and flowers from chilling damage. Plant J 2004; 40: 474–87.

Park EJ, Jeknic Z, Chen THH. Exogenous application of glycinebetaine increases chilling tolerance in tomato plants. Plant Cell Physiol 2006; 47: 706–14.

Park EJ, Jeknic Z, Pino MT, Murata N, Chen THH. Glycinebetaine accumulation is more effective in chloroplasts than in the cytosol for protecting transgenic tomato plants against abiotic stress. Plant Cell Environ 2007; 30: 994–1005.

Park EJ, Lee WY, Kurepin LV, Zhang R, Janzen L, Pharis RP. Plant hormone-assisted early family selection in *Pinus densiflora* via a retrospective approach. Tree Physiol 2015; 35: 86–94.

Peleg Z, Blumwald E. Hormone balance and abiotic stress tolerance in crop plants. Curr Opin Plant Biol 2011; 14: 290–295.

Prasad KVSK, Saradhi PP. Enhanced tolerance to photoinhibition in transgenic plants through targeting of glycinebetaine biosynthesis into the chloroplasts. Plant Sci 2004; 166: 1197–212.

Qaderi MM, Kurepin LV, Reid DM. Effects of temperature and watering regime on growth, gas exchange and abscisic acid content of canola (*Brassica napus*) seedlings. Environ Exp Bot 2012; 75: 107–13.

Qin X, Zeevaart JAD. Overexpression of a 9-cis-epoxycarotenoid dioxygenase gene in *Nicotiana plumbaginifolia* increases abscisic acid and phaseic acid levels and enhances drought tolerance. Plant Physiol 2002; 128: 544–51.

Quan R, Shang M, Zhang H, Zhao Y, Zhang J. Improved chilling tolerance by transformation with betA gene for the enhancement of glycinebetaine synthesis in maize. Plant Sci 2004a; 166: 141–9.

Quan R, Shang M, Zhang H, Zhao Y, Zhang J. Engineering of enhanced glycine betaine synthesis improves drought tolerance in maize. Plant Biotech J 2004b; 2: 477–86.

Rahman S, Miyake H, Takeoka Y. Effects of exogenous glycinebetaine on growth and ultrastructure of salt-stressed rice seedlings (*Oryza sativa* L.). Plant Prod Sci 2002; 5: 33–44.

Rajagopal S, Carpentier R. Retardation of photo-induced changes in Photosystem I submembrane particles by glycinebetaine and sucrose. Photosyn Res 2003; 78: 77–85.

Rajasekaran LR, Kriedemann PE, Aspinall D, Paleg LG. Physiological significance of proline and glycinebetaine: Maintaining photosynthesis during NaCl stress in wheat. Photosynthetica 1997; 34: 357–66.

Rapacz M, Waligórski P, Janowiak F. ABA and gibberellin-like substances during prehardening, cold acclimation, de- and reacclimation of oilseed rape. Acta Physiol Plant 2003; 25: 151–161.

Rathinasabapathi B, Burnet M, Russel BL, Gage DA, Liao P-C, Nye GJ, Scott P, Golbeck JH, Hanson AD. Choline monooxygenase, an unusual iron-sulfur enzyme catalyzing the first step of glycine betaine synthesis in plants: Prosthetic group characterization and cDNA cloning. Proc Natl Acad Sci USA 1997; 94: 3454–3458.

Rhodes D, Hanson AD. Quaternary ammonium and tertiary sulfonium compounds in higher plants. Ann Rev Plant Physiol Plant Mol Biol 1993; 44: 357–84.

Sahu BB, Shaw BP. Isolation, identification and expression analysis of salt-induced genes in *Suaeda maritima*, a natural halophyte using PCR-based suppression subtractive hybridization. BMC Plant Biol 2009; 9: 69.

Sakamoto A, Murata N. Metabolic engineering of rice leading to biosynthesis of glycinebetaine and tolerance to salt and cold. Plant Mol Biol 1998; 38: 1011–9.

Sakamoto A, Murata N. The role of glycinebetaine in the protection of plants from stress: clue from transgenic plants. Plant Cell Environ 2002; 25: 63–71.

Saneoka H, Nagasaka C, Hahn DT, Yang WJ, Premachandra GS, Joly RJ, Rhodes D. Salt tolerance of glycinebetaine-deficient and -containing maize lines. Plant Physiol 1995; 107: 631–8.

Saneoka H, Ishiguro S, Moghaieb REA. Effect of salinity and abscisic acid on accumulation of glycinebetaine and betaine aldehyde dehydrogenase mRNA in Sorghum leaves (*Sorghum bicolor*). J Plant Physiol 2001; 158: 853–9.

Scheller H, Haldrup A. Photoinhibition of photosystem I. Planta 2005; 221: 5–8.

Schwartz SH, Zeevaart JAD. Abscisic acid biosynthesis and metabolism. In: Davies PJ, Ed. Plant Hormones: Biosynthesis, Signal Transduction and Action! (revised third edition). Springer Dordrecht Heidelberg London New York, 2010; pp. 137–55.

Scott IM, Clarke SM, Wood JE, Mur LAJ. Salicylate accumulation inhibits growth at chilling temperature in Arabidopsis. Plant Physiol 2004; 135: 1040–1049.

Shahbaz M, Zia B. Does exogenous application of glycinebetaine through rooting medium alter rice (*Oryza sativa* L.) mineral nutrient status under saline conditions? J Appl Bot Food Qual 2011; 84: 54–60.

Shirasawa K, Takabe T, Takabe T, Kishitani S. Accumulation of glycinebetaine in rice plants that overexpress choline monooxygenase from spinach and evaluation of their tolerance to abiotic stress. Ann Bot 2006; 98: 565–71.

Sinclair TR, Muchow RC, Ludlow MM, Leach GJ, Lawn RJ, Foale MA. Field and model analysis of the effect of water deficits on carbon and nitrogen accumulation by soybean, cowpea and black gram. Field Crop Res 1987; 17: 121–40.

Sonoike K. Photoinhibition of photosystem I: its physiological significance in the chilling sensitivity of plants. Plant Cell Physiol 1996; 37: 239–247

Takahashi W, Oishi H, Ebina M, Komatsu T, Takamizo T. Production of transgenic Italian ryegrass expressing the betaine aldehyde dehydrogenase gene of zoysiagrass. Breed Sci 2010; 60: 279–85.

Takhtajan AL. Outline of the classification of flowering plants (Magnoliophyta). Bot Rev 1980; 46: 225–359.

Thompson AJ, Jackson AC, Symonds RC, Mulholland BJ, Dadswell AR, Blake PS, Burbidge A, Taylor IB. Ectopic expression of a tomato 9-*cis*-epoxycarotenoid dioxygenase gene causes over-production of abscisic acid. Plant J 2000; 23: 363–74.

Turnbull MH, Pharis RP, Kurepin LV, Sarfati M, Mander LN, Kelly D. Flowering in snow tussock (*Chionochloa* spp.) is influenced by temperature and hormonal cues. Funct Plant Biol 2012; 39: 38–50.

Wahid A, Shabbir A. Induction of heat stress tolerance in barley seedlings by pre-sowing seed treatment with glycinebetaine. Plant Growth Regul 2005; 46: 133–41.

Walton LJ, Kurepin LV, Reid DM, Chinnappa CC. Stem and leaf growth of alpine sun and prairie shade ecotypes of *Stellaria longipes* under different photoperiods: role of ethylene. Botany 2006; 84: 1496–1502.

Walton LJ, Kurepin LV, Reid DM, Chinnappa CC. - Narrow-band light regulation of ethylene and gibberellin levels in hydroponically-grown *Helianthus annuus* hypocotyls and roots. Plant Growth Regul 2010; 61: 53–59.

Walton LJ, Kurepin LV, Yeung EC, Shah S, Emery RJN, Reid DM, Pharis RP. Ethylene involvement in silique and seed development of canola (*Brassica napus* L.). Plant Physiol Biochem 2012; 58: 142–50.

Wang LJ, Huang WD, Liu YP, Zhan JC. Changes in salicylic and abscisic acid contents during heat treatment and their effect on thermotolerance of grape plants. Russ J Plant Physiol 2005; 52: 516–520.

Wang GP, Li F, Zhang J, Zhao MR, Hui Z, Wang W. Over-accumulation of glycine betaine enhances tolerance of the photosynthetic apparatus to drought and heat stress in wheat. Photosynthetica 2010; 48: 30–41.

Weisz PR, Denison RF, Sinclair TR. Response to drought stress of nitrogen fixation (acetylene reduction) rates by field-grown soybeans. Plant Physiol 1985; 78: 525–30.

Wilson KE, Ivanov AG, Öquist G, Grodzinski B, Sarhan F, Huner NPA. Energy balance, organellar redox status and acclimation to environmental stress. Can J Bot 2006; 84 1355–1370.

Wu L, Zu X, Wang X, Sun A, Zhang J, Wang S, Chen Y. Comparative proteomic analysis of the effects of salicylic acid and abscisic acid on maize (*Zea mays* L.) leaves. Plant Mol Biol Rep 2013; 31: 507–16.

Wyn Jones RG, Storey R, Leigh RA, Ahmad N, Pollard A. A hypothesis on cytoplasmic osmoregulation. In: Marre E, Cifferi O, Eds. Regulation of Cell Membrane Activities in Plants. Elsevier, Amsterdam, 1977; pp. 121–36.

Xing W, Rajashekar CB. Glycine betaine involvement in freezing tolerance and water stress in *Arabidopsis thaliana*. Environ Exp Bot 2001; 46: 21–8.

Yalpani N, Enyedi AJ, Leon J, Raskin I. Ultraviolet light and ozone stimulate accumulation of salicylic acid, pathogenesis related proteins and virus resistance in tobacco. Planta 1994; 193: 372–376.

Yang X, Liang Z, Lu C. Genetic engineering of the biosynthesis of glycinebetaine enhances photosynthesis against high temperature stress in transgenic tobacco plants. Plant Physiol 2005; 138: 2299–309.

Yang X, Liang Z, Wen X, Lu C. Genetic engineering of the biosynthesis of glycinebetaine leads to increased tolerance of photosynthesis to salt stress in transgenic tobacco plants. Plant Mol Biol 2008; 66: 73–86.

Yu Y-B, Adams DO, Yanf SF. Inhibition of ethylene production by 2, 4-dinitrophenol and high temperature. Plant Physiol 1980; 66: 286–90.

Zaman M, Ghani A, Kurepin LV, Pharis RP, Khan S, Smith TJ. Improving ryegrass-clover pasture dry matter yield and urea efficiency with gibberellic acid. J Sci Food Agricul 2014; 94: 2521–2528.

Zaman M, Kurepin LV, Catto W, Pharis RP. Enhancing crop yield with the use of N-based fertilizers co-applied with plant hormones or growth regulators. J Sci Food Agric 2015; 95: 1777–1785.

Zaman M, Kurepin LV, Catto W, Pharis RP. Evaluating the use of plant hormones and biostimulators in forage pastures to enhance shoot dry biomass production by perennial ryegrass (*Lolium perenne* L.). J Sci Food Agric 2016; 96: 715–726.

Zhang Z, Huang R. Enhanced tolerance to freezing in tobacco and tomato overexpressing transcription factor TERF2/LeERF2 is modulated by ethylene biosynthesis. Plant Mol Biol 2010; 73: 241–9.

Zhang LX, Li SX, Liang ZS. Differential plant growth and osmotic effects of two maize (*Zea mays* L.) cultivars to exogenous glycinebetaine application under drought stress. Plant Growth Regul 2009; 58: 297–305.

Zhang L, Gao M, Hu J, Zhang X, Wang K, Ashraf M. Modulation role of abscisic acid (ABA) on growth, water relations and glycinebetaine metabolism in two maize (*Zea mays* L.) cultivars under drought stress. Int J Mol Sci 2012; 13: 3189–202.

Zhao X-X, Ma Q-Q, Liang C, Fang Y, Wang YQ, Wang W. Effect of glycinebetaine on function of thylakoid membranes in wheat flag leaves under drought stress. Biol Plant 2007; 51: 584–8.

Photosynthetic Responses Under Harmful and Changing Environment: Practical Aspects in Crop Research

10

Marek Zivcak, Katarina Olsovska, and Marian Brestic

Summary

Climate change at global and regional scales as well as increased needs for crop production is predicted, emphasizing the urgent need for introduction of crops with enhanced productivity and tolerance to unfavorable abiotic conditions. Unlike to previous breeding strategies in main crops, the future gain in yield potential can be obtained probably only through an increase of photosynthetic productivity in optimum as well as in stress conditions. This fact emphasizes the importance of photosynthetic measurements, especially those based on non-invasive techniques, useful in real selection of crop genotypes. Photosynthetic responses at the leaf or canopy level can be well characterized by measurements of gas exchange and chlorophyll fluorescence which help to identify different sensitive components and important protective mechanisms within the photosynthetic apparatus. This chapter focuses on actual experiences with the photosynthetic measurements in strategic crops, applications and limits of ecophysiological methodology in detection of crop photosynthetic productivity as well as in the screening for improved drought and heat stress tolerance. The chapter also outlines the perspectives of photosynthesis research at the crop plant level, especially the application of photosynthetic methods for a high-throughput crop phenotyping.

Keywords

Photosynthesis • Stress • Drought • High temperature • Chlorophyll fluorescence • Phenotyping

Contents

10.1	Introduction	204
10.2	Photosynthesis and Productivity in Crop Plants	204
10.2.1	Gas Exchange Analyses of Photosynthetic Limitations	207
10.2.2	Mesophyll Limitations to Photosynthesis	209
10.2.3	The Chlorophyll Fluorescence Methods in Crop Research	210
10.2.3.1	Saturation Pulse Method	211
10.2.3.2	Analysis of Fast Chlorophyll Fluorescence Induction	213
10.2.3.3	Chlorophyll Fluorescence Imaging	215
10.2.4	Photosynthetic Data in Context of Plant and Canopy	216
10.3	The Photosynthesis and Abiotic Stress Factors	218
10.3.1	Limitation of Photosynthesis by Drought	218
10.3.2	Limitation of Photosynthesis by High Temperature	224
10.4	Future Challenges: The Central Role of Plant Phenotyping	230
10.5	Concluding Remarks	233
References		234

M. Zivcak • K. Olsovska • M. Brestic (✉)
Department of Plant Physiology, Slovak University of Agriculture, Tr. A. Hlinku 2, Nitra 94901, Slovakia
e-mail: marian.brestic@uniag.sk

10.1 Introduction

Crop production is a result of many processes, which take place at the chloroplast, leaves and canopy level, respectively. In order to increase production of strategic crops in the world it will be crucial either to extend their growing season or to improve efficiency of light conversion by plants. While the extension of the growing season will only be possible in regard to the cultivated crop species as well as local impacts of climate changes influencing the development of new progressive farming technologies, improvement of light conversion efficiency seems to be one of the few promising ways based on the increased photosynthetic rate per unit leaf area.

Besides the fact that there are many limiting factors operating within the photosynthetic machinery, which determine the photosynthetic rate under different environmental conditions, there are also many theoretical and field studies confirming that improvement of net photosynthetic rate does not ensure the same level of biomass production, as it depends also on both sink capacity of crops and regulation of photosynthates transport and allocation into the generative organs.

As traditional breeding and selection of crops for higher yields has already maximized many of easily reachable parameters, such as harvest index, crop architecture or plant growth cycle, and, by contrast, has not resulted so much in increased total biomass production, future work on the photosynthesis and productivity improvement remains a big challenge ensuring the food, feed or fuel production in a changing environment.

Up to 70% of global crop production is lost due to inappropriate environmental conditions that mostly inhibit the process of photosynthetic energy conversion and photosynthates production. Therefore a considerable effort has been paid to determine mechanisms of photosynthetic acclimation to different environmental factors.

In the chapter we summarize present knowledge on photosynthetic reactions of crop plants in response to fluctuating environment with a more detailed view on different sensitive sites within the photosynthetic apparatus as related to photosynthetic potential and abiotic stress factors, which may negatively influence crop productivity and yield.

10.2 Photosynthesis and Productivity in Crop Plants

Photosynthesis is the crucial process necessary for plant production. Improvement of photosynthesis can enhance food security in the following decades since we can expect an increase in world population (Evans 2013). The relationships among photosynthesis, growth and crop yield are not simple and has been subject of discussion for many years (Demetriades-Shah et al. 1992; Monteith 1994). Plants differ in the efficiency of conversion of photosynthetic intermediates into biomass as well as in partitioning of biomass into harvestable parts of crops producing crop yield. Therefore it is not surprising that the leaf photosynthetic rate was found to be weakly correlated with crop productivity when different genotypes were compared. Such conclusions led to the broadly extended opinion that the increase of leaf photosynthesis would not lead to improvement of crop yield (Long et al. 2006; Flood et al. 2011). The typical example was the classical study of Evans and Dunstone (1970) who showed that the leaf photosynthesis in modern genotypes of bread wheat is typically lower than the same parameter measured in leaves of wild ancestors. Similarly, comparisons of wheat varieties released over time, growing in the same field trials under favorable conditions indicated that the total aboveground biomass of new and old genotypes is similar despite increasing trend of grain yields (Austin et al. 1980, 1989; Evans 2013). Moreover, it was shown that the rate of photosynthetic assimilation can be limited by the capacity of the sink (i.e. ability to use photosynthates). For example, in grain crops, the major sink capacity is determined by the seed number and genetic predisposition for the seed size. Lower capacity of sinks leads to down-regulation of photosynthetic capacity (Peet and Kramer 1980). Influential studies have shown that crop yield seems to be more limited by the capacity of the sink than by photosynthetic capacity in non- stressed conditions (Borrás et al. 2004; Reynolds et al. 2005). Therefore, the possibility to increase crop yield through an increase of photosynthetic rate was considered unrealistic.

Anyway, Long and coauthors (2006) argued that this scientific dogma was broken up after the release of results of experiments with increased CO_2 under different conditions (Mitchell et al. 1999; Bender et al. 1999; Drake et al. 1997; Kimball 1983; Ainsworth et al. 2002; Ainsworth and Long 2005). For example, the series of experiments with spring wheat under ambient and elevated (doubled) concentration of CO_2 in open-top chambers in well-watered conditions showed the 50% increase of the flag leaf photosynthetic rate and 35% increase of grain yield (Mitchell et al. 1999; Bender et al. 1999). However, in other experiments increase of photosynthetic rate by 30% was associated with only 10% increase of relative growth rate. This discrepancy can be attributed to the fact that the enhanced availability of carbohydrates exceeded the ability of crop plants to utilize them due to the inherent limitations of growth capacity or nutrient insufficiencies (Kirschbaum 2011). C4 crops in similar conditions show little or no increase in photosynthetic rate or in yield (Ghannoum et al. 2001; Long et al. 2004, 2005). This meets the expectation that C4 photosynthesis is CO_2-saturated in natural conditions (Ghannoum et al. 2001). In addition, analysis of photosynthetic rate in Australian bread wheat genotypes with different date of release indicated that selection for higher grain yield led to unconscious selection for higher photosynthetic rate (Watanabe et al. 1994). Similarly, our results support this idea, comparing the photosynthetic and the leaf parameters of high-yielding wheat (*Triticum aestivum* L.) genotype with a wild relative species of the same group (*Aegilops cylindrica* Host.), which is used in advanced breeding programs as a gene donor in wheat breeding (Schoenenberger et al. 2005) (Fig. 10.1). The data indicate that a wild relative species of the same botanical group, having almost the same vegetation period and canopy development as the observed genotype of winter wheat, is much less productive, it has smaller and thicker leaves. Moreover, the leaves of *Aegilops* demonstrated significantly lower metabolic activity, as shown by the rate of CO_2 assimilation and mitochondrial respiration. Flood et al. (2011) reviewed the studies analyzing releases of rice and wheat genotypes through several decades, which show that yield increases were obtained through the increases in harvest index until the 1980s; in ongoing decades the increase in photosynthesis and accumulation of plant biomass have become more important. This well documents a transition from sink limitation to source limitation in major C3 crops (Shearman et al. 2005; Hubbart et al. 2007) providing strong arguments for the improvement of photosynthesis.

These and many other research findings strongly support the hypothesis that a sustained increase in the leaf photosynthetic rate can lead to the increase in total production of biomass that would be necessary to gain further increase in yield (Long et al. 2006; Parry et al. 2007; Zhu et al. 2010; Evans 2013). Moreover, some reports have shown that a higher leaf photosynthesis often correlates with a higher invasive potential and fecundity (Arntz and Delph 2001; Zou et al. 2007; Mozdzer and Zieman 2010; Arntz et al. 2000).

In this regard, the effect of CO_2 increase on yield increase suggest that similar yield gain can be obtained by improvement of efficiency of CO_2 fixation in metabolic pathways of photosynthesis. It can be achieved by an increase of regeneration of the ribulose biphosphate (RuBP), more efficient carboxylation catalysis or indirectly by decreased photorespiration, as ~30% of carbon fixed in C3 pathway is lost in the process of photorespiration (Monteith and Moss 1977). In addition, the decrease of losses caused by mitochondrial respiration is also regarded; however, this process is essential for cell metabolism and its decrease might have negative side effects

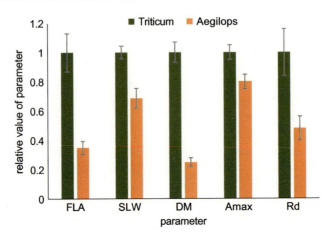

Fig. 10.1 Comparison of the growth characteristics and the photosynthetic parameters measured in flag leaves (*the upper*) of field grown wheat (*Triticum aestivum* L. cv. Astella) and Aegilops (*Aegilops cylindrica* Host.). The parameters are shown in relative units compared to the average values recorded in wheat after anthesis (the average values +/− standard errors from 6 measurements are presented). The parameters presented are: the flag leaf area (*FLA*); the specific leaf weight, i.e. the leaf dry mass per leaf area unit (*SLW*); the maximum steady state photosynthesis at ambient CO_2 level (A_{max}); mitochondrial respiration measured in the dark (R_d). The original figure is based on unpublished data of the authors. The photosynthetic measurements were done at saturating light (PAR ~1000 µmol m^{-2} s^{-1}) and high stomatal conductance (indicating stomata fully open)

(Long et al. 2006). Despite the fact that the relationship between the photosynthesis at the leaf level and productivity is very complex, some examples have shown a clear relationship (Flood et al. 2011, etc.).

Recently there are different attempts of employing genetic tools for improvement of photosynthetic efficiency. The first one is a conversion of a C3 to a C4 metabolism in major crop species, such as rice or wheat (Long et al. 2006; Sheehy et al. 2007; Hibberd et al. 2008; Gowik and Westhoff 2011). Up to now, the success of the conversion of C3 crops into C4 has been limited to the expression of enzymes from the C4 cells in the mesophyll of C3 plants. However, it was shown that simple expression of C4 enzymes into cells of C3 crops doesn't lead to improvement of energetic efficiency (von Caemmerer 2003). An alternative route involving a simple 'genetic switch' inducing the formation of Kranz anatomy (Surridge 2002) seemed to be more promising, but the efforts have not been crowned with the success, yet. In addition, Evans (2013) suggests a simpler approach by transferring bicarbonate transporters from cyanobacteria into chloroplasts, thus preventing CO_2 leakage (Price et al. 2008, 2013; Zarzycki et al. 2013; Meyer and Griffiths 2013).

Another principal approach is either an increase of the catalytic efficiency or specificity factor of the key photosynthetic enzyme Rubisco (Parry et al. 2007), which became a part of breeding program in wheat (Reynolds et al. 2011). Zhu et al. (2010) indicated that the majority of plant species still have a quantity and qualitative parameters of Rubisco adjusted to the CO_2 concentrations in pre-industrial era. It means that C3 plants have generally the excess of Rubisco and are prone to be RuBP-limited. Anyway, the increase of catalytic efficiency would increase the rate of photosynthetic fixation without a need of additional Rubisco. This would be beneficial especially in high light conditions, where Rubisco enzymatic activity limits the photosynthetic rate. In contrary, an increase in specificity factor would increase net CO_2 uptake in low light conditions, where the electron transport rate limits the photosynthesis; the photosynthetic electron transport could be directed into carboxylation process instead of oxygenation.

Unfortunately, it was shown that increase in specificity factor leads to the decrease in catalytic activity of Rubisco (Bainbridge et al. 1995). It has contradictory effects at the canopy level; higher specificity factor would increase photosynthetic rate in low light conditions, while the concomitant decline in the catalytic efficiency will lead to the decrease of photosynthesis of well-illuminated leaves (Zhu et al. 2004). Therefore, instead of increasing one of this enzymatic properties, replacing of a high catalytic activity of Rubisco with a form with high specificity factor during the acclimation to shade could be beneficial. Thus, plants would show a high catalytic activity of Rubisco in the sun-exposed leaves (upper leaves) and a high specificity factor of Rubisco in the leaves inside the canopy (Long et al. 2006).

In addition to changes of Rubisco properties, the electron transport (RuBP) limitation can also be partially eliminated. Two limiting points in the RuBP regeneration were identified: sedoheptulose-1,7-bisphosphatase (SbPase) catalyzing one of the steps in the Calvin cycle and the cytochrome b6/f complex in the thylakoid membrane playing essential role in regulation of the electron transport chain (Price et al. 1998; Harrison et al. 2001; Raines 2003). The upregulation of SbPase led to the increase of electron transport and decreased RuBP limitation (Lefebvre et al. 2005). However, this might be important mostly in light limited environments and low canopy levels (Long et al. 2006).

The previous paragraphs clearly indicate that the measurements of photosynthetic productivity still play and will play an important role in the crop research and improvement. There is a high number of examples of genetic variation in photosynthetic traits in crops and wild species. Moreover, the interaction of photosynthetic phenotypes with environmental factors is also well described. However, the genetic variation in plant photosynthesis is a largely unexplored, resulting in insufficiently utilized crop genetic resources (Flood et al. 2011).

10.2.1 Gas Exchange Analyses of Photosynthetic Limitations

The technical development brought new possibilities for laboratory and especially for field measurements. Introduction of portable infrared CO_2 analyzers provided a tool for selection of crop varieties based on the rates of leaf photosynthesis and transpiration (Long et al. 1996). Despite the technical development, the method is relatively time consuming. Therefore, in most of the cases field records of gas exchange measurements have been limited on single leaf records of the light-saturated assimilation rate, often at a single phenological stage (Long et al. 2006, Long 1998). Although this can provide important information, the correlation with overall plant photosynthetic production can be poor. It is because even half of the crop carbon fixation may be held by leaves in sub-optimal light conditions. The leaves of lower positions have different biochemical properties and physiological manifestations than upper leaves (Long 1993). Moreover, an increase in leaf area can result to lower investment per leaf area unit (Beadle and Long 1985; Evans 1993). Therefore the complex measurements at different canopy levels are needed to evaluate more precisely the photosynthetic performance of crops.

Gas exchange measurements can provide a valuable information, including kinetic parameters related to the activity of Rubisco enzyme and photosynthetic electron transport. For this purpose, the special protocol called A/Ci curve was developed. The analysis of the A/Ci curve is based on advanced mathematical models, derived from the biochemical model of the steady-state C3 photosynthesis introduced originally by Farquhar et al. (1980) (later updated by other authors). It enables quantification of photosynthetic limitations, i.e. whether the photosynthetic rate is limited by the Rubisco catalytic efficiency ($V_{c,max}$) or by insufficient rate of regeneration of RuBP (J_{max}).

An example of A/Ci curve recorded in modern bread wheat and *Aegilops* (the wild relative of wheat) is presented in Fig. 10.2, indicating even

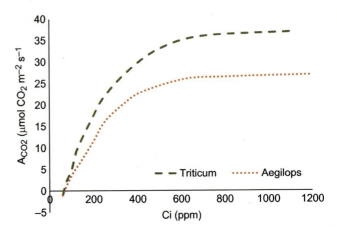

Fig. 10.2 An example of A/Ci curve measured in flag leaves (*the upper*) of field grown wheat (*Triticum aestivum* L. cv. Astella) and Aegilops (*Aegilops cylindrica* Host.). The original figure based on the unpublished data of the authors

visually the lower photosynthetic productivity in the leaves of wild species. The steeper initial slope in wheat sample indicates even visually the higher carboxylation efficiency. The protocols based on gradually increasing (or decreasing) CO_2 concentration in measuring chamber are needed to obtain the A/Ci; these can be further used in analyses using the Farquhar model of C3 photosynthesis (Farquhar et al. 1980) providing important kinetic parameters, such as $V_{c,max}$, J_{max}, mentioned before, as well as for estimation of mesophyll limitation of CO_2 diffusion inside the leaf (discussed below). The analysis was further developed by employing the simultaneous measurements of gas exchange and chlorophyll fluorescence, making the analysis of A/Ci curves more precise. Anyway, for the correct estimation of photosynthetic parameters, the precisely realized measurements are expected. Mistakes can occur especially if the measurements are carried out in stress conditions, when the stomata closes. The potential sources of errors are well described e.g. by Yin et al. (2009). Moreover, the measurements are time consuming and devices are expensive; therefore it is only hardly useful for obtaining the statistically relevant datasets necessary for screening of larger collections of genotypes.

Another valuable protocol of gas exchange records is the light response curve, providing the information on the maximum quantum yield of carboxylation, light compensation point, the level of light saturation and fully light saturated steady-state photosynthetic rate. The photosynthetic light response curve enables to determine some important issues of the intrinsic limitations of photosynthetic assimilation (Skillman et al. 2011). In the Fig. 10.3 we see that in the dark-exposed leaves net CO_2 assimilation is negative due to mitochondrial respiration (R_D). Although rates of R_D decrease in light-exposed samples, the mitochondrial respiratory pathway is not completely stopped (Atkin et al. 1998). The real rate of CO_2 assimilation by plant, leaf or cell represents net balance between the dissimilation (mitochondrial respiration, photorespiration, etc.) and carbon assimilation (i.e., gross photosynthetic carbon fixation). The rate of respiration varies across growing conditions and is strongly variable also among species or among individual tissues in the same plant (Griffin et al. 2001). It is frequently emphasized that the mitochondrial respiration determines the net carbon gain, becoming a subject of interest for crop improvement programs (Skillman et al. 2011).

The slope of the steeply-increasing linear part of the light-response curve (Fig. 10.3), represents an estimate of the quantum yield (Φ_{ACO_2}), i.e. it serves as a measure of the maximum efficiency by which the light energy is converted into assimilates by the photosynthetic machinery of the leaf.

As it is evident from the figure, there can be some variation in the Φ_{ACO_2} among different

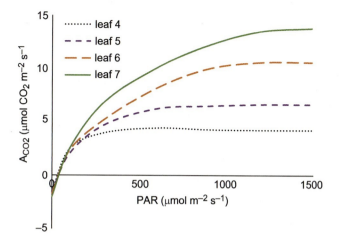

Fig. 10.3 An example of light response curves, i.e. relationships between the CO_2 assimilation rate measured by infra-red analyzer and incident photosynthetically active radtiation (*PAR*), measured on different leaf positions in barley (*Hordeum vulgare* L., cv. Kompakt) in post-anthesis growth stage. The leaves are numbered in the order in which they appeared. The leaf 7 represents a penultimate leaf (the second leaf from the *top*, well exposed to sun), the leaf 4 (an older leaf, *below* others) was almost completely shaded inside the canopy. The measurements were done at leaf temperature 20 °C and air CO_2 content of 370 ppm

samples; such a variation was commonly found across the different C3-species (Ehleringer and Björkman 1977). The maximum Φ_{ACO_2} measured in normal O_2 and CO_2 atmospheric concentrations is much lower compared to the maximum potential Φ_{ACO_2} determined under low O_2 concentrations. This inefficiency is caused mainly by the photorespiratory activity, which can be responsible for relatively high energy losses. Therefore, photorespiration might be an important target for improvement of photosynthetic performance in C3 plants (Skillman 2008; Skillman et al. 2011), as mentioned above. The maximum net photosynthetic capacity (A_{max}) can be assessed as a steady-state rate of CO_2 assimilation (alternatively also as O_2 release) at saturation light level (Fig. 10.3). A_{max} values quantified in the same conditions are quite variable across species and they also vary in the same species depending on the growing conditions and physiological status of a sample (e.g., Skillman et al. 2005; Skillman et al. 2011). While high quantum yield (Φ_{ACO_2}) is necessary for high photosynthetic rate in light limited conditions, the light-saturated level (A_{max}) characterizes light responses of the leaf in high light (Farquhar et al. 1980; Collatz et al. 1992).

10.2.2 Mesophyll Limitations to Photosynthesis

Today's understanging mesophyll conductance (g_m) is increasingly important not only in terms of efforts to maximize crop productivity and yield (Price et al. 2013), but also understand the coupled processes taking place between the vegetation and atmosphere (Ballantyne et al. 2011) as well as role of evolution in plant physiological processes (Griffiths and Helliker 2013).

Recent photosynthesis models used for prediction of crop yields or plant responses to climate change are based on two main regulation and limiting factors, such as stomatal diffusive conductance (g_s) (Medrano et al. 2002; Chaves et al. 2003) and metabolic capacity of photosynthetic apparatus (Wullschleger 1993; Farquhar et al. 2001). However, the knowledge of last two decades has shown that there is a third player in the photosynthesis regulation, called mesophyll conductance for CO_2 flow (g_m), which may affect the photosynthetic carbon gain (Loreto et al. 1992), and should be included into existing models (Sharkey et al. 2007).

Internal leaf conductance to CO_2 diffusion (g_m) is known to be finite, which means that

CO_2 concentration measured in sub-stomatal cavities (c_i) is substantially higher than CO_2 concentration inside the chloroplasts (c_c). So, this concentration gradient for CO_2 in the leaf mesophyll introduces another diffusive limitation to photosynthesis in addition to the stomatal conductance (g_s) (Bernacchi et al. 2002; Flexas and Medrano 2002).

The mesophyll conductance (g_m) includes several individual conductances: firstly, CO_2 is dilluted in the mesophyll cell walls, then diffuses through plasmalema into cytosol; after that it enters chloroplasts through the chloroplastic envelope with the next movement to Rubisco enzyme. Some of the authors also mention the CO_2 originating from photorespiration, which may supply the CO_2 diffusion and use in the mesophyll (Pinelli and Loreto 2003; Kebeish et al. 2007; Tholen and Zhu 2011; Tholen et al. 2012).

There are some indications of a strong co-regulation of g_s and g_m especially under conditions of drought and salinity, so that the sum of the stomatal and mesophyll diffusion limits (resistances), and not the metabolic impairment sets the limit for photosynthesis (Bernacchi et al. 2002; Flexas et al. 2004). However, the mechanisms that down-regulate g_m under such conditions are still to be investigated and confirmed. Likely mechanisms might be regulation activity of carbonic anhydrase, a protein regulator (Gillon and Yakir 2000) and aquaporines, a common agent for both water and CO_2 molecules transport through the membranes (Martin and Ruiz-Torrez 1992; Uehlein et al. 2003; Flexas et al. 2006; Kaldenhoff 2012).

The A/c_i curve analyses have shown that stomatal limitation to photosynthesis has been found in most of analysed plant species under water stress conditions as prevailing (Martin and Ruiz-Torrez 1992; Escalona et al. 1999), however, there are some reports showing: (i) the appearance of both non-stomatal (e.g. metabolic) and stomatal limitations at the same time during the early stages of water stress; (ii) that the changes of g_m can be as fast as those of g_s (Centritto et al. 2003), and (iii) a higher mesophyll limitation compared to the stomatal resistance (Tezara et al. 1999; Lawlor and Cornic 2002).

The results of Flexas et al. (1999) and others support the fact that metabolism impairment of photosynthesis does not occur at mild to moderate water stress conditions, as plants show very fast recovery of A_{CO_2} (less than 1 day), but at more severe water stress, the Rubisco, RuBP and TPU factors mostly restrict the biochemical capacity to assimilate CO_2, even if they are indirect factors, mediated by stomatal closure and reduced CO_2 concentration inside the leaf. As g_m makes a concentration gradient for CO_2 inside the leaf, the estimates of photosynthetic limitations from A/ci curves may lead to incorrect interpretations of photosynthetic responses of plants to the abiotic factors. Measurements of A/c_c curves show that while the photosynthetic activity is a function of diffusive limits, photosynthetic capacity is preserved under water and salinity stress (Flexas et al. 2004; Niinemets et al. 2009; Hu et al. 2010).

Future research in the topic of g_m lays on analyses of interspecific differences of g_m and its responses to environental factors as well as the study of structural, physiological and molecular mechanisms underlying the regulation of g_m. Incorporation of g_m and its variability into the existing photosynthetic models will improve the prediction ability and open a new promising way for biotechnological improvement of crop photosynthesis, water use efficiency and yield (Flexas et al. 2008; Parry et al. 2011).

10.2.3 The Chlorophyll Fluorescence Methods in Crop Research

The previous paragraphs clearly indicated that measurements of photosynthetic performance will be an important part of future strategy of crop improvement. Individual leaves of crops have different photosynthetic properties and the values of measured parameters strongly vary with time. In this respect, an extensive measurement campaign is necessary to obtain a dataset necessary for a comprehensive description of canopy photosynthesis, which is often not feasible. To be used in the process of high throughput screening, the method should be reproducible, fast and non-invasive (Flood et al. 2011). The measurements of gas exchange are non-invasive and, thanks to the possibility of precisely controlled light intensity, leaf temperature, air humidity and CO_2 content, they are also reproducible. Anyway, field portable systems that measure photosynthetic CO_2 assimilation are slow, a fact which severely limits the ability to survey variation in photosynthetic characters between cultivars or within germplasm collections (Evans 2013). Therefore, there are attempts to substitute them by other, more efficient methods. The most advanced are chlorophyll fluorescence measurements realized either as conventional single-point measurements or as screens of emitted chlorophyll fluorescence from the larger leaf areas, processed in a pixel scale, generally known as the chlorophyll fluorescence imaging. A wide range of relevant photosynthetic parameters can be derived from the fluorescence data, including photosynthetic electron transport rate, which can serve to estimate the photosynthetic rate, analogically to measurements of CO_2 fixation (Flood et al. 2011).

The several techniques of chlorophyll fluorescence are broadly used now with well developed and commercially available technical equipments. The methods based on variable chlorophyll a fluorescence measurements differ in the manner by which the photochemistry is saturated. The most frequently used are the saturation pulse analysis method using pulse amplitude modulation (PAM) fluorometers (Schreiber et al. 1986; Schreiber 2004) and the direct fluorescence recording represented by the method fast chlorophyll a fluorescence kinetics analysis (Strasser and Govindjee 1991, 1992).

10.2.3.1 Saturation Pulse Method

The saturation pulse method has been recently the most frequently used and generally accepted chlorophyll fluorescence technique. Thanks to this method, the measurements of the correct fluorescence yield through quenching analysis using modulated fluorescence and saturation pulses became possible (Bradbury and Baker 1981; Quick and Horton 1984; Schreiber et al. 1986; Schreiber 2004). The majority of chlorophyll fluorescence parameters, which can be found in literature, can be determined from five basic values taken from a record of slow chlorophyll fluorescence kinetics (F_m, F_0, F_m', F_s', F_0') (Rohacek et al. 2008; Baker 2008). The most frequently used parameters are:

– *The maximum quantum yield of PS II photochemistry*, F_v/F_m, the most frequently used parameter, often applied as the indicator of photoinhibition or other kind of injury caused to the PS II complexes (Rohacek et al. 2008). It quantifies the maximum photochemical efficiency (capacity) of open PS II reaction centers. It is almost constant for many different plant species when measured under non-stressed conditions and equals to 0.832 (Bjorkman and Demmig 1987). For stressed and/or damaged plants, F_v/F_m is strongly reduced.

– *Effective quantum yield of photochemical energy conversion in PS II*, Φ_{PSII} (Fq'/Fm'; $\Delta F'/Fm'$). Assessment of Φ_{PSII} does not require previous dark adaptation of the sample. Therefore, it is often used for field investigations. It quantifies the efficiency of the electron transport, as well as a fraction of photons absorbed in PS II antennae and utilized in the PS II photochemistry. If the photochemical and biochemical processes of photosynthesis are equilibrated under non-stress conditions, Φ_{PSII} is often correlated with the quantum yield of CO_2 fixation or the rate of photorespiration (Genty et al. 1989; Rohacek et al. 2008).

– *Non-photochemical quenching of chlorophyll fluorescence*, NPQ, often used as an indicator of the excess-radiant energy dissipation to heat in the PS II antennae. The extent of NPQ is linearly correlated to xanthophyll deepoxidation through the xanthophyll cycle. NPQ reflects also the decrease of the light-harvesting antenna size, PS II inactivation,

etc. (Bilger and Bjorkman 1990; Rohacek et al. 2008).
– Electron transport rate (ETR) as a function of the quantum yield and illumination. The most frequently used formula is: ETR = $0.84 \times 0.5 \times I \times \Phi_{PSII}$. However, this calculation must be used carefully, especially in conditions of plant stress. If the chlorophyll content decrease occurs, the absorbance change must be considered (Baker 2008).

Similarly to the gas exchange records, the chlorophyll fluorescence measurements can be done as single records or by using different protocols (Brestic and Zivcak 2013). The most simple and fast approach is the application of a single saturation pulse on dark adapted sample. The obtained value of Fv/Fm parameter provides only very basic and in the most cases insufficient information about the physiological status of the photosynthetic apparatus. The better approach useful for plant screening of larger collections is use of the chlorophyll fluorescence measurements without previous dark adaptation (used e.g. by Kuckenberg et al. 2009; Morales et al. 2012). Although the parameters that need previous dark adaptation cannot be calculated, it is possible to measure the actual quantum yield of PS II (Φ_{PSII}) and hence the electron transport rate (ETR) using the light intensity provided by the device or measured value of the photosynthetic active radiation incident on a leaf, if the fluorescence was measured under external (e.g. solar) irradiation. Such an approach enables to make more measurements in a short time. Kuckenberg et al. (2009) reported more heterogeneous data measured without previous dark adaptation compared to the dark adapted samples, but the data from both approaches showed the same tendency. In more detailed studies, the slower, automated measuring protocols can be employed (in detail, see chapter of Brestic and Zivcak 2013):

Slow induction curve with recovery is regularly measured after dark adaptation period (usually 15–20 min), followed by the period of actinic light and the next dark (recovery) period. The first phase can serve for assessment of photosynthetic induction (startup), while the recovery period enables to recognize the major constituents of non-photochemical quenching (Lichtenthaler et al. 2005a).

Slow and rapid light response curves represent plots of fluorescence parameters related to the graduated light intensities. The time of steps with particular light intensities should be enough to get the steady-state at each light level. Recently, the curves have been frequently measured simultaneously with other parameters, e.g. gas exchange (CO_2 or O_2). As this approach is very time-consuming, the more efficient version called "rapid light curves" is used more frequently. They can be used for assessment of the physiological flexibility of photosynthetic apparatus to the rapid changes in light relations, similar to environmental conditions (Schreiber et al. 1997; White and Critchley 1999; Ralph and Gademann 2005; Guarini and Moritz 2009), providing a detailed eco-physiological information on performance of plant photochemistry, which is strongly determined by environmental conditions and plant physiological status (Wing and Patterson 1993; Kubler and Raven 1996; Hewson et al. 2001; Seddon and Cheshire 2001).

Measurement of relative fluorescence decrease represents less frequent protocol for estimation of the physiological status of samples to obtain the simple *Rfd* parameter (the relative fluorescence decrease ratio). For the calculation of this parameter, saturation light pulses are not needed, as it uses the values of fluorescence at defined points measured during the dark-to-light induction at strong actinic light. According to the methodical paper of Lichtenthaler et al. (2005a), this irradiance-induced chlorophyll fluorescence induction kinetic is characterized by fast increase of chlorophyll fluorescence from the initial F_0 level (also termed base fluorescence) to a local maximum fluorescence level, plateau,

F_P, within 0.1–0.2 s, where the F_p is recorded. Then, after startup of the photosynthetic machinery, the fluorescence within 3–5 min decreases to a much lower steady state Fs. The greater this Chl fluorescence decrease F_d ($F_d = F_p - F_s$), the higher the net photosynthetic rate (A_{CO_2}) of the leaf examined (Lichtenthaler and Rinderle 1988; Lichtenthaler and Miehé 1997; Lichtenthaler and Babani 2004; Lichtenthaler et al. 2005a; Brestic and Zivcak 2013). Rfd ratio is calculated as: $Rfd = F_d/F_s$. The RFd-values being measured at the saturated irradiance of photosynthesis exhibited a highly significant linear correlation to A_{CO_2} as shown in Lichtenthaler et al. (2005a).

10.2.3.2 Analysis of Fast Chlorophyll Fluorescence Induction

As the methods based on saturation pulse analysis are relatively time-consuming, a big effort has been applied to develop a more efficient way of measurements of photosynthetic performance and environmental effect. In the last decades, the exponential increase of the studies applying the fast fluorescence kinetics can be observed. Chlorophyll fluorescence induction represents a plot of measured fluorescence intensity as a function of time of continuous illumination (Fig. 10.4).

Such a curve recorded under continuous light has a fast (less than one second) exponential phase, and a slow decay phase (duration of few min). The rise has a typical polyphasic shape, well evident when the curve is plotted on the logarithmic time scale, or if the individual steps are plotted separately, in different time resolution (Fig. 10.4). The shape of OJIP-transient is sometimes denoted as a 'fingerprint' of a sample of a given physiological status; any deviation of the curve indicate photochemical changes at the thylakoid membrane level. The analysis of OJIP curve taking the theoretical assumptions and probabilities derives different photosynthetic parameters for a dark adapted state of the photosynthetic systems (Strasser et al. 2000, 2004; Stirbet and Govindjee 2011). The nomenclature for 'OJIP' is as followed: O for origin or $Fo = F_0$ level measured at 50 μs (or less) after illumination, J and I represent intermediate states measured after 2 ms and 30 ms, respectively, and P is the peak or $F_P = F_m$ (maximal fluorescence). This is valid only if a sufficient light intensity is used. In heat-stressed samples, another peak arise between F_0 and F_J at app. 300 μs, which is usually called K-step (Guisse et al. 1995; Srivastava et al. 1997; Strasser et al. 2000); therefore some authors call the fast chlorophyll fluorescence induction the OKJIP-curve or transient. The OJIP curve from F_0 to F_m is correlated with the primary photochemical reactions of PS II (Duysens and Sweers 1963) and the fluorescence yield is controlled by a PS II acceptor quencher (the primary quinone acceptor, Q_A) (Van Gorkom 1986). Thus, the OJIP transient can be used for estimation of the photochemical quantum yield of PS II photochemistry, and electron transport properties. The OJIP fluorescence curve analysis can be used to monitor the effect of various biotic and abiotic stresses, and photosynthetic mutations affecting the structure and function of the photosynthetic apparatus (Strasser et al. 2004).

There are several groups of parameters derived from the fluorescence rise. In addition to the basic fluorescence values and fundamental parameters, such as Fo, Fm, Fv/Fm (similar to the saturation pulse method), there is also a group of parameters derived from the JIP-test, introduced by Strasser and coauthors (1995, 2000, 2004, 2010), reviewed well by Stirbet and Govindjee (2011). We can divide it into the fluorescence parameters derived from the data extracted from OJIP transient and the biophysical parameters calculated using the previous group of fluorescence parameters (Strasser et al. 2010). In plant stress research there are several possible ways of interpreting the data. A multiparametric approach is based on the visualization of data e.g. by spider plots or pipeline models. On the other hand, the model offers the integrative parameters enabling simple assessment of the status and vitality of the photosynthetic apparatus, which are sensitive and created mostly for

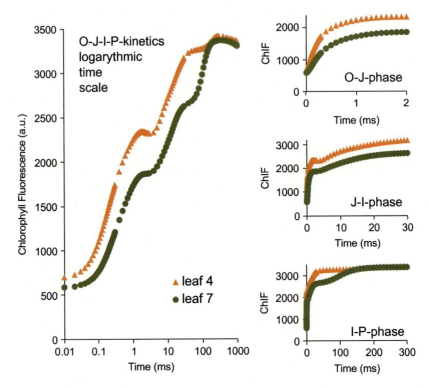

Fig. 10.4 Examples of O-J-I-P-curves recorded in two different leaves of barley (*Hordeum vulgare* L. cv. Kompakt) in post-anthesis stage. The leaves are numbered in the order in which they appeared. The leaf 7 represents a penultimate leaf (the second leaf from the *top*, well exposed to sun), the leaf 4 (an older leaf, *below* the others) was almost completely shaded inside the canopy. The main graph (*left*) shows the entire O-J-I-P-kinetics plotted on a logarithmic time scale. The small graphs (*right*) show individual phases plotted on a regular time scale: O-J phase in time 0–2 ms, J-I phase in time 2–30 ms, and I-P phase in time 30–300 ms. Unpublished data of the authors

possible practical applications in pre-screening or selection in research and breeding programs (Brestic and Zivcak 2013), including special applications such as assessment of nutrition status (Kalaji et al. 2014a, b) or toxic effects (Shaw et al. 2014; Kalaji et al. 2014b).

From numerous JIP-test parameters, for practical applications in the crop research Performance Index (PI) was introduced (Strasser et al. 2000). This complex parameter integrates several independent structural and functional properties of the photochemistry, reflecting the functionality of both photosystems and providing a quantitative information on the current state of plant performance under stress conditions (Strasser et al. 2004). An example of the comparison based on PI values in field grown winter wheat genotypes is presented in Fig. 10.5.

The average data (based on the highest number measurements across the vegetation period, creating statistically relevant datasets) show significant differences among genotypes in PI values. More specifically, the lowest values were found in historical local landraces, while new modern genotypes of the same provenance had usually higher PI values. This suggests, at least, that the structural and functional properties in modern genotypes differ from the old landmarks. However, the current level of knowledge does not entitle us to draw further conclusions about photosynthetic performance based on the fast chlorophyll fluorescence only. Even usefulness of the fast chlorophyll fluorescence for leaf photosynthetic performance testing could be proven in the future, more probably, the method will remain the tool for assessment of

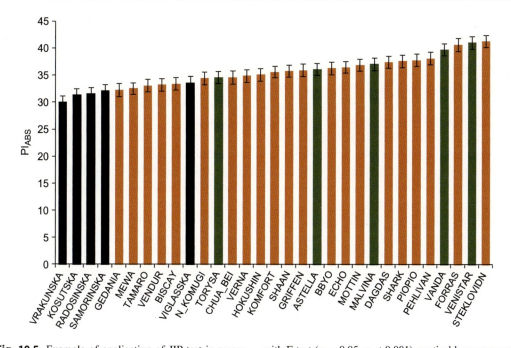

Fig. 10.5 Example of application of JIP-test in assessment of leaf properties in wheat genotypes. Average values of Performance Index (PI_{abs}) recorded in 31 wheat genotypes. The data represent mean values for several subsequent measurements done from April to June on leaves of wheat. Data were compared using ANOVA with F-test ($\alpha = 0.05$, $p < 0.001$), vertical bars represent standard error. Genotypes are ranked according to mean values. *Black* columns represent old landraces of Slovak origin, the *green* columns represent modern wheat varieties of Slovak origin (original figure based on the results published by Brestic et al. 2012)

the stress effects on the photosynthetic functions mostly.

In this respect, Stirbet and Govindjee (2011) postulated that the availability of user-friendly portable fluorometers for high-frequency record of OJIP-transient and the useful software for the analysis of experimental data, make the JIP test derived from the fast chlorophyll fluorescence attractive even for users without a deep knowledge on photochemical processes at the thylakoid membrane level. As we have mentioned above, the small and portable devices allow efficient data records even in the field conditions. The chlorophyll fluorescence induction kinetics contains a valuable information about the photochemical efficiency of primary conversion of incident light energy, electron transport events, and related regulatory processes. These issues can be deciphered using advanced mathematical models based on the analysis of fluorescence curves, providing a large number of the fluorescence parameters. However, the JIP test, in this context, is nothing more than a systematic method, to be used as a practical tool, to obtain quick information on the various possibilities of effects on photosynthesis, particularly on PSII, and to a limited extent on PSI. Till now, it is not sufficiently confirmed that the measurements could be used to obtain detailed information on the entire system, without direct measurements of electron transport, and overall photosynthesis rates (Stirbet and Govindjee 2011) Nevertheless, despite the large number of partial results, the potential of the method for the use in practice is, in fact, almost unrealized.

10.2.3.3 Chlorophyll Fluorescence Imaging

One of the most useful innovations of the chlorophyll fluorescence technique has been the development of chlorophyll fluorescence imaging, which involves advancements in the technology of light emission, imaging detectors, and rapid data handling (Nedbal and Whitmarsh 2004;

Gorbe and Calatayud 2012). Fluorescence imaging devices have been constructed for their use either at the microscopic level (Oxborough and Baker 1997; Rolfe and Scholes 1995), at the plant, leaf and organ level (Omasa et al. 1987; Calatayud et al. 2006) or for remote sensing of chlorophyll fluorescence (Calatayud et al. 2006; Gorbe and Calatayud 2012). Fluorescence imaging represents an efficient tool for observing the fluorescence emission pattern of sub-cellular structures, cells, tissues, leaves or other plant organs, or whole plants, providing precise visual information about the plant stress effects (Calatayud et al. 2006). The fluorescence imaging method enables the observation of the temporal and spatial heterogeneities of photosynthetic processes over the relatively large observed area; such heterogeneities occur as a result of internal and/or environmental factors (Nedbal and Whitmarsh 2004). They can be just barely (or not at all) detected through the conventional point measurements by non-imaging chlorophyll fluorescence (Ellenson and Amundson 1982; Oxborough and Baker 1997; Omasa and Takayama 2003).

An example of chlorophyll fluorescence imaging record confirming the unequal spatial distribution of photosynthetic performance within the crop canopy can be seen in Fig. 10.6.

It is evident that the newest leaves in upper positions, which are exposed to the maximum light intensity have a higher photosynthetic rate (expressed in photosynthetic electron transport units). Moreover, the figure also indicates the unequal distribution (heterogeneity) of photosynthetic rate across the leaf. It is particularly important as far as considering the dominant use of point measurements of photosynthesis. Here, we can see that the results of point measurements are similar to the full area results in upper leaves, where the photosynthetic rate was distributed more-or-less homogenically; this was not the case of lower leaf positions where the point measurements done in the middle of a leaf led to overestimation of the leaf photosynthetic electron transport rate. This example clearly documents the advantages of chlorophyll fluorescence imaging compared to the common point measurements. The standard, commercially available instruments for chlorophyll fluorescence imaging enable to record the automated protocols, such as light response curves (Fig. 10.6, right below), analogically to those presented in the gas exchange measurements (Fig. 10.2).

One of the possible fields for future applications of chlorophyll fluorescence imaging is the plant breeding, especial the process of selection of useful parental germplasm or progeny. Harbinson et al. (2012) stated that the recent advanced studies with chlorophyll fluorescence imaging are focused on high-throughput screening of genotypes tolerant to abiotic and biotic stress factors. Recently, the assessment of stress tolerance or disease resistance in breeding programs is based mostly on a visual scoring by evaluators, i.e. it is strongly subjective and prone to error; moreover, it is time-consuming. In this regard, an important advantage of chlorophyll fluorescence imaging is that it can be applied to screen simultaneously a large number of samples. Moreover, fluorescence imaging can be integrated into robots for automatic measurements (Baker and Rosenqvist 2004; Charlie et al. 2007; Gorbe and Calatayud 2012).

10.2.4 Photosynthetic Data in Context of Plant and Canopy

Any increase in photosynthetic process at the chloroplast or leaf level will only bring substantial practical benefits if it confers an improvement at the level of the crop canopy. The rate of crop photosynthesis is typically measured by enclosing part of a single leaf in measuring head of gazometer, but to understand growth, the daily integral of photosynthetic uptake by the whole plant or canopy and its further allocation need to be considered (Evans 2013). The variation in photosynthetic data depends on the unit of measurement chosen (Flood et al. 2011). The crop plant photosynthesis is most frequently quantified as the assimilation rate per leaf area unit, as the activity of the photosynthetic process is it enables to compare results and easily

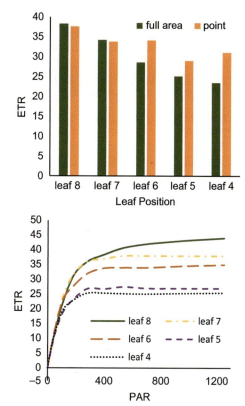

Fig. 10.6 An example of chlorophyll fluorescence imaging screen of different leaf positions in wheat (*Triticum aestivum* L.). The figure *left* shows a two-dimensional distribution of recorded values of electron transport rate of PS II photochemistry (*ETR*), recorded at leaf illuminated by blue actinic light with intensity ~200 µmol m^{-2} s^{-1} for 5 min. Figure right above shows the comparison of average values recorded for full screened leaf area with the results obtained if the point measurement in the *middle* part of the leaf was used. The figure right below shows the result of rapid light curves performed at the same levels. Individual ETR values were obtained after 20 s at each light intensities, the curve was plotted using 12 levels of PAR measured in different leaf positions in barley (*Hordeum vulgare* L., cv. Kompakt) in post-anthesis stage. The leaves are numbered in the order in which they appeared. The leaf 8 represents the flag leaf (first leaf from the *top*, well exposed to sun), the leaf 4 (an older leaf, *below* the others) was almost completely shaded by younger leaves. Measurements were performed by Maxi-Imaging-PAM (Walz, Germany), unpublished data of the authors

evaluate the leaf photosynthetic activity. For studies at the canopy level, it needs to be integrated to give the rate per plant or unit area of the crop. For this purpose, parameters like leaf area index (LAI) as well as those describing the photosynthetic variation between the leaves and canopy architecture are needed. The photosynthetic activity at the leaf level is combined with the architecture and plant morphology, which are non-photosynthetic in their character. Although leaf area index and the measures of plant architecture are important traits in crop research, they have to be considered separately to the leaf photosynthetic capacity expressed per unit of leaf area. Moreover, in the major crop plants, almost optimal crop architecture and LAI were achieved by previous breeding, and thus potential for further improvement is almost depleted (Long et al. 2006; Parry et al. 2011).

More specifically, at high irradiance, the rate of CO_2 assimilation is limited by the photosynthetic capacity. The loss of efficiency at high irradiance by a single leaf can be reduced in a plant canopy by distributing light capture between leaves. Thus, a larger leaf area would operate with a greater efficiency at moderate irradiances and lower leaf area operating less efficiently at high irradiance. To achieve this, leaves at the top of canopy need to be held more erected (Song et al. 2013a; Evans 2013).

The sample size represents one of the most frequent complications, in terms of the interpretation of photosynthetic measurements. In many cases it is not feasible to realize measurements more than just on one leaf in one time only. The rate of photosynthesis is, however, rather variable depending on growth phase, leaf age, irradiation level, etc. A lack of correlation between the photosynthetic capacity and plant yield, which was often reported in the past, might be caused by generalization of results of the photosynthetic rate per unit leaf area in a portion of one leaf for the whole crop canopy. To obtain true results, the sufficient representative dataset across the leaf positions must be recorded during whole season; the necessary representative sample size will vary depending on factors such as canopy architecture and leaf age distribution (Flood et al. 2011).

In addition to the classical physiological measurements at the plant scale, the remote-sensing approaches offer potential to estimate crop productivity (Evans 2013). The most promising data were introduced by measurements of leaf reflectance spectroscopy in trees, which enabled to estimate key determinants of photosynthetic capacity ($V_{c,max}$ and J_{max}). This approach applied successfully in broadleaf trees (Serbin et al. 2012) represents a promising strategy for developing remote sensing method for advanced analysis of canopy photosynthetic metabolism at broad scales.

10.3 The Photosynthesis and Abiotic Stress Factors

As it was shown above, the increase in crop yield potential can be achieved by increasing the maximum capacity of the source through improvement of maximum efficiency of carboxylation process in leaves. However, in a global range, the yields of crops are far away from the yield potential, which is caused mainly by adverse environmental conditions (Araus et al. 2002). Therefore, another way of crop yield improvement is an increase of crop resistance or tolerance to stress factors. In the next part of chapter we will aim at presenting the effects of main abiotic stresses on crop photosynthesis and highlighting of some examples of the application of photosynthetic techniques in screening for improved plant stress tolerance.

10.3.1 Limitation of Photosynthesis by Drought

Drought is the main environmental factor limiting plant productivity in a global range. Drought stress leads to stomatal closure and reduced rate of transpiration, a decrease in the water content in plant tissues, inhibition of plant growth and decline in photosynthesis. It is associated with accumulation of abscisic acid (ABA), being the main stress signal molecule as well as other stress related compounds such as compatible osmolytes (sorbitol, mannitol, proline) or radical scavenging compounds (e.g. ascorbate, glutathione, α-tocopherol) (Yordanov et al. 2003). Plants also undergo changes at different level of organization, starting with ultrastructural changes of the chloroplast (Vassileva et al. 2012), through anatomical to morphological alterations. At the level of whole plant, the drought stress effects are mostly manifested as a decline in growth and photosynthesis, and is connected with changes in the metabolism of nitrogen and carbon. The plant responses are complex as they reflect the influences of stress over time and space, and responses are distributed at all levels of plant organization. Under field conditions, the one stress-related changes might by modified by the superimposition of other stresses; either synergistically or antagonistically (Blum 1996; Cornic and Massacci 1996; Yordanov et al. 2003).

The knowledge of the biophysical, biochemical, and physiological bases for a decrease of photosynthesis in plants experienced to water deficits is necessary for improvement of plant stress responses. Drought-induced effects on leaf photosynthesis and, in the next step, on plant performance have been discussed over the decades (Genty et al. 1987; Sharkey and Seemann 1989; Cornic and Briantais 1991; Brestic et al. 1995; Lawlor and Tezara 2009). Results confirmed by several research groups have shown that a decrease of leaf photosynthetic rate caused by moderate water stress is mostly the result of stomatal activity and the stomatal conductance and photosynthesis are mutually co-regulated. The stomata close progressively with the drought progress, followed by parallel decreases of net photosynthesis (Medrano et al. 2002). The photosynthetic techniques are frequently employed in drought stress studies in crop plants, as it is documented in Table 10.1,

Table 10.1 The overview of the most recent studies employing the photosynthetic techniques in drought stress studies in crop plants

Crops	Applied photosynthetic techniques	References
Wheat	Gas exchange, multispectral fluorescence	Burling et al. (2013)
	Gas exchange, chlorophyll fluorescence slow	Zivcak et al. (2013), Hou et al. (2013) and Fábián et al. (2013)
	Gas exchange, WUE	Li et al. (2010), Bencze et al. (2011), Deák et al. (2011), Akhkha et al. (2011), Karim et al. (2012) and Yasir et al. (2013)
	Gas exchange, chlorophyll fluorescence (fast)	Roohi et al. (2013) and Li et al. (2010)
	Chlorophyll fluorescence (fast)	Zivcak et al. (2008a, b, c), Zivcak et al. (2009a, b) and Kovačevič et al. (2013)
Barley	Gas exchange, chlorophyll fluorescence (slow, fast)	Wójcik-Jagła et al. (2013)
	Gas exchange, WUE	Bencze et al. (2011)
	Fast chlorophyll fluorescence	Repkova et al. (2008)
	Gas exchange, fast chlorophyll fluorescence	Roohi et al. (2013) and Rapacz et al. (2010)
	Fast chlorophyll a fluorescence	Jedmowski et al. (2013) and Ashoub et al. (2013)
Rice	Gas exchange	Ishizaki et al. (2013) and Farooq et al. (2008, 2009)
	Chlorophyll fluorescence (slow)	Do et al. (2013)
	Chlorophyll fluorescence fast	Redillas et al. (2011) and Phung et al. (2011)
Maize	Gas exchange	Wang et al. (2008), Markelz et al. (2011), Anjum et al. (2011), Barnaby et al. (2013), Zhang et al. (2013) and Nguyen et al. (2013)
	Slow chlorophyll fluorescence	Fan et al. (2013)
Soybean	Gas exchange, chlorophyll fluorescence (slow)	de Souza et al. (2013)
	Gas exchange, WUE	Li et al. (2013) and Anjum et al. (2013)
Sorghum	Gas exchange, fast chlorophyll fluorescence	Zegada-Lizarazu and Monti (2013)
Bean	Gas exchange, chlorophyll fluorescence (slow)	Ramalho et al. (2014) and Santos et al. (2009)
Tomato	Gas exchange, fast chlorophyll fluorescence	Giannakoula and Ilias (2013)
Sugar beet	Fast chlorophyll fluorescence with P700	Ceppi et al. (2012)
Tobacco	Gas exchange, chlorophyll fluorescence slow	Deeba et al. (2012)
Cotton	Gas exchange, chlorophyll fluorescence, imaging	Massacci et al. (2008)
	Gas exchange, A/Ci curve	Kuppu et al. (2013)

referencing the most recent contributions in this area.

The most useful method for observation of changes in photosynthetic activity in drought stress are the gas exchange records, mostly the non-invasive measurements of CO_2 assimilation and transpiration by infra-red analyzers, as described above. An example of photosynthetic induction record in non-stressed and severely stressed wheat plants is shown in Fig. 10.7, where the lower CO_2 assimilation in steady-state (after 30 min on light) was associated with low internal CO_2 concentration, indicating the stomatal limitation of the net photosynthetic rate.

The leaf CO_2 assimilation rate is down-regulated at mild drought stress even before the leaf water content falls down due to the soil water shortage (Gollan et al. 1986; Davies and Zhang 1991) or in response to a drop in humidity of atmosphere (Bunce 1981). The proportion of photosynthetic stomatal effect depends on the severity of drought stress. Under mild water deficit the stomatal closure is a first event, followed by changes of photosynthetic reactions (Cornic and Briantais 1991). Under field conditions, stomatal regulation of transpiration was shown as a primary event in plant response to water deficit leading to decrease of CO_2 uptake by the leaves (Chaves 1991; Cornic and Massacci 1996; Chaves et al. 2002; Ditmarova et al. 2010). The level of photosynthetic limitation of steady-state photosynthesis caused by drought-induced stomata closure can be evaluated from the relationship between stomata conductance and CO_2 assimilation, shown in Fig. 10.8a, indicating direct linear relationship between g_s and A_{CO_2} below the threshold level of stomatal conductance. On the other hand, the range of stomatal

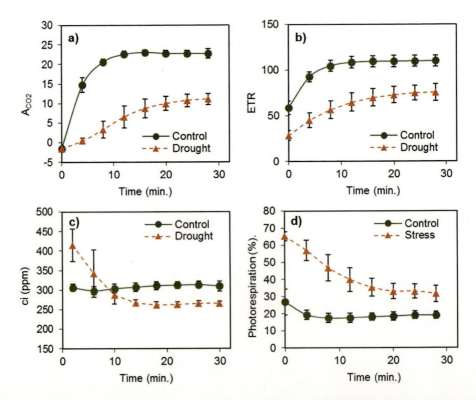

Fig. 10.7 The effect of severe drought stress in wheat leaves on induction of assimilation process (A_{CO_2}) and related parameters: Electron transport rate (*ETR*), internal CO_2 content in leaf (ci), and the proportion of electron transport spent in the process of photorespiration. The measurements were carried out after switching-on actinic light (after 20 min in darkness); the point in time 0 represents the record at the moment of switching-on light. The figure is composed using the data published by Zivcak et al. (2013)

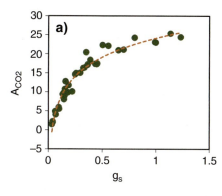
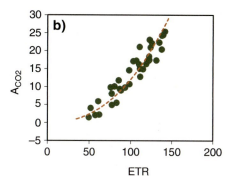

Fig. 10.8 Relationship between the stomatal conductance (**a**) or electron transport rate derived from chlorophyll fluorescence measurements (**b**) and CO_2 assimilation. Measurements were performed during the prolonged drought stress in winter wheat. The simultaneous measurements of gas exchange and chlorophyll fluorescence was done by gas exchange system Licor 6400 (Licor, USA). The figure is composed from the data published by Zivcak et al. (2013)

conductance, which is sufficient to reach the maximum photosynthetic rate is rather wide, indicating possible inefficiency in water use when stomata fully open.

In addition to the lower stomatal conductance, the opening stomata to get steady-state levels took much longer time in drought stressed plants compared to non-stressed Zivcak et al. (2013). It was shown that stomatal closure occurs in response either to a decrease in leaf turgor and/or water potential, or to low air humidity (Maroco et al. 1997). Responses of stomata correlate better with the soil water content than with water content in leaf. It indicates stomatal responses to "non-hydraulic" chemical signals (Yordanov et al. 2003), mainly to abscisic acid (ABA) synthesized in the roots in response to decrease of water content in soil (Davies and Zang 1991). This is comfirmed by an early drought-induced stomatal closure before any change in leaf water status is detected (Medrano et al. 2002). Some role in stomata responses play also the circadian rhythm (Chaves et al. 2002). Lawlor (2002) has shown that variations in cell carbon metabolism are frequently occurring at the beginning of drought stress. The stomatal function in drought-tolerant species is controlled to remain some carbon fixation running even in stress conditions; hence, the water use efficiency increases. Moreover, when water deficit is relieved, they open stomata more rapidly. Some studies have shown (Faver et al. 1996; Herppich and Peckmann 1997) that in severe drought stress, photosynthesis depends more on the carbon fixation capacity determined by enzymatic properties than by decreased conductance for CO_2 diffusion.

The non-stomatal responses of photosynthetic assimilation (PS II energy conversion, the dark reaction of Rubisco carbon fixation) have been shown to be resistant to water deficits (Genty et al. 1987; Chaves 1991). The strong drought-induced reductions of Rubisco activity has also been reported (Maroco et al. 2002; Parry et al. 2002) but a lot of studies have observed negligible effect of drought (Lal et al. 1996). Flagella et al. (1998) has shown that the quantum yield of PS II in relation to Calvin cycle metabolism was reduced only under extreme water shortage. However, Lauer and Boyer (1992) identified a metabolic damage of photosynthetic machinery by assessment of intercellular CO_2 partial pressure measured directly in the leaves. An example of such observation can be seen in Fig. 10.9.

It is evident that the decrease of leaf relative water content below 70% was associated with the decrease of carboxylation efficiency, as a function of both increase of photorespiration rate and metabolic limitations mentioned above. Interestingly, the diurnal changes in carboxylation

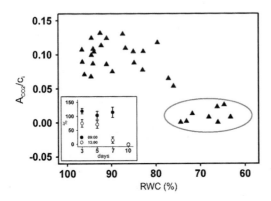

Fig. 10.9 The decrease of carboxylation efficiency (expressed as the ratio of CO_2 assimilation rate and internal CO_2 content) caused by drought stress in wheat plants. The small plot shows the differences in carboxylation efficiency recorded in two different times during the day (Zivcak 2006)

efficiency are also evident (small insertion in Fig. 10.9), probably as a function of leaf temperature and/or changes in leaf water content.

Low activity of biochemical pathways under drought stress may lead to down-regulation of photochemical reaction by decreasing consumption of photosynthetic intermediates (Santos et al. 2009). It was concluded (Nogues and Baker 2000) that in C3 under conditions with reduced CO_2 assimilation due to stomata closure, the Rubisco-driven reduction of O_2 in the process of photorespiration plays a protective role by partial utilization of excessive excitation energy converted to NADPH and ATP. However, the photorespiration itself is not sufficient to dissipate the full excess of energy absorbed by PS II antennae therefore the photoprotective responses towards the dissipation of light energy in PSII are up-regulated.

Tezara et al. (1999) indicated that the low ATP synthesis cause limitation of electron transport, leading to decrease of ribulose-1,5-bisphosphate (RuBP) supply. Similarly, Lawlor and Cornic (2002) postulated that both the decrease in ATP content and related imbalance in the redox status substantially affect plant cell metabolism. They argued that impaired metabolism, associated with a shortage of ATP, limits RuBP synthesis without negative effects on enzymes of carbon reduction cycle including Rubisco. In conditions of mild drought stress, the cycles of photosynthetic Calvin cycle and photorespiration serve as the main PSII electron sinks (Cornic and Fresneau 2002). Activity and responses at the level of PS II were not substantially changed during early phases of water deficit. The CO_2 content in the chloroplasts decreases after stomatal closure in drying leaves, leading (in C3 plants) to increase of RuBP oxygenation, where O_2 plays a role of electron acceptor; the extent of this activity depends on the intensity of incident light. The estimation of partitioning of the electron transport destination between Calvin cycle and photorespiration (Fig. 10.7d) by the method of Epron et al. (1995) confirm the significant increase of photorespiration in drought stressed plants, being the main factor decreasing the net photosynthetic efficiency (indictated e.g. as A/ci ratio in Fig. 10.9). Similarly, increasing yield of CO_2 assimilation per unit of electron transport rate, indicated from the exponential shape of relationship between A_{CO_2} and ETR (Fig. 10.8b) suggest that the losses of energy by photorespiration increased with drought-induced decrease in CO_2 assimilation.

According to Girardi et al. (1996) long-term drought reduction in water content led to negative changes in PSII structures. The rest of PS II core complex was functional, but they observed changes in organization and stoichiometry. Similarly, Yordanov et al. (2003) reported that drought stress induced accumulation of PS II-inactive reaction centers in bean; this effect was more serious in drought-sensitive cultivars compared to drought-tolerant varieties.

The genotypic differences in stomatal behavior and photochemical drought resistance can be subscribed partially also to osmotic adjustment. It represents the mechanism important for maintenance of plant cell turgor and hence it is important for delay of full stomata closure (Zhang et al. 1999). Results of correlation analyses (Table 10.2) show significant positive correlation between capacity for osmotic adjustment and values of stomatal conductance during drought

Table 10.2 Results of correlation analyses between capacity for osmotic adjustment and selected parameters in winter wheat genotypes

Parameter[a]	r	p
Stomatal conductance	0.590	0.019
Net assimilation rate	0.789	0.035
Grain yield	0.510	0.053

Zivcak et al. (2009b)
[a]r correlation index, p probability value

stress (Zivcak et al. 2009b). The same trend was confirmed also for a relationship of osmotic adjustment and net assimilation rate. It indicates that a higher capacity for the accumulation of osmotically active compounds is beneficial for the elongation of active period of CO_2 assimilation under drought conditions.

A main role often attributed to the osmotic adjustment and especially to free proline is to protect the subcellular structures within the photosynthetic apparatus. Proline plays a role in the regulation of cellular redox potential, signaling and recovery from stress conditions (Ashraf and Fooladad 2006). The effects of drought stress on primary photosynthetic processes were evaluated using analysis of rapid chlorophyll fluorescence kinetics with the Performance Index as a parameter useful for determination of drought stress effects (Zivcak et al. 2008a).

Stress physiologists are particularly interested in the process of photosynthesis, because it is a very good sensitive indicator of the overall fitness of plants. One of the first responses of plants to harmful environment is the decrease in the rate of photosynthesis and inhibition of several molecular mechanisms. In this context chlorophyll fluorescence analysis has become a very good tool for estimating various photosynthetic parameters and for optimization and control of crop photosynthesis in the field conditions (Brestic and Zivcak 2013). However, the gas exchange method, even being a very efficient tool for observing the drought effect on photosynthetic process, is too slow and labor-consuming and its usefulness for practical applications in screening for drought tolerance in crops is limited (as we have documented before). Therefore the chlorophyll fluorescence methods are the permanent object of interest of many stress physiologists.

The effect of drought stress on chlorophyll fluorescence parameters depends on the degree of water deficit. It is generally accepted, that mild to moderate drought stress decrease the photosynthetic rate mainly due to stomata closure, whilst the metabolic processes remain almost unaffected (Cornic and Massacci 1996). The critical leaf relative water content is app. 70%; below this value, the non-stomatal limitation of photosynthesis increases its importance. This phenomenon is reflected by the values of chlorophyll fluorescence and calculated fluorescence parameters.

One of the most frequently used fluorescence parameters in plant physiological research, including drought stress research, is the maximum quantum yield of PS II photochemistry (Fv/Fm). It is mostly because this parameter is very easy to measure and it is generally well accepted measure of photosynthetic status. However, this parameter is highly insensitive to stomatal effects neither to any other effects occurring in moderate drought stress recorded during quick dehydration of wheat leaf in darkness. This parameter was shown to be insensitive to changes in leaf photosynthetic capacity even if the chlorophyll content was significantly reduced, e.g. due to insufficient nutrient supply (Zivcak et al. 2014a, b). It is obvious, that Fv/Fm values are extremely stable and they start to decrease at the level that can be entitled as the lethal level. If the drought stress run in field conditions, the water deficit acts to leaf much longer and hence the decrease of Fv/Fm starts at something higher values of relative water content. Even here, it is obvious that the Fv/Fm values keep high and start its decline below 70% of relative water content in leaf; anyway we didn't observe the strong decrease even at 50% of relative water content (Zivcak et al. 2008a). As it was reported in many other studies, any decrease of Fv/Fm cannot be attributed to drought stress at the physiologically relevant level; however, the Fv/Fm measurements during drought stress make sense, as they can draw attention to the effects of co-occurring stresses (heat stress, photoinhibition, etc.) or to the early phases of leaf senescence (Brestic and Zivcak 2013).

Anyway, both measurements of the slow and the fast chlorophyll fluorescence kinetics were shown to be sensitive to drought stress (Fracheboud and Leipner 2003; Oukarroum et al. 2007, 2009; Zivcak et al. 2008a, b, c, 2009a, b). The decrease of effective PS II quantum yield (Φ_{PSII}) and ETR in drought stressed leaves compared to well-hydrated is mainly due to lack of CO_2 inside the leaf (closed stomata). In C3 leaves, the decrease of Φ_{PSII} was correlated with the net assimilation rate, but the correlation was not linear as the increased photorespiration efficiently consumes part of electrons flowing within the linear electron transport chain (as it can be seen in Fig. 10.8b). Anyway, the measurement of the slow fluorescence kinetics and calculation of quantum yields and electron transport rate seems to be useful for determination of the drought stress effects, reflecting both stomatal and non-stomatal effects; however, such measurements during drought stress cannot be directly related directly to CO_2 assimilation (Baker 2008). Similarly, Lichtenthaler et al. (2005a) reported the relative fluorescence decrease ratio (R_{fd}) being more sensitive and better correlating with photosynthetic assimilation compared to PS II quantum yield or ETR. A very promising approach to assess the effects of drought stress, as well as protective mechanisms at the level of photochemical processes is the simultaneous measurement of slow chlorophyll fluorescence with redox changes of photosystem I (Zivcak et al. 2014c), as its proper activity is crucial for photoprotection (Brestic et al. 2014, Zivcak et al. 2015).

The gradual decrease of parameter Performance Index (PI_{ABS}), derived from fast chlorophyll fluorescence kinetics and the JIP-test (Strasser et al. 2000) during dehydration indicates the drought induced changes at the PS II electron acceptor side. Such a decrease was observed in laboratory conditions (Zivcak et al. 2009a, b), even in natural conditions during slowly advancing drought stress (Zivcak et al. 2008a). The decrease of Performance Index is associated with changes of chlorophyll fluorescence transients. The mentioned results, as well as the high number of other scientific papers describe the possibilities, but also the limits of the chlorophyll fluorescence method in study of drought stress (see e.g. Brestic and Zivcak 2013). An example of experimental results comparing sensitivity of different parameters (Zivcak et al. 2009a, b) is shown in Fig. 10.10.

Similarly to the results published by others (e.g. Oukarroum et al. (2007, 2009), the Performance Index represents the parameter useful for screening of the drought stress effects on photosynthetic apparatus. Moreover, it enables to recognize the genotypic differences in drought stress effects. Despite some partial results indicate positive correlation (Zivcak et al. 2008b, c), there is still a lack of information on a direct relationship between the drought tolerance evaluated by chlorophyll fluorescence parameters and crop yield. Nevertheless, one cannot exclude that further research will bring the desired results, as the method meets the demands for simplicity, speed and non-invasiveness mentioned before.

10.3.2 Limitation of Photosynthesis by High Temperature

Description of general mechanisms for adjustment of photosynthesis to temperature in higher plants is very difficult for three reasons: (i) unequal strategies in growth and development, (ii) inherent genetic diversity and (iii) responses of organisms to temperature changes rather than to absolute temperature (Falk et al. 1996). In addition to the general heat tolerance, there can be recognized also the acquired stress tolerance to temperature extremes. It represents the complex trait depending on different factors. We can define it as enhanced ability to survive a severe temperature stress induced by exposure to a non-lethal (mild) temperature (or even some other abiotic or biotic) stress (Sung et al. 2003). Plant responses to the external environmental changes can be classified within two groups. Adaptations express stable genotypic responses to long-term changes while acclimations cause

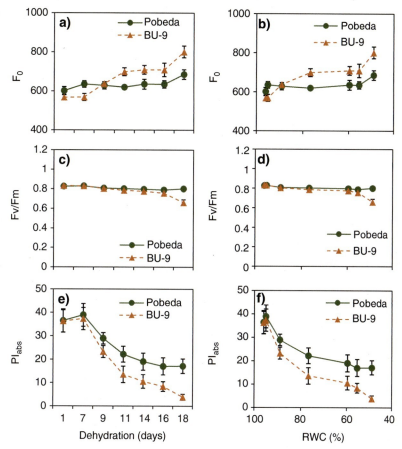

Fig. 10.10 An example of application of fast chlorophyll fluorescence kinetics in comparison of drought stress responses and drought tolerance in two genotypes of winter wheat. The values of basal fluorescence – F_0 (**a**, **b**), the maximum quantum yield of PSII photochemistry – F_v/F_m (**c**, **d**) and performance index (PI_{abs}) are compared related to duration of dehydration (*left*) and leaf relative water content – RWC (*right*). Figure presents the data published by Zivcak et al. (2009a)

phenotypic changes within a single life cycle without apparent genetic changes (Falk et al. 1996).

The sensitivity of photosynthetic reactions to high temperatures is generally well known. Even short exposure of plants to moderate temperature (from 35 to 40 °C) can lead to inhibition of photosynthesis in various (but not in all) plant species (Crafts-Brandner and Salvucci 2002; Sinsawat et al. 2004). After heat stress, photosynthetic electron transport was lowered due to low PSII thermostability. Unlike the drought stress, the stomatal conductance and internal CO_2 content in plant tissues appears to play only a minor role in the temperature limitation of photosynthesis. This fact can be observed well by A/Ci curves based on gas exchange analyses, where the metabolic limitation can be also recognized (Fig. 10.11).

Havaux (1996) has shown that PS II is damaged by severe high temperature stress, i.e. higher than 45 °C, while the significant decrease of CO_2 assimilation occurs at moderate heat stress. Recently, the most of authors believe that the decrease of CO_2 assimilation is associated with the inhibition of Rubisco activation via a direct effect of temperature on Rubisco activase (Feller et al. 1998; Law and Crafts-Brandner 1999; Crafts-Brandner and Salvucci 2002; Salvucci and Crafts-Brandner 2004). Anyway, it is also generally accepted that the photochemical reactions on the level of thylakoids are endangered by heat stress effects (directly or indirectly) (Wise et al. 2004; Wahid et al. 2007). PS II is highly susceptible to high temperature, and its activity is significantly impaired under heat stress conditions (Bukhov et al. 1999; Camejo et al. 2005; Datko et al. 2008).

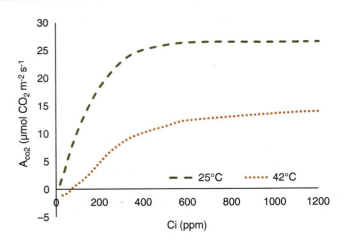

Fig. 10.11 An example of A/Ci curves recorded in wheat leaves at leaf temperature 25 °C (grown at normal temperature level) and at leaf temperature 42 °C in plants pre-exposed to the same high air temperature for 8 h prior to the measurements (Based on unpublished experimental data of the authors)

An important role can play the properties of thylakoid membranes causing heat susceptibility of the system (Sharkey and Zhang 2010). Moreover, the protein components of thylakoid membranes have several weak spots. Heat exposure can lead to dissociation of manganese-stabilizing protein at PS II reaction center complex and the release of Mn atoms (Yamane et al. 1998). Heat stress may also lead to the damage of other components in the reaction center, e.g. major protein units (D1, D2) or other proteins (De Las Rivas and Barber 1997). In barley, heat pulses led to damage of the PS II and to the decline of photosynthetic capacity (measured as O_2 release) and hence, to a limited electron transport (Toth et al. 2005). Sharkova (2001) has shown that high temperatures in high light conditions led to damage of different sites of PS II. It was associated with reinforcing of mechanisms recovering the PSII function. Impaired PS II units are degraded and then synthesized *de novo* to maintain the PS II activity (Wahid et al. 2007). In addition to PSII, Cytochrome b559 (Cytb559) and plastoquinone (PQ) are also affected by heat (Mathur et al. 2014).

Under high temperature conditions, photosystem I (PS I), chloroplast envelopes, and stromal enzymes are thermostable and the PS I driven cyclic electron flow, contributing to thylakoid proton gradient, is activated (Bukhov et al. 1999). However, in heat susceptible plants, the decrease of cyclic electron flow associated with PSI damage was also observed (Essemine et al. 2011; Brestic et al. 2016). Photosynthetic apparatus is sensitive to high temperature; anyway, the thermotolerance can be improved by exposure to moderately high temperature. Havaux (1993) observed that exposition of potato plants at 35 °C for 20 min significantly increased the PS II thermostability. Similar effect observed Brestic et al. (2012) in field grown wheat after the air temperature exceeded 30 °C. The rapid acclimation occured probably thanks to the accumulation of the xanthophylls, especially zeaxanthin (Havaux and Tardy 1996; Ye et al. 2000), which stabilize the lipids of the thylakoid membrane.

Heat tolerance is associated also with the enhanced PSI activity (Zhang and Sharkey 2009), which reflects its quantum yield (Harbinson and Hedley 1989; Datko et al. 2008). The cyclic electron flow around PS I is strongly stimulated by high temperature, and plants lacking this mechanism were significantly more heat-susceptible (Zhang and Sharkey 2009). The cyclic electron transport have several pathways, which can be stimulated to increase or maintain the proton gradient across the thylakoid membrane without increase of NADPH production (Shikanai 2007; Rumeau et al. 2007). This is especially important when the heat stress induce leakiness of thylakoid membrane (Bukhov et al. 1999). Thanks to this mechanism, the ATP levels are not affected by heat as much as expected (Schrader et al. 2004). Hence, the PS II fluorescence measurements may lead to overestimation of the PSII electron transport rate measured at high temperature.

Although many of the physiological mechanisms of crop tolerance to high temperature are well described, further research is needed to uncover the physiological basis of phenotypic flexibility, which enables plant to withstand high temperature. Chlorophyll fluorescence has been used as an early, *in vivo*, indicator of many types of plant stress including temperature stress (Brestic and Zivcak 2013). The chlorophyll fluorescence parameters correlated well with other physiological determinants under high temperature conditions (Wahid et al. 2007). Yamada et al. (1996) found correlation between the maximum quantum yield of PS II photochemistry (Fv/Fm) or the basal fluorescence (F_0) and the heat tolerance in tropical fruits. Gautam et al. (2014) demonstrated use of chlorophyll a fluorescence measurement as an effective tool to screen plants grown in tropical zone for their heat stress tolerance. In many cases the chlorophyll fluorescence may be a more reliable measurement of photosynthesis than CO_2 exchange, which can be low due to closed stomata and not by heat itself (Ierna 2007).

The high temperature effects on PS II and the thermostability at the PS II level has been often evaluated by the basic parameters, such as F_0, Fm and Fv/Fm. Anyway, minimum and maximum fluorescence values are variable, depending on the optical and physiological properties of the sample. Hence, it causes high variability even in non-stressed samples and can become a source of mistakes. The use of the Fv/Fm ratio diminished the risk of error; however, the correct values of Fv/Fm are obtained only in that case, when F_0 is measured in open reaction centers (Q_A fully oxidized) and Fm in closed reaction centers (Q_A fully reduced). In heated samples, the F_0 can slightly grow up due to process of chlororespiration leading to partial reduction of Q_A in the dark, which is enhanced by high temperature (Sazanov et al. 1998). Moreover, if the electron supply by the donor side of PSII is limited due to highly impaired oxygen evolving complex, the electron transport is low and there is the risk that the saturation flash can be insufficient to reach Fm value. Thus, in heat stressed samples the maximum fluorescence Fm can be underestimated and Fv/Fm value does not refer to the maximum quantum yield of PSII photochemistry (Toth et al. 2007; Chen and Cheng 2009). Both the overestimation of F_0 and underestimation of Fm lead to Fv/Fm underestimation showing the temperature effect more severe than it is (Brestic and Zivcak 2013).

Some studies have shown, that the parameters derived from the fast chlorophyll fluorescence kinetics are more sensitive to heat than the common fluorescence parameters such as Fv/Fm. General effects of heat stress on fast fluorescence kinetics are the decrease of maximum fluorescence Fm, increase of minimum fluorescence F_0 and an appearance of additional peak at ~0.3 ms, known as the K- step (Srivastava et al. 1997). The effect of different temperature levels on OJIP-transient of chlorophyll fluorescence induction can be seen in Fig. 10.12.

This is usually associated with a decrease in J-I phase and by the final rise of fluorescence in I-P-phase. The K-step is an indicator of the decrease of capacity of the PSII donor side caused by down-regulation of electron transport between oxygen evolving complex and PSII reaction centre (Guisse et al. 1995; Srivastava et al. 1997). As this effect can be invisible at moderate heat stress if the chlorophyll fluorescence kinetics is shown in common way, the better form is the plot of relative variable fluorescence $W = V_t/V_J$ from 0.05 to 2 ms. The level of this parameter at 0.3 ms ($W_K = V_K/V_J$ value) is very stable (~ 0.50) in non-stressed samples, while in heat treated samples the values increase gradually, in some cases reaching several times higher values in comparison with non-stressed controls (Brestic et al. 2012; Brestic and Zivcak 2013).

This effect is caused only by high temperature, for example even sub-lethal level of drought stress doesn't induce the K-step (Brestic and Zivcak 2013). The effects of exposition at different temperature levels (30 min in darkness) is presented in Fig. 10.13.

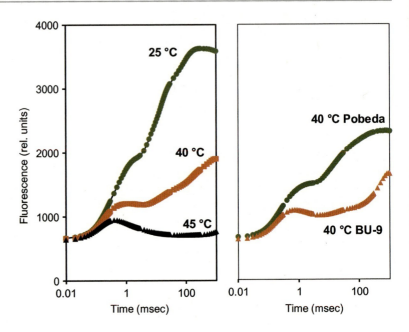

Fig. 10.12 An example of the fast chlorophyll fluorescence kinetics recorded after exposition of leaf segments at different temperature levels for 30 min (the termotest described by Brestic et al. (2012). Figure *left* shows average curves for set of wheat genotypes measured after exposition at 25, 40 and 45 °C, as published by Zivcak et al. (2009a). The figure *right* shows the different level of PSII thermostability observed in two wheat genotypes (2008c)

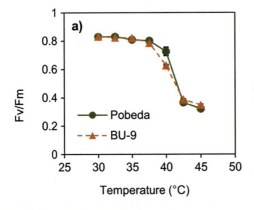
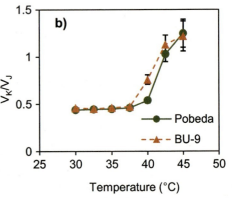

Fig. 10.13 An example of parameters derived from the fast chlorophyll fluorescence kinetics recorded after the exposition of leaf segments at different temperature levels for 30 min. (**a**) The maximum quantum yield of PSII photochemistry – Fv/Fm. (**b**). The ratio of relative variable fluorescence in K step (time 0.3 ms). Based on the data published by Zivcak et al. (2009a)

Although the resistance of PSII photochemistry to high temperatures can be evaluated by exposition at one threshold temperature level, the more efficient way is the testing at wider temperature scale, as it enables to estimate the critical temperature level. Analogically, the PS II thermostability, as a part of the overall leaf heat tolerance, has been determined frequently by the dependence of basal fluorescence and temperature (F_0-T curve), introduced by Schreiber and Berry (1977) (Fig. 10.14).

The method is based on a continuous increase of sample temperature and the regular record of F_0 values. The critical temperature T_C is the temperature at which the F_0 starts a steep increase. The values of critical temperature range in broad scale, mostly between 42 and 50 °C (Dreyer et al. 2001; Robakowski et al. 2002; Froux et al. 2004). Similarly, the method analogous to the continuous F_0 measurements can be used, which is the exposition of leaf samples at several temperature levels (graduated

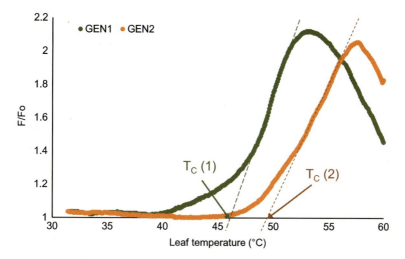

Fig. 10.14 Example of estimation of critical temperature in leaves of two wheat genotypes. The method is based on a continuous record of chlorophyll fluorescence in continuously heated leaves. The rate of heating was app. 1 °C per 1 min. Measurements were made by Maxi-Imaging-PAM (Walz, Germany). The original figure is based on unpublished data provided by the authors, the improved method was previously presented by Brestic and Zivcak (2013)

temperature approach) with the fresh sample used at each level (Brestic et al. 2012); such measurements give more complex information about the heat effects on PS II photochemistry, but it enables also estimate the critical temperature for F_0 parameter (but even for any other measured parameter) similarly as in previous technique (Brestic and Zivcak 2013). The slight increase of F_0 in moderate temperatures might be the result of chlororespiration (Sazanov et al. 1998) or by other processes causing partial re-reduction of fully oxidized Q_A during dark heat treatment. However, beyond the threshold level (in our case app. 42 °C and more) we observed the start of the exponential F_0 increase; this observation is consistent with critical temperatures published for annual species grown at moderate temperatures up to 25 °C (Havaux et al. 1990; Froux et al. 2004). The critical temperature can be stated also using any heat sensitive parameter of fast fluorescence kinetics, if the measurements were done at graduated levels of high temperature (Zivcak et al. 2013).

Critical temperature based on Fo level represents the temperature, at which the serious disorganization of structure and loss of main functions occurs (Yamane et al. 1998). Hence, it is temperature mostly out of range of "physiologically relevant" temperature, as in most cases leaves in a field are not heated up to 50 °C or more during their life. On the other hand, the occurrence of K-step even at temperature 4–5 °C lower than a steep F_0 increase was found (Brestic et al. 2012; Zivcak et al. 2013) and the V_K/V_j values in non-treated plants are almost stable. Moreover, a higher capacity to increase thermostability at the K-step level than at F_0 level was observed (Brestic et al. 2012).

In addition, by the analysis of OJIP-transient, the responses to mild heat stress between 35 and 40 °C can be also observed, especially the decrease of active reaction centers, followed by limitation of the donor side indicated by the appearance of K-step (Fig. 10.15).

The physiological meaning of the K-step increase is clear and it represents probably the first (and hence the critical) irreversible heat effect inactivating the PS II reaction centers. For all these reasons the application of fast chlorophyll a fluorescence measurements aimed at the K-step determination can be considered to be an efficient approach of testing PS II heat

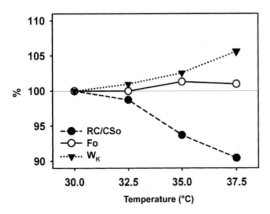

Fig. 10.15 The early heat stress responses recorded by the changes in fast chlorophyll fluorescence kinetics in leaves of winter wheat. The first response to heat is a decrease of number of fully cative reaction centers, evident eve at 35 °C. The donor side inhibition appeared mostly at temperature over 37 °C. The Fo component was insensitive to moderate heat stress (Zivcak 2006)

thermostability and plasticity in plants (Brestic and Zivcak 2013) as shown also by other studies (Oukarroum et al. 2012); moreover it has potential of practical use in plant screening, as this method is fast and reliable. The record of fast chlorophyll a fluorescence transient in dark adapted samples represent an efficient tool for assessment of adverse effects on PS II photochemistry (Strasser et al. 2004).

Numerous research papers deal with the application of photosynthetic methods in heat stress responses in crop plants; the most recent papers are listed in Table 10.3.

In general, the last years brought promising data towards practical applications of photosynthetic techniques, mostly chlorophyll fluorescence in screening and breeding new genotypes characterized by higher heat tolerance. The challenges in this area come with new technical applications, like chlorophyll fluorescence imaging and its modifications, as well as simultaneous measurement of chlorophyll fluorescence (or even fluorescence imaging) with other physiological parameters, can that verify fluorescence measurements and bring more complex and useful information about the mechanisms contributing to stress tolerance.

10.4 Future Challenges: The Central Role of Plant Phenotyping

Plant phenotyping has been introduced as a new discipline in plant sciences, using non- or minimally-invasive sensors to measure plant performance and uncover genetic determinants of enhanced agronomic traits (Rascher et al. 2011). Complex traits, such as plant and canopy growth, accumulation of biomass and harvestable yield are products of the interplay of environmental and genetic factors during vegetation season of crops. Different external factors modulate the gene expression and influence markedly the plant phenotype. Reactions can be generalized for individual traits and environmental factors using advanced statistical tools, requiring a high number of experimental data (Poorter et al. 2010; Rascher et al. 2011).

Phenotyping can be done under laboratory conditions, in greenhouses or in the field. The term phenotyping covers many aspects associated with selection of useful germplasm and, in fact, there is often big difference between approaches employed in field conditions and in laboratory. In laboratory, external conditions may be easily regulated. In contrast, field trials are grown in highly variable, fluctuating conditions (Rascher and Nedbal 2006; Schurr et al. 2006; Mittler and Blumwald 2010). The specific part of phenotyping represent so called automate high throughput phenotyping, which is more frequently applied in technical platforms in controlled conditions; however, there have been introduced also advanced field automated phenotyping platforms. The new technologies employed in the high-throughput phenotyping are done in a line with monitoring environmental factors, and they provide highly valuable datasets for statistical approaches (Rascher et al. 2009). Exponentially increasing number of papers have been published dealing with the crop phenotyping, including general concepts as well as practical outputs of field and laboratory approaches; the most recent papers are listed in Table 10.4.

Although plant phenotyping can be aimed at different traits (e.g. growth patterns of

Table 10.3 The list of the most recent papers dealing with heat stress and photosynthethesis

Crops	Applied photosynthetic methods	References
Wheat	Gas exchange	Tian et al. (2012), Almeselmani et al. (2012) and Fang et al. (2013)
	Gas exchange, chlorophyll fluorescence (slow)	Tian et al. (2012), Dias et al. (2011), Almeselmani et al. (2012), Amirjani (2012), Ushakova et al. (2013), Shanmugan et al. (2013), Wang et al. (2014) and Wang et al. (2011)
	Chlorophyll fluorescence (slow)	Chauhan et al. (2012)
	Chlorophyll fluorescence (fast)	Brestic et al. (2013), Olsovska et al. (2013), Shefazadeh et al. (2012), Brestic et al. (2012), Oukarroum et al. (2012), Zivcak et al. (2008a, b, c) and Mathur et al. (2011a, b)
	Chlorophyll fluorescence (slow and fast)	Sharma et al. (2012)
Barley	Gas exchange, chlorophyll fluorescence	Kaňa et al. (2008)
	Fast chlorophyll fluorescence	Kalaji et al. (2011), Janeczko et al. (2011) and Oukarroum et al. (2012)
	Fast chlorophyll fluorescence with P700	Oukarroum et al. (2009)
	Slow chlorophyll fluorescence with P700	Datko et al. (2008)
Rice	Gas exchange	Thussagunpanit et al. (2012)
	Gas exchange, chlorophyll fluorescence (JIP-test)	Restrepo-Diaz and Garces-Varon (2013)
	Gas exchange, A/Ci curves	Scafaro et al. (2012)
	Gas exchange, chlorophyll fluorescence (slow),	Chandrakala et al. (2013) and Song et al. (2013b)
Maize	Gas exchange, WUE	Suwa et al. (2010)
	Gas exchange, slow chlorophyll fluorescence	Xu et al. (2011)
Soybean	Gas exchange, chlorophyll fluorescence (slow)	Herzog and Chai-Arree (2012)
	O_2 evolution, chlorophyll fluorescence (fast)	Li et al. (2009)
	gas exchange	Djanaguiraman et al. (2011)
Sorghum	Gas exchange, fast chlorophyll fluorescence	Yan et al. (2012) and Yan et al. (2011)
	Gas exchange, fast chlorophyll fluorescence, P700	Yan et al. (2013a, b)
Potato	Gas exchange	Aien et al. (2011)
Sunflower	Gas exchange	Kalyar et al. (2013)
Oilseed rape	Gas exchange	Qaderii et al. (2012)
Bean	Gas exchange, chlorophyll fluorescence (slow)	Huve et al. (2011) and González-Cruz and Pastenes (2012)
	Chlorophyll fluorescence (slow)	Ribeiro et al. (2008)
	Fast chlorophyll fluorescence	Stefanov et al. (2011)
Tomato	Gas exchange, chlorophyll fluorescence with PSI	Qi et al. (2013)
Jer. artichoke	Gas exchange, fast chlorophyll fluorescence, P700	Yan et al. (2013a, b)
Mulberry	Gas exchange with chlorophyll fluorescence	Yu et al. (2013)

(continued)

Table 10.3 (continued)

Crops	Applied photosynthetic methods	References
Eggplant	*Gas exchange, chlorophyll fluorescence (slow)*	Wu et al. (2014)
Melon	*Gas exchange, chlorophyll fluorescence (slow)*	Zhang et al. (2013)
Tall fescue	*Gas exchange (A/Ci curves), Rubisco activity*	Yu et al. (2014)
Jatropha	*Gas exchange, chlorophyll fluorescence (slow)*	Silva et al. (2010)
Tobacco	*Gas exchange, fast chlorophyll fluorescence*	Frolec et al. (2010) and Tan et al. (2011)
Cotton	*Gas exchange, chlorophyll fluorescence (slow)*	Snider et al. (2009), Hozain et al. (2012) and Carmo-Silva et al. (2012)

Table 10.4 The overview of recent papers dealing with photosynthetic traits in crop phenotyping

Topic	Crops	References
Crop phenotyping (general reviews)	General	Montes et al. (2007), Furbank and tester (2011), Tuberosa (2012), Fiorani et al. (2012), Pieruschka and Poorter (2012), Dhondt et al. (2013) and Cobb et al. (2013)
Field phenotyping	General	Cativelli et al. (2008), Jones et al. (2009), Tuberosa (2012), Cabrera-Bosquet et al. (2012), Fiorani and Schurr (2013), Costa et al. (2013), Sadras et al. (2013) and Araus and Cairns (2014)
	Wheat	Munns et al. (2010), Richards et al. (2010), Reynolds et al. (2009), Saint Pierre et al. (2012), Rebetzke et al. (2012) and Passioura (2012)
	Barley	Munns et al. (2010)
	Maize	Masuka et al. (2012), Ciampitti et al. (2012) and Araus et al. (2012)
	Rice	Kamoshita et al. (2008), Serraj et al. (2011) and Fischer et al. (2012)
	Soybean	Mengistu et al. (2011)
	Bean	Rascher et al. (2011) and Beebe et al. (2013)
	Potato	Prashar et al. (2013)
Automated phenotyping (laboratory)	General	Granier and Tardieu (2009), Berger et al. (2010), Kolukisaoglu and Thurow (2010), Pereyra-Irujo et al. (2012), Harbinson et al. (2012) and Brien et al. (2013)
	Wheat	Golzarian et al. (2011) and Hackl et al. (2013)
	Barley	Cseri et al. (2013)
	Rice	Feng et al. (2013)
	Maize	Fuad-Hassan et al. (2008), Chapuis et al. (2012) and Dignat et al. (2013)
	Soybean	Vollmann et al. (2011)
	Bean	Rascher et al. (2011)

aboveground biomass or roots), an important part of crop phenotyping programs is focused on improvement of the photosynthetic efficiency under optimal or stress conditions. In addition to different approaches aimed at improvement of light use efficiency, mentioned above (Parry et al. 2003; Long et al. 2006; Hibberd et al. 2008; Murchie et al. 2009; Rascher et al. 2010; Whitney et al. 2011; Gowik and Westhoff 2011), also the phenotyping programs aimed at improvement of water use efficiency (Gilbert et al. 2011) or plant stress tolerance use photosynthetic parameters.

Among numerous methods, the quantification of chlorophyll fluorescence represents a robust approach to determine the photosynthetic light use efficiency (Maxwell and Johnson 2000; Baker 2008). However, when using the common fluorescence methods mentioned above, the leaves need to be excited actively by high light intensities, thus reducing applicability of these methods for fluorescence imaging of larger areas under natural light. The possible alternative method represents the LIFT, i.e. the laser induced fluorescence transient that was previously applied to determine the chlorophyll fluorescence parameters for detection of stress effects from a distance of 50 m (Ananyev et al. 2005; Kolber et al. 2005; Rascher and Pieruschka 2008; Pieruschka et al. 2010). Another approach represents the passive recording of the chlorophyll fluorescence (Fs) emmited by naturally iluminated plants, which has overcome the problem of required active excitation (Moya et al. 2004; Liu et al. 2005; Alonso et al. 2008; Malenovsky et al. 2009; Meroni et al. 2009). It has been experimentally documented that Fs correlates well with leaf photosynthetic efficiency and with the stress induced limitation of the photosynthetic electron transport. In this respect, it may be used to quantify photosynthetic light use efficiency (Rosema et al. 1998; Flexas et al. 2000; Meroni and Colombo 2006; Damm et al. 2010). However, when we take into account the 3-D structure of canopy, the Fs signal has not a simple interpretation as it depends also on the light-leaf interaction determined by the orientation of the different canopy elements. Anyway, recent results of the combination of Fs with the 3-D parameterisation of plant morphological data, which enable to obtain satisfactory results (Rascher et al. 2011).

In addition to the imaging methods based on chlorophyll fluorescence, mapping crops can be done also by rapidly developing imaging spectroscopy. While these methods in the past were based on a limited number of spectral bands, now it is possible to record a continuous light spectrum (Ustin et al. 2004; Rascher et al. 2007). Imaging spectroscopy deals with the reflected radiation to obtain information about different plant and canopy properties. The value of canopy reflectance in the visible spectra (400–700 nm) depends on the absorbance of light by the plant photosynthetic pigments. Near infrared band (700–1100 nm) is influenced mainly by light absorption by the internal plant tissues. The shortwave infrared band (1100–2500 nm) depends on the water content in plants (Curran 1989; Rascher et al. 2010, 2011). Moreover, an important advance represents the possibility to derive the natural light-induced chlorophyll fluorescence data from imaging spectroscopy records (Malenovsky et al. 2009; Meroni et al. 2009; Rascher et al. 2009).

The spatio-temporal changes of metabolic and structural components of crop canopy result in a variable interactions with incident radiation. In most of cases, the research studies have been concentrated on a few spectral bands. Most frequently, the normalized difference vegetation index (NDVI) has been applied, as it correlates well with the leaf area index (LAI) and chlorophyll content. However, the methods of spectral reflectance have been improved significantly and it one may expect that the improved methods enriched in different information will be employed in the automated crop phenotyping in a broad extent (Rascher et al. 2011).

10.5 Concluding Remarks

A system process of the improvement of main crop species has led to attending the harvest index nearby to the theoretical maxima; therefore, the only way to reach another gain of yield potential in the future genotypes have to be associated with an increase in the production of crop biomass through the improvement of photosynthetic efficiency. Moreover, the decrease of photosynthetic production caused by abiotic stress factors represents the main reason why yields in a global range are far away from the yield potential of main crops. In this respect, the application of non-invasive methods for assessment of crop photosynthesis is gaining a greater practical importance. The most important parameters can be obtained from gas exchange method, especially from the measurements of A/c_i curves, which enable to estimate the main photosynthetic limitations, including stomatal, enzymatic and mesophyll limitations. However, the method has a limited practical application in the screening of useful germplasm in genotype collections, as the measurements are very time-consuming. Therefore, attention has been paid also to chlorophyll fluorescence, which is fast, more efficient and potentially useful in plant testing. More than in the study of improved photosynthetic production, chlorophyll fluorescence corresponds better with the screening for improved stress tolerance, as it has been documented by numerous studies. Analyses of the fast chlorophyll fluorescence kinetics has a big potential in the future because of its simplicity, large screening capacity and low time requirements. Especially promising are the results directed to the application of fast chlorophyll fluorescence in the screening of photosynthetic apparatus for high temperature tolerance.

The future challenges lie mainly in the application of photosynthetic methods for a high-throughput crop phenotyping. For automated systems, imaging methods seem to be the most useful. In addition to the chlorophyll fluorescence imaging, methods based on the canopy reflectance open a path for efficient development of large-scale models useful in practical applications. In the context of above-mentioned, for a model parameterization of individual crops and specific environmental conditions, the extensive involvement of "classical" photosynthetic methods will be necessary.

Acknowledgements This work was supported by grants APVV-15-0721 and the research project of the Scientific Grant Agency of Slovak Republic VEGA-1-0923-16.

References

Aien A, Khetarpal S, Pal M (2011) Photosynthetic characteristics of potato cultivars grown under high temperature. Am-Eurs J Agric Environ Sci 11:633–639

Ainsworth EA, Long SP (2005) What have we learned from 15 years of free-air CO_2 enrichment (FACE)? A meta-analytic review of the responses of photosynthesis, canopy. New Phytol 165:351–371

Ainsworth EA, Davey PA, Bernacchi CJ, Dermody OC, Heaton EA, Moore DJ, Morgan PB, Naidu SL, Ra HSY, Zhu XG, Curtis PS, Long SP (2002) A meta-analysis of elevated [CO_2] effects on soybean (*Glycine max*) physiology, growth and yield. Glob Change Biol 8:695–709

Akhkha A, Boutraa T, Alhejely A (2011) The rates of photo-synthesis, chlorophyll content, dark respiration, proline and abscicic acid (ABA) in wheat (*Triticum durum*) under water deficit conditions. Internat J Agric Biol 13:215–221

Almeselmani M, Deshmukh PS, Chinnusamy V (2012) Effects of prolonged high temperature stress on respiration, photosynthesis and gene expression in wheat (*Triticum aestivum* L.) varieties differing in their thermotolerance. Plant Stress 6:25–32

Alonso L, Gomez-Chova L, Vila-Frances J, Amoros-Lopez J, Guanter L, Calpe J, Moreno J (2008) Improved Fraunhofer line discrimination method for vegetation fluorescence quantification. IEEE Geosci Remote Sens Lett 5:620–624

Amirjani M (2012) Estimation of wheat responses to "high" heat stress. Amer-Euro J Sustain Agri 6:222–233

Ananyev G, Kolber ZS, Klimov D, Falkowski PG, Berry JA, Rascher U, Martin R, Osmond CB (2005) Remote sensing of heterogeneity in photosynthetic efficiency, electron transport and dissipation of excess light in *Populus deltoides* stands under ambient and elevated CO_2 concentrations, and in a tropical forest canopy, using a new laser-induced fluorescence transient device. Glob Change Biol 11:1195–1206

Anjum SA, Wang LC, Farooq M, Hussain M, Xue LL, Zou CM (2011) Brassinolide application improves the drought tolerance in maize through modulation of enzymatic antioxidants and leaf gas exchange. J Agron Crop Sci 197:177–185

Anjum SA, Ehsanullah LX, Wang L, Farrukh M, Saleem CJH (2013) Exogenous benzoic acid (BZA) treatment can induce drought tolerance in soybean plants by improving gas-exchange and chlorophyll contents. Austral J Crop Sci 7:555

Araus JL, Cairns JE (2014) Field high-throughput phenotyping: the new crop breeding frontier. Trends Plant Sci 19:52–61

Araus JL, Slafer GA, Reynolds MP, Royo C (2002) Plant breeding and drought in C3 cereals: what should we breed for? Ann Bot 89:925–940

Araus JL, Serret MD, Edmeades GO (2012) Phenotyping maize for adaptation to drought. Front Physiol 3:305

Arntz M, Delph L (2001) Pattern and process: evidence for the evolution of photosynthetic traits in natural populations. Oecologia 127:455–467

Arntz AM, DeLucia EH, Jordan N (2000) From fluorescence to fitness: variation in photosynthetic rate affects fecundity and survivorship. Ecology 81:2567–2576

Ashoub A, Beckhaus T, Berberich T, Karas M, Brüggemann W (2013) Comparative analysis of barley leaf proteome as affected by drought stress. Planta 237:771–781

Ashraf M, Foolad MR (2006) Roles of glycine betaine and proline in improving plant abiotic stress resistance. Environ Exp Bot 59:206–216

Atkin OK, Evans JR, Siebke K (1998) Relationship between the inhibition of leaf respiration by light and enhancement of leaf dark respiration following light treatment. Aust J Plant Physiol 24:437–443

Austin R, Bingham J, Blackwell R, Evans L, Ford M, Morgan C, Taylor M (1980) Genetic improvements in winter wheat yields since 1900 and associated physiological changes. J Agr Sci 94:675–689

Austin RB, Ford MA, Morgan CL (1989) Genetic improvement in the yield of winter wheat: a further evaluation. J Agr Sci 112:295–301

Bainbridge G, Madgwick P, Parmar S, Mitchell R, Paul M, Pitts J, Keys AJ, Parry MAJ (1995) Engineering Rubisco to change its catalytic properties. J Exp Bot 46:1269–1276

Baker NR (2008) Chlorophyll fluorescence: a probe of photosynthesis in vivo. Annu Rev Plant Biol 59:659–668

Baker NR, Rosenqvist E (2004) Applications of chlorophyll fluorescence can improve crop production strategies: an examination of future possibilities. J Exp Bot 403:1607–21

Ballantyne AP, Miller JB, Tans PP, White JWC (2011) Testing biosphere models with atmospheric observations: novel applications of isotopes in CO_2. Biogeosciences 8:3093–3106

Barnaby JY, Kim M, Bauchan G, Bunce J, Reddy V, Sicher RC (2013) Drought responses of foliar metabolites in three maize hybrids differing in water stress tolerance. PloS One 8:77145

Beadle CL, Long SP (1985) Photosynthesis – is it limiting to biomass production? Biomass 8:119–168

Beebe SE, Rao IM, Blair MW, Acosta-Gallegos JA (2013) Phenotyping common beans for adaptation to drought. Front Physiol 4:35

Bencze S, Bamberger Z, Janda T, Balla K, Bedő Z, Veisz O (2011) Drought tolerance in cereals in terms of water retention, photosynthesis and antioxidant enzyme activities. Centr Europ J Biol 6:376–387

Bender J, Hertstein U, Black CR (1999) Growth and yield responses of spring wheat to increasing carbon dioxide, ozone and physiological stresses: a statistical analysis 'ESPACE-wheat' results. Eur J Agron 10:185–195

Berger B, Parent B, Tester M (2010) High-throughput shoot imaging to study drought responses. J Exp Bot 61:3519–3528

Bernacchi CJ, Portis AR, Nakano H, von Caemmerer S, Long SP (2002) Temperature response of mesophyll conductance: implications for the determination of Rubisco enzyme kinetics and for limitations to photosynthesis in vivo. Plant Physiol 130:1992–1998

Bilger W, Bjorkman O (1990) Role of the xanthophyll cycle in photoprotection elucidated by measurements of light-induced absorbance changes, fluorescence and photosynthesis in leaves of *Hedera canariensis*. Photosynth Res 25:173–185

Björkman O, Demmig B (1987) Photon yield of O_2 evolution and chlorophyll fluorescence characteristics at 77 K among vascular plants of diverse origins. Planta 170:489–504

Blum A (1996) Crop responses to drought and the interpretation of adaptation. Plant Growth Regul 20:135–148

Borrás L, Slafer GA, Otegui ME (2004) Seed dry weight response to source-sink manipulations in wheat, maize and soybean: a quantitative reappraisal. Field Crop Res 86:131–146

Bradbury M, Baker NR (1981) Analysis of the slow phases of the in vivo chlorophyll fluorescence induction curve. Changes in the redox state of photosystem II electronacceptors and fluorescence emission from photosystems I and II. Biochim Biophys Acta 635:542–55

Brestic M, Zivcak M (2013) PS II fluorescence techniques for measurements of drought and high temperature stress signal in crop plants: protocols and applications. In: Rout GR, Das AB (eds) Molecular Stress Physiology of Plants. Springer, Dordrecht, pp 87–131

Brestic M, Cornic G, Fryer MJ, Baker NR (1995) Does photorespiration protect the photosyn-thetic apparatus in French bean leaves from photoinhibition during drought stress? Planta 196:450–457

Brestic M, Zivcak M, Kalaji HM, Allakhverdiev SI, Carpentier R (2012) Photosystem II ther-mostability in situ: environmentally induced acclimation and genotype-specific reactions in *Triticum aestivum* L. Plant Physiol Biochem 57:93–105

Brestic M, Zivcak M, Olsovska K, Repkova J (2013) Involvement of chlorophyll a fluorescence analyses for identification of sensitiveness of the photosynthetic apparatus to high temperature in selected wheat genotypes. Photosynthesis Research for Food, Fuel and the Future. Springer, Berlin Heidelberg, pp 510–513

Brestic M, Zivcak M, Olsovska K, Shao HB, Kalaji HM, Allakhverdiev SI (2014) Reduced glutamine synthetase activity plays a role in control of photosynthetic responses to high light in barley leaves. Plant Physiol Biochem 81:74–83

Brestic M, Zivcak M, Kunderlikova K, Sytar O, Shao H, Kalaji HM, Allakhverdiev SI (2015) Low PSI content limits the photoprotection of PSI and PSII in early growth stages of chlorophyll b-deficient wheat mutant lines. Photosynth Res 125:151–166

Brestic M, Zivcak M, Kunderlikova K, Allakhverdiev SI (2016) High temperature specifically affects the photoprotective responses of chlorophyll b-defficient wheat mutant lines. Photosynth Res (in press). doi:10.1007/s11120-016-0249-7

Brien CJ, Berger B, Rabie H, Tester M (2013) Accounting for variation in designing greenhouse experiments with special reference to greenhouses containing plants on conveyor systems. Plant Methods 9:5

Bukhov NG, Wiese C, Neimanis S, Heber U (1999) Heat sensitivity of chloroplasts and leaves: Leakage of protons from thylakoids and reversible activation of cyclic electron transport. Photosynth Res 59:81–93

Bunce JA (1981) Comparative responses of leaf conductance to humidity in single attached leaves. J Exp Bot 32:629–634

Bürling K, Cerovic ZG, Cornic G, Ducruet JM, Noga G, Hunsche M (2013) Fluorescence-based sensing of drought-induced stress in the vegetative phase of four contrasting wheat genotypes. Environ Exp Bot 89:51–59

Cabrera-Bosquet L, Crossa J, von Zitzewitz J, Serret MD, Araus JL (2012) High-throughput Phenotyping and Genomic Selection: The Frontiers of Crop Breeding Converge. J Integr Plant Biol 54:312–320

Calatayud A, Roca D, Martínez PF (2006) Spatial-temporal variations in rose leaves under water stress conditions studied by chlorophyll fluorescence imaging. Plant Physiol Biochem 44:564–573

Camejo D, Rodriguez P, Morales MA, Dell'Amico JM, Torrecillas A, Alarcon JJ (2005) High temperature effects on photosynthetic activity of two tomato cultivars with different heat susceptibility. J Plant Physiol 162:281–289

Carmo-Silva AE, Gore MA, Andrade-Sanchez P, French AN, Hunsaker DJ, Salvucci ME (2012) Decreased CO_2 availability and inactivation of Rubisco limit photosynthesis in cotton plants under heat and drought stress in the field. Environ Exp Bot 83:1–11

Cattivelli L, Rizza F, Badeck FW, Mazzucotelli E, Mastrangelo AM, Francia E, Mare C, Tondelli A, Stanca AM (2008) Drought tolerance improvement in crop plants: an integrated view from breeding to genomics. Field Crop Res 105:1–14

Centritto M, Loreto F, Chartzoulakis K (2003) The use of low CO_2 to estimate diffusional and non-diffusional limitations of photosynthetic capacity of salt stressed olive saplings. Plant Cell Environ 26:585–594

Ceppi MG, Oukarroum A, Çiçek N, Strasser RJ, Schansker G (2012) The IP amplitude of the fluorescence rise *OJIP* is sensitive to changes in the photosystem I content of leaves: a study on plants exposed to magnesium and sulfate deficiencies, drought stress and salt stress. Physiol Plantarum 144:277–288

Chaerle L, Leinonen I, Jones HG, van der Straeten D (2007) Display Settings: Monitoring and screening plant populations with combined thermal and chlorophyll fluorescence imaging. J Exp Bot 58:773–84

Chandrakala JU, Chaturvedi AK, Ramesh KV, Rai P, Khetarpal S, Pal M (2013) Acclimation response of signalling molecules for high temperature stress on photosynthetic characteristics in rice genotypes. Indian J Plant Phys 18:142–150

Chapuis R, Delluc C, Debeuf R, Tardieu F, Welcker C (2012) Resiliences to water deficit in a phenotyping platform and in the field: How related are they in maize? Eur J Agron 42:59–67

Chauhan H, Khurana N, Nijhavan A, Khurana JP, Khurana P (2012) The wheat chloroplastic small heat shock protein (sHSP26) is involved in seed maturation and germination and imparts tolerance to heat stress. Plant Cell Environ 35:1912–1931

Chaves MM (1991) Effects of water deficits on carbon assimilation. J Exp Bot 42:1–16

Chaves MM, Pereira JS, Maroco J, Rodrigues ML, Ricardo CP, Osorio ML, Carvalho I, Faria T, Pinheiro C (2002) How plants cope with water stress in the field. Photosynthesis and growth. Annals Bot 89:907–916

Chaves MM, Maroco JP, Pereira JS (2003) Understanding plant responses to drought – from genes to the whole plant. Funct Plant Biol 30:239–264

Chen LS, Cheng L (2009) Photosystem 2 is more tolerant to high temperature in apple (*Malus domestica* Borkh) leaves than in fruit peel. Photosynthetica 47:112–120

Ciampitti IA, Zhang H, Friedemann P, Vyn TJ (2012) Potential physiological frameworks for mid-season field phenotyping of final plant nitrogen uptake, nitrogen use efficiency, and grain yield in maize. Crop Sci 52:2728–2742

Cobb JN, DeClerck G, Greenberg A, Clark R, McCouch S (2013) Next-generation phenotyping: requirements and strategies for enhancing our understanding of genotype–phenotype relationships and its relevance to crop improvement. Theor Appl Genet 126:867–887

Collatz GJ, Ribascarbo M, Berry JA (1992) Coupled photosynthesis-stomatal conductance model for leaves of C4 plants. Aust J Plant Physiol 19:519–538

Cornic G, Briantais JM (1991) Partitioning of photosynthetic electron flow between CO_2 and O_2 reduction in a C3 leaf (*Phaseolus vulgaris* L.) at different CO_2 concentrations and during water stress. Planta 183:178–184

Cornic G, Fresneau C (2002) Photosynthetic carbon reduction and carbon oxidation cycles are the main electron sinks for photosystem 2 activity during a mild drought. Annals Bot 89:887–894

Cornic G, Massacci A (1996) Leaf photosynthesis under drought stress. In: Baker NR (ed) Photosynthesis and the Environment. Kluwer Academic Publishers, Dordrecht, pp 347–366

Costa JM, Grant OM, Chaves MM (2013) Thermography to explore plant–environment interactions. J Exp Bot 64:3937–3949

Crafts-Brandner SJ, Salvucci ME (2002) Sensitivity of photosynthesis in a C4 plant, maize to heat stress. Plant Physiol 129:1773–1780

Cseri A, Sass L, Torjek O, Pauk J, Vass I, Dudits D (2013) Monitoring drought responses of barley genotypes with semi-robotic phenotyping platform and association analysis between recorded traits and allelic variants of some stress genes. Austral J Crop Sci 7:1560–1570

Curran PJ (1989) Remote-sensing of foliar chemistry. Remote Sens Environ 30:271–278

Damm A, Elbers J, Erler E, Gioli B, Hamdi K, Hutjes R, Kosvancova M, Meroni M, Miglietta F, Moersch A, Moreno J, Schickling A, Sonnenschein R, Udelhoven T, van der Linden S, Hostert P, Rascher U (2010) Remote sensing of sun induced fluorescence to improve modeling of diurnal courses of gross primary production (GPP). Glob Change Biol 16:171–186

Datko M, Zivcak M, Brestic M (2008) Proteomic analysis of barley (*Hordeum vulgare* L.) leaves as affected by high temperature treatment. In: Allen JF, Gantt E, Goldbeck JH, Osmond B (eds) Photosynthesis. Energy from the sun: 14th International congress on photosynthesis. Springer, Dordrecht, pp 1523–1527

Davies WJ, Zhang J (1991) Root signals and the regulation of growth and development of plant in drying soil. Annu Rev Plant Physiol Plant Mol Biol 42:55–76

de Las Rivas J, Barber J (1997) Structure and thermal stability of photosystem II reaction centers studied by infrared spectroscopy. Biochemistry 36:8897–8903

de Souza TC, Magalhães PC, de Castro EM, de Albuquerque PEP, Marabesi MA (2013) The influence of ABA on water relation, photosynthesis parameters, and chlorophyll fluorescence under drought conditions in two maize hybrids with contrasting drought resistance. Acta Physiol Plant 35:515–527

Deák C, Jäger K, Fábián A, Nagy V, Albert Z, Miskó A, Barnabás B, Papp I (2011) Investigation of physiological responses and leaf morphological traits of wheat genotypes with contrasting drought stress tolerance. Acta Biol Szeged 55:69–71

Deeba F, Pandey AK, Ranjan S, Mishra A, Singh R, Sharma YK, Shirke PA, Pandey V (2012) Physiological and proteomic responses of cotton (*Gossypium herbaceum* L.) to drought stress. Plant Physiol Bioch 53:6–18

Demetriades-Shah TH, Fuchs M, Kanemasu ET, Flitcroft I (1992) A note of caution concerning the relationship between cumulated intercepted solar radiation and crop growth. Agr Forest Meteorol 58:193–207

Dhondt S, Wuyts N, Inzé D (2013) Cell to whole-plant phenotyping: the best is yet to come. Trends Plant Sci 18:428–439

Dias AS, Semedo J, Ramalho JC, Lidon FC (2011) Bread and durum wheat under heat stress: a comparative study on the photosynthetic performance. J Agron Crop Sci 197:50–56

Dignat G, Welcker C, Sawkins M, Ribaut J M, Tardieu F (2013) The growths of leaves, shoots, roots and reproductive organs partly share their genetic control in maize plants. Plant Cell Environ 36:1105–1119

Ditmarova L, Kurjak D, Palmroth S, Kmet J, Strelcova K (2010) Physiological responses of Norway spruce (*Picea abies*) seedlings to drought stress. Tree Physiol 30:205–213

Djanaguiraman M, Prasad PVV, Al-Khatib K (2011) Ethylene perception inhibitor 1-MCP decreases oxidative damage of leaves through enhanced antioxidant defense mechanisms in soybean plants grown under high temperature stress. Environ Exp Bot 71:215–223

Do PT, Degenkolbe T, Erban A, Heyer AG, Kopka J, Köhl KI, Zuther E (2013) Dissecting rice polyamine metabolism under controlled long-term drought stress. PloS One 8:60325

Drake BG, Gonzalez-Meler M, Long SP (1997) More efficient plants: a consequence of rising atmospheric CO_2? Ann Rev Plant Physiol 48:609–639

Dreyer E, Le Roux X, Montpied P, Daudet FA, Masson F (2001) Temperature response of leaf photosynthetic capacity in seedlings from seven temperate tree species. Tree Physiol 21:223–232

Duysens LMN, Sweers HE (1963) Mechanism of the two photochemical reactions in algae as studied by means of fluorescence. In: Japanese Society of Plant Physiologists (ed) Studies on Microalgae and Photosynthetic Bacteria, University of Tokyo Press, Tokyo

Ehleringer J, Björkman O (1977) Quantum yields for CO_2 uptake in C3 and C4 plants. Plant Physiol 59:86–90

Ellenson JL, Amundson RG (1982) Delayed light imaging for the early detection of plant stress. Science 215:1104–1106

Epron D, Godard G, Cornic G, Genty B (1995) Limitation of net CO_2 assimilation rate by internal resistances to CO_2 transfer in the leaves of two tree species (*Fagus sylvatica* and *Castanea sativa* Mill.). Plant Cell Environ 18:43–51

Escalona JM, Flexas J, Medrano H (1999) Stomatal and non-stomatal limitations to photosynthesis under water stress in field-grown grapevine. Aust J Plant Physiol 26:421–433

Essemine J, Govindachary S, Ammar S, Bouzid S, Carpentier R (2011) Abolition of photosystem I cyclic electron flow in *Arabidopsis thaliana* following thermal-stress. Plant Physiol Bioch 49:235–243.

Evans LT (1993) Crop Evolution, Adaptation and Yield. Cambridge University Press, Cambridge, 500 p.

Evans JR (2013) Improving photosynthesis. Plant Physiol 162:1780–1793

Evans LT, Dunstone RL (1970) Some physiological aspects of evolution in wheat. Aust J Biol Sci 23:725–741

Fábián A, Jäger K, Barnabás B (2013) Developmental stage dependency of the effect of drought stress on photosynthesis in winter wheat (*Triticum aestivum* L.) varieties. Acta Agron Hung 61:13–21

Falk S, Maxwell DP, Laudenbach DE, Huner NPA (1996) Photosynthetic adjustment to temperature. In: Baker NR (ed) Photosynthesis and the Environment. Advances in Photosynthesis and Respiration, Volume 5. Kluwer, Dordrecht, pp 367–385

Fan X, Huang G, Zhang L, Deng T, Li Y (2013) Adaptability and recovery capability of two maize inbred-line foundation genotypes, following treatment with progressive water-deficit stress and stress recovery. Agr Sci Finland 4:389–398

Fang S, Su H, Liu W, Tan K, Ren S (2013) Infrared warming reduced winter wheat yields and some physiological parameters, which were mitigated by irrigation and worsened by delayed sowing. PloS One 8:67518

Farooq M, Basra SMA, Wahid A, Cheema ZA, Cheema MA, Khaliq A (2008) Physiological role of exogenously applied glycinebetaine to improve drought tolerance in fine grain aromatic rice (*Oryza sativa* L.). J Agron Crop Sci 194:325–333

Farooq M, Wahid A, Basra SMA (2009) Improving water relations and gas exchange with brassinosteroids in rice under drought stress. J Agron Crop Sci 195:262–269

Farquhar GD, von Caemmerer S, Berry JA (1980) A biochemical model of photosynthetic CO_2 assimilation in leaves of C3 species. Planta 149:78–90

Farquhar GD, von Caemmerer S, Berry JA (2001) Models of photosynthesis. Plant Physiol 125:42–45

Faver KL, Gerik TJ, Thaxton PM, El-Zik KM (1996) Late season water stress in cotton: II. Leaf gas exchange and assimilation capacity. Crop Sci 36:922–928

Feller U, Crafts-Brandner SJ, Salvucci E (1998) Moderately high temperatures inhibit ribulose-1,5-bisphosphate carboxylase/oxygenase activase-mediated activation of Rubisco. Plant Physiol 116:539–546

Feng H, Jiang N, Huang C, Fang W, Yang W, Chen G, Liu Q (2013) A hyperspectral imaging system for an accurate prediction of the above-ground biomass of individual rice plants. Rev Sci Instrum 84:095107

Fiorani F, Schurr U (2013) Future Scenarios for Plant Phenotyping. Annu Rev Plant Biol 64:267–291

Fiorani F, Rascher U, Jahnke S, Schurr U (2012) Imaging plants dynamics in heterogenic environments. Curr Opin Biotech 23:227–235

Fischer KS, Fukai S, Kumar A, Leung H, Jongdee B (2012) Field phenotyping strategies and breeding for adaptation of rice to drought. Front Physiol 3:282

Flagella Z, Campanile RG, Stoppelli MC, de Caro A, di Fonzo N (1998) Drought tolerance of photosynthetic electron transport under CO_2 enriched and normal air in cereal species. Physiol Plantarum 104:753–759

Flexas J, Medrano H (2002) Drought-inhibition of photosynthesis in C3 plants: stomatal and non-stomatal limitations revisited. Ann Bot 89:183–189

Flexas J, Escalona JM, Medrano H (1999) Water stress induces different levels of photosynthesis and electron transport rate regulation in grapevines. Plant Cell Environ 22:39–48

Flexas J, Briantais JM, Cerovic Z, Medrano H, Moya I (2000) Steady-state and maximum chlorophyll fluorescence responses to water stress in grapevine leaves: a new remote sensing system. Remote Sens Environ 73:283–297

Flexas J, Bota J, Loreto F, Cornic G, Sharkey TD (2004) Diffusive and metabolic limitations to photosynthesis under drought and salinity in C(3) plants. Plant Biol 6:269–279

Flexas J, Ribas-Carbo M, Hanson DT, Bota J, Otto B, Cifre J, McDowell N, Medrano H, Kaldenhoff R (2006) Tobacco aquaporin NtAQP1 is involved in mesophyll conductance to CO_2 *in vivo*. The Plant J 48:427–439

Flexas J, Ribas-Carbo M, Diaz-Espej A, Galmes J, Medrano H (2008) Mesophyll conductance to CO_2: current knowledge and future prospects. Plant Cell Environ 31:602–621

Flood PJ, Harbinson J, Aarts MG (2011) Natural genetic variation in plant photosynthesis. Trends Plant Sci 16:327–335

Fracheboud Y, Leipner J (2003) The application of chlorophyll fluorescence to study light, temperature, and drought stress. In: DeEll JR, Toivonen PMA (eds) Practical Applications of Chlorophyll Fluorescence in Plant Biology. Kluwer Academic Publishers, Dordrecht, pp 125–150

Frolec J, Řebíček J, Lazár D, Nauš J (2010) Impact of two different types of heat stress on chloroplast movement and fluorescence signal of tobacco leaves. Plant Cell Rep 29:705–714

Froux F, Ducrey M, Epron D, Dreyer E (2004) Seasonal variations and acclimation potential of the thermostability of photochemistry in four Mediterranean conifers. Ann For Sci 61:235–241

Fuad-Hassan A, Tardieu F, Turc O (2008) Drought-induced changes in anthesis-silking interval are related to silk expansion: a spatio-temporal growth analysis in maize plants subjected to soil water deficit. Plant Cell Environ 31:1349–1360

Furbank RT, Tester M (2011) Phenomics–technologies to relieve the phenotyping bottleneck. Trends Plant Sci 16:635–644

Gautam A, Agrawal D, SaiPrasad SV, Jajoo A (2014) A quick method to screen high and low yielding wheat cultivars exposed to high temperature. Physiol Mol Biol Plants, 20: 533–537.

Genty B, Briantais JM, da Silva JB. (1987) Effects of drought on primary photosynthet-ic processes of cotton leaves. Plant Physiol 83:360–364

Genty B, Briantais JM, Baker NR (1989) The relationship between the quantum yield of photosynthetic electron transport and quenching of chlorophyll fluorescence. Biochim Biophys Acta 990:87–92

Ghannoum O, von Caemmerer S, Conroy JP (2001) Plant water use efficiency of 17 Australian NAD-ME and NADP-ME C-4 grasses at ambient and elevated CO_2 partial pressure. Aust J Plant Physiol 28:1207–1217

Giannakoula AE, Ilias IF (2013) The effect of water stress and salinity on growth and physiology of tomato (*Lycopersicon esculentum* Mil.). Arch Biol Sci 65:611–620

Gilbert ME, Zwieniecki MA, Holbrook NM (2011) Independent variation in photosynthetic capacity and stomatal conductance leads to differences in intrinsic water use efficiency in 11 soybean genotypes before and during mild drought. J Exp Bot 62:2875–2887

Gillon JS, Yakir D (2000) Internal conductance to CO_2 diffusion and $C^{18}OO$ discrimination in C3 leaves. Plant Physiol 123:201–2013

Girardi MT, Cona B, Geiken B, Kucera T, Masojidek J, Matoo AK (1996) Longterm drought stress induces structural and functional reorganization of photosystem II. Planta 199:118–125

Gollan Y, Passioura JB, Munns R. (1986) Soil water status affects the stomatal conductance of fully turgid wheat and sunflower leaves. Aust J Plant Physiol 48:575–579

Golzarian MR, Frick RA, Rajendran K, Berger B, Roy S, Tester M, Lun DS (2011) Accurate inference of shoot biomass from high-throughput images of cereal plants. Plant Methods 7:2

González-Cruz J, Pastenes C (2012) Water-stress-induced thermotolerance of photosynthesis in bean (*Phaseolus vulgaris* L.) plants: The possible involvement of lipid composition and xanthophyll cycle pigments. Environ Exp Bot 77:127–140

Gorbe E, Calatayud A (2012) Applications of chlorophyll fluorescence imaging technique in horticultural research: A review. Sci Hortic 138:24–35

Gowik U, Westhoff P (2011) The path from C3 to C4 photosynthesis. Plant Physiol 155:56–63

Granier C, Tardieu F (2009) Multi-scale phenotyping of leaf expansion in response to environmental changes: the whole is more than the sum of parts. Plant Cell Environ 32:1175–1184

Griffin KL, Anderson OR, Gastrich MD, Lewis JD, Lin GH, Schuster W, Seemann JR, Tissue DT, Turnbull MH, Whitehead D (2001) Plant growth in elevated CO_2 alters mitochondrial number and chloroplast fine structure. Proc Natl Acad Sci USA 98:2473–2478

Griffiths H, Helliker BR (2013) Mesophyll conductance: internal insights of leaf carbon exchange. Plant Cell Environ 36:733–735

Guarini JM, Moritz C (2009) Modelling the dynamics of the electron transport rate measured by PAM fluorimetry during Rapid Light Curve experiments. Photosynthetica 47:206–214

Guisse B, Srivastava A, Strasser RJ (1995) The polyphasic rise of the chlorophyll a fluorescence (O-K-J-I-P) in heat stressed leaves. Arch Sci Geneve 48:147–160

Hackl H, Mistele B, Hu Y, Schmidhalter U (2013) Spectral assessments of wheat plants grown in pots and containers under saline conditions. Funct Plant Biol 40:409–424

Harbinson J, Hedley CL (1989) The kinetic of P700 reduction in leaves: a novel in situ probe of thylakoid functioning. Plant Cell Environ 12:357–369

Harbinson J, Prinzenberg AE, Kruijer W, Aarts MG (2012) High throughput with chlorophyll fluorescence imaging and its use in crop improvement. Curr Opin Biotech 23:221–6

Harrison EP, Olcer H, Lloyd JC, Long SP, Raines CA (2001) Small decreases in SBPase cause a linear decline in the apparent RuBP regeneration rate, but do not affect Rubisco carboxylation capacity. J Exp Bot 52:1779–1784

Havaux M (1993) Characterization of thermal damage to the photosynthetic electron transport system in potato leaves. Plant Sci 94:19–33

Havaux M (1996) Short-term responses of photosystem I to heat stress. Induction of a PS II-independent electron transport through PS I fed by stromal components. Photosynth Res 47:85–97

Havaux M, Tardy F (1996) Temperature-dependent adjustment of the thermal stability of photosystem II in vivo: possible involvement of xanthophyll-cycle pigments. Planta 198:324–333

Havaux M, Strasser RJ, Greppin H (1990) In vivo photoregulation of photochemical and non-photochemical deactivation of photosystem II in intact plant leaves. Plant Physiol Bio 28:735–746

Herppich WB, Peckmann K (1997) Responses of gas exchange, photosynthesis, nocturnal acid accumulation and water relations of *Aptenia cordifolia* to short-term drought and rewatering. J Plant Physiol 150:467–474

Herzog H, Chai-Arree W (2012) Gas exchange of five warm-season grain legumes and their susceptibility to heat stress. J Agron Crop Sci 198:466–474

Hewson I, O'Neil JM, Dennison WC (2001) Virus-like particles associated with Lyngbya ma-juscula (Cyanophyta; Oscillatoriacea) bloom decline in Moreton Bay, Australia. Aquat Microb Ecol 25:207–213

Hibberd JM, Sheehy JE, Langdale JA (2008) Using C_4 photosynthesis to increase the yield of rice-rationale and feasibility. Curr Opin Plant Biol 11:228–231

Hou X, Li R, Jia Z, Han Q (2013) Rotational tillage improves photosynthesis of winter wheat during reproductive growth stages in a semiarid region. Agron J 105:215–221

Hozain M, Abdelmageed H, Lee J, Kang M, Fokar M, Allen RD, Holaday AS (2012) Expression of *AtSAP5* in cotton up-regulates putative stress-responsive genes and improves the tolerance to rapidly developing water deficit and moderate heat stress. J Plant Physiol 169:1261–1270

Hu L, Wang Z, Huang B (2010) Diffusion limitations and metabolic factors associated with inhibition and recovery of photosynthesis from drought stress in a C3 perennial grass species. Physiol Plantarum 139:93–106

Hubbart S, Peng S, Horton P, Chen Y, Murchie EH (2007) Trends in leaf photosynthesis in historical rice varieties developed in the Philippines since 1966. J Exp Bot 58:3429–3438

Hüve K, Bichele I, Rasulov B, Niinemets U (2011) When it is too hot for photosynthesis: heat-induced instability of photosynthesis in relation to respiratory burst, cell permeability changes and H_2O_2 formation. Plant Cell Environ 34:113–126

Ierna A (2007) Characterization of potato genotypes by chlorophyll fluorescence during plant aging in a Mediterranean environment. Photosynthetica 45:568–575

Ishizaki T, Maruyama K, Obara M, Fukutani A, Yamaguchi-Shinozaki K, Ito Y, Kumashiro T (2013) Expression of *Arabidopsis DREB1C* improves survival, growth, and yield of upland New Rice for Africa (NERICA) under drought. Mol Breeding 31:255–264

Janeczko A, Oklešťková J, Pociecha E, Kościelniak J, Mirek M (2011) Physiological effects and transport of 24-epibrassinolide in heat-stressed barley. Acta Physiol Plant 33:1249–1259

Jedmowski C, Ashoub A, Brüggemann W (2013) Reactions of Egyptian landraces of *Hordeum vulgare* and *Sorghum bicolor* to drought stress, evaluated by the *OJIP* fluorescence transient analysis. Acta Physiol Plant 35:345–354

Jones HG, Serraj R, Loveys BR, Xiong L, Wheaton A, Price AH (2009) Thermal infrared imaging of crop canopies for the remote diagnosis and quantification of plant responses to water stress in the field. Funct Plant Biol 36:978–989

Kalaji HM, Bosa K, Koscielniak J, Hossain Z (2011) Chlorophyll a fluorescence—a useful tool for the early detection of temperature stress in spring barley (*Hordeum vulgare* L.). OMICS 15:95–934

Kalaji HM, Schansker G, Ladle RJ, Goltsev V, Bosa K, Allakhverdiev SI (2014a) Frequently asked questions about in vivo chlorophyll fluorescence: practical issues. Photosynth Res 122:121–158

Kalaji HM, Oukarroum A, Alexandrov V, Kouzmanova M, Brestic M, Zivcak M, Samborska IA, Cetner MD, Allakhverdiev SI, Goltsev V (2014b) Identification of nutrient deficiency in maize and tomato plants by in vivo chlorophyll a fluorescence measurements. Plant Physiol Bioch 81:16–25

Kaldenhoff R (2012) Mechanisms underlying CO_2 diffusion in leaves. Curr Opin Plant Biol 15:276–281

Kalyar T, Rauf S, Teixeira da Silva JA, Iqbal Z (2013) Variation in leaf orientation and its related traits in sunflower (*Helianthus annuus* L.) breeding populations under high temperature. Field Crop Res 150:91–98

Kamoshita A, Babu R, Boopathi N, Fukai S (2008) Phenotypic and genotypic analysis of drought-resistance traits for development of rice cultivars adapted for rainfed environments. Field Crop Res 109:1–23

Kaňa R, Kotabova E, Prášil O (2008) Acceleration of plastoquinone pool reduction by alternative pathways precedes a decrease in photosynthetic CO_2 assimilation in preheated barley leaves. Physiol Plantarum 133:794–806

Karim M, Zhang YQ, Zhao RR, Chen XP, Zhang FS, Zou CQ (2012) Alleviation of drought stress in winter wheat by late foliar application of zinc, boron, and manganese. J Plant Nutr Soil Sci 175:142–151

Kebeish R, Niessen M, Thiruveedhi K, Bari R, Hirsch HJ, Rosenkranz R, Staebler N, Schoenfeld B, Kreuzaler F, Peterhaenzel C (2007) Chloroplastic photorespiratory bypass increases photosynthesis and biomass production in *Arabidopsis thaliana*. Nat Biotechnol 25:593–599

Kimball BA (1983) Carbon-dioxide and agricultural yield – an assemblage and analysis of 430 prior observations. Agron J 75:779–788

Kirschbaum MU (2011) Does enhanced photosynthesis enhance growth? Lessons learned from CO_2 enrichment studies. Plant Physiol 155:117–124

Kolber Z, Klimov D, Ananyev G, Rascher U, Berry JA, Osmond CB (2005) Measuring photosynthetic parameters at a distance: laser induced fluorescence transient (LIFT) method for remote measurements of PSII in terrestrial vegetation. Photosynth Res 84:121–129

Kolukisaoglu Ü, Thurow K (2010) Future and frontiers of automated screening in plant sciences. Plant Sci 178:476–484

Kovačević J, Kovačević M, Cesar V, Drezner G, Lalić A, Lepeduš H, Kovačević V (2013) Photosynthetic efficiency and quantitative reaction of bread winter wheat to mild short-term drought conditions. Turk J Agric Forest 37:385–393

Kubler JE, Raven J (1996) Nonequilibrium rates of photosynthesis and respiration under dynamic light supply. J Phycol 32:963–969

Kuckenberg J, Tartachnyk I, Noga G (2009) Temporal and spatial changes of chlorophyll fluorescence as a basis for early and precise detection of leaf rust and powdery mildew infections in wheat leaves. Precision Agric 10:34–44

Kuppu S, Mishra N, Hu R, Sun L, Zhu X, Shen G, Blumwald E, Payton P, Zhang H (2013) Water-deficit inducible expression of a cytokinin biosynthetic gene *IPT* improves drought tolerance in cotton. PloS One 8:64190

Lal A, Ku MSB, Edwards GE (1996) Analysis of inhibition of photosynthesis due to water-stress in the C3 species *Hordeum vulgare* and *Vicia faba* – electron-transport, CO_2 fixation and carboxylation capacity. Photosynth Res 49:57–69

Lauer MJ, Boyer JS (1992) Internal CO_2 measured directly in leaves. Plant Physiol 98:1310–1316

Law RD, Crafts-Brandner SJ (1999) Inhibition and acclimation of photosynthesis to heat stress is closely correlated with activation of ribulose-1,5-bisphosphate Carboxylase/Oxygenase. Plant Physiol 120:173–181

Lawlor DW (2002) Limitation to photosynthesis in water-stressed leaves: Stomatal metabolism and the role of ATP. Annals Hort 89:871–885

Lawlor DW, Cornic G (2002) Photosynthetic carbon assimilation and associated metabolism in relation to water deficits in higher plants. Plant Cell Environ 25:275–294

Lawlor DW, Tezara W (2009) Causes of decreased photosynthetic rate and metabolic capacity in water-deficient leaf cells: a critical evaluation of mechanisms and integration processes. Ann Bot 103:561–579

Lefebvre S, Lawson T, Zakhleniuk OV, Lloyd JC, Raines CA (2005) Increased sedoheptulose-1,7-

bisphosphatase activity in transgenic tobacco plants stimulates photosynthesis and growth from an early stage in development. Plant Physiol 138:451–460

Li P, Cheng L, Gao H, Jiang C, Peng T (2009) Heterogeneous behavior of PSII in soybean (*Glycine max*) leaves with identical PSII photochemistry efficiency under different high temperature treatments. J Plant Physiol 166:1607–1615

Li X, Shen X, Li J, Eneji AE, Li Z, Tian X, Duan L (2010) Coronatine alleviates water deficiency stress on winter wheat seedlings. J Integr Plant Biol 52:616–625

Li D, Liu H, Qiao Y, Wang Y, Cai Z, Dong B, Liu M (2013) Effects of elevated CO_2 on the growth, seed yield, and water use efficiency of soybean (*Glycine max* L. Merr.) under drought stress. Agr Water Manage 129:105–112

Lichtenthaler HK, Babani F (2004) Light adaptation and senescence of the photosynthetic apparatus: changes in pigment composition, chlorophyll fluorescence parameters and photo-synthetic activity. In: Papageorgiou GC, Govindjee (eds) Chlorophyll a fluorescence: A Signature of Photosynthesis. Advances in Photosynthesis and Respiration, Volume 19. Springer, Dordrecht, pp 713–736

Lichtenthaler HK, Miehé JA (1997) Fluorescence imaging as a diagnostic tool for plant stress. Trends Plant Sci 2:316–320

Lichtenthaler HK, Rinderle U (1988) Role of chlorophyll fluorescence in the detection of stress conditions in plants. CRC Cr Rev Anal Chem 19:29–85

Lichtenthaler HK, Buschmann C, Knapp M (2005a) How to correctly determine the different chlorophyll fluorescence parameters and the chlorophyll fluorescence decrease ratio RFd of leaves with the PAM fluorometer. Photosynthetica 43:379–393

Liu LY, Zhang YJ, Wang JH, Zhao CJ (2005) Detecting solar-induced chlorophyll fluorescence from field radiance spectra based on the Fraunhofer line principle. IEEE Trans Geosci Remote Sensing 43:827–832

Long SP (1993) The significance of light-limited photosynthesis to crop canopy carbon gain and productivity – a theoretical analysis. In Abrol YP, Mohanty P, Govindjee (eds) Photosynthesis: Photoreactions to Plant Productivity. Oxford & IBH Publishing, New Delhi, pp 547–560

Long SP (1998) Rubisco, the key to improved crop production for a world population of more than eight billion people? In: Waterlow JC, Armstron DG, Fowdenand L, Riley R (eds) Feeding a World Population of More Than Eight Billion People – A Challenge to Science. Oxford University Press, New York, pp 124–136

Long SP, Farage PK, Garcia RL (1996) Measurement of leaf and canopy photosynthetic CO_2 exchange in the field. J Exp Bot 47:1629–1642

Long SP, Ainsworth EA, Rogers A, Ort DR (2004) Rising atmospheric carbon dioxide: plants face their future. Annu Rev Plant Phys 55:591–628

Long SP, Ainsworth EA, Leakey ADB, Morgan PB (2005) Global food insecurity. Treatment of major food crops with elevated carbon dioxide or ozone under large-scale fully open-air conditions suggests recent models may have overestimated future yields. Philos Trans R Soc Lond B Biol Sci 360:2011–2020

Long SP, Zhu XG, Naidu SL, Ort DR (2006) Can Improvement in Photosynthesis Increase Crop Yields? Plant Cell Environ 29:315–330

Loreto F, Harley PC, Di Marco G, Sharkey TD (1992) Estimation of mesophyll conductance to CO_2 flux by three different methods. Plant Physiol 98:1437–1443

Malenovsky Z, Mishra KB, Zemek F, Rascher U, Nedbal L (2009) Scientific and technical challenges in remote sensing of plant canopy reflectance and fluorescence. J Exp Bot 60:2987–3004

Markelz RC, Strellner RS, Leakey AD (2011) Impairment of C4 photosynthesis by drought is exacerbated by limiting nitrogen and ameliorated by elevated [CO_2] in maize. J Exp Bot 62:3235–3246

Maroco JP, Pereira JS, Chaves MM (1997) Stomatal responses of leaf-to-air vapour pressure deficit in Sahelian species. Aust J Plant Physiol 24:381–387

Maroco JP, Rodriges ML, Lopes C, Chaves MM (2002) Limitation to leaf photosynthesis in grapevine under drought – metabolic and modeling approaches. Funct Plant Biol 29:1–9

Martin B, Ruiz-Torres NA (1992) Effect of water-deficit stress on photosynthesis, its components and components limitations, and on water use efficiency in wheat (*Triticum aestivum* L.). Plant Physiol 100:733–739

Massacci A, Nabiev SM, Pietrosanti L, Nematov SK, Chernikova TN, Thor K, Leipner J (2008) Response of the photosynthetic apparatus of cotton (*Gossypium hirsutum*) to the onset of drought stress under field conditions studied by gas-exchange analysis and chlorophyll fluorescence imaging. Plant Physiol Bioch 46:189–195

Masuka B, Araus JL, Das B, Sonder K, Cairns JE (2012) Phenotyping for Abiotic Stress Tolerance in Maize. J Integr Plant Biol 54:238–249

Mathur S, Allakhverdiev SI, Jajoo A (2011a) Analysis of high temperature stress on the dynamics of antenna size and reducing side heterogeneity of Photosystem II in wheat leaves (*Triticum aestivum*). Biochym Biophys Acta 1807:22–29

Mathur S, Jajoo A, Mehta P, Bharti S (2011b) Analysis of elevated temperature-induced inhibition of photosystem II using chlorophyll *a* fluorescence induction kinetics in wheat leaves (*Triticum aestivum*). Plant Biol 13:1–6

Mathur S, Agrawal D, Jajoo A (2014) Photosynthesis: response to high temperature stress. J Photochem Photobiol B Biol 137:116–126

Maxwell K, Johnson GN (2000) Chlorophyll fluorescence-a practical guide. J Exp Bot 51:659–668

Medrano H, Escalona JM, Bota J, Gulias J, Flexas J (2002) Regulation of photosynthesis of C3 plants in response to progressive drought: stomatal conductance as a reference parameter. Ann Bot 89:895–905

Mengistu A, Bond J, Mian R, Nelson R, Shannon G, Wrather A (2011) Identification of soybean accessions resistant to by field screening, molecular markers, and phenotyping. Crop Sci 51:1101–1109

Meroni M, Colombo R (2006) Leaf level detection of solar induced chlorophyll fluorescence by means of a subnanometer resolution spectroradiometer. Remote Sens Environ 103:438–448

Meroni M, Rossini M, Guanter L, Alonso L, Rascher U, Colombo R, Moreno J (2009) Remote sensing of solar-induced chlorophyll fluorescence: review of methods and applications. Remote Sens Environ 113:2037–2051

Meyer M, Griffiths H (2013) Origins and diversity of eukaryotic CO_2-concentrating mechanisms: lessons for the future. J Exp Bot 64:769–786

Mitchell RAC, Black CR, Burkart S, Burke JI, Donnelly A, De Temmmerman L, Fangmeier A, Mulholland BJ, Theobald JC, van Oijen M (1999) Photosynthetic responses in spring wheat grown under elevated CO_2 concentrations and stress conditions in the European, multiple-site experiment 'ESPACE-wheat'. Eur J Agron 10:205–214

Mittler R, Blumwald E (2010) Genetic engineering for modern agriculture: challenges and perspectives. Annu Rev Plant Biol 61:443–462

Monteith JL (1994) Validity of the correlation between intercepted radiation and biomass. Agr Forest Meteorol 68:213–220

Monteith JL, Moss CJ (1977) Climate and the efficiency of crop production in Britain. Philos Trans R Soc Lond B 281:277–294

Montes JM, Melchinger AE, Reif JC (2007) Novel throughput phenotyping platforms in plant genetic studies. Trends Plant Sci 12:433–436

Moya I, Camenen L, Evain S, Goulas Y, Cerovic ZG, Latouche G, Flexas J, Ounis A (2004) A new instrument for passive remote sensing: 1. Measurements of sunlight-induced chlorophyll fluorescence. Remote Sens Environ 91:186–197

Mozdzer TJ, Zieman JC (2010) Ecophysiological differences between genetic lineages facilitate the invasion of non-native *Phragmites australis* in North American Atlantic coast wetlands. J Ecol 98:451–458

Munns R, James RA, Sirault XRR, Furbank RT, Jones HG (2010) New phenotyping methods for screening wheat and barley for beneficial responses to water deficit. J Exp Bot 61:3499–3507

Murchie EH, Pinto M, Horton P (2009) Agriculture and the new challenges for photosynthesis research. New Phytol 181:532–552

Nedbal L, Whitmarsh J (2004) Chlorophyll fluorescence imaging of leaves and fruits. In: Papageorgiou GC, Govindjee (eds) Chlorophyll a Fluorescence: A Signature of Photosynthesis. Advances in Photosynthesis and Respiration, Volume 19. Springer, Dordrecht, pp 389–407

Nguyen TX, Nguyen T, Alameldin H, Goheen B, Loescher W, Sticklen M (2013) Transgene pyramiding of the *HVA1* and *mtlD* in T3 maize (*Zea mays* L.) plants confers drought and salt tolerance, along with an increase in crop biomass. Int J Agron 2013:10

Niinemets U, Diaz-Espejo A, Flexas J, Galmes J, Warren C (2009) Importance of mesophyll diffusion conductance in estimation of plant photosynthesis in the field. J Exp Bot 60:2271–2282

Nogués S, Baker NR (2000) Effects of drought on photosynthesis in Mediterranean plants grown under enhanced UV-B radiation. J Exp Bot 51:1309–1317

Olšovská K, Živčák M, Hunková E, Dreveňáková P (2013) Assessment of the photosynthesis-related traits and high temperature resistance in tetraploid wheat (Triticum L.) genotypes. J Centr Europ Agric 14:289–302

Omasa K, Takayama K (2003) Simultaneous measurement of stomatal conductance, non-photochemical quenching, and photochemical yield of photosystem II in intact leaves by thermal and chlorophyll fluorescence imaging. Plant Cell Physiol 44:1290–1300

Omasa K, Shimazaki KI, Aiga I, Larcher W, Onoe M (1987) Image analysis of chlorophyll fluorescence transients for diagnosing the photosynthetic system of attached leaves. Plant Physiol 84:748–752

Oukarroum G, Schansker, Strasser RJ (2009) Drought stress effects on photosystem I content and photosystem II thermotolerance analyzed using Chl a fluorescence kinetics in barley varieties differing in their drought tolerance. Physiol Plantarum 137:188–199

Oukarroum A, El Madidi S, Schansker G, Strasser RJ (2007) Probing the responses of barley cultivars (*Hordeum vulgare* L.) by chlorophyll a fluorescence OLKJIP under drought stress and re-watering. Environ Exp Bot 60:438–446

Oukarroum A, El Madidi S, Strasser RJ (2012) Exogenous glycine betaine and proline play a protective role in heat-stressed barley leaves (*Hordeum vulgare* L.): A chlorophyll a fluorescence study. Plant Biosyst 146:1037–1043

Oxborough K, Baker NR (1997) Resolving chlorophyll a fluorescence images of photosynthetic efficiency into photochemical and non-photochemical components – calculation of qP and Fv'/Fm' without measuring Fo'. Photosynth Res 54:135–142

Parry M, Andraloje PJ, Khan S, Lea PJ, Keys A (2002) Rubisco activity: effect of drought stress. Annals of Bot 89:833–639

Parry MAJ, Andralojc PJ, Mitchell RAC, Madgwick PJ, Keys AJ (2003) Manipulation of Rubisco: the amount, activity, function and regulation. J Exp Bot 54:1321–1333

Parry MAJ, Madgwick PJ, Carvalho JFC, Andralojc PJ (2007) Prospects for increasing photosynthesis by overcoming the limitations of Rubisco. J Agr Sci 145:31–43

Parry MAJ, Reynolds M, Salvucci ME, Reines C, Andralojc PJ, Zhu XG, Price GD, Condon AG, Furbank RT (2011) Raising yield potential of wheat. II. Increasing photosynthetic capacity and efficiency. J Exp Bot 62:453–467

Passioura JB (2012) Phenotyping for drought tolerance in grain crops: when is it useful to breeders? Funct Plant Biol 39:851–859

Peet MM, Kramer PJ (1980) Effects of decreasing source-sink ratio in soybeans on photosynthesis, photorespiration, transpiration and yield. Plant Cell Environ 3:201–206

Pereyra-Irujo GA, Gasco ED, Peirone LS, Aguirrezábal LA (2012) GlyPh: a low-cost platform for phenotyping plant growth and water use. Funct Plant Biol 39:905–913

Phung TH, Jung HI, Park JH, Kim JG, Back K, Jung S (2011) Porphyrin biosynthesis control under water stress: sustained porphyrin status correlates with drought tolerance in transgenic rice. Plant Physiol 157:1746–1764

Pieruschka R, Poorter H (2012) Phenotyping plants: genes, phenes and machines. Funct Plant Biol 39:813–820

Pieruschka R, Klimov D, Kolber ZS, Berry JA (2010) Monitoring of cold and light stress impact on photosynthesis by using the laser induced fluorescence transient (LIFT) approach. Funct Plant Biol 37:395–402

Pinelli P, Loreto F (2003) $^{12}CO_2$ emission from different metabolic pathways measured in illuminated and darkened C3 and C4 leaves at low, atmospheric and elevated CO_2 concentration. J Exp Bot 54:1761–1769

Poorter H, Niinemets Ü, Walter A, Fiorani F, Schurr U (2010) A method to construct dose–response curves for a wide range of environmental factors and plant traits by means of a meta-analysis of phenotypic data. J Exp Bot 61:2043–2055

Prashar A, Yildiz J, McNicol JW, Bryan GJ, Jones HG (2013) Infra-red thermography for high throughput field phenotyping in Solanum tuberosum. PloS One 8:65816

Price GD, von Caemmerer S, Evans JR, Siebke K, Anderson JM, Badger MR (1998) Photosynthesis is strongly reduced by antisense suppression of chloroplastic cytochrome bf complex in transgenic tobacco. Aust J Plant Physiol 25:445–452

Price GD, Badger MR, Woodger FJ, Long BM (2008) Advances in understanding the cyanobacterial CO_2-concentrating-mechanism (CCM): functional components, Ci transporters, diversity, genetic regulation and prospects for engineering into plants. J Exp Bot 59:1441–1461

Price GD, Pengelly JJL, Forster B, Du J, Whitney SM, von Caemmerer S, Badger MR, Howitt SM, Evans JR (2013) The cyanobacterial CCM as a source of genes for improving photosynthetic CO_2 fixation in crop species. J Exp Bot 64:753–768

Qaderi MM, Kurepin LV, Reid DM (2012) Effects of temperature and watering regime on growth, gas exchange and abscisic acid content of canola (Brassica napus) seedlings. Environ Exp Bot 75:107–113

Qi M, Liu Y, Li T (2013) Nano-TiO_2 Improve the Photosynthesis of Tomato Leaves under Mild Heat Stress. Biol Trace Elem Res 156:1–6

Quick WP, Horton P (1984) Studies on the induction of chlorophyll fuorescence in barley protoplasts. II. Resolution of fluorescence quenching by redox state and the transthylakoid pH gradient. Philos Trans R Soc Lond B Biol Sci 220:371–382

Raines CA (2003) The Calvin cycle revisited. Photosynth Res 75:1–10

Ralph PJ, Gademann R (2005) Rapid Light Curve: A powerful tool to assess photosynthetic activity. Aquat Bot 82:222–237

Ramalho JC, Zlatev ZS, Leitão AE, Pais IP, Fortunato AS, Lidon FC (2014) Moderate water stress causes different stomatal and non-stomatal changes in the photosynthetic functioning of Phaseolus vulgaris L. genotypes. Plant Biology 16:133–146

Rapacz M, Kościelniak J, Jurczyk B, Adamska A, Wójcik M (2010) Different patterns of physiological and molecular response to drought in seedlings of malt- and feed-type barleys (Hordeum vulgare). J Agron Crop Sci 196:9–19

Rascher U, Nedbal L (2006) Dynamics of photosynthesis in fluctuating light – commentary. Curr Opin Plant Biol 9:671–678

Rascher U, Pieruschka R (2008) Spatio-temporal variations of photosynthesis – the potential of optical remote sensing to better understand and scale light use efficiency and stresses of plant ecosystems. Precis Agric 9:355–366

Rascher U, Nichol CJ, Small C, Hendricks L (2007) Monitoring spatiotemporal dynamics of photosynthesis with a portable hyperspectral imaging system. Photogramm Eng Rem S 73:45–56

Rascher U, Agati G, Alonso L, Cecchi G, Champagne S, Colombo R, Damm A, Daumard F, de Miguel E, Fernandez G, Franch B, Franke J, Gerbig C, Gioli B, Gómez JA, Goulas Y, Guanter L, Gutiérrez-de-la-Cámara Ó, Hamdi K, Hostert P, Jiménez M, Kosvancova M, Lognoli D, Meroni M, Miglietta F, Moersch A, Moreno J, Moya I, Neininger B, Okujeni A, Ounis A, Palombi L, Raimondi V, Schickling A, Sobrino JA, Stellmes M, Toci G, Toscano P, Udelhoven T, van der Linden S, Zaldei A (2009) CEFLES2: the remote sensing component to quantify photosynthetic efficiency from the leaf to the region by measuring sun-induced fluorescence in the oxygen absorption bands. Biogeosciences 6:1181–1198

Rascher U, Biskup B, Leakey ADB, McGrath JM, Ainsworth EA (2010) Altered physiological function, not structure, drives increased radiation use efficiency of soybean grown at elevated CO_2. Photosynth Res 105:15–25

Rascher U, Blossfeld S, Fiorani F, Jahnke S, Jansen M, Kuhn AJ et al (2011) Non-invasive approaches for phenotyping of enhanced performance traits in bean. Funct Plant Biol 38:968–983

Rebetzke GJ, Chenu K, Biddulph B, Moeller C, Deery DM, Rattey AR, Bennet D, Barrett-Lennard G, Mayer JE (2012) A multisite managed environment facility for targeted trait and germplasm phenotyping. Funct Plant Biol 40:1–13

Redillas MC, Strasser RJ, Jeong JS, Kim YS, Kim JK (2011) The use of JIP test to evaluate drought-tolerance of transgenic rice overexpressing *OsNAC10*. Plant Biotechnol Rep 5:169–175

Repkova J, Brestic M, Zivcak M (2008) Bioindication of barley leaves vulnerability in conditions of water deficit. Cereal Res Commun 36:1747–1750

Restrepo-Diaz H, Garces-Varon G (2013) Response of rice plants to heat stress during initiation of panicle primordia or grain-filling phases. J Stress Physiol Biochem 9:318–325

Reynolds MP, Pellegrineschi A, Skovmand B (2005) Sink-limitation to yield and biomass: a summary of some investigations in spring wheat. Ann Appl Biol 146:39–49

Reynolds M, Manes Y, Izanloo A, Langridge P (2009) Phenotyping approaches for physiological breeding and gene discovery in wheat. Ann Appl Biol 155:309–320

Reynolds M, Bonnett D, Chapman SC, Furbank RT, Manès Y, Mather DE, Parry MA (2011) Raising yield potential of wheat. I. Overview of a consortium approach and breeding strategies. J Exp Bot 62:439–452

Ribeiro RV, Santos MG, Machado EC, Oliveira RF (2008) Photochemical heat-shock response in common bean leaves as affected by previous water deficit. Russ J Plant Physiol 55:350–358

Richards RA, Rebetzke GJ, Watt M, Condon AT, Spielmeyer W, Dolferus R (2010) Breeding for improved water productivity in temperate cereals: phenotyping, quantitative trait loci, markers and the selection environment. Funct Plant Biol 37:85–97

Robakowski P, Montpied P, Dreyer E (2002) Temperature response of photosynthesis of silver fir (*Abies alba* Mill.) seedlings. Ann For Sci 59:159–166

Rohacek K, Soukupova J, Bartak M (2008) Chlorophyll fluorescence: A wonderful tool to study plant physiology and plant stress. In: Benoit Schoefs (ed) Plant Cell Compartments – Selected Topics, Trivandrum, India

Rolfe SA, Scholes JD (1995) Quantitative imaging of chlorophyll fluorescence. New Phytol 131:69–79

Roohi E, Tahmasebi-Sarvestani Z, Modarres-Sanavy SAM, Siosemardeh A (2013) Comparative study on the effect of soil water stress on photosynthetic function of triticale, bread wheat, and barley. J Agr Sci Tech 15:215–225

Rosema A, Snel JFH, Zahn H, Buurmeijer WF, van Hove LWA (1998) The relation between laser-induced chlorophyll fluorescence and photosynthesis. Remote Sens Environ 65:143–154

Rumeau D, Peltier G, Cournac L (2007) Chlororespiration and cyclic electron flow around PS I during photosynthesis and plant stress response. Plant Cell Environ 30:1041–1051

Sadras VO, Rebetzke GJ, Edmeades GO (2013) The phenotype and the components of phenotypic variance of crop traits. Field Crop Res 154:255–259

Saint Pierre C, Crossa JL, Bonnett D, Yamaguchi-Shinozaki K, Reynolds MP (2012) Phenotyping transgenic wheat for drought resistance. J Exp Bot 63:1799–1808

Salvucci ME, Crafts-Brandner SJ (2004) Inhibition of photosynthesis by heat stress: the acti-vation state of Rubisco as a limiting factor in photosynthesis. Physiologia Plant 120:179–186

Santos MG, Ribeiro RV, Machado EC, Pimentel C (2009) Photosynthetic parameters and leaf water potential of five common bean genotypes under mild water deficit. Biol Plantarum 53:229–236

Sazanov LA, Burrows PA, Nixon PJ (1998) The chloroplast Ndh complex mediates the dark reduction of the plastoquinone pool in response to heat stress in tobacco leaves. FEBS Lett 429:115–118

Scafaro AP, Yamori W, Carmo-Silva AE, Salvucci ME, von Caemmerer S, Atwell BJ (2012) Rubisco activity is associated with photosynthetic thermotolerance in a wild rice (*Oryza meridionalis*). Physiologia Plant 146:99–109

Schoenenberger N, Felber F, Savova-Bianchi D, Guadagnuolo R (2005) Introgression of wheat DNA markers from A, B and D genomes in early generation progeny of *Aegilops cylindrica* Host× *Triticum aestivum* L. hybrids. Theor Appl Genet 111:1338–1346

Schrader SM, Wise RR, Wacholtz WF, Ort DR, Sharkey TD (2004) Thylakoid membrane responses to moderately high leaf temperature in Pima cotton. Plant Cell Environ 27:725–735

Schreiber U (2004) Pulse-amplitude-modulation (PAM) fluorometry and saturation pulse method: an overview. In: Papageorgiou GC, Govindjee (eds) Chlorophyll a Fluorescence: A Signature of Photosynthesis. Advances in Photosynthesis and Respiration, Volume 19. Springer, Dordrecht, pp 279–319

Schreiber U, Berry JA (1977) Heat-induced changes of chlorophyll fluorescence in intact leaves correlated with damage of the photosynthetic apparatus. Planta 136:233–238

Schreiber U, Schliwa U, Bilger W (1986) Continuous recording of photochemical and non-photochemical chlorophyll fluorescence quenching with a new type of modulation fluorometer. Photosynth Res 10:51–62

Schreiber U, Gademann R, Ralph PJ, Larkum AWD (1997) Assessment of photosynthetic performance of Prochloron in Lissoclinum patella in hospite by chlorophyll fluorescence measurements. Plant Cell Physiol 38:945–951

Schurr U, Walter A, Rascher U (2006) Functional dynamics of plant growth and photosynthesis – from steady-state to dynamics – from homogeneity to heterogeneity. Plant Cell Environ 29:340–352

Seddon S, Cheshire AC (2001) Photosynthetic response of Amphibolis antarctica and Posidonia australis to temperature and dessication using chlorophyll fluorescence. Mar Ecol Progr Ser 220: 119–130

Serbin SP, Dillaway DN, Kruger EL, Townsend PA (2012) Leaf optical properties reflect variation in photosynthetic metabolism and its sensitivity to temperature. J Exp Bot 63:489–502

Serraj R, McNally KL, Slamet-Loedin I, Kohli A, Haefele SM, Atlin G, Kumar A (2011) Drought resistance improvement in rice: an integrated genetic and resource management strategy. Plant Prod Sci 14:1–14

Shanmugam S, Kjær KH, Ottosen CO, Rosenqvist E, Kumari Sharma D, Wollenweber B (2013) The alleviating effect of elevated CO_2 on heat stress susceptibility of two wheat (*Triticum aestivum* L.) cultivars. J Agron Crop Sci 199:340–350

Sharkey TD, Seeman JR. (1989) Mild water stress effects on carbon-reduction-cycle intermediates, ribulose bisphosphate carboxylase activity, and spatial homogeneity of photosynthesis in intact leaves. Plant Physiol 89:1060–1065

Sharkey TD, Zhang R (2010) High temperature effects on electron and proton circuits of photosynthesis. J Integr Plant Biol 52:712–722

Sharkey TD, Bernacchi CJ, Farquhar GD, Singsaas EL (2007) Fitting photosynthetic carbon dioxide response curves for C(3) leaves. Plant Cell Environ 30:1035–1040

Sharkova VE (2001) The effect of heat shock on the capacity of wheat plants to restore their photosynthetic electron transport after photoinhibition or repeated heating. Russ J Plant Physiol 48:793–797

Sharma DK, Andersen SB, Ottosen CO, Rosenqvist E (2012) Phenotyping of wheat cultivars for heat tolerance using chlorophyll *a* fluorescence. Funct Plant Biol 39:936–947

Shaw AK, Ghosh S, Kalaji HM, Bosa K, Brestic M, Zivcak M, Hossain Z (2014) Nano-CuO stress induced modulation of antioxidative defense and photosynthetic performance of Syrian barley (*Hordeum vulgare* L.). Environ Exp Bot 102:37–47

Shearman VJ, Sylvester-Bradley R, Scott RK, Foulkes MJ (2005) Physiological processes associated with wheat yield progress in the UK. Crop Sci 45:175–185

Sheehy JE, Ferrer AB, Mitchell PL, Elmido-Mabilangan A, Pablico P, Dionora MJA (2007) How the rice crop works and why it needs a new engine. Charting new pathways to C. International Rice Institute 4:3–26

Shefazadeh MK, Mohammadi M, Karimizadeh R. (2012) Genotypic difference for heat tolerance traits under real field conditions. J Food Agric Envir 10:484–487

Shikanai T (2007) Cyclic electron transport around photosystem I: Genetic approaches. Annu Rev Plant Biol 58:199–217

Silva EN, Ferreira-Silva SL, Fontenele ADV, Ribeiro RV, Viégas RA, Silveira JAG (2010) Photosynthetic changes and protective mechanisms against oxidative damage subjected to isolated and combined drought and heat stresses in *Jatropha curcas* plants. J Plant Physiol 167:1157–1164

Sinsawat V, Leipner J, Stamp P, Fracheboud Y (2004) Effect of heat stress on the photosynthetic apparatus in maize (*Zea mays* L.) grown at control or high temperature. Environ Exp Bot 52:123–129

Skillman JB (2008) Quantum yield variation across the three pathways of photosynthesis. J Exp Bot 59:1647–1661

Skillman JB, Garcia M, Virgo A, Winter K (2005) Growth irradiance effects on photosynthesis and growth in two co-occurring shade-tolerant neotropical perennials of contrasting photosynthetic pathways. Am J Bot 92:1811–1819

Skillman JB, Griffin KL, Earll S, Kusama M (2011) Photosynthetic productivity: can plants do better? In: Moreno-Piraján JC (ed) Thermodynamics – Systems in Equilibrium and Non-Equilibrium

Snider JL, Oosterhuis DM, Skulman BW, Kawakami EM (2009) Heat stress-induced limitations to reproductive success in *Gossypium hirsutum*. Physiol Plantarum 137:125–138

Song L, Yue L, Zhao H, Hou M (2013a) Protection effect of nitric oxide on photosynthesis in rice under heat stress. Acta Physiol Plant 35:3323–3333

Song Q, Zhang G, Zhu XG (2013b) Optimal crop canopy architecture to maximise canopy photosynthetic CO_2 uptake under elevated CO_2: atheoretical study using a mechanistic model of canopy photosynthesis. Funct Plant Biol 40:108–124

Srivastava B, Guisse H, Greppin H, Strasser RJ (1997) Regulation of antenna structure and electron transport in photosystem II of *Pisum sativum* under elevated temperature probed by the fast polyphasic chlorophyll a fluorescence transient: OKJIP. Biochim Biophys Acta 1320:95–106

Stefanov D, Petkova V, Denev ID (2011) Screening for heat tolerance in common bean (*Phaseolus vulgaris* L.) lines and cultivars using *JIP*-test. Sci Hortic-Amsterdam 128:1–6

Stirbet A, Govindjee (2011) On the relation between the Kautsky effect (chlorophyll a fluorescence induction) and Photosystem II: Basics and applications of the OJIP fluorescence transient. J Photochem Photobio B 104:236–257

Strasser RJ, Govindjee (1992) On the OJIP fluorescence transients in leaves and D1 mutants of *Chlamydomonas reinhardtii*. In: Murata N (ed) Research in Photosynthesis. Kluwer Academic Publishers, Dordrecht, Volume II, pp 29–32

Strasser RJ, Govindjee (1991) The Fo and the OJIP fluorescence rise in higher plants and algae. In: Argyroudi-Akoyunoglou JH (ed) Regulation of Chloroplast Biogenesis. Plenum Press, New York, pp 423–426

Strasser BJ, Strasser RJ (1995) Measuring fast fluorescence transients to address environmental questions: The JIP-test. In: Mathis P (ed) Photosynthesis: from Light to Biosphere. Kluwer Academic Publishers, Dordrecht, pp 977–980

Strasser RJ, Srivastava A, Tsimilli-Michael M (2000) The fluorescence transient as a tool to characterize and screen photosynthetic samples. In: Yunus M, Pathre U, Mohanty P (eds) Probing Photosynthesis: Mechanism, Regulation and Adaptation. Taylor and Francis, London, pp 445–483

Strasser RJ, Srivastava A, Tsimilli-Michael M (2004) Analysis of the chlorophyll a fluorescence transient. In: Papageorgiou GC, Govindjee (eds) Chlorophyll a Fluorescence: A Signature of Photosynthesis. Advances in Photosynthesis and Respiration, Volume 19. Kluwer Academic Publishers, The Netherlands, pp 321–362

Strasser RJ, Tsimili-Michael M, Qiang S, Goltsev V (2010) Simultaneous in vivo recording of prompt and delayed fluorescence and 820-nm reflection changes during drying and after rehydratation of the resurrection plant *Haberlea rhodopensis*. Biochim Biophys Acta 1797:1313–1326

Sung DY, Kaplan F, Lee KJ, Guy CHL (2003) Acquired tolerance to environmental extrems. Trends Pl Sci 8:179–187

Surridge C (2002) The rice squad. Nature 416:576–578

Suwa R, Hakata H, Hara H, El-Shemy HA, Adu-Gyamfi JJ, Nguyen NT, Kanai S, Lightfoot DA, Mohapatra PK, Fujita K (2010) High temperature effects on photosynthate partitioning and sugar metabolism during ear expansion in maize (*Zea mays* L.) genotypes. Plant Physiol Bioch 48:124–130

Tan W, Brestič M, Olšovská K, Yang X (2011) Photosynthesis is improved by exogenous calcium in heat-stressed tobacco plants. J Plant Physiol 168:2063–2071

Tezara W, Mitchell VJ, Driscoll SD, Lawlor DW (1999) Water stress inhibits plant photosynthesis by decreasing coupling factors and ATP. Nature 401:914–917

Tholen D, Zhu XG (2011) The mechanistic basis of internal conductance: a theoretical analysis of mesophyll cell photosynthesis and CO_2 diffusion. Plant Physiol 156:90–105

Tholen D, Ethier G, Genty B, Pepin S, Zhu X (2012) Variable mesophyll conductance revisited: theoretical background and experimental implications. Plant Cell Environ 35:2087–2103

Thussagunpanit J, Jutamanee K, Chai-Arree W, Kaveeta L (2012) Increasing photosynthetic efficiency and pollen germination with 24-Epibrassinolide in rice (*Oryza sativa* L.) under heat stress. Thai J Bot 4:135–143

Tian Y, Chen J, Chen C, Deng A, Song Z, Zheng C, Hoogmoed W, Zhang W (2012) Warming impacts on winter wheat phenophase and grain yield under field conditions in Yangtze Delta Plain, China. Field Crops Res 134:193–199

Toth SZ, Schansker G, Kissimon J, Kovacs L, Garab G, Strasser RJ (2005) Biophysical studies of photosystem II-related recovery processes after a heat pulse in barley seedlings (*Hordeum vulgare* L.). J Plant Physiol 162:181–194

Toth SZ, Schansker G, Strasser RJ (2007) A non-invasive assay of the plastoquinone pool redox state based on the OJIP-transient. Photosynth Res 93:193–203

Tuberosa R (2012) Phenotyping for drought tolerance of crops in the genomics era. Front Physiol 3:347

Uehlein N, Lovisolo C, Siefritz F, Kaldenhoff R (2003) The tobacco aquaporin NtAQP1 is a membrane CO_2 pore with physiological functions. Nature 425:734–737

Ushakova SA, Tikhomirov AA, Shikhov VN, Gros JB, Golovko TK, Dal'ke IV, Zakhozhii IG (2013) Tolerance of wheat and lettuce plants grown on human mineralized waste to high temperature stress. Adv Space Res 51:2075–2083

Ustin SL, Roberts DA, Gamon JA, Asner GP, Green RO (2004) Using imaging spectroscopy to study ecosystem processes and properties. Bioscience 54:523–534

van Gorkom HJ (1986) Fluorescence measurements in the study of photosystem II electron transport. In: Govindjee, Amesz J, Fork DC (eds) Light Emission by Plants and Bacteria. Academic Press, New York, pp 267–289

Vassileva V, Demirevska K, Simova-Stoilova L, Petrova T, Tsenov N, Feller U (2012) Long-term field drought affects leaf protein pattern and chloroplast ultrastructure of winter wheat in a cultivar-specific manner. J Agron Crop Sci 198:104–117

Vollmann J, Walter H, Sato T, Schweiger P (2011) Digital image analysis and chlorophyll metering for phenotyping the effects of nodulation in soybean. Comput Electron Agr 75:190–195

von Caemmerer S (2003) C4 photosynthesis in a single C3 cell is theoretically inefficient but may ameliorate internal CO_2 diffusion limitations of C3 leaves. Plant Cell Environ 26:1191–1197

Wahid A, Gelani S, Ashraf M, Foolad MR (2007) Heat tolerance in plants: an overview. Environ Exp Bot 61:199–223

Wang B, Li Z, Eneji AE, Tian X, Zhai Z, Li J, Duan L (2008) Effects of coronatine on growth, gas exchange traits, chlorophyll content, antioxidant enzymes and lipid peroxidation in maize (*Zea mays* L.) seedlings under simulated drought stress. Plant Production Sci 11:283–290

Wang X, Cai J, Jiang D, Liu F, Dai T, Cao W (2011) Pre-anthesis high-temperature acclimation alleviates damage to the flag leaf caused by post-anthesis heat stress in wheat. J Plant Physiol 168:585–593

Wang X, Cai J, Liu F, Dai T, Cao W, Wollenweber B, Jiang D (2014) Multiple heat priming enhances thermo-tolerance to a later high temperature stress *via* improving subcellular antioxidant activities in wheat seedlings. Plant Physiol Bioch 76:185–192

Watanabe N, Evans JR, Chow WS (1994) Changes in the photosynthetic properties of Australian wheat cultivars over the last century. Aust J Plant Physiol 21:169–183

White AJ, Critchley C (1999) Rapid Light Curves: A new fluorescence method to assess the state of the photosynthetic apparatus. Photosynth Res 59:63–72

Whitney SM, Houtz RL, Alonso H (2011) Advancing our understanding and capacity to engineer Nature's CO_2-sequestering enzyme, Rubisco. Plant Physiol 155:27–35

Wing SR, Patterson MR (1993) Effects of wave-induced light flecks in the intertidal zone on photosynthesis in the macroalgae *Postelsia palmaeformis* and *Hedophyllum sessile* (Phaeophyceae). Mar Biol 116:519–52

Wise RR, Olson AJ, Schrader SM, Sharkey TD (2004) Electron transport is the functional limitation of photosynthesis in field-grown Pima cotton plants at high temperature. Plant Cell Environ 27:717–724

Wójcik-Jagła M, Rapacz M, Tyrka M, Kościelniak J, Crissy K, Żmuda K (2013) Comparative QTL analysis of early short-time drought tolerance in Polish fodder and malting spring barleys. Theor Appl Genet 126:3021–3034

Wu X, Yao X, Chen J, Zhu Z, Zhang H, Zha D (2014) Brassinosteroids protect photosynthesis and antioxidant system of eggplant seedlings from high-temperature stress. Acta Physiol Plant 36:251–261

Wullschleger SD (1993) Biochemical limitations to carbon assimilation in C3 plants—a retrospective analysis of the A/C_i curves from 109 species. J Exp Bot 44:907–920

Xu Z, Zhou G, Han G, Li Y (2011) Photosynthetic potential and its association with lipid peroxidation in response to high temperature at different leaf ages in maize. J Plant Growth Regul 30:41–50

Yamada M, Hidaka T, Fukamachi H (1996) Heat tolerance in leaves of tropical fruit crops as measured by chlorophyll fluorescence. Sci Hortic 67:39–48

Yamane Y, Kashino Y, Koike H, Satoh K (1998) Effects of high temperatures on the photosynthetic systems in spinach: oxygen-evolving activities, fluorescence characteristics and the denaturation process. Photosynth Res 57:51–59

Yan K, Chen P, Shao H, Zhang L, Xu G (2011) Effects of short-term high temperature on photosynthesis and photosystem II performance in sorghum. J Agron Crop Sci 197:400–408

Yan K, Chen P, Shao H, Zhao S, Zhang L, Xu G, Sun J (2012) Responses of photosynthesis and photosystem II to higher temperature and salt stress in sorghum. J Agron Crop Sci 198:218–225

Yan K, Chen P, Shao H, Shao C, Zhao S, Brestič M (2013a) Dissection of photosynthetic electron transport process in sweet sorghum under heat stress. PloS One 8:62100

Yan K, Chen P, Shao H, Zhao S (2013b) Characterization of photosynthetic electron transport chain in bioenergy crop Jerusalem artichoke (*Helianthus tuberosus* L.) under heat stress for sustainable cultivation. Ind Crop Prod 50:809–815

Yasir TA, Min D, Chen X, Condon AG, Hu YG (2013) The association of carbon isotope discrimination (Δ) with gas exchange parameters and yield traits in Chinese bread wheat cultivars under two water regimes. Agr Water Manage 119:111–120

Ye L, Gao HY, Zou Q (2000) Responses of the antioxidant systems and xanthophyll cycle in *Phaseolus vulgaris* to the combined stress of high irradiance and high temperature. Photosynthetica 38:205–210

Yin X, Struik PC, Romero P, Harbinson J, Evers JB, van der Putten PEL, Vos J (2009) Using combined measurements of gas exchange and chlorophyll fluorescence to estimate parameters of a biochemical C3 photosynthesis model: A critical appraisal and a new integrated approach applied to leaves in a wheat (*Triticum aestivum*) canopy. Plant Cell Environ 32:448–464

Yordanov I, Velikova V, Tsonev T (2003) Plant responses to drought and stress tolerance. Bulg J Plant Physiol, special issue 187–206

Yu C, Huang S, Hu X, Deng W, Xiong C, Ye C, Li Y, Peng B (2013) Changes in photosynthesis, chlorophyll fluorescence, and antioxidant enzymes of mulberry (*Morus* ssp.) in response to salinity and high-temperature stress. Biologia 68:404–413

Yu J, Yang Z, Jespersen D, Huang B (2014) Photosynthesis and protein metabolism associated with elevated CO_2 mitigation of heat stress damages in tall fescue. Environ Exp Bot 99:75–85

Zarzycki J, Axen SD, Kinney JN, Kerfeld CA (2013) Cyanobacterial-based approaches to improving photosynthesis in plants. J Exp Bot 64:787–798

Zegada-Lizarazu W, Monti A (2013) Photosynthetic response of sweet sorghum to drought and re-watering at different growth stages. Physiol Plantarum 149:56–66

Zhang R, Sharkey TD (2009) Photosynthetic electron transport and proton flux under moderate heat stress. Photosynth Res 100:29–43

Zhang JX, Nguyen HT, Blum A (1999) Genetic analysis of osmotic adjustment in crop plants. J Exp Bot 50:291–302

Zhang YP, Zhu XH, Ding HD, Yang SJ, Chen YY (2013) Foliar application of 24-epibrassinolide alleviates high-temperature-induced inhibition of photosynthesis in seedlings of two melon cultivars. Photosynthetica 51:341–349

Zhu XG, Portis AR, Long SP (2004) Would transformation of C3 crop plants with foreign Rubisco increase productivity? A computational analysis extrapolating from kinetic properties to canopy photosynthesis. Plant Cell Environ 27:155–165

Zhu XG, Long SP, Ort DR (2010) Improving photosynthetic efficiency for greater yield. Ann Rev Plant Biol 61:235–261

Zivcak M (2006) Application of physiological reaction diversity in screening of wheat genotypes for drought and high temperature tolerance. Dissertation, Slovak University of Agriculture.

Zivcak M, Brestic M, Olsovska, K (2008a) Application of photosynthetic parameters in screening of wheat (*Triticum aestivum* L.) genotypes for imroved drought and high tem-perature tolerance. In: Allen JF, Gantt E, Goldbeck JH, Osmond B (eds) Photosynthesis. Energy from the sun: 14th International congress on photosynthesis. Springer, Dordrecht, pp 1247–1250.

Zivcak M, Brestic M, Olsovska, K, Slamka P (2008b) Performance index as a sensitive indicator of water stress in *Triticum aestivum*. Plant Soil Environ 54:133–139

Zivcak M, Brestic M, Olsovska, K (2008c) Physiological parameters useful in screening for improved tolerance to drought in winter wheat (*Triticum aestivum* L.). Cereal Res Commun 36:1943–1946

Zivcak M, Brestic M, Olsovska, K (2009a) Application of chlorophyll fluorescence for screening wheat (*Triticum aestivum* L.) genotype susceptibility to drught and high temperature. Vagos 82:82–87

Zivcak M, Repkova J, Olsovska K, Brestic M (2009b) Osmotic adjustment in winter wheat varieties and its importance as a mechanism of drought tolerance. Cereal Res Commun 37:569–572

Zivcak M, Olsovska K, Brestic M, Slabbert MM (2013) Critical temperature derived from the selected chlorophyll a fluorescence parameters of indigenous vegetable species of South Africa treated with high temperature. In: Lu C (ed) Photosynthesis: Research for Food, Fuel and Future–15th International Conference on Photosynthesis, Springer, Dordrecht, pp 628–632

Živčák M, Brestič M, Balátová Z, Dreveňáková P, Olšovská K, Kalaji HM, Yang X, Allakhverdiev SI (2013) Photosynthetic electron transport and specific photoprotective responses in wheat leaves under drought stress. Photosynth Res 117:529–546

Živčák M, Olšovská K, Slamka P, Galambošová J, Rataj V, Shao HB, Brestič M (2014a) Application of chlorophyll fluorescence performance indices to assess the wheat photosynthetic functions influenced by nitrogen deficiency. Plant Soil Environ 60:210–215

Živčák M, Olšovská K, Slamka P, Galambošová J, Rataj V, Shao HB, Kalaji HM, Brestič M (2014b) Measurements of chlorophyll fluorescence in different leaf positions may detect nitrogen deficiency in wheat. Zemdirbyste-Agriculture 101:437–444

Zivcak M, Kalaji HM, Shao HB, Olsovska K, Brestic M (2014c) Photosynthetic proton and electron transport in wheat leaves under prolonged moderate drought stress. J Photochem Photobiol B Biol 137:107–115

Zivcak M, Brestic M, Kunderlikova K, Sytar O, Allakhverdiev SI (2015) Repetitive light pulse-induced photoinhibition of photosystem I severely affects CO_2 assimilation and photoprotection in wheat leaves. Photosynth Res 126:449–463

Zou J, Rogers WE, Siemann E (2007) Differences in morphological and physiological traits between native and invasive populations of *Sapium sebiferum*. Funct Ecol 21:721–730

Effects of Environmental Pollutants Polycyclic Aromatic Hydrocarbons (PAH) on Photosynthetic Processes

Anjana Jajoo

Summary

Increasing pollution of the environment has become an important problem of the present era. Polycyclic aromatic hydrocarbons (PAHs) are widely known as anthropogenic pollutants harmful to plants, animals and humans. Plants are an integral component of the terrestrial ecosystem and have ability to take up, transform and accumulate environmental pollutants including PAHs. It has been shown that PAHs influence the biochemical and physiological processes in plants, just similar to other toxic organic compounds, i.e. herbicides. They not only change the processes of energetic metabolism, but also change mechanisms associated with plant growth and development. In this chapter we shall be discussing the effects of PAH on plant growth, particularly the photosynthetic apparatus. A comprehensive and updated knowledge of the effects of various PAHs including naphthalene, anthracene, pyrene, fluoranthene on the photosynthetic mechanisms has been presented and discussed.

A. Jajoo (✉)
School of Life Science, Devi Ahilya University, Indore 452017, MP, India
e-mail: anjanajajoo@hotmail.com

Keywords

Polycyclic aromatic hydrocarbons (PAHs) • Photosynthesis • Plant growth • Photosystem II • Phototoxicity

Contents

11.1 Introduction 249
11.2 Structure and Properties of PAH 250
11.3 Effects of PAHs on Plant Growth 251
11.4 Effects of PAHs on Photosynthesis 252
11.5 Effects of PAHs on Photosystem II Heterogeneity 253
11.6 Photo-Toxicity of PAH 253
11.7 Phytoremediation Methods 255
11.8 Conclusions and Future Prospective 256
References 257

11.1 Introduction

In the last few decades, industrial activities carried out by human beings, involving partial combustion of fossil fuels, oil spills etc. has led to accumulation of various harmful chemical substances. Of the numerous environmental pollutants, polycyclic aromatic hydrocarbons (PAHs) are amongst the most toxic and highly persistent pollutants (Burritt 2008). PAHs are extremely harmful compounds that can disturb several physiological processes in plant and thus ultimately inhibiting crop yield. This problem is of great significance since PAHs are commonly present in water, air, soil, food and living organisms. As much as 90% of the PAHs accumulate in the soil due to their hydrophobic character which favours rapid association with soil solid particles and their permeation to bottom sediments (Bałdyga et al. 2005).

Polycyclic aromatic hydrocarbons (PAHs), a prevalent group of toxic and mutagenic contaminants, are of environmental concern because they are both ubiquitous and highly hydrophobic. They readily accumulate in the lipoprotein membranes of both aquatic and terrestrial organisms. Hydrophobicity also limits bioavailability of PAHs and these chemicals are usually sequestered to sediment and particulate phases in aquatic environments. Some polycyclic aromatic hydrocarbons are harmful mainly due to their cytostatic and immunostatic effects, genotoxic properties and carcinogenic products of their transformation.

11.2 Structure and Properties of PAH

The basic chemical structure of PAHs consists of carbon (C) and hydrogen atom (H) arranged in the form of two or more fused benzene rings in linear, angular or cluster arrangements (Wilson and Jones 1993). The stability of PAHs is indicated by the ring arrangement, linear being the most unstable and angular the most stable.

anthracene   *phenanthrene*   *pyrene*

The reactivity of PAHs is dependent on the number and organization of the condensed rings (Sverdrup et al. 2003). Polycyclic aromatic hydrocarbons are classified as Persistent, Bio-acumulative and Toxic (PBT) compounds.

PAHs are relatively inert compounds and are resistant to the degradation processes. The reactivity of PAHs is influenced by several factors like temperature, light, oxygen (O_2), ozone (O_3) and other chemical reagents.

Soil is a crucial environmental factor involved in the translocation and the phytotoxic effects of organic pollutants in plants (Desalme et al. 2011). Much of PAHs released into atmosphere eventually reach the soil by direct deposition or by deposition on vegetation. Forest and prairie fires or decomposition of dead plants are important natural sources of PAHs for soils. The industrial emission and usage of sewage sludges are also important soil sources of PAHs.

Plants have the ability to take up PAHs from the environment. Four major pathways include passive and active uptake through root system, gaseous and particulate deposition to aboveground shoots and direct contact between soil and plant tissues (Collins et al. 2006; Vácha et al. 2010). It was shown that the transfer of PAHs into plants is influenced mainly by chemical characteristics of the substances (the number and position of aromatic nuclei); by soil characteristics (content and quality of soil organic matter) and by plant characteristics (plant species and plant bodies). More than 50% of total atmospheric PAHs are adsorbed into soils, eluted to deeper layers and then taken up by plants. More than 20 toxic PAHs have been identified amongst those which are commonly found in the environment and they include anthracene, pyrene, fluoranthene, naphthalene and phenanthrene.

The study of influence of selected PAHs on higher plants is very important because plants are dominant component of terrestrial ecosystems with ability to uptake PAHs from the environment. For these reasons, plants are used as early indicators of environment pollution. The evaluation of biological parameters can allow for faster and economical preferable pollution information than complex chemical analyses. Therefore, some plant species may be used for detoxification of the environment.

11.3 Effects of PAHs on Plant Growth

Increasing fluoranthene (FLT) concentration inhibited the energy of germination and the germination rate of seeds of all plant species. In none of the plant species, FLT affect the uptake of water by the seeds (Kummerová et al. 2006, 2012). Seedling growth was a far more sensitive endpoint than seed emergence for all substances (Sverdrup et al. 2003). The addition of PAHs, especially higher level (400 mg kg^{-1}), to soils resulted in some adverse effects on rice growth such as a decrease in biomass and chlorophyll content and an increase of water content and the chlorophyll a/b ratio. Certain responses of rice, including enhancement of soluble protein contents and inducement of SOD activities, indicated that rice possessed some resistance to stress due to the presence of PAHs. Among the parameters examined, water content and SOD activity were the most sensitive indicators. Inoculation with PAH-degrading bacteria significantly promoted biomass and the photosynthetic effectiveness of rice (Li et al. 2008).

PAHs have been shown to induce oxidative stress, reduce growth, and cause leaf deformation as well as tissue necrosis. Some of the physiological systems that participate in the PAH-induced stress response in Arabidopsis have been identified (Weisman et al. 2010). Microarray results have revealed numerous perturbations in signalling and metabolic pathways that regulate reactive oxygen species (ROS) and responses related to pathogen defence. A number of glutathione S-transferases that may tag xenobiotics for transport to the vacuole were upregulated. Comparative microarray analyses indicated that the phenanthrene response was closely related to other ROS conditions, including pathogen defence conditions. The ethylene-inducible transgenic reporters were activated by phenanthrene. Mutant experiments showed that PAH inhibits growth through an ethylene-independent pathway, as PAH-treated ethylene-insensitive *etr1–4* mutants exhibited a greater growth reduction than WT. Further, phenanthrene-treated, constitutive ethylene signalling mutants had longer roots than the untreated control plants, indicating that the PAH inhibits parts of the ethylene signalling pathway (Weisman et al. 2010).

In fact, many PAH-induced responses involve interference with membrane-mediated physiological and biochemical processes; these include membrane permeability, enzymatic dysfunction, and photosynthesis (Duxbury et al. 1997). Because PAHs partition primarily into thylakoids, they could disrupt the ordered arrangement of chloroplasts and interfere with electron transport and photosynthesis. Modified PAHs, being more hydrophilic, reactive, and electrophilic, may bind covalently to cellular structures such as proteins, DNA, and RNA, interfering with metabolism, enzyme activity, and growth.

The PAH stress response was studied in Arabidopsis (*Arabidopsis thaliana*) exposed to the three-ring aromatic compound, phenanthrene (Alkio et al. 2005). Morphological symptoms of PAH stress were growth reduction of the root and shoot, deformed trichomes, reduced root hairs, chlorosis, late flowering, and the appearance of white spots, which later developed into necrotic lesions. At the tissue and cellular levels, plants experienced oxidative stress as indicated by localized H_2O_2 production and cell death. These findings show that (i) Arabidopsis takes up phenanthrene, suggesting possible degradation in plants, (ii) a PAH response in plants and animals may share similar stress mechanisms, since in animal cells detoxification of PAHs also results in oxidative stress, and (iii) plant specific defence mechanisms contribute to PAH stress response in Arabidopsis. It has been proven that PAHs influence the biochemical and physiological processes in plant, just similar to other toxic organic compounds such as herbicides. They not only change the processes of energetic metabolism but also mechanisms associated with plant growth and development.

11.4 Effects of PAHs on Photosynthesis

Photosynthesis is one of the important metabolic processes in plants which is highly sensitive to stress and at the same time is directly correlated with biomass. Any stress which results in inhibition of this process ultimately leads to reduced crop yield. Photosynthesis is susceptible to inhibition from many environmental contaminants, including herbicides, metals, and organic contaminants (Marwood et al. 2001, 2003). Photosynthetic electron transport is a universal feature of higher plants, algae, and cyanobacteria.

It is also known that PAHs affect primary and secondary processes involved in photosynthesis. The multifaceted action of PAHs observed in treated plants suggests that hydrophobic PAHs accumulate in thylakoid membranes (Duxbury et al. 1997) and may induce conformational changes in their structures, thus causing disturbance of electron transport at both donor and acceptor site of PS II (Aksmann and Tukaj 2008; Kummerová et al. 2008; Tomar et al. 2015). Some PAH, like anthracence, inhibit photosynthetic activities of green alga due to lowering of quantum efficiency of electron trapping and transport and impairment of PS II and PS I. Lower activity of oxygen electron complex (OEC) is observed in anthracene treated algal cells (Aksmann and Tukaj 2008). It is also reported that anthracene inhibits carbon fixation, i.e. it affects net photosynthesis (Huang et al. 1997). Other PAHs like phenanthrene and pyrene also decrease net photosynthesis in higher plants. Phenanthrene and pyrene exposure caused a decrease in growth, photosynthetic pigment contents, stomatal conductance, maximal quantum yield, effective quantum yield of PSII and photochemical quenching coefficient (Ahammed et al. 2012).

In higher plants, the exposure to PAHs cause diminished biomass accumulation, chlorosis and inhibition of photosynthesis (Huang et al. 1996; Oguntimehin et al. 2008). PAHs cause a decrease in photosynthetic pigment content by direct interaction with the pigment molecules or by inhibition of their synthesis (Kummerová et al. 2006; Oguntimehin et al. 2010). Images from Transmission electron microscopy have revealed that PAHs can cause gross deformation in chloroplast and may lead to oxidation stress (Liu et al. 2009). PAH has been found to enhance production of reactive oxygen species (ROS) in higher plants (Oguntimehin et al. 2008). Several studies have shown inhibitory effects of PAH on donor and acceptor sides of Photosystem II (PS II) including oxygen evolving complex (OEC) and electron transport chain (Aksmann and Tukaj 2008; Kummerová et al. 2008). Fluoranthene was found to inhibit both light as well as dark reactions of photosynthesis in wheat (Tomar and Jajoo 2014).

By measuring the polyphasic rise of fluorescence transients to evaluate the modifications in PSII photochemistry in PAHs exposed wheat (Triticum aestivum), effects of fluoranthene on energy trapping and utilization rather than on other components of photosynthetic electron transport chain have been proposed (Tomar and Jajoo 2013a).

Several parameters including growth, net photosynthesis rate, production of ethylene, ethane and carbon dioxide, and the content of cytokinins etc. of in vitro cultivated pea plants responded sensitively to PAH (fluoranthene) in the environment (Kummerová et al. 2010). Decrease in the net photosynthesis rate and CO_2 production, increase in ethylene and ethane production and lowering in cytokinin content induced by FLT, especially its higher concentrations, reflected negatively on the growth of plants (Kummerová et al. 2010). BN PAGE analysis of thylakoid membrane proteins in Arabidopsis grown in PAH showed no major change in the composition of protein complexes (unpublished data).It will be of interest to investigate the effects of various PAH differing in their aromaticity (2-ring, 3-ring, 4- ring PAHs) on photosynthetic activity.

Recently, a study of prevalent polycyclic aromatic hydrocarbon fluoranthene (FLT) on wheat plants (Triticum aestivum) has been carried out and considerable attention has been paid to the effect of this compound on the primary reactions of photosynthesis (Tomar and Jajoo 2013a). Chl *a* fluorescence is used as an indicator of the toxicity of FLT in wheat. FLT has the potential to have impact on the primary productivity of crops via inhibition of primary processes of photosynthesis. It is proposed that FLT inhibits quantum efficiency of energy absorption and trapping by converting active PS II reaction centers into inactive centers (heat sink). The method and the interpretations can be applied to assess the effects of environmental pollutants on a wide range of samples including higher plants, algae etc.

11.5 Effects of PAHs on Photosystem II Heterogeneity

As compared to other protein pigment complexes participating in light reaction, photosystem II (PSII) shows diverse nature in structural and functional aspects and this is termed as PS II heterogeneity. Most important observation is that the extent and nature of PS II heterogeneity varies under different physiological conditions (Lavergne and Briantais 1996) i.e. salinity, osmotic, temperature and pH stress (Mehta et al. 2010; Mathur et al. 2011a, b; Tongra et al. 2011).Two major types of PS II heterogeneity have been identified: reducing side heterogeneity and antenna heterogeneity. It is possible to estimate heterogeneity of PS II by measuring Chlorophyll *a* fluorescence. It is a rapid method for measuring photosynthetic electron transport from plants *in vivo* and can be used to calculate several parameters that describe the efficiency of a photochemical process within the photosynthetic apparatus, especially PS II.

Reducing side heterogeneity of PS II is based on the electron transport properties on the reducing side of the reaction centers. PS II in green plants occur in two distinct forms: centers with efficient electron transport from Q_A to Q_B are known as Q_B reducing type, while centers that are photochemically competent but unable to transfer electron from Q_A to Q_B are known as Q_B non-reducing type (Graan and Ort 1986). In such centers Q_A^- can only be reoxidized by a back reaction with the donor side of PS II. Q_B reducing centers are so called 'active centers' while Q_B non-reducing centers are termed as 'inactive centers (Fig. 11.1).

Antenna heterogeneity of PS II is related to heterogeneity in the antenna size as well as their energetic connectivity between PS IIs. Analysis of the biphasic data obtained in DCMU treated plant material suggested the presence of three distinct populations of PS II centers in the chloroplast, termed as PS IIα, PS IIβ and PS IIγ. The characteristics of these are listed in Fig. 11.1.

Tomar and Jajoo (2013b) have described inhibitory response of Fluoranthene on PSII heterogeneity in crop plant wheat (Table 11.1). With FLT treatment the fractions of Q_B-non-reducing centers increased and Q_B-reducing centers decreased which implies that the active centers were converted into inactive centers. In case of antenna heterogeneity, increase in FLT concentration caused change in the relative amount of α, β and γ centers. The active α centers were converted into inactive β and γ centers. Decrease in Q_B-reducing center and alteration in α, β and γ center is associated with down regulation of photochemical activity and damage of photosynthetic apparatus. Measurement of PSII heterogeneity in early developmental stages under stress condition may prove to be an excellent biomarker to assess tolerance to toxicity.

11.6 Photo-Toxicity of PAH

Photo-induced processes involving these chemicals are environmentally relevant, and are observed as increased toxicity of PAHs in the presence of simulated solar radiation (SSR) and natural sunlight. One of the most important factors is solar radiation, especially the ultraviolet wavelengths (UV-A 320–400 nm and UV-B 290–320 nm). Interestingly, PAHs strongly absorb solar ultraviolet (UV) radiation, resulting in photo-modification (generally oxidation) of the carbon skeleton (Duxbury et al. 1997). The absorption of UV-photons causes excitation of the PAH molecule and leads to triplet-state

> Photosystem II heterogeneity
> 1. Reducing side heterogeneity: (i) Q_B reducing centres
>   (ii) Q_B non-reducing centres
> 2. Antenna heterogeneity: (i) Connectivity- Grouped, Ungrouped
>   (ii) Antenna size – Alpha (α), Beta (β), Gamma (γ) centres
>
> PS II α:
> - Dominant form, localized in the grana partition regions.
> - Exhibits a large light harvesting antenna (210-250 Chl *a* and *b*).
> - Excited states transfer between PS II units is possible as reflected in a sigmoidal fluorescence rise when measured with DCMU.
>
> PS II β:
> - Mainly located in the stromal region of thylakoid membranes.
> - Exhibit small light harvesting antenna (100-120 Chl *a* and *b*).
> - No excited states transfer between PS II occurs as reflected in an exponential fluorescence rise when measured with DCMU.
>
> PS II γ:
> - Antenna size of centre is supposed to be very small as compared to alpha (α) centres

Fig. 11.1 Types of photosystem heterogeneity

Table 11.1 Changes in reducing side and antenna size heterogeneity of PS II in 25 days FLT exposed wheat plants

Concentration	% of Q_B non-reducing centers	% of Q_B reducing centers	α centres	β centres	γ centres
Control	21 ± 1	79 ± 1	69 ± 1	25 ± 1	6 ± 1
5 µM	24 ± 1	76 ± 2	57 ± 2	33 ± 2	10 ± 1
10 µM	25 ± 1	75 ± 1	55 ± 1	36 ± 1	9 ± 1
25 µM	28 ± 2	72 ± 1	52 ± 1	38 ± 1	10 ± 1
50 µM	30 ± 1	70 ± 1	52 ± 2	40 ± 2	8 ± 1
75 µM	34 ± 1	66 ± 1	50 ± 1	41 ± 1	9 ± 1
100 µM	38 ± 2	62 ± 2	44 ± 2	42 ± 2	14 ± 1

formation. PAHs are highly efficient at promoting formation of singlet oxygen ($1O_2$) and other reactive oxygen species which can cause oxidative damage in biological systems.

Photo-induced toxicity of PAHs is derived from two photochemical processes: photosensitization and photo-modification. In a photosensitization reaction, intracellular singlet-state oxygen ($1O_2$) and other active oxygen species are generated, which can cause oxidative damage in biological systems. During photo-modification, PAHs are structurally altered and form a complex mixture of photoproducts which may be more toxic than parent compounds (Ankley et al. 1999). Since these photoproducts are more polar, their increased solubility and bioavailability cause them to be more toxic (Tomar et al. 2015; Mcconkey et al. 1997).Interestingly, some PAH photo-oxidation products are similar to those produced by biological oxidation via cytochrome P450 (Huang et al. 1997).

When PAHs are photo-modified, many of the resultant products are quinones. Such chemicals partition into thylakoid membranes where they potentially can block photosynthesis, especially where plastoquinone is used as an electron acceptor or donor. However, little information is currently available on the route(s) by which PAHs or their environmental photoproducts can interfere with photosynthetic electron transport. Both photosystem I (PSI) and PSII were found to be inhibited, with the primary site of action being PSI. Photo-modified fluoranthene was found to exert more harmful effects as compared to intact fluoranthene by inhibiting growth and photosynthetic parameters in wheat (Tomar and Jajoo 2015).

Generation of reactive oxygen intermediates (ROI) such as $^-O_2$, H_2O_2, and ^-OH is known to be a major mechanism of damage in biological systems. Photo-sensitization is an overproduction of reactive oxygen species (ROS) inside cells treated with PAHs and exposed to natural or artificial sunlight. After absorption of energy, especially UV radiation, PAHs in the excited state can transfer their energy to molecular oxygen to generate ROS including singlet oxygen, superoxide radical and hydrogen peroxide. ROS can be also generated when light excited photosynthetic pigments cannot transfer electrons due to inhibition of photosynthetic electron chain by PAHs located inside chloroplast. Oxidative stress is often suggested as one of the PAHs toxicity mechanisms (Aksmann and Tukaj 2008; Tukaj and Aksmann 2007). Due to oxidative stress in pants, chloroplast and mitochondria in treated plants undergo gross deformation, and cellular structures collapses as revealed by Transmission electron microscopy (Liu et al. 2009).

Photo-induced toxicity of PAHs could be based in the formation of intracellular singlet oxygen and other ROI that lead to biological damage. Phyto-toxicities appear to vary, depending on the particular PAH and plant species (Hwang et al. 2003). The production of reduced and excited species of ROI in chloroplasts has been reviewed (Asada 2006). The reaction centers of PSI and PSII in chloroplast thylakoids are the major sites of ROI generation. Superoxide dismutase (SOD) constitutes the first line of defence against ROI within a cell (Chen and Schopfer 1999). Peroxidases (PERO, hydrogen peroxide oxidoreductase) are widely found in plants, and oxidize a vast array of compounds in the presence of hydrogen peroxide (H_2O_2) (Upham and Jahnke 1986). Mannitol (MANN) is produced in some plants and is recognized as a potent ROI quencher. It was shown to scavenge hydroxyl radicals (^-OH) generated by cell-free oxidant systems (Muratova et al. 2003).

Oguntimehin et al. (2010) showed that peroxidase (PERO), superoxide dismutase (SOD) and mannitol (MANN) were all effective scavengers of ROI. MANN was found to be most effective and scavenged ^-OH, the most lethal of the ROI.

Role of polyamines has been suggested in the PAH induced oxidative damage. Recovery appeared, at least in part, due to increased synthesis of PAs, achieved via increased activities of the enzymes arginine decarboxylase (ADC) and S-adenosylmethionine decarboxylase (SAMDC). Chemical inhibition of these enzymes inhibited plant recovery, while treatment with PAs aided recovery. Finally, as chloroplasts and the plasma membrane are to be key targets for PHEN-induced damage, the potential roles of PAs in protecting these cellular components has been suggested (Burritt 2008).

11.7 Phytoremediation Methods

The effectiveness of vegetation for the bioremediation of PAH-contaminated soils has previously been studied (Muratova et al. 2003, 2009; Joner et al. 2002; Yoshitomi and Shann 2001). It has been shown that phyto-remediation of PAH-contaminated soils may be enhanced by plant root exudates, which stimulate microbial PAH degradation in rhizosphere. Studies by Rentz et al. (Rentz et al. 2004) and Kamath et al. (Kamath et al. 2004) observed repression of PAH-catabolic genes by root exudates and demonstrated greater PAH degradation by bacteria grown on root products compared to PAH, supporting the notion that prolific microbial growth provides improved degradation in the rhizosphere. Consequently, the mechanisms responsible for PAH degradation in the rhizosphere requires deeper understanding.

Various plant species have been shown to accumulate and tolerate PAHs and half of the PAHs released from the contamination resources can be absorbed and removed by plants. Amaranth, a selenium accumulator, lowered phenanthrene and pyrene amounts by 87.85–94.03% and 46.89–76.57%, respectively, in contaminated soils. Trees such as poplar and jack pine, grasses such as rye, oat, wheat, and maize, as well as agricultural crops including sunflower, soybean, pea and carrot, have been shown to remove PAH and crude oil pollutants from the environment (Liu et al. 2009). Surprisingly, inhibitory effects of PAHs on photosynthetic performance were not found to be related to their aromaticity (Jajoo et al. 2014).

There is a lack of information related to the interaction between higher plants and PAH-degrading bacteria. It was reported that inoculation with the *Pseudomonas putida* strain PCL1444 isolated from the rhizosphere of a grass species (*Lolium multiflorum* cv.Barmultra) from a PAH-contaminated site protected grass seeds and seedlings against high naphthalene concentrations (Li et al. 2008). *Mycobacterium vanbaalenii* PYR-1 is capable of degrading a wide range of high-molecular-weight PAHs, including fluoranthene. A combination of metabolomic, genomic, and proteomic technologies were used to investigate fluoranthene degradation in this strain (Kweon et al. 2007). The strain operates multiple pathways for fluoranthene degradation. Many genome-predicted proteins were identified, and more detailed roles were suggested with respect to the degradation of fluoranthene. The presence of *S. densiflora* significantly increases phenanthrene degradation in soils (Gómez et al. 2011).

To improve phyto-remediation processes, multiple techniques that comprise different aspects of contaminant removal from soils have been combined. Using creosote as a test contaminant, a multi-process phyto-remediation system composed of physical (volatilization), photochemical (photo-oxidation) and microbial remediation, and phyto-remediation (plant-assisted remediation) processes was developed (Huang et al. 2004).The techniques applied to realize these processes were land-farming (aeration and light exposure),introduction of contaminant degrading bacteria, plant growth promoting rhizobacteria (PGPR), and plant growth of contaminant tolerant tall fescue (*Festuca arundinacea*). Over a 4-month period, the average efficiency of removal of 16 priority PAHs by the multiprocess remediation system was twice that of land-farming, 50% more than bioremediation alone, and 45% more than phyto-remediation by itself. Importantly, the multi-process system was capable of removing most of the highly hydrophobic, soil bound PAHs from soil. The key elements for successful phyto-remediation were the use of plant species that have the ability to proliferate in the presence of high levels of contaminants and strains of PGPR that increase plant tolerance to contaminants and accelerate plant growth in heavily contaminated soils. The synergistic use of these approaches resulted in rapid and massive biomass accumulation of plant tissue in contaminated soil, putatively providing more active metabolic processes, leading to more rapid and more complete removal of PAHs.

11.8 Conclusions and Future Prospective

Polycyclic aromatic hydrocarbons (PAH) being prevalent environmental pollutants deserve special attention by plant physiologists and biotechnologists. Although some research has been done unraveling the harmful effects of PAH on plant growth and Photosynthesis, mechanisms underlying the physiological effects need to be explored. Molecular aspects of effects of PAH should be investigated to understand its mechanism of action and thereby to plan strategies to protect from them.

Acknowledgements AJ would thank Department of Science and Technology (DST) India for the project (INT/RUS/RFBR/P-173). Rupal Singh Tomar is thanked for her help during the preparation of this manuscript.

References

Ahammed, G.J.; Yuan, H.L.; Ogweno, J.O.; Zhou, Y.H.; Xia, X.J.; Mao, W.H.; Shi, K.; Yu, J.Q. Brassino steroid alleviates phenanthrene and pyrene phytotoxicity by increasing detoxification activity and photosynthesis in tomato. Chemosphere, 2012, 86, 546–555.

Aksmann, A.; Tukaj, Z. Intact anthracene inhibits photosynthesis in algal cells: a fluorescence induction study on *Chlamydomonas reinhardtii* cw92 strain. Chemosphere, 2008, 74, 26–32.

Alkio, M.; Tabuchi, T.M.; Wang, X.C.; Colon-Carmona, A. Stress responses to polycyclic aromatic hydrocarbons in *Arabidopsis* include growth inhibition and hypersensitive response-like symptoms. J. Exp. Bot., 2005, 56, 2983–2994.

Ankley, G.; Mount, D.; Erickson, R.; Diamond, S.; Burkhard, L.; Sibley, P.; Cook, P. In: 9th Annual meeting of SETAC-Europe, Phototoxic polycyclic aromatic hydrocarbon in sediments: a model based approach for assessing risk. Leipzig, Germany, 1999.

Asada, K. Production and scavenging of reactive oxygen species in chloroplasts and their functions. Plant Physiol., 2006, 141, 391–396.

Bałdyga, B.; Wieczorek, J.; Smoczyński, S.; Wieczorek, Z.; Smoczyńska, K. Pea plant response to anthracene present in soil. Pollut. J. Environ. Stud., 2005, 14, 397–401.

Burritt, D. J. The polycyclic aromatic hydrocarbon phenanthrene causes oxidative stress and alters polyamine metabolism in the aquatic liverwort *Riccia fluitans* L. Plant Cell Environ., 2008, 31, 1416–1431.

Chen, S.; Schopfer, P. Hydroxyl-radical production in physiological reactions. A novel functions of peroxidase. Eur. J. Biochem., 1999, 260, 726–735.

Collins, C.; Martin, I.; Fryer, M. Principal pathways for plant uptake of organic chemicals. Environment Agency, Rio House, Bristol, England, 2006.

Desalme, D.; Binet, P.; Bernard, N.; Gilbert, D.; Toussaint, M.L.; Chiapusio, G. Atmospheric phenanthrene transfer and effects on two grassland species and their root symbionts: A microcosm study. Environ. Exp. Bot., 2011, 71, 146–151.

Duxbury, C.L.; Dixon, D.G.; Greenberg, B.M. Effects of simulated solar radiation on the bioaccumulation of polycyclic aromatic hydrocarbons by the duckweed *Lemna gibba*. Environ. Toxicol. Chem., 1997, 16, 1739–1748.

Gómez, S.R.; Andrades-Moreno, L.; Parra, R.; Valera-Burgos, J.; Real, M.; Mateos-Naranjo, E.; Cox, L.; Cornejo, J. *Spartina densiflora* demonstrates high tolerance to phenanthrene in soil and reduces it concentration. Mar. Pollut. Bull., 2011, 62, 1800–1808.

Graan, T.; Ort, D.R. Detection of oxygen-evolving photosystem II centers inactive in plastoquinone reduction. Biochim. Biophys. Acta, 1986, 852, 320–330.

Huang, X.D.; Lorelei, F.; Zeiler, D.; Dixon, G.; Greenberg B.M. Photoinduced toxicity of PAHs to the foliar region of *Brassica napus* (canola) and *Cuumbis sativus* (cucumber) in simulated solar radition. Ecotoxicol. Environ. Saf., 1996, 35, 190–197.

Huang, X.D.; McConkey, B.J.; Babu, T.S.; Greenberg, B.M. Mechanisms of photoinduced toxicity of photomodified anthracene to plants: inhibition of photosynthesis in the aquatic higher plant *Lemna gibba* (duckweed). Environ. Toxicol. Chem., 1997, 16, 1707–1715.

Huang, X.D.; El-Alawi, Y.; Penrose, D.M.; Glick, B.R.; Greenberg, B.M. A multi-process phytoremediation system for removal of polycyclic aromatic hydrocarbons from contaminated soils. Environ. Pollut., 2004, 130, 465–476.

Hwang, H.M.; Wade, T.; Sericano, J.L. Concentrations and source characterization of polycyclic aromatic hydrocarbons in pine needles from Korea, Mexico, and United States. Atmos. Environ., 2003, 37, 2259–2267.

Jajoo, A.; Mekala, N.R.; Tomar, R.S.; Grieco, M.; Tikkanen, M.; Aro, E-M. Inhibitory effects of polyciclic aromatic hydrocarbons (PAHs) on photosynthetic performance are not related to their aromaticity, J. Photochem. Photobiol. B:Biol., 2014, 137, 151–155.

Joner, E.J.; Corgié, S.C.; Amellal, N.; Leyval, C. Nutritional contributions to degradation of polyciclic aromatic hydrocarbons in a stimulated rhizosphere. Soil Biol. Biochem., 2002, 34, 859–864.

Kamath, R.; Schnoor, J.L.; Alvarez, P.J.J. Effects of plant derived substrates on expression of catabolic genes using a nah-lux reporter. Environ. Sci. Tech., 2004, 38, 1740–1745.

Kummerová, M.; Barták, M.; Dubová, J.; Třiska, J.; Zubrová, E.; Zezulka, Š. Inhibitory effect of fluoranthene on photosynthetic processes in lichens detected by chlorophyll fluorescence. Ecotoxicology, 2006, 15, 121–131.

Kummerová, M.; Vanová, L.; Krulová, J.; Zezulká, S. The use of physiological characteristics for comparison of organic compounds phytotoxicity. Chemosphere, 2008, 71, 2050-2059.

Kummerová, M.; Váňová, L.; Fišerová, H.; Klemš, M.; Zezulka, Š.; Krulová, J. Understanding the effect of organic pollutant fluoranthene on pea in vitro using cytokinins, ethylene, ethane and carbon dioxide as indicators. Plant Growth Regul., 2010, 61, 161–174.

Kummerová, M.; Zezulka, Š.; Váňová, L.; Fišerová, H. Effect of organic pollutant treatment on the growth of pea and maize seedlings. Cent. Eur. J. Biol., 2012, 7, 159–166.

Kweon, O.; Kim S.J.; Jones, R.C.; Freeman, J.P.; Adjei, M.D.; Edmondson, R.D.; Cerniglia, C.E. A polyomic approach to elucidate the fluoranthene-degradative pathway in *Mycobacterium vanbaalenii* PYR-1. J. Bacteriol., 2007, 189, 4635–4647.

Lavergne, J.; Briantais, J.M. Photosystem II heterogeneity, In: Oxygeneic Photosynthesis: The light reactions; Ort, R.D.; Yocum, C.F. Eds.; Kluwer Publishers, Dordrecht, The Netherlands, 1996; pp. 265–287.

Li, J. H.; Gao, Y.; Wu, S.C.; Cheung, K.C.; Wang, X.R.; Wong, M. H. Physiological and Biochemical Responses of Rice (*Oryza sativa* L.) to Phenanthrene and Pyrene. Int. J. Phytorem., 2008, 10, 106–118.

Liu, H.; Weisman, D.; Ye, Y.B.; Cui, B.; Huang, Y.H.; Colon-Carmona, A.; Wang, Z.H. An oxidative stress response to polycyclic aromatic hydrocarbon exposure is rapid and complex in *Arabidopsis thaliana*. Plant Sci., 2009, 17, 6357–6382.

Marwood, C.A.; Solomon, K.R.; Greenberg, B.M. Chlorophyll fluorescence as a bioindicator of effects on growth in aquatic macrophytes from mixtures of polycyclic aromatic hydrocarbons. Environ. Toxicol. Chem., 2001, 20, 890–898.

Marwood, C.A.; Jim, K.T.; Bestari, R.; Gensemer, W.; Solomon, K.R.; Greenberg, B. M. Creosote toxicity to photosynthesis and plant growth in aquatic microcosms. Environ. Toxicol. Chem., 2003, 22, 1075–1085.

Mathur, S.; Jajoo, A.; Mehta, P.; Bharti, S. Analysis of elevated temperature-induced inhibition of photosystem II by using chlorophyll *a* fluorescence induction kinetics in wheat leaves (*Triticum aestivum*). Plant Biol., 2011a, 13, 1–6.

Mathur, S.; Allakhverdiev, S.I.; Jajoo, A. Analysis of the temperature stress on the dynamic of antenna size and reducing side heterogeneity of photosystem II in wheat leaves (*Triticum aestivum*). Biochim. Biophys. Acta, 2011b, 1807, 22–29.

Mcconkey, B.J.; Duxbury, C.L.; Dixon, D.G.; Greenberg, B.M. Toxicity of a PAH photooxidation product to the bacteria *Photobacterium phosphoreum* and the duckweed *Lemna gibba*: Effects of phenanthrene and its primary photoproducts, phenantrene quinone. Environ. Toxicol. Chem., 1997, 16, 892–899.

Mehta, P.; Allakhverdiev, S.I.; Jajoo, A. Characterization of photosystem II heterogeneity in response to high salt stress in wheat leaves (*Triticum aestivum*). Photosynth. Res., 2010, 105, 249–255.

Muratova, A.Y.; Turkovskaya, O.V.; Huebner, T.; Kuschk, P. Study of the efficacy of alfalfa and reed in the phytoremediation of hydrocarbon polluted soil. Appl. Biochem. Microbiol., 2003, 39, 599–605.

Muratova, A.Y.; Kapitonova, V.V.; Chernyshova, M.P.; Turkovskaya O.V.; Enzymatic activity of alfalfa in a phenanthrene-contaminated environment. World Agr. Sci. Eng. Tech., 2009, 58, 569–574.

Oguntimehin, I.; Sakugawa, H. Fluoranthene fumigation and exogenous scavenging of reactive oxygen intermediates (ROI) in evergreen Japanese red pine seedlings (*Pinus Densiflora* Sieb. et. Zucc.). Chemosphere, 2008, 72, 747–754.

Oguntimehin, I.; Nakatani, N.; Sukugawa, H. Phytotoxicities of fluoranthene and phenanthrene deposited on needle surfaces of the evergreen conifer, Japanese red pine (*Pinus densiflora* Sieb. et Zucc.). Environ. Pollut., 2008,154, 264–271.

Oguntimehin, I.; Eissa, F.; Sakugawa, H. Negative effects of fluoranthene on the eco-physiology of tomato plants (*Lycopersicon esculentum Mill*). Chemosphere, 2010, 78, 877–884.

Rentz, J.A.; Alvarez, P.J.J.; Schnoor, J.L. Repression of *Pseudomonas putida* phenanthrene-degrading activity by plant root extracts and exudates. Environ. Microbiol., 2004, 6, 574–583.

Sverdrup, L.E.; Krogh, P.H.; Nielsen, T.; Kjaer, C.; Stenersen, J. Toxicity of eight polycyclic aromatic compounds to red clover (*Trifolium pratense*), ryegrass (*Lolium perenne*) and mustard (*Sinapsis alba*). Chemosphere,2003, 53, 993–1003.

Tomar, R.S.; Jajoo, A. A quick investigation of the detrimental effects of environmental pollutant polycyclic aromatic hydrocarbon fluoranthene on the photosynthetic efficiency of wheat (*Triticum aestivum*). Ecotoxicology, 2013a, DOI 10.1007/s10646-013-1118-1.

Tomar, R.S.; Jajoo, A. Alteration in PSII heterogeneity under the influence of polycyclic aromatic hydrocarbon (fluoranthene) in wheat leaves (*Triticum aestivum*). Plant Sci., 2013b, 209, 58–63

Tomar, R.S.; Jajoo, A. Fluranthene, a polycyclic aromatic hydrocarbon, inhibits light as well as dark reactions of photosynthesis in wheat (*Triticum aestivum*), Ecotoxico. Environ. Safety, 2014, 109,110–115.

Tomar, R.S.; Jajoo, A. Photomodified fluranthene exerts more harmful effects as compared to intact fluoranthene by inhibiting growth and photosynthetic processes in wheat, Ecotoxico. Environ. Safety, 2015, 122, 31–36.

Tomar, R.S.; Sharma, A.; Jajoo, A. Assessment of phytotoxicity of anthracene in soybean (*Glycine max*) with a quick method of chlorophyll fluorescence. Plant Biol., 2015, 17, 870–876.

Tongra, T.; Mehta, P.; Mathur, S.; Agrawal, D.; Bharti, S.; Los, D.A.; Allakhverdiev, S.I.; Jajoo, A. Computational analysis of fluorescence induction curves in intact spinach leaves treated at different pH. Biosystems, 2011,103, 158–163

Tukaj, Z.; Aksmann A. Toxic effect of anthraquinone and phenanthrene quinone upon *Scenedesmus* strains (green algae) at low and elevated concentration of CO_2. Chemosphere,2007, 66, 480–487.

Upham, B.L.; Jahnke, L.S. Photooxidative reactions in chloroplast thylakoids. Evidence for a Fenton-type reaction by superoxide or ascorbate. Photosynth. Res., 1986, 8, 235–247.

Vácha, R.; Čechmánková, J.; Skála, J. Polycyclic aromatic hydrocarbons in soil and selected plants. Plant Soil Environ., 2010, 56, 434–443.

Weisman, D.; Alkio, M.; Colón-Carmona, A. Transcriptional responses to polycyclic aromatic hydrocarbon-induced stress in *Arabidopsis thaliana*. BMC Plant Biol., 2010, 10, 59–71.

Wilson, S.C.; Jones, K.C. Bioremediation of soil contaminated with polynuclear aromatic hydrocarbons (PAHs): A review. Environ. Pollut., 1993, 81, 229–249.

Yoshitomi, K.J.; Shann, J.R. Corn (*Zea mays L*) root exudates and their impact on 14C-Pyrene minerdization. Soil Biol. Biochem., 2001, 33, 1769–1776.

Chlorophyll Fluorescence for High-Throughput Screening of Plants During Abiotic Stress, Aging, and Genetic Perturbation

Krishna Nath, James P. O'Donnell, and Yan Lu

Summary

Chlorophyll (Chl) is nature's gift to oxygenic photosynthetic organisms which capture solar radiation and convert it into chemical energy to drive the whole process of photosynthesis for proper growth and development of plants. Understanding the responses of photosynthetic apparatus in crop plants under various stress conditions has become a major target for many research programs. In this chapter, we describe the principal of Chl fluorescence and the recent advances in the application of Chl fluorescence. Chl fluorescence measurement is one of the most useful, cost-effective, and non-invasive tools to measure efficiency of photosystem II photochemistry. Incorporated with improved imaging and computer technologies, it can be utilized on a small or large scale for examination of photosynthetic performance, stress tolerance, and aging. Further advancements are being made to develop efficient more tools to apply Chl fluorescence measurement for large-scale high-throughput photosynthesis phenotyping, forestry and crop management.

Keywords

Chlorophyll fluorescence • High-throughput screening • Photosynthesis • Abiotic stress • Aging • Large-scale phenomics studies • Forestry and crop management

Contents

12.1	Introduction	261
12.2	The Basic Principle of Chl Fluorescence	262
12.3	Chlorophyll Fluorescence as a Tool to Study Abiotic Stresses	263
12.4	Chlorophyll Fluorescence as an Approach to Investigate the Aging Process	266
12.5	Chlorophyll Fluorescence as a Biomarker in Large-Scale Phenomics Studies	268
12.6	The Application of Chl Fluorescence in Forestry and Crop Management	269
12.7	Concluding Remarks	269
References		270

12.1 Introduction

Plants provide food, fiber, shelters, and fuel to humans and other animals. In this regard, photosynthesis is directly related to the existence of life on Earth. Photosynthesis involves a series of oxidation-reduction reactions in which solar radiation is captured and converted to chemical energy by plants and some other organisms (Arnon 1959;

K. Nath (✉) • J.P. O'Donnell • Y. Lu
Department of Biological Sciences, Western Michigan University, Kalamazoo, MI 49008, USA
e-mail: Krishnanath@gmail.com

Björkman and Demmig-Adams 1995; Eberhard et al. 2008). The net process of photosynthesis is the temporary capture of light energy in the chemical bonds of ATP and NADPH during the light reactions and the permanent conversion of captured energy in long-term energy storage molecules such as glucose during the dark reactions (Allen 1975; Whitmarsh 1999; Baker 2008; Eberhard et al. 2008; Araújo et al. 2014). Finally, this stored energy is utilized in cellular metabolism for growth and development of all living organisms, because almost all non-photosynthetic organisms consume energy from photosynthetic organisms (Whitmarsh 1999). Therefore, photosynthesis could be considered the ultimate source of cellular energy.

When a photon strikes a chlorophyll (Chl) molecule, the Chl molecule becomes excited by absorbing energy in the form of visible light (photons) (Müller et al. 2001). In leaves, photosynthetic light absorption is carried out by Chl-binding light harvesting complexes (LHCs) that are associated with Photosystem I (PSI) and Photosystem II (PSII) (Müller et al. 2001; Björn et al. 2009). Light absorbed by Chl molecules is further transferred from one pigment molecule to another within the LHC and eventually to a reaction center (RC) (Huner et al. 1998). Some of the light energy absorbed by Chl molecules can be re-emitted as fluorescence; this process is called Chl fluorescence (Flexas et al. 2000, 2002, Maxwell 2000; Freedman et al. 2002; Zarco-Tejada et al. 2003; Soukupová et al. 2008; Rascher et al. 2010; Porcar-Castell 2011; Garbulsky et al. 2013). At ambient temperature, PSII is the primary contributor to Chl fluorescence (Müller et al. 2001). Therefore, Chl fluorescence is an important indicator for the activity of PSII (Baker 2008; Li et al. 2010; Stirbet and Govindjee 2011; Zivcak et al. 2014). The measurement of Chl fluorescence is simple and straightforward. Among different photosynthetic complexes, PSII is highly sensitive to environmental stresses, such as drought, heat, high light, and aging (Saibo et al. 2009; Sperdouli and Moustakas 2011; Nath et al. 2013a, c). For these reasons, Chl fluorescence is considered as an efficient and non-invasive technique to monitor the responses of PSII to environmental, developmental, and genetic perturbations (Baker 2008; Finkel 2009; Furbank and Tester 2011; Lu et al. 2011b; Araus and Cairns 2014). In this chapter, we summarize the principle of Chl fluorescence and recent advancements in its application to different aspects of plant science including; the responses of PSII to abiotic stresses, aging, and large-scale phenotypic screening of genetic variants.

12.2 The Basic Principle of Chl Fluorescence

Chl molecules are present in the photosynthetic antenna in thylakoid membranes where absorbed light energy is used to perform photosynthesis (Govindjee 1995; Maxwell 2000). Absorption of light results in the excitation of Chl from its ground state (^{1}Chl) to its singlet excited state (^{1}Chl*), which can be returned to the original ground state, ^{1}Chl, through several internal decay pathways (Fig. 12.1). The excitation energy of ^{1}Chl* is then transferred to reaction centers to drive photosynthetic reactions. The remaining energy that is not used by photosynthesis can be re-emitted as fluorescence or released by thermal dissipation processes. Because the yield of fluorescence is the reciprocal of thermal dissipation and photochemistry processes, the release of heat is called non-photochemical quenching (NPQ) and the energy emitted as fluorescence is known as photochemical quenching (qP). However, ^{1}Chl* may also decay via triplet state of Chl (^{3}Chl*) and produce singlet oxygen (^{1}O$_2^*$), a cause of photo-oxidative damage to plants tissues and photosynthetic complexes, particularly PSII (Havaux and Niyogi 1999; Jahns and Holzwarth 2012). The excessive light energy dissipated by qP and NPQ can be considered as two important protective mechanisms that compete with *each other* and minimize the harmful process which produces highly reactive ^{1}O$_2^*$.

In addition to understanding the role of NPQ as a protective mechanism under adverse environmental conditions, it is important to understand how plants maintain their photosynthetic activity under extreme environments. To monitor photosynthetic performance in algae and higher

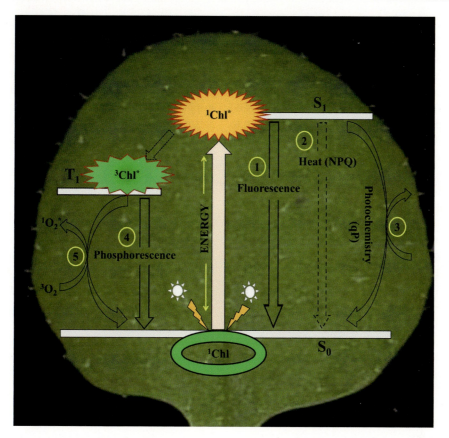

Fig. 12.1 Model for the possible fates of light energy in the de-activation pathway of excited $^1Chl^*$. Light energy absorbed by Chl molecules associated with PSII can be utilized in the subsequent process of photosynthesis. When a leaf is exposed to light, photons absorbed by Chl molecules may cause excitation of an electron (e^-) in Chl molecules. At the same time, the ^1Chl molecule is excited into $^1Chl^*$ from ground state (S_0) to excited state (S_1). From there, $^1Chl^*$ has several routes to get back to the ground state (^1Chl) such as, fluorescence (①), thermal dissipation processes (NPQ, ②), and photochemical processes (qP, ③). Chl fluorescence, heat dissipation, and photochemistry are in direct competition for the excitation energy. In addition to the three above mentioned fates, $^1Chl^*$ can be transformed into $^3Chl^*$ and return to its ground state by phosphorescence (④). $^3Chl^*$ has a relatively long life-time and it can produce highly reactive oxygen species $^1O_2^*$, which may cause photo-oxidative damages (⑤). The yield of $^3Chl^*$ and fluorescence depends on the yield of qP and NPQ, and thus they may help minimize the production of $^1O_2^*$ during excessive light illumination

plants, several photosynthetic parameters, such as photochemical efficiency of PSII, electron transport, excitation pressure, and performance index (PI_{ABS}) can be measured with conventional or advanced Chl fluorescence imaging systems (Maxwell 2000; Baker 2008; Stirbet and Govindjee 2011; Zivcak et al. 2014). Among these instruments, conventional pulse-amplitude-modulated (PAM) and imaging-PAM (I-PAM) are most widely used (Baker 2008). Chl fluorescence can be utilized in different research areas such as photosynthesis, phenomics, stress biology, developmental biology, and aging biology. Additional applications of Chl fluorescence include remote sensing in forestry, agriculture, and crop biology.

12.3 Chlorophyll Fluorescence as a Tool to Study Abiotic Stresses

The responses of photosynthetic apparatus to abiotic stress conditions are complex and involve the interplay of stress responses occurring at different sites of the photosynthetic machinery (Eberhard et al. 2008; Takahashi and Murata 2008; Kreslavski et al. 2013). Changes in the environmental conditions can result in imbalance between energy supplied by photochemistry and energy consumed by metabolism. In order to survive under such fluctuating environmental conditions, photosynthetic organisms have to balance between photochemistry and carbon metabolism (Huner et al. 1998). Algae and higher plants have developed the ability to tolerate a series of abiotic stresses such as low or high light, low or high temperature, flood or drought, the presence of heavy metals or mineral toxicity, excess of salts, deficiency of oxygen (hypoxia) in the soil, ultraviolet radiation, and oxidative stress to allow survival in adverse conditions (Vranova et al. 2002; Foyer and Noctor 2005; Mittler 2006; Gill and Tuteja 2010; Nath et al. 2013a). Photosynthetic efficiency of crop plants is known to be tightly associated with stress tolerance (Mittler 2006; Rascher et al. 2010). Abiotic stresses have adverse effects on plant growth, for example, reduction in biomass and crop yield (Furbank and Tester 2011). Among these stress conditions, high light and high temperature are most critical, because they directly influence the photosynthetic efficiency and the repair cycle of PSII (Sharkey 2005; Takahashi and Murata 2008; Tikkanen et al. 2008; Nath et al. 2013a; Kreslavski et al. 2013). In response to abiotic stresses, plants have developed mechanisms that lead to suppression of many physiological functions (Apel and Hirt 2004; Mittler 2006; Nath et al. 2013a). To understand how plants cope with severe environmental conditions, versatile and efficient techniques are needed to monitor photosynthetic performance. Chl fluorescence was developed as a tool to study photosynthesis. Several methods are available to detect signals from stressed leaf samples, for example, microarray analysis of stress-inducible genes, proteomics analysis, ultra-thin transmission electron microscopy of chloroplasts, and high performance liquid chromatography analysis of pigments. However, these approaches generally require expensive instrumentation and long sample preparation and measurement time (Cortese et al. 2009; Miller and Tang 2009; Slonim et al. 2009). In contrast to these methods, Chl fluorescence analysis is very sensitive to changes in thylakoid membranes or perturbations in plant cells in given environmental conditions and it does not involve sample preparation procedure such as homogenization of leaf tissues (Baker et al. 2007; Baker 2008; Li et al. 2010). Therefore, Chl fluorescence has been widely used as a non-invasive tool to monitor photosynthetic performance (Baker 2008). At the onset of an abiotic stress condition, a readily visible physiological stress response is often absent (Chaerle et al. 2007; Mishra et al. 2011; Murchie and Lawson 2013; Rousseau et al. 2013). However, Chl fluorescence can detect the initial stress response of a plant before the morphological stress response becomes visible. Therefore, Chl fluorescence is an efficient tool to examine the responses of photosynthetic organisms to environmental stresses. Recent advances in computerized digital imaging systems make it more convenient to perform Chl fluorescence measurements, allowing collection of a massive amount of data in a very short time (Baker 2008; Eberhard et al. 2008; Rousseau et al. 2013). Modern instruments for Chl fluorescence imaging are equipped with high-resolution cameras and a variety of light sources (Murchie and Lawson 2013). Coupling of such advanced instrumentation systems with other measurement techniques and software is important for Chl fluorescence measurements in given environmental conditions. For example, Chl fluorescence imaging systems equipped with large light-emitting diodes are perfect for large-scale phenomics studies, but they are not suitable in a field condition due to the relatively big size. In contrast, Chl fluorescence imaging systems equipped with standard fiber optic systems are compact and portable. Therefore, they are suitable for field measurements.

Chl fluorescence can be used to detect short-term effects of abiotic stresses. For example, we treated cherry (*Prunus avium* L) leaves at 50 °C for 5 min and measured Chl fluorescence before and after the treatment (Fig. 12.2a). Leaf tips

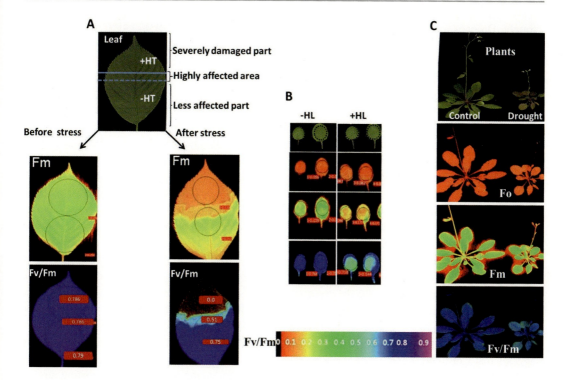

Fig. 12.2 Analysis of Chl fluorescence under short-term and long-term abiotic stresses. (**a**) Chl fluorescence of a control cherry leaf before and after a short-term (5-min) high-temperature treatment at 50 °C. +HT, high-temperature-treated; −HT, non-treated. (**b**) Chl fluorescence of *A. thaliana* leaves before and after a short-term (5-min) high-light treatment at 4000 μmol photons m^{-2} s^{-1}. +HL, high-light-treated; −HL, non-treated. (**c**) Chl fluorescence of a control *A. thaliana* plant and a plant under continuous drought stress for 7 days. *Black circles* in (**a**) and (**b**) represent areas of interest selected for generating numeric values for Chl fluorescence parameters Fo, Fm, and Fv/Fm. All Chl parameters were measured with the M-series I-PAM system (Heinz Walz GmbH, Effeltrich, Germany)

treated with high temperature (+HT) did not show visible phenotypic signs when compared to the non-treated basal portion of the same leaf (−HT), but dramatic changes were observed in maximum Chl fluorescence (Fm) and maximum photochemical efficiency of PSII (Fv/Fm) (Fig. 12.2a right). As shown in Fig. 12.2a, before the 5-min high temperature treatment, there was no detectable difference in Fv/Fm between the leaf tip and the basal portion of the same leaf. Substantial changes in Fv/Fm were seen after the treatment: the Fv/Fm ratio was ~0.5 at the junction of the treated and untreated portion of the leaf (between the full blue line and the dotted blue line); the Fv/Fm ratio at the untreated base of the leaf was ~0.76. There was a dramatic change in the Fv/Fm ratio after the 5-min high temperature treatment to the leaf tip. These results demonstrate that Chl fluorescence can be easily utilized to screen for high-temperature-sensitive plants.

To investigate the effect of high light stress on *Arabidopsis thaliana* leaves, we treated the 3rd and 4th leaves of *A. thaliana* plants under 4000 μmol photons m^{-2} s^{-1} for 5 min and measured Chl fluorescence before and after the treatment (Fig. 12.2b). As shown in Fig. 12.2b, Fm and Fv/Fm were significantly reduced in high-light-treated areas as compared to untreated leaves. These data suggest that Chl fluorescence can be used to screen for high-light-sensitive plants.

Chl fluorescence can also be used to monitor long-term effects of abiotic stresses, such as drought. Drought stress is and will continue to be one of the major limitations placed on plant growth and crop yield (Khush 2001; Sperdouli and Moustakas 2011; Ashraf and Harris 2013). Therefore, it is important to screen for drought-tolerant plants using efficient methods, such as Chl fluorescence. To test whether Chl fluorescence can be employed to monitor plant responses to drought, we simulated long-term drought stress by blocking the regular watering cycle, and monitored the responses of plants along with the changing Chl fluorescence parameters (Fig. 12.2c). This experiment allowed us to identify specific Chl fluorescence parameters as the indicators of drought stress during long-term drought conditions. Interestingly, false-color images of Chl fluorescence showed that a number of Chl parameters, Fv/Fm in particular, were dramatically altered in plants under drought stress (Fig. 12.2c). This suggests that Chl fluorescence parameter Fv/Fm can be utilized as a biological marker to screen drought-tolerant plants.

12.4 Chlorophyll Fluorescence as an Approach to Investigate the Aging Process

Plants often change their visible phenotypes and photosynthetic characteristics throughout their developmental stages. As the primary organ for photosynthesis, leaves go through a series of developmental, metabolic, and physiological transitions throughout their lifespan, which culminates in senescence and eventually cell death (Pennell and Lamb 1997; Woo et al. 2013). Leaf senescence is a developmental process that is controlled by an array of internal and external factors such as age of leaf, levels of plant growth hormones, transcription factors, environmental conditions, and pathogen infection (Guo and Gan 2005; Lim et al. 2007; Park et al. 2007; Breeze et al. 2011; Woo et al. 2013).

One visible symptom of leaf senescence is color change. Changes in leaf color could be caused by preferential degradation of Chl or accumulation of other pigments during senescence (Matile et al. 1992). The gradual loss of Chl is normally accompanied by a decrease in photosynthetic activity during senescence (Park et al. 2007; Sakuraba et al. 2012; Nath et al. 2013b). During leaf senescence, most macromolecules such as lipids, nucleic acids, and proteins are degraded, which results in sharp decreases in photosynthetic activity (Oh 2003; Kusaba et al. 2007, 2013; Hörtensteiner 2009; Nath et al. 2013b). It is well known that leaf senescence is initiated by chloroplast degeneration (Gepstein 2004), which leads to dramatic imbalances in photostasis, redox status, and sugar levels (Mohapatra et al. 2013; Woo et al. 2013). However, the relocation of nutrients from senesced organs to young organs is not well understood and there is little data available regarding this process (Lim et al. 2007; Woo et al. 2013). It is believed that nutrients released by catabolic activities during senescence are translocated from source to sink, *i.e.*, from old organs to young and actively growing organs such as new buds, young leaves, and developing seeds and fruits (Guo and Gan 2005; Yang and Ohlrogge 2008; Breeze et al. 2011; Pottier et al. 2014). Successful nutrient translocation requires crosstalk between photosynthetic functions and plant development.

The aging of plants, especially leaf senescence, is one of the major limitations of crop yields (Kusaba et al. 2013; Woo et al. 2013). One potential approach to increase crop yield is delaying senescence thereby increasing the duration of active photosynthesis (Dohleman and Long 2009). Significant progress have been made in the attempts to extend the duration of active photosynthesis in crop plants via genetic engineering of Chl biosynthetic pathways (Buchanan-Wollaston 1997; Woo et al. 2001, 2004; Park et al. 2007; Balazadeh et al. 2007; Rauf et al. 2013). In general, there are two types of mutations that could result in delayed senescence and extended life in plants. One of them is "functional stay-green type senescence mutants", where leaf senescence is initiated as scheduled but proceeds slowly. These mutants stay green longer and retain their photosynthetic activity at higher levels than wild-type plants (Thomas and

Howarth 2000; Thomas et al. 2002). Alterations in physiological, molecular and signaling pathways, such as slower metabolism, suppression of ethylene, abscisic acid, brassinosteroid, and strigolactone signal transduction, and activation of cytokinin signaling can result in stay-green phenotypes and prolong the lifespan of the plants (Woo et al. 2001; Oh 2003; Kim et al. 2009; Kusaba et al. 2013). More than a dozen *Oresara* mutants (*Oresara*, a Korean word for "long-lived", hereinafter referred to as Ore, have been identified (Oh et al. 1997; Woo et al. 2001, 2004; Oh 2003; Lim et al. 2007). For example, an ORE1 transcription factor has been found to control the senescence process by regulating the expression of downstream genes including senescence genes (Kim et al. 2009; Rauf et al. 2013). ORE9 was identified as an F-box motif containing protein that limits leaf longevity by removing target proteins which are essential for delaying leaf senescence (Oh 2003). ORE12 was found to encode a cytokinin receptor and a gain-of-function mutant of ORE12 demonstrated prolonged lifespan (Kim et al. 2006). The above mentioned mutants showed delayed senescence because they retain their greenness and maximum photochemical efficiency of PSII longer than wild-type plants.

The second type of mutants with delayed senescence is "non-functional", where mutant plants retain green color due to the aggregation of Chl-containing LHCs. These mutants are not functionally active and their photosynthetic activity becomes impaired much earlier than that in wild-type plants (Oh 2003). This type of mutant is usually caused by defects in the degradation pathway of Chl (Thomas and Howarth 2000; Thomas et al. 2002), and is very useful to study the disassembly process of Chl-binding protein complexes in PSI and PSII (Oh 2003). One representative "non-functional stay-green type mutant" is *ore10*. Deletion of the *ORE10* gene causes the aggregation of Chl-containing LHCs during senescence. In contrast to functional stay-green type senescence mutants, such as *ore1*, *ore9* and *ore12*, non-functional stay-green type mutants, such as *ore10* fail to maintain their functionality, for example their maximum photochemical efficiency of PSII. Their Fv/Fm value decreases much earlier than that in wild-types but the mutants appear green during senescence. Therefore, they are called "earlier than wild-type progenies". However, because they look green during senescence, they are also known as "non-functional mutants" (Oh 2003).

The dramatic changes in photosynthetic activity and the gradual loss of Chl content during senescence can be monitored by Chl fluorescence measurements (Kusaba et al. 2007; Park et al. 2007; Nath et al. 2013b). For example, we determined the Chl content and monitored the Chl fluorescence parameters in 7–35 days old leaves (Fig. 12.3). Leaves reached their maximum photochemical efficiency of PSII (i.e., Fv/Fm) at maturity, coinciding with the highest amount of

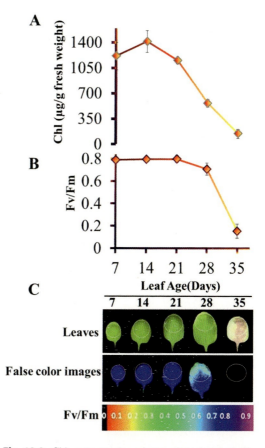

Fig. 12.3 Chl content and maximum photochemical efficiency (Fv/Fm) of PSII as functions of leaf age. Visible symptoms and the false color images of Fv/Fm of aging leaves are shown as well. Maximum photochemical efficiency of PSII (Fv/Fm) was measured with the M-series I-PAM system (Heinz Walz GmbH, Effeltrich, Germany)

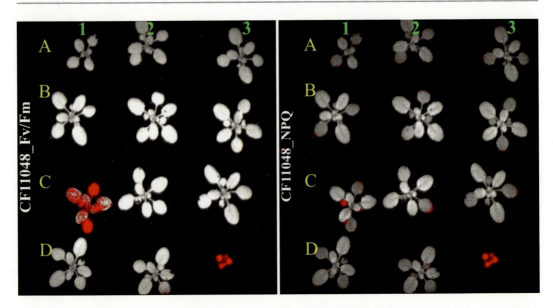

Fig. 12.4 Chl fluorescence assay to identify mutants with reduced photosynthetic parameters. False-colored images of maximum photochemical efficiency of PSII (Fv/Fm, *left*) and NPQ (*right*) are shown

Chl. This was followed by a gradual decline of the Chl content during senescence (Fig. 12.4). These data indicate that photosynthetic activity is tightly associated with Chl contents.

12.5 Chlorophyll Fluorescence as a Biomarker in Large-Scale Phenomics Studies

In modern agricultural research, it is important to identify desirable traits and apply them into crop plants. The search for desirable traits among large populations of genetic variants could benefit from rapid and high-throughput screening methods (Oxborough and Baker 1997; Baker et al. 2007; Baker 2008; Araus and Cairns 2014). For example, the collaborative Chloroplast Phenomics project (http://www.plastid.msu.edu) at Michigan State University and Western Michigan University has successfully utilized Chl fluorescence kinetics measurement to screen ~5200 *A. thaliana* insertional mutants for changes in photosynthetic parameters Fv/Fm and NPQ (Lu et al. 2011a). The use of 3-week-old *A. thaliana* plants in small pots allows simultaneous measurement of Chl fluorescence parameters in 12 plants in the same sub-flat (Fig. 12.4). The *npq1* mutant, in which the biosynthetic pathway for zeaxanthin is blocked, was used as a positive control. The identification of positive controls such as the *npq1* mutant in a double-blind and high-throughput method demonstrated the power of Chl fluorescence in large-scale phenomics studies. In addition to the confirmation of known photosynthetic mutants, the Chloroplast Phenomics project also identified >90 novel genes with potential roles in photosynthesis (http://bioinfo.bch.msu.edu/2010_LIMS). Select mutants are subject to detailed physiological and biochemical studies to understand the molecular functions of the corresponding genes. Although there are increasing interests in high-throughput phenotyping platforms (Araus and Cairns 2014), large-scale data acquisition and process continues to be a daunting task. The development of a laboratory information management system (LIMS) and a standardized laboratory operational workflow ensured the accuracy and ease of large-scale data acquisition (Lu et al. 2011b). In addition to the Chloroplast Phenomics Project, a number of other plant phenomics facilities, for example, the Australian Plant Phenomics Facility (http://

www.plantphenomics.org.au/) and the Michigan State University Center for Advanced Algal and Plant Phenotyping (http://www.prl.msu.edu/caapp), have emerged recently.

Although it is fast and straightforward to generate Chl fluorescence images, the extraction of numeric values for photosynthetic parameters from Chl fluorescence images is still a slow and tedious process. Because one has to manually select areas of interest from individual leaves and then export the fluorescence values for the areas of interest. Fortunately, the recently developed computer vision algorithms demonstrate the potential to speed up this process substantially (Tessmer et al. 2013). The current challenges of imaging systems and computer vision algorithms include: (a) the need to train established algorithms in the presence of emerging new leaves; (b) the necessity of implementing three-dimensional imaging systems and vision algorithms suitable for taller plants such as tobacco, tomato, and bean (Tessmer et al. 2013). Ultimately, the three-dimensional Chl fluorescence imaging technology could be applied to investigate the impacts of herbicides on a whole plant (Konishi et al. 2009).

12.6 The Application of Chl Fluorescence in Forestry and Crop Management

There is a growing interest in utilizing Chl fluorescence in remote sensing to understand the heterogeneity and dynamics of the global vegetation physiology (Gamon et al. 1990; Baret and Guyot 1991; Flexas et al. 2000; Malenovsky et al. 2009; Meroni et al. 2009; Rascher et al. 2009; Damm et al. 2010; Garbulsky et al. 2013; Zarco-Tejada et al. 2013). Remote sensing of plant canopy could be accomplished by using airborne multispectral sensors that detect both reflectance and fluorescence from vegetation canopies (Malenovsky et al. 2009). These remote sensing instruments have been employed to monitor the diurnal and seasonal changes in photosynthesis (Zarco-Tejada et al. 2003; Dobrowski et al. 2005; Malenovsky et al. 2009; Lausch et al. 2013). During Chl fluorescence measurements of vegetation canopies in field conditions, manipulation of actinic light intensity is technically challenging. Therefore, steady state Chl fluorescence is often utilized to track changes in photosynthetic activity of forest and field vegetation (Flexas et al. 2000, 2002; Freedman et al. 2002; Zarco-Tejada et al. 2003; Soukupová et al. 2008; Rascher et al. 2010; Porcar-Castell 2011; Garbulsky et al. 2013). The challenges of remote sensing include: (a) the need to extract simple parameters from complex steady-state Chl fluorescence signals; (b) the requirement of scaling up steady-state Chl fluorescence signals from individual leaves to heterogeneous canopies; (c) the necessity to increase the sensitivity of space-borne remote sensors; and (d) the need for accurate algorithms for downstream data processing (see Malenovsky et al. 2009 for a detailed review on this topic). Recently, Konishi et al. (2009) combined Chl fluorescence analysis of photosynthetic parameters with three-dimensional imaging and modeling technology to study spatiotemporal effects of herbicides on a whole melon (*Cucumis melo* L) plant (Omasa et al. 2009). This study illustrated the power of combined use of Chl fluorescence and other advanced technologies. The approach can also be used to investigate spatiotemporal effects of environmental, developmental, and genetic perturbations and spatiotemporal responses of photosynthetic apparatus to perturbations.

12.7 Concluding Remarks

Chl fluorescence analysis is an efficient and non-destructive method to determine photosynthetic parameters. When coupled with methods that lead to changes in the environment developmental stages, and/or genetic backgrounds, Chl fluorescence can be used to monitor the responses of photosynthetic apparatus to environmental, developmental, and/or genetic perturbations. In this chapter, we discussed the applications and challenges of Chl fluorescence in plant abiotic stress responses, aging, large-scale phenomics studies of genetic variants, and forestry and crop management. Overall, it is quick and easy to generate data with the Chl fluorescence method. Further improvements in instrument sensitivity and downstream data processing algorithms will make Chl fluorescence even more useful, especially in large-scale phenomics studies, remote sensing, forestry and crop management.

Acknowledgments This work was supported by U.S. National Science Foundation Grant MCB-1244008.

References

Allen JF (1975) Oxygen reduction and optimum production of ATP in photosynthesis. Nature 256:599–600. doi: 10.1038/256599a0

Apel K, Hirt H (2004) Reactive oxygen species: metabolism, oxidative stress, and signal transduction. Annu Rev Plant Biol 55:373–399.

Araújo WL, Nunes-Nesi A, Fernie AR (2014) On the role of plant mitochondrial metabolism and its impact on photosynthesis in both optimal and sub-optimal growth conditions. Photosynth Res 119:141–156.

Araus JL, Cairns JE (2014) Field high-throughput phenotyping: the new crop breeding frontier. Trends Plant Sci 19:52–61. doi: 10.1016/j.tplants.2013.09.008

Arnon DI (1959) Conversion of light into chemical energy in photosynthesis. Nature 184:10–20.

Ashraf M, Harris PJC (2013) Photosynthesis under stressful environments: An overview. Photosynthetica 51:163–190.

Baker NR (2008) Chlorophyll Fluorescence: A Probe of Photosynthesis In Vivo. Annu Rev Plant Biol 59:89–113. doi: 10.1146/annurev.arplant.59.032607.092759

Baker NR, Harbinson J, Kramer DM (2007) Determining the limitations and regulation of photosynthetic energy transduction in leaves. Plant, cell & Environ 30:1107–1125.

Balazadeh S, Riaño-Pachón DM, Mueller-Roeber B (2008) Transcription factors regulating leaf senescence In *Arabidopsis thaliana*. Plant Biol 10:63–75. doi: 10.1111/plb.2008.10.issue-s1

Baret F, Guyot G (1991) Potentials and limits of vegetation indices for LAI and APAR assessment. Remote Sens Environ 35:161–173.

Björkman O, Demmig-Adams B (1995) Regulation of photosynthetic light energy capture, conversion, and dissipation in leaves of higher plants. In: Ecophysiology of photosynthesis. Springer, pp 17–47

Björn LO, Papageorgiou GC, Blankenship RE, Govindjee (2009) A viewpoint: Why chlorophyll a? Photosynth Res 99:85–98. doi: 10.1007/s11120-008-9395-x

Breeze E, Harrison E, McHattie S, et al (2011) High-Resolution Temporal Profiling of Transcripts during Arabidopsis Leaf Senescence Reveals a Distinct Chronology of Processes and Regulation. Plant Cell 23:873–894. doi: 10.1105/tpc.111.083345

Buchanan-Wollaston V (1997) The molecular biology of leaf senescence. J Exp Bot 48:181–199.

Chaerle L, Leinonen I, Jones HG, Van Der Straeten D (2007) Monitoring and screening plant populations with combined thermal and chlorophyll fluorescence imaging. J Exp Bot 58:773–784.

Cortese K, Diaspro A, Tacchetti C (2009) Advanced Correlative Light/Electron Microscopy: Current Methods and New Developments Using Tokuyasu Cryosections. J Histochem Cytochem 57:1103–1112. doi: 10.1369/jhc.2009.954214

Damm A, Elbers JAN, Erler A, et al (2010) Remote sensing of sun-induced fluorescence to improve modeling of diurnal courses of gross primary production (GPP). Glob Chang Biol 16:171–186.

Dobrowski SZ, Pushnik JC, Zarco-Tejada PJ, Ustin SL (2005) Simple reflectance indices track heat and water stress-induced changes in steady-state chlorophyll fluorescence at the canopy scale. Remote Sens Environ 97:403–414.

Dohleman FG, Long SP (2009) More productive than maize in the Midwest: how does Miscanthus do it? Plant Physiol 150:2104–2115.

Eberhard S, Finazzi G, Wollman F-A (2008) The Dynamics of Photosynthesis. Annu Rev Genet 42:463–515. doi: 10.1146/annurev.genet.42.110807.091452

Finkel E (2009) With "Phenomics," Plant Scientists Hope to Shift Breeding Into Overdrive. Science 325:380–381.

Flexas J, Briantais J-M, Cerovic Z, et al (2000) Steady-state and maximum chlorophyll fluorescence responses to water stress in grapevine leaves: a new remote sensing system. Remote Sens Environ 73:283–297.

Flexas J, Escalona JM, Evain S, et al (2002) Steady-state chlorophyll fluorescence (Fs) measurements as a tool to follow variations of net CO2 assimilation and stomatal conductance during water-stress in C3 plants. Physiol Plant 114:231–240.

Foyer CH, Noctor G (2005) Oxidant and antioxidant signalling in plants: a re-evaluation of the concept of oxidative stress in a physiological context. Plant, Cell & Environ 28:1056–1071.

Freedman A, Cavender-Bares J, Kebabian PL, et al (2002) Remote sensing of solar-excited plant fluorescence as a measure of photosynthetic rate. Photosynthetica 40:127–132.

Furbank RT, Tester M (2011) Phenomics – technologies to relieve the phenotyping bottleneck. Trends Plant Sci. 16:635–644.

Gamon JA, Field CB, Bilger W, et al (1990) Remote sensing of the xanthophyll cycle and chlorophyll fluorescence in sunflower leaves and canopies. Oecologia 85:1–7.

Garbulsky MF, Filella I, Verger A, Peñuelas J (2013) Photosynthetic light use efficiency from satellite sensors: From global to Mediterranean vegetation.

Gepstein S (2004) Leaf senescence-not just awear and tear'phenomenon. Genome Biol 5:212–212.

Gill SS, Tuteja N (2010) Reactive oxygen species and antioxidant machinery in abiotic stress tolerance in crop plants. Plant Physiol Biochem 48:909–930.

Govindjee (1995) 63 Years since Kautsky Chlorphyll -a fluorescence. Aust J Plant Physiol 131–160.

Guo Y, Gan S (2005) Leaf Senescence: Signals, Execution, and Regulation. Elsevier

Havaux M, Niyogi KK (1999) The violaxanthin cycle protects plants from photooxidative damage by more than one mechanism. Proc Natl Acad Sci 8762–8767.

Hörtensteiner S (2009) Stay-green regulates chlorophyll and chlorophyll-binding protein degradation during senescence. Trends Plant Sci 14:155–162. doi: 10.1016/j.tplants.2009.01.002

Huner NPA, Öquist G, Sarhan F (1998) Energy balance and acclimation to light and cold. Trends Plant Sci. 3:224–230.

Jahns P, Holzwarth AR (2012) The role of the xanthophyll cycle and of lutein in photoprotection of photosystem II. Biochim et Biophys Acta 1817:182–193. doi: 10.1016/j.bbabio.2011.04.012

Khush GS (2001) Green revolution: the way forward. Nat Rev Genet 2:815–822.

Kim HJ, Ryu H, Hong SH, et al (2006) Cytokinin-mediated control of leaf longevity by AHK3 through phosphorylation of ARR2 in Arabidopsis. Proc Natl Acad Sci United States Am 103:814–819.

Kim JH, Woo HR, Kim J, et al (2009) Trifurcate feed-forward regulation of age-dependent cell death involving miR164 in Arabidopsis. Science 323:1053–1057.

Konishi A, Eguchi A, Hosoi F, Omasa K (2009) 3D monitoring spatio–temporal effects of herbicide on a whole plant using combined range and chlorophyll a fluorescence imaging. Funct Plant Biol 36:874–879.

Kreslavski VD, Zorina AA, Los DA, Fomina IR, and Allakhverdiev SI (2013) Molecular Mechanisms of StressResistance of Photosynthetic Machinery. Mol. Stress Physiol. Plants 21–50.

Kusaba M, Ito H, Morita R, et al (2007) Rice NON-YELLOW COLORING1 Is Involved in Light-Harvesting Complex II and Grana Degradation during Leaf Senescence. Plant Cell 19:1362–1375. doi: 10.1105/tpc.106.042911

Kusaba M, Tanaka A, Tanaka R (2013) Stay-green plants: what do they tell us about the molecular mechanism of leaf senescence. Photosynth Res 117:221–234.

Lausch A, Pause M, Merbach I, et al (2013) A new multiscale approach for monitoring vegetation using remote sensing-based indicators in laboratory, field, and landscape. Environ Monit Assess 185:1215–1235.

Li J, Pandeya D, Nath K, et al (2010) ZEBRA-NECROSIS, a thylakoid-bound protein, is critical for the photoprotection of developing chloroplasts during early leaf development. Plant J 62:713–725.

Lim PO, Kim HJ, Nam HG (2007) Leaf senescence. Annu Rev Plant Biol 58:115–136. doi: 10.1146/annurev.arplant.57.032905.105316

Lu Y, Hall DA, Last RL (2011a) A small zinc finger thylakoid protein plays a role in maintenance of photosystem II in *Arabidopsis thaliana*. Plant Cell 1861–1875.

Lu Y, Savage LJ, Larson MD, et al (2011b) Chloroplast 2010: a database for large-scale phenotypic screening of Arabidopsis mutants. Plant Physiol 155:1589–1600.

Malenovsky Z, Mishra KB, Zemek F, et al (2009) Scientific and technical challenges in remote sensing of plant canopy reflectance and fluorescence. J Exp Bot 60:2987–3004.

Matile P, Schellenberg M, Peisker C (1992) Production and release of a chlorophyll catabolite in isolated senescent chloroplasts. Planta 187:230–235.

Maxwell K (2000) Chlorophyll fluorescence–a practical guide. J Exp Bot 51:659–668. doi: 10.1093/jexbot/51.345.659

Meroni M, Rossini M, Guanter L, et al (2009) Remote sensing of solar-induced chlorophyll fluorescence: Review of methods and applications. Remote Sens Environ 113:2037–2051.

Miller MB, Tang Y-W (2009) Basic concepts of microarrays and potential applications in clinical microbiology. Clin Microbiol Rev 22:611–633.

Mishra A, Mishra KB, Höermiller II, et al (2011) Chlorophyll fluorescence emission as a reporter on cold tolerance in *Arabidopsis thaliana* accessions.

Mittler R (2006) Abiotic stress, the field environment and stress combination. Trends Plant Sci 11:15–19.

Mohapatra PK, Joshi P, Ramaswamy NK, et al (2013) Damage of photosynthetic apparatus in the senescing basal leaf of *Arabidopsis thaliana*: A plausible mechanism of inactivation of reaction center II. Plant Physiol Biochem 62:116–121.

Müller P, Li X-P, Niyogi KK (2001) Non-photochemical quenching. A response to excess light energy. Plant Physiol 125:1558–1566.

Murchie EH, Lawson T (2013) Chlorophyll fluorescence analysis: a guide to good practice and understanding some new applications. J Exp Bot 64:3983–3998.

Nath K, Jajoo A, Poudyal RS, et al (2013a) Towards a critical understanding of the photosystem II repair mechanism and its regulation during stress conditions. FEBS Lett 587:3372–3381. doi: 10.1016/j.febslet.2013.09.015

Nath K, Phee B-K, Jeong S, et al (2013b) Age-dependent changes in the functions and compositions of photosynthetic complexes in the thylakoid membranes of *Arabidopsis thaliana*. Photosynth. Res. 117:547–556.

Nath K, Poudyal RS, Eom J-S, et al (2013c) Loss-of-function of OsSTN8 suppresses the photosystem II core protein phosphorylation and interferes with the photosystem II repair mechanism in rice (*Oryza sativa*). Plant J 76:675–686. doi: 10.1111/tpj.2013.76.issue-4

Oh M-H (2003) Increased Stability of LHCII by Aggregate Formation during Dark-Induced Leaf Senescence in the Arabidopsis Mutant, ore10. Plant Cell Physiol. 44:1368–1377.

Oh SA, Park J-H, Lee GI, et al (1997) Identification of three genetic loci controlling leaf senescence in *Arabidopsis thaliana*. Plant J 12:527–535.

Omasa K, Konishi A, Tamura H, Hosoi F (2009) 3D confocal laser scanning microscopy for the analysis of chlorophyll fluorescence parameters of chloroplasts in intact leaf tissues. Plant cell Physiol 50:90–105.

Oxborough K, Baker NR (1997) An instrument capable of imaging chlorophyll a fluorescence from intact leaves at very low irradiance and at cellular and subcellular levels of organization. Plant, Cell & Environ 20:1473–1483.

Park S-Y, Yu J-W, Park J-S, et al (2007) The Senescence-Induced Staygreen Protein Regulates Chlorophyll Degradation. Plant Cell 19:1649–1664. doi: 10.1105/tpc.106.044891

Pennell RI, Lamb C (1997) Programmed cell death in plants.

Porcar-Castell A (2011) A high-resolution portrait of the annual dynamics of photochemical and non-photochemical quenching in needles of *Pinus sylvestris*. Physiol Plant 143:139–153.

Pottier M, Masclaux-Daubresse C, Yoshimoto K, Thomine S (2014) Autophagy as a possible mechanism for micronutrient remobilization from leaves to seeds. Front Plant Sci. doi: 10.3389/fpls.2014.00011

Rascher U, Agati G, Alonso L, et al (2009) CEFLES2: the remote sensing component to quantify photosynthetic efficiency from the leaf to the region by measuring sun-induced fluorescence in the oxygen absorption bands.

Rascher U, Damm A, van der Linden S, et al (2010) Sensing of photosynthetic activity of crops. In: Precision Crop Protection-the Challenge and Use of Heterogeneity. Springer, pp 87–99

Rauf M, Arif M, Dortay H, et al (2013) ORE1 balances leaf senescence against maintenance by antagonizing G2-like-mediated transcription. EMBO reports – Issue 14:382–388. doi: 10.1038/embor.2013.24

Rousseau C, Belin E, Bove E, et al (2013) High throughput quantitative phenotyping of plant resistance using chlorophyll fluorescence image analysis. Plant Methods 9:17.

Saibo NJM, Lourenco T, Oliveira MM (2009) Transcription factors and regulation of photosynthetic and related metabolism under environmental stresses. Ann. Bot. 103:609–622.

Sakuraba Y, Schelbert S, Park S-Y, et al (2012) STAY-GREEN and Chlorophyll Catabolic Enzymes Interact at Light-Harvesting Complex II for Chlorophyll Detoxification during Leaf Senescence in Arabidopsis. Plant Cell 24:507–518. doi: 10.1105/tpc.111.089474

Sharkey TD (2005) Effects of moderate heat stress on photosynthesis: importance of thylakoid reactions, rubisco deactivation, reactive oxygen species, and thermotolerance provided by isoprene. Plant, Cell & Environ 28:269–277.

Slonim DK, Yanai I, Troyanskaya OG (2009) Getting Started in Gene Expression Microarray Analysis. PLoS Comput Biol 5:1000543. doi: 10.1371/journal.pcbi.1000543

Soukupová J, Cséfalvay L, Urban O, et al (2008) Annual variation of the steady-state chlorophyll fluorescence emission of evergreen plants in temperate zone. Funct Plant Biol 35:63–76.

Sperdouli I, Moustakas M (2011) Spatio-temporal heterogeneity in *Arabidopsis thaliana* leaves under drought stress. Plant. Biol. doi: 10.1111/j.1438-8677.2011.00473.x

Stirbet A, Govindjee (2011) On the relation between the Kautsky effect (chlorophyll a fluorescence induction) and Photosystem II: Basics and applications of the OJIP fluorescence transient. J Photochem Photobiol B: Biol 104:236–257. doi: 10.1016/j.jphotobiol.2010.12.010

Takahashi S, Murata N (2008) How do environmental stresses accelerate photoinhibition? Trends Plant Sci 13:178–182. doi: 10.1016/j.tplants.2008.01.005

Tessmer OL, Jiao Y, Cruz JA, et al (2013) Functional approach to high-throughput plant growth analysis. BMC Syst Biol 7:1–13.

Thomas H, Howarth CJ (2000) Five ways to stay green. J Exp Bot 51:329–337.

Thomas H, Ougham H, Canter P, Donnison I (2002) What stay-green mutants tell us about nitrogen remobilization in leaf senescence. J Exp Bot 53:801–808.

Tikkanen M, Nurmi M, Kangasjärvi S, Aro E-M (2008) Core protein phosphorylation facilitates the repair of photodamaged photosystem II at high light. Biochim et Biophys Acta 1777:1432–1437.

Vranova E, Inze D, Van Breusegem F (2002) Signal transduction during oxidative stress. J Exp Bot 53:1227–1236.

Whitmarsh J (1999) The photosynthetic process. In: Concepts in Photobiology. Springer, pp 11–51

Woo HR, Chung KM, Park J-H, et al (2001) ORE9, an F-box protein that regulates leaf senescence in Arabidopsis. Plant Cell 1779–1790.

Woo HR, Kim JH, Nam HG, Lim PO (2004) The delayed leaf senescence mutants of Arabidopsis, ore1, ore3, and ore9 are tolerant to oxidative stress. Plant Cell Physiol 45:923–932.

Woo HR, Kim HJ, Nam HG, Lim PO (2012) Plant leaf senescence and death – regulation by multiple layers of control and implications for aging in general. J cell Sci 126:4823–4833. doi: 10.1242/jcs.109116

Yang Z, Ohlrogge JB (2008) Turnover of fatty acids during natural senescence of Arabidopsis, Brachypodium, and switchgrass and in Arabidopsis ?-oxidation mutants. Plant Physiol 150:1981–1989.

Zarco-Tejada PJ, Pushnik JC, Dobrowski S, Ustin SL (2003) Steady-state chlorophyll a fluorescence detection from canopy derivative reflectance and double-peak red-edge effects. Remote Sens Environ 283–294.

Zarco-Tejada PJ, Morales A, Testi L, Villalobos FJ (2013) Spatio-temporal patterns of chlorophyll fluorescence and physiological and structural indices acquired from hyperspectral imagery as compared with carbon fluxes measured with eddy covariance. Remote Sens Environ 133:102–115.

Zivcak M, Brestic M, Kalaji HM, Govindjee (2014) Photosynthetic responses of sun- and shade-grown barley leaves to high light: is the lower PSII connectivity in shade leaves associated with protection against excess of light? Photosynth. Res. 119:339–354.

Adaptation to Low Temperature in a Photoautotrophic Antarctic Psychrophile, *Chlamydomonas sp.* UWO 241

Beth Szyszka, Alexander G. Ivanov, and Norman P.A. Hüner

Summary

Permanent cold environments account for a large portion of the Earth. These environments are inhabited by various micro-organisms that have often adapted to unique combinations of selection pressures. Therefore, there is considerable interest in understanding survival strategies utilized by extremophiles to exist in such harsh environments. This chapter summarizes common adaptive mechanisms of psychrophilic organisms with focus on the unique photosynthetic characteristics of a unicellular green microalga, *Chlamydomonas sp.* UWO241. *Chlamydomonas sp.* UWO 241 was isolated from perennially, ice-covered Lake Bonney, Antarctica where it has adapted to constant low temperatures and high salinity. A unique characteristic of this algal strain is its inability to undergo state transitions combined with its high rates of photosystem I cyclic electron transport. Consequently, in contrast to mesophilic green algal species such as *Chlamydomonas reinhardtii* which undergo state transitions, *Chlamydomonas sp.* UWO241 does not phosphorylate LHCII polypeptides. Rather, the Antarctic psychrophile exhibits a unique, light-dependent, thylakoid polypeptide phosphorylation profile associated with a photosystem I supercomplex which also contains the cytochrome b_6/f complex. The stability of this photosystem I supercomplex in *Chlamydomonas sp.* UWO 241 is sensitive to its phosphorylation status as well as high salt concentrations. The role of the photosystem I supercomplex and its phosphorylation status in the regulation of photosystem I cyclic electron transport is discussed. We suggest that *Chlamydomonas sp.* UWO 241 should be considered a model system to study psychrophily and adaptation to low temperature in eukaryotic photoautrophs.

Keywords

Cold adaptation • Cold acclimation • Photoautotrophs • *Chlamydomonas* • Psychrophile • Protein phosphorylation • PSI-Cyt b_6/f supercomplex

B. Szyszka • N.P.A. Hüner (✉)
Department of Biology and the Biotron Center for Experimental Climate Change Research, Western University, London, ON, Canada N6A 5B7
e-mail: nhuner@uwo.ca

A.G. Ivanov
Department of Biology and the Biotron Center for Experimental Climate Change Research, Western University, London, ON, Canada N6A 5B7

Institute of Biophysics and Biomedical Engineering, Bulgarian Academy of Sciences, Acad. G. Bonchev Street, Bl. 21, 1113 Sofia, Bulgaria

Contents

13.1	Introduction	276
13.2	Ultrastructure and Cell Morphology	277
13.3	Membrane Fatty Acid Composition	277
13.4	Cold-Adapted Enzymes	278
13.5	Energy Metabolism	280
13.6	Photosynthesis at Low Temperatures	281
13.7	Acclimation to Light and Low Temperatures	282
13.8	*Chlamydomonas* sp. UWO241	285
13.8.1	Natural Habitat of UWO241	285
13.8.2	Growth of UWO241	285
13.8.2.1	Temperature	285
13.8.2.2	Salt	286
13.9	Adaptation of UWO241 to Low Temperature	286
13.10	Photosynthetic Electron Transport	288
13.10.1	Structure of PETC	288
13.10.2	Function of PETC	288
13.10.3	Acclimation to Temperature and Irradiance	289
13.10.4	Acclimation to Light Quality	290
13.10.5	State Transitions	291
13.10.6	Thylakoid Polypeptide Phosphorylation Profile	291
13.10.7	PsbP-Like Proteins	292
13.11	A Model of Cyclic Electron Flow in *Chlamydomonas* sp. UWO241	293
13.12	Conclusions	295
References		296

13.1 Introduction

The majority of the Earth's biosphere is characterized by cold ecosystems, which for the most part, remain largely unexplored. These permanently cold environments consist of polar and alpine habitats, as well as the ocean depths, where 90% of the volume is below 5 °C (Feller and Gerday 2003). In addition, constant low temperatures of these habitats are frequently accompanied by additional abiotic stresses, including desiccation, osmotic stress, as well as immense fluctuations in salinity, irradiance, photoperiod and nutrient availability (Eicken 1992). Consequently, while plants and animals exhibit a rather limited diversity in such extreme environments, microorganisms contribute to the majority of the ecosystem biomass, including bacteria, yeasts, fungi, *Archaea*, protists and cyanobacteria (Feller and Gerday 2003). These microorganisms are often "polyextremophiles", which are able to tolerate several extreme conditions simultaneously. Despite the large, low-temperature coverage of the Earth, relatively little is known about the microorganisms that colonize these areas.

Cold-adapted microorganisms which have successfully inhabited low temperature regions are classified as being either "psychrophilic" or "psychrotrophic". Psychrophiles are defined as organisms having an optimal growth temperature below 15 °C but do not grow and survive above temperatures of 20 °C (Feller and Gerday 2003). Alternatively, psychrotrophic, or cold-tolerant organisms, do not share this cold requirement, and although they are capable of growth and survival temperatures below 15 °C, their optimal growth temperatures are generally above 18 °C. Thus, psychrophilic or "cold-loving" organisms are considered to be truly adapted to grow only at low temperatures, whereas psychrotolerant organisms are usually "eurythermal", meaning they are able to grow over a wide range of temperatures.

Organisms can respond to their specific low-temperature environments by either using short-term mechanisms of acclimation or long-term adaptation. Adaptation refers to the combined genetic traits that have been altered by natural selection/selective pressures, which are heritable and maintained over many generations. In contrast, acclimation describes short-term adjustments to phenotypic and physiological traits which occur within a single lifetime (Hüner et al. 1998). Survival in cold environments necessitates that cellular metabolic rates are balanced with growth and maintenance requirements, and therefore, the energy produced must be sufficient to fulfill all energy demands. Habitats characterized by decreased temperatures pose several major challenges to microbial metabolism, growth and survival, including decreased membrane fluidity, reduced biochemical activity and cold-denaturation of proteins. Therefore, psychrophilic microorganisms exhibit a complex array of mechanisms to compensate for the metabolic challenges associated with cold environments. Remarkably, psychrophiles have evolved to not only survive, but to thrive at temperatures as low as −20 °C (Thomas and Dieckmann 2002; D'Amico et al. 2006). Adaptive mechanisms employed by these organisms have been observed on many organizational levels which range from modifications in cell morphology, to membrane lipid and fatty acid composition to alterations in the structure and function of enzymes.

13.2 Ultrastructure and Cell Morphology

During short periods of cold acclimation, several temperate/mesophilic green algal species exhibit an increase in the number and individual size of both starch grains and oil droplets (Hatano et al. 1982; Nagao et al. 2008). In addition, short-term cold-acclimation induces an increase in the cytoplasmic volume of algal cells, with a concomitant decrease in the vacuole size (Nagao et al. 2008). These structural characteristics have also been observed among cold-adapted algae permanently inhabiting Antarctica (Pocock et al. 2004; Hu et al. 2008; Chen 2012; Chen et al. 2012). Another common trait of Antarctic algal species is that they exhibit thicker cell walls, and are often described to exist in multi-cellular structures, surrounded by a mucilaginous sheath (Pocock et al. 2004; Hu et al. 2008; Chen 2012). It is not clear whether these groups of cells represent temporarily aggregated, cohering cells or reproductive structures involved in sexual or asexual reproduction.

Although some green algae exhibit morphological variability under certain growth conditions, relatively few studies have investigated this phenomenon. Therefore, little is known about the potential advantage of membrane-bound multicellular structures, over motile unicells. However, at low temperatures, Antarctic bacteria are known to produce large amounts of extracellular exopolysaccharides (Mancuso Nichols et al. 2004). These polysaccharides prevent intracellular freezing, provide protection against cold-denaturation of extracellular enzymes, aid in nutrient trapping, and alter adhesive properties of cells (Mancuso Nichols et al. 2004). Thus, formation of mucus-embedded multi-cellular structures in Antarctic algae may provide a more favorable microenvironment under low temperature conditions.

13.3 Membrane Fatty Acid Composition

Biological membranes function as physical barriers, regulate the movement of solutes and ions, and provide a matrix for protein complexes as well as signal transduction pathways (Hazel 1995). As temperatures decline, membrane lipid molecules become packed more tightly, and membranes become increasingly rigid. Such reduction of membrane fluidity can have adverse effects on the physical properties and the overall membrane function (Hazel 1995; Sinensky 1974; Los et al. 2013). Psychrophilic and psychrotolerant organisms can regulate the degree of membrane fluidity by adjusting fatty acid composition of their membrane lipids (Los et al. 2013; Los and Murata 2004; Morgan-Kiss et al. 2006a). Biosynthesis of unsaturated fatty acids in response to low temperature in order to modulate membrane fluidity is called "homeoviscous acclimation" (Sinensky 1974). Fatty acid desaturases (FADs) regulate the biosynthesis of unsaturated fatty acids and have been shown to be cold-inducible in a wide range of heterotrophic as well as photoautotrophic organisms (Los et al. 2013; Chen and Thelen 2013). A change in membrane fluidity has been proposed to act as a temperature sensor in photoautotrophic micro-organisms (Wada et al. 1993; Nishida and Murata 1996; Murata and Los 1997; Los and Murata 2002). In the cyanobacterium, *Synechosystis*, cold induces more than 50 genes (Inaba et al. 2003) of which about 50% are regulated by the HIK33-Rre, two component sensing-signalling system (Los et al. 2013). HIK33 is the transmembrane protein with histidine kinase activity whose sensor domain is autophosphorylated at a specific His residue upon low temperature-induced decreases in membrane fluidity. The phosphate group of the sensor domain is transferred to an aspartate residue on Rre, the protein response regulator. This causes a conformational change in Rre which allows this transcription factor to bind to the promoters that activate the expression of FADs such as *desA*, *desB*, *desC* and *desD* in *Synechocystis* (Los et al. 2013; Los and Murata 2002). As a consequence of the activation of these FAD genes, the level of fatty acid unsaturation increases which enhances membrane fluidity at low temperature.

For photoautotrophic organisms, sunlight represents the ultimate source of energy which is transformed into chemical energy (ATP) and reducing power (NADPH) by the photosynthetic electron transport system localized to thylakoid membranes. The major chloroplast thylakoid lipids include mono- and digalactosidediacylglycerides (MGDG and DGDG), the sulfolipid, sulfoquinovosyldiacylglyceride (SQDG) and the phospholipid, phosphatidyldiacylglyceride (PG). MGDG and DGDG are the major lipid species of thylakoid membranes and constitute about 40 and 30 mol % respectively whereas SQDG and PG constitute 5 and 10 mol % respectively of the total thylakoid lipids. The primary fatty acids associated with these lipids usually include the 16 and 18 carbon saturated fatty acids, palmitic (16:0) and stearic acid (18:0) respectively combined with their cis-unsaturated forms, 16:3, as well as 18:1, 18:2 and 18:3 which results in array of lipid molecular species associated with thylakoid membranes (Murata and Siegenthaler 1998). The content of 18:1, 18:2 and 18:3 in MGDG and DGDG represents approximately 95% of the fatty acids associated with these major thylakoid lipids even at optimal growth temperatures due to the high protein/lipid ratio characteristic of thylakoid membranes (Murata and Siegenthaler 1998). Thus, thylakoid membranes isolated from cold acclimated, cold tolerant winter rye exhibited minimal (2–5%) changes in thylakoid lipid unsaturation compared with thylakoids isolated from non-acclimated plants grown to the same developmental state as the cold acclimated plants. However, the major change in fatty acid composition was reflected in a significant decrease in the unique Δ^3-transhexadecenoic acid (Δ^3-trans16:1) esterified at the sn2 position of PG with no change in the fluidity of the thylakoid membrane between cold acclimated and nonacclimated rye as determined by either electron spin resonance or differential scanning calorimetry (Hüner et al. 1987). The decrease in Δ^3-trans16:1 in PG was correlated with a predisposition for stabilization of the monomeric form of LHCII whereas high levels of this fatty acid predisposed the stabilization of oligomeric LHCII under controlled growth (Hüner et al. 1987, 1993; Dubacq and Tremolieres 1983; Remy et al. 1982; Krol et al. 1988; Krupa et al. 1992; Tremolieres and Siegenthaler 1998) as well as natural field conditions (Gray et al. 2005). Furthermore, the extent to which low growth temperature modulates Δ^3-trans16:1 content in PG and the oligomeric state of LHCII are excellent predictors of the freezing tolerance of cereals (Hüner et al. 1989).

13.4 Cold-Adapted Enzymes

Temperature reflects the kinetic energy of molecules – the higher the temperature the higher the kinetic energy of molecules. Biology is characterized by specialized proteins called enzymes that catalyze biochemical reactions within cells. In general, the ability of enzymes to catalyze high rates of chemical bond formation to convert reactants to products is the ability of enzymes to reduce the activation energy for this conversion. To lower the activation energy, most biochemical reactions require a conformational change in the enzyme catalyzing the reaction to optimize the structure of active site for maximal reaction rates. This is considered to be the rate limiting step for most biochemical reactions and is very temperature sensitive (Hochachka and Somero 2002). Consequently, low temperature imposes a thermodynamic constraint on the rate of enzyme catalyzed reactions. Furthermore, the increased medium viscosity at low temperatures further contributes to reducing biochemical reaction rates (Demchenko et al. 1989). Thus, cold-adapted organisms must compensate for both of these temperature-dependent thermodynamic constraints in order to maintain metabolic homeostasis at low temperature.

Catalytic performance is traditionally measured as k_{cat}, the rate that substrate is converted to product per active site of enzyme (k_{cat} = rate of reaction/[enzyme], assuming one active site per enzyme molecule). To maintain appropriate enzyme reaction rates at low temperatures, psychrophiles are characterized by cold-adapted enzymes which can exhibit k_{cat} that can be fivefold to tenfold higher at low temperature than the same enzyme from mesophilic and thermophilic organisms (Feller and Gerday 2003; Hochachka and Somero 2002; Van den Burg 2003). This enhanced catalytic performance of cold adapted enzymes at low temperature is achieved through enhanced protein flexibility especially near the active site (Feller and Gerday 2003; Thomas and Dieckmann 2002; Hochachka and Somero 2002). Increased protein flexibility of cold adapted enzymes reflects changes in the primary structure which is consistent with a general lower thermal stability and higher heat sensitivity than the same enzymes from mesophilic species (Feller and Gerday 2003; Van den Burg 2003). However, the increased protein flexibility exhibited by cold adapted enzymes reduces the activation energy for enzyme catalysis which leads to increased k_{cat} at low temperature (Hochachka and Somero 2002).

Although the increased protein flexibility of cold adapted enzymes reflects differences in primary structure of cold adapted versus warm adapted enzymes, an increased k_{cat} for a biochemical reaction can also be achieved by increasing the amount of enzyme present in a cellular compartment since enzyme activity is the product of k_{cat} and enzyme concentration. Increasing enzyme concentration would also help to overcome limitations imposed by the lower rates of diffusion through an aqueous solution at low temperature. Rubisco (ribulose-1,5-bisphosphate carboxylase/oxygenase), a key enzyme in photosynthesis, is the most abundant protein in nature and present in the chloroplast stroma at mM concentrations but catalyzes the fixation of CO_2 with a relatively low k_{cat} (Spreitzer and Salvucci 2002).

In contrast to the typical response of psychrophilic enzymes discussed above, Devos et al. (1998) reported that the carboxylase activity of Rubisco from psychrophilic unicellular green algae appears to be reduced at low temperatures compared to the enzyme from mesophilic species and that the psychrophilic and mesophilic forms of Rubisco exhibit similar optimal temperature for maximal activities (Devos et al. 1998). Although similar trends were obtained for enzyme activity measured either in crude extracts or the enzyme purified by sucrose gradient purification, the psychrophilic species appeared to exhibit significantly higher amounts of Rubisco per cell (Devos et al. 1998). It is interesting to note that Devos et al. (1998) did not verify their *in vitro* results by *in vivo* measurements of maximum light and CO_2 saturated rates of photosynthesis which reflect the maximum Rubisco carboxylation rates (Devos et al. 1998). Pocock et al. (2007) examined the temperature dependence of the maximum light- and CO_2-saturated rates of photosynthesis in the psychrophile, *Chlamydomonas sp.* UWO241, with that of the model mesophile, *Chlamydomonas reinhardtii* (Pocock et al. 2007). The psychrophilic form of Rubisco exhibited maximum activity at 10 °C which progressively decreased with an increase in temperature with minimal activity at 45 °C. In contrast, the mesophilic form of the enzyme exhibited a typical temperature response curve for mesophilic enzymes with minimal activity at 10 °C that was threefold lower than the psychrophilic form measured at the same temperature with a maximum between 25 °C and 30 °C. Subsequently, the rates of photosynthesis decreased to minimal levels at 45 °C (Pocock et al. 2007). Since the maximum light- and CO_2-saturated rates of photosynthesis reflect Rubisco activity,

Pocock et al. (2007) conclude that, in contrast to Devos et al. (1998), Rubisco in the psychrophile *Chlamydomonas sp.* UWO241 exhibits a photosynthetic activity temperature profile consistent with cold adaptation relative to the mesophilic enzyme (Pocock et al. 2007).

13.5 Energy Metabolism

As described above, rigidification of membranes and decreased reaction rates represent challenges to the management of biochemical processes at low temperatures. However, under low physiological growth temperatures where biological processes for most microorganisms become non-functional, psychrophiles must generate adequate energy levels to maintain metabolic homeostasis. Photosynthesis and respiration represent the two primary processes involved in energy metabolism in photoautotrophic psychrophiles. These two metabolic processes are coupled since light energy is converted into reductants and ATP for the biosynthesis of triose-P intermediates, via the Calvin-Benson Cycle, that are subsequently exported from the chloroplast to the cytosol either for conversion into sucrose or oxidation via glycolysis and aerobic respiration in the matrix of the mitochondria (Hopkins and Hüner 2009). Hydrolysis of ATP by mitochondria in the cytoplasm is the major energy source for metabolism, growth and development. Thus, cellular concentrations of ATP and adenylates provide information on cellular metabolic states (Bott and Kaplan 1985). Cold-adapted organisms generally exhibit elevated concentrations of adenylate compounds – key molecules of energy metabolism, compared to mesophilic species (Napolitano and Shain 2004).

In both mesophilic and thermophilic organisms, growth rates and ATP levels decline with decreasing temperatures. Therefore, at higher growth temperatures, higher growth rates correspond to increased energy supply, as measured by total adenylate levels (Napolitano and Shain 2004, 2005). In contrast, psychrophilic microorganisms exhibit an inverse relationship between adenylate pool size and growth temperature (Napolitano and Shain 2004, 2005). Therefore, the presence of high levels of ATP and total adenylate pool size at low temperature may represent an additional adaptive mechanism utilized by psychrophiles to offset decreased rates of biochemical reactions at low temperatures (Napolitano and Shain 2004, 2005). It has been suggested that this adaptation likely results from adjustments to components involved in ATP synthesis, either due to structural modifications or alterations in the cellular environment (e.g. pH) (Portner et al. 1999).

This is consistent with both adaptation and acclimation to low temperature in photoautotrophs. Pocock et al. (2007) reported that the maximum absolute light- and CO_2-saturated *in vivo* rates of photosynthesis observed for the psychrophile, *Chlamydomonas sp.* UWO241, at its optimal growth temperature (10 °C) was equal to that of the mesophile, *Chlamydomonas reinhardtii*, at its optimal growth temperature (25–30 °C) (Pocock et al. 2007). The low temperature adaptation of maximum photosynthetic rates was matched by low temperature adaptation of rates of respiration which ensures metabolic homeostasis at low temperature as confirmed by the temperature for maximum growth.

Dahal et al. (2012a, b, c) reported that winter cereals acclimated to growth at 5 °C exhibited significantly higher rates of light- and CO_2 saturated rates photosynthesis at all temperatures between 5 and 20 °C, comparable rates between 25 and 30 °C and similar inhibition of photosynthetic rates between 30 and 45 °C (Dahal et al. 2012a). These enhanced rates of CO_2 assimilation at low temperature are matched by increased levels of expression and higher activities of the regulatory enzymes, fructose bisphosphatase and sucrose-P synthase involved in cytosolic sucrose biosynthesis (Hurry et al. 1995a; Strand et al. 1997, 1999; Savitch et al. 2000, 2005; Stitt and Hurry 2002). Cold acclimation of winter cultivars typically results in increased concentrations of Rubisco and other enzymes involved in photosynthetic carbon metabolism (Dahal et al. 2012a; Hurry et al. 1995a; Strand et al. 1997, 1999).

Furthermore, enhanced rates of respiration are characteristic of cold acclimated plants relative to the warm acclimated state (Dahal et al. 2012a; Hurry et al. 1995b; Atkin et al. 2005; Armstrong et al. 2008; Ruelland et al. 2009). Thus, the adjustments in photosynthesis and respiration are consistent with the establishment of metabolic homeostasis and optimal growth rates during acclimation as well as adaptation to cold temperatures in photoautotrophs.

13.6 Photosynthesis at Low Temperatures

In addition to low temperature, photoautotrophs must deal with a unique and critical challenge to survival at low temperature that is not faced by heterotrophs. Photosynthetic organisms transform absorbed visible electromagnetic radiation into electrons (NADPH) and chemical energy (ATP) as their primary energy source for all subsequent cellular processes necessary for the establishment of metabolic homeostasis, growth and development. However, one must distinguish between the roles of light quality versus light intensity in the regulation of the photosynthetic apparatus. Pogson and co-workers (2008) differentiate sensing/signalling associated with changes in light quality through photoreceptors involved in photomorphogenesis as "biogenic signals" versus sensing/signalling associated with changes in light intensity through photosynthetic electron transport in mature chloroplasts as "operational signals" (Pogson et al. 2008; Estavillo et al. 2011). For example, biogenic light signals are involved in chloroplast biogenesis and govern the proper biosynthesis and assembly of thylakoid membranes and their constituent membrane protein complexes whereas light energy as an operational signal is required to ensure energy balance and cellular homeostasis in an environment where light intensity, temperature, water and nutrient availability constantly fluctuate (Hüner et al. 1998, 2012, 2013a; Wilson et al. 2006; Kurepin et al. 2013). The former is governed by changes in light quality which is sensed through photoreceptors such as phytochrome and cryptochrome which regulate chloroplast biogenesis and photomorphogenesis (Pogson et al. 2008). In contrast, the latter is sensed through alterations in the redox state of the photosynthetic electron transport chain (PETC) which, in turn, directly regulates chloroplast gene expression in addition to nuclear gene expression through a retrograde signalling process (Nott et al. 2006; Fernandez and Strand 2008; Jung and Chory 2010; Foyer et al. 2012). It has been proposed that chloroplast operational signals are associated with PETC (Foyer et al. 2012), specifically, the redox state of the plastoquinone pool (Allen 1993; Escoubas et al. 1995; Maxwell et al. 1995a; Pfannschmidt 2003; Rosso et al. 2009), singlet oxygen generated by PSII (Apel and Hirt 2004; Aluru et al. 2009), redox intermediates associated with the acceptor side of PSI (Dietz 2003; Dietz and Pfannschmidt 2011; Brautigam et al. 2009) as well as intermediates of the chlorophyll and heme biosynthetic pathway (Lermontova and Grimm 2006; Luo et al. 2012; Woodson et al. 2011). Identification of intermediates of the signal transduction pathway between the chloroplast and the nucleus remains a challenge (Jung and Chory 2010) but may include H_2O_2 (Apel and Hirt 2004; Karpinski et al. 1999), GUN1 and Mg-protoporphyrin IX (Fernandez and Strand 2008) as well as the phosponucleotide, 3-′-phosphoadenosine-5′-phosphate (Estavillo et al. 2013). Noctor and co-workers (2013) have suggested that glutathione acts as an important cellular "redox gatekeeper" involved in redox homeostasis and signalling associated with plant stress (Noctor et al. 2013).

The operational signals generated by chloroplasts is a consequence of the potential imbalance between the primary, photophysical and photochemical processes involved light absorption, energy transfer within light harvesting complexes and charge separation in the reaction centers that generate electrons and the biochemical processes that utilize these electrons for the reduction of C, N and S (Hüner and Grodzinski 2011). The photophysical and photochemical processes are temperature-insensitive, in contrast to downstream metabolic reactions that consume photochemically formed energy products (NADPH and ATP), which are temperature-dependent (Hüner and Grodzinski 2011). Since the rate of PSI photo-oxidation-reduction is not considered limiting during steady-state photosynthesis, the rate of PSII photochemistry is considered to be significantly slower than PSI, in part, due to the diffusion limited oxidation of plastoquinol by the Cyt b_6/f complex (Haehnel 1984; Mitchell et al. 1990). A balance between the photophysical and photochemical processes (energy source) and the biochemical processes (energy sinks) that utilize the energy is termed photostasis (Hüner et al. 2003a). Photostasis may be represented by the following equation (Falkowski and Chen 2003a):

$$\sigma_{PSII} \times E_k = \tau^{-1}$$

where σ_{PSII} is the effective cross section area of photosystem II, E_k is the irradiance (I) at which the maximum photochemical yield of PSII balances photosynthetic capacity, and τ^{-1} is the rate at which photosynthetic electrons are consumed by downstream metabolic sinks (Falkowski and Chen 2003a).

Thus, an imbalance in energy budget occurs whenever the rate of energy absorbed through PSII and the rate of electron flux into photosynthetic electron transport exceeds the metabolic sink capacity, i.e. whenever $\sigma_{PSII} \times E_k > \tau^{-1}$. This may occur as a consequence of increased growth irradiance to exceed E_k, or lowering growth temperature, which causes a decrease in τ^{-1} (Hüner et al. 1998, 2003b, 2013b). Exposure to excess light may lead to a reduced ability to utilize the increased amount of energy trapped by photochemistry, whereas exposure to low temperatures results in the downregulation of metabolic processes downstream of photochemistry (Hüner et al. 1998). Therefore, conditions of high light and cold temperatures result in increased energy input or decreased energy utilization, respectively, which leads to increased excitation pressure (Hüner et al. 1998, 2003a). High excitation pressure occurs when components of the photosynthetic electron transport chain, and the plastoquinone pool (PQ) become over-reduced, as electrons are generated faster by PSII than they can be consumed by the metabolic sinks (Hüner et al. 1998). In turn, a balance can be restored either by increasing the rate of energy utilization and storage (τ^{-1}) by the metabolic sinks or by decreasing the energy input ($\sigma_{PSII} \times I$), by reducing either the physical size and/or the effective, functional absorption cross section of PSII (Hüner et al. 2003b, 2013b).

13.7 Acclimation to Light and Low Temperatures

To maintain photostasis, photosynthetic organisms integrate a variety of short-term and long-term acclimation mechanisms (Anderson et al. 1995). On a short time scale of minutes, organisms can reduce the efficiency of energy transfer to PSII by diverting energy away from PSII in favour of PSI through state transitions (Fig. 13.1) (Kargul and Barber 2008; Rochaix 2011). During exposure to conditions that cause an over-reduction of the plastoquinone (PQ) pool, plastoquinol (PQH$_2$) binds to the Q (o) site of the cytochrome b_6/f complex. Upon docking of plastoquinol, a membrane bound protein kinase, STT7 in *Chlamydomonas reinhardtii*, is activated, which phosphorylates LHCII proteins. Recently, it was reported that the STN8 protein kinase appears to be the primary thylakoid protein kinase that regulates the phosphorylation of the PSII core proteins whereas STN7 governs the phosphorylation of Lhcb1, Lchb2 and Lhcb4 in *Arabidopsis thaliana* (Rochaix 2011; Bonardi et al. 2005; Tikkanen et al. 2008, 2012; Fristedt et al. 2009; Wunder et al. 2013). The orthologous thylakoid protein kinases in *Chlamydomonas reinhardtii* are STT8 and STT7 (Fig. 13.1) (Rochaix 2011; Depège et al. 2003; Vener 2006).

Upon phosphorylation, major LHCII trimers dissociate from photosystem II and physically migrate to couple with photosystem I (state II) (Finazzi and Forti 2004). Therefore, this transition to state II results in an increase of PSI

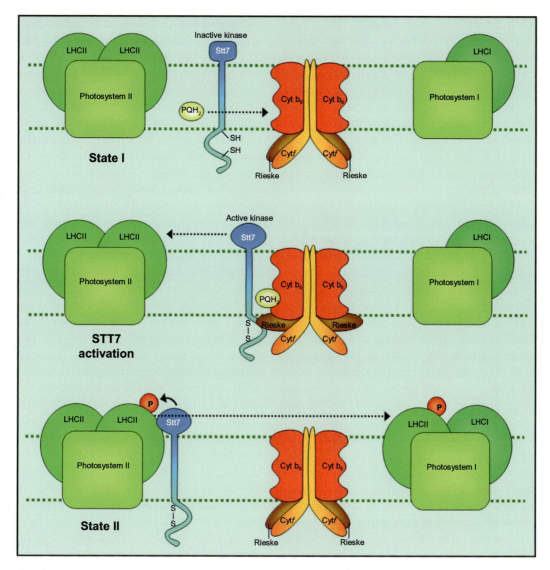

Fig. 13.1 A hypothetical model for state transitions from state I to state II in green algae. Transition from state I to state II occurs when the redox state of the plastoquinone (PQ) pool is reduced. Docking of a plastoquinol (PQH_2) to the Q(o) site of Cyt b_6 causes the Rieske protein to shift from a distal to proximal position, relative to the thylakoid membrane. This movement results in the interaction of Rieske with the Stt7 kinase, and its subsequent activation. The active STT7 kinase interacts with PSII, leading to the phosphorylation of LHCII proteins. Phosphorylated LHCII proteins dissociate from PSII and physically migrate to dock with PSI. The transition from state II to state I occurs upon LHCII dephosphorylation by a phosphatase, resulting in LHCII detachment from PSI and re-association with PSII (not shown)

antenna size at the expense of the PSII antenna (Finazzi and Forti 2004). In contrast, when the plastoquinone pool becomes increasingly oxidized, thylakoid phosphatases such as PPH1/TAP38 remove the phosphate group, and the mobile fraction of the PSII antenna becomes associated with PSII (state I) (Rochaix 2011). Alternatively, the photosynthetic apparatus can be protected from excess absorbed light by dissipation of the excess energy as heat by non-photochemical quenching (NPQ) via the xanthophyll cycle (Hüner et al. 2003a; Demmig-Adams et al.

1999; Horton et al. 2008; Li et al. 2009). Both state transitions and NPQ are rapid and require no de novo biosynthesis of either new metabolites or proteins and can be activated by either a change in irradiance or temperature (Adams et al. 1995; Krol et al. 1999). Since both processes are regulated by the redox state of the plastoquinone pool, state transitions and NPQ are governed by excitation pressure (Wilson et al. 2006; Rochaix 2011; Wilson and Hüner 2000).

In contrast, long-term acclimation to excess absorbed light due to exposure to either low temperature or high light is a consequence of alterations in gene expression and translation. Typically, in many single cell photosynthetic micro-organisms such as *Dunaliella tertiolecta* (Escoubas et al. 1995; Falkowski and Chen 2003b), *Chlorella vulgaris* (Hüner et al. 1998; Maxwell et al. 1995b; Wilson et al. 2003) and *Dunaliella salina* (Maxwell et al. 1995a; Masuda et al. 2003), this is reflected in alterations in the abundance of light-harvesting antenna polypeptides and/or adjustments in PSI:PSII stoichiometry to balance the excitation light energy absorbed by the two photosystems (Melis et al. 1996; Fujita 1997; Yamazaki et al. 2005). Coordinated regulation of chlorophyll (Chl) biosynthesis and Chl binding LHC abundance is likely an important characteristic of acclimation to excitation pressure in green algae such as *Chlorella vulgaris*. The majority of Chl b is bound to LHC polypeptides and is required both for the assembly and function of LHCs (Peter and Thornber 1991). Masuda et al. (Masuda et al. 2003) demonstrated that changes in the levels of *CAO* transcripts encoding chlorophyll a oxidase, the enzyme catalyzing the conversion of Chl a to Chl b, occur concomitant with changes in *Lhcb* transcript abundance during acclimation to high light intensity in *D. salina* (Masuda et al. 2003). Furthermore, the use of site specific inhibitors of the PETC demonstrated that the redox state of the PQ pool regulates both *CAO* and *Lhcb* transcript abundance. This is consistent with previous work demonstrating regulation of *Lhcb* transcription rates by the redox state of the PQ pool in *D. tertiolecta* during photoacclimation (Escoubas et al. 1995). Similarly, parallel decreases in *Lhcb* and *CAO* transcript abundance have also been observed in green algae during salt stress (Chen et al. 2010). These studies are consistent with acclimation to high excitation pressure as both high light and high salt stress have the potential create imbalances in energy flow between light energy absorption and consumption by increasing the influx of electrons into intersystem electron transport or through reduction of growth rates, respectively (Hüner et al. 1998). Reduction of the PQ pool through environmentally induced high excitation pressure is proposed to act as a source of retrograde signals which affect a decrease in the transcript abundance of nuclear encoded genes, *CAO* and *Lhcb*, which in turn, co-ordinately decreases LHCII and Chl abundance. This results in a pale-yellow green phenotype. Conversely, oxidation of the PQ pool signals for increases *CAO* and *Lhc* transcript abundance thereby coordinating an increase in LHCII polypeptide and Chl accumulation which results in the typical dark green phenotype.

Thus, it appears that *Chlorella vulgaris* inherently exhibits a limited capacity to adjust photosynthetic carbon metabolism (τ^{-1}) in response to changes in excitation pressure (Savitch et al. 1996). Consequently, *Chlorella vulgaris* maintains photostasis by reducing σ_{PSII} coupled with an increase in xanthophyll cycle activity which stimulates NPQ to balance energy absorbed with energy consumption by the metabolic sinks (τ^{-1}). This strategy to maintain photostasis allows organisms to survive at the expense of a decrease in photosynthetic light use efficiency. On the other hand, some marine phytoplankton avoid exposure to excess light by changing their vertical position in the water column (Falkowski 1983).

Alternatively, winter cereals (Hüner et al. 1993, 1998, Savitch et al. 2002; Ensminger et al. 2006; Dahal et al. 2013a, b; c), *Brassica napus* (Hurry et al. 1995a; Savitch et al. 2005; Dahal et al. 2012b), *Arabidopsis thaliana* (Strand et al. 1999, 2003; Savitch et al. 2001) and the native Antarctic species, *Colobanthus quitensis* (Kunth) Bartl. (Bravo et al. 2007) maintain photostasis at low temperature by enhancing photosynthetic capacity. This is the result of an increased capacity for photosynthetic electron consumption by increasing the mRNA and protein levels of Calvin cycle enzymes such as Rubisco as well as cytosolic enzymes involved in sucrose biosynthesis (Hüner et al. 1998; Strand et al. 1999, 2003; Savitch et al. 2005). This is, in turn, coupled to increased rates of export from source leaves to metabolic sinks (Leonardos et al. 2003). Thus, there is no requirement for these plants to adjust σ_{PSII} or enhance NPQ since the enhanced sink capacity (τ^{-1}) appears to be sufficient in these plants to maintain photostasis at low temperature. Consequently, the enhanced sink capacity induced by cold acclimation results in a significant increase in photosynthetic light use efficiency which is translated into increased biomass production and enhanced seed yield in winter cereals (Dahal et al. 2013a, b).

13.8 *Chlamydomonas* sp. UWO241

Although cold acclimation, low temperature stress and freezing tolerance have been studied in myriad mesophilic photoautotrophs, we know relatively little about adaptation of true psychrophilic photoautotrophs to cold environments (Hüner et al. 1993). For the past two decades, *Chlamydomonas sp.* UWO241 has been the most studied psychrophilic green alga to date, and consequently can be considered the model organism for the study of psychrophily in photoautrophs (Morgan-Kiss et al. 2006a).

13.8.1 Natural Habitat of UWO241

Chlamydomonas sp. UWO241 originates from the east lobe of Lake Bonney in Taylor Valley, Antarctica – one of the coldest and driest deserts of our planet – with annual air temperatures average of -20 °C, and annual precipitation is lower than 10 cm (Neale and Pricu 1995; Priscu 1998). Psychrophilic UWO241 was isolated from the deepest trophic zone, situated at 17 m below the permanent ice-cover of this unique lake. The presence of a thick (3–4.5 m) ice cover limits gas exchange between lake water and the atmosphere and prevents vertical mixing within the water column, due to the absence of wind turbulence (Spigel and Priscu 1996). Mixing occurs predominantly via molecular diffusion and the vertical mixing time for Lake Bonney is estimated to be approximately 50,000 years (Moorhead et al. 1999).

Vertical stratification in Lake Bonney results from strong salinity gradients, reaching levels up to ten times that of seawater at maximum lake depths (Spigel and Priscu 1996; Priscu et al. 1998). Stable salinities of approximately 700 mM are characteristic at the depth where UWO241 was isolated, which is above that of seawater (545 mM) (Priscu 1998). The light environment at the depth where UWO241 naturally exists is characterized by low light (<50 μmol photons m^{-2} s^{-1}) during austral summer and a light spectral distribution that is heavily biased to the blue region of the visible spectrum (450–550 nm) (Lizotte and Priscu 1994; Lizotte et al. 1996).

A recent study of the distribution of microbial eukaryotes using 18S rRNA libraries revealed that Lake Bonney is dominated by photosynthetic protists, with the majority being related to flagellated strains (Bielewicz et al. 2011). Both lobes of the lake were vertically stratified with dominating populations of a cryptophyte at shallow depths (6–10 m), a haptophyte at mid-depths (13 m) and various chlorophytes, including UWO241 residing in the deepest photic zone (15–20 m) (Bielewicz et al. 2011). Thus, UWO241 is a psychrophile that has adapted to an extreme but stable growth regime that is characterized by low temperatures, high salinity and extreme shade in a narrow spectral range distribution.

13.8.2 Growth of UWO241

13.8.2.1 Temperature
Cultures of UWO241 exhibit an optimal growth temperature of 8 °C and exponential growth up to temperatures of 16 °C. However, at 20 °C or

higher, growth of UWO241 is inhibited which confirms that UWO241 is an obligate psychrophile adapted to growth at low temperatures (Morgan et al. 1998a). To address the inability UWO241 to grow at temperatures above 16 °C, a recent study examined the effects of supra-optimal temperatures on the physiology of UWO241 (Possmayer et al. 2011). Using the membrane impermeable SYTOX green assay to stain cellular DNA, it was determined that exposure of UWO241 cells to 24 °C, results in cell death with a half-time of 34.9 h (Possmayer et al. 2011). Surprisingly, cell death occurred independently of light, as dark incubation of cells showed a comparable half-time of 43.7 h (Possmayer et al. 2011). This suggests that absorption of excess light and formation of reactive oxygen species in the chloroplast of UWO241 play relatively minor roles in cell death (Possmayer et al. 2011). A 12 h shift of UWO241 from 10 °C to 24 °C caused reductions in light-saturated rates of photosynthesis and respiration and stimulated excitation pressure and altered energy partitioning. This was associated with a decrease in transcript abundances for LHCII polypeptides and ferredoxin but dramatic increases in the transcript abundance for heat shock proteins (HSPs) (Possmayer et al. 2011). All of these effects were reversible within 24–48 h of recovery at 10 °C, indicating that UWO241 exhibits significant physiological plasticity at a temperature that is lethal to this psychrophile (Possmayer et al. 2011). However, this molecular and physiological plasticity is time dependent at supra-optimal temperatures.

13.8.2.2 Salt

Although UWO241 has adapted to a salinity of 700 mM, it has been demonstrated that UWO241 is not halophilic, but halotolerant (Pocock et al. 2011). This psychrophilic strain exhibits the ability to grow under a wider salinity range of 10 mM to its upper critical salinity limit of 1300 mM at 8 °C, compared to a mesophilic strain that was unable to grow at salt concentrations above 100 mM at 24 °C (Pocock et al. 2011). However, UWO241 exhibited decreased growth rates as salinity increased, with a 35% reduction in growth rate at 700 mM compared to 10 mM NaCl (Pocock et al. 2011). Therefore, despite the hypersaline environment of Lake Bonney, high salt is not an absolute requirement for growth of UWO241, but rather a condition that is tolerated by this psychrophile.

13.9 Adaptation of UWO241 to Low Temperature

Detailed microscopy studies (Pocock et al. 2004) revealed that *Chlamydomonas sp.* UWO241 cells exist as motile, biflagellate single cells of approximately 10–15 µm in length, as well as non-motile membrane-bound palmelloids, consisting of 16–32 cells that measure approximately 30 µm (Fig. 13.2). It is not clear if these palmelloids are a product of mitosis or whether they represent meiotic reproductive structures. However, it has been observed that cell morphology in UWO241 is regulated by growth temperature. Increasing growth temperature between 8 and 16 °C caused a significant shift from predominantly motile, single cells to palmelloids. Thus, this shift in cell morphology may play a role in cold adaptation. Similar to other Antarctic algae, UWO241 cells secrete ice-binding proteins (IBPs), which function extracellularly to alter the structure of surrounding ice and increase freezing tolerance (Raymond and Morgan-Kiss 2013).

Analysis of the membrane lipid composition in UWO241 exhibited significantly higher levels of unsaturated fatty acids compared to the mesophilic model organism, *Chlamydomonas reinhardtii*, with an unsaturation index of 2.74, compared to 1.90, respectively (Morgan-Kiss et al. 2002a). In addition, UWO241 membranes revealed the presence of several unique polyunsaturated fatty acids, with altered unsaturated bond positions that occurred close to the lipid head group (Morgan-Kiss et al. 2002a, 2006b). As chloroplasts comprise more than 80% of the total cell membranes, the majority of these fatty acids represent galactololipids (MGDG, DGDG, SQDG and PG) of the photosynthetic thylakoid membranes (Morgan-Kiss et al. 2002a, 2006b).

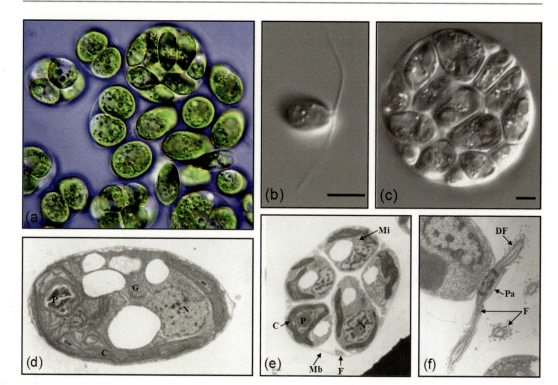

Fig. 13.2 Light microscope images (**a–c**) of *Chlamydomonas* sp. UWO241. Cells exist as motile, flagellated single cells (**a**) and palmelloids (**b**). Magnification 100×. Scale bars, 5 μm. Electron micrographs (**d–f**) showing a single cell (**d**), a palmelloid (**e**) and flagella (**f**). Note the pyrenoids (*P*), the membrane (*Mb*) surrounding the flagellated daughter cells in the colony, the flat apical papilla (*Pa*), and the vesicular outpocketings of the flagellar membrane (*DF*). *C* chloroplast, *N* nucleus, *G* Golgi apparatus, *Mi* mitochondrion, *F* flagellum. Scale bars, 1.0 mm. Panels **b–f** are reproduced, with permission, from Pocock et al. (2004) (Journal of Phycology)

In a mechanism used to overcome reduced catalytic efficiency at low temperatures, UWO241 exhibits high concentrations of the large unit of Rubisco, compared to a mesophilic control, *C. raudensis* SAG 49.72 (Dolhi et al. 2013). However, despite these increased levels, Rubisco activity was 30% lower in UWO241 compared to a mesophile, when cells were grown under their optimal growth temperatures of 8 °C and 29 °C (Dolhi et al. 2013). However, the *in vitro* results for Rubisco activity are inconsistent with the temperature dependence for light and CO_2 saturated rates of photosynthetic gas exchange which indicate that photosynthetic rates for UWO241 are maximum at the optimal low growth temperature and minimum at non-permissive, high temperatures (Pocock et al. 2007). Furthermore, *in vivo* photosynthetic rates for UWO241 at its optimal growth temperature (8 °C) were comparable to those of the mesophile, *Chlamydomonas reinhardtii*, at its optimal growth temperature (29 °C) (Pocock et al. 2007).

In addition, growth under similar optimal conditions showed that UWO241 significantly increased concentrations of two major (α and β) subunits of the CF_1 complex of the chloroplast ATP synthase compared to the mesophilic control, *C. reinhardtii* (Morgan et al. 1998b). An increase in chloroplast ATP synthase is likely an adaptive strategy to maintain elevated adenylate pools, which are required for ATP-dependent biochemical reactions occurring at low temperatures (Napolitano and Shain 2004,

2005). Increased levels of ATP in UWO241 may also be necessary to actively pump sodium across cell membranes in hypersaline growth environments, as observed for the halotolerant model, *Dunaliella salina* (Liska et al. 2004).

13.10 Photosynthetic Electron Transport

13.10.1 Structure of PETC

Adaptation of *C. raudensis* UWO241 to its unique environment has resulted in the evolution of distinct structural features in its photosynthetic apparatus. Previous studies have revealed that this Antarctic psychrophile exhibits altered overall stoichiometry of PSII/PSI (Morgan et al. 1998a; Szyszka et al. 2007). This reflects the reduced abundance of both PSI and LHCI, resulting in a smaller functional absorptive cross section of this photosystem (σ_{PSI}), compared to *C. reinhardtii* (Morgan et al. 1998a). Conversely, the absorptive cross section of PSII (σ_{PSII}) was larger compared to *C. reinhardtii* and a higher proportion of LHCII proteins were found in an oligomeric, rather than a monomeric state (Morgan et al. 1998a). As chlorophyll b is present exclusively bound to light harvesting complex II (LHCII) proteins, high levels of LHCII also result in characteristically low chlorophyll a/b ratios (~1.8-2.0) in UWO241, compared to typical values of ~3.0 in *C. reinhardtii* (Pocock et al. 2004; Morgan et al. 1998a). Relatively high levels of PSII, LHCII and increased PSII cross section most likely allow for more efficient use of light and reflect an adaptation of extreme shade conditions (Neale and Pricu 1995; Morgan et al. 1998a).

In addition to altered stoichiometry of photosystems, UWO241 shows higher levels of the Cyt b_6/f complex compared to mesophilic algae under the same growth conditions (Morgan et al. 1998a; Szyszka et al. 2007). Based on SDS-PAGE, the cytochrome *f* protein of UWO241 has a 7-kDa lower apparent molecular mass (34 kDa) compared to that of *C. reinhardtii* (41 kDa) (Morgan-Kiss et al. 2002b). This mass difference does not affect the ability of the psychrophilic cytochrome *f* to bind the heme factor; however, this covalently bound heme is significantly less stable to high temperature than that of *C. reinhardtii*, which likely reflects an adaptive protein modification towards cold environments (Gudynaite-Savitch et al. 2006). The amino acid sequence of the *petA* gene, encoding the UWO241 cytochrome *f* protein was found to be 79% identical to that of *C. reinhardtii*, with a similar calculated molecular mass, despite the observed apparent molecular mass variance based on SDS-PAGE (Gudynaite-Savitch et al. 2006). Notably, the UWO241 cytochrome *f* sequence revealed the presence of three cysteine residues (C21, C24, and C261), compared to the two cysteine residues (C21 and C24) that are usually observed in other photosynthetic organisms and involved in heme binding (Gudynaite-Savitch et al. 2006). The additional C261 residue of UWO241 cytochrome *f* originates from a transmembrane helix and is not implicated in heme binding (Gudynaite-Savitch et al. 2006). The role of C261 remains to be elucidated.

13.10.2 Function of PETC

In addition to structural differences, the photosynthetic electron transport chain of UWO241 reveals major functional changes (Szyszka et al. 2007; Morgan-Kiss et al. 2002b). Two inhibitors of electron transport, DCMU (3-(3,4-dichlorophenyl)-1,1-dimethylurea) and DBMIB (2,5-dibromo-3-methyl-6-isopropyl-p-benzoquinone) were used to examine the redox state of the photosynthetic electron transport chain of UWO241. DCMU competes for the plastoquinone binding site of PSII, blocking the transfer of electrons from PSII to the PQ pool, while DBMIB occupies the Q (o) site of the Cyt b_6/f complex, thereby preventing the oxidation of PQ (Trebst 1980). Spectroscopic studies assessing the rates of intersystem electron transport through the light dependent reduction of P_{700}^+ revealed that intersystem electron transport in UWO241 appears to be less sensitive to the presence of DCMU and DBMIB than that of

C. reinhardtii. These results are consistent with the fact that UWO241 is locked in state I and is unable to undergo a state transition. Thus, it appears that the redox status of the intersystem electron transport of UWO241 is predisposed to favour the oxidized state compared to that of *C. reinhardtii* (Morgan-Kiss et al. 2002b).

In addition to production of both ATP plus NADPH through linear electron flow (LEF) between PSII and PSI, cyclic electron flow (CEF) recycles photosynthetic electrons solely around PSI to produce ATP, with no net production of NADPH (Finazzi et al. 1999; Eberhard et al. 2008; Cardol et al. 2011). This pathway functions to adjust the appropriate ratio of ATP/NADPH needed for carbon fixation as well as other cellular processes which require ATP (Lucker and Kramer 2013a). An increase in the ratio of CEF/LEF has been observed under environmental conditions associated with increased ATP demand, including, low temperatures (Clarke and Johnson 2001), low CO_2 (Golding and Johnson 2003; Lucker and Kramer 2013b), drought (Golding and Johnson 2003) and high light (Munekage et al. 2004). Several reports have suggested that CEF is activated by state I-II transitions (Iwai et al. 2010; Wollman 2001a). However, recent studies demonstrated that CEF occurs independently of state transitions and that the kinetics of state transitions are considerably slower than the rapid kinetics observed for CEF activation (Lucker and Kramer 2013a, b; Takahashi et al. 2013). *Chlamydomonas sp.* UWO241 is locked in state I (Morgan-Kiss et al. 2002b; Gudynaite-Savitch et al. 2006; Takizawa et al. 2009) and maintains rates of cyclic electron flow that are up to twofold greater than that observed for the mesophiles, *C. reinhardtii* and *C. raudensis* SAG 49.72 (Fig. 13.3) (Morgan-Kiss et al. 2002b; Szyszka-Mroz et al. 2015).

Cyclic electron flow also plays a role in photoprotection through dissipation of excess excitation energy and the down regulation of photosystem II (Golding and Johnson 2003; Heber and Walker 1992; Munekage et al. 2002). CEF-dependent generation of a pH gradient (ΔpH) across the thylakoid membrane results from electron transfer from PSII to the Cyt b_6/f complex. Acidification of the thylakoid lumen increases non-photochemical quenching (NPQ), thereby reducing the efficiency of PSII, PSII photodamage, and electron transfer from PSII to PSI (Golding and Johnson 2003; Heber and Walker 1992). Therefore, CEF may exert regulatory control over PSII activity and electron transport rates through the associated changes in lumenal pH. This is called photosynthetic control (Foyer et al. 2012).

A comparison of energy partitioning of UWO241 with the mesophile *C. raudensis* SAG 49.72 at their optimal growth temperatures of 8 °C and 28 °C, respectively, showed that the psychrophile exhibits a 30% reduction in the efficiency of PSII with a nearly twofold higher level of non-photochemical quenching (Szyszka et al. 2007). Consistently, UWO241 maintains lower epoxidation states of xanthophyll pigments (violaxanthin, antheraxanthin and zeaxanthin) compared to the mesophilic species, *C. raudensis* SAG 49.72 (Szyszka et al. 2007), which are known to play an important role in energy dissipation and photoprotection under conditions of high light. Higher levels of NPQ and lower epoxidation states could be associated with higher rates of cyclic electron flow in UWO241, since this electron pathway results in the generation of a proton gradient.

As a consequence of the differences in both structure and function of the photosynthetic apparatus, UWO241 maintains comparable photosynthetic capacity, PSII excitation pressure and thus, energy balance to that of mesophilic algae when grown under their respective optimal growth conditions (Morgan et al. 1998a; Szyszka et al. 2007).

13.10.3 Acclimation to Temperature and Irradiance

Recently, the capacity of UWO241 to acclimate to different steady state temperature and light growth regimes was examined and compared to that of the mesophile, *C. raudensis* SAG 49.72 (Szyszka et al. 2007). Although the psychrophile

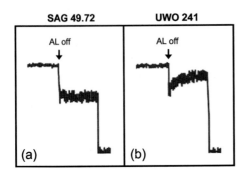

Fig. 13.3 The post-illumination chlorophyll fluorescence transient of the (**a**) mesophilic SAG 49.72 and (**b**) psychrophilic UWO 241 *Chlamydomonas* strains. Consistent with higher rates of cyclic electron flow observed in UWO 241, this Antarctic psychrophile exhibits an increase in fluorescence following actinic light illumination (AL of 250 µmol photons m^{-2} s^{-1}), due to reduction of the plastoquinone pool in the dark. (**c**) Effects of antimycin A on the half times for P700$^+$ reduction after turning off the FR light in a mesophilic SAG 49.72 and psychrophilic UWO 241 strain of *Chlamydomonas* grown under steady state growth temperatures of 28 °C and 5 °C, respectively, and a common temperature of 16 °C for both strains. Antimycin A, an inhibitor of cyclic electron flow, lowered the rate of P700$^+$ re-reduction after turning off the FR light (t^{red}½) by 748% in UWO 241, compared to only 257% in SAG 49.72, confirming the greater activity of CEF in UWO 241 cells

retained the capacity to acclimate to these various conditions, the mechanism employed to maintain photostasis appears to be quite different. UWO241 exhibits alterations in the partitioning of excess excitation energy in response to both elevated growth temperatures and increased growth irradiance (Szyszka et al. 2007). While *C. raudensis* SAG 49.72 favoured energy partitioning through typical down regulatory processes associated with the xanthophyll cycle and antenna quenching, UWO241 favoured energy partitioning through other constitutive processes involved in energy dissipation, which most likely reflect PSII reaction centre quenching (Szyszka et al. 2007).

Although the xanthophyll cycle does not appear to be the primary process employed for dissipation of excess light energy in UWO241 at low temperature, this psychrophile does maintain a fully functional xanthophyll cycle, and exhibits adjustments of epoxidation states in response to changes in irradiance and temperature (Pocock et al. 2007; Szyszka et al. 2007). Despite the adaptation of UWO241 to a cold environment, surprisingly, this Antarctic psychrophile exhibits greater susceptibility to low temperature photoinhibition of PSII compared to the mesophilic *C. reinhardtii* (Pocock et al. 2007).

However, to compensate for this sensitivity to low temperature-induced photoinhibition, UWO241 displays an unusually rapid rate of recovery from photoinhibition at 8 °C, as a result of a unique D1 repair cycle that operates maximally at low temperatures (Pocock et al. 2007).

13.10.4 Acclimation to Light Quality

Adjustments of PSII:PSI stoichiometry are controlled by the redox state of the PQ pool and represents a mechanism for maintaining maximum efficiency of photosynthetic electron transport during long-term acclimation to light intensity and light quality (Melis et al. 1996; Fujita 1997; Yamazaki et al. 2005; Falkowski et al. 1981; Miskiewicz et al. 2002). Under blue light, PSII is preferentially excited and the intersystem electron transport components are mostly reduced, whereas under conditions enriched in red light absorbed mainly by PSI, the components of the intersystem electron pool are mostly in an oxidized state (Melis et al. 1996). Studies with UWO241 have demonstrated that UWO241 is able to adjust photosystem stoichiometry in response to growth temperature, irradiance and light quality (Szyszka et al. 2007;

Morgan-Kiss et al. 2005). However, *Chlamydomonas sp.* UWO241 is unable to grow under red light illumination exclusively even though the light is absorbed by Chl a and Chl b. Exposure to such a light environment inhibits growth and photosynthetic rates and results in a fourfold increase in excitation pressure with concomitant increases in nonphotochemical quenching (Morgan-Kiss et al. 2005).

13.10.5 State Transitions

Despite the ability of UWO241 to adjust photosystem stoichiometry, one of the most striking characteristics of UWO241 is that this psychrophile lacks the capacity to redistribute light energy among photosystem I and photosystem II through a process called state transitions (Morgan-Kiss et al. 2002b). Thus, *Chlamydomonas sp.* UWO241 is the first natural variant deficient in the state transition response (Morgan-Kiss et al. 2002b). Consequently, the ability to undergo state transitions is not an absolute requirement for survival in green algae. However, the presence of the thylakoid STT7 kinase has been detected in UWO241 at comparable levels to that observed for *C. reinhardtii* and *C. raudensis* SAG 49.72 using immunodetection with specific STT7 antibodies (Szyszka and Hüner, unpublished).

Activation of the STT7 kinase is initiated by its interaction with the Cyt b_6/f complex, which occurs due to a rotation of the Rieske Fe-S protein from a distal to proximal position upon binding of plastoquinol to the Q(o) site of cytochrome b_6 (Fig. 13.1) (Zito et al. 1999; Finazzi et al. 2001; Wollman 2001b). Since STT7 kinase activation depends on the structure of the Cyt b_6/f complex, the observed 7 kDa difference in apparent molecular mass of cytochrome *f* in UWO241 compared to that of *C. reinhardtii*, and its possible impairment of state transitions were assessed (Gudynaite-Savitch et al. 2006). The substitution of cytochrome *f* in a *petA* deletion mutant of the mesophilic *C. reinhardtii* with the *petA* from UWO241 clearly showed that the altered structure of psychrophilic cytochrome *f* is not responsible for its inability of UWO241 to undergo state transitions (Gudynaite-Savitch et al. 2006).

Another requirement for state transitions has recently been established by Zer et al. (Zer et al. 1999) who demonstrated that in order for phosphorylation to occur, LCHII polypeptides must undergo a light-induced conformational change in order to expose the phosphorylation site to the STT7 kinase (Zer et al. 1999). Although it is not known whether LHCII proteins of the Antarctic psychrophile have the capacity to undergo such conformational changes, non-denaturing and SDS-PAGE analyses have revealed differences in both the structure of LHCII complexes, as well as variation in the apparent molecular masses of several individual LHCII proteins, compared to *C. reinhardtii* and *C. raudensis* SAG 49.72 (Morgan et al. 1998a; Szyszka et al. 2007).

13.10.6 Thylakoid Polypeptide Phosphorylation Profile

Phosphorylation of light harvesting complex II (LHCII) polypeptides is essential in the regulation of state transitions (Fig. 13.1) (Depège et al. 2003; Finazzi et al. 2001; Zer et al. 1999). Consistent with the inability to perform state transitions, UWO241 does not phosphorylate LHCII proteins (25–40 kDa) (Fig. 13.4) in response to either growth temperature, irradiance or light quality (Szyszka et al. 2007; Morgan-Kiss et al. 2002b; Morgan-Kiss et al. 2005). In contrast to the typical phosphorylation patterns observed for photosynthetic organisms, UWO241 exhibits a light dependent phosphorylation of high molecular mass polypeptides (70 kDa, >115 kDa), as well as a small 17 kDa protein (Fig. 13.3) (Szyszka et al. 2007; Morgan-Kiss et al. 2002b; Szyszka-Mroz et al. 2015).

Recently, the nature of the altered thylakoid membrane phosphorylation pattern exhibited by the Antarctic psychrophile was examined through the identification of its unique phosphoproteins (Szyszka-Mroz et al. 2015). Two dimensional blue native PAGE of thylakoid membrane complexes demonstrated that

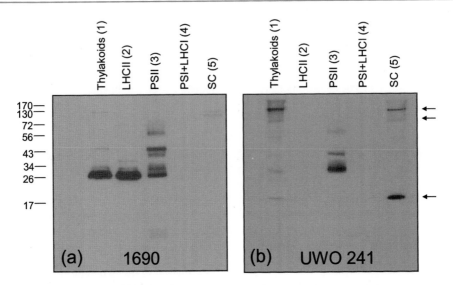

Fig. 13.4 Differential phosphorylation profiles of thylakoid proteins in the model organism (**a**) *Chlamydomonas reinhardtii* 1690 and (**b**) *Chlamydomonas* sp. UWO 241. Thylakoid membranes and purified complexes were separated on SDS-PAGE and immunoblotted with phospho-threonine antibodies. In contrast to *C. reinhardtii*, UWO 241 exhibits the absence of LHCII phosphorylation (**b**, lane 2), and instead, phosphorylation of subunits associated with a high density supercomplex (**b**, lane 5). *Arrows* indicate the novel phosphorylated protein bands in UWO 241. Molecular masses are indicated on the left (kDa)

UWO241 preferentially phosphorylates proteins associated with a large ~1000 kDa pigment-protein supercomplex. Subsequent isolation of this supercomplex using sucrose density gradients revealed that it contains components of both PSI and the Cyt b_6/f complex and thus, most likely functions in CEF. The stability of this supercomplex was dependent upon the phosphorylation state of the thylakoid membrane. Treatment of purified thylakoids with the phosphatase inhibitor, NaF, stabilized while treatment with staurosporine, a kinase inhibitor, destabilized the PSI-Cyt b_6/f supercomplex. The abundance of this supercomplex was also dependent on the concentration of NaCl. The stability of the PSI-Cyt b_6/f supercomplex was significantly increased in high-salt (700 mM NaCl) grown cells compared to low salt (70 mM NaCl) grown cells. The physical stability of this supercomplex was correlated with functional measurements of CEF by P700 photo-oxidation and its dark relaxation kinetics, which indicated that high-salt grown UWO241 with a stable PSI-Cyt b_6/f supercomplex exhibited rates of CEF that were 2.5 times faster than low-salt grown cells which exhibit minimal levels of this complex (Fig. 13.4).

Separation of individual subunits of this PSI-Cyt b_6/f supercomplex by isoelectric focusing, led to the identification of 2 major phosphorylated proteins, including a 17 kDa PsbP-like protein and a 70 kDa ATP-dependent zinc metalloprotease FtsH. The UWO241 PsbP-like protein showed 70.6% amino acid sequence identity relative to the authentic PsbP protein associated with the oxygen evolving complex of PSII in *Chlamydomonas reinhardtii* (Szyszka-Mroz et al. 2015).

13.10.7 PsbP-Like Proteins

Authentic PsbP proteins normally are lumenal PSII subunits associated with the oxygen-evolving complex (OEC), and are not associated with PSI. However, in addition to the authentic *PsbP* genes, recent genomic and proteomic studies have revealed the existence of a family of chloroplast PsbP proteins, categorized as either PsbP-Like (PPL) proteins or PsbP-domain (PPD) proteins (Ifuku et al. 2010). While these proteins share significant sequence and structural homologies, their functions are diverse and unrelated to that of authentic PsbP proteins (Ishihara et al. 2007). Interestingly, several recent studies have demonstrated that the PsbP-like 2 protein (PPL2) was not associated with PSII, but instead, required for the accumulation of chloroplast NDH complex in *Arabidopsis* (Peng et al. 2009; Suorsa et al. 2009). Furthermore, an *Arabidopsis* mutant of the PsbP-domain 5 protein (PPD5) displayed decreased levels of NADPH dehydrogenase (NDH) activity (Roose et al. 2011). The PsbP-domain protein 1 (PPD1) interacts with PsaA and PsaB of PSI and is thought to be essential for the assembly of this complex (Liu et al. 2012).

Based on these findings, together with the recent evidence from UWO241, it was concluded that the unique thylakoid phosphoproteins of this psychrophile are subunits of a PSI- Cyt b_6/f supercomplex that governs CEF. Since stability of the PSI-Cyt b_6/f supercomplex is dependent on the phosphorylation status of the thylakoid membrane, this supercomplex likely functions in the dynamic regulation between linear and cyclic electron transport in UWO241. Therefore, we suggest that the balance of excitation energy between the two photosystems may be maintained through regulation of the CEF pathway, rather than regulation of state transitions in the Antarctic psychrophile.

13.11 A Model of Cyclic Electron Flow in *Chlamydomonas* sp. UWO241

Until recently, the composition of the thylakoid membrane-enclosed lumen has been poorly characterized. Thus, the function of this chloroplast compartment has been considered to play a limited role in electron transport and photosynthesis. A recent proteomic analysis had identified over 80 proteins in the lumen of *Arabidopsis*, for which only approximately half have been assigned a putative function, including isomerases, m-type thioredoxins, chaperones, peroxidases and proteases (Schlicher and Soll 1996; Kieselbach et al. 2000; Adam 2001; Peltier et al. 2002). Two-dimensional isoelectrophoresis of lumenal proteins also revealed the expression of a surprising number of paralogs (Peltier et al. 2002). In addition to several plastocyanin, PsbO (OEC33) and PsbQ (OEC16) paralogs, the authors discovered seven weakly related PsbP (OEC23)-like paralogs (Peltier et al. 2002).

In addition, several reports have shown evidence for nucleotide-dependent processes in the chloroplast lumen [reviewed in (Spetea et al. 2004)], including the presence of a nucleoside diphosphate kinase (NDPKIII), a trans-thylakoid membrane nucleotide transport system, and the capacity of lumenal PsbO (OEC33) to bind GTP (Spetea et al. 2004; Thuswaldner et al. 2007; Yin et al. 2010). Several studies have revealed the existence of a thylakoid ATP/ADP carrier (TAAC) that delivers stromal ATP to the thylakoid lumen in exchange for ADP (Spetea et al. 2004; Thuswaldner et al. 2007; Yin et al. 2010). Spetea et al. (Spetea et al. 2004) demonstrated the presence of a 17 kDa nucleoside diphosphate kinase III (NDPKIII) in the thylakoid lumen, which catalyzes the transfer of a phosphate group from ATP to GDP, in the synthesis of GTP, a nucleotide known to alter the conformation of a wide array of GTP-binding proteins (GTPases) (Spetea et al. 2004). These same authors demonstrated the light and DCMU sensitive high-affinity binding of GTP to the PSII-associated, lumenal PsbO (OEC33) protein, suggesting a correlation to photosynthetic electron transport as well as the additional function of PsbO as a GTPase involved in signal transduction (Spetea et al. 2004). Subsequent studies showed that GTP binding to PsbO induces changes in the structure of this protein and stimulates the dissociation of PsbO from PSII (Lundin et al. 2007a). Furthermore, in *Arabidopsis,* PsbO is encoded by two isoforms, PsbO1 and PsbO2 with seemingly different functions. While PsbO1 supports oxygen evolution, PsbO2 regulates the phosphorylation state and turnover of the D1 protein (Lundin et al. 2007b, 2008). Small plant GTPases act as important molecular switches in plant signaling and exhibit significant diversity in both structure and function (Yang 2002). These G-proteins are typically between 18 and 33 kDa, require Mg^{2+} ions for activity, exist as both soluble and membrane-associated forms and are activated by a guanine nucleotide exchange factor (GEF) (Lundin et al. 2007a; Leipe et al. 2002).

Coincidently, the crystal structure of authentic PsbP from *Nicotiana tabacum* also suggests a novel function for this protein in GTP-regulated metabolism, as PsbP exhibits strong structural similarity to Mog1p, (Ifuku et al. 2004). Analysis of PsbP structural homology indicated that the folding of this protein is very similar to that of Mog1p, a 24 kDa regulatory protein which interacts with a multifunctional Ran, a small GTPase of the RAS superfamily (Ifuku et al. 2004; Baker et al. 2001). Studies in *Saccharomyces cerevisiae* have shown that Mog1p is a novel factor involved in regulating the nucleotide state of Ran (Ifuku et al. 2004). Upon formation of a complex between Mog1p and a Ran GTPase, Mog1p stimulates the release of bound GDP or GTP from Ran (Ifuku et al. 2004). It has also been suggested that Mog1p may function in the binding of a kinase involved in signal transduction of osmotic response genes (Lu et al. 2004).

Based on our recent studies of a PSI-Cyt b_6/f supercomplex, together with established functions of several related proteins, we propose a model for cyclic electron flow in *Chlamydomonas sp.* UWO241 (Fig. 13.5). The major subunits identified in this supercomplex include photosystem I proteins, cytochrome b_6, cytochrome f, a 17 kDa PsbP-like protein, an ADP-ribosylation factor associated with the ARF family of GTPases, a PSI assembly protein associated with the Ycf4 superfamily in *Chlamydomonas*, an ATP-dependent FtsH metalloprotease, heat shock protein 70 and an adenine nucleotide (ATP/ATP) translocator. Both FtsH and the PsbP-like protein appear to be the primary phosphorylated subunits of this supercomplex.

We propose that under conditions associated with increased ATP demand (i.e. low temperatures, high irradiance, high salt), phosphorylation of a lumenal PsbP-like protein in UWO241 may increase its affinity towards photosystem I. Increased levels of PSI bound PsbP-like proteins may promote their interaction with membrane-associated small Ras GTPase proteins. Binding of a PsbP-like protein with an inactive GTPase-GDP could induce a conformational change to remove the GDP nucleotide group. This would facilitate the guanine nucleotide exchange of GTPase, possibly via a lumenal nucleoside diphosphate kinase (NDPK) in the activation of GTPase-GTP, which in turn, would play a role in regulating GTPase effector proteins. These proteins may include some of the subunits found associated with the PSI-Cyt b_6/f supercomplex, including heat shock protein 70 (HSP70), ATP-dependent metalloprotease (FtsH) and Ycf4, which have previously been identified to function as chaperones or assembly proteins and are therefore, likely involved in maintaining the stability of this supercomplex. Thus, phosphorylation of a PsbP-like protein may subsequently activate a GTPase together with its effector proteins to ultimately assemble the structure of the PSI-Cyt b_6/f supercomplex and potentially signal other GTPase-dependent events in the lumen.

Conversely, during conditions when ATP demand is reduced (i.e. low salt, low light), dephosphorylated PsbP-like proteins may dissociate from PSI and subsequently localize to the thylakoid lumen. Decreased PsbP-like protein interaction with GTPase may result in the inactive GTPase-GDP form and favour disassembly of the PSI-Cyt b_6/f supercomplex, resulting in lower levels of CEF.

In addition, previous studies have demonstrated that a hypersaline environment can result in NaCl-induced phosphorylation of several thylakoid membrane proteins by activating the corresponding kinase (Liu and Shen 2004). Therefore, under conditions of high-salt, increased levels of phosphorylated PSI-Cyt b_6/f supercomplex subunits could further enhance CEF in UWO241.

The chloroplast ATP/ADP transporter functions to supply ATP to the lumen in exchange for ADP, which can subsequently be converted to GTP. It has been demonstrated that

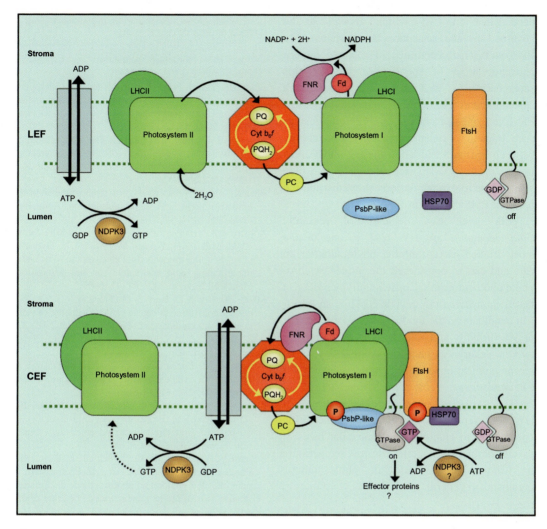

Fig. 13.5 A hypothetical model of linear electron flow (*top*) and the PSI-Cyt b6/f supercomplex (*bottom*) involved in cyclic electron flow in the psychrophilic *Chlamydomonas* sp. UWO 241

lumenal GTP enhances D1 degradation by altering the proteolytic system (Spetea et al. 2000). Therefore, enhanced D1 photoinhibition and repair observed in UWO241 at low temperatures may be associated with increased levels of lumenal GTP resulting from higher rates of CEF, in addition to increased adenylate pools.

13.12 Conclusions

Relatively little is known about cold-adapted microorganisms that dominate the large, low-temperature portion of the Earth, in particular, the psychrophilic primary producers that depend on photoautotrophic metabolism. While UWO241 exhibits some typical cold adapted characteristics with respect to cell morphology, individual proteins and thylakoid membrane lipids and fatty acids, this psychrophile has revealed some unique photosynthetic features. The major distinguishing feature between UWO241 and mesophilic green algae is its inability to undergo state transitions. This appears to be due to the unique composition and organization of the electron transport chain. Compared to similar green algae from temperate environments, UWO241 exhibits a greater dependence on CEF which appears to be governed by the phosphorylation status of a PSI-Cyt b_6/f supercomplex. The novel protein phosphorylation of specific subunits of this supercomplex as well as the salt concentration appear to control the stability and function of this PSI-Cyt b_6/f supercomplex. Therefore, just as state transitions balance the excitation energy between PSII and PSI when these photosystems operate in concert with each other, CEF may play a prominent role in regulating energy distribution between PSII and PSI in UWO241.

With over two decades of research, *Chlamydomonas sp.* UWO241 is one of the most studied psychrophilic phytoplankton to date and can be considered the model psychrophilic photoautotroph. Consequently, we suggest that *Chlamydomonas sp.* UWO241 is an excellent candidate as the model system for the study of psychrophily and adaptation to low temperature in eukaryotic photoautotrophs. Given the recent publication of the genome sequence for *Chlamydomonas reinhardtii* (Merchant et al. 2007), comparison with genome sequence of UWO241 represents an exceptional scientific opportunity to elucidate the molecular basis of psychrophily and adaptation to low temperature in *Chlamydomonas* and in photosynthetic organisms in general.

Acknowledgements BS is the recipient of an NSERC Postgraduate Fellowship. NPAH gratefully acknowledges the long-term financial support from the NSERC Discovery Grants programme as well as funding from the Canada Foundation for Innovation and the Canada Research Chairs programme.

Conflict of Interest We declare no conflict of interest.

References

Adam Z. Chloroplast proteases and their role in photosynthetic regulation. In: Aro EM, Andersson B, Eds. Advances in Photosynthesis and Respiration. Dordrecht, Kluwer, 2001; pp. 265-276.

Adams III WW, Hoehn A, Demmig-Adams B. Chilling temperatures and the xanthophyll cycle. A comparison of warm-grown and overwintering spinach. Aust J Plant Physiol 1995; 22: 75-85.

Allen JF. Redox control of gene expression and the function of chloroplast genomes- an hypothesis. Photosynth Res 1993; 36: 95-102.

Aluru MR, Zola J, Foudree A, Rodermel SR. Chloroplast photooxidation-induced transcriptome reprogramming in *Arabidopsis* immutans white leaf sectors. Plant Physiol 2009; 150: 904-23.

Anderson JM, Chow WS, Park YI. The grand design of photosynthesis: acclimation of the photosynthetic apparatus to environmental cues. Photosynth Res 1995; 46: 129-39.

Apel K, Hirt H. Reactive oxygen species: metabolism, oxidative stress, and signal transduction. Annu Rev Plant Biol 2004; 55: 373-99.

Armstrong AF, Badger MR, Day DA, Barlett MM, Smith PMC, Millar AH, Whelan JM, Atkin OK. Dynamic changes in the mitochondrial electron transport chain underpinning cold acclimation of leaf respiration. Plant Cell Environ 2008; 31: 1156-69.

Atkin OK, Bruhn D, Hurry VM, Tjoelker MG. The hot and the cold: unravelling the variable response of plant respiration to temperature. Funct Plant Biol 2005; 32: 87-105.

Baker RP, Harreman MT, Eccleston JF, Corbett AH, Stewart M. Interaction between Ran and Mog1 is required for efficient nuclear protein import. J Biol Chem 2001; 276: 41255-62.

Bielewicz S, Bell EM, Kong W, Friedberg I, Priscu JC, Morgan-Kiss RM. Protist diversity in a permanently ice-covered Antarctic lake during polar night transition. ISME J 2011; 5: 1559-64.

Bonardi V, Pesaresi P, Becker T, Schleiff E, Wagner R, Pfannschmidt T, Jahns P, Leister D. Photosystem II core phosphorylation and photosynthetic acclimation require two different protein kinases. Nature 2005; 437: 1179-82.

Bott TL, Kaplan LA. Bacterial biomass, metabolic state, and activity in stream sediments: relation to environmental variables and multiple assay comparisons. Appl Environ Microbiol 1985; 50: 508-22.

Brautigam K, Dietzel L, Kleine T, Stroher E, Wormuth D, Dietz KJ, Radke D, Wirtz M, Hell R, Dormann P, Nunes-Nesi A, Schauer N, Fernie AR, Oliver SN, Geigenberger P, Leister D, Pfannschmidt T. Dynamic plastid redox signals integrate gene expression and metabolism to induce distinct metabolic states in photosynthetic acclimation in *Arabidopsis*. Plant Cell 2009; 21: 2715-32.

Bravo LA, Saavedra-Mella FA, Vera F, Guerra A, Cavieres LA, Ivanov AG, Hüner NPA, Corcuera LJ. Effect of cold acclimation on the photosynthetic performance of two ecotypes of *Colobanthus quitensis* (Kunth) Bartl. J Exp Bot 2007; 58: 3581-90.

Cardol P, Forti G, Finazzi G. Regulation of electron transport in microalgae. Biochim Biophys Acta 2011; 1807: 912-18.

Chen Z, HE C, Hu H. Temperature responses of growth, photosynthesis, fatty acid and nitrate reductase in Antarctic and temperate *Stichococcus*. Extremophiles 2012; 16: 127–33.

Chen M, Thelen JJ. Acyl-lipid desaturase 2 is required for chilling and freezing tolerance in *Arabidopsis*. Plant Cell 2013; 25: 1430–44.

Chen WM, Jin N, Shi Y, Su YQ, Fei BJ, Li W, Qiao DR, Cao Y. Coordinate expression of light-harvesting chlorophyll a/b gene family of photosystem II and chlorophyll a oxygenase gene regulated by salt-induced phosphorylation in *Dunaliella salina*. Photosynthetica 2010; 48: 355–60.

Chen Z, Gong Y, Fang X, Hu H. *Scenedesmus sp*. NJ-1 isolated from Antarctica: a suitable renewable lipid source for biodiesel production. World J Microbiol Biotechnol 2012; 28: 3219–25.

Clarke JE, Johnson GN. In vivo temperature dependence of cyclic and pseudocyclic electron transport in barley. Planta 2001; 212: 808–16.

D'Amico S, Collins T, Marx JC, Feller G, Gerday C. Psychrophilic microorganisms: challenges for life. EMBO Rep 2006; 7: 385–89.

Dahal K, Kane K, Gadapati W, Webb E, Savitch LV, Singh J, Sharma P, Sarhan F, Longstaffe FJ, Grodzinski B, Hüner NPA. The effects of phenotypic plasticity on photosynthetic performance in winter rye, winter wheat and *Brassica napus*. Physiol Plant 2012a; 144: 169–88.

Dahal K, Gadapati W, Savitch L, Singh J, Hüner NPA. Cold acclimation and BnCBF17-over-expression enhance photosynthetic performance and energy conversion efficiency during long-term growth of *Brassica napus* under elevated CO_2 conditions. Planta 2012b; 236: 1639–52.

Dahal K, Kane K, Gadapati W, Webb E, Savitch LV, Singh J, Sharma P, Sarhan F, Longstaffe FJ, Grodzinski B, Hüner NPA. The effects of phenotypic plasticity on photosynthetic performance in winter rye, winter wheat and *Brassica napus*. Physiol Plant 2012c; 144: 169–88.

Dahal K, Knowles V, Plaxton WC, Hüner NPA. Enhancement of photosynthetic performance, water use efficiency and grain yield under long-term growth at elevated CO_2 in wheat and rye is growth temperature and cultivar dependent. Environ Exp Bot 2013a; 106: 207–220.

Dahal K, Kane K, Sarhan F, Savitch LV, Singh J, Grodzinski B, Hüner NPA. C-Repeat transcription factors as targets for the maintenance of crop yield under suboptimal growth conditions. In: Pessarkli M, Ed. Handbook of Plant and Crop Physiology (3rd ed), Florida, Taylor & Francis Books, 2013b; pp. 313–32.

Demchenko AP, Ruskyn OI, Saburova EA. Kinetics of the lactate dehydrogenase reaction in high-viscosity media. Biochim Biophys Acta 1989; 998: 196–03.

Demmig-Adams B, Adams WW, Ebbert V, Logan BA. Ecophysiology of the xanthophyll cycle. In: Frank HA, Young AJ, Britton G, Cogdell RJ, Eds.
Advances in Photosynthesis. The Photochemistry of Carotenoids. Dordrecht, Kluwer Academic, 1999; pp. 245–269.

Depège N, Bellafiore S, Rochaix JD. Role of chloroplast protein kinase Stt7 in LHCII phosphorylation and state transition in *Chlamydomonas*. Science 2003; 299: 1572–5.

Devos N, Ingouff M, Loppes R, Matagne RF. Rubisco adaptation to low temperatures: A comparative study in psychrophilic and mesophilic unicellular algae. J Phycol 1998; 34: 655–60.

Dietz KJ. Plant peroxiredoxins. Annu Rev Plant Biol 2003; 54: 93–107.

Dietz KJ, Pfannschmidt T. Novel regulators in photosynthetic redox control of plant metabolism and gene expression. Plant Physiol 2011; 155: 1477–85.

Dolhi JM, Maxwell DP, Morgan-Kiss RM. Review: the Antarctic *Chlamydomonas raudensis*: an emerging model for cold adaptation of photosynthesis. Extremophiles 2013; 17:711–22.

Dubacq JP, Tremolieres A. Occurence and function of phosphatidylglycerol containing delta3-transhexadecenoic acid in photosynthetic lamellae. Physiol Veg 1983; 21: 293–12.

Eberhard S, Finazzi G, Wollman FA. The dynamics of photosynthesis. Annu Rev Genet 2008; 42: 463–15.

Eicken H. The role of sea ice in structuring Antarctic ecosystems. Polar Biol 1992; 12: 3–13.

Ensminger I, Busch F, Hüner NPA. Photostasis and cold acclimation: sensing low temperature through photosynthesis. Physiol Plant 2006; 126: 28–44.

Escoubas JM, Lomas M, LaRoche J, Falkowski PG. Light intensity regulates cab gene transcription via the redox state of the plastoquinone pool in the green alga, *Dunaliella tertiolecta*. Proc Natl Acad Sci USA 1995; 92: 10237–41.

Estavillo GM, Crisp PA, Pornsiriwong W, Wirtz M, Collinge D, Carrie C, Giraud E, Whelan J, David P, Javot H, Brearley C, Hell R, Marin E, Pogson BJ. Evidence for a SAL1-PAP chloroplast retrograde pathway that functions in drought and high light signaling in *Arabidopsis*. Plant Cell 2011; 23: 3992–12.

Estavillo GM, Chan KX, Phua SY, Pogson BJ. Reconsidering the nature and mode of action of metabolite retrograde signals from the chloroplast. Front Plant Sci 2013; 3: 300.

Falkowski PG. Light–shade adaptation and vertical mixing of marine phytoplankton: a comparative field study. J Mar Res 1983; 41: 215–37.

Falkowski PG, Chen YB. Photoacclimation of light harvesting systems in eukaryotic algae. In: Green BR, Parson WW, Eds. Advances in photosynthesis and respiration Light harvesting antennas in photosynthesis. Dordrecht, Kluwer Academic, 2003a; pp. 423–47.

Falkowski PG, Chen YB. Photoacclimation of light harvesting systems in eukaryotic algae. Advances in Photosynthesis and Respiration. Light Harvesting Antennas in Photosynthesis (Green BR and Parson

WW, eds). Kluwer Academic Publishers, Dordrecht 2003b; 13: 423–447.

Falkowski PG, Owens TG, Ley AC, Mauzerall DC. Effects of growth irradiance levels on the ratio of reaction centers in two species of marine phytoplankton. Plant Physiol 1981; 68: 969–73.

Feller G, Gerday C. Psychrophilic enzymes: hot topics in cold adaptation. Nat Rev Microbiol 2003; 1: 200–208.

Fernandez AP, Strand A. Retrograde signaling and plant stress: plastid signals initiate cellular stress responses. Curr Opin Plant Biol 2008; 11: 509–13.

Finazzi G, Forti G. Metabolic flexibility of the green alga *C. reinhardtii* as revealed by the link between state transitions and cyclic electron flow. Photosynth Res 2004; 82: 327–38.

Finazzi G, Furia A, Barbagallo RP, Forti G. State transitions, cyclic and linear electron transport and photophosphorylation in *Chlamydomonas reinhardtii*. Biochim Biophys Acta 1999; 1413: 117–29.

Finazzi G, Zito F, Barbagallo RP, Wollman FA. Contrasted effects of inhibitors of cyt b_6/f complex on state transitions in *Chlamydomonas reinhardtii*: the role of Qo site occupancy in LHCII kinase activation. J Biol Chem 2001; 276: 9770–74.

Foyer CH, Neukermans J, Queval G, Noctor G, Harbinson J. Photosynthetic control of electron transport and the regulation of gene expression. J Exp Bot 2012; 63: 1637–61.

Fristedt R, Willig A, Granath P, Crevecoeur M, Rochaix JD, Vener AV. Phosphorylation of photosystem II controls functional macroscopic folding of photosynthetic membranes in *Arabidopsis*. Plant Cell 2009; 21: 3950–64.

Fujita Y. A study on the dynamic features of photosystem stoichiometry - accomplishments and problems for future studies. Photosynth Res 1997; 53: 83–93.

Golding AJ, Johnson GN. Down-regulation of linear and activation of cyclic electron transport during drought. Planta 2003; 218: 107–14.

Gray GR, Ivanov AG, Krol M, Williams JP, Kahn MU, Myscich EG, Hüner NPA. Temperature and light modulate the trans-delta3-hexadecenoic acid content of phosphatidylglycerol: light-harvesting complex II organization and non-photochemical quenching. Plant Cell Physiol 2005; 46: 1272–82.

Gudynaite-Savitch L, Gretes M, Morgan-Kiss R, Savitch LV, Simmonds J, Kohalmi SE, Hüner NPA. Cytochrome *f* from the Antarctic psychrophile, *Chlamydomonas raudensis* UWO 241: structure, sequence, and complementation in the mesophile *Chlamydomonas reinhardtii*. Mol Genet Genom 2006; 275: 387–98.

Haehnel W. Photosynthetic electron transport in higher plants. Annu Rev Plant Physiol 1984; 35: 659–93.

Hatano S, Kabata K, Yoshimoto M, Sadakane H. Accumulation of free fatty acids during hardening of *Chlorella ellipsoidea*. Plant Physiol 1982; 70: 1173–77.

Hazel JR. Thermal adaptation in biological membranes: is homeoviscous adaptation the explanation? Annu Rev Physiol 1995; 57: 19–42.

Heber U, Walker D. Concerning a dual function of coupled cyclic electron transport in leaves. Plant Physiol 1992; 100: 1621–26.

Hochachka PW, Somero GN, Ed. Biochemical Adaptation. Mechanisms and processes in physiological evolution. New York: Oxford University Press Inc 2002; pp. 290–450.

Hopkins WG, Hüner NPA, Ed. Introduction to Plant Physiology, 4th edition. New York: Wiley and Sons 2009; pp. 77–90.

Horton P, Johnson MP, Perez-Bueno ML, Kiss AZ, Ruban AV. Photosynthetic acclimation: does the dynamic structure and macro-organisation of photosystem II in higher plant grana membranes regulate light harvesting states? FEBS J 2008; 275: 1069–79.

Hu H, Li H, Xu X. Alternative cold response modes in *Chlorella* (Chlorophyta, Trebouxiophyceae) from Antarctica. Phycologia 2008; 47: 28–34.

Hüner NPA, Grodzinski B. Photosynthesis and photoautotrophy. In: Murray Moo-Young, Ed. Comprehensive Biotechnology (second edition) Amsterdam, Elsevier, 2011; pp. 315–22.

Hüner NPA, Krol M, Williams JP, Maissan E, Low P, Roberts D, Thompson JE. Low temperature development induces a specific decrease in trans-$\Delta 3$-hexadecenoic acid content which influences LHCII organization. Plant Physiol 1987; 84: 12–18.

Hüner NPA, Williams JP, Maissan EE, Myscich EG, Krol M, Laroche A, Singh J. Low temperaure-induced decrease in trans-delta3-hexadecenoic acid content is correlated with freezing tolerance in cereals. Plant Physiol 1989; 89: 144–50.

Hüner NPA, Öquist G, Hurry VM, Krol M, Falk S, Griffith M. Photosynthesis, photoinhibition and low temperature acclimation in cold tolerant plants. Photosynth Res 1993; 37: 19–39.

Hüner NPA, Öquist G, Sarhan F. Energy balance and acclimation to light and cold. Trends Plant Sci 1998; 3: 224–30.

Hüner NPA, Öquist G, Melis A. Photostasis in plants, green algae and cyanobacteria: the role of light harvesting antenna complexes. In: Green BR, Parson WW, Eds. Advances in Photosynthesis and Respiration Light Harvesting Antennas in Photosynthesis. Dordrecht, Kluwer Academic, 2003a; pp. 401–421.

Hüner NPA, Öquist G, Melis A. 2003b. Photostasis in plants, green algae and cyanobacteria: the role of light harvesting antenna complexes. In: Green BR, Parson WW, Eds. Advances in photosynthesis and respiration Light harvesting antennas in photosynthesis. Dordrecht, Kluwer Academic, 2003; pp. 401–21.

Hüner NPA, Dahal K, Hollis L, Bode R, Rosso D, Krol M, Ivanov AG. Chloroplast redox imbalance governs phenotypic plasticity: the "grand design of photosynthesis" revisited. Front Plant Sci 2012; 3: 255.

Hüner NPA, Bode R, Dahal K, Busch FA, Possmayer M, Szyszka B, Rosso D, Ensminger I, Krol M, Ivanov AG, Maxwell DP. Shedding some light on cold acclimation, cold adaptation, and phenotypic plasticity. Botany 2013a; 91: 127–36.

Hüner NPA, Bode R, Dahal K, Busch FA, Possmayer M, Szyszka B, Rosso D, Ensminger I, Krol M, Ivanov AG, Maxwell DP. Shedding some light on cold acclimation, cold adaptation, and phenotypic plasticity. Botany 2013b; 91: 127–36.

Hurry VM, Keerberg O, Pärnik T, Gardeström P, Öquist G. Cold-hardening results in increased activity of enzymes involved in carbon metabolism in leaves of winter rye (*Secale cereale L*). Planta 1995a; 195: 554–62.

Hurry V, Tobiaeson M, Krömer S, Gardeström P, Öquist G. Mitochondria contribute to increased photosynthetic capacity of leaves of winter rye (*Secale cereale L.*) following cold-hardening. Plant Cell Environ 1995b; 18: 69–76.

Ifuku K, Nakatsu T, Kato H, Sato F. Crystal structure of the PsbP protein of photosystem II from *Nicotiana tabacum*. EMBO Rep 2004; 5: 362–7.

Ifuku K, Ishihara S, Sato F. J. Integr. Molecular functions of oxygen-evolving complex family proteins in photosynthetic electron flow. Plant Biol 2010; 52: 723–34.

Inaba M, Suzuki I, Szalontai B, Kanesaki Y, Los DA, Hayashi H, Murata N. Gene engineered rigidification of membrane lipids enhances the cold inducibility of gene expression in *Synechocystis*. J Biol Chem 2003; 278: 12191–98.

Ishihara S, Takabayashi A, Ido K, Endo T, Ifuku K, Sato F. Distinct functions for the two PsbP-like proteins PPL1 and PPL2 in the chloroplast thylakoid lumen of *Arabidopsis*. Plant Physiol 2007; 145: 668–79.

Iwai M, Takizawa K, Tokutsu R, Okamuro A, Takahashi Y, Minagawa J. Isolation of the elusive supercomplex that drives cyclic electron flow in photosynthesis. Nature 2010; 464: 1210–13.

Jung H-S, Chory J. Signaling between chloroplasts and the nucleus: can a systems biology approach bring clarity to a complex and highly regulated pathway? Plant Physiol 2010; 152: 453–59.

Kargul J, Barber J. Photosynthetic acclimation: structural reorganisation of light harvesting antenna; role of redox-dependent phosphorylation of major and minor chlorophyll a/b binding proteins. FEBS J 2008; 275: 1056–68.

Karpinski S, Reynolds H, Karpinska B, Wingsle G, Creissen G, Mullineaux P Systemic signaling and acclimation in response to excess excitation energy in *Arabidopsis*. Science 1999; 284: 654–7.

Kieselbach T, Bystedt M, Hynds P, Robinson C, Schröder WP. A peroxidase homologue and novel plastocyanin located by proteomics to the *Arabidopsis* chloroplast thylakoid lumen. FEBS Lett 2000; 480: 271–6.

Krol M, Hüner NPA, Williams JP, Maissan E. Chloroplast biogenesis at cold hardening temperatures. Kinetics of trans-3-hexadecenoic acid accumulation and the assembly of LHCII. Photosynth Res 1988; 15: 115–32.

Krol M, Ivanov AG, Jansson S, Kloppstech K, Hüner NPA. Greening under high light or cold temperature affects the level of xanthophyll-cycle pigments, early light-inducible proteins, and light-harvesting polypeptides in wild-type barley and the chlorina f2 mutant. Plant Physiol 1999; 120: 193–04.

Krupa Z, Williams JP, Hüner N. The role of acyl lipids in reconstitution of lipid-depleted light harvesting complex II from cold hardened and nonhardened rye. Plant Physiol 1992; 100: 931–38.

Kurepin L, Dahal K, Savitch L, Singh J, Bode R, Ivanov AG, Hurry V, Hüner NPA. Role of CBFs as integrators of chloroplast redox, phytochrome and plant hormone signaling during cold acclimation. Int J Mol Sci 2013; 14: 12729–63.

Leipe DD, Wolf YI, Koonin EV, Aravind L. Classification and evolution of P-loop GTPases and related ATPases. J Mol Biol 2002; 317: 41–72.

Leonardos ED, Savitch LV, Hüner NPA, Öquist G, Grodzinski B. Daily photosynthetic and C-export patterns in winter wheat leaves during cold stress and acclimation. Physiol Plant 2003; 117: 521–31.

Lermontova I, Grimm B. Reduced activity of plastid protoporphyrinogen oxidase causes attenuated photodynamic damage during high-light compared to low-light exposure. Plant J 2006; 48: 499–10.

Li Z, Ahn TK, Avenson TJ, Ballottari M, Cruz JA, Kramer DM, Bassi R, Fleming GR, Keasling JD, Niyogi KK. Lutein accumulation in the absence of zeaxanthin restores nonphotochemical quenching in the *Arabidopsis thaliana* npq1 mutant. Plant Cell 2009; 21: 1798–12.

Liska AJ, Shevchenko A, Pick U, Katz A. Enhanced photosynthesis and redox energy production contribute to salinity tolerance in *Dunaliella* as revealed by homology-based proteomics. Plant Physiol 2004; 136: 2806–17.

Liu XD, Shen YG. NaCl-induced phosphorylation of light harvesting chlorophyll a/b proteins in thylakoid membranes from the halotolerant green alga, *Dunaliella salina*. FEBS Lett 2004; 569: 337–40.

Liu J, Yang H, Lu Q, Wen X, Chen F, Peng L, Zhang L, Lu C. PsbP-domain protein1, a nuclear-encoded thylakoid lumenal protein, is essential for photosystem I assembly in *Arabidopsis*. Plant Cell 2012; 24: 4992–06.

Lizotte MP, Priscu JC. Natural fluorescence and quantum yields in vertically stationary phytoplankton from perennially ice-covered lakes. Limnol Oceanogr 1994; 39: 1399–410.

Lizotte MP, Sharp TR, Priscu JC. Phytoplankton dynamics in the stratified water column of Lake Bonney, Antarctica: biomass and productivity during the winter-spring transition. Polar Biol 1996; 16: 155–162.

Los DA, Murata N. Sensing and responses to low temperature in cyanobacteria. In: Storey KB, Storey JM, Eds.

Sensing, Signalling and Cell Adaptation. Amsterdam, Elsevier Science BV, 2002; pp. 139–53.

Los DA, Murata N. Membrane fluidity and its roles in the perception of environmental signals. Biochim Biophys Acta 2004; 1666:142–57.

Los D, Mironov K, Allakhverdiev S. Regulatory role of membrane fluidity in gene expression and physiological functions. Photosynth Res 2013; 116: 489–509.

Lu JM, Deschenes RJ, Fassler JS. Role for the Ran binding protein, Mog1p, in *Saccharomyces cerevisiae* SLN1-SKN7 signal transduction. Eukaryot Cell 2004; 3:1544–56.

Lucker B, Kramer D. Regulation of cyclic electron flow in *Chlamydomonas reinhardtii* under fluctuating carbon availability. Photosynth Res 2013a; 117: 449–59.

Lucker B, Kramer DM. Regulation of cyclic electron flow in *Chlamydomonas reinhardtii* under fluctuating carbon availability. Photosynth Res 2013b; 117: 449–59.

Lundin B, Thuswaldner S, Shutova T, Eshaghi S, Samuelsson G, Barber J, Andersson B, Spetea C. Subsequent events to GTP binding by the plant PsbO protein: structural changes, GTP hydrolysis and dissociation from the photosystem II complex. Biochim Biophys Acta. 2007a; 1767: 500–8.

Lundin B, Hansson M, Schoefs B, Vener AV, Spetea C. The *Arabidopsis* PsbO2 protein regulates dephosphorylation and turnover of the photosystem II reaction centre D1 protein. Plant J 2007b; 49: 528–39.

Lundin B, Nurmi M, Rojas-Stuetz M, Aro EM, Adamska I, Spetea C. Towards understanding the functional difference between the two PsbO isoforms in *Arabidopsis thaliana* – insights from phenotypic analyses of psbo knockout mutants. Photosynth Res 2008; 98: 405–14.

Luo T, Fan T, Liu Y, Rothbart M, Yu J, Zhou S, Grimm B, Luo M. Thioredoxin redox regulates ATPase activity of magnesium chelatase CHLI subunit and modulates redox-mediated signalling in tetrapyrrole biosynthesis and homeostasis of reactive oxygen species in pea plants. Plant Physiol 2012; 159: 118–30.

Mancuso Nichols CA, Garon S, Bowman JP, Raguénès G, Guézennec J. Production of exopolysaccharides by Antarctic marine bacterial isolates. J Appl Microbiol 2004; 96: 1057–66.

Masuda T, Tanaka A, Melis A. Chlorophyll antenna size adjustments by irradiance in *Dunaliella salina* involve coordinate regulation of chlorophyll a oxygenase (CAO) and Lhcb gene expression. Plant Mol Biol 2003; 51: 757–71.

Maxwell DP, Laudenbach DE, Hüner NPA. Redox regulation of light-harvesting complex II and cab mRNA abundance in *Dunaliella salina*. Plant Physiol 1995a; 109: 787–95.

Maxwell DP, Falk S, Hüner NPA. Photosystem II excitation pressure and development of resistance to photoinhibition I. LHCII abundance and zeaxanthin content in *Chlorella vulgaris*. Plant Physiol 1995b; 107: 687–94.

Melis A, Murakami A, Nemson JA, Aizawa K, Ohki K, Fujita Y. Chromatic regulation in *Chlamydomonas reinhardtii* alters photosystem stoichiometry and improves the quantum efficiency of photosynthesis. Photosynth Res 1996; 47: 253–65.

Merchant SS, Prochnik SE, Vallon O, Harris EH, Karpowicz SJ, Witman GB, Terry A, Salamov A, Fritz-Laylin LK, Marechal-Drouard L, Marshall WF, Qu L-H, Nelson DR, Sanderfoot AA, Spalding MH, Kapitonov VV, Ren Q, Ferris P, Lindquist E, Shapiro H, Lucas SM, Grimwood J, Schmutz J, Cardol P, Cerutti H, Chanfreau G, Chen C-L, Cognat V, Croft MT, Dent R, Dutcher S, Fernandez E, Fukuzawa H, Gonzalez-Ballester D, Gonzalez-Halphen D, Hallmann A, Hanikenne M, Hippler M, Inwood W, Jabbari K, Kalanon M, Kuras R, Lefebvre PA, Lemaire SD, Lobanov AV, Lohr M, Manuell A, Meier I, Mets L, Mittag M, Mittelmeier T, Moroney JV, Moseley J, Napoli C, Nedelcu AM, Niyogi K, Novoselov SV, Paulsen IT, Pazour G, Purton S, Ral J-P, Riano-Pachon DM, Riekhof W, Rymarquis L, Schroda M, Stern D, Umen J, Willows R, Wilson N, Zimmer SL, Allmer J, Balk J, Bisova K, Chen C-J, Elias M, Gendler K, Hauser C, Lamb MR, Ledford H, Long JC, Minagawa J, Page MD, Pan J, Pootakham W, Roje S, Rose A, Stahlberg E, Terauchi AM, Yang P, Ball S, Bowler C, Dieckmann CL, Gladyshev VN, Green P, Jorgensen R, Mayfield S, Mueller-Roeber B, Rajamani S, Sayre RT, Brokstein P, Dubchak I, Goodstein D, Hornick L, Huang YW, Jhaveri J, Luo Y, Martinez D, Ngau WCA, Otillar B, Poliakov A, Porter A, Szajkowski L, Werner G, Zhou K, Grigoriev IV, Rokhsar DS, Grossman AR. The *Chlamydomonas* genome reveals the evolution of key animal and plant functions. Science 2007; 318: 245–50.

Miskiewicz E, Ivanov AG, Hüner NPA. Stoichiometry of the photosynthetic apparatus and phycobilisome structure of the cyanobacterium *Plectonema boryanum* UTEX 485 are regulated by both light and temperature. Plant Physiol 2002; 130: 1414–25.

Mitchell R, Spillmann A, Haehnel W. Plastoquinol diffusion in linear photosynthetic electron transport. Biophys J 1990; 58: 1011–24.

Moorhead DL, Doran PT, Fountain AG, Lyons WB, McKnight DM, Priscu JC, Virginia RA, Wall DH. Ecological legacies: impacts on ecosystems of the McMurdo Dry Valleys. BioScience 1999; 49: 1009–19.

Morgan RM, Ivanov AG, Priscu JC, Maxwell DP, Hüner NPA. Structure and composition of the photochemical apparatus of the antarctic green alga, *Chlamydomonas subcaudata*. Photosynth Res 1998a; 56: 303–14.

Morgan RM, Ivanov AG, Priscu JC, Maxwell DP, Hüner NPA. Structure and composition of the photochemical apparatus of the Antarctic green alga, *Chlamydomonas subcaudata*. Photosynth Res 1998b; 56: 303–14.

Morgan-Kiss R, Ivanov AG, Williams J, Khan M, Hüner NPA. Differential thermal effects on the energy distribution between photosystem II and photosystem I in thylakoid membranes of a psychrophilic and a mesophilic alga. Biochim Biophys Acta 2002a; 1561: 251–65.

Morgan-Kiss RM, Ivanov AG, Hüner NPA. The Antarctic psychrophile, *Chlamydomonas subcaudata*, is deficient in state I-state II transitions. Planta. 2002b; 214: 435–45.

Morgan-Kiss RM, Ivanov AG, Pocock T, Krol M, Gudynaite-Savitch L, Hüner NPA. The Antarctic psychrophile, *Chlamydomonas raudensis* Ettl (UWO241) (Chlorophyceae, Chlorophyta) exhibits a limited capacity to photoacclimate to red light. J Phycol 2005; 41: 791–00.

Morgan-Kiss RM, Priscu JC, Pocock T, Gudynaite-Savitch L, Hüner NP. Adaptation and acclimation of photosynthetic microorganisms to permanently cold environments. Microbiol Mol Biol R 2006a; 70: 222–52.

Morgan-Kiss R, Priscu JC, Pocock T, Gudynaite-Savitch-L, Hüner NPA. Adaptation and acclimation of photosynthetic microorganisms to permanently cold environments. Microbiol Mol Biol Rev 2006b; 70: 222–52.

Munekage Y, Hojo M, Meurer J, Endo T, Tasaka M, Shikanai T. PGR5 is involved in cyclic electron flow around photosystem I and is essential for photoprotection in *Arabidopsis*. Cell 2002; 110: 361–71.

Munekage Y, Hashimoto M, Miyake C, Tomizawa K, Endo T, Tasaka M, Shikanai T. Cyclic electron flow around photosystem I is essential for photosynthesis. Nature 2004; 429: 579–82.

Murata N, Los DA. Membrane fluidity and temperature perception. Plant Physiol 1997; 115: 875–79.

Murata N, Siegenthaler PA. Lipids in photosynthesis: an overview. In: Siegenthaler P-A, Ed. Advances in Photosynthesis. Lipids in Photosynthesis: Structure Function and Genetics. Dordrecht, Kluwer Academic, 1998; pp. 1–20.

Nagao M, Matsui K, Uemura M. *Klebsormidium flaccidum*, a charophycean green alga, exhibits cold acclimation that is closely associated with compatible solute accumulation and ultrastructural changes. Plant Cell Environ 2008; 31: 872–85.

Napolitano MJ, Shain DH. Four kingdoms on glacier ice: convergent energetic processes boost energy levels as temperatures fall. Proc Biol Sci 2004; 271: S273-6.

Napolitano MJ, Shain DH. Distinctions in adenylate metabolism among organisms inhabiting temperature extremes. Extremophiles 2005; 9: 93–8.

Neale PJ, Pricu JC. The photosynthetic apparatus of phytoplankton from a perennially ice-covered Antarctic lake: acclimation to an extreme shade environment. Plant Cell Physiol 1995; 36: 253–63.

Nishida I, Murata N. Chilling sensitivity in plants and cyanobacteria: the crucial contribution of membrane lipids. Annu Rev Plant Physiol Plant Mol Biol 1996; 47: 541–68.

Noctor G, Mhamdi A, Queval G, Foyer CH. Regulating the redox gatekeeper: vacuolar sequestration puts glutathione disulfide in its place. Plant Physiol 2013; 163: 665–71.

Nott A, Jung H-S, Koussevitzky S, Chory J. Plastid-to-nucleus retrograde signaling. Annu Rev Plant Biol 2006; 57: 739–59.

Peltier JB, Emanuelsson O, Kalume DE, Ytterberg J, Friso G, Rudella A, Liberles DA, Söderberg L, Roepstorff P, von Heijne G, van Wijk KJ. Central functions of the lumenal and peripheral thylakoid proteome of *Arabidopsis* determined by experimentation and genome-wide prediction. Plant Cell 2002; 14: 211–36.

Peng L, Fukao Y, Fujiwara M, Takami T, Shikanai T. Efficient operation of NAD(P)H dehydrogenase requires supercomplex formation with photosystem I via minor LHCI in *Arabidopsis*. Plant Cell 2009; 21: 3623–40.

Peter GF, Thornber JP. Biochemical composition and organization of higher plant photosystem II light-harvesting pigment-proteins. J Biol Chem 1991; 266: 16745–54.

Pfannschmidt T. Chloroplast redox signals: how photosynthesis controls its own genes. Trends Plant Sci 2003; 8: 33–41.

Pocock T, Lachance MA, Pröschold T, Priscu JC, Kim SS, Hüner NPA. Identification of a psychrophilic green alga from Lake Bonney Antarctica: *Chlamydomonas raudensis* Ettl. (UWO 241) Chlorophyceae. J Phycol 2004; 40:1138–48.

Pocock TH, Koziak A, Rosso D, Falk S, Hüner NPA. *Chlamydomonas raudensis* (UWO241), chlorophyceae, exhibits the capacity for rapid D1 repair in response to chronic photoinhibition at low temperature. J Phycol 2007; 43: 924–36.

Pocock T, Vetterli A, Falk S. Evidence for phenotypic plasticity in the Antarctic extremophile *Chlamydomonas raudensis* Ettl. UWO 241. J Exp Bot 2011; 62: 1169–77.

Pogson BJ, Woo NS, Förster B, Small ID. Plastid signalling to the nucleus and beyond. Trends Plant Sci 2008; 13: 602–09.

Portner HO, Peck LS, Zielinski S, Conway LZ. Intracellular pH and energy metabolism in the highly stenothermal Antarctic bivalve *Limopsis marionensis* as a function of ambient temperature. Polar Biol 1999; 22: 17–30.

Possmayer M, Berardi G, Beall BFN, Trick CG, Hüner NPA, Maxwell DP. Plasticity of the psychrophilic green alga *Chlamydomonas raudensis* (UWO 241) (Chlorophyta) to supraoptimal temperature stress. J Phycol 2011; 47:1098–09.

Priscu JC, Ed. Ecosystem Dynamics in a Polar Desert: The McMurdo Dry Valleys, Antarctica. Vol. 72, Antarctic Research Series. Washington: AGU Press 1998.

Priscu JC, Fritsen CH, Adams EE, Giovannoni SJ, Paerl HW, McKay CP, Doran PT, Gordon DA, Lanoil BD,

Pinckney JL. Perennial Antarctic lake ice: an oasis for life in a polar desert. Science 1998; 280: 2095–98.

Raymond JA, Morgan-Kiss R. Separate origins of ice-binding proteins in Antarctic *Chlamydomonas* species. PLoS One 2013; 8: e59186.

Remy R, Tremolieres A, Duval JC, Ambard-Bretteville F, Dubacq JP. Study of the supramolecular organization of the oligomeric light harvesting chlorophyll-protein (LHCP): conversion of the oligomeric form into the monomeric form by phospholipase A2 and reconsitution with liposomes. FEBS Lett 1982; 137: 271–75.

Rochaix JD. Regulation of photosynthetic electron transport. Biochim Biophys Acta 2011; 1807: 375–83.

Roose JL, Frankel LK, Bricker TM. Developmental defects in mutants of the PsbP domain protein 5 in *Arabidopsis thaliana*. PLoS One 2011; 6: e28624.

Rosso D, Bode R, Li W, Krol M, Saccon D, Wang S, Schillaci LA, Rodermel SR, Maxwell DP, Hüner NPA. Photosynthetic redox imbalance governs leaf sectoring in the *Arabidopsis thaliana* variegation mutants immutans, spotty, var1, and var2. Plant Cell 2009; 21: 3473–92.

Ruelland E, Vaultier MN, Zachowski A, Hurry V. Cold signalling and cold acclimation in plants. Adv Bot Res 2009; 49: 35–150.

Savitch LV, Maxwell DP, Hüner NPA. Photosystem II excitation pressure and photosynthetic carbon metabolism in *Chlorella vulgaris*. Plant Physiol 1996; 111: 127–36.

Savitch LV, Harney T, Hüner NPA. Sucrose metabolism in spring and winter wheat in response to high irradiance, cold stress and cold acclimation. Physiol Plant 2000; 108: 270–78.

Savitch LV, Barker-Astrom J, Ivanov AG, Hurry V, Öquist G, Hüner NPA, Gardestrom P. Cold acclimation of *Arabidopsis thaliana* results in incomplete recovery of photosynthetic capacity which is associated with an increased reduction of the chloroplast stroma. Planta 2001; 214: 295–01.

Savitch LV, Leonardos ED, Krol M, Jansson S, Grodzinski B, Hüner NPA, Öquist G. Two different strategies for light utilization in photosynthesis in relation to growth and cold acclimation. Plant Cell Environ 2002; 25: 761–71.

Savitch LV, Allard G, Seki M, Robert LS, Tinker NA, Hüner NPA, Shinozaki K, Singh J. The effect of overexpression of two Brassica CBF/DREB1-like transcription factors on photosynthetic capacity and freezing tolerance in *Brassica napus*. Plant Cell Physiol 2005; 46: 1525–39.

Schlicher T, Soll J. Molecular chaperones are present in the thylakoid lumen of pea chloroplasts. FEBS Lett 1996; 379: 302–04.

Sinensky M. Homeoviscous adaptation-a homeostatic process that regulates the viscosity of membrane lipids in *Escherichia coli*. Proc Natl Acad Sci USA. 1974; 71: 522–25.

Spetea C, Keren N, Hundal T, Doan JM, Ohad I, Andersson B. GTP enhances the degradation of the photosystem II D1 protein irrespective of its conformational heterogeneity at the Q(B) site. J Biol Chem 2000; 275: 7205–11.

Spetea C, Hundal T, Lundin B, Heddad M, Adamska I, Andersson B. Multiple evidence for nucleotide metabolism in the chloroplast thylakoid lumen. Proc Natl Acad Sci USA 2004; 101: 1409–14.

Spigel RH, Priscu JC. Evolution of temperature and salt structure of Lake Bonney, a chemically stratified Antarctic lake. Hydrobiologia 1996; 321: 177–90.

Spreitzer RJ, Salvucci ME. Rubisco: Structure, regulatory interactions, and possibilities for a better enzyme. Annu Rev Plant Biol 2002; 53: 449–75.

Stitt M, Hurry V. A plant for all seasons: alterations in photosynthetic carbon metabolism during cold acclimation in *Arabidopsis*. Curr Opin Plant Biol 2002; 5: 199–206.

Strand A, Hurry V, Gustafsson P, Gardestrom P. Development of *Arabidopsis thaliana* leaves at low temperature releases the suppression of photosynthesis and photosynthetic expression despite the accumulation of soluble carbohydrates. Plant J 1997; 12: 605–14.

Strand A, Hurry V, Henkes S, Hüner NPA, Gustafsson P, Gardestrom P, Stitt M. Acclimation of *Arabidopsis* leaves developing at low temperatures. Increasing cytoplasmic volume accompanies increased activities of enzymes in the Calvin cycle and in the sucrose-biosynthesis pathway. Plant Physiol 1999; 119: 1387–97.

Strand A, Foyer CH, Gustafsson P, Gardestrom P, Hurry V. Altering flux through the sucrose biosynthesis pathway in transgenic *Arabidopsis thaliana* modifies photosynthetic acclimation at low temperatures and the development of freezing tolerance. Plant Cell Environ 2003; 26: 523–35.

Suorsa M, Sirpiö S, Aro EM. Towards characterization of the chloroplast NAD(P)H dehydrogenase complex. Mol Plant 2009; 2: 1127–40.

Szyszka B, Ivanov AG, Hüner NPA. Psychrophily is associated with differential energy partitioning, photosystem stoichiometry and polypeptide phosphorylation in *Chlamydomonas raudensis*. Biochim Biophys Acta 2007; 1767: 789–00.

Szyszka-Mroz B, Pittock P, Ivanov AG, Lajoie G, Hüner NPA. The Antarctic psychrophile, *Chlamydomonas sp*. UWO 241, preferentially phosphorylates a PSI-cytochrome b6/f supercomplex. Plant Physiol 2015; 169: 717–736.

Takahashi H, Clowez S, Wollman FA, Vallon O, Rappaport F. Cyclic electron flow is redox-controlled but independent of state transition. Nature Commun 2013; 4 DOI: 10.1038/ncomms2954.

Takizawa K, Takahashi S, Hüner NPA, Minagawa J. Salinity affects the photoacclimation of *Chlamyomonas raudensis* Ettl UWO241. Photosynth Res 2009; 99: 195–03.

Thomas DN, Dieckmann GS. Antarctic sea ice – a habitat for extremophiles. Science 2002; 295: 641–6644.

Thuswaldner S, Lagerstedt JO, Rojas-Stütz M, Bouhidel K, Der C, Leborgne-Castel N, Mishra A, Marty F, Schoefs B, Adamska I, Persson BL, Spetea C. Identification, expression, and functional analyses of a thylakoid ATP/ADP carrier from *Arabidopsis*. J Biol Chem 2007; 282: 8848–59.

Tikkanen M, Nurmi M, Kangasjarvi S, Aro E-M. Core protein phosphorylation facilitates the repair of photodamaged photosystem II at high light. Biochim Biophys Acta 2008; 1777: 1432–37.

Tikkanen M, Gollan P, Suorsa M, Kangasjarvi S, Aro EM. STN7 operates in retrograde signalling through controlling redox balance in the electron transfer chain. Front Plant Sci 2012; 3: 277.

Trebst A. Inhibitors in electron flow: tools for the functional and structural localization of carriers and energy conservation sites. Methods Enzymol 1980; 69: 675-15.

Tremolieres A, Siegenthaler PA. Reconstitution of photosynthetic structures and activities with lipids. In: Siegenthaler PA, Ed. Advances in Photosynthesis and Respiration. Lipids in Photosynthesis: Structure, Function and Genetics. Dordrecht, Kluwer Academic, 1998; pp. 175–89.

Van den Burg B. Extremophiles as a source for novel enzymes. Curr Opin Microbiol 2003; 6: 213–8.

Vener AV. Phosphorylation of thylakoid proteins. In: Demmig-Adams B, Adams III, WW, Mattoo AK, Eds. Advances in Photosynthesis and Respration. Photoprotection, Photoinhibition Gene Regulation and Environment. Dordrecht, Kluwer Academic, 2006; pp. 107–126.

Wada H, Gombos Z, Sakamoto T, Murata N. Role of lipids in low temperature adaptation. In: Yamomoto HY, Smith CM, Eds. Current Topics in Plant Physiology: Photosynthetic Responses to the Environment. Amer Soc Plant Physiol 1993; 8: 78–87.

Wilson KE, Hüner NPA. The role of growth rate, redox-state of the plastoquinone pool and the trans-thylakoid ΔpH in photoacclimation of *Chlorella vulgaris* to growth irradiance and temperature. Planta 2000; 212: 93-02.

Wilson KE, Krol M, Hüner NPA. Temperature-induced greening of *Chlorella vulgaris*. The role of the cellular energy balance and zeaxanthin-dependent nonphotochemical quenching. Planta 2003; 217: 616–27.

Wilson KE, Ivanov AG, Öquist G, Grodzinski B, Sarhan F, Hüner NPA. Energy balance, organellar redox status and acclimation to environmental stress. Can J Bot 2006; 84: 1355–70.

Wollman FA. State transitions reveal the dynamics and flexibility of the photosynthetic apparatus. EMBO J 2001a; 20: 3623–30.

Wollman FA. State transitions reveal the dynamics and flexibility of the photosynthetic apparatus. EMBO J 2001b; 20: 3623–30.

Woodson JD, Perez-Ruiz JM, Chory J. Heme synthesis by plastid ferrochelatase I regulates nuclear gene expression in plants. Curr Biol 2011; 21: 897–03.

Wunder T, Liu Q, Aseeva E, Bonardi V, Leister D, Pribil M. Control of STN7 transcript abundance and transient STN7 dimerisation are involved in the regulation of STN7 activity. Planta 2013; 237: 541–58.

Yamazaki JY, Suzuki T, Maruta E, Kamimura Y. The stoichiometry and antenna size of the two photosystems in marine green algae, *Bryopsis maxima* and *Ulva pertusa*, in relation to the light environment of their natural habitat. J Exp Bot 2005; 56: 1517–23.

Yang Z. Small GTPases versatile signaling switches in plants. Plant Cell 2002; 14: S375-S388.

Yin L, Lundin B, Bertrand M, Nurmi M, Solymosi K, Kangasjärvi S, Aro EM, Schoefs B, Spetea C. Role of thylakoid ATP/ADP carrier in photoinhibition and photoprotection of photosystem II in *Arabidopsis*. Plant Physiol 2010; 153: 666–77.

Zer H, Vink M, Keren N, Dilly-Hartwig HG, Paulsen H, Herrmann RG, Andersson B, Ohad I. Regulation of thylakoid protein phosphorylation at the substrate level: reversible light-induced conformational changes expose the phosphorylation site of LHCII. FEBS J 1999; 96: 8277–82.

Zito F, Finazzi G, Delosme R, Nitschke W, Picot D, Wollman FA. The Qo site of cytochrome b$_6$/f complexes controls the activation of the LHCII kinase. EMBO J 1999; 18: 2961–69.

Nitric Oxide Mediated Effects on Chloroplasts

14

Amarendra N. Misra, Ranjeet Singh, Meena Misra, Radka Vladkova, Anelia G. Dobrikova, and Emilia L. Apostolova

Summary

Nitric oxide (NO) is emerging as a signaling molecule in plants. Its metabolism, site and mode of action in chloroplasts are still not clear. Chloroplasts are emerging as an alternative site for NO synthesis in plants. However, exogenous NO donors show direct evidence on the action of this molecule on chloroplasts under stress as well non-stress conditions. Nitric oxide is also implicated in the development and senescence of the organelle. The effects of NO on chloroplasts, particularly on photosynthetic and antioxidative processes are described. The target sites and probable sites of action are enumerated.

Keywords

Nitric Oxide • Chloroplast • Electron transport • Photosynthesis • Photosystems • Photophosphorylation • Greening • Senescence • Development • Abiotic stress

Contents

14.1	Introduction	306
14.2	Sources of Nitric Oxide in Plants	307
14.3	Effect of Nitric Oxide on Photosynthetic Pigment Dynamics	307
14.4	Interaction of Nitric Oxide with Oxygen Evolving Complex	308
14.5	Effect of Nitric Oxide on Photosynthetic Electron Transport	309
14.6	Effect of Nitric Oxide on the Photophosphorylation in Chloroplasts	311
14.7	Ameliorating Effect of Nitric Oxide on Photosynthetic Stress Responses	311
14.7.1	Nitric Oxide Under Osmotic Stress	311
14.7.2	Nitric Oxide Under Temperature Stress	312
14.7.3	Nitric Oxide Under High Light Stress and UV Radiation	312
14.7.4	Nitric Oxide Under Heavy Metals	313
14.7.5	Nitric Oxide Under Herbicides	314

A.N. Misra (✉) • M. Misra
Centre for Life Sciences, School of Natural Sciences, Central University of Jharkhand, Ratu Lohardaga Road, Brambe, Ranchi 435020, India
e-mail: misraan@yahoo.co.uk

R. Singh
Centre for Life Sciences, School of Natural Sciences, Central University of Jharkhand, Ratu Lohardaga Road, Brambe, Ranchi 435020, India

Biotechnology Department, Beej Sheetal Research Pvt. Ltd., Mantha Choufuli, Jalna 431203, India

R. Vladkova • A.G. Dobrikova • E.L. Apostolova
Institute of Biophysics and Biomedical Engineering, Bulgarian Academy of Sciences, Acad. G. Bonchev Str., Bl. 21, Sofia 1113, Bulgaria

14.8 Conclusion 315
References 316

Abbreviations

PSI	photosystem I
PSII	photosystem II
LHCII	light-harvesting chlorophyll a/b complex of PSII
GSNO	S-nitrosoglutathione
GSSG	glutathione disulphide
NO	nitric oxide
qL	coefficient of photochemical fluorescence quenching assuming interconnected PSII antennae
qP	coefficient of photochemical fluorescence quenching assuming non-interconnected PSII antennae
NPQ	non-photochemical quenching
Rubisco	ribulose-1,5-bisphosphate carboxylase
PTIO	2-phenyl-4,4,5,5-tetramentyl-imidazoline-1-oxyl-3-oxide
NOS	nitric oxide synthase
L-NNA	Nω-nitro-L-arginine
SNP	sodium nitropruside
OEC	oxygen-evolving complex
CPTIO	2-(4-carboxyphenyl)-4,4,5,5-tetramethylimidazoline-l-oxyl-3-oxide

14.1 Introduction

Nitric oxide (NO) is a gaseous molecule with a signaling role in plant growth, development and responses to environmental changes (Neill et al. 2008; Palavan-Unsal and Arisan 2009; Misra et al. 2010a, b, 2011, 2012). The effects of NO in plants can be direct or through intermediate effector molecules regulating cellular metabolism (Krasylenko et al. 2010). Nitric oxide action is achieved also by modifying the redox state of the cell and can modulate the activity of proteins, through reversible reactions with functional groups such as thiols and heme. It is well known that iron is a necessary element for synthesis and development of chloroplast and NO plays an important role in the distribution of iron in the chloroplasts in plant leaves (Sun et al. 2007). Nitric oxide, a highly unstable free radical, has been described both as a cytotoxin and a cytoprotectant in plants, as well (Beligni and Lamattina 1999a, 2001a). This signal molecule appears to take part in the regulation of cellular redox homeostasis, acting either as an oxidant or as an antioxidant (Stamler et al. 1992). However, at lower concentrations, NO promotes normal plant growth and development (Beligni and Lamattina 2001b). Nitric oxide stimulates leaf expansion, prevents etiolation, retards leaf senescence and induces stomatal closure (Leshem et al. 1998; Beligni and Lamattina 2000; García-Mata and Lamattina 2001). When applied at relatively high doses to plants, NO clearly perturbs normal metabolism and reduces the net photosynthesis in leaves of oats and alfalfa (Hill and Bennett 1970). Nitric oxide in concentrations above optimal (above 10^{-6} M) inhibits the expansion of leaf lamina, increases the viscosity of simulated thylakoid lipid monolayers and potentially impairs photosynthetic electron transport (Leshem et al. 1998; Leterrier et al. 2012).

Chloroplasts are highly specialized semiautonomous photosynthesizing organelles found in green plants. There is wide diversity in chloroplast structure, function and adaptation. The chloroplasts encode a large number of their own RNAs and proteins, in addition to that synthesized by the nuclear genes, economizing the cellular energy demand for its structural organization. Chloroplast develops from a progenitor known as proplastid accompanied with the coordinated regulation of plastid and nuclear-encoded genes (Baumgartner et al. 1989; Dilnawaz et al. 2001; Joshi et al. 2013). The chloroplasts are the only organelle that supports autotrophy in plants through its role in photosynthesis and also sustains life on earth. The process involves coordination between the primary photochemical processes in the thylakoid membrane and reduction of CO_2 in the stroma of chloroplasts. The photochemical process includes solar energy trapping, photolysis of water, electron transfer and generation of reductants for the reduction of CO_2 to carbohydrates in the stroma – the soluble fraction of chloroplasts. The thylakoid membrane has four main multi-subunit protein complexes: photosystem II (PSII), photosystem I (PSI), cytochrome b_6f and ATPase (Nelson and Yocum 2006).

Studies on the effect of NO on chloroplasts are crucial for understanding its role in green plants. In addition, chloroplasts are reported to be one of the several cellular sites for the synthesis of endogenous nitric oxide (Guo and Crawford 2005; Jasid et al. 2006; Galatro et al. 2013; Tewari et al. 2013). However, till date there is a lack of a precise report on the effect of the NO on the regulation of different physiological, biochemical and molecular processes in chloroplasts. In this review, we consolidate the up-to-date studies on the effect of NO on chloroplasts. An emphasis will be given in this chapter for the effect of NO on the photochemical efficiency in chloroplasts under physiological conditions and abiotic stress.

14.2 Sources of Nitric Oxide in Plants

The nitric oxide production in plant cells is compartmentalized and is mediated through several different pathways (Gupta et al. 2011). It has been shown to produce NO from nitrite (Desikan et al. 2002), from L – arginine by NOS like activity (Guo et al. 2003) and from S-nitrosoglutathione decomposition (Jasid et al. 2006) in chloroplasts. Non-enzymatic production of NO from nitrite involving plastid pigments such as carotenoids has also been reported (Cooney et al. 1994). Interestingly, NO synthesis in response to iron, elicitors, high temperatures, salinity or osmotic stress is first detected in chloroplasts using NO-sensitive diaminofluorescein probes (Foissner et al. 2000; Gould et al. 2003; Arnaud et al. 2006). In spite of several ifs and buts, these results corroborate the hypothesis that plastids are key players in the control of NO levels in plant cells. Nitric oxide originates in chloroplasts through the reduction of nitrite to NO and/or through nitric-oxide synthases (NOS) like activities (NOA) mediated NO biosynthetic pathway using arginine as a precursor molecule.

14.3 Effect of Nitric Oxide on Photosynthetic Pigment Dynamics

Nitric oxide improves the accumulation of chlorophylls (Chls) and even imitates red light responses in greening leaves (Beligni and Lamattina 2000). It is well known that the photosynthetic pigments, Chls in particular, are visible markers for chloroplast development and senescence in leaves (Misra and Biswal 1980, 1982; Misra and Misra 1986, 1987; Biswal et al. 2001; Dilnawaz et al. 2001). The synthesis of Chls and plastid proteins is intricately connected and is essential for the stability of Chl-protein complexes *in vivo* (Dilnawaz et al. 2001; Neill et al. 2003; Joshi et al. 2013). Seedlings grown in darkness develop etioplasts from proplastids, which ultimately transform into well organized chloroplasts in light (Misra and Misra 1987; Joshi et al. 2013). But, genes for nuclear-encoded Chl *a/b* – binding antennae and plastid-encoded Chl *a* – binding polypeptides are obligatorily dependent on incidence of light only. These photo-regulatory processes have several light receptors such as phytochrome and cryptochrome (Pogson and Albrecht 2011; Lepistö and Rintamäki 2012).

Nitric oxide donor sodium nitroprusside (SNP) enhanced Chl synthesis and accumulation of light-harvesting chlorophyll *a/b* complex of PSII (LHCII) and PSIA/B, primary photochemistry of PSII and effective quantum yield of PSII of the developing chloroplasts in greening of barley leaves (Zhang et al. 2006). Nitric oxide scavenger PTIO (2-phenyl-4,4,5,5-tetramentylimidazoline-1-oxyl-3-oxide) or NOS inhibitor L-NNA (nitro-nitro-L-arginine) retarded the greening process. Moreover, sodium ferrocyanide, an analog of SNP, nitrite and nitrate etc. do not have any effect on the greening process, suggesting a positive role of NO in the greening process (Zhang et al. 2006). The endogenous NO content of greening leaves also increased in parallel with the greening (measured by Chl accumulation) of leaves indicating a direct role of plastid NO in leaf greening (Zhang et al. 2006). Leaf senescence is associated with the symptoms of Chl degradation through various enzymatic and oxidative processes in the green cells (Biswal et al. 2001; Misra et al. 2006). Endogenous and exogenous NO at lower concentrations delayed leaf senescence, but at higher concentrations accelerated leaf senescence (Leshem et al. 1997; Guo and Crawford 2005; Mishina et al. 2007; Selcukcam and Cevahir 2008; Prochazkova and Wilhelmova 2011).

14.4 Interaction of Nitric Oxide with Oxygen Evolving Complex

Photosystem II is one of the sites of action for NO in chloroplasts (Wodala et al. 2008). At the electron donor side of PSII it acts at the oxygen-evolving complex (OEC). This complex is a part of the PSII, which is a multi-subunit chlorophyll-protein complex that uses light energy to oxidize water and form molecular oxygen, with a concomitant reduction of plastoquinone to plastoquinol (Debus 1992; Britt 1996). The functional conformation of the Mn cluster is expected to be maintained by a 33 kDa hydrophilic protein subunit of OEC attached to the luminal side of the D1/D2 heterodimer. During the oxidation of two water molecules to one oxygen molecule and protons, the OEC cycles through five intermediate redox states termed S_0–S_4 (Fig. 14.1). Dark adapted photosynthetic apparatus contains S_0 and S_1 states. The most reduced state is S_0, while S_1, S_2 and S_3 represent higher oxidation states and molecular oxygen being evolved during the transition from S_4 to S_0 states (Haumann et al. 2005; Penner-Hahn and Yocum 2005).

It is suggested that NO interacts with Mn cluster of PSII and leads to rapid destabilization of the excited states of OEC (Schansker and Petrouleas 1998). Studies with PSII and the di-manganese catalase have shown a similar mode of interaction of NO with the different oxidation states of the Mn clusters (Ioannidis et al. 2000). It is also suggested by Ioannidis et al. (2000) that one-electron reduction of the cluster occurs followed by release of NO_2^- as described below:

$$Mn^n + NO \rightarrow Mn^{n-1} + NO^+;$$
$$NO^+ + OH^- \rightarrow NO_2^- + H^+$$

Schansker et al. (2002) studied the oxygen oscillation patterns of PSII-enriched membranes and observed shift of the maximum flash-induced oxygen yield from flash 3 to flash 6/7 in the NO-treated samples. Considering these observations, the authors suggested the reduction of Mn cluster to the S_{-2} state by NO, which is assigned to the formation of Mn(II)-Mn(III) dimer. During catalysis the enzyme appears to cycle between the states Mn(II)-Mn(II) and Mn(III)-Mn(III) (Khangulov et al. 1990; Waldo and Penner-Hahn 1995).

Ioannidis et al. (2000) proposed a rapid interaction of NO with S_3 state of the OEC. This is explained by a metallo-radical characteristic of the S_3 state. A probable role of Tyr Y_D in oxidizing of Mn complex to the lower oxidizing state S_0 than the S_1 state was proposed by Styring and Rutherford (1987). Sanakis et al. (1997) proposed the formation of a Tyr-NO species which can act as an electron donor to PSII. An iminoxyl

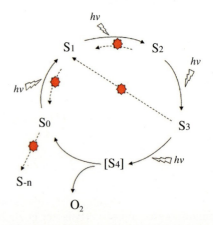

Fig. 14.1 Schematic presentation of the Kok's model of oxygen evolution, consisting of four stabile states (S_0–S_3) and one transient state (S_4). The possible sites of action of NO on the cycle are shown by *asterisks*. The back reduction of S-states by NO and possible over-reduction to S_{-n} states are shown by *dashed arrows*

radical is formed upon light-induced oxidation of this species, which is the first example of a chemical modification of one of the tyrosines of PSII to produce a photochemically active species. Our recent *in vitro* study demonstrated that the exogenous NO donor SNP (above 5 µM) has a clearly pronounced damaging effect on the primary oxygen-evolving reactions at the electron donor side of photosynthetic apparatus (Vladkova et al. 2011). In addition, our investigation for influence of exogenous NO donor SNP on isolated thylakoid membranes also revealed a dramatic increase of PSII population in the most reduced S_0 state and an increase of the turnover time of the oxygen-evolving centers, i.e. delayed the process of the electron donor capturing or S_i states turnover (Vladkova et al. 2011).

14.5 Effect of Nitric Oxide on Photosynthetic Electron Transport

One of the important reactions of NO in biology is interaction with metal complexes (Wink and Mitchell 1998). Because NO possesses an unpaired electron, it has high affinity to transition metals to form metal-nitrosyl complexes (Wink and Mitchell 1998). Due to this reason, proteins containing transition metal ions in either heme or non-heme complexes can be the potential targets for NO (Wink and Mitchell 1998). Nitric oxide is able to influence the photosynthetic electron transport chain directly by binding to such non-heme iron in the core complex of the PSII (Wodala et al. 2008). The important binding sites of NO in PSII are the non-heme iron between Q_A and Q_B binding sites (Diner and Petrouleas 1990; Petrouleas and Diner 1990) and Y_D, the Tyr residue of D2 protein (Sanakis et al. 1997). Electron paramagnetic resonance (EPR) studies confirmed the NO binding to non-heme iron and that NO competes with bicarbonate for its binding (Diner and Petrouleas 1990). Formate, an anion which also competes with bicarbonate, binds simultaneously with NO (Diner and Petrouleas 1990) besides the other anions like fluoride (Sanakis et al. 1999).

Experiments with isolated thylakoids indicated that NO binding slows down the rate of electron transfer between Q_A and Q_B (Diner and Petrouleas 1990). Binding of NO to the $Q_AFe^{2+}Q_B$ complex is facilitated in the presence of reduced Q_A acceptor, as this reduction weakens the bond between bicarbonate and iron (Goussias et al. 2002). Nitric oxide binding to PSII can also decrease the rate of electron transport on the donor side as well, since *in vitro* experiments have proven that NO interacts with the $Y_D\bullet$ tyrosine residue and the OEC. The latter is reduced to the S_{-2} state by NO, as shown by oxygen electrode, fluorescence and EPR measurements (Schansker et al. 2002). Measurements in the presence of DCMU demonstrated that NO induces inhibition of Q_A^- recombination with the S_2 state of the OEC. This donor side inhibition of electron transport may sufficiently be accounted by the reduction of either the OEC, or the $Y_D\bullet$ residue by NO. To the contrary, our recent results showed that the NO donor SNP is probably the only NO donor which stimulates the electron transport through PSII at sub-µmolar concentrations (Vladkova et al. 2011). Nitric oxide interacts with the tyrosine residue of the D2 protein (Sanakis et al. 1997) and the resulting $Y_D\bullet$–NO couple has a decreased redox potential low enough to become a more efficient electron donor in isolated thylakoid membranes than the immediate redox-active tyrosine residue (Y_Z) located on the D1 protein. The probable binding sites and sites of action in thylakoid membranes are summarized in Fig. 14.2.

Chlorophyll fluorescence studies have provided contradictory effects of NO on chloroplasts *in vivo*. However, these results depend on the used NO donor. In the leaves, NO derived from S-nitroso-N-acetylpenicillinamine (SNAP) showed no effect on the maximum quantum efficiency of photosynthesis, but that from SNP and S-nitrosoglutathione (GSNO) decreased this parameter (Takahashi and Yamasaki 2002; Yang et al. 2004; Wodala 2006; Wodala et al. 2008). All the NO donors induced a decrease in effective quantum efficiency, which is related to photochemical

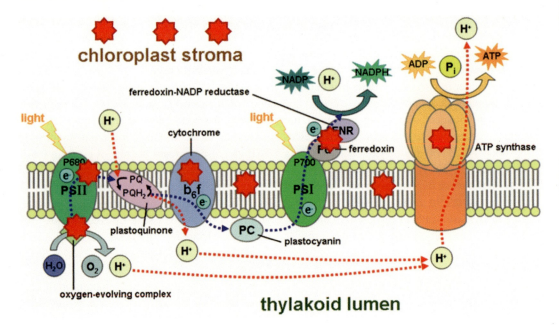

Fig. 14.2 Summary and schematic presentation of the probable binding of NO to different components of PSII, cytochrome b_6f, PSI, ATPase complexes. The nitrosylation of membrane lipids and of both thylakoid and stromal polypeptides are shown as red colored *circular asterisks*. These binding and nitrosylation affect photosynthetic processes

quenching (qP). These studies indicated that NO increases the proportion of closed PSII reaction centers in intact leaves (Wodala 2006; Wodala et al. 2008). However, all these results are not uniform and unequivocal. Wodala et al. (2008) suggested that the different chemical properties of NO donors and the different experimental conditions generate conflicting experimental results *in vivo*. Fast Chl fluorescence induction kinetics of GSNO-treated leaf disks confirmed significant donor and acceptor side inhibition of electron transport (Wodala et al. 2008).

It has been found that NO influences the non-photochemical quenching (NPQ). Besides reducing steady-state NPQ values, NO changes the amplitude and kinetics of an NPQ transient, which resembles reaction-center NPQ described by Finazzi et al. (2004). Reaction-center NPQ arises upon the onset of illumination of dark-adapted leaves and, at low light intensities, it is relaxed rapidly after a few min of illumination. On the basis of its fast relaxation and ΔpH-dependency, Finazzi et al. (2004) showed that reaction-center NPQ is caused by the rapid and transient over acidification of the thylakoid lumen, which is created by the immediate onset of the photochemistry. In addition, they suggested that the ΔpH may be further increased by cyclic and pseudo-cyclic electron transport (Mehler-reaction) and explain the relaxation of this transient form of NPQ by the activation of the carbon fixation apparatus, which decreases ΔpH and redox pressure. Although a potential effect of NO on Calvin cycle activation would account for changes in this NPQ transient, steady-state NPQ values below control values indicate that NO does not decrease the maximum rate of the Calvin cycle (Finazzi et al. 2004).

Photosynthetic studies on stomata of peeled epidermal strips respond to exogenous NO by instantly decreasing photochemical fluorescence quenching coefficients (qP and qL), the operating quantum efficiency of PSII, and NPQ to close to zero. However, NO effect *in vivo* is reversible. The reversible inhibition by NO of the electron transport rate could be restored by bicarbonate, a compound known to compete with NO for one of the two coordination sites of the non-heme iron (II) in the $Q_A Fe^{2+} Q_B$ complex (Ordog et al. 2013).

14.6 Effect of Nitric Oxide on the Photophosphorylation in Chloroplasts

Previous studies revealed that NO donor SNAP inhibits the linear electron transport rate and light-induced pH formation (ΔpH) across thylakoid membrane, and decreased the rate of ATP synthesis (Takahashi and Yamasaki 2002). The inhibitory effect of NO on the photophosphorylation can be prevented by a supplemental high concentration of bicarbonate. It has been reported that high concentrations of bicarbonate enable the bound NO to liberate from the reaction center and recover electron transport activity (Diner and Petrouleas 1990). Thus, sensitivity of photosynthesis against NO would depend on a local concentration of bicarbonate within thylakoid membranes. It is plausible that the decreased rate of electron transport, due to the NO-induced dissociation of bicarbonate from the specific sites of thylakoids, is involved in the mechanism of the inhibitory effect of NO on the photophosphorylation. The overall effect of NO on the electron transport and the photophosphorylation as well its resultant effect on the stromal enzymes is summarized and shown in Fig. 14.2.

14.7 Ameliorating Effect of Nitric Oxide on Photosynthetic Stress Responses

Plants produce substantial amounts of NO in their natural environments (Wilson et al. 2008). In recent years, there has been increasing evidence that NO is involved in regulating, if not all, many key physiological processes in plants under normal and stress conditions. NO is reported to ameliorate several stress responses in plants (García-Mata and Lamattina 2001; 2002; Arasimowicz and Floryszak 2007; Krasylenko et al. 2010; Misra et al. 2011). Abiotic stresses, such as drought, high and low temperature, salinity, heavy metals, UV-B and oxidative stress is reported to induce NO production in plants (Shi et al. 2005; Arasimowicz and Floryszak 2007; Qiao and Fan 2008; Misra et al. 2011). Stressors form free radicals and other oxidants, resulting in increased level of reactive oxygen species (ROS) in plant cells (Mittler 2002; Qiao and Fan 2008). NO eliminates the superoxide radicals and as a signal molecule interacts with plant hormones and ROS (Laxalt et al. 1997; Zhao et al. 2004; Qiao and Fan 2008). In addition to its signal roles, NO may also function as a regulator of gene expression (Kopyra and Gwozdz 2004; Qiao and Fan 2008). A large amount of NO may combine with O^{2-} to form peroxynitrite ($ONOO^-$), which has been reported to damage lipids, proteins and nucleic acids (Yamasaki et al. 1999).

14.7.1 Nitric Oxide Under Osmotic Stress

Osmotic stress is one of the major abiotic factors limiting crop productivity and natural status of the environment, affecting functions of the plants (Misra et al. 2001). Drought, high salinity and freezing impose osmotic stress on plants. Plants respond to this stress in part by modulating gene expression, which eventually leads to the restoration of cellular homeostasis, detoxification of ROS and recovery of growth (Xiong and Zhu 2002). One of the most important responses of the plants under osmotic stress is increased synthesis of abscisic acid (ABA) (Neill et al. 2008). Nitric oxide maintains leaf water content by regulating ABA-induced stomatal closure during osmotic stress. But the physiological role of NO-induced ABA accumulation remains unknown. Reactive oxygen species are one of the main damaging compounds that are produced during stress and osmotic stress in particular (Beligni and Lamattina 1999a, b). The protective action of NO against oxidative damage can be explained by two mechanisms. Firstly, NO operates as a signaling molecule, which activates cellular antioxidant enzymes (Huang et al. 2002; Shi et al. 2005; Zhao et al. 2008). Secondly, NO might detoxify ROS directly. Mutant analysis of genetically modified *Arabidopsis* showed evidences that NO is an endogenous regulator of the tolerance of the plants to salt stress (Martinez et al. 2000; Zhao et al. 2004, 2007). These authors showed that *Atnos 1* mutant with decreased NO production due to a T-DNA insertion in *AtNOS* gene was hypersensitive to oxidative stress induced by NaCl (Martinez et al. 2000; Zhao et al. 2004, 2007).

14.7.2 Nitric Oxide Under Temperature Stress

Extreme temperatures either high or low (chilling) are stressful to plants affecting photosynthesis in plants (Misra and Misra 1986; Misra et al. 1997). Heat stress can also cause an overproduction of ROS, which could be involved in triggering defence responses against potentially damaging temperatures (Suzuki and Mittler 2006; Volkov et al. 2006; Kotak et al. 2007; Locato et al. 2008). A significant rise in NO production under heat stress in alfalfa sprouts (Leshem et al. 1998) and in tobacco leaf cells (Gould et al. 2003) are reported. Conversely, pea plants exposed at 38 °C for 4 h reduced the NO content of leaves, but it was found that the S-nitrosothiol (SNOs) content increases threefold and that the protein nitrosylation is enhanced (Corpas et al. 2008a). Protein nitrosylation can cause an inhibition of the activities of photosynthetic enzymes such as carbonic anhydrase and of ferredoxin-NADP reductase (Chaki et al. 2011). In *Arabidopsis*, several mutants have been identified to have impairment in the *GSNOR1* gene, showing the involvement of this gene in the mechanism of response against heat stress. Thus, the mutant HOT5 (sensitive to hot temperatures) showed that GSNOR modulates the intracellular level of SNOs, enabling thermo-tolerance, as well as the regulation of plant growth and development (Lee et al. 2008). In calluses of reed, the exogenous application of SNP or ABA elevated thermo-tolerance by alleviating ion leakage, lipid peroxidation and growth suppression induced by heat stress (45 °C for 2 h). On the other hand, exogenous ABA notably activated NOS activity and increased NO release, maintaining the heat tolerance (Song et al. 2008). Studies so far unequivocally report an increase in NO content and increase in thermo-tolerance of plants.

Chilling (below 4 °C) or cold (below 10 °C) stress is also detrimental for plant systems (Misra et al. 1997). A proteomic analysis showed that the chilling stress induced S-nitrosylation of Rubisco, which correlated with the inhibition of photosynthesis (Abat and Deswal 2009). Pea plants exposed to low temperature (8 °C for 48 h) showed an enhanced activity of L-arginine NOS and GSNOR, and an increase in the content of SNOs (Sharma et al. 2005). Low temperature caused an imbalance of the ROS and reactive nitrogen species metabolism in leaves, triggering a rise in the lipid oxidation and the protein tyrosine nitration, which indicates an induction of oxidative and nitrosative stress (Corpas et al. 2008b; Airaki et al. 2012). Similar responses and an increase in NO content have been reported in *Arabidopsis thaliana* exposed to chilling (4 °C for 1–4 h) stress or during cold acclimation (Zhao et al. 2009; Cantrel et al. 2011). It is most probable that temperature stress induced increase in NO and stress ameliorating effect of NO could be due to the antioxidative action of NO (Desikan et al. 2002).

14.7.3 Nitric Oxide Under High Light Stress and UV Radiation

High light causes photoinhibition of photosynthesis and generates ROS in green plants (Misra et al. 1997). Protective role of NO during photoinhibition has been reported in higher plants. Pronounced increases of NO production are found in tall fescue leaves after exposure to high-light stress (Xu et al. 2010). Nitric oxide might act as a signaling molecule to enhance antioxidant enzyme activities, further protecting against injuries caused by high light stress (Xu et al. 2010). However, *Chlamydomonas reinhardtii* cells under very high light conditions induces $^1O_2$ (singlet oxygen) accumulation due to a decrease in the $^1O_2$ scavenging capacity caused by NO-mediated inhibition of carotenoid synthesis and PSII electron transport, which in turn leads to oxidative damage and cell death (Chang et al. 2013).

The UV-B radiation (280–320 nm) clearly affects plant growth and usually also induces oxidative stress. Moreover, UV-B triggered a rise in ROS widely distributed in chloroplasts and mesophyll cells, causing cell damage. It has been observed that apocynin reduces UV-B-induced oxidative damage because it reduces the Chl breakdown caused by H_2O_2, and this is correlated with NO production mediated by an enhanced NOS activity (Tossi et al. 2009). In the case of maize leaves, the UV-B irradiation provoked a simultaneous rise in the concentration of ABA, H_2O_2 and NO in leaves. These authors also reported that the accumulation of endogenous NO is ABA-dependent and is responsible for tolerance to high doses of UV-B radiation (Tossi et al. 2012). Exogenous NO partially alleviated the UV-B effect characterized by a decrease in Chl contents and oxidative damage to the thylakoid membrane in bean seedlings (Shi et al. 2005). Zhang et al. (2009) suggested that under UV-B stress, NO production is mediated by H_2O_2 through the enhancement of NOS activity. It is well established that H_2O_2 induces NO synthesis and accumulation and vice versa (cf. Mazid et al. 2011).

The other possible explanation of the protective action of NO on the photosynthesis under UV and temperature stress can be either due to the modification of the OEC after interaction with NO (for SNP treated thylakoid membranes), which leads to a strong increase of the most reduced S_0 state in the dark or an increase of PSII open centers (Vladkova et al. 2011). It is well established that the most sensitive component to UV and temperature stress in photosynthetic membranes is LHCII-PSII supercomplex (Ivanova et al. 2008; Dankov et al. 2009; Apostolova and Dobrikova 2010; Dobrikova et al. 2013). Apostolova et al. (2006) showed that there is a relationship between the organization of PSII complex and oxidation state of the Mn clusters of the OEC. On the other hand, it has been demonstrated that UV radiation causes increase of PSII centers in the S_0 state, as one of the reasons for the UV-induced inhibition of the oxygen evolution is the direct absorption of UV light by Mn ions in Mn(III) and Mn (IV) oxidation states (see in Ivanova et al. 2008; Dobrikova et al. 2013). Therefore, it could be assumed that modification of the oxido-reduction states (i.e. more reduced states) of the Mn cluster in OEC after NO interaction is also a possible reason for protection of photosynthetic apparatus under abiotic stress.

14.7.4 Nitric Oxide Under Heavy Metals

The sensitivity of the photosynthetic processes to heavy metals is studied extensively (Arellano et al. 1995; Barón et al. 1995; Boucher and Carpentier 1999; Prasad 2004; Rouillon et al. 2006). One of the primary sources of metal toxicity in chloroplasts is through the generation of ROS, which affects chloroplast structure and function, as in other abiotic stresses. Recently, Saxena and Shekhawat (2013) reported the role of NO in heavy metal tolerance in plants. As reported for other abiotic stresses, NO has both beneficial and harmful effects during heavy metal stress, depending on the concentration and location of NO in the plant cells. Nitric oxide decreases the harmful effects of the ROS generated by heavy metal stress, interacts with other target molecules and regulates the expression of stress responsive genes (Saxena and Shekhawat 2013).

Recently it has been shown that the metalloid arsenic (As) also triggers the NO and S-nitrosoglutathione (GSNO) metabolism in Arabidopsis (Leterrier et al. 2012). Arsenic-treated seedlings showed a significant decrease in growth and an increase in lipid oxidation due to an alteration in antioxidant enzymes with a significant increase in NO content, protein tyrosine nitration and also the S-nitrosoglutathione reductase (GSNOR) activity, which reduced the glutathione and S-nitrosoglutathione content. Talukdar (2013) reported that the exogenous addition of NO significantly reversed the As-induced oxidative stress in *Phaseolus vulgaris* seedlings, maintaining H_2O_2 in a certain level through balanced alterations of antioxidant enzyme activities. The role of SNP donated NO in the process of amelioration has ultimately been manifested by significant decrease of the membrane damage and improvement of growth performance in plants grown on As + SNP medium. Nitric oxide synthesis inhibitor PTIO accelerates As-induced oxidative damage, which clearly demonstrates the role of NO in ameliorating metal stress in plants. Yu et al. (2013) studied the effect of Cd-induced stress in cucumber seedlings. Both leaf pigments and net photosynthesis, and antioxidant activity decreased after Cd treatment, which could be reversed by the application of exogenous NO (100 μM SNP). Similarly, Srivastava and Dubey (2012) showed ameliorating effect of NO on the ROS scavenging machinery in Mn-induced oxidative stress in rice seedlings.

14.7.5 Nitric Oxide Under Herbicides

The widespread use of herbicides in agriculture has resulted in increasing pollution of soil and water with these toxic compounds. Herbicides are one of the major abiotic stressors as that of salinity, drought, temperature extremes, flooding, toxic metals, high light intensity and UV-radiation. All of these are the major causes of yield loss in cultivated crops worldwide and pose major threats to agriculture and food security (Rodríguez et al. 2005). The primary site of action of many herbicides is chloroplast, besides the mitochondria. As many other abiotic stressors (Qiao and Fan 2008; Misra et al. 2010a, b, 2011), the herbicides also elicit the production of NO (Klepper 1979; Mallick et al. 2000; Sakihama et al. 2002). Exogenous NO has been shown to reduce herbicide toxicity by its protective effects on chloroplast membrane and by retarding herbicide induced loss of Chl (Beligni and Lamattina 1999a, b; Hung et al. 2002).

One of the most widely used herbicides is atrazine, which belongs to the triazine group of chemicals. The primary site of atrazine action is blocking the PSII electron transport via binding to the Q_B-binding site on the D1 polypeptide of PSII reaction center and inhibition of light-driven electron transport from Q_A to Q_B in PSII (Trebst 1987; Draber et al. 1991). A high sensitivity of the photosynthetic apparatus to atrazine is well documented (Qian et al. 2009; Vladkova et al. 2009; Apostolova et al. 2011; Rashkov et al. 2012). Qian et al. (2009) have shown that in unicellular green algae *Chlorella vulgaris*, atrazine (100 μg/L) or glufosinate (10 mg/L) with low concentrations of NO donor SNP (10–20 μM) significantly decreased herbicide induced ROS generation and membrane peroxidation and increased the chlorophyll content of leaves.

Other widely used herbicide in agriculture is paraquat (also known as methyl viologen), which belongs to the bipyridinium herbicides. This herbicide exerts its toxic effects by catalyzing the electron transfer from PSI to molecular oxygen, producing oxygen radicals that cause lipid peroxidation and membrane damage (Cha et al. 1982). Hung et al. (2002) have evaluated the protective effect of NO against paraquat toxicity of rice leaves. They showed that NO-donors (PBN, N-tert-butyl-α-phenylnitrone, SNP, sodium nitropruside and SIN-1, 3-morpholinosydnonimine), as well as the ascorbic acid and $NaNO_2$ are effective in reducing paraquat toxicity in rice leaves, most likely mediated through an increase in antioxidant enzyme activities and decrease in lipid peroxidation.

Recently, Sood et al. (2012) revealed the effects of the exogenous NO (donor SNP) on the paraquat treated *Azolla microphylla*. The authors results suggested that SNP released NO can work both as cytoprotective and cytotoxic in concentration dependent manner and involvement of NO in protecting *Azolla* against paraquat toxicity. Paraquat (8 µM) alone increased the activities of antioxidant enzymes SOD, CAT, GPX, APX and the amount of H_2O_2. The supplementation of SNP (8–100 µM) suppressed the activities of antioxidant enzymes and the amount of H_2O_2 compared to paraquat alone. The addition of NO scavengers along with NO donor in paraquat treated fronds neutralized the effect of exogenously supplied NO, indicating that NO can effectively protect *Azolla* against paraquat toxicity by quenching ROS. Higher SNP concentration (200 µM) is reported to reverse the effect of NO.

Diquat, like as the paraquat is a bipyridinium herbicide and serves as an artificial electron acceptor of PSI (Beligni and Lamattina 1999a, b), influencing also the level of NO in the plants. Beligni and Lamattina (2002) reported that diquat triggered lipid peroxidation, ribulose-1,5-biphosphate carboxylase/oxygenase (Rubisco) and D1 protein loss. Supplementation of NO-donors SNP and S-nitroso-N-acetylpenicillamine greatly reduced lipid peroxidation, rapid protein turn-over and mRNA breakdown caused by the application of a high dose of diquat to potato leaf pieces or isolated chloroplasts. Moreover, diquat caused an increase in the rate of photosynthetic electron transport in isolated chloroplasts and NO restored it back to the control levels (Beligni and Lamattina 2002).

Lactofen is a diphenylether herbicide, which inhibits Chl biosynthesis by blocking the enzyme protoporphyrinogen-IX oxidase activity, which catalyses the oxidation of protoporphyrinogen-IX to protoporphyrin-IX (proto-IX) (Matringe et al. 1989). Accumulation of protoporphyrins leads to ROS generation causing oxidative stress in the plants. SNP donated NO was able to scavenge ROS generated by the lactofen action in soybean plants, avoiding the photosynthetic pigment breakdown, but the lipid peroxidation was not completely prevented (Ferreira et al. 2010). Later, Ferreira et al. (2011) demonstrated that the lactofen-induced morphological and physiological alterations in soybean leaves are reduced with NO.

14.8 Conclusion

This review clearly shows NO effects on the chloroplast structure and function under physiological and stress conditions. Many investigations revealed the role of the nitric oxide as a key signal molecule in plants. It has also been shown that NO participates under abiotic stress. Nitric oxide production increases in the plants as a response to abiotic stress (Qiao and Fan 2008). Under stress, NO enhances activities of the antioxidant enzymes (Shi et al. 2005; Neill et al. 2008) and as a signal molecule interacts with plant hormones, and affects physiological processes (Laxalt et al. 1997; Zhao et al. 2004; Misra et al. 2006; Qiao and Fan 2008; Misra et al. 2011). The other possibility of the protection of NO on the photosynthesis under abiotic stress can be due to the direct action on the thylakoid membranes, which leads to an increase of the open PSII centers and a modification of the OEC (Vladkova et al. 2011). These changes influence the structure of the LHCII-PSII supercomplex in response to stress factors (Ivanova et al. 2008; Dankov et al. 2009; Apostolova and Dobrikova 2010). On the other hand, it has been shown that there is relationship between the organization of the PSII complex and the oxidation state of the Mn cluster in the OEC (Apostolova et al. 2006). It has been proposed that the modification of the oxido-reduction state of the Mn cluster after NO interaction is a possible reason for the protection of the photosynthetic apparatus under abiotic stress.

Acknowledgements This is a part of the UGC MRP No. F. 36-302/2008 and DBT BUILDER project No. BT/PR9028/INF/22/193/2013 to ANM and is the result of International cooperation grants BIn-01/07 of the NSF of Bulgaria and project Grant No. INT/BULGARIA/B70/06 DST, India. MM acknowledges the UGC, India grant [No. F.15-14/11 (SA-II)] of PDF for Women.

References

Abat JK, Deswal R (2009) Differential modulation of S-nitrosoproteome of *Brassica juncea* by low temperature: change in S-nitrosylation of Rubisco is responsible for the inactivation of its carboxylase activity. Proteomics 9:4368–4380

Airaki M, Leterrier M, Mateos RM, Valderrama R, Chaki M, Barroso JB, del Río LA, Palma JM, Corpas FJ (2012) Metabolism of reactive oxygen species and reactive nitrogen species in pepper (*Capsicum annuum* L.) plants under low temperature stress. Plant Cell Environ 35:281–295

Apostolova EL, Dobrikova AG (2010) Effect of high temperature and UV-A radiation on the photosystem II. In: Pessarakli M (ed) Handbook of Plant and Crop Stress, 3rd ed. CRC Press, BocaRaton, pp 577–591

Apostolova EL, Dobrikova AG, Ivanova PI, Petkanchin IB, Taneva SG (2006) Relationship between the organization of the PSII supercomplex and the functions of the photosynthetic apparatus. J. Photochem Photobiol B 83:114–122

Apostolova EL, Dobrikova AG, Rashkov GD, Dankov KG, Vladkova RS, Misra AN (2011) Prolonged sensitivity of immobilized thylakoid membranes in cross-linked matrix to atrazine. Sens Acuat B 156:140–146

Arasimowicz M, Floryszak J (2007) Nitric oxide as a bioactive signaling molecule in plant stress response. Plant Science 172:876–887

Arellano JB, Lázaro JJ, López-Gorgé J, Barón M (1995) The donor side of Photosystem II as the copper-inhibitory binding site. Photosynth Res 45:127–134

Arnaud N, Murgia I, Boucherez J, Briat JF, Cellier F, Gaymard F (2006) An iron-induced nitric oxide burst precedes ubiquitin-dependent protein degradation for Arabidopsis AtFer1 ferritin gene expression. J Biol Chem 281:23579–23588

Barón M, Arellano JB, López-Gorgé J (1995) Copper and photosystem II: A controversial relationship. Physiol Plant 94:174–180

Baumgartner BJ, Rapp JC, Mullet JE (1989) Plastid transcription activity and DNA copy number increase early in Barley chloroplast development. Plant Physiol 89:1011–1018

Beligni MV, Lamattina L (1999a) Is nitric oxide toxic or protective? Trends in Plant Science 4:299–300

Beligni MV, Lamattina L (1999b) Nitric oxide counteracts cytotoxic processes mediated by reactive oxygen species in plant tissues. Planta 208:337–344

Beligni MV, Lamattina L (2000) Nitric oxide stimulates seed germination and deetiolation, and inhibits hypocotyls elongation, three light-inducible responses in plants. Planta 210:215–221

Beligni MV, Lamattina L (2001a) Nitric oxide in plants: the history is just beginning. Plant Cell Environ 24:267–278

Beligni MV, Lamattina L (2001b) Nitric oxide: A non-traditional regulator of plant growth. Trends in Plant Sci 6:508–509

Beligni MV, Lamattina L (2002) Nitric oxide interferes with plant photooxidative stress by detoxifying reactive oxygen species. Plant Cell Environ 25:737–748

Biswal AK, Dilnawaz F, David KAV, Ramaswamy NK, Misra AN (2001) Increase in the intensity of thermoluminescence Q-band during leaf ageing is due to a block in the electron transfer from Q_A to Q_B. Luminescence 16:309–313

Boucher N, Carpentier R (1999) Heat-stress stimulation of oxygen uptake by Photosystem I involves the reduction of superoxide radicals by specific electron donors. Photosynth Res 59:167–174

Britt RD (1996) Oxygen evolution, In Ort DR, Yocum CF (eds) Advances in Photosynthesis: Oxygenic Photosynthesis, The Light Reactions. Kluwer Academic Publishers, Dordrecht, The Netherlands, pp 137–164

Cantrel C, Vazquez T, Puyaubert J, Rezé N, Lesch M, Kaiser WM, Dutilleul C, Guillas I, Zachowski A, Baudouin E (2011) Nitric oxide participates in cold-responsive phosphosphingolipid formation and gene expression in *Arabidopsis thaliana*. New Phytol 189: 415–427

Cha LS, McRae DG, Thompson JE (1982) Light-dependence of paraquat-initiated membrane deterioration in bean plants. Evidence for the involvement of superoxide. Physiol Plant 56:492–499

Chaki M, Valderrama R, Fernández-Ocãna A, Carreras M, Gómez-Rodríguez V, Pedrajas JR, Begara FJ, Morales JC, Sánchez-Calvo B, Luque F, Leterrier M, Corpas FJ, Barroso JB (2011) Mechanical wounding induces anitrosative stress by down-regulation of GSNO reductase and an increase in S-nitrosothiols in sunflower (*Helianthus annuus*) seedlings. J Exp Bot 62:1803–1813

Chang HL, Hsu YT, Kang CY, Lee TM (2013) Nitric Oxide down-regulation of carotenoid synthesis and photosystem II activity in relation to very high light-induced singlet oxygen production and oxidative stress in *Chlamydomonas reinhardtii*. Plant Cell Physiol 58:1296–1315

Cooney RV, Harwood PJ, Custer LJ, Franke AA (1994) Light-mediated conversion of nitrogen dioxide to nitric oxide by carotenoids. Environ Health Perspect 102:460–462

Corpas FJ, Chaki M, Fernández-Ocãna A, Valderrama R, Palma JM, Carreras A, Begara FJ, Morales JC, Airaki M, del Río LA, Barroso JB (2008a) Metabolism of reactive nitrogen species in pea plants under abiotic stress conditions. Plant Cell Physiol 49:1711–1722

Corpas FJ, del Río LA, Barroso JB (2008b) Post-translational modifications mediated by reactive nitrogen species: Nitrosative stress responses or components of signal transduction pathways? Plant Signaling Behaviour 3:301–303

Dankov K, Taneva S, Apostolova EL (2009) Freeze-thaw damage of photosynthetic apparatus. Effect of the organization of LHCII-PSII supercomplex. Comp Rend Acad Bulg Sci 62:1103–1110

Debus RJ (1992) The manganese and calcium ions of photosynthetic oxygen evolution. Biochim Biophys Acta 1102:269–352

Desikan R, Griffiths R, Hancock J, Neill S (2002) New role for an old enzyme: nitrate reductase-mediated nitric oxide generation is required for abscisic acid-induced stomatal closure in *Arabidopsis thaliana*. Proc Nat Acad Sci, USA 99:16314–16318

Dilnawaz F, Mahapatra P, Misra M, Ramaswamy NK, Misra AN (2001) The distinctive pattern of photosystem II activity, photosynthetic pigment accumulation and ribulose-1, 5-bisphosphate carboxylase/oxygenase content of chloroplasts along the axis of primary wheat leaf lamina. Photosynthetica 39:557–563

Diner BA, Petrouleas V (1990) Formation by NO of nitrosyl adducts of redox components of the Photosystem II reaction center. II: Evidence that HCO_3^-/CO_2 binds to the acceptor-side non-heme iron. Biochim Biophys Acta 1015:141–149

Dobrikova AG, Krasteva V, Apostolova EL (2013) Damage and protection of the photosynthetic apparatus from UV-B radiation I. Effect of ascorbate. J Plant Physiol 170:251–257

Draber W, Tietjen K, Kluth JF, Trebst A (1991) Herbicides in photosynthesis research. Angew Chem 30:1621–1633

Ferreira LC, Cataneo AC, Remaeh LMR, Corniani N, Fumis TDF, Souza YAD, Scavroni J, Soares BJA (2010) Nitric oxide reduces oxidative stress generated by lactofen in soybean plants. Pest Biochem Physiol 97:47–54

Ferreira LC, Cataneo AC, Remaeh LMR, Búfalo J, Scavroni J, Andréo-Souza Y, Cechin I, Soares BJA (2011) Morphological and physiological alterations induced by lactofen in soybean leaves are reduced with nitric oxide. Planta Daninha 29:837–847

Finazzi G, Johnson GN, Dall.osto L, Joliot P, Wollman FA, Bassi RA (2004) Zeaxanthin-independent non photochemical quenching mechanism localized in the photosystem II core complex. Proc Natl Acad Sci, USA 101:12375–12380

Foissner I, Wendehenne D, Langebartels C, Durner J (2000) In vivo imaging of an elicitor-induced nitric oxide burst in tobacco. Plant J 23:817–824

Galatro A, Puntarulo S, Guiamet JJ, Simontacchi M (2013) Chloroplast functionality has a positive effect on nitric oxide level in soybean cotyledons. Plant Physiol Biochem 66:26–33

García-Mata C, Lamattina L (2001) Nitric oxide induces stomatal closure and enhances the adaptive plant responses against drought stress. Plant Physiol 126:1196–1204

García-Mata C, Lamattina L (2002) Nitric oxide and abscisic acid cross talk in guard cells. Plant Physiol 128:790–792

Gould KS, Lamotte O, Klinger A, Pugin A, Wendehenne D (2003) Nitric oxide production in tobacco leaf cells: A generalized stress response? Plant Cell Environ 26:1851–1862

Goussias C, Deligiannakis Y, Sanakisk Y, Ioannidis N, Petrouleas V (2002) Probing subtle coordination changes in the iron-quinone complex of photosystem II during charges separation by the use of NO. Biochemistry 41:15212–15223

Guo FQ, Crawford NM (2005) Arabidopsis nitric oxide synthase 1 is targeted to mitochondria and protects against oxidative damage and dark-induced senescence. Plant Cell 17:3436–3450

Guo FQ, Okamoto M, Crawford NM (2003) Identification of a plant nitric oxide synthase gene involved in hormonal signaling. Science 302:100–103

Gupta KJ, Fernie AR, Kaiser WM, van Dongen JT (2011) On the origins of nitric oxide. Trends in Plant Science 16:160–168

Haumann M, Liebisch P, Müller C (2005) Photosynthetic O_2 formation tracked by time-resolved X-ray experiments. Science 310:1019–1021

Hill AC, Bennett JH (1970) Inhibition of apparent photosynthesis by nitrogen oxides. Atmospheric Environ 4:341–348

Huang X, Rad U, Durner J (2002) Nitric oxide induces transcriptional activation of the nitric oxide-tolerant alternative oxidase in Arabidopsis suspension cells. Planta 215:914–923

Hung KT, Chang CJ, Kao CH (2002) Paraquat toxicity is reduced by nitric oxide in rice leaves. J Plant Physiol 159:159–166

Ioannidis N, Schansker G, Barynin VV, Petrouleas V (2000) Interaction of nitric oxide with the oxygen evolving complex of photosystem II and manganese catalase: A comparative study. J Biol Inorg Chem 5:354–363

Ivanova PI, Dobrikova AG, Taneva SG, Apostolova EL (2008) Sensitivity of the photosynthetic apparatus to UV-A radiation: a role of light-harvesting complex II – photosystem II supercomplex organization, Radiat Environ Biophys 47:169–177

Jasid S, Simontacchi M, Bartoli CG, Puntarulo S (2006) Chloroplasts as a nitric oxide cellular source. Effect of reactive nitrogen species on chloroplastic lipids and proteins. Plant Physiol 142:1246–1255

Joshi P, Misra AN, Nayak L, Biswal B (2013) Response of mature, developing and senescing chloroplast to environmental stress. In: Advances in Photosynthesis and Respiration, Volume 36. Springer, Dordrecht, pp 641–668

Khangulov SV, Barynin VV, Antonyuk-Barynina SV (1990) Manganese containing catalase from Thermus thermophilus peroxide-induced redox transformation of manganese ions in the presence of specific inhibitor of catalase activity. Biochim Biophys Acta 1020:25–33

Klepper LA (1979) Nitric oxide (NO) and nitrogen dioxide (NO2) emissions from herbicide-treated soybean plants. Atmosph Environ 13:537–542

Kopyra M, Gwozdz EA (2004) The role of nitric oxide in plant growth regulation and responses to abiotic stress. Acta Physiol Plant 26:459–472

Kotak S, Larkindale J, Lee U, von Koskull-Doring P, Vierling E, Scharf KD (2007) Complexity of the heat stress response in plants. Curr Opin Plant Biol 10:310–316

Krasylenko YA, Yemets AI, Blume YB (2010) Functional role of nitric oxide in plants. Russian J Plant Physiol 57:451–461

Laxalt AM, Beligni MV, Lamattina L (1997) Nitric oxide preserves the level of chlorophyll in potato leaves infected by *Phytophthora infestans*. European J Plant Pathol 103:643–651

Lee U, Wie C, Fernández M, Feelisch BO, Vierling E (2008) Modulation of nitrosative stress by S-nitrosoglutathione reductase is critical for thermotolerance and plant growth in Arabidopsis, Plant Cell 80:786–802

Lepistö A, Rintamäki E (2012) Coordination of plastid and light signaling pathways upon development of Arabidopsis leaves under various photoperiods. Mol Plant 5:799–816

Leshem YY, Haramaty E, Iluz D, Malik Z, Sofer Y, Roitman L, Leshem Y (1997) Effect of stress nitric oxide (NO): Interaction between chlorophyll fluorescence, galactolipid fluidity and lipoxygenase activity. Plant Physiol Biochem 35:573–579

Leshem YY, Wills RBH, Ku VVV (1998) Evidence for the function of the free radical gas – nitric oxide (NO•) – as an endogenous maturation and senescence regulating factor in higher plants. Plant Physiol Biochem 36:825–833

Leterrier M, Airaki M, Palma J, Chaki M, Corpas FJ (2012) Arsenic triggers the nitric oxide (NO) and S-nitrosoglutathione (GSNO) metabolism in Arabidopsis. Environ Pol 166:136–143

Locato V, Gadaleta C, De Gara L, De Pinto MC (2008) Production of reactive species and modulation of antioxidant network in response to heat shock: A critical balance for cell fate. Plant Cell Environ 31:1606–1619

Mallick N, Mohn FH, Rai L, Soeder CJ (2000) Impact of physiological stresses on nitric oxide formation by green alga, *Scenedesmus obliquus*. J Microbiol Biotechnol 10:300–306

Martinez GR, Mascio PD, Bonini MG, Augusto O, Briviba K, Sies H (2000) Peroxynitrite does not decompose to singlet oxygen (1gO2) and nitroxyl (NO-). Proc Nat Acad Sci, USA 97:10307–10312

Matringe M, Camadro JM, Labbe P, Scalla R (1989) Protoporphyrinogen oxidase as a molecule target for diphenyl ether herbicides. Biochem J 260:231–235

Mazid M, Khan TA, Mohammad F (2011) Role of Nitric oxide in regulation of H_2O_2 mediating tolerance of plants to abiotic stress: A synergistic signalling approach. J Stress Physiol Biochem 7:34–74

Mishina TE, Lamb C, Zeier J (2007) Expression of a nitric oxide degrading enzyme induces a senescence programme in Arabidopsis. Plant Cell Environ 30:39–52

Misra AN, Biswal UC (1980) Effect of phytohormones on the chlorophyll degradation during aging of chloroplasts in vivo and in vitro. Protoplasma 105:1–8

Misra AN, Biswal UC (1982) Differential changes in the electron transport properties of chloroplasts during aging of attached and detached leaves, and of isolated chloroplasts. Plant Cell Environ 5:27–30

Misra AN, Misra M (1986) Effect of temperature on senescing rice leaves. I. Photoelectron transport activity of chloroplasts. Plant Science 46:1–4

Misra AN, Misra M (1987) Effect of age and rehydration on greening of wheat leaves. Plant Cell Physiol 28:47–51

Misra AN, Ramaswamy NK, Desai TS (1997) Thermoluminescence studies on photoinhibition of pothos leaf discs at chilling, room and high temperature. J Photochem. Photobiol B: Biology 38:164–168

Misra AN, Srivastava A,; Strasser RJ (2001) Utilization of fast chlorophyll a fluorescence technique in assessing the salt/ion sensitivity of mung bean and Brassica seedlings. J Plant Physiol 158:1173–1181

Misra AN, Latowski D, Strzalka K (2006) The xanthophylls cycle activity in kidney bean and cabbage leaves under salinity stress. Russian J Plant Physiol 53:102–109

Misra AN, Misra M, Singh R (2010a) Nitric oxide biochemistry, mode of action and signalling in plants. J Med Plants Res 4:2729–2749

Misra AN, Misra M, Singh R (2010b) Nitric oxide: An ubiquitous signaling molecule with diverse role in plants. African J Plant Sci 5:57–74

Misra AN, Misra M, Singh R (2011) Nitric oxide ameliorates stress responses in plants. Plant Soil Env 57:95–100

Misra AN, Misra M, Singh R (2012) Nitric oxide signaling during senescence and programmed cell death in leaves. In: Ekinci D (ed) Chemical Biology, Intech Open. pp 159–186

Mittler R (2002) Oxidative stress, antioxidants and stress tolerance, Trends Plant Sci 7:405–410

Neill S, Desikan R, Hancock J (2003) Nitric oxide signalling in plants. New Phytol 159:11–35

Neill S, Barros R, Bright J, Desikan R, Hancock J, Harrison J, Morris P, Ribeiro D, Wilson I (2008) Nitric oxide, strimal closure, and abiotic stress. J Exp Bot 59:165–176

Nelson N, Yocum CF (2006) Structure and function of Photosystems I and II. Annu Rev Plant Biol 57:521–565

Ordog A, Wodala B, Rozsavolgyi T, Tari I, Horvath F (2013) Regulation of guard cell photosynthetic electron transport by nitric oxide. J Exp Bot 64:1357–1366.

Palavan-Unsal N, Arisan D (2009) Nitric oxide signaling in plants. Bot Res 75:203–229

Penner-Hahn JE, Yocum CF (2005) The photosynthesis "oxygen clock" gets a new number. Science 310:982–983

Petrouleas V, Diner BA (1990) Formation by NO of nitrosyl adducts of redox components of the

Photosystem II reaction center. I. NO binds to the acceptor-side non-heme iron. Biochim Biophys Acta 1015:131–140

Pogson BJ, Albrecht V (2011) Genetic dissection of chloroplast biogenesis and development: an overview. Plant Physiol 155:1545–1551

Prasad MNV (ed.) Heavy metal stress in plants: From biomolecules to ecosystems,. 2004. Springer-Verlag, Berlin-Heidelberg, pp 1–391

Prochazkova D, Wilhelmova N (2011) Nitric oxide, reactive nitrogen species and associated enzymes during plant senescence. Nitric Oxide 24:61–65

Qian H, Chen W, Li J, Wang J, Zhou Z, Liu W, Fu W (2009) The effect of exogenous nitric oxide on alleviating herbicide damage in *Chlorella vulgaris*. Aqua Toxicol 92:250–257

Qiao W, Fan LM (2008) Nitric oxide signaling in plant responses to abiotic streses. J Integrative Plant Biol 50:1238–1246

Rashkov GD, Dobrikova AG, Pouneva ID, Misra AN, Apostolova EL (2012) Sensitivity of *Chlorella vulgaris* to herbicides. Possibility of using it as a biological receptor in biosensors. Sens Actuat: B. Chemical 161:151–155

Rodríguez M, Canales E, Borrás-Hidalgo O (2005) Molecular aspects of abiotic stress in plants. Biotecnol Aplic 22:1–10

Rouillon R, Piletsky SA, Breton F, Piletska EV, Carpentier R (2006) Photosystem II biosensors for Heavy Metals Monitoring. In: Giardi MT, Piletska E (eds) Biotechnological Applications of Photosynthetic Proteins: Biochips, Biosensors and Biodevices. Springer US, pp 166–174

Sakihama Y, Nakamura S, Yamasaki H (2002) Nitric oxide production mediated by nitrate reductase in the green alga *Chlamydomonas reinhardtii*: an alternative NO production pathway in photosynthetic organisms. Plant Cell Physiol 43:290–297

Sanakis Y, Goussias C, Mason RP, Petrouleas V (1997) NO interacts with the tyrosine radical YD• of photosystem II to form an iminoxyl radical, Biochemistry 36:1411–1417

Sanakis Y, Petasis D, Petrouleas V, Hendrich M (1999) Simultaneous binding of fluoride and NO to the non-heme iron of Photosystem II: Quantitative EPR evidence for a weak exchange interaction between the semiquinone QA – and the iron-nitrosyl complex. J Am Chem Soc 121:9155–9164

Saxena I, Shekhawat GS (2013) Nitric oxide (NO) in alleviation of heavy metal induced phytotoxicity and its role in protein nitration. Nitric Oxide 32:13–20

Schansker G, Goussias C, Petrouleas V, Rutherford AW (2002) Reduction of the Mn cluster of the water-oxidizing enzyme by nitric oxide: Formation of an S_{-2} state. Biochemistry 41:3057–3064

Schansker G, Petrouleas V (1998) In: Garab G (ed) Photosynthesis: mechanisms and effects. Kluwer, Dordrecht., Vol. 2, pp 1319–1322

Selcukcam EC, Cevahir OG (2008) Investigation on the relationship between senescence and nitric oxide in sunflower (*Helianthus annuus* L.) seedlings. Pak J Bot 40:1993–2004

Sharma P, Sharma N, Deswal R (2005) The molecular biology of the low-temperature response in plants. Bioessays 27:1048–1059

Shi SY, Wang G, Wang YD, Zhang LG, Zhang LX (2005) Protective effect of nitric oxide against oxidative stress under ultraviolet-B radiation. Nitric Oxide 13:1–9

Song L, Ding W, Shen J, Zhang Z, Bi Y, Zhang L (2008) Nitric oxide mediates abscisic acid induced thermotolerance in the calluses from two ecotypes of reed under heat stress. Plant Sci 175:826–832

Sood A, Kalra C, Pabbi S, Uniyal PL (2012) Differential responses of hydrogen peroxide, lipid peroxidation and antioxidant enzymes in *Azolla microphylla* exposed to paraquat and nitric oxide. Biologia 67:1119–1128

Srivastava S, Dubey RS (2012) Nitric oxide alleviates manganese toxicity by preventing oxidative stress in excised rice leaves, Acta Physiol Plant 34:819–825

Stamler JS, Singel DJ, Loscalzo J (1992) Biochemistry of nitric oxide and its redox-activated forms. Science 258:1898–1902

Styring S, Rutherford AW (1987) In the oxygen-evolving complex of photosystem II the S0-state is oxidized to the S1-state by D+ (Signal II slow). Biochemistry 26:2401–2405

Sun J, Jiang H, Xu Y, Li H, Wu X, Xie Q, Li C (2007) The CCCH-type zinc finger proteins AtSZF1 and AtSZF2 regulate salt stress responses in Arabidopsis, Plant Cell Physiol 48:1148–1158

Suzuki N, Mittler R (2006) Reactive oxygen species and temperature stresses: a delicate balance between signaling and destruction. Physiol Plant 126:45–51

Takahashi S, Yamasaki H (2002) Reversible inhibition of photophosphorylation in chloroplasts by nitric oxide. FEBS Lett 512:145–148

Talukdar D (2013) Arsenic-induced oxidative stress in the common bean legume, *Phaseolus vulgaris* L. seedlings and its amelioration by exogenous nitric oxide. Physiol Mol Biol Plants 19:69–79.

Tewari RK, Prommer J, Watanabe M (2013) Endogenous nitric oxide generation in protoplast chloroplasts, Plant Cell Rep 32:31–44

Tossi V, Lamattina L, Cassia R (2009) An increase in the concentration of abscisic acid is critical for nitric oxide-mediated plant adaptive responses to UV-B irradiation. New Phytol 181:871–879

Tossi V, Cassia R, Bruzzone S, Zocchi E, Lamattina L (2012) ABA says NO to UV-B: a universal response? Trends in Plant Sci 17:510–517

Trebst A (1987) The three-dimensional structure of the herbicide binding niche on the reaction center polypeptides of photosystem II. Z Naturforsch 42c:742–750

Vladkova R, Ivanova P, Krasteva V, Misra AN, Apostolova E (2009) Assessment of chlorophyll fluorescence and photosynthetic oxygen evolution parameters in development of biosensors for detection of Q$_B$ binding herbicides. Compt rend Acad bulg Sci 62:355–360

Vladkova R, Dobrikova AG, Singh R, Misra AN, Apostolova E (2011) Photoelectron transport ability of chloroplast thylakoid membranes treated with NO donor SNP: Changes in flash oxygen evolution and chlorophyll fluorescence. Nitric Oxide 24:84–90

Volkov RA, Panchuk II, Mullineaux PM, Schöfl F (2006) Heat stress-induced H2O2 is required for effective expression of heat shock genes in Arabidopsis. Plant Mol Biol 61:733–746

Waldo GS, Penner-Hahn JE (1995) Mechanism of Manganese Catalase Peroxide Disproportionation – Determination of Manganese Oxidation-States During Turnover. Biochemistry 34:1507–1512

Wilson ID, Neill SJ, Hancock JT (2008) Nitric oxide synthesis and signalling in plants. Plant Cell Environ 31:622–631

Wink DA, Mitchell JB (1998) Chemical biology of nitric oxide: insights into regulatory, cytotoxic, and cytoprotective mechanisms of nitric oxide. Free Rad Biol Med 25:434–456

Wodala B (2006) Combined effects of nitric oxide and cyanide on the photosynthetic electron transport of intact leaves. Acta Physiol Szeg 50:185–188

Wodala B, Deák Z, Vass I, Erdei L, Altorjay I, Horváth F (2008) In vivo target sites of NO in photosynthetic electron transport as studied by chlorophyll fluorescence in pea leaves. Plant Physiol 146:1920–1927

Xiong L, Zhu JK (2002) Molecular and genetic aspects of plant responses to osmotic stress. Plant Cell Envron 25:131–139

Xu YF, Sun XL, Jin J-W, Zhou H (2010) Protective roles of nitric oxide on antioxidant systems in tall fescue leaves under high-light stress. Afr J Biotechnol 9:300–306

Yamasaki H, Sakihama Y, Takahashi S (1999) An altenative pathway for nitric oxide production in plants: new features of an old enzyme. Trends in Plant Sci 4:128–139

Yang JD, Zhao HL, Zhang TH, Yun JF (2004) Effects of exogenous nitric oxide on photochemical activity of photosystem II in potato leaf tissue under non-stress condition. Acta Sin 46:1009–1014

Yu L, Gao R, Shi Q, Wang X, Wei M, Yang F (2013) Exogenous application of sodium nitroprusside alleviated cadmium induced chlorosis, photosynthesis inhibition and oxidative stress in cucumber. Pak J Bot 45:813–819

Zhang L, Wang Y, Zhao L, Shi S, Zhang L (2006) Involvement of nitric oxide in light-mediated greening of barley seedlings. J Plant Physiol 163:818–826

Zhang L, Zhou S, Xuan Y, Sun M, Zhao L (2009) Protective effect of nitric oxide against oxidative damage in Arabidopsis leaves under ultraviolet-B irradiation. J Plant Biol 52:135–140

Zhao L, Zhang F, Guo J, Yang Y, Li B, Zhang L (2004) Nitric oxide functions as a signal in salt resistance in the calluses from two ecotypes of reed. Plant Physiol 134:849–857

Zhao M, Zhao X, Wu Y, Zhang L (2007) Enhanced sensitivity to oxidative stress in an Arabidopsis nitric oxide synthase mutant. J Plant Physiol 164:737–745

Zhao L, He J, Wang X, Zhang L (2008) Nitric oxide protects against polyethylene glycol-induced oxidative damage in two ecotypes of reed suspension cultures. J Plant Physiol 165:182–191

Zhao MG, Chen L, Zhang LL, Zhang WH (2009) Nitric reductase-dependent nitric oxide production is involved in cold acclimation and freezing tolerance in *Arabidopsis*. Plant Physiol 151:755–767

Nanostructured Mn Oxide/Carboxylic Acid or Amine Functionalized Carbon Nanotubes as Water-Oxidizing Composites in Artificial Photosynthesis

15

Mohammad Mahdi Najafpour, Saeideh Salimi, Małgorzata Hołyńska, Fahime Rahimi, Mojtaba Tavahodi, Tatsuya Tomo, and Suleyman I. Allakhverdiev

Summary

Herein, we report on nano-sized Mn oxide/carboxylic acid or amine-functionalized carbon nanotubes as water-oxidizing composites in artificial photosynthesis. The composites are synthesized by simple procedures and characterized by different methods. The water-oxidizing activities of these composites are also considered in the presence of cerium (IV) ammonium nitrate and Ru(bpy)$_3^{3+}$ in a photochemical reaction. Some composites are efficient Mn-based catalysts with turnover frequency (TOF, mmolO$_2$/(mol Mn·s)) of more than 1.

Keywords

Water oxidation • Water splitting • Manganese • Manganese oxide • Artificial photosynthesis • Carbon nanotube • Nanostructure • Composite • Cerium

Contents

15.1 Introduction 321
15.2 Experiments 322
15.3 Results and Discussion 324
15.4 Conclusions 329
References 330

15.1 Introduction

The finding of an efficient, cheap and environmentally friendly water-oxidizing or reducing catalyst is highly desirable for artificial photosynthetic systems (Jafari et al. 2016). As it is

M.M. Najafpour (✉)
Department of Chemistry, Institute for Advanced Studies in Basic Sciences (IASBS), Zanjan 45137-66731, Iran

Center of Climate Change and Global Warming, Institute for Advanced Studies in Basic Sciences (IASBS), Zanjan 45137-66731, Iran
e-mail: mmnajafpour@iasbs.ac.ir

S. Salimi • F. Rahimi • M. Tavahodi
Department of Chemistry, Institute for Advanced Studies in Basic Sciences (IASBS), Zanjan 45137-66731, Iran

M. Hołyńska
Fachbereich Chemie und Wissenschaftliches Zentrum für Materialwissenschaften (WZMW), Philipps-Universität Marburg, Hans-Meerwein-Straße, D-35032 Marburg, Germany

T. Tomo
Faculty of Science, Department of Biology, Tokyo University of Science, Kagurazaka 1-3, Shinjuku-ku, Tokyo 162-8601, Japan

S.I. Allakhverdiev (✉)
Controlled Photobiosynthesis Laboratory, Institute of Plant Physiology, Russian Academy of Sciences, Botanicheskaya Street 35, Moscow 127276, Russia

Institute of Basic Biological Problems, Russian Academy of Sciences, Pushchino, Moscow 142290, Russia

Faculty of Biology, Department of Plant Physiology, M.V. Lomonosov Moscow State University, Leninskie Gory 1-12, Moscow 119991, Russia
e-mail: suleyman.allakhverdiev@gmail.com

© Springer International Publishing AG 2017
H.J.M. Hou et al. (eds.), *Photosynthesis: Structures, Mechanisms, and Applications*,
DOI 10.1007/978-3-319-48873-8_15

known, water oxidation is a bottleneck for water splitting. Among different compounds, Mn oxides are very promising because they are cheap and environmentally friendly. Glikman and Shcheglova indicated that Mn oxides can catalyze water-oxidation in the presence of cerium(IV) salts as an oxidant (Glikman and Shcheglova 1968). The electrochemical water oxidation by Mn dioxide was considered by Morita's group (Morita et al. 1977). Harriman in 1988 showed that among Mn oxides, Mn(III) oxide is an efficient catalyst (Harriman et al. 1988). After these groups, many Mn oxides were reported for water oxidation and under different conditions. An active MnO_x/glassy carbon catalyst for water oxidation and oxygen reduction obtained by atomic layer deposition was reported (Pickrahn et al. 2012). Strasser and Behrens used an incipient wetness impregnation method for the preparation of MnO_x/carboxylic acid electrocatalysts for efficient water splitting (Mette et al. 2012). Besides the use of the two methods, the MnO_x/carbon nanotubes (CNTs) sample was obtained by conventional impregnation and was reported as a promising catalytic anode material for water electrolysis at neutral pH, showing a high activity and stability (Mette et al. 2012). In 2014, nano-sized Mn oxide/CNT, graphene and graphene oxide as water-oxidizing compounds in artificial photosynthesis were reported (Najafpour et al. 2014a). Mn oxides on other nanocarbon structures, such as nanodiamond and C_{60}, show good water-oxidizing activity toward water oxidation. However, the decomposition of these carbon nanostructure was also reported (Najafpour et al. 2014b, c). The composites were synthesized by different simple procedures and characterized by a number of methods. The water-oxidizing activities of these composites were also considered in the presence of cerium(IV) ammonium nitrate (Ce(IV)) as a usual oxidant (Parent et al. 2013). Some composites were efficient catalysts toward water oxidation.

MnO_2/multi wall CNTs were synthesized by coating oxidized multi-walled CNTs with MnO_2 via simple immersion of the multi-wall CNTs in $KMnO_4$ solution. This catalyst, comprising the outer region of catalytic MnO_2 and the inner region of highly conductive the multi-wall CNTs, showed enhanced photocatalytic activity toward water oxidation.

Herein, we report on the catalytic activity of different nanostructured Mn oxide/carboxylic acid or amine functionalized CNTs toward water oxidation in the presence of Ce(IV) and photo-produced $Ru(bpy)_3^{2+}$.

15.2 Experiments

Water oxidation experiments in the presence of Ce(IV) were performed using an HQ40d portable dissolved oxygen-meter connected to an oxygen monitor with digital readout at 25 °C. In a typical run, the instrument readout was calibrated against air-saturated distilled water stirred continuously with a magnetic stirrer in an air-tight reactor. After ensuring a constant baseline reading, water in the reactor was replaced with a Ce(IV) solution. Without the catalyst, Ce (IV) was stable under these conditions and oxygen evolution was not observed. After deaeration of the Ce(IV) solution with argon, Mn oxides as several small particles were added and oxygen evolution was recorded with the oxygen meter under stirring (Scheme 15.1). The formation of oxygen was followed and the oxygen formation rates per Mn site were obtained from linear fits of the data by the initial rate. Electrochemical water oxidation was performed with the use of a setup shown in Scheme 15.1.

Photochemical water oxidation experiments were performed in a flask containing 10 mL of aqueous buffer (Na_2SiF_6-$NaHCO_3$, 0.022–0.028 M) with pH held at 5.8, Na_2SO_4 (150 mg), $K_2S_2O_8$ (800 mg), $[Ru(bpy)_3]$$Cl_2 \cdot 6H_2O$ (15 mg), and the catalyst **1** (20 mg). After deaeration of the solution by Ar, the reactor was irradiated with 4 LED (each one 10 W) around flask in a home-made device.

For Fabrication of modified electrodes in electrochemical experiments, 30 μL of the composites suspension were dripped on the FTO electrode surface and dried at room temperature. Eventually, 10 μL of 0.5 wt % Nafion solution was deposited onto the center of the

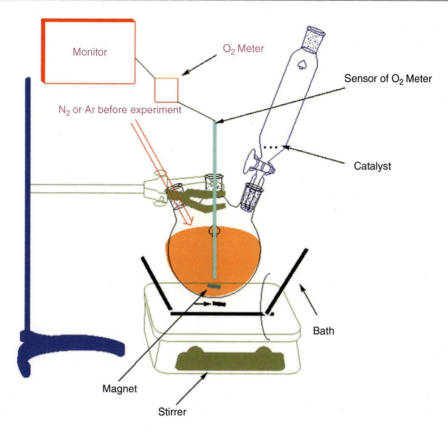

Scheme 15.1 Setup for a water-oxidation experiment

modified electrode. A three-electrode system was applied for the investigation of the electrochemical properties of the modified electrodes by cyclic voltammetry. Electrochemical experiments were performed using an EmStat3+ device from the Palm Sens Company (the Netherlands). In this case, a conventional three electrodes setup were used in which a FTO electrode or FTO electrode modified with the composites, an Ag|AgCl|KCl$_{sat}$ electrode and a Pt rod served as the working, reference and auxiliary electrodes, respectively.

A-1 Nanostructured Mn oxide/amine functionalized CNT (A-1) was synthesized with a simple method by mixing and stirring of KMnO$_4$ (50 mL, 12.6 mM) and the amine-functionalized CNT (500 mg) for 24 h at room temperature. The solid was washed with water to remove KMnO$_4$ and dried at 60 °C.

A-2 Nanostructured Mn oxide/amine-functionalized CNT (A-2) was synthesized by a simple method by mixing and stirring of KMnO$_4$ (50 mL, 12.6 mM) and amine-functionalized CNT (500 mg) for 10 h under hydrothermal conditions at 120 °C. The solid was washed with water to remove KMnO$_4$ and dried at 60 °C.

C-1 Nanostructured Mn oxide/carboxylic acid CNT (C-1) was synthesized by mixing and stirring of KMnO$_4$ (50 mL, 12.6 mM) and the carboxylic acid CNT (100 mg) for 24 h at room temperature. The solid was washed with water to remove KMnO$_4$ and dried at 60 °C.

C-2 Nanostructured Mn oxide/carboxylic acid CNT (C-2) was synthesized with a simple method by mixing and stirring of KMnO$_4$ (50 mL, 12.6 mM) and amine-functionalized CNT (100 mg) for 10 h under hydrothermal conditions at 120 °C.

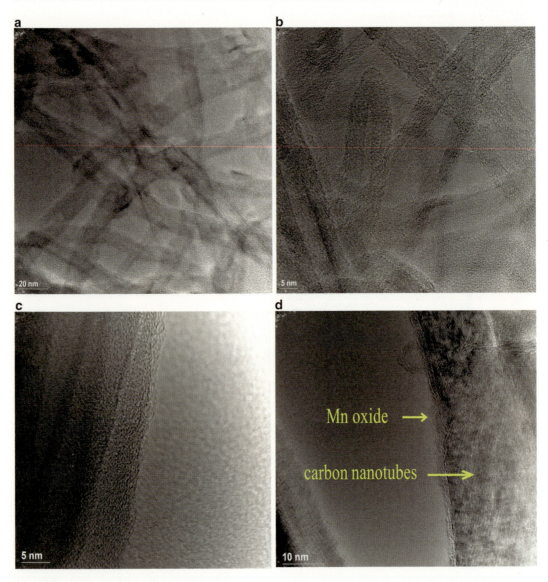

Fig. 15.1 TEM images of carboxylic acid (**a, b**) or amine (**c**) functionalized carbon nanotube and A-2 (**d**)

The solid was washed with water to remove $KMnO_4$ and dried at 60 °C.

15.3 Results and Discussion

The composites were synthesized by two different methods and under different conditions. In contrast to other supports, such as gold, platinum or silica, carbon nanostructures react with $KMnO_4$ even at room temperature. The amounts of Mn (%) in A-1, A-2, C-1 and C-2 are 6.6, 8.8, 8.7 and 7.0%, respectively, which shows that the method is promising to synthesize Mn oxide composites with CNTs. TEM images of CNTs and their composites with Mn oxides show very similar structure, which prove that there are no significant changes in the morphology. TEM images are shown in Fig. 15.1. These images for CNT clearly show CNTs with diameters at about 5–10 nm and lengths of 0.5–1.5 μm.

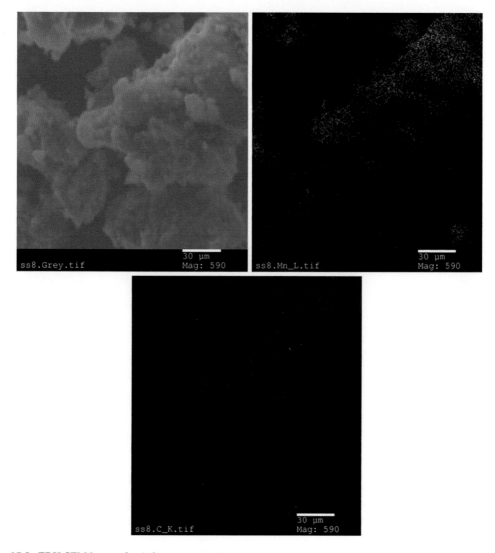

Fig. 15.2 EDX-SEM images for A-2

EDX-SEM images of the material are shown in Fig. 15.2. The images show C, Mn and K on the surface of the composite.

Cyclic voltammograms (CVs) of the materials on FTO electrode show no specific peaks, but Mn oxide/CNT composites display peaks for Mn(II)/Mn(III) oxidation and Mn(III)/Mn(IV) oxidation at 0.5 and 0.8 V (Ag|AgCl), which are clear in A-1 and A-2. Mn oxide shows also increased charge capacity compared to Mn oxide (Fig. 15.3).

Turnover frequencies (TOFs) of different compounds for water oxidation catalyzed by the composites are shown in Table 15.1.

In comparison to TOFs displayed by other catalysts, Mn oxide/amine-functionalized CNTs composites are among good catalysts toward water oxidation in the presence of Ce (IV) (Table 15.2). However, Mn oxide/ CNTs functionalized with carboxylic acids are not good catalysts in the presence of Ce (IV) (Table 15.2). However, in the presence of Ru(bpy)$_3^{3+}$, the results are different and Mn oxide/ CNTs functionalized with carboxylic acids are more active toward water oxidation (Fig. 15.4, Table 15.1). We speculated that the effect may be related to the interaction of

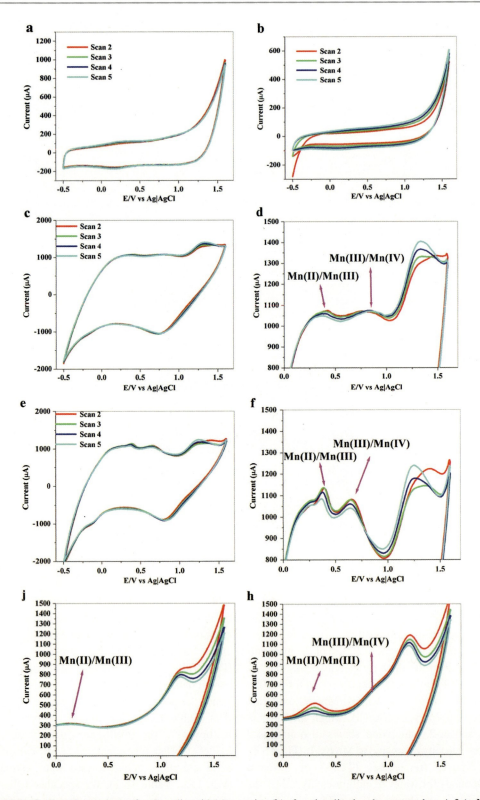

Fig. 15.3 Cyclic voltamograms of carboxylic acid (**a**) or amine (**b**) functionalized carbon nanotubes, A-2 (**c,d**) and A-1 (**e,f**), C-2 (**g,h**) and C-1 (**i,j**)

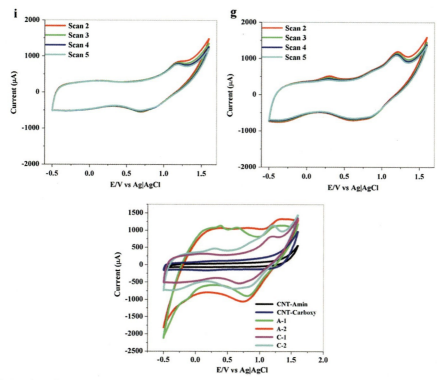

Fig. 15.3 (continued)

Table 15.1 Oxygen evolution by the composites in the presence of oxidants

Compound	Oxidant	TOF (mmol O_2/(mol Mn))
Carbon nanotubes functionalized with carboxylic acids	Ce(IV)	0
Amine-functionalized carbon nanotubes	Ce(IV)	0
Carbon nanotubes functionalized with carboxylic acids	Ru(bpy)$_3^{3+}$	0
Amine-functionalized carbon nanotubes	Ru(bpy)$_3^{3+}$	0
A-1	Ce(IV)	1.3
A-2	Ce(IV)	1.7
C-1	Ce(IV)	0.2
C-2	Ce(IV)	0.07
A-1	Ru(bpy)$_3^{3+}$	0.3
A-2	Ru(bpy)$_3^{3+}$	0.1
C-1	Ru(bpy)$_3^{3+}$	0.2
C-2	Ru(bpy)$_3^{3+}$	0.5

negatively charged carboxylate groups with positive Ru(bpy)$_3^{3+}$ ions. In contrast to Ru(bpy)$_3^{3+}$, the interaction of negative Ce(IV)(NO$_3$)$_6^{2-}$ with negative carboxylate groups on CNTs significantly inhibit from interaction Mn oxide/CNTs and Ce(IV)(NO$_3$)$_6^{2-}$.

In addition to efficiency, the stability of the catalyst is also an important issue for applications on industrial scale. The self-healing is a strategy to increase the stability of the catalyst. The Najafpour's group reported a self-healing reaction of Mn oxides in the presence of Ce(IV), which remakes the Mn oxide phase from MnO$_4^-$. In this mechanism, MnO$_4^-$ ions are produced as the decomposition product of the reaction of Mn oxide and Ce(IV).

Table 15.2 Oxygen evolution by different Mn compounds in the presence of oxidants

Compound	Oxidant	TOF (mmol O_2/(mol Mn))
Ca-Mn oxide	Ce(IV)	3.0
Nano-scale Mn oxide within NaY zeolite	Ce(IV)	2.62
Layered Mn-calcium oxide	Ce(IV)	2.2
Nanolayered Mn oxide/CNT,G or GO	Ce(IV)	0.5–2.6
Layered Mn-Al, Zn, K, Cd and Mg oxide	Ce(IV)	0.8–2.2
Mn oxide/amine- functionalized carbon nanotubes	**Ce(IV)**	**1.3–1.7**
$CaMn_2O_4 \cdot H_2O$	Ce(IV)	0.54
Amorphous Mn Oxides	$Ru(bpy)_3^{3+}$	0.06
	Ce(IV)	0.52
$CaMn_2O_4 \cdot 4H_2O$	Ce(IV)	0.32
Mn oxide nanoclusters	$Ru(bpy)_3^{3+}$	0.28
Mn oxide-coated montmorillonite	Ce(IV)	0.22
Mn oxide/carboxylic acid functionalized carbon nanotubes	**Ce(IV)**	**0.07–0.2**
Nano-sized α-Mn_2O_3	Ce(IV)	0.15
Octahedral Molecular Sieves	$Ru(bpy)_3^{3+}$	0.11
	Ce(IV)	0.05
MnO_2 (colloid)	Ce(IV)	0.09
α-MnO_2 nanowires	$Ru(bpy)_3^{3+}$	0.059
$CaMn_3O_6$	Ce(IV)	0.046
$CaMn_4O_8$	Ce(IV)	0.035
α-MnO_2 nanotubes	$Ru(bpy)_3^{3+}$	0.035
Mn_2O_3	Ce(IV)	0.027
β-MnO_2 nanowires	$Ru(bpy)_3^{3+}$	0.02
$Ca_2Mn_3O_8$	Ce(IV)	0.016
$CaMnO_3$	Ce(IV)	0.012
Nano-sized λ-MnO_2	$Ru(bpy)_3^{3+}$	0.03
Bulk α-MnO_2	$Ru(bpy)_3^{3+}$	0.01
Mn Complexes	Ce(IV)	0.01–0.6
PSII	Sunlight	$100–400 \times 10^3$

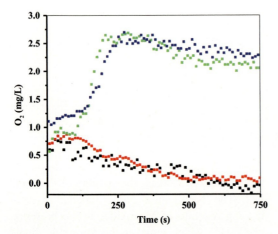

Fig. 15.4 The water oxidation in the presence of carboxylic acid (*red*) or amine (*black*) functionalized carbon nanotube, A-1 (*blue*) and C-2 (*green*) in the presence of $Ru(bpy)_3Cl_3/K_2S_2O_8$ and LED (for details see text) at 30 °C. For detail see experimental section

This is in accordance with the following observations (Scheme 15.2) (Najafpour et al. 2013; Najafpour 2015a, b):

- The MnO_4^- ions reacted with compounds such as Mn oxide or supports at the end of the reaction when the concentration of Ce (IV) decreased.
- The MnO_4^- ions formation in the reaction of Mn(II) nitrate and Ce(IV) is observed (Scheme 15.2).
- The amounts of the MnO_4^- ions in the absence of Mn oxide showed no change during 100 h of the experiment indicating that Mn oxide is necessary for the reduction of MnO_4^-.

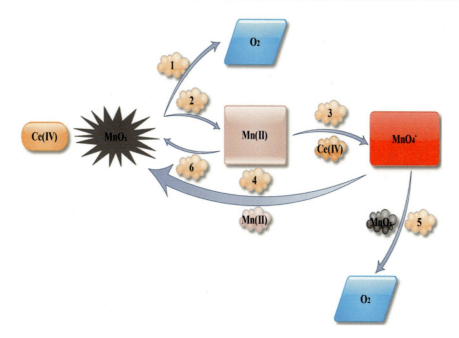

Scheme 15.2 Self-healing process in water oxidation catalysed by Mn oxides in the presence of Ce(IV). *1*: Oxygen evolution was detected with an oxygen meter. The origin of oxygen is water. *2*: Mn(II) was detected by EPR. *3*: the MnO_4^- ions formation could be detected by UV-Vis in a reaction of Mn(II) and Ce(IV). *4*: It is known that in the reaction of Mn(II) and MnO_4^- at different pH values Mn oxide is produced. *5*: The MnO_4^- ions in the presence of Mn oxide oxidize water. In this reaction, the MnO_4^- ions are reduced to Mn oxide. *6*: Mn(II) in the presence of Ce(IV) forms Mn oxide. In a typical experiment, the reaction of $MnSO_4$ in the presence of Ce (IV) (1.0 M) yields MnO_2 that can be detected by XRD. Images and captions are taken (Najafpour 2015b) (Reprinted with permission from Najafpour 2015b. Copyright (2015) by American Chemical Society)

In accordance with all these results, a mechanism was proposed for the self-healing of Mn oxide in the presence of Ce(IV) (Scheme 15.2).

Mn oxide/CNT composites undergo a very similar self-healing process: Mn oxide/CNT after the reaction with Ce(IV) forms MnO_4^- ions (Fig. 15.5). MnO_4^- ions are not a catalyst for water oxidation in the presence of Ce(IV) and oxygen evolution occurs on the surface of Mn oxide. These MnO_4^- ions disappear completely after conversion of Ce(IV) to Ce(III). Our experiments showed that the MnO_4^- ions react not only with the Mn oxide/CNT composite, but also with CNT to remake the catalyst. In other words, the reaction of MnO_4^- with CNT is also the method to synthesize these composites.

15.4 Conclusions

In accordance with the presented results we concluded that:

- The reaction of MnO_4^- ions with CNTs is a promising procedure to synthesize of water-oxidizing composites. In this case, a composite with the TOF $(mmolO_2/(mol\ Mn \cdot s)) > 1$ can be obtained. Thus, the synthesis of carboxylic acid or amine-functionalized CNTs as efficient water-oxidizing composites with very simple methods is possible.
- Amine-functionalized CNTs as support are better than CNTs functionalized with carboxylic acids, at least, under herein reported conditions.

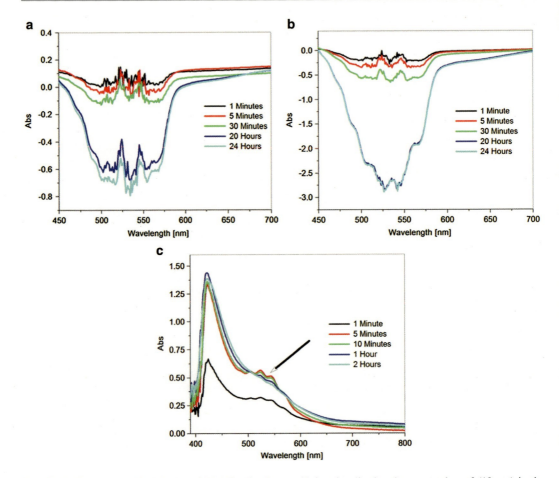

Fig. 15.5 The reaction of a solution of KMnO$_4$ (3 mL, 0.5 mM) and Mn oxide on amine (**a**) or carboxylic acid (**b**) functionalized carbon nanotubes (10 mg) at different times. Electronic spectra of Mn oxide phase on carboxylic acid functionalized carbon nanotubes of (10 mg) in the presence of Ce(IV) (3.5 mL, 0.05 M) after different times (**c**). The *arrows* show KMnO$_4$ formation

Acknowledgements MMN, SS, FR and MT are grateful to the Institute for Advanced Studies in Basic Sciences and the National Elite Foundation for financial support. MH acknowledges Prof. Dr. S. Dehnen and Prof. Dr. Florian Kraus for generous support. This work was supported by Grants-in-aid for Scientific Research from the Ministry of Education of Japan (Grant No: 26220801 to TT), and and by the Russian Science Foundation (Grant No: 15-14-30007). All authors thank Dr. Leonid Kurepin and Professor Harvey Hou for their helpful comments.

References

Glikman TS, Shcheglova IS (1968) The catalytic oxidation of water by quadrivalent cerium ions. Kinet. Katal 9:461–480

Harriman A, Pickering IJ, Thomas JM, Christensen PA (1988) Metal oxides as heterogeneous catalysts for oxygen evolution under photochemical conditions. J. Chem. Soc. Faraday Trans. 1 F 84:2795–2806

Jafari T, Moharreri E, Shirazi Amin A, Miao R, Song W, Suib SL (2016) Photocatalytic Water Splitting—The Untamed Dream: A Review of Recent Advances. Molecules. 9:900–929.

Mette K, Bergmann A, Tessonnier JP, Havecker M, Yao L, Ressler T, Schlögl R, Strasser P, Behrens M (2012) Nanostructured manganese oxide supported on carbon nanotubes for electrocatalytic water splitting. ChemCatChem 4:851–862

Morita M, Iwakura C, Tamura H (1977) The anodic characteristics of manganese dioxide electrodes prepared by thermal decomposition of manganese nitrate. Electrochim. Acta 22:325–328

Najafpour MM, Rahimi F, Fathollahzadeh M, Haghighi B, Hołyńska M, Tomo T, Allakhverdiev SI

(2014a) Nanostructured manganese oxide/carbon nanotubes, graphene and graphene oxide as water-oxidizing composites in artificial photosynthesis. Dalton Trans. 43:10866–10876

Najafpour MM, Abasi M, Tomo T, Allakhverdiev SI (2014b) Mn oxide/nanodiamond composite: a new water-oxidizing catalyst for water oxidation. RSC Adv 4:37613–37619

Najafpour MM, Abasi M, Tomo T, Allakhverdiev SI (2014c) Nanolayered manganese oxide/C_{60} composite: a good water-oxidizing catalyst for artificial photosynthetic systems. Dalton Trans 43:12058–12064

Najafpour MM, Kompany-Zareh M, Zahraei A, Sedigh DJ, Jaccard H, Khoshkam M, Britt RD, Casey WH (2013) Mechanism, decomposition pathway and new evidence for self-healing of manganese oxides as efficient water oxidizing catalysts: new insights. Dalton Trans. 42:14603–14611

Najafpour MM, Khoshkam M, Sedigh DJ, Zahraei A, Kompany-Zareh M (2015a) Self-healing for nanolayered manganese oxides in the presence of cerium (IV) ammonium nitrate: new findings. New J Chem 39:2547–2550

Najafpour MM, Fekete M, Sedigh GJ, Aro EM, Carpentier R, Eaton-Rye JJ, Nishihara H, Shen JR, Allakhverdiev SI, Spiccia L (2015b) Damage management in water-oxidizing catalysts: From photosystem II to nanosized metal oxides. ACS Catal 5:1499–1512

Parent AR, Crabtree RH, Brudvig GW (2013) Comparison of primary oxidants for water-oxidation catalysis. Chem Soc Rev 42: 2247–2252

Pickrahn KL, Park SW, Gorlin Y, Lee HBR, Jaramillo TF, Bent SF (2012) Active MnO_x electrocatalysts prepared by atomic layer deposition for oxygen evolution and oxygen reduction reactions. Adv. Energy Mater 2:1269–1277

Self-Healing in Nano-sized Manganese-Based Water-Oxidizing Catalysts

16

Mohammad Mahdi Najafpour, Seyed Esmael Balaghi, Moayad Hossaini Sadr, Behzad Soltani, Davood Jafarian Sedigh, and Suleyman I. Allakhverdiev

Summary

Water splitting is considered as a method to storage of renewable energies to hydrogen. The water-oxidation reaction in water splitting is an efficiency-limiting process for water splitting, and thus, there has been notable progress to find highly efficient water-oxidizing catalysts made from cost-effective and earth-abundant elements. In addition to efficiency, the stability of the water-oxidizing compounds is very important. Herein we focus on self-healing in manganese-based water-oxidizing catalysts in artificial photosynthetic systems.

Keywords

Water oxidation • Water splitting • Manganese • Manganese oxide • Artificial photosynthesis • Carbon nanotube • Nanostructure • Composite • Cerium • Mechanism

Contents

16.1 Introduction 333
16.2 Self-Healing in Water-Oxidizing Catalysts 334
16.3 Manganese Based Water-Oxidizing Catalyst 336
References 340

16.1 Introduction

Hydrogen as a fuel is proposed as a solution for the problems of the global warming and depletion of fossil fuels and a good way to hydrogen production is water splitting (Barber 2009; Bockris

M.M. Najafpour (✉)
Department of Chemistry, Institute for Advanced Studies in Basic Sciences (IASBS), Zanjan 45137-66731, Iran

Center of Climate Change and Global Warming, Institute for Advanced Studies in Basic Sciences (IASBS), Zanjan 45137-66731, Iran
e-mail: mmnajafpour@iasbs.ac.ir

S.E. Balaghi • M.H. Sadr (✉) • B. Soltani
Faculty of Basic Sciences, Department of Chemistry, Azarbaijan Shahid Madani University, Tabriz, Iran
e-mail: mp.sadr@gmail.com

D.J. Sedigh
Department of Chemistry, Institute for Advanced Studies in Basic Sciences (IASBS), Zanjan 45137-66731, Iran

S.I. Allakhverdiev (✉)
Controlled Photobiosynthesis Laboratory, Institute of Plant Physiology, Russian Academy of Sciences, Botanicheskaya Street 35, Moscow 127276, Russia

Institute of Basic Biological Problems, Russian Academy of Sciences, Pushchino, Moscow 142290, Russia

Faculty of Biology, Department of Plant Physiology, M.V. Lomonosov Moscow State University, Leninskie Gory 1–12, Moscow 119991, Russia
e-mail: suleyman.allakhverdiev@gmail.com

© Springer International Publishing AG 2017
H.J.M. Hou et al. (eds.), *Photosynthesis: Structures, Mechanisms, and Applications*,
DOI 10.1007/978-3-319-48873-8_16

1975; Pace 2005). There are a number of methods for water splitting, and water electrolysis is one of the best methods (Scheme 16.1) (Rasten et al. 2003; Takenaka et al. 1982; Zeng and Zhang 2010). It is possible to convert energy from Sun, wind, ocean currents, tides or wave to electricity.

One big problem in water splitting is water oxidation, which is involving multi-electron transfer (Scheme 16.1) (Nocera 2012):

$$2H_2O \rightarrow 4H^+ + O_2 + 4e^- \quad (16.1)$$

Thus, the development of novel water oxidizing catalysts is necessary (Blakemore et al. 2015; Frey et al. 2014; Hocking et al. 2011; Kärkäs et al. 2014). Platinum and iridium are widely used for water oxidation in modern technology (Liu and Wang 2012). On the other hand, cyanobacteria, algae, and higher plants manage to use abundant, non-toxic transition metals for the same purpose (Fig. 16.1) (Najafpour et al. 2015b).

Recently, there has been significant progress in improving the activities of catalysts for water oxidation (Blakemore et al. 2015; Frey et al. 2014; Hocking et al. 2011; Kärkäs et al. 2014; Liu and Wang 2012; Nocera 2012). However, the stability of these catalysts is also necessary for commercial application. To solve the stability problem we have, at least, two ways. The first strategy is using a very stable catalyst with a very high turnover number, and the second is using a catalyst with self-healing (Najafpour et al. 2015a). The term 'self-healing' or 'self-repair' means self-recovery of the compound following damage caused by the external environment or internal stresses (Najafpour et al. 2015a). As discussed by Nosonovsky (Nosonovsky and Bhushan 2009), healing is performed by shifting the system away from the thermodynamic equilibrium, which causes a restoring thermodynamic force to drive the system back to equilibrium (Scheme 16.2) (Nosonovsky and Bhushan 2009). The force drives the healing process. Shifting the system away from equilibrium can be achieved by placing it in a metastable state, so that the rupture breaks the fragile metastable equilibrium, and the system drives to the new most stable state (Scheme 16.2). The metastability can be achieved by energy for example heating or redox potential.

16.2 Self-Healing in Water-Oxidizing Catalysts

There are a few observed self-healing phenomena in compounds:

The compounds require no external intervention to restore the damage, and other compounds are capable of nonautonomous self-healing that requires an external trigger to start the self-healing process (Amendola and Meneghetti 2009; Ghosh 2009; Zheludkevich 2009). These self-healing are important to increase the robustness and extend the life of a compound when in cases where repair or replacement of compounds is economically detrimental, dangerous, or impossible (Amendola and Meneghetti 2009; Ghosh 2009; Zheludkevich 2009). Nanomaterials are prone to both fast degradation and structural damage and self-healing can be more important for them. A few strategies were used in self-healing compounds (Amendola and Meneghetti 2009; Ghosh 2009; Zheludkevich 2009). One strategy is using epoxy that can serve as a healing agent that stored within brittle macrocapsules embedded into the matrix. When damage occurs, the capsules fracture and the healing agent is released and propagates into the crack due to capillarity. In the next step, the healing agent reacts with the catalyst in the matrix, which starts the cross-linking reaction and hardening of the epoxy that seals the crack (Amendola and Meneghetti 2009; Ghosh 2009; Zheludkevich 2009). Another strategy involves thermoplastic polymers with various ways of incorporating the healing agent into the material that needs heating to initiate healing

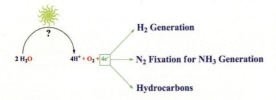

Scheme 16.1 Water oxidation to form cheap electrons

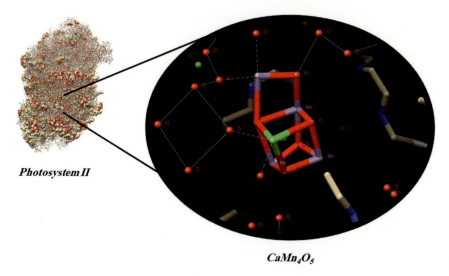

Fig. 16.1 The Mn-Ca cluster (Ca:*green*; Mn:*blue*; O: *red*) cluster and surrounding amino acids. Amino acid residues in PSII are involved in proton, water and oxygen transfer

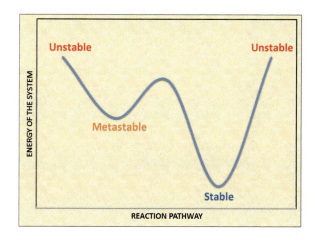

Scheme 16.2 Possible reaction pathways (marked by *arrows*) related to the energy state of a system. Stable, metastable and unstable points are indicated

(Amendola and Meneghetti 2009; Ghosh 2009; Zheludkevich 2009). These two strategies are used for polymers.

For metallic systems, there are three main methods for self-healing. The first is the formation of precipitates at the defect sites that immobilize further growth until failure. The second is using an alloy matrix with a microfiber or wires with ability to recover their original shape after some deformation has occurred (Amendola and Meneghetti 2009; Ghosh 2009; Zheludkevich 2009). This process usually needs heating above the phase transformation temperature. The third approach is to use a healing agent such as an alloy with a low melting temperature embedded into a metallic solder matrix.

The compounds that catalyse multi-electron reactions are prone to structural rearrangement, and instability, during turnover as discussed by Nocera (Lutterman et al. 2009; Surendranath et al. 2012). Nature used a self-healing for the WOC as an enzyme that performs a four-electron reaction (Najafpour et al. 2015a). As water-oxidizing catalysts are very significant for energy science, the designs of catalysts that repair themselves is a necessary issue in industrial application. In water oxidation, a self-healing mechanism in *in-situ* for cobalt oxides under water oxidation was first mentioned by Kanan and Nocera (Lutterman et al. 2009; Surendranath et al. 2012). In that experiment, an amorphous cobalt oxide catalyst film (CoCF) was electrodeposited from a solution containing cobalt and phosphate ions at neutral pH. The fate of the Co- and phosphate ions during electrodeposition and catalyst operation was examined when the oxide film (Co-Pi) was prepared from a radio labelled solution containing ^{57}Co and ^{32}P isotopes. When a potential bias of 1.3 V vs. normal hydrogen electrode (NHE) was applied no film dissolution was observed until the potential bias was removed (Lutterman et al. 2009; Surendranath et al. 2012). This process could be reversed when a buffering electrolyte was present. It has been described as self-healing of the catalyst film, which entails the re-deposition of the leached Co(II) ions at operational potentials as an amorphous phosphate-rich Co(III) oxide catalyst (Lutterman et al. 2009; Surendranath et al. 2012).

16.3 Manganese Based Water-Oxidizing Catalyst

A tetranuclear Mn cluster with diarylphosphinate ligand was reported as an efficient catalyst for water oxidation. $[Mn_4O_4L_6]^+$ (L: diarylphosphinate) embedded into a Nafion shows to be robust and to function for up to 3 days with minor loss in activity (Dismukes et al. 2009). A few mechanisms were proposed for water oxidation by the complex. In 2011, Hocking et al. proposed a mechanism for water oxidation by the $[Mn_4O_4L_6]^+$ compound (Parent et al. 2013). The group demonstrated that in situ cycling between the Mn(II) photoreduced product and an oxidized, disordered Mn(III/IV) oxide phase most likely forms the basis of the observed water oxidation catalysis. To proof the cycling, the group used K-edge X-ray absorption spectroscopy (XAS) for $[Mn_4O_4L_6]^+$ both in acetonitrile and embedded in a Nafion film coated on an electrode (Parent et al. 2013).

A shift of the X-ray absorption near edge structure (XANES) peak to a lower energy related to a Mn(II) was also observed immediately upon loading of $[Mn_4O_4L_6]^+$ into Nafion (Parent et al. 2013). The extended X-ray absorption fine structure (EXAFS) and its Fourier transform was consistent with a Mn(II) inner sphere occupied by six oxygen donors with no ordered second sphere (Parent et al. 2013). Electro-oxidation of the film shifted the XANES rising edge to a higher energy expected for layered Mn oxide. Thus, the group concluded that $[Mn_4O_4L_6]^+$ may not be responsible for the prolonged water oxidation catalysis. The protonated ligand was detected clearly on the Nafion film by HNMR and PNMR that show decoordination of ligand from $[Mn_4O_4L_6]^+$ (Parent et al. 2013). High resolution TEM studies for both the oxidized and reduced states of Mn in the Mn compound show that nanoparticles (1–2 nm diameter) were identified in the TEM image of the oxidized films from both Mn(II) and $[Mn_4O_4L_6]^+$ (Parent et al. 2013), whereas no nanoparticles were found for the reduced state (Fig. 16.2).

Najafpour's group introduced new mechanism for self-healing of Mn oxide in the presence of Ce(IV) (Najafpour et al. 2012a, b, 2013a, b, 2015a, c). From Pourbaix diagram (Fig. 16.3) (Hocking et al. 2014), we understand that in the absence of any more activation energy all species that place above the water oxidation line, such as MnO_4^-, should cause water oxidation. However, water-oxidation reactions are usually slow by the compound kinetically and in high enough potential both water oxidation and MnO_4^- reactions take place. MnO_4^- and similar species cause a long-term stability problem and separate Mn ions from catalysts and decrease activity of these compounds toward water oxidation.

Recently, the Najafpour's group reported more details about self-healing of Mn oxides (Najafpour et al. 2013b). The group fund that in the reaction of these Mn oxides with Ce(IV), MnO_4^- is produced that changes in concentration with time. All MnO_4^- was consumed in the

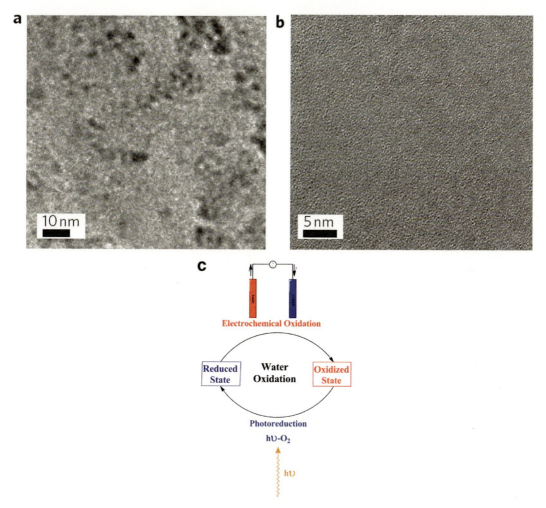

Fig. 16.2 Self-healing cycle for Mn oxide proposed by Spiccia and co-workers. High-resolution transmission electron microscopy images: Nafion film showing presence of manganese-oxide nanoparticles formed on electro-oxidation of $[Mn_4O_4L_6]^+$ at 1.0 V (versus Ag/AgCl) (**a**). Nafion film examined after introduction of $[Mn_4O_4L_6]^+$ into the film (**b**). Images and captions are from (Hocking et al. 2011) (Reprinted with permission from Hocking et al. 2011. Copyright (2011) by Nature Publishing Group. A proposed mechanism for self-healing by the system (**c**))

end of reaction when Ce(IV) was completely consumed, suggesting the operation of a self-healing mechanism (Najafpour et al. 2015c). Regarding another experiment, a MnO_4^- solution in the presence of K-layered oxide showed a linear reduction in concentration of MnO_4^- (Najafpour et al. 2013b). On the other hand, MnO_4^- formation was observed by the reaction of Mn(II) nitrate (0.5 mM) and Ce(IV) (0.2–2 M) (Najafpour et al. 2013b). Without any Mn oxide, the amounts of the MnO_4^- showed no change during 100 hours of experiment indicating that Mn(II) or Mn(III) in Mn oxide is necessary for reduction of MnO_4^- (Najafpour et al. 2013b). Using the multivariate curve resolution-alternative least squares, spectroscopic data was analyzed and the concentration profiles of cerium (IV) ammonium nitrate (Ce(IV)) and MnO_4^- during the reaction under different conditions of water oxidation were reported (Najafpour et al. 2015c). As shown in Fig. 16.4, the results showed that the concentration profile of Ce (IV) (red line) is in descending manner and is expected since it is an initial oxidant and is

Fig. 16.3 Pourbaix diagram for Mn ion

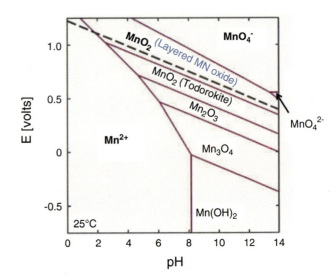

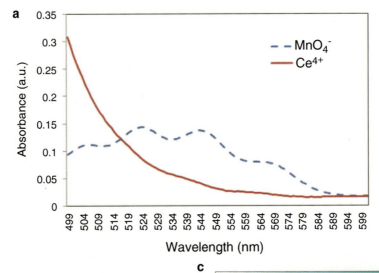

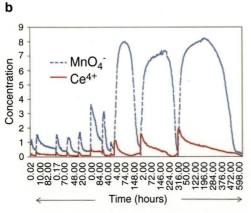

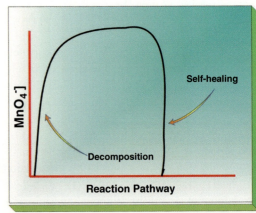

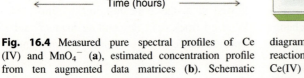

Fig. 16.4 Measured pure spectral profiles of Ce (IV) and MnO_4^- (**a**), estimated concentration profile from ten augmented data matrices (**b**). Schematic diagram to show the decomposition and self-healing reactions for nanolayered Mn oxide in the presence of Ce(IV) (**c**)

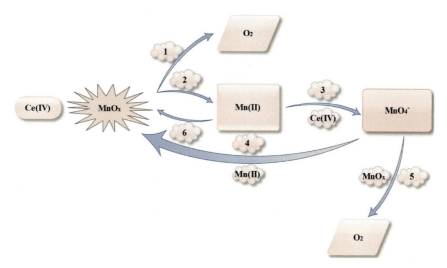

Scheme 16.3 Self-healing in water oxidation by Mn oxides in the presence of Ce(IV). *1*: Oxygen evolution was detected by an oxygen meter. The origin of oxygen is water. *2*: Mn(II) was detected by EPR (see text). *3*: MnO_4^- formation could be detected by UV-Vis spectrophotometry in reacting Mn(II) and Ce(IV). *4*: It is known that in the reaction of Mn(II) and MnO_4^- at different pH values, Mn oxide is produced. *5*: MnO_4^- in the presence of Mn oxide oxidizes water. In the reaction, MnO_4^- reduces to Mn oxide. *6*: Mn(II) in the presence of Ce (IV) form Mn oxide. In a typical experiment, the reaction of $MnSO_4$ in the presence of Ce(IV) (1.0 M), forms MnO_2 that can be detected by XRD. Images and captions are taken (Najafpour et al. 2015a) (Reprinted with permission from (Najafpour et al. 2015b). Copyright (2015a) by American Chemical Society)

consumed during the water oxidation. On the other hand, MnO_4^- is produced and consumed during the reaction time (Najafpour et al. 2015c). As shown in Fig. 16.4, the concentration profile of MnO_4^- (blue line) is similar to an intermediate during water oxidation. It is also clear that the rate of MnO_4^- formation is decreased when the initial concentration of Ce(IV) is decreased (Najafpour et al. 2015c). When the concentration of Ce(IV) is low, the concentration of MnO_4^- cannot be clearly observed, which shows, in low concentration of Ce(IV), MnO_4^- formation is not favorable (Najafpour et al. 2015c).

Regarding this experiments, the group suggest that (Najafpour et al. 2015c):

(a) MnO_4^- may be formed by oxidizing leaked Mn(II) from Mn oxide by Ce(IV).
(b) Mn oxide is responsible for reduction of MnO_4^-. At least two reactions are possible for reduction of MnO_4^- in the presence of Mn oxide: (i) The first proposed reaction is the reaction of MnO_4^- with Mn(II) or Mn(III) ions on the surface of Mn oxide (Scheme 16.3). (ii) It is also possible that MnO_4^-, as an oxidant, and Mn oxide, as water-oxidizing catalyst, react to oxidize water (Scheme 16.3) (Najafpour et al. 2015c).

In Nature, the self-healing processes are usually performed by replacing the 'outdated' components but that is difficult to mimic in synthetic systems (Urban 2012). However, the systems introduced here, similar to biological systems, use decomposed and 'outdated' components (Mn(II) or MnO_4^-) for self-healing.

Mn(II) as a labile specie leads to degradation of the catalyst in the water-oxidation reaction because the population of the electrons in the antibonding orbitals of metal oxides, e_g^* and t_{2g}^* correlates with stability of the catalyst (Huynh et al. 2014). Regarding this issue, the

Fig. 16.5 Decomposition and self-healing of the PSII-WOC (during photodeactivation) or metal oxides (during catalytic operation), could be summarized by this simple schematic image. In this image, each structure contains only a few components. The attachment or removal of these components results in self-healing (*green arrow*) or decomposition (*red arrows*) (Images and captions are from Najafpour et al. 2015a. Reprinted with permission from Najafpour et al. 2015a. Copyright (2015) by American Chemical Society. A proposed mechanism for self-healing by the system (**c**))

decomposition of Mn oxides in acidic condition is lower than other late first-row metal oxides with higher d-electron counts. So the MnO$_x$ is stable under the water oxidation conditions up to 4 M HCl due to self-healing (Huynh et al. 2014).

Self-healing mechanisms for biological and artificial compounds are usually different but a few similarities were reported among different water-oxidizing catalysts in self-healing reaction (Najafpour et al. 2015a). Such systems can be considered to be composed of a few simple "building blocks" (Najafpour et al. 2015a):

1. Mn-Ca cluster in Photosystem II: Mn, Ca and O
2. Mn oxides: primarily Mn and O
3. Co oxides: primarily Co and O

Both decomposition (red arrows in Fig. 16.5) and self-healing (green arrows in Fig. 16.5) reactions can occur for these structures. The detachment and reattachment of these components is related to the decomposition and self-healing processes, respectively.

Acknowledgments MMN and DJS are grateful to the Institute for Advanced Studies in Basic Sciences and the National Elite Foundation for financial support. MHS, BS and SEB are grateful from Azarbaijan Shahid Madani University. This work was supported by the Russian Science Foundation (Grant №14-14-00039). The authors thank Dr. Wu Xu and Professor Harvey Hou for their comments.

References

Amendola V, Meneghetti M (2009) Self-healing at the nanoscale. Nanoscale 1:74–88.
Barber J (2009) Photosynthetic energy conversion: natural and artificial. Chem Soc Rev 38:185–196.
Blakemore JD, Crabtree RH, Brudvig GW (2015) Molecular catalysts for water oxidation. Chem Rev 115:12974–13005.
Bockris JO (1975) Energy: the solar-hydrogen alternative.
Dismukes GC, Brimblecombe R, Felton GA, Pryadun RS, Sheats JE, Spiccia L, Swiegers GF (2009) Development of bioinspired mn4o4– cubane water oxidation catalysts: Lessons from photosynthesis. Acc Chem Res 42:1935–1943.
Frey CE, Wiechen M, Kurz P (2014) Water-oxidation catalysis by synthetic manganese oxides–systematic variations of the calcium birnessite theme. Dalton Trans 43:4370–4379.
Ghosh SK (2009) Self-healing materials: fundamentals, design strategies, and applications. John Wiley & Sons.
Hocking RK, Brimblecombe R, Chang L-Y, Singh A, Cheah MH, Glover C, Casey WH, Spiccia L (2011) Water-oxidation catalysis by manganese in a geochemical-like cycle. Nat Chem 3:461–466.
Hocking RK, Malaeb R, Gates WP, Patti AF, Chang SL, Devlin G, Macfarlane DR, Spiccia L (2014) Formation of a Nanoparticulate Birnessite-Like Phase in Purported Molecular Water Oxidation Catalyst Systems. ChemCatChem 6:2028–2038.
Huynh M, Bediako DK, Nocera DG (2014) A functionally stable manganese oxide oxygen evolution catalyst in acid. J Am Chem Soc 136:6002–6010.
KäRkäS MD, Verho O, Johnston EV, Åkermark BR (2014) Artificial photosynthesis: molecular systems for catalytic water oxidation. Chem Rev 114:11863–12001.
Liu X, Wang F (2012) Transition metal complexes that catalyze oxygen formation from water: 1979–2010. Coord Chem Rev 256:1115–1136.

Lutterman DA, Surendranath Y, Nocera DG (2009) A self-healing oxygen-evolving catalyst. J Am Chem Soc 131:3838–3839.

Najafpour MM, Rahimi F, Amini M, Nayeri S, Bagherzadeh M (2012a) A very simple method to synthesize nano-sized manganese oxide: an efficient catalyst for water oxidation and epoxidation of olefins. Dalton Trans 41:11026–11031.

Najafpour MM, Rahimi F, Aro E-M, Lee C-H, Allakhverdiev SI (2012b) Nano-sized manganese oxides as biomimetic catalysts for water oxidation in artificial photosynthesis: a review. J R Soc Interface 9:2383–2395.

Najafpour MM, Kompany-Zareh M, Zahraei A, Sedigh DJ, Jaccard H, Khoshkam M, Britt RD, Casey WH (2013a) Mechanism, decomposition pathway and new evidence for self-healing of manganese oxides as efficient water oxidizing catalysts: new insights. Dalton Trans 42:14603–14611.

Najafpour MM, Sedigh DJ, Pashaei B, Nayeri S (2013b) Water oxidation by nano-layered manganese oxides in the presence of cerium (IV) ammonium nitrate: important factors and a proposed self-repair mechanism. New J Chem 37:2448–2459.

Najafpour MM, Fekete M, Sedigh DJ, Aro E-M, Carpentier R, Eaton-Rye JJ, Nishihara H, Shen J-R, Allakhverdiev SI, Spiccia L (2015a) Damage Management in Water-Oxidizing Catalysts: From Photosystem II to Nanosized Metal Oxides. ACS Catal 5:1499–1512.

Najafpour MM, Ghobadi MZ, Larkum AW, Shen J-R, Allakhverdiev SI (2015b) The biological water-oxidizing complex at the nano–bio interface. Trends Plant Sci 20:559–568.

Najafpour MM, Khoshkam M, Sedigh DJ, Zahraei A, Kompany-Zareh M (2015c) Self-healing for nanolayered manganese oxides in the presence of cerium (IV) ammonium nitrate: new findings. New J Chem 39:2547–2550.

Nocera DG (2012) The artificial leaf. Acc Chem Res 45:767–776.

Nosonovsky M, Bhushan B (2009) Thermodynamics of surface degradation, self-organization and self-healing for biomimetic surfaces. Philos Trans R Soc A 367:1607–1627.

Pace RJ (2005) An integrated artificial photosynthesis model. Artificial photosynthesis: From basic biology to industrial application:13–34.

Parent AR, Crabtree RH, Brudvig GW (2013) Comparison of primary oxidants for water-oxidation catalysis. Chem Soc Rev 42:2247–2252.

Rasten E, Hagen G, Tunold R (2003) Electrocatalysis in water electrolysis with solid polymer electrolyte. Electrochim Acta 48:3945–3952.

Surendranath Y, Lutterman DA, Liu Y, Nocera DG (2012) Nucleation, growth, and repair of a cobalt-based oxygen evolving catalyst. J Am Chem Soc 134:6326–6336.

Takenaka H, Torikai E, Kawami Y, Wakabayashi N (1982) Solid polymer electrolyte water electrolysis. Int J Hydrogen Energy 7:397–403.

Urban MW (2012) Dynamic materials: the chemistry of self-healing. Nat Chem 4:80–82.

Zeng K, Zhang D (2010) Recent progress in alkaline water electrolysis for hydrogen production and applications. Prog Energy Combust Sci 36:307–326.

Zheludkevich M (2009) Self-healing anticorrosion coatings. Wiley-VCH Verlag GmbH & Co., KGaA: Weinheim, Germany.

17

A Robust PS II Mimic: Using Manganese/Tungsten Oxide Nanostructures for Photo Water Splitting

Harvey J.M. Hou

Summary

Photosystem II is able to catalyze water-splitting reaction to achieve energy storage on the large scale at room temperature and neutral pH in green plants, algae, and cyanobacteria. The three-dimensional structure of photosystem II with oxygen-evolving activity has been determined at an atomic level, which provides a thorough image with the specific position of each atom in the Mn_4CaO_5 cluster. These advancements have significantly enhanced our understanding of the mechanisms of water splitting in photosynthesis and offered a unique opportunity for solar fuel production. Inspired by the natural photosynthesis, great progresses in using earth abundant elements based artificial catalytic systems have been made to achieve artificial catalysis in photo water splitting. In this chapter, I describe a robust PS II mimic containing manganese/tungsten oxide nanostructure to accomplish the photo water splitting chemistry. The synthesis, structural characterization, photo water splitting activity, and possible mechanism of the manganese/tungsten oxide system are presented and discussed. This PS II mimic shows a compelling working principle by combining the active catalysts in water splitting with semiconductor hetero-nanostructures for effective solar energy harnessing and is highly likely to offer novel technology for transforming the solar energy into our future energy systems.

Keywords

Photosystem II • Manganese • Tungsten oxide • Semiconductor • Nanomaterial • Water oxidation • Water splitting • Catalysis

Contents

17.1 Introduction 343
17.2 Manganese Oxo Dimer and Manganese Oxo Oligomer 344
17.3 Design, Synthesis, and Structural Characterization of Manganese/Tungsten Oxide Nanostructure 348
17.4 Catalytic Activity of Manganese/Tungsten Oxide Nanostructure 350
17.5 Mechanism of the Manganese/Tungsten Oxide System 353
17.6 Conclusions 355
References 356

17.1 Introduction

Natural photosynthesis has supported the life on our planet over several billion years by storing the solar energy and generating oxygen

H.J.M. Hou (✉)
Department of Physical Sciences, Alabama State University, Montgomery, AL 36104, USA
e-mail: hhou@alasu.edu

molecules via water splitting reaction. Photosystem II is able to catalyze a water-splitting reaction to liberate electrons and achieve energy storage on the large scale at room temperature and neutral pH in green plants, algae, and cyanobacteria (Blankenship 2002; Diner and Rappaport 2002; Brudvig 2008; Barber 2009). To address the global energy crisis now, photosynthesis is a unique and an excellent example for design and mimicry for solar energy utilization and renewable fuel production on the large scale via water splitting chemistry (Lewis and Nocera 2006; Blankenship et al. 2011; Najafpour et al. 2016; Moore 2016).

In photosynthetic organisms, photons are captured by light-harvesting antenna complexes, and the light energy is efficiently transferred to reaction centers by the antenna systems via well-structured protein complexes such as the phycobilisome megacomplex (Liu et al. 2013). The three-dimensional structure of photosystem II with oxygen-evolving activity has been determined at an atomic level, which provides a thorough image with the specific position of each atom in the Mn_4CaO_5 cluster (Ferreira et al. 2004; Loll et al. 2005; Umena et al. 2011). Recently intense femtosecond X-ray pulses were used for simultaneous X-ray diffraction and X-ray emission spectroscopy of microcrystals of PS II at room temperature and revealed electronic structure of Mn_4CaO_5 cluster on S_1 and S_2 states (Kern et al. 2013). These advancements have significantly enhanced our understanding of the mechanisms of water splitting in photosynthesis and offered a unique opportunity for solar fuel production.

17.2 Manganese Oxo Dimer and Manganese Oxo Oligomer

Inspired by natural photosynthesis, great progresses have been made in using earth abundant materials to achieve photoinduced charge separation and water oxidation (Blakemore et al. 2015; Li et al. 2015). The design and synthesis of functional oxygen evolving catalysts using manganese-containing compounds was successfully accomplished to mimic natural photosynthesis (Bolton 1996; Meyer 2008; Brimblecombe et al. 2008, 2010; Sala et al. 2009; Youngblood et al. 2009; Kudo and Miseki 2009; Hou 2010; Rivalta et al. 2012; Young et al. 2012, 2015; He et al. 2013; Najapour et al. 2013). One well known example is a Mn-oxo tetrameric compound, $[Mn_4O_4](dpp)_6]$, developed by Dismukes and co-workers (Ruettinger and Dismukes 1997; Ruettinger et al. 2000). The important feature of the compound is its cubical Mn_4O_4 core. Recently the Mn-oxo tetramer/Nafion system has been developed by deposition of cubium into the Nafion membrane and is able to efficiently photo oxidize water to oxygen gas as a supramolecular model system (Brimblecombe et al. 2009; Dismukes et al. 2009).

Another exceptional example is Brudvig Mn-oxo dimer, $[OH_2(terpy)Mn(O)_2Mn(terpy) OH_2](NO_3)_3$, which contains the key structure of the two terminal water molecules for water oxidation (Limburg et al. 1999, 2001; Cady et al. 2010). The Mn-oxo dimer as a PS II functional model is able to oxidize water to produce dioxygen in the presence of chemical oxidants such as oxone or Ce(IV) ion (Cady et al. 2008). The water oxidation mechanism of Mn-oxo dimer is well studied and proposed that the active catalytic species is Mn(V) = O or Mn(V)-oxo radical, which is capable of releasing oxygen and close the catalytic cycle. The Mn-oxo dimer compound can be directly deposited on the surface of TiO_2 nanoparticles and forms Mn-oxo tetrameric structure, which is able to oxidized water in the present of Ce(IV) ion (Li et al. 2009; McNamara et al. 2009). The Mn-oxo dimer immobilized in Nafion membranes is able to catalyze the photo reaction of water oxidation (Young et al. 2011).

However, the Mn-oxo dimer compound seems unstable in aqueous solution and may decompose to form precipitate after exposed to room temperature for several days. Figure 17.1 shows the effect of the aqueous solution temperatures on the UV-Vis spectra of Mn-oxo dimer (Zhang et al. 2011). The elevated temperature causes substantial changes in three regions of 280 nm,

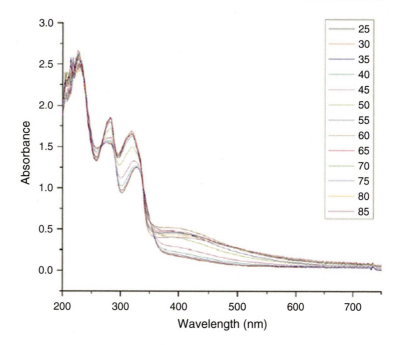

Fig. 17.1 UV-Vis spectra of the Mn-oxo dimer compound taken at the elevated temperatures shows the decomposition reaction of the compound (Zhang et al. 2011, reproduced with permission from Elsevier)

310–340 nm, and 400–440 nm. As Mn(IV/IV)-oxo dimer species has strong absorption peak at 400–440 nm (Limburg et al. 2001), the increase in absorption at these wavelengths upon heating may be due to the conversion of Mn(III/IV)-oxo dimer to Mn(IV/IV) dimer. The increase at 280 nm and 310–340 nm is likely due to the dissociation of the terpy ligand from the Mn center ions.

The absorbance at 400 nm indicates a two-step process in the decomposion of Mn-oxo dimer compound as shown in Fig. 17.2 (Zhang et al. 2011). The first phase occurred at temperature below 60 °C, possibly due to formation of an intermediate with a valence change in Mn ion. The intermediate species is likely a Mn(IV/IV)-oxo compound. The second phase happened at above 60 °C accompanied by the precipitation of the intermediate to form a more stable solid species, which may be a Mn-oxo oligomer material.

To confirm the valence change in Mn center ion in the compound, EPR spectra of frozen solution of the Mn-oxo dimer collected over time during the heating process are recorded in Fig. 17.3 (Zhang et al. 2011). The native Mn (III/IV)-oxo dimer sample in acetate buffer shows a 16-line signal in the range of 2800–4100 G, which is characteristic for the Mn(III/IV) mixed-valence species. When the solution was heated at 75 °C, the 16-line EPR signal was decreased by a factor of 90% within 10 min. This indicates that the Mn(III/IV)-oxo dimer is converted by heating into an EPR silent species, such as the Mn(IV/IV)-oxo dimer suggested by the UV–visible spectrophotometric data.

To further characterize the thermal transformation of the Mn(III/IV)-oxo dimer, we performed a kinetic study at 60 °C using UV–Vis spectrometry to measure the spectral changes at different time intervals. A solution of the Mn (III/IV)-oxo dimer was placed in a reactor at 60 °C controlled by a circulated water bath (Fig. 17.4) (Zhang et al. 2011). The conversion of the Mn(III/IV)-oxo dimer into the intermediate with an increased absorbance at 400 nm took approximately 8 min at a transformation temperature of 60 °C. A fit of the kinetic trace of absorbance at 400 nm revealed that the half time of transformation of the Mn(III/IV)-oxo dimer was 3.5 ± 0.5 min in the first fast step

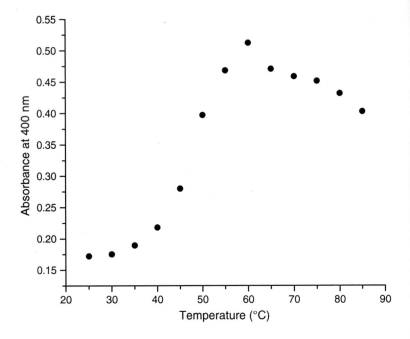

Fig. 17.2 The absorbance at 400 nm of the Mn-oxo dimer compound at the elevated temperatures reveals the two-step mechanism of decomposition reaction of the compound (Zhang et al. 2011, reproduced with permission from Elsevier)

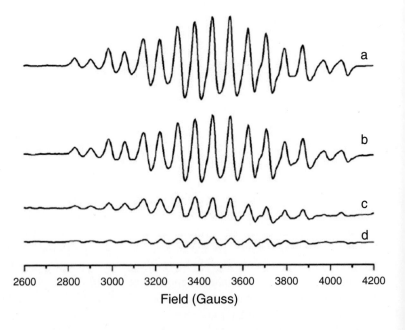

Fig. 17.3 EPR spectra of Mn-oxo dimer compound after heating at 75 °C at times of 0, 2, 6, and 10 min shows the valence change of Mn ion in the compound (Zhang et al. 2011, reproduced with permission from Elsevier)

and 19 ± 4 min in the following slow step. The fit of the kinetic trace of absorbance at 400 nm was temperature dependent. The rate constants of the two steps of decomposition of the Mn(III/IV)-oxo dimer at the different temperatures were used to determine the activation energy (Ea) based on Arrhenius plots. The activation energies for step 1 and step 2 were 68 ± 10 kJ/mol, and 82 ± 16 kJ/mol, respectively.

We would expect to see the change in water oxidation activity when the Mn-oxo dimer is heated as the active species, Mn-oxo mix valence

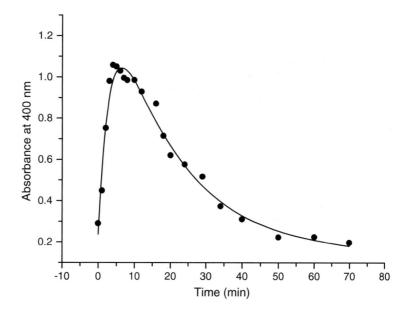

Fig. 17.4 Kinetic curve of the thermal induced decomposition reaction of the Mn-oxo dimer compound monitored at 400 nm at 60 °C reveals the lifetime of 3.0 min for the first step and the lifetime of 19 min for the second step (Zhang et al. 2011, reproduced with permission from Elsevier)

compound, is decomposed to produced a Mn-oxo oligomer complex. Figure 17.5 shows the effect of elevated temperatures on the initial rate of O_2 evolution catalyzed by the Mn(III/IV)-oxo dimer. The oxygen evolution rates of solutions of the Mn(III/IV)-oxo dimer were measured at room temperature in the presence of oxone in acetate buffer solution. We found that the catalytic activity of the Mn(III/IV)-oxo dimer contains two phases during the heating process in the range of 25–85 °C. In the first step in the temperature range of 25 °C to 60 °C, the O_2 evolution rate measured at 22 °C showed a small decline and then an increase.

As the Mn(III/IV)-oxo dimer was heated, we noticed a change in the color of the solution of the Mn(III/IV)-oxo dimer from light green to light brown. In addition, a brown precipitate was formed during the heating. The UV–Vis spectrophotometric and EPR data demonstrated that the Mn(III/IV)-oxo dimer decomposed. As the active catalytic species has previously been shown to be the Mn(III/IV)-oxo dimer (Limburg et al. 2001), one would anticipate the activity would decrease after decomposition of Mn (III/IV)-oxo dimer. Unexpectedly, Fig. 17.5 showed an increase in catalytic rate after heating to 60 °C (Zhang et al. 2011). We propose that heating the solution of Mn(III/IV)-oxo dimer produces a novel Mn-containing complex, which has higher activity to catalyze the O_2 evolving reaction in the presence of oxone. Comparing the O_2 evolving reaction catalyzed by the supernatant aliquot and the precipitate fraction in the reaction, we observe that both the supernatant and precipitate per Mn showed higher activity than the native Mn(III/IV)-oxo dimer solution. It is likely that two types of Mn-containing catalysts were formed: one in solid and one in aqueous form (Hou 2011).

We propose a two-step mechanism for decomposition of the Mn(III/IV)-oxo dimer under elevated temperature between 25 and 85 °C. The fast step involving a valence change in Mn has a lifetime of 3.5 ± 0.5 min and an activation energy of 68 ± 10 kJ/mol. In this first reaction step, the Mn(III/IV)-oxo dimer may have disproportionated into Mn(II)–terpy complex, Mn(IV/IV)-oxo dimer, and an unknown water-soluble species (I) with high catalytic activity. The following slow step with a lifetime of 19 ± 4 min leads to a transformation of the intermediate species to a highly active Mn-containing precipitate. The activation energy of the second step is 82 ± 16 kJ/mol. In the second reaction step, an additional unknown

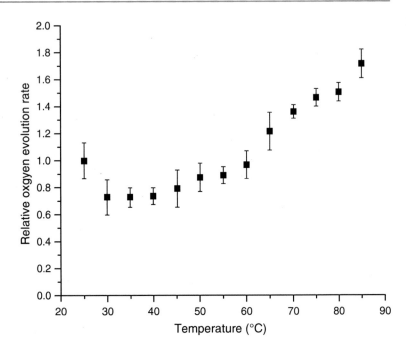

Fig. 17.5 The formation of the novel Mn-oxo oligomer material with high active oxygen evolving activity after heating the Mn-oxo dimer compound (Zhang et al. 2011, reproduced with permission from Elsevier)

water-soluble species (P) with high catalytic activity was also presented. The novel Mn-containing water splitting catalyst, $[(Mn_2O_2(terpy)_2(H_2O)_2]_x$, is a Mn-oxo oligomer solid material. The Mn-oxo oligomer is thermally stable and may be coupled with a semiconductor for efficient photo-water splitting.

17.3 Design, Synthesis, and Structural Characterization of Manganese/Tungsten Oxide Nanostructure

Inspired by natural photosynthesis, great progresses in using earth abundant element based artificial catalytic systems have been made to achieve artificial catalysis in photo water splitting (Nocera 2012; He et al. 2013; Najapour et al. 2013). In particular, semiconductor material is proved to be an excellent candidate to achieve this purpose (Fujishima and Honda 1972). Take the example of Mn-oxo oligomer complex and semiconductor materials such as tungsten oxide, Fig. 17.6 illustrates the proposed working mechanism of the semiconductor/catalyst system (Liu et al. 2011). The sunlight is absorbed by tungsten oxide semiconductor and causes the charge separation to produce electrons and holes. The holes receive electrons from the Mn-oxo oligomer complex, which is the precipitate of Brudvig catalyst (Mn-oxo dimer) under thermal conditions (Liu et al. 2011). The Mn-oxo compound has a high catalytic activity to extract electrons from water to generate oxygen gas. The electrons are transferred to the cathode by an electric wire to produce hydrogen gas. The key feature of the design is the ultrathin layer of Mn-oxo compound (2–5 nm). The design will take the advantage of high catalytic activity and minimize the electron resistance in electron transfer from water to Mn compound and tungsten oxide.

We choose ALD to prepare WO_3 because of the following advantages: (1) a high degree of control over the resulting materials; (2) excellent step coverage to yield conformal coatings; and (3) process versatility to tailor the composition of the deposit. WO_3 was studied because it is one of the most researched compounds for water splitting. The widely available literature makes

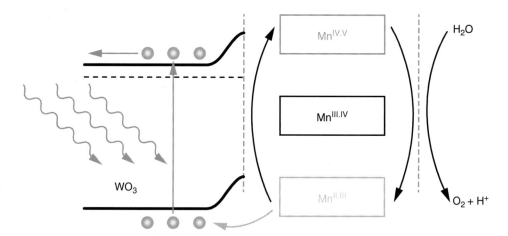

Fig. 17.6 Design of a catalytic system containing semiconductor tungsten oxide and manganese-oxo oligomer complex in photo water splitting mimicking photosynthesis (Liu et al. 2011, reproduced with permission from Wiley-VCH)

it easy to compare our results with existing reports and thus allows us to test the power of the heteronanostructure design. To avoid the production of corrosive byproducts during the ALD process and to ensure the reaction occurs in the true ALD regime, we used $(tBuN)_2(Me_2N)_2W$ as tungsten precursor and H_2O as oxygen precursor.

The goal was to verify that the growth indeed takes place in the ALD regime. The dependence of the growth rate on the precursor pulse times and on the substrate temperature unambiguously confirms this. In addition, the excellent linear dependence of the deposition thickness on the number of precursor pulses supports the ALD growth mechanism and shows the extent of control we can achieve. That a long H_2O pulse time is necessary to initiate growth is a key finding of this work. Despite intentional strengthening of the oxidative conditions, as-grown WO_3 exhibited a tinted color, indicating the existence of oxygen deficiencies, which was then corrected by an annealing step in O_2 at 550 °C. The crystalline nature of the product is shown in the high resolution (HR) TEM image in Fig. 17.7 and XRD data in Fig. 17.8 (Liu et al. 2011). The as-grown tungsten oxide film was annealed at 550 °C in O_2 to achieve the desired stoichiometry and crystallinity. Consistent with the TEM result, formation of monoclinic tungsten oxide was confirmed by XRD after annealing. The uniformity and good coverage around the nanonet branches show that this deposition technique is suitable for the creation of heteronanostructures (Lin et al. 2009).

Different Mn-oxo dimer solution concentrations, different deposition temperatures and different deposition times were studied. The general trend was that more concentrated dimer solutions, lower temperatures and longer times tend to produce thicker film. The coating thickness was measured by TEM and spectroscopic ellipsometer. Important to the H_2O solar splitting functionalities, films thicker than 5 nm are undesired because charge transfer through the coating becomes hindered. Films thinner than 1 nm also exhibit detrimental effects because they fail to provide adequate protection for the WO_3. We identified the following optimum conditions to yield a continuous coating of 2 nm. Energy dispersive spectroscopy (EDS) and X-ray photoelectron spectroscopy (XPS) confirmed the presence of Mn element in the amorphous layer.

The surface of WO_3 was coated with the Mn-oxo oligomer catalyst by thermal treatment of a solution of the Mn-oxo dimer compound at 75 °C as shown in Fig. 17.9 (Liu et al. 2011). This step was typically 5 min and did not cause noticeable colorization of WO_3. No measurable differences were observed in the absorption spectra of WO_3 before and after this deposition, that

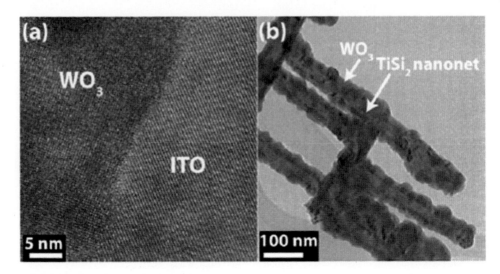

Fig. 17.7 Structure of ALD-grown semiconductor tungsten oxide on ITO (**a**) and on a TiS$_2$ nanonet (**b**) showing the formation of uniformed tungsten oxide and viable approach of ALD to prepare heteronanostructures (Liu et al. 2011, reproduced with permission from Wiley-VCH)

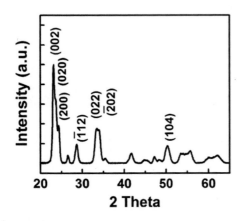

Fig. 17.8 XRD pattern of tungsten oxide after annealing shows that the crystal structure is identified as monoclinic tungsten oxide material (Liu et al. 2011, reproduced with permission from Wiley-VCH)

is, solvation on the WO$_3$ surface was insignificant. This observation also suggests that the catalyst poses no appreciable competition to WO$_3$ in light absorption, which is an extremely important feature, because light absorbed by the catalyst would be wasted. WO$_3$ can be protected in non-acidic solutions by depositing other materials such as TiO$_2$ by, for example, ALD. However, deposition of the Mn catalyst is preferred for at least three reasons: (1) deposition is straightforward, (2) the coating does not compete with WO$_3$ in light absorption, and (3) the catalyst facilitates hole transfer from the semiconductor to the solution.

17.4 Catalytic Activity of Manganese/Tungsten Oxide Nanostructure

As shown in Fig. 17.10, the working hypothesis is that the light radiation is absorbed by the WO$_3$ semiconductor and causes a charge separation to produce electrons and holes. The electrons may be transferred to the cathode by an electric wire to produce hydrogen gas in a photoelectrochemical cell. Donating electrons to the holes, the Mn-oxo oligomer is able to extract electrons from water to evolve O$_2$ following the catalytic mechanism described (Chou et al. 2012). A Pt mesh was used as counter electrode, and the reference electrode was Ag/AgCl in 1 m KCl solution. The electrolyte solution was 1 m KCl with HCl added to adjust the pH from 2 to 7. A CHI 600C potentiostat was used. The voltage was swept between 0 and 1.3 V (vs. RHE) at a rate of 10 mV/s. The light source was a 150 W Newport Mercury lamp, and the intensity was adjusted to 100 mW/cm^2.

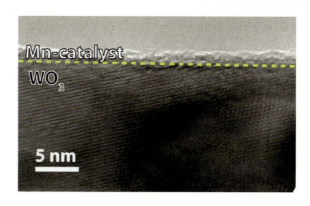

Fig. 17.9 TEM image and EDS analysis reveal the thin coating of the Mn-oxo oligomer material on the surface of tungsten oxide. The Mn-oxo dimer concentration, 1 mM; deposition temperature: 75 °C; and deposition time: 5 min (Liu et al. 2011, reproduced with permission from Wiley-VCH)

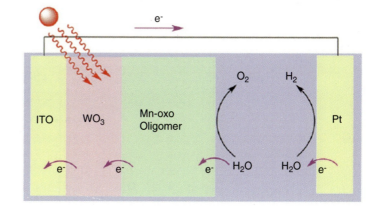

Fig. 17.10 An experiment set-up of the PS II mimic using manganese-oxo complex/tungsten oxide for solar water splitting (Chou et al. 2012, reproduced with permission from Elsevier)

To quantify the amount of O_2 and H_2 generated by the WO_3/Mn catalyst electrode, we conducted photocatalytic experiments with GC analysis. An HP 5890 GC instrument equipped with an HP-Plot MoleSeive column was used for this experiment. The inject and the detector temperatures were set at 100 °C. Helium was used as the carrier gas for oxygen measurements, and nitrogen was used for hydrogen. The flow rate of the carrier gases was 5.4 ml/min. For the stability test, the reaction vesicle was purged with N_2 every 7 h while all other test parameters were kept constant. When the vessel was purged with N_2 and the experiment was restarted, O_2 was produced at the same rate as in the original experiment in Fig. 17.11, which shows the robustness of the catalytic system in efficient photo water splitting.

An apparent difference is observed for the electrodes with and without the Mn catalyst as shown in Fig. 17.12 (Liu et al. 2011). When the catalyst is present, the amount of O_2 increases with time, following a linear relationship for up to 5 h, after which the rate of O_2 generation slows down. Without the Mn catalyst, the amount of O_2 measured was only approximately 50% of that with the Mn catalyst after 3 h. Thereafter, the electrode ceased to function, showed obvious colorization, and eventually peeled off from the ITO support. Better stability was observed when solutions with lower pH were used, and no obvious colorization was seen when WO_3 was tested in solutions of pH 2 for up to a day. The protecting effect of the Mn catalyst was more pronounced when the electrodes were tested in less acidic solutions. At pH 7, WO_3 without the

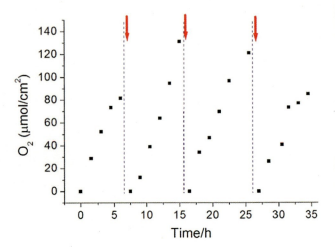

Fig. 17.11 The robustness of the catalytic system containing manganese-oxo oligomer complex and tungsten oxide in photo water splitting. The *arrows* show where the reaction vesicle is purged by inert gas nitrogen (Liu et al. 2011, reproduced with permission from Wiley-VCH)

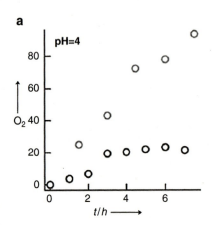
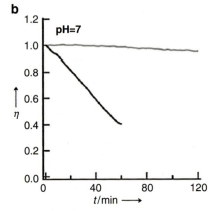

Fig. 17.12 Stability of the catalytic systems containing tungsten oxide (*gray symbols*) and manganese-oxo oligomer complex/tungsten oxide (*black symbols*) in photo water splitting at different pH aqueous solutions (**a**) pH 4.0 and (**b**) pH 7.0 (Liu et al. 2011, reproduced with permission from Wiley-VCH)

Mn catalyst decayed more quickly than at pH 4, whereas approximately 4% performance degradation was observed up to 2 h. In contrast, it took more than 19 h in the Mn/WO$_3$ case for the efficiency to drop to 50% of the initial value.

The pH dependence of photocurrent over the range of pH 2 to 8 is shown in Fig. 17.13) (Chou et al. 2012). The upper panel showed the photocurrents under light and dark at different pH when the polarization is at 1.0 V for 10 min. The dark signal is almost zero, and the light-induced current is dramatically different and around 1.2–1.4 mA. Almost the same photocurrents were observed over the pH range from 2 to 8, indicating that the performance of Mn-oxo oligomer complex/WO$_3$ system in water splitting is virtually independent of pH values. The lower panel shows the photocurrent in different pH buffers when the applied potential scans from 0 to 1 V at the rate of 10 mV/min. The steady photocurrents over the pH 2 to 8 supported the idea that Mn-oxo oligomer complex/WO$_3$ is functional over the wide range of pH values. We concluded that Mn-oxo catalytic material is a robust catalyst over the range of pH 2 to pH 8.

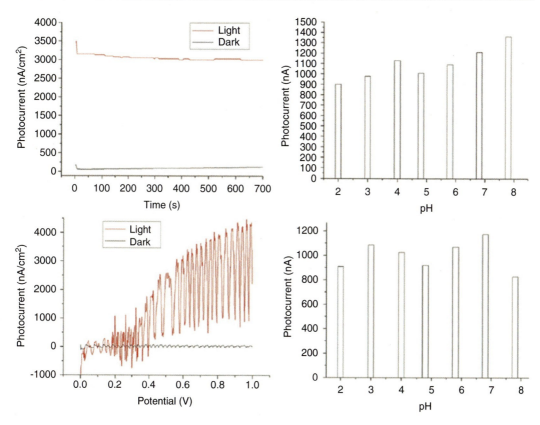

Fig. 17.13 The pH dependence experiments show the stability of the catalytic system in aqueous solutions ranging of pH 2.0 to 8.0 (Chou et al. 2012, reproduced with permission from Elsevier)

17.5 Mechanism of the Manganese/Tungsten Oxide System

In nature photosynthetic water splitting involves transfer of electrons and protons from water molecule initiated by a light-driven step. The reactions are catalyzed by a Mn_4CaO_5 complex bound to PS II that undergo a cycle containing five intermediate S-states (S_0–S_4), leading to dioxygen formation (Joliot et al. 1969; Kok et al. 1970; Hoganson and Babcock 1997; Yano et al. 2006). The Mn-oxo dimer compound is a functional model for PS II OEC complex, Brudvig and co-workers proposed a catalytic mechanism for oxygen evolution by the Mn-oxo dimer which involves oxidation of Mn to Mn(V) = O species (Limburg et al. 1999; McEvoy et al. 2005).

With an energy gap between the conduction and valence bands, semiconductors represent an appealing candidate to effectively absorb photons and transform the optical energy into free charges (electrons and holes). It has been demonstrated that these charges can be readily utilized for water splitting. Theoretical calculations have shown that the power conversion efficiency of using semiconductor for water photo-splitting can be as high as that of solid-state solar cells. Combining semiconductor nanomaterials with the Mn-oxo catalyst as proposed here overcomes a key challenge in using semiconductor directly, which is the low catalytic activity of semiconductors. The low reactivity often leads to a high overpotential and results in significant reduction in the overall energy conversion efficiency (Fig. 17.14).

As shown in Fig. 17.15, the W content is almost unchanged after 19 h of reaction (left panel) (Chou et al. 2012). It supported the idea that Mn plays a protective role in keeping WO_3

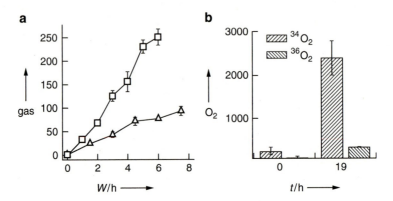

Fig. 17.14 Photocatalytic experiments confirm that the measured oxygen gas is the direct product of complete water splitting. (**a**) The rate of hydrogen gas production is approximately twice that of oxygen gas. (**b**) Isotopic labeling experiments verify that oxygen atoms in oxygen gas come from water (Liu et al. 2011, reproduced with permission from Wiley-VCH)

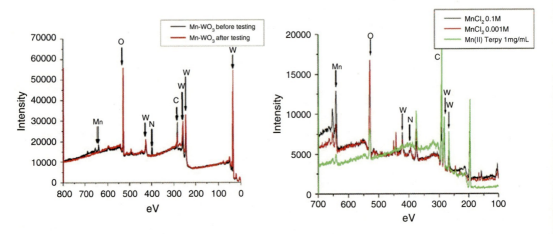

Fig. 17.15 XPS analysis of the catalytic system containing manganese-oxo oligomer complex and tungsten oxide in the absence (*left panel*) and presence (*right*) of manganese (II) ions in solutions

active in the water splitting cycle. In contrast, Mn on the surface of WO₃ was absent after 19 h. This observation suggested that Mn is likely dissolved in the aqueous solution by forming of Mn(II) ion during the catalytic cycle of water oxidation. To confirm this hypothesis, we added different amounts of Mn(II) ion (1.0 mM and 0.10 M) and Mn-terpy complex in the aqueous solution. After 19 h of reaction, the photoanode was washed three times and subjected to XPS analysis. In right panel, a Mn signal was observed in all three cases. This observation indicated that Mn ions in aqueous solution may be oxidized to high valent species and deposited on the surface of WO₃ during the catalytic cycle of water splitting.

Addition of Mn(II) in the aqueous solution prevents the disappearing of the Mn peak after testing. A question arises if it is necessary that the Mn-oxo dimer to be used as a precursor, or if just using Mn(II) ion would lead to the formation of the same catalyst? We conducted the photocurrent measurements using WO₃/ITO electrode in the aqueous solution containing MnCl₂ or Mn (II)-terpy species under identical conditions. No catalytic photocurrent was observed, and no Mn species on the WO₃ electrode was found by XPS, indicating that Mn-oxo dimer as a precursor is required for the formation of Mn-oxo oligmer/WO$_3$ catalyst.

The complete structure of the catalyst is unknown. We have tried to optimize the experimental conditions in order to obtain the crystals of the compound. However the crystallization has not been successful so far. We did use EPR, X-ray absorption, electrochemistry, FTIR, and elemental analysis to characterize the catalyst. The experimental data of EPR and FTIR data suggest that the catalyst is not MnO$_2$. The stoichiometry of elemental analysis indicates the significant amount of N and O, matching the composition of terpyridine ligand. The preliminary X-ray absorption spectroscopic data showed the Mn-Mn distance is somewhat similar to that of Mn-oxo mix valent dimer. We propose that the catalyst is an Mn-oxo oligomer with terpyridine ligand bound possibly to Mn.

We propose a working model of Mn-oxo oligmer/WO$_3$ for photo water splitting in Fig. 17.16 (Chou et al. 2012). The semiconductor tungsten oxide serves as the light harvesting system and reaction center, which absorbs photons and generates electrons and holes. The holes in WO$_3$ are filled by the electrons of Mn-oxo oligomer complex via electron transfer steps. The Mn-oxo oligomer complex is oxidized to form an Mn(V) = O species which is attacked by water molecules. The O-O bond is formed together with the reduction of Mn(V) = O to a Mn(II)-terpy species via proton-coupled electron transfer reaction and closes the water splitting cycle. The Mn(II)-terpy species is released to the aqueous phase. It is very likely that the exchange of oxygen atoms among the solid phase would involve major structural rearrangement and probably destruction of the complex. The Mn-oxo oligomer complex or similar catalytic form is regenerated by the oxidation of Mn (II)-terpy.

The nature of the active species in the catalytic cycle of water splitting is under debate. We proposed that the Mn(III/V) might be the active species in our model. However, our experimental data cannot rule out the possible involvement of Mn(II) or Mn-oxide as active species. The terpyridine is likely a proton shuttle. This is supported by the pH experiment. The ions in the phosphate and acetate buffers (pH 4–8) may act as proton acceptors, where in HCl solution (pH 2–4) this is not the case. The effect of HCl and acetate/phosphate buffers on photocurrent is insignificant; suggesting the proton shuttle might be terpyridine.

17.6 Conclusions

To face the global energy crisis today, efficiently and expensively converting solar energy to electricity by splitting water is one of the most pressing issues. Several promising and exciting catalytic systems using earth abundant elements via visible light-driven water splitting reaction have been developed recently (Yin et al. 2010; Han et al. 2012; Reece et al. 2012; Birkner et al. 2013; Alibabaei et al. 2013; Mayer et al. 2013). The major challenges in this area include capabilities to design, make, and study novel

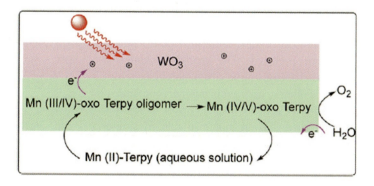

Fig. 17.16 Proposed water splitting cycle of catalytic system containing tungsten oxide and manganese-oxo oligomer complex (Chou et al. 2012, reproduced with permission from Elsevier)

materials that can perform this reaction with meaningful efficiency and at low cost (Cook et al. 2010; Concepcion et al. 2009, Han et al. 2012). The case study presented in this chapter will contribute significantly to this goal.

Light is absorbed by WO_3 to generate electrons and holes. The built-in field in WO_3 helps concentrate electrons away from the solid/liquid interface to be collected by the supporting substrate, which is indium tin oxide (ITO). Holes are driven by the built-in field toward the solid/liquid interface, where they transfer to the solution to oxidize H_2O. This schematic neglects the potential influence of the Mn catalyst coating on the electronic energy of WO_3 because of the thinness of the former. The oxidation process is mediated by the catalyst, which we suggest works in a fashion similar to the oxo-bridged Mn catalyst, that is, mixed-valent MnIII and MnIV are oxidized by the photogenerated holes from the semiconductor, and the product of the oxidation process is reduced by H_2O to produce O_2. A distinguishing feature of the WO_3/Mn catalyst system is that the presence of the catalyst facilitates charge transfer from the semiconductor to the solution, and the effect is more obvious when the charge density is high.

The ALD growth of tungsten oxide without production of corrosive byproducts has not been reported elsewhere, and the synthetic technique makes it easy to form heteronanostructures. The manganese catalyst derived from the oxo-bridged Mn dimer is easy to prepare and exhibits good stability and catalytic properties. When interfaced with tungsten oxide, it acts as a protecting layer without adverse effect on the water-splitting properties. To the best of our knowledge, this is the first time that tungsten oxide photoelectrodes stable in neutral solution has been prepared. The manganese/tungsten oxide heteronanostrutrue design combines multiple components, each with unique complementary and critical functions, and offers integration of properties that are not available in single-component materials. The versatility of the method will find applications in numerous areas where the availability of materials is the limiting factor.

Acknowledgements The author thanks the support from the Alabama State University. He is grateful to his collaborators Dr. Gary Brudvig at Yale University and Dr. Dunwei Wang at Boston College.

Conflict of Interest The author does not have conflict of interest.

References

Alibabaei, L., Brennaman, M.K., Norris, M.R., Kalanyan, B., Song, W., Losego, M.D., Concepcion, J.J., Binstead, R.A., Parsons, G.N., and Meyer, T.J. (2013) Solar water splitting in a molecular photoelectrochemical cell, Proc. Natl. Acad. Sci. U S A, in press (Doi: 10.1073/pnas.1319628110)

Barber, J. (2009). Photosynthetic energy conversion: natural and artificial. Chem. Soc. Rev., 38, 185–196.

Birkner, N., Nayeri, S.N., Pashaei, B., Najafpour, M.M., Casey, W.H., Navrotsky A. (2013) Energetic basis of catalytic activity of layered nanophase calcium manganese oxides for water oxidation. Proc. Natl. Acad. Sci. U S A, 110, 8801–8806

Blakemore, J.D., Crabtree, R.H. and Brudvig, G.W. (2015) Molecular Catalysts for Water Oxidation. Chem. Rev. 115, in press.

Blankenship R.E. (2002) Molecular Mechanisms of Photosynthesis, Blackwell Science

Blankenship, R. E., Tiede, D. M., Barber, J., Brudvig, G. W., Fleming, G., Ghirardi, M., et al. (2011). Comparing Photosynthetic and Photovoltaic Efficiencies and Recognizing the Potential for Improvement. Science, 332, 805–809.

Bolton, J. R. (1996). Solar photoproduction of hydrogen: A review. Solar Energy, 57, 37–50.

Brimblecombe, R., Swiegers, G. F., Dismukes, G. C., & Spiccia, L. (2008). Sustained water oxidation photocatalysis by a bioinspired manganese cluster, Angew. Chem. Int. Ed., 47, 7335–7338.

Brimblecombe, R., Dismukes, G. C., Swiegers, G. F., & Spiccia, L. (2009). Molecular water-oxidation catalysts for photoelectrochemical cells. Dalton Trans., 43, 9374–9384.

Brimblecombe, R., Koo, A., Dismukes, G. C., Swiegers, G. F., & Spiccia, L. (2010). Solar-driven Water Oxidation by a Bio-inspired Manganese Molecular Catalyst. J. Am. Chem. Soc., 132, 2892–2894.

Brudvig, G. W. (2008). Water oxidation chemistry of photosystem II. Philos. Trans. R. Soc., B, 363, 1211–1219.

Cady, C. W., Crabtree, R. H., & Brudvig, G. W. (2008). Functional models for the oxygen-evolving complex of photosystem II. Coord. Chem. Rev., 252, 444–455.

Cady, C. W., Shinopoulos, K. E., Crabtree, R. H., & Brudvig, G. W. (2010) [(H$_2$O)(terpy)Mn(μ-O)$_2$Mn (terpy)(OH$_2$)](NO$_3$)$_3$ (terpy = 2,2′:6,2″-terpyridine) and its relevance to the oxygen-evolving complex of photosystem II examined through pH dependent cyclic voltammetry. Dalton Trans., 39, 3985–3989.

Chou, L.-Y., Liu, R., He, W., Geh, N., Lin, Y., Hou, E. Y. F., et al. (2012). Direct Oxygen and Hydrogen Production by Water Splitting Using a Robust Bioinspired Manganese-oxo Oligomer Complex/Tungsten Oxide Catalytic System. Int. J. Hydrogen Energy, 37, 8889–8896.

Concepcion, J. J., Jurss, J. W., Brennaman, M. K., Hoertz, P. G., Patrocinio, A. O. T., Murakami Iha, N. Y., et al. (2009). Making Oxygen with Ruthenium Complexes. Acc. Chem. Res., 42, 1954–1965.

Cook, T. R., Dogutan, D. K., Reece, S. Y., Surendranath, Y., Teets, T. S., & Nocera Daniel, G. (2010). Solar energy supply and storage for the legacy and nonlegacy worlds. Chem. Rev., 110, 6474–6502.

Diner, B. A., & Rappaport, F. (2002). Structure, dynamics, and energetics of the primary photochemistry of photosystem II of oxygenic photosynthesis. Annu Rev Plant Biol, 53, 551–580.

Dismukes, G. C., Brimblecombe, R., Felton, G. A. N., Pryadun, R. S., Sheats, J. E., Spiccia, L., et al. (2009). Development of bioinspired Mn_4O_4-cubane water oxidation catalysts: lessons from photosynthesis. Acc. Chem. Res., 42, 1935–1943.

Ferreira, K. N., Iverson, T. M., Maghlaoui, K., Barber, J., & Iwata, S. (2004). Architecture of the photosynthetic oxygen-evolving center. Science, 303, 1831–1838.

Fujishima, A., & Honda, K. (1972). Electrochemical Photolysis of Water at a Semiconductor Electrode. Nature, 238, 37–38.

Han, Z., Qiu, F., Eisenberg, R., Holland, R.L., Krauss, T.D. (2012) Robust Photogeneration of H_2 in Water Using Semiconductor Nanocrystals and a Nickel Catalyst. Science, 338, 1321–1324

He, W., Zhao, K.H., and Hou, H.J.M. (2013) Toward solar fuel production using manganese/semiconductor systems to mimic photosynthesis. NanoPhotoBioSciences, 1, 63–78

Hoganson, C. W., & Babcock, G. T. (1997). A metalloradical mechanism for the generation of oxygen from water in photosynthesis. Science 277, 1953–1956.

Hou, H. J. M. (2010). Structural and mechanistic aspects of Mn-oxo and Co-based compounds in water oxidation catalysis and potential application in solar fuel production. J. Integr. Plant Biol, 52, 704–711.

Hou, H. J. M. (2011). Manganese-based materials inspired by photosynthesis for water-splitting. Materials, 4, 1693–1704.

Joliot P., Barbieri G., Chabaud R. (1969) Un nouveau modele des centres photochimiques du systém II. Photochem Photobiol, 10: 309–329.

Kern, J., Alonso-Mori, R., Tran, R., Hattne, J., Gildea, R. J., Echols, N., Glöckner, C., et al (2013) Simultaneous Femtosecond X-ray Spectroscopy and Diffraction of Photosystem II at Room Temperature. Science, 340, 491–495

Kok, B., Forbush, B., & McGloin, M. (1970). Cooperation of charges in photosynthetic O2 evolution-I. A linear four step mechanism. Photochem. Photobiol., 11(6), 457–475.

Kudo, A., & Miseki, Y. (2009). Heterogeneous photocatalyst materials for water splitting. Chem. Soc. Rev. 38, 253–278

Lewis, N. S., & Nocera, D. G. (2006). Powering the planet: chemical challenges in solar energy utilization. Proc. Natl. Acad. Sci. U S A, 103, 15729–15735.

Li, G., Sproviero, E. M., Snoeberger, R. C., III, Iguchi, N., Blakemore, J. D., Crabtree, R. H., et al. (2009). Deposition of an oxomanganese water oxidation catalyst on TiO_2 nanoparticles: computational modeling, assembly and characterization. Energy Environ. Sci., 2, 230–238.

Li, W., Sheehan, S.W., He, D., He, Y., Yao, X., Grimm, R.L., Brudvig, G.W., Wang, D. (2015) Hematite-Based Solar Water Splitting in Acidic Solutions: Functionalization by Mono- and Multi-layers of Ir Oxygen-evolution Catalysts. Angew. Chem. Int. Ed., 54, 11428–11432

Limburg, J., Vrettos, J. S., Liable-Sands, L. M., Rheingold, A. L., Crabtree, R. H., & Brudvig, G. W. (1999). A functional model for O-O bond formation by the O_2-evolving complex in photosystem II. Science, 283, 1524–1527.

Limburg, J., Vrettos, J. S., Chen, H., de Paula, J. C., Crabtree, R. H., & Brudvig, G. W. (2001). Characterization of the O_2-evolving reaction catalyzed by [(terpy)(H_2O)Mn(III)(O)$_2$Mn(IV)(OH_2)(terpy)](NO_3)$_3$ (terpy = 2,2′:6,2″-terpyridine). J. Am. Chem. Soc., 123, 423–430.

Lin, Y., Zhou, S., Liu, X., Sheehan, S., & Wang, D. (2009). TiO_2/$TiSi_2$ Heterostructures for High-Efficiency Photoelectrochemical H_2O Splitting. J. Am. Chem. Soc., 131, 2772–2773.

Liu, R., Lin, Y., Chou, L.-Y., Sheehan, S. W., He, W., Zhang, F., et al. (2011). Water splitting by tungsten oxide prepared by atomic layer deposition and decoraed with an oxygen-evolving catalyst. Angew. Chem. Int. Ed., 50, 499–502.

Liu, H., Zhang H., Niedzwiedzki, D.M., Prado, M., He G., Gross M.L., and Blankenship R.E. (2013) Phycobilisomes Supply Excitations to Both Photosystems in a Megacomplex in Cyanobacteria, Science, 342, 1104–1107

Loll, B., Kern, J., Saenger, W., Zouni, A., & Biesiadka, J. (2005). Towards complete cofactor arrangement in the 3.0 A resolution structure of photosystem II. Nature, 438, 1040–1044.

Mayer, M.T., Lin, Y., Yuan, G., Wang D. (2013) "Forming Junctions at the Nanoscale for Improved Water Splitting by Seminconductor Materials: Case Studies of Hematite," Acc. Chem. Res. 2013, 46, 1558–1566

McEvoy, J. P., Gascon, J. A., Batista, V. S., & Brudvig, G. W. (2005). The mechanism of photosynthetic water splitting. Photochem Photobiol Sci, 4, 940–949.

McNamara, W. R., Snoeberger, R. C., III, Li, G., Richter, C., Allen, L. J., Milot, R. L., et al. (2009). Hydroxamate anchors for water-stable attachment to TiO$_2$ nanoparticles. Energy Environ. Sci., 2, 1173–1175.

Meyer, T. J. (2008). The art of splitting water. Nature, 451, 778–779.

Moore, G.F. (2016) Concluding Remarks and Future Perspectives: Looking Back and Moving Forward, In: "*Photosynthesis: Structures, Mechanisms, and Applications,*" Harvey J.M. Hou, Madhi M. Najafpour, Gary F. Moore and Suleyman Allakhverdiev Eds., Springer, Chapter 1, in press

Najafpour, M.M., Abasi, M., and Allakhverdiev, S.I. (2013) Recent Proposed Mechanisms for Biological Water Oxidation, NanoPhotoBioSciences, 1, 79–92

Najafpour, M.M., Hou, H.J.M., and Allakhverdiev, S.I. (2016) Photosynthesis: Natural Nanomachines Toward Energy and Food Production, In: "*Photosynthesis: Structures, Mechanisms, and Applications,*" Harvey J.M. Hou, Madhi M. Najafpour, Gary F. Moore and Suleyman Allakhverdiev Eds., Springer, Chapter 1, in press

Nocera, D. G. (2012). The Artificial Leaf. Acc. Chem. Res., 45, 767–776.

Reece, S.Y., Hamel, J.A., Sung, K., Jarvi, T.D., Esswein, A.J., Pijpers J.J.H., Nocera, D.G. (2012) Wireless Solar Water Splitting Using Silicon-Based Semiconductors and Earth-Abundant Catalysts, Science, 334, 645–648

Rivalta, I., Brudvig Gary, W., & Batista, V. S. (2012). Oxomanganese complexes for natural and artificial photosynthesis. Curr. Opinion Chem. Biol., 16, 11–18.

Ruettinger, W., & Dismukes, G. C. (1997). Synthetic Water-Oxidation Catalysts for Artificial Photosynthetic Water Oxidation. Chem. Rev., 97(1), 1–24.

Ruettinger, W., Yagi, M., Wolf, K., Bernasek, S., & Dismukes, G. C. (2000). O2 Evolution from the Manganese-Oxo Cubane Core Mn$_4$O$_4^{6+}$: A Molecular Mimic of the Photosynthetic Water Oxidation Enzyme. J. Am. Chem. Soc., 122, 10353–10357.

Sala, X., Romero, I., Rodriguez, M., Escriche, L., & Llobet, A. (2009). Molecular catalysts that oxidize water to dioxygen., Angew. Chem. Int. Ed., 48, 2842–2852.

Umena, Y., Kawakami, K., Shen, J. R., & Kamiya, N. (2011). Crystal structure of oxygen-evolving photosystem II at a resolution of 1.9 A. Nature, 473, 55–61.

Yano, J., Kern, J., Sauer, K., Latimer, M. J., Pushkar, Y., Biesiadka, J., et al. (2006). Where water is oxidized to dioxygen: structure of the photosynthetic Mn4Ca cluster. Science, 314, 821–825.

Yin, Q., Tan, J. M., Besson, C., Geletii, Y. V., Musaev, D. G., Kuznetsov, A. E., et al. (2010). A Fast Soluble Carbon-Free Molecular Water Oxidation Catalyst Based on Abundant Metals. Science, 328, 342–345.

Young, K. J., Gao, Y., & Brudvig Gary, W. (2011). Photocatalytic Water Oxidation Using Manganese Compounds Immobilized in Nafion Polymer Membranes. Australian J. Chem., 64, 1219–1226.

Young, K. J., Martini, L. A., Milot, R. L., Snoeberger, R., Batisa, V. S., Schmuttenmaer, C., et al. (2012). Light-driven Water Oxidation for Solar Fuels. Coord Chem Rev, 256, 2503–2520.

Young, K.J., Brennan, B.J., Tagore, R. and Brudvig, G.W. (2015) Photosynthetic Water Oxidation: Insights from Manganese Model Chemistry. Accts. Chem. Res. 48, 567–574

Youngblood, W. J., Lee, S.-H. A., Maeda, K., & Mallouk, T. E. (2009). Visible Light Water Splitting using Dye-Sensitized Oxide Semiconductors. Acc. Chem. Res., 42, 1966–1973.

Zhang, F., Cady, C. W., Brudvig Gary, W., & Hou, H. J. M. (2011). Thermal Stability of [Mn(III)(O)$_2$Mn(IV)(H$_2$O)$_2$(Terpy)$_2$](NO$_3$)$_3$ (Terpy = 2,2′:6′,2″-terpyridine) in aqueous solution. Inorg. Chim. Acta, 366, 128–133.

Time-Resolved EPR in Artificial Photosynthesis

18

Art van der Est and Prashanth K. Poddutoori

Summary

Time-resolved electron paramagnetic resonance (TREPR) methods often play an important role in characterizing artificial photosynthetic systems. The radical pairs and triplet states generated in such systems are spin polarized because of the initial correlation of the electron spins and the spin selectivity of the electronic relaxation and electron transfer. The polarization makes the TREPR signals of the states fundamentally different from those of equilibrium systems and makes it possible to extract information about the geometry of the radical pairs and about the pathway and kinetics of electronic relaxation and electron transfer. In this chapter, we give an overview of the different types of TREPR experiments that can be performed on artificial photosynthetic complexes and the different polarization patterns that are observed. This is followed by a summary of recent results on a several selected systems, which illustrate the strengths and weakness of the technique.

Keywords

Electron transfer • Spin polarization • Light-induced radical pairs • Molecular triplet states

Contents

18.1	**Introduction**	360
18.1.1	Transient Paramagnetic Species in Natural Photosynthesis	360
18.1.2	Differences Between Natural and Artificial Photosynthesis	360
18.2	**Time-Resolved EPR Methods**	362
18.2.1	Transient EPR	362
18.2.2	Pulsed EPR	362
18.3	**Quasi-Static Polarization Patterns**	363
18.3.1	Spin-Polarized, Weakly-Coupled Radical Pairs	364
18.3.2	Strongly-Coupled Radical Pairs	368
18.3.3	Sequential Radical Pairs	370
18.3.4	Triplet States	372
18.4	**Time Dependent Effects**	374
18.4.1	Electron Transfer	374
18.4.2	Quantum Beats	375
18.4.3	Out of Phase Echo Modulation	375

A. van der Est (✉)
Department of Chemistry, Brock University, 1812 Sir Isaac Brock Way, St. Catharines, ON L2S 3A1, Canada

Freiburg Institute of Advanced Studies (FRIAS), Albert-Ludwigs-Universität Freiburg, Albertstr. 19, D-19104 Freiburg, Germany
e-mail: avde@brocku.ca

P.K. Poddutoori
Department of Chemistry, Brock University, 1812 Sir Isaac Brock Way, St. Catharines, ON L2S 3A1, Canada

© Springer International Publishing AG 2017
H.J.M. Hou et al. (eds.), *Photosynthesis: Structures, Mechanisms, and Applications*,
DOI 10.1007/978-3-319-48873-8_18

18.5	**Recent Results**	376
18.5.1	Early Results on Donor-Acceptor Complexes	376
18.5.2	Sequential Electron Transfer in Triads	377
18.5.3	Quantum Beats	378
18.5.4	Echo Modulations	379
18.5.5	Polymer-Fullerene Blends	380
18.6	**Concluding Remarks**	382
References		382

18.1 Introduction

18.1.1 Transient Paramagnetic Species in Natural Photosynthesis

In oxygenic photosynthesis, the absorption of light leads to electron transfer through several proteins embedded in the thylakoid membrane (Fig. 18.1). This process results in the oxidation of water in the lumen and the reduction of NADP in the stroma. The initial steps take place in Photosystems I and II in which a series of sequential radical pairs are generated as the electron is transferred along the electron transfer chain. These radical pairs have been studied extensively using time-resolved electron paramagnetic resonance (TREPR) spectroscopy (Möbius 1997; van der Est 2001, 2009; Bittl and Zech 2001; Stehlik 2006; Thurnauer et al. 2004). A key feature of natural photosynthetic systems is the extremely high quantum yield of charge separation, which is achieved by optimizing the properties of the cofactors such that the rates of forward electron transfer are orders of magnitude faster than competing processes such as charge recombination or intersystem crossing. As a consequence of the fast electron transfer, the radical pairs are produced in spin states that are non-eigenstates of the spin system and the population distribution is far from equilibrium. This has a profound effect on the TREPR signals observed from photosynthetic reaction centers and allows details of the structure and dynamics of the system to be determined. TREPR methods and their application to natural photosynthesis have been summarized in detail in a number of review articles and book chapters (Möbius 1997; van der Est 2001, 2009; Bittl and Zech 2001; Stehlik 2006; Thurnauer et al. 2004).

The goal of artificial photosynthesis is to mimic the features of the natural systems and TREPR methods have been used widely to study artificial photosynthetic complexes. Despite the goal of mimicking the natural systems, the artificial complexes often display significant differences, which have important consequences for their time-resolved EPR spectra (Forbes et al. 2013).

18.1.2 Differences Between Natural and Artificial Photosynthesis

The term artificial photosynthesis refers to a wide range of complexes all of which mimic natural photosynthesis to some degree. These systems can be broadly grouped into the two classes shown in Fig. 18.2 based on whether the catalysis is homogeneous or heterogeneous. In photocells (Fig. 18.2a) a photosensitizer injects electrons into the conduction band of a semiconductor and activates heterogeneous catalysis at the electrode surface. In homogeneous biomimetic systems (Fig. 18.2b), synthetic analogues of the components of the photosynthetic reaction centers are coupled together to act as a shuttle for electrons between two catalysts that should perform oxidative and reductive chemistry in solution. The arrangement, shown in Fig. 18.2b, is only one of many possible schemes of this type and the position of the chromophore in the electron transfer chain, the nature of the bridges, donors and acceptors, the addition of antenna chromophores and many other factors can be varied. The vast majority of TREPR studies involve the characterization of such complexes, normally in the absence the redox catalysts, and thus, the primary focus of this chapter will be in this area.

In natural photosynthesis, the protein plays an essential role in tuning the properties of the cofactors and high quantum efficiency electron transfer has been achieved primarily through optimization of the protein-cofactor interactions (Allen and Williams 2014). In artificial donor-

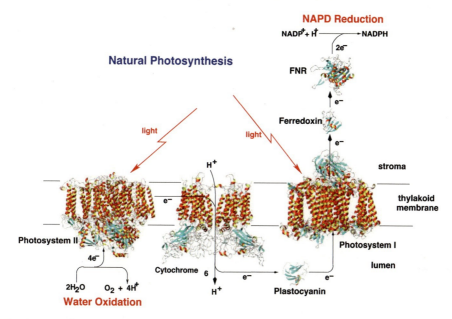

Fig. 18.1 Schematic diagram of the electron transport chain in oxygenic photosynthesis. The structures of the proteins have been generated from the following protein databank files using the program molmol (Koradi et al. 1996): Photosytem I 1JB0 (Jordan et al. 2001); Photosystem II 2AXT (Loll et al. 2005) Cytochrome b_6f 1VF5(Kurisu et al. 2003); Plastocyanin 1JXD (Bertini et al. 2001); Ferredoxin 1FXI(Tsukihara et al. 1990); Ferredoxin-NADP$^+$ reductase (*FNR*) 1FNB (Bruns and Karplus 1995)

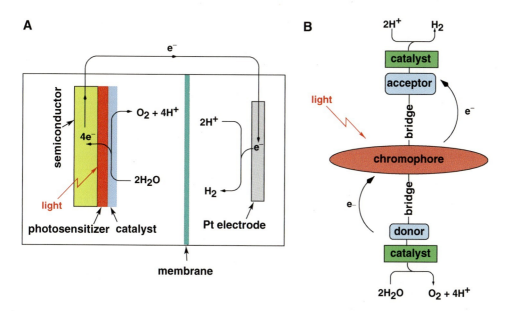

Fig. 18.2 Different possible schemes for artificial photosynthesis. (**a**). Photocell in which a photo-anode is used to split water and drive the reduction of protons at the cathode. (**b**). Donor-acceptor triad coupled to two catalysts to drive the oxidation of water and reduction of protons

acceptor (D-A) complexes, the solvent and bridging groups between the donor and acceptor play the role of the protein. Because it is difficult to achieve the same level of control found in the natural photosystems, intersystem crossing in the donor and back electron transfer often compete with forward electron transfer and the lifetime of the final charge-separated state is usually shorter. The covalently bound bridging groups also generally provide stronger electronic coupling between the electron donors and acceptors than is found in a protein matrix in which the cofactors are non-covalently bound. In addition, partially ordered solvents are sometimes used to stabilize the charge separation. All of these factors lead to a wider variety of paramagnetic states and TREPR spectra for synthetic donor acceptor complexes compared to the natural photosystems.

In the following section, an overview of the basic principles of TREPR will be given, followed by a description of the spectra of the different species observed in artificial photosynthetic systems.

18.2 Time-Resolved EPR Methods

18.2.1 Transient EPR

The most straightforward TREPR experiment is transient EPR. A detailed description of the method can be found in Forbes et al. (2013) and only a brief overview is given here. As shown in Fig. 18.3, a laser flash with a pulse length of typically 10 ns or less is applied to the sample while it is irradiated with continuous microwaves in a static magnetic field. The microwave absorption is measured as a function of time giving a transient response. The magnetic field is then stepped over a range of values and a transient is collected at each field position. The resulting collection of transients forms a time/magnetic field dataset and transient EPR spectra can be generated by plotting the signal amplitude in a chosen time window as a function of the magnetic field.

If lock-in detection with 100 kHz field modulation is used, the response time is on the order of 100 μs. However, with so-called direct detection, the response time can be reduced to several tens of nanoseconds, and is limited by the Fourier broadening at short times. The faster response time comes at the cost of much lower sensitivity but this loss in sensitivity is compensated for by the strong spin polarization observed at short times. The polarization arises from the spin selectivity of the photoreactions and gives rise to both absorptive (A) or emissive (E) signal contributions. An important feature of TREPR studies is the analysis of the patterns of the absorptive and emissive polarization. In most spectroscopic experiments the absorption coefficient, the concentration and the temperature determine the intensity. However, for TREPR spectra the pathway by which the paramagnetic state was generated is the most important factor. Thus, the same state can have different spectra if it can be generated by different mechanisms. For example, singlet and triplet electron transfer can be distinguished by the polarization pattern of the resulting radical pair.

18.2.2 Pulsed EPR

Pulsed EPR techniques can also be used to measure light-induced paramagnetic species (Schweiger and Jeschke 2001). Such experiments are technically more challenging than transient EPR measurements but they provide additional information and allow the spin system to be manipulated by shifting population between the spin states. They can be grouped into two general types illustrated in Fig. 18.4. In the field-swept echo experiment (Fig. 18.4a) a two-pulse sequence is used to generate a spin echo (other echo sequences can also be used) and the height of the echo is recorded over a range of magnetic field positions to give a spectrum. In systems undergoing photochemical reactions, the delay after the laser flash at which the microwave pulses are applied can be varied to obtain the spectrum at different times during the reaction. In addition, the delay between the microwave pulses can be chosen to suppress or emphasize signal contributions from different species. To obtain a field-swept echo spectrum

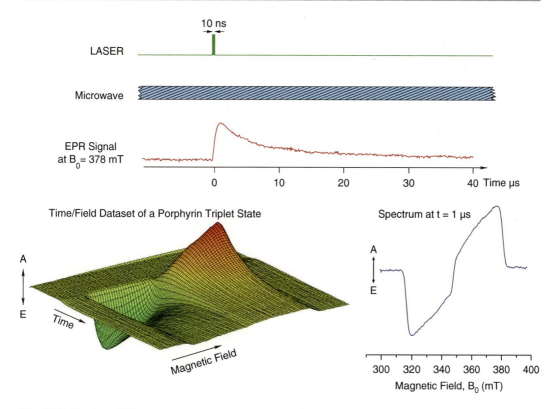

Fig. 18.3 Transient EPR. Continuous microwaves are used to monitor the EPR signal response of the sample to a short laser flash at a fixed magnetic field strength. The field is stepped over a region to create a time/field dataset from which transient EPR spectra can be extracted

corresponding to the allowed single-quantum transitions of the spin system, the excitation bandwidths of the pulses should be small compared to the width of the spectrum. For molecular triplet states this is essentially always the case. However, for weakly coupled radical pairs, which have narrow spectral widths, the excitation bandwidth must be kept small.

Echo modulation measurements are the other main class of pulse EPR experiments. They are illustrated in Fig. 18.4b using the out-of-phase echo modulation of a light-induced radical pair as an example. In an echo modulation experiment the height of the echo is measured as a function of the spacing between the pulses at a fixed magnetic field. The height of the echo is modulated by the weak interactions of the spin system. In the sequence shown in Fig. 18.4b only one pulse spacing is varied and hence a single modulation decay curve is obtained. For more complex pulse sequences multiple delays can be varied giving multidimensional modulation datasets.

In the case of a weakly coupled singlet-born radical pair, the echo is phase-shifted by 90° compared to that of a stable radical and shows deep modulations due to the coupling between the two spins. The values of the coupling constants can be obtained either by fitting calculated modulation curves to the experimental data or from the positions of the features in the Fourier transform of the modulation curve.

18.3 Quasi-Static Polarization Patterns

The EPR signals from a photoreaction depend on both the kinetics of the reaction and the spin dynamics of the light-induced paramagnetic species. In general, the analysis of TREPR data requires that the time dependent spin density

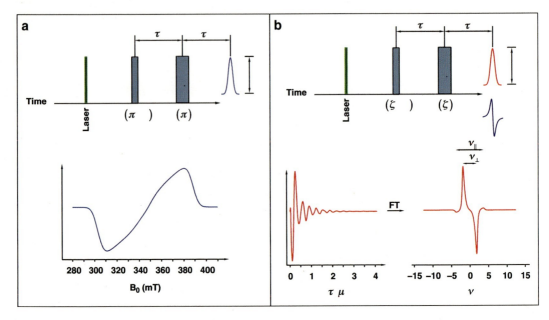

Fig. 18.4 Pulsed EPR experiments. (**a**) Field Swept Echo. After a laser flash, which generates a paramagnetic species, the two microwave pulses generate an echo. The field-swept echo spectrum is generated by measuring the echo amplitude as a function of the magnetic field. (**b**) Out-of-phase echo modulation. For a singlet-born radical pair the echo is phase shifted by 90° and is the amplitude of the echo is modulated strongly as a function of the spacing between the pulses by the spin-spin coupling

matrix of the system be calculated (Schweiger and Jeschke 2001). However, if the microwave field is weak (i.e. in the linear response regime) and the signals are measured after the decay of any coherence effects (i.e. if off-diagonal elements of the density matrix can be ignored), the spin polarized transient EPR spectrum or field-swept echo spectrum can be calculated from the transitions between energy levels of the static spin Hamiltonian of the system.

In the light-induced radical pairs generated by electron transfer in D-A complexes, the strength of the spin-spin coupling has a strong effect on the observed TREPR spectra. There are two components to the coupling and they depend on the structure of the complex and the distance between the electrons. The dipolar coupling falls off with r^{-3} and the exchange coupling between the electrons can be written as:

$$J = J_0 \exp(-\beta(r - r_0)) \quad (18.1)$$

where r_0 is the van der Waals radius and the attenuation factor β depends on the bridge between the donor and acceptor (Schubert et al.

2015). Thus, within the range of distances found in D-A complexes large variations in the strength of the coupling occur. For a system of two coupled $S = 1/2$ spins, there are two distinct regimes. If the spin-spin coupling is much larger than the difference of the Larmor precession frequencies, the eigenstates can be separated into singlet and triplet manifolds and the spin system is strongly coupled. In contrast, if the coupling is small compared to the difference in the precession frequencies, the spin system is weakly coupled.

18.3.1 Spin-Polarized, Weakly-Coupled Radical Pairs

Weakly coupled radical pairs generated from a singlet precursor have been described extensively in the literature because their importance in photosynthetic reaction centers (Kandrashkin et al. 1998; Norris et al. 1990; Closs et al. 1987; Hore et al. 1987; Stehlik et al. 1989; Angerhofer and Bittl 1996; Kandrashkin and van der Est 2001, 2007; Kamlowski et al. 1998; Savitsky

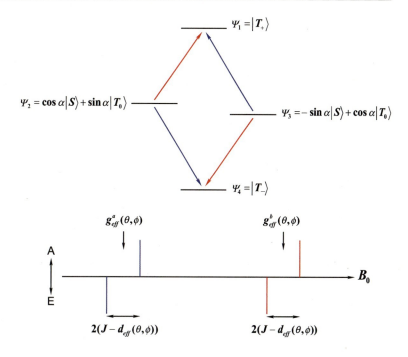

Fig. 18.5 Energy level diagram and stick spectrum of a singlet-born weakly coupled radical pair

et al. 2007, 2013). The electron spin polarization in such systems can be described using the spin correlated radical pair (SCRP) model (Forbes et al. 2013; Kandrashkin et al. 1998; Closs et al. 1987; Hore et al. 1987; Stehlik et al. 1989; Buckley et al. 1987). As shown in Fig. 18.5, there are four spin energy levels two of which are pure triplet states and two which have both singlet and triplet character. If the radical pair is formed from a pure singlet state only the energy levels with singlet character are populated and thus there are two absorptive and two emissive transitions. Electron transfer from a triplet precursor, also gives two absorptive and two emissive transitions but the pattern of emission and absorption is different. For a fixed orientation of the radical pair in the magnetic field, the spectrum (Fig. 18.5, bottom) consists of two antiphase doublets. The splitting of the doublets is determined by the spin-spin coupling and each of the doublets is centered at the effective g-value of the respective radicals. Both the splitting and the positions of the lines are orientation dependent.

The observed spectrum also depends on the motion of the donor-acceptor complex. Several different situations arise depending on the solvent used and the temperature. At low temperature in frozen solution, the system can be treated as being in the rigid limit, while in liquid or nematic solution, rapid motion occurs that leads to averaging of the interactions. In the rigid limit, the sum over all possible orientations of the radical pair in the magnetic field must be taken. In natural photosynthetic reactions centers, the tumbling of the protein complexes is slow even in liquid solution so that the rigid limit applies to both frozen and liquid solution samples. In contrast, in donor-acceptor mimics, the rigid limit only applies below the freezing point of the solvent.

In the experimental spectrum, the lines are normally broadened by unresolved hyperfine couplings and if the inhomogenous linewidth is larger than the spin-spin coupling the antiphase doublet for each radical of a singlet-born radical pair is given by (Kandrashkin and van der Est 2001, 2007):

$$I_i = \frac{2(J - d_{eff}(\theta,\phi))}{\sqrt{2\pi}\Delta\omega_i^3}(\omega_{eff}(\theta,\phi) - \omega_0)_i$$
$$\exp\left(-\frac{(\omega_{eff}(\theta,\phi) - \omega_0)_i^2}{2\Delta\omega_i^2}\right)$$
(18.2)

where i refers to the two radicals and the total spectrum is $I_{total} = I_{donor} + I_{acceptor}$. $\Delta\omega_i$ is the inhomogeneous linewidth of each radical, $\omega_{eff} = \hbar^{-1} g_{eff} \beta B_0$ is the resonance frequency and ω_0 is the microwave frequency. The orientation dependences of $d(\theta,\phi)$ and $g_{eff}(\theta,\phi)$ makes the polarization pattern dependent on the relative orientations of the principal axes of the two g-tensors and the vector from one radical to the other that defines the dipolar coupling. Thus, the spectra can be used to determine the geometry of the radical pair. However, if the exchange coupling is negligible, additional information is required to obtain a unique solution for the geometry (Kandrashkin and van der Est 2001, 2007). This is because the spin-spin coupling influences only the amplitude of the antiphase doublet described by Eq. 18.2 and not its shape. As a result, the absolute amplitude of the spectrum must be known if the spin-spin coupling and geometry is to be determined uniquely. Determination of the geometry also requires that the nature of the precursor (singlet, triplet or mixed singlet-triplet) be known. In the natural photosystems, the highly efficient forward electron transfer ensures that the initial state of the first observable radical pair is a pure singlet. In artificial donor-acceptor complexes, triplet electron transfer and/or significant singlet-triplet mixing can occur.

Figure 18.6 illustrates the sensitivity of radical pair powder spectra to geometry of the radical pair and the nature of the precursor. The black spectra (a, d) are for a singlet precursor and the red and blue spectra (b, c, e, f) for a triplet precursor with different population distributions of the spin states. The spectra on the (a–c) were calculated with the dipolar coupling axis parallel to the x-axis of the g-tensor of radical 1 and in the spectra on the right (d–f) the g-tensor has been rotated by 90° so that the y-axis is parallel to the dipolar coupling axis. For the singlet precursor spectra, (a, d) the sign of the polarization on the low-field end of the spectrum is sensitive to the orientation of the dipolar coupling vector relative to the x-axis of the g-tensor of radical 1. For a triplet precursor, however, the polarization pattern depends on the spin selectivity of the intersystem crossing by which the triplet state is formed (spin selective intersystem crossing is discussed in Sect. 18.3.4) and either sign (absorptive or emissive) is possible on the low field end of the spectrum. A characteristic feature of triplet state formation by intersystem crossing is that it also creates net polarization of the spin system, which is generally emissive (Salikhov et al. 1984). Thus, the spectra of radical pairs generated from a triplet precursor typically have net emissive polarization as shown in spectra c, d, e and f in Fig. 18.6

When the tumbling of the molecules is fast, e.g. liquid solution at room temperature, the orientation dependent terms are replaced by their average values. An important consequence of this is that the spectrum no longer depends on the dipolar coupling since its average value is zero. As a result the spectra no longer depend on the internal geometry of the radical pair. Moreover, if the exchange coupling is negligible and the radical pair is formed from a singlet precursor the absorptive and emissive lines cancel each other completely and no spectrum is observed. However, in most donor acceptor complexes the exchange coupling is sufficiently large that this situation does not arise but the radical pair signals are sometimes weak because the cancellation.

Liquid crystalline solvents in the nematic phase are often used for studying donor-acceptor complexes because the solvent dynamics stabilize the radical pair states (Hasharoni and Levanon 1995; Wiederrecht et al. 1999a, 1997.) In such solvents, the motion of the molecules is rapid but they have a non-zero average alignment relative to a direction known as the director. In the absence of any external fields the director varies randomly and there is no macroscopic

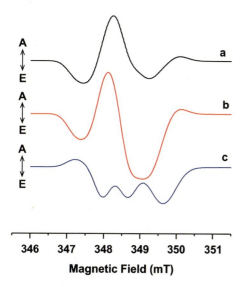
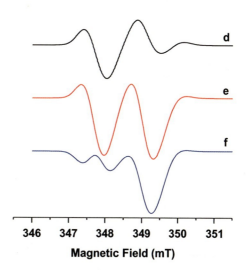

Fig. 18.6 Calculated spin-polarized EPR spectra of a weakly coupled radical pair in the rigid limit. The spectra illustrate the effect of the relative orientation of the donor and acceptor and the influence of the precursor spin state. *a, b, c*: dipolar-coupling axis parallel to the *x*–axis of radical 1. *d, e, f*: Dipolar-coupling axis parallel to the *y*–axis of radical 1. *a, d*: singlet precursor. *b, e*: triplet precursor populated according to $S_{\Omega,z}^2 - \frac{1}{3}\vec{S}^2$ in the molecular frame. *c, f*: triplet precursor populated according to $S_{\Omega,x}^2 - S_{\Omega,y}^2$ in the molecular frame. Principal g-values of the two radicals of the radical pair: radical 1 $g_{xx} = 2.014$, $g_{yy} = 2.008$, $g_{zz} = 2.002$; radical 2 $g_{xx} = g_{yy} = g_{zz} = 2.002$. Dipolar coupling constant $= -0.6$ mT. Exchange coupling $= 0$

alignment of the solvent. However, in the magnetic field of an EPR spectrometer, the director becomes aligned and in most nematic phases is parallel to the magnetic field. When a solute is dissolved in a nematic phase it also becomes partially ordered largely as a result of short-range interactions with the solvent (van der Est et al. 1987; Burnell and De Lange 1998). The ordering of a solute molecule is described by its order matrix, the elements of which are given by:

$$S_{ij} = \left\langle \frac{3}{2}\cos\theta_{iZ}\cos\theta_{jZ} - \frac{1}{2}\delta_{ij} \right\rangle \quad (18.3)$$

where *i* and *j* refer to a set of axes fixed in the molecule and Z is the direction of the magnetic field. Determining the principal axes of the order matrix in a molecule of low symmetry is not trivial but since the ordering is the result of short-range interactions, the principal *z*-axis of rod-like molecules is along the rod axis and for extended planar molecules it is perpendicular to the plane.

For a radical pair in a rapidly tumbling donor-acceptor complex in a nematic liquid crystal, the dipolar coupling averages to:

$$d_{eff} = D\left\langle \frac{3}{2}\cos^2\theta - \frac{1}{2} \right\rangle \quad (18.4)$$

where θ is the angle between the vector and the magnetic field and the angled brackets mean the weighted average over all orientations. The term in angled brackets is the order parameter of the dipolar coupling vector. The g-values of the radicals are also averaged:

$$g_{eff} = g_{iso} + \frac{2}{3}\sum_{i=x,y,z} g_{ii}S_{ii} \quad (18.5)$$

where $S_{ii}, i = x,y,z$ are the order parameters of the principal axes of the g-tensor. The order parameters are difficult to predict accurately for donor-acceptor complexes but approximate values can be estimated from the structure. In general the bridging groups define the long axis

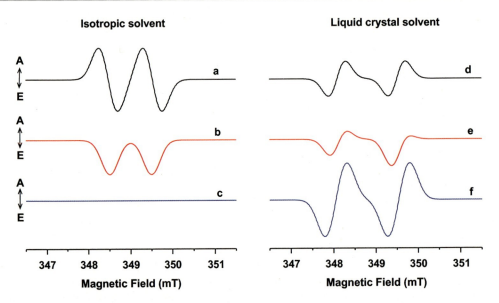

Fig. 18.7 Calculated spin-polarized EPR spectra of a weakly coupled radical pair undergoing rapid motion. The spectra illustrate the influence of the precursor spin state and the effect of partial ordering when the exchange and dipolar coupling constants are of the same sign. *Left*: isotropic motional averaging. *Right*: anisotropic averaging such that the dipolar-coupling axis has an order parameter of 0.4. The magnetic parameters and geometry of the radical pair are the same as in Fig. 18.6 except that $J = -0.17$ mT has been assumed for spectra *a*, *b*, *d* and *e*. Spectra *a* and *d* singlet precursor; *b* and *e* triplet precursor with populated according to $S_{\Omega,z}^2 - \frac{1}{3}\vec{S}^2$; *c* and *f* exchange coupling $J = 0$, singlet precursor

of the molecule (Fig. 18.2b) and the direction of the dipolar coupling. Since the ordering is determined by short-range interactions, orientations of the molecule with its long axis parallel to the director and the magnetic field are more probable than orientations with the long axis perpendicular to the field. Thus, the order parameter of the dipolar-coupling axis is generally positive.

Several calculated spectra of rapidly tumbling radical pairs are shown in Fig. 18.7. On the left, the motion is assumed to be isotropic, while on the right anisotropic motion as would be observed in a nematic liquid crystal has been assumed. The exchange coupling J has been taken to be negative. In the case of isotropic motion, the negative sign of J results in A/E polarization for each of the antiphase doublets when the precursor is a pure singlet state (a). For a triplet prescursor the sign of the multiplet polarization depends on the orientation of the complex in the magnetic field. For rapid motion, the multiplet polarization averages to zero and only the emissive polarization remains (b). If the exchange coupling is negligible (c) no spectrum is observed as discussed above. In an anisotropic solvent the spectra show significant differences. If the exchange and dipolar coupling constants have the same sign, then the splitting of the lines in the antiphase doublet, $2(J–d)$, can be of opposite sign in isotropic and anisotropic solvents, which results in an inversion of the sign of the polarization. With a triplet precursor the multiplet polarization does not average to zero in an anisotropic solvent and an E/A/E/A pattern with net emission is obtained for the choice of couplings used (e). Similarly, when the exchange coupling is negligible (f) the spectrum does not average to zero because the average dipolar coupling is non-zero.

18.3.2 Strongly-Coupled Radical Pairs

In donor-acceptor dyads it is common for the spin-spin coupling to be larger than the difference of the resonance frequencies of the two

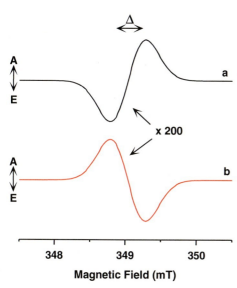
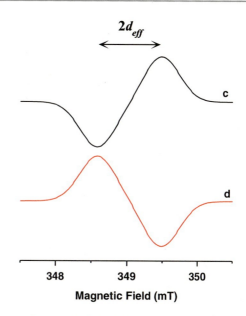

Fig. 18.8 Calculated spectra of a strongly coupled radical pair. *a, b*: isotropic solution *c, d* liquid crystalline solution with S = 0.6 for the dipolar coupling axis. *a, c* singlet precursor or triplet precursor with excess population in T_0. *b, d*: triplet precursor with excess population in T_+/T_-. Parameters: $g_1 = 2.0023$, $g_2 = 2.0030$, $J = -2.0$ mT, $D = -1.5$ mT, $\Delta\omega = 0.5$ mT

radicals. Under these conditions, the radical pair is strongly coupled and its spin states are nearly pure singlet and triplet states. Since the singlet state is EPR silent, the spectrum of such a radical pair is that of the triplet state, which gives an antiphase doublet. The sign of the polarization pattern and the amount of net polarization depend on how the radical pair was formed and what type of solvent is used (Fig. 18.8). In an isotropic solvent, the dipolar coupling is averaged to zero but an antiphase doublet is still observed (Fig. 18.8a, b). The reason for this is that the resonance positions of the two transitions in a triplet state are not identical and the difference between them is approximately $\left(\omega_{eff}^D - \omega_{eff}^A\right)^2/4J$ (van der Est and Poddutoori 2013). For a molecular triplet state, the exchange coupling J is large and this difference is negligible but for a radical pair, the value of J can be sufficiently small that the two peaks are observable despite the fact that the dipolar coupling is averaged. In a liquid crystalline environment, the average dipolar coupling is no longer zero and the separation between the lines is determined by d_{eff} (Fig. 18.8c, d) In either case, if the separation is much smaller than inhomogeneous broadening of the lines, the observed splitting of the antiphase doublet is determined by the linewidth, $\Delta\omega$, of the two overlapping lines and the intensity of the resulting pattern is determined by the separation, $\left(\omega_{eff}^D - \omega_{eff}^A\right)^2/4J$ or $2d_{eff}$. Since the g-factors of most organic radicals do not differ strongly from the free electron value, $\left(\omega_{eff}^D - \omega_{eff}^A\right)$ is small and the separation of the lines in an isotropic solvent is typically much smaller than in a liquid crystal and hence the intensity is much weaker. Often, in an isotropic solvent the observed splitting is determined by the linewidth, while in a nematic liquid crystal it is due to the dipolar coupling as shown in Fig. 18.8. The sign of the dipolar-coupling constant in a radical pair is negative, and hence the sign of the pattern E/A or A/E, can be used to determine the population distribution. An E/A pattern (Fig. 18.8a, c) implies that the T_0 level of the radical pair is preferentially populated. An A/E pattern (Fig. 18.8b, d) implies excess

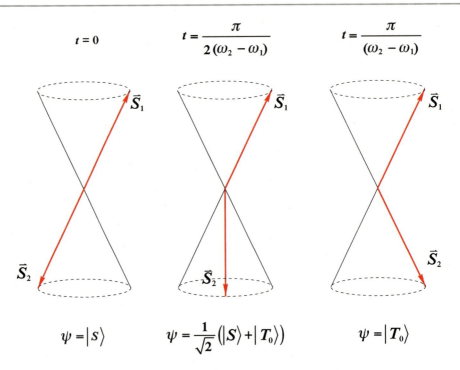

Fig. 18.9 Schematic representation of singlet-triplet mixing

population in T_+ and T_-, which is only possible from triplet electron transfer.

18.3.3 Sequential Radical Pairs

In D-A triads, a series of radical pairs can be produced by any of the following reactions, depending on where the chromophore is located in the electron transfer chain:

$$D \to^{h\nu} D^* \to D^+ A_1^- \to D^+ A_2^-$$
$$A \to^{h\nu} A^* \to D_1^+ A^- \to D_2^+ A^-$$
$$I \to^{h\nu} I^* \to I^+ A^- \to D^+ A^-$$
$$I \to^{h\nu} I^* \to D^+ I^- \to D^+ A^-$$

Here, I refers to a chromophore in the middle of the triad. In all cases, two sequential radical pairs are produced and if the lifetime of the first radical pair is long enough to allow singlet-triplet mixing to occur, the spin polarized TREPR spectrum of the second radical pair is affected. The mechanism of singlet-triplet mixing is illustrated in Fig. 18.9. Initially, the spin system is in a pure singlet state, in which the two spin vectors are antiparallel to one another. However, because the two spins in the radical pair are in different environments their precession frequencies differ. Thus, spin 2 will precess with a frequency ($\omega_1 - \omega_2$) in a frame of reference rotating with the precession frequency of spin 1. As a result of the different precession frequencies, the spin system oscillates between the S and T_0 states.

The effect of singlet-triplet mixing on the spin polarization in sequential electron transfer has been described theoretically in detail (Kandrashkin et al. 1998, 2002; Norris et al. 1990; Hore 1996; Wang et al. 1992; Tang et al. 1996) and involves calculating the evolution of the density matrix. This is a rather involved process since the time at which the electron hops from one acceptor to the next is a statistical process and the integral over the ensemble must be taken. However, the polarization is usually observed at times that are long compared to the electron transfer lifetime and the decay of coherence effects. Under these conditions, Kandrashkin et al. (1998, 2002, 2007) were able derive analytical expressions for the spin polarization, which are extremely useful for

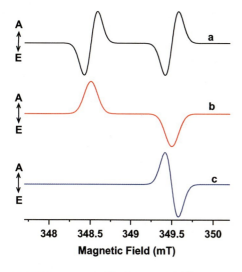
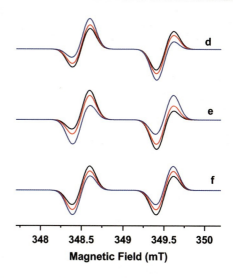

Fig. 18.10 The effect of ST-mixing on the TREPR spectra of the secondary radical pair in sequential electron transfer. *a*: singlet contribution *b*: additional polarization due the g-factor difference in the precursor radical pair; *c*: additional polarization resulting from inhomogenous hyperfine broadening in the precursor; *d*: Influence of the precursor lifetime, *black* $\tau = 1$ ns, *black* $\tau = 2$ ns, *blue* $\tau = 3$ ns; *e*: Influence of the spin-spin coupling in the precursor, *black* $J = -1.0$ mT, *red* $J = -0.5$ mT, *blue* $J = 0.5$ mT; *f*: Influence of the g-factor difference in the precursor, *black* $g_{donor} - g_{acceptor} = -0.0027$, *red* $g_{donor} - g_{acceptor} = 0.0$, *blue* $g_{donor} - g_{acceptor} = 0.0027$.

understanding the observed polarization patterns. If the electron transfer is initiated from a singlet state, the singlet triplet mixing in each radical pair in the series generates additional polarization in the subsequent radical pairs. The polarization of a given radical pair can be broken down into three contributions: (i) singlet polarization, *i.e.* the polarization that would be seen without any singlet triplet mixing, (ii) a contribution that arises from the singlet-triplet mixing due differences in the g-values of the two radicals (iii) polarization associated with inhomogenous broadening in cases in which only one of the two unpaired electrons is transferred. These three contributions are illustrated in Fig. 18.10, spectra a, b and c, respectively. The relative intensities of these three contributions can be written as functions of the electron transfer rates and magnetic parameters of the radical pairs. The singlet contribution (spectrum a) is proportional to:

$$I^s \propto \frac{2(J_2 - d_{eff,2})}{\Delta\omega} \quad (18.6)$$

where the subscript 2 refers to the secondary radical pair. The additional polarization arising from singlet-triplet mixing due to the difference in the g-values of the radicals in the precursor (spectrum b) is given by:

$$I^z \propto \frac{\beta B_0}{h} \frac{(2J_1 + d_{1,eff})(g_D - g_A)_1}{k_1^2} \quad (18.7)$$

where k_1 is the decay rate of the precursor, which is usually dominated by the forward electron transfer rate. The contribution from the inhomogeous line broadening (spectrum c) is:

$$I^h \propto \frac{(J_1 + d_{eff,1})\Delta\omega}{k_1^2} \quad (18.8)$$

The spectra on the right of Fig. 18.10 show how the polarization pattern of the secondary radical pair changes as the lifetime and magnetic parameters of the precursor change. The observed changes are primarily a result of changes in the contribution described by Eq. 18.8. In Fig. 18.10d, the lifetime of the precursor has been varied. In accordance with Eq. 18.8, the net polarization of the two radicals increases as the lifetime is increased from 1 ns (Fig. 18.10d, black spectrum) to 3 ns

(Fig. 18.10d, blue spectrum). In Fig. 18.10e, the spin-spin coupling in the precursor has been varied. The sign of the net polarization depends on the sign of the coupling in the precursor and in the observed radical pair. In the spectra in Fig. 18.10 the dipolar coupling has been set to zero and the value of J in the secondary pair is positive. When the value of J in precursor is also positive (Fig. 18.10e, blue spectrum) the radical with resonances at lower field shows net emission and the higher field antiphase doublet shows net absorption. When the sign of J in the precursor is negative (Fig. 18.10e, red and black spectra) the pattern is reversed and the low field antiphase doublet shows net absorption. The effect of the difference in the g-values of the radicals in the precursor is shown in Fig. 18.10f. If the two radicals in the precursor have the same g-factor then no net polarization is observed in the secondary radical pair (Fig. 18.10f, red spectrum). When the sign of the g-factor difference is the same in the precursor and secondary radical pair, and J is negative in the precursor and positive in the secondary radical pair (Fig. 18.10f, black spectrum), the low field doublet has net absorption and the high field doublet has net emission. When the sign of the g-factor difference is opposite the sign of the net polarization in each doublet is reversed (Fig. 18.10f, blue spectrum). Thus, with sufficient information about the g-factors of the radicals and the sign of the coupling it is possible to deduce, the approximate lifetime of the precursor from such spectra.

18.3.4 Triplet States

In both natural and artificial photosynthetic systems, molecular triplet states can also be formed, either by intersystem crossing from the excited singlet state of a chromophore or by charge recombination. Because both of these processes are spin selective, the resulting triplet state is spin polarized. However, the selectivity of the two processes is different and hence the TREPR spectra differ (Budil and Thurnauer 1991; Thurnauer 1979). This can be a very useful tool in characterizing the photo-physics of D-A complexes, especially when used in combination with transient optical methods.

Figure 18.11 illustrates the two different pathways by which a molecular triplet state can be formed and how they result in different population distributions in the spin sublevels. Charge recombination is shown in Fig. 18.11a and intersystem crossing in Fig. 18.11b. In both cases, singlet-triplet interconversion must occur but the mechanism is different. As discussed above, when the charge-separated state D^+A^- is formed, mixing between S and T_0 occurs as the spins precess and as a result recombination to the triplet state can occur. Because the singlet-triplet mixing in the radical pair is only between S and

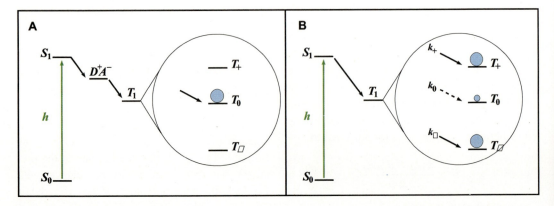

Fig. 18.11 Different possible pathways by which a molecular triplet state can be formed. (**a**): Radical pair recombination. (**b**): Intersystem crossing

T_0, only the T_0 sublevel of the triplet state is populated.

In the case of intersystem crossing, the presence of spin-orbit coupling results in mixing of the singlet and triplet states. Because this interaction is governed by molecular symmetry, while the wave functions of the spin states are determined largely by the Zeeman interaction with the external field, the intersystem crossing rates to the three triplet sublevels differ and depend on the orientation of the molecule in the field. The probability of intersystem crossing to a given sublevels can be written as a linear combination of probabilities associated with each of the three principal axes in the molecule (Forbes et al. 2013; Levanon 1987):

$$p_{i=+,0,-} = |c_{ix}|^2 p_x + |c_{iy}|^2 p_y + |c_{iz}|^2 p_z \quad (18.9)$$

The probabilities p_x, p_y and p_z depend on the strength of the spin-orbit coupling in the molecular x, y and z directions. The coefficients c_{ix}, c_{iy}, and c_{iz} depend on the orientation of the molecule and can be obtained by writing the triplet wavefunctions in terms of the zero-field wavefunctions:

$$\psi_{i=+,0,-} = c_{ix}\psi_x + c_{iy}\psi_y + c_{iz}\psi_z \quad (18.10)$$

Because, the absolute amplitude of the spin polarization is usually not known only the ratios of the probabilities can be determined. Moreover, since their sum equals one, there are only two independent probabilities. This description also does not take net polarization of the spin system into account.

An more elegant way to describe the population distribution is to use the traceless diagonal part of the density matrix $\Delta\rho$, which represents the differences in the populations of the spin levels and can be expanded in terms of the matrix representations of the spin operators (Kandrashkin et al. 2006a). It may also be written as the sum of a multiplet polarization contribution $\Delta\rho_M$ with equal amounts of absorption and emission and a net polarization contribution $\Delta\rho_N$ with either pure emission or absorption. The mutliplet polarization is invariant to inversion and thus can be described by the even powers of the spin operators (Salikhov et al. 1984). In the case of radical pair recombination the multiplet polarization can be written:

$$\Delta\rho_M^{RP} \propto S_z^2 - \frac{1}{3}\vec{S}^2 \quad (18.11)$$

For intersystem crossing the multiplet polarization represents the differences in the intersystem crossing rates that arise from the traceless anisotropy of the spin-orbit coupling, and two contributions are needed (Kandrashkin et al. 2006a, b):

$$\Delta\rho_M^{axial} \propto (1 - 3\cos^2\theta)\left(S_z^2 - \frac{1}{3}\vec{S}^2\right)$$
$$\Delta\rho_M^{non-axial} \propto \sin^2\theta \cos 2\phi \left(S_z^2 - \frac{1}{3}\vec{S}^2\right)$$
$$(18.12)$$

where θ and ϕ describe the orientation of the molecule in the magnetic field. The important difference between the two cases is that the polarization generated by radical pair recombination does not depend on the orientation of the molecule while the polarization generated during intersystem crossing does. Any measured polarization pattern can be reproduced as a linear combination of these contributions:

$$\Delta\rho \propto \kappa_M^{axial} \Delta\rho_M^{axial} + \kappa_M^{non-axial} \Delta\rho_M^{non-axial} + \kappa_N \Delta\rho_N \quad (18.13)$$

The parameters κ_M^{axial}, $\kappa_M^{non-axial}$ can be related to the probabilities p_x, p_y and p_z such that $\kappa_M^{axial} = 1.0$, $\kappa_M^{non-axial} = 0.0$ corresponds to p_x:p_y:$p_z = 0$:0:1; $\kappa_M^{axial} = -1.0$, $\kappa_M^{non-axial} = 1.0$ corresponds to p_x:p_y:$p_z = 0$:1:0 and $\kappa_M^{axial} = -1.0$, $\kappa_M^{non-axial} = -1.0$ corresponds to p_x:p_y:$p_z = 0$:0:1.

Figure 18.12 shows examples of these contributions to the polarization pattern for a free-base porphyrin. The spectra a and b correspond to $\Delta\rho_M^{axial}$ and $\Delta\rho_M^{non-axial}$, respectively and are the contributions to the multiplet polarization from intersystem crossing. Spectrum c is the pattern generated by radical pair recombination and spectrum d is the net polarization. Spectrum

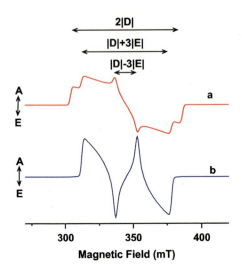
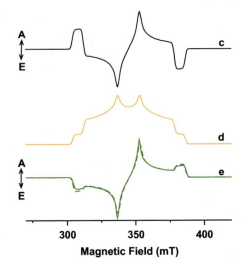

Fig. 18.12 Spin polarization patterns for molecular triplet states. *a*: multiplet contribution $\Delta\rho_M^{axial}$ (Eq. 18.8); *b*: multiplet contribution $\Delta\rho_M^{non-axial}$ (Eq. 18.8); *c*: radical pair recombination contribution $\Delta\rho_M^{RP}$ (Eq. 18.7); *d*: net polarization contribution proportional to $-S_z$; e simulation of the polarization pattern of a free-base porphyrin with $\kappa_{axial} : \kappa_{non-axial} : \kappa_{net} = -1.0 : 0.47 : 0$. *solid line*: experimental spectrum, *dashed line*: simulation. For all of the spectra $D = 38.4$ mT, $E = 7.8$ mT and $g = 2.0023$ e, dashed line, is a weighted sum of spectra a, b and d that reproduces the experimental spectrum (solid line).

18.4 Time Dependent Effects

The time traces in a TREPR experiment also contain a significant amount of information. However, extracting this information is more complicated than the analysis of the spin polarization patterns.

18.4.1 Electron Transfer

TREPR data can be used to determine electron transfer rates but the range of accessible lifetimes is limited. The theoretical lower limit of the time resolution is determined by Fourier broadening and for an experiment performed at 9 GHz (X-band) this limit is on the order of 1–10 ns. In practice, the bandwidth of the resonator and the detection results in a response time that is several tens of nanoseconds or more, depending on the instrument. The sensitivity of such instruments is usually not high enough to detect the equilibrium Boltzmann polarization and hence the range of available lifetimes is also limited by T_1 relaxation. Nonetheless, in favorable cases, lifetimes in the range of ~50 ns to several tens of microseconds are accessible. Usually, the decay of the TREPR signal of a radical pair is determined by a combination of charge recombination and spin relaxation and it is difficult to distinguish the two effects using the EPR data alone. Hence, in general it is necessary to use both optical and EPR methods to obtain a clear picture of the kinetics. The main advantage of TREPR methods for measuring kinetics is that in contrast to optical methods, only paramagnetic species are observed and there are no overlapping signals from diamagnetic excited states or the ground state. In addition, triplet states and radical pairs are easily distinguished and the spin selectivity of the electron transfer can be studied. An example of this is illustrated in Fig. 18.13. Electron transfer with predominantly singlet character generates a weakly coupled radical pair with the population distribution shown on the left. This distribution results in the polarization pattern shown under the energy level diagram. Charge recombination from states with singlet character is faster than from those with

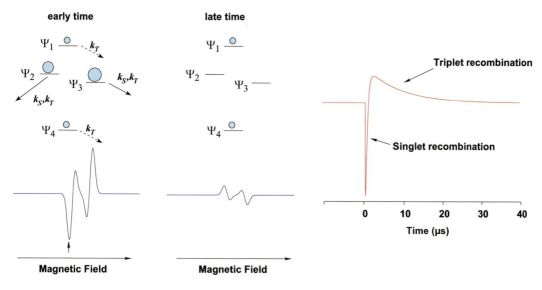

Fig. 18.13 Example of spin selective charge recombination of a radical pair and its effect on the associated TREPR signals

pure triplet character. Thus states Ψ_2 and Ψ_3 are depopulated relative to Ψ_1 and Ψ_4. As a result the polarization pattern becomes weaker and inverts. A time trace corresponding to the field position marked with an arrow under the spectrum on the left is shown on the right of Fig. 18.13. The initial population distribution results in a strong emissive signal at this field position. As the singlet recombination occurs the signal rapidly inverts and then decays slowly as the states with triplet character are depopulated. From such time traces, the singlet and triplet backreaction rates can be determined. In most cases, spin relaxation between the pure triplet states (Ψ_1 and Ψ_4) and the mixed singlet/triplet states (Ψ_2 and Ψ_3) is expected to be the rate-limiting step in triplet recombination.

18.4.2 Quantum Beats

If a radical pair is generated on a time scale such that no significant singlet-triplet mixing occurs, its initial state is a pure singlet as shown in Fig. 18.9 and singlet-triplet mixing occurs due to the subsequent precession of the spins. This motion of the spin system modulates the intensity of the EPR transitions. The intensity of the transitions shown in Fig. 18.5 are determined by the product of the transition probability and the population difference. Because ψ_1 and ψ_4 are pure triplet sates, the transition probability is proportional to the triplet character of the mixed states ψ_2 and ψ_3 Thus, as the spin system oscillates between a pure singlet and pure triplet state the intensity of the EPR transitions oscillate. Such oscillations are referred to as quantum beats (Bittl and Kothe 1991; Salikhov et al. 1990) and they can be observed at short times following the laser flash. At longer time, they decay as the relative phase of the oscillations in different parts of the sample becomes random. Typically, phase relaxation times are on the order of 100 ns or less, which makes the quantum beats difficult to observe. Because they depend on the rate of singlet-triplet mixing they can be analyzed to obtain the magnetic parameters of the spin system. However, this information can also be obtained from the analysis of the spin polarization pattern, which is much easier to measure.

18.4.3 Out of Phase Echo Modulation

Early observation of the electron spin echo signals from photosynthetic samples showed that the echo was phase shifted by 90° compared to the echo from a stable radical (Thurnauer et al. 1979, 1982; Thurnaur and Norris 1980; Thurnauer and Clark 1984). The origin of the phase shift was later shown to be a result of the correlation between the electron spins and that the amplitude of the echo is modulated by the spin-spin interactions (Salikhov et al. 1992; Tang et al. 1994). This provides an elegant way of measuring the spin-spin coupling and it has been widely used to characterize the natural photosynthetic systems (Bittl and Zech 2001; Angerhofer and Bittl 1996; Savitsky et al. 2007, 2013; Bittl and Kothe 1991; Borovykh et al. 2002; Bittl and Zech 1997; Zech et al. 1996). For many artificial complexes the out-of-phase echo is not observed or the modulation decays too rapidly to determine the spin-spin coupling. The absence of an out-of-phase echo indicates a lack the correlation between the spins, which can occur as a result of dephasing of singlet-triplet mixing in the primary radical pair or when triplet electron transfer is the dominant mechanism. As a result there are far fewer reports of the use of out-of-phase echo modulation experiments on D-A complexes.

18.5 Recent Results

18.5.1 Early Results on Donor-Acceptor Complexes

The early results of TREPR experiments on donor-acceptor complexes for artificial photosynthesis are summarized in a number of review articles (Levanon and Möbius 1997; Savitsky and Moebius 2006; Gust et al. 2001; Wasielewski 1992, 2006; Verhoeven 2006). In these initial studies, the goal was primarily to mimic various aspects of the natural photosystems. Some of the first systems to be investigated were the porphyrin-quinone dyads and triads studied by Möbius and Kurreck (Lendzian et al. 1991; Hasharoni et al. 1993; Batchelor et al. 1995; Kay et al. 1995; Elger et al. 1998; Fuhs et al. 2000; Wiehe et al. 2001)

that were designed to mimic the chlorophyll donor and quinone acceptors of photosynthetic reaction centers. While most of these complexes displayed light induced electron transfer from the porphyrin to the quinone, the TREPR spectra were very different from those of the natural systems because of differences in the strength of the spin-spin coupling, relaxation dynamics and electron transfer pathway. The widest array of D-A systems has been studied by Wasielewski and co-workers (Wiederrecht et al. 1999a, 1997, 1999b; Wasielewski 1992, 2006; Wasielewski et al. 1988, 1990, 1991, 1993, 1995; Hasharoni et al. 1995, 1996; Laukenmann et al. 1995; van der Est et al. 1996; Levanon et al. 1998; Heinen et al. 2002; Shaakov et al. 2003; Dance et al. 2006, 2008a, b; Jakob et al. 2006; Tauber et al. 2006; Ahrens et al. 2007; Mi et al. 2009; Carmieli et al. 2009; Giacobbe et al. 2009; Scott et al. 2009; Miura et al. 2010; Wilson et al. 2010; Miura and Wasielewski 2011; Ls et al. 2012; Colvin et al. 2012, 2013). Using the charge transfer bands of compounds such as 4-aminonaphthalene-1,8-dicarboximide and 3,5-dimethyl-4-(9-anthracenyl) julolidine to initiate electron transfer, efficient, long-lived charge separation could be achieved by attaching secondary donors such as aniline, tetrathiafulvalene (TTF) and tetramethylbenzobisdioxole (BDX) and secondary acceptors such naphthalenediimide (NDI) and pyromellitic diimide (PI). An important feature of these dyads and triads is that long-lived radical pair formation occurs in both liquid and frozen solution, which greatly facilitates TREPR measurements. The low reorganization energy of fullerene (C_{60}) as an electron acceptor also promotes long-lived charge separation as first shown by Gust and Moore (Liddell et al. 1994; Carbonera et al. 1998). Using these systems it has been possible to reproduce all of the characteristic TREPR signatures seen in photosynthetic reaction centers and obtain important information about the electron transfer. In the following sections we highlight a number of recent examples and summarize the TREPR studies on organic bulk heterojunction solar cells.

18.5.2 Sequential Electron Transfer in Triads

As discussed in Sect 18.3.3, in a sequential electron transfer reaction the spin dynamics that occurs during the lifetime of the primary radical pair affects the polarization pattern of the subsequent radical pairs. We have investigated this phenomenon in a series of triads based on aluminum(III) porphyrin (AlPor) in which an acceptor (NDI or C_{60}) and secondary donor (TTF) are attached on opposite faces of the porphyrin as shown in Fig. 18.14 (van der Est and Poddutoori 2013; Poddutoori et al. 2013, 2015). Excitation of the AlPor leads to transfer of the excited electron in the LUMO of the porphyrin to the LUMO of the acceptor as well as hole transfer from the HOMO of the porphyrin to the HOMO of the donor. This reaction creates the radical pair $TTF^{•+}NDI^{•-}$ or $TTF^{•+}C_{60}^{•-}$ depending on which acceptor is used.

The transient EPR spectra of the triads are shown in Fig. 18.15. In both cases, the spectrum consists of two antiphase doublets with an E/A/E/A polarization pattern. The low field doublet is due to $TTF^{•+}$, while the high-field doublet arises from the reduced acceptor, either $NDI^{•-}$ or $C_{60}^{•-}$. The antiphase doublets are not symmetric and in each doublet one of the two peaks is considerably stronger than the other. With NDI as the acceptor (Fig. 18.15, left) the low-field doublet due to $TTF^{•+}$ shows net emissive polarization and the $NDI^{•-}$ doublet at higher field has net absorption. With C_{60} as the acceptor, the situation is reversed (Fig. 18.15, right). The E/A/E/A pattern is consistent with singlet electron transfer and negative spin-spin coupling. However, pure singlet electron transfer leads to symmetric antiphase doublets as shown in Fig. 18.10. The net polarization of each doublet is due to singlet-triplet mixing during the lifetime of the precursors. The simulations of the spectra (red curves) take this mixing into account (van der Est and Poddutoori 2013; Poddutoori et al. 2013, 2015). The different sign of the net polarization in the two cases can be easily understood as the result of a change in the sign of the g-factor difference in the precursor state. As shown in Eq. 18.7, the sign of the additional polarization generated by singlet-triplet mixing depends on the sign of $g_D - g_A$ in the precursor state. For the radical pair $AlPor^{•+}NDI^{•-}$ the g-factor difference is negative, while for $TTF^{•+}AlPor^{•-}$ and $AlPor^{•+}C_{60}^{•-}$ it is positive. Thus, the data are consistent with the following electron transfer sequences in the two triads:

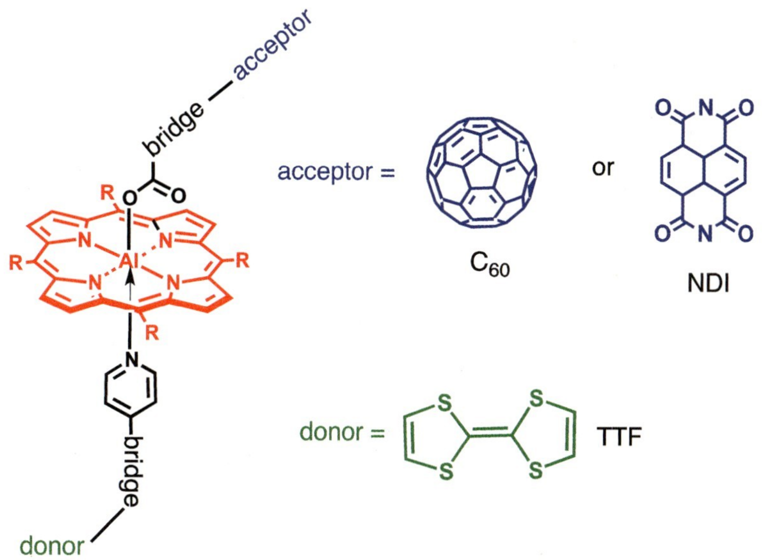

Fig. 18.14 Structure of axial aluminum(III) porphyrin-based D-A triads

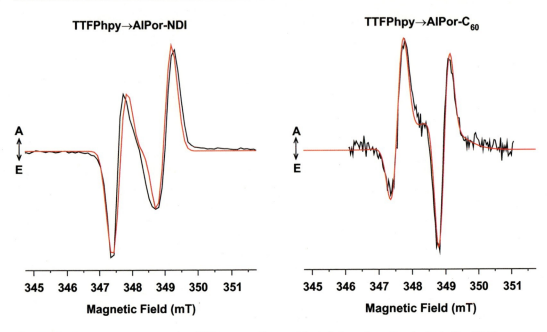

Fig. 18.15 Room temperature transient EPR spectra of two axial triads from the series shown in Fig. 18.14

$$AlPor \rightarrow AlPor^* \rightarrow AlPor^{\cdot+}NDI^{\cdot-}$$
$$\rightarrow TTF^{\cdot+}NDI^{\cdot-}$$

and

$$AlPor \rightarrow AlPor^* \rightarrow TTF^{\cdot+}AlPor^{\cdot-}$$
$$\rightarrow TTF^{\cdot+}C_{60}^{\cdot-}$$

In the latter case, the EPR data are also consistent with electron transfer to C_{60} being the initial step, however, fluorescence quenching data suggest that electron donation by TTF occurs first. In principle, the lifetime of the precursor state can be estimated from the net polarization of the antiphase doublets, however, this requires that the strength of the spin-spin coupling in the precursor is known. For the spectra shown in Fig. 18.15, a lifetime of about 1 ns is obtained if the spin-spin coupling is assumed to be ~2 mT but this combination of values for the lifetime and coupling is not unique.

18.5.3 Quantum Beats

As discussed in Sect. 18.4.2, the strong correlation between the electrons when a radical pair is generated from a pure singlet state produces coherence effects in the TREPR signals. These quantum beat oscillations are not easy to observe because of the rapid de-phasing of the spins. This is particularly true when the measurements must be performed at room temperature. In most D-A complexes charge separation that is long lived enough to be observed by TREPR requires stabilization of the radical pair states by solvent reorganization. As a result, the radical pairs cannot be observed at low temperature and hence there are very few reports of quantum beat measurements. However, there are some complexes in which this phenomenon can be studied (Laukenmann et al. 1995; Krzyaniak et al. 2015). The first observation of quantum beat oscillations in a D-A complex was reported by Kothe, Norris, Wasielewski and co-workers (Laukenmann et al. 1995). More recently, a new optical method of indirect observation of the singlet-triplet coherence has reported by Wasielewski and co-workers (Kobr et al. 2012; Krzyaniak et al. 2015). They designed a complex in which two electron-transfer steps can be driven by two short laser pulses at different wavelengths. The first pulse generates the primary radical pair, which is then irradiated to drive the secondary electron transfer and create the secondary radical pair. In a first study it was shown that the secondary radical pair is spin-correlated and generated initially in a singlet state (Kobr et al. 2012). In a very recent report it was then shown that by varying the delay between the two laser flashes changes in the TREPR spectrum of the secondary radical pair are observed that are consistent with coherent singlet-triplet mixing in the primary radical pair.

18.5.4 Echo Modulations

The values of the spin-spin couplings are crucial for understanding the relationship between molecular structure and the efficiency of electron transfer. This is because the exchange coupling can be directly related to the electronic coupling between the donor and acceptor and the dipolar coupling provides information about the distance between the separated electrons. However, as discussed in Sect. 18.3.1, if the coupling is smaller than the inhomogeneous linewidth it cannot be obtained unambiguously from the TREPR spectrum. However, the out-of-phase echo experiment described in Sect. 18.2.2 provides an elegant and accurate way of the determining the couplings. Again, however, this experiment is difficult to perform at room temperature because of short T_2 relaxation. Carmieli et al. (2009) recently reported out-of-phase echo modulation results on the complexes shown in Fig. 18.16, which can be measured at low temperature. In complexes 1 and 2 excitation of the charge transfer band between 6ANI and PI and subsequent donation by BDX or TTF results in a long-lived radical pair. In complex 3, the charge transfer band between DMJ and An is excited and the charge separation is stabilized by electron transfer to NI.

The out-of-phase echo modulation curves for the three complexes and their Fourier transforms are shown in Fig. 18.17. It is immediately apparent that the modulation frequency for complex 3 is much lower than for complexes 1 and 2 and that this results in a narrower Fourier transform spectrum. This lower frequency is the result of a significantly larger distance between the spins in complex 3. Closer inspection of the curves for complexes 1 and 2 shows that there are subtle differences between them despite the fact that the complexes are virtually identical. In 2 the perpendicular component of the coupling is smaller than in 1 but the parallel component is larger and the sign of the $\nu_\parallel$ and $\nu'_\parallel$ signal components are opposite in 1 and 2. These differences are a result of a difference in the sign of J in the two complexes. This is a surprising result, given

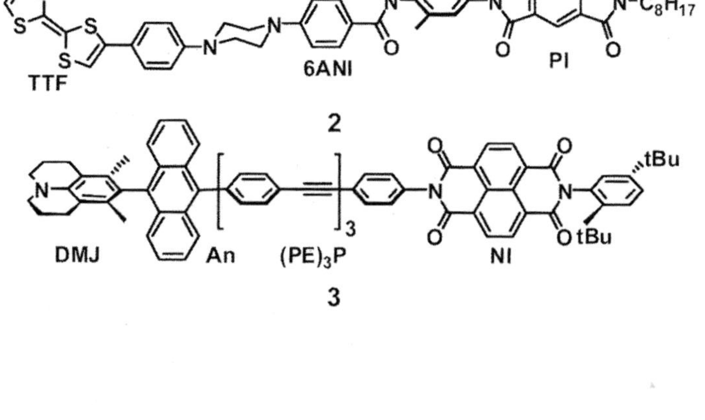

Fig. 18.16 Structures of donor acceptor complexes for out-of-phase echo modulation experiments (Reprinted with permission from (Carmieli et al. 2009). Copyright (2009) American Chemical Society)

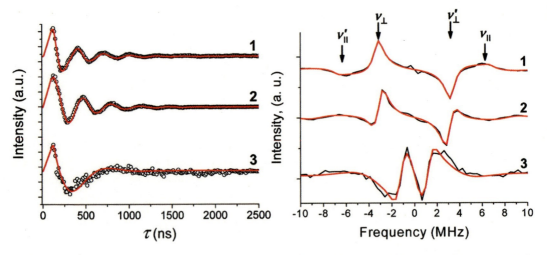

Fig. 18.17 Out-of-phase echo modulation curves of the light-induced radical pairs in the donor-acceptor complexes shown in Fig. 18.16 (Reprinted with permission from (Carmieli et al. 2009). Copyright (2009) American Chemical Society)

that the two complexes differ only in the nature of the donor. Although the origin of the difference remains unclear, the results provide a good demonstration of the use of the out-of-phase echo modulation to accurately determine the dipolar and exchange contributions to the coupling.

18.5.5 Polymer-Fullerene Blends

TREPR methods have also been applied to the study of charge separation in organic bulk heterojunction solar cells. These systems consist of blends of a conductive polymer, usually a substituted polythiophene, and an electron acceptor usually a substituted fullerene.

Light excitation of the polymer leads to charge separation at the polymer/fullerene junction with good yield and a long lifetime and if the heterojunction is sandwiched between two electrodes of a closed circuit a photocurrent is generated (Sariciftci et al. 1992). For such systems, the factors that allow the Coulomb attraction between the separated charges to be overcome is not well understood and this has motivated a number of TREPR studies (Da Ros et al. 1999; Pasimeni et al. 2001a, b; Behrends et al. 2012; Kobori et al. 2013; Miura et al. 2013; Kraffert et al. 2014; Lukina et al. 2015; Niklas et al. 2015). The structures of some of the fullerenes and polythiophenes that have been used in these studies are shown in Fig. 18.18. Pasimeni and co-workers (Da Ros et al. 1999; Pasimeni et al. 2001a, b) reported the first TREPR spectra on a blend of N-mTEGFP and ST6 (Figs. 18.18 and 18.19). The E/A/E/A polarization is indicative of a spin correlated radical pair. The authors were able to simulate the spectrum (Fig. 18.19 dotted curve) assuming a singlet-born radical pair with a dipolar coupling constant of $D = -121 \pm 4$ μT, which corresponds to a distance of 28.4 ± 0.3 Å between the radicals. Because a distribution of geometries for the radical pair is expected, a random distribution of orientations of the dipolar coupling vector relative to the g-tensors was taken and the principal axes of the g-tensors of the two radicals were assumed to be collinear. No out-of-phase echo was detectable for these samples (Pasimeni et al. 2001b). Behrends et al. (2012; Kraffert et al. 2014) obtained a virtually identical spectrum from a blend of PCBM and P3HT (Fig. 18.18) and showed that it could be simulated using purely isotropic spin-spin coupling and argued that it is not possible to obtain a unique value for the coupling. As discussed in Sect. 18.3.1, the geometry and spin-spin coupling can only be determined if the absolute intensity

N-mTEGFP
(R=triethylene glycol mono methylether)

PCBM

ST6 (R = SCH_3, n=6)

P3HT (R = hexyl)

P3DDT (R = dodecyl)

PTB7 (R=2-ethylhexyl)

PCDTBT (R=octyl)

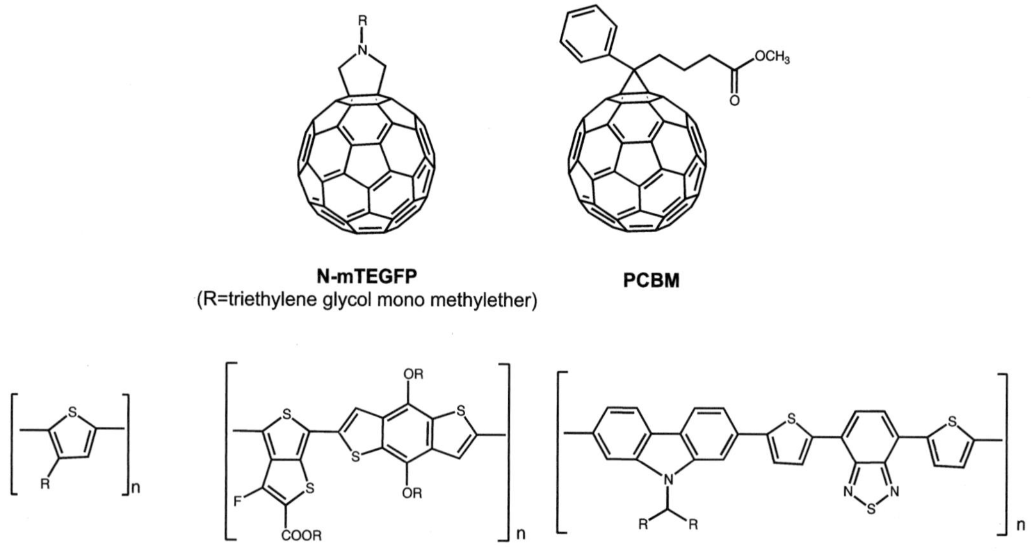

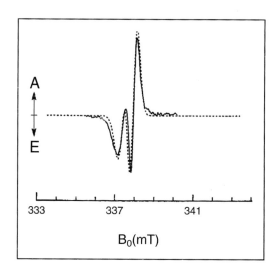

Fig. 18.18 Structures of several polythiophenes and fullerene derivates used in organic bulk heterojunction solar cells

Fig. 18.19 Transient EPR spectrum of a sexithiophene – fulleropyrrolidine blend. *Solid line*: experimental spectrum, *Dashed line*: simulation (Reprinted from (Pasimeni et al. 2001b), with permission from Elsevier)

of the spectrum is known and this is generally not the case for TREPR data. They also showed that the polarization decays to a purely absorptive pattern at late times that was the same as that observed by steady-state EPR, suggesting that it was due to separated polarons.

Kobori and co-workers (Kobori et al. 2013; Miura et al. 2013) studied the same P3HT-PCBM blend and obtained similar time-dependent spectra. In contrast to Behrends et al. (2012; Kraffert et al. 2014) they argued that the absorptive spectrum observed at late time was broader than the steady-state spectrum and therefore assigned it to the coupled radical pair after relaxation of the initial spin polarization. They were able to reproduce the experimental spectra as a function of time using the stochastic Liouville equation to take the relaxation into account. To calculate the spectra, they assumed a single average geometry for the $P3HT^{\bullet+}PCBM^{\bullet-}$ radical pair and estimated the dipolar coupling from the width of the spectrum. The exchange coupling and geometry were obtained by fitting to the experimental data. When the side chains on the polythiophene were varied and the polymer blend was annealed, different values for the exchange and dipolar couplings and geometries were obtained (Miura et al. 2013). From these data, the distance dependence of the exchange coupling was calculated and a value of $\beta = 0.2$ Å^{-1} for the attenuation factor of the electronic coupling was estimated.

Niklas et al. (2015) compared the TREPR spectra of the blends of PCBM with the three polythiophenes P3HT, PTB7 and PCDTBT shown in Fig. 18.18. These blends show similar TREPR spectra but differing degrees of net polarization of the two radicals occur. For the PTB7-containing blend a spin-polarized spectrum of the triplet state of PTB7 is observed in addition to the radical pair. The polarization pattern of the triplet state showed that it is due to radical pair recombination (see Sect. 18.3.4). The intensity of the triplet state increased substantially in experiments performed at 130 GHz as a result of the faster singlet-triplet mixing at higher magnetic field. Together the polarization patterns of the radical pairs and the appearance of the triplet state spectrum clearly indicate that a series of sequential radical pairs is formed and that the observed pair is the secondary pair.

In the analysis by Kobori et al. (2013) an average geometry was assumed but the width of the distribution was not evaluated, although the relaxation times they obtained suggest a relatively broad distribution. Out-of-phase echo studies provide a more accurate measure of the spin-spin coupling and the width of the distribution. Lukina et al. (2015) recently reported such experiments and found that although a light-induced out-of-phase echo from P3HT$^+$PCBM$^-$ radical pair is observed it is not modulated. They rationalized the absence of modulation as a result of delocalization of the hole on the polythiophene chain resulting in a distribution of spin-spin couplings, which leads to damping of the echo modulation. The echo decay curve could be simulated with a simple model in which the hole on the polythiophene is distributed over a distance spanning about 4 nm.

18.6 Concluding Remarks

As can be seen from the examples presented here TREPR experiments provide important information both for the characterization and design of artificial photosynthetic D-A complexes. As for any spectroscopic technique, there are limitations. The most important of these for TREPR is the comparitively slow inherent response time which is on the order of 10 ns. However as discussed above, the spin dynamics occuring on a shorter time scale can have an influence on the observed data and it is possible to deduce properties of the states that cannot be observed directly. The organic bulk heterojunction studies illustrate how the complexity of the data analysis increases with the complexity of the systems. However, they also show that by combining data from various experiments it is possible to gain useful insights even in disordered amorphous systems.

Acknowledgments This work was supported by a Discovery Grant from the Natural Sciences and Engineering Council Canada. AvdE wishes to thank Stefan Weber and the members of his research group and the members of FRIAS in Freiburg for their great hospitality during the writing of this manuscript.

References

Ahrens MJ, Kelley RF, Dance ZE, Wasielewski MR (2007) Photoinduced Charge Separation in Self-Assembled Cofacial Pentamers of Zinc-5, 10, 15, 20-Tetrakis (Perylenediimide) Porphyrin. Phys Chem Chem Phys 9 (12):1469–1478

Allen JP, Williams JC (2014) Energetics of Cofactors in Photosynthetic Complexes: Relationship Between Protein–Cofactor Interactions and Midpoint Potentials. In: Golbeck JH, van der Est A (eds) The Biophysics of Photosynthesis. Biophysics for the Life Sciences. Springer, New York, pp 275–295

Angerhofer A, Bittl R (1996) Radicals and Radical Pairs in Photosynthesis. Photochem Photobiol 63 (1):11–38

Batchelor SN, Sun L, Möbius K, Kurreck H (1995) Time-resolved EPR Studies of Covalently Linked Porphyrin—Crown Ether—Quinones, Dissolved in Liquid Crystals. Magn Reson Chem 33 (13):S28–S33

Behrends J, Sperlich A, Schnegg A, Biskup T, Teutloff C, Lips K, Dyakonov V, Bittl R (2012) Direct detection of photoinduced charge transfer complexes in polymer fullerene blends. Phys Rev B 85 (12):125206

Bertini I, Bryant DA, Ciurli S, Dikiy A, Fernández CO, Luchinat C, Safarov N, Vila AJ, Zhao J (2001) Backbone Dynamics of Plastocyanin in Both Oxidation States Solution Structure of the Reduced Form and

Comparison with the Oxidized State. J Biol Chem 276 (50):47217–47226

Bittl R, Kothe G (1991) Transient EPR of Radical Pairs in Photosynthetic Reaction Centers: Prediction of Quantum Beats. Chem Phys Lett 177 (6):547–553

Bittl R, Zech SG (1997) Pulsed EPR Study of Spin-Coupled Radical Pairs in Photosynthetic Reaction Centers: Measurement of the Distance Between and in Photosystem I and between and in Bacterial Reaction Centers. J Phys Chem B 101 (8):1429–1436

Bittl R, Zech SG (2001) Pulsed EPR Spectroscopy on Short-Lived Intermediates in Photosystem I. Biochim Biophys Acta 1507 (1):194–211

Borovykh I, Kulik L, Dzuba S, Hoff A (2002) Out-of-phase Stimulated Electron Spin-Echo Appearing in the Evolution of Spin-Correlated Photosynthetic Triplet-Radical Pairs. J Phys Chem B 106 (46):12066–12071

Bruns CM, Karplus AP (1995) Refined Crystal Structure of Spinach Ferredoxin Reductase at 1.7 Å Resolution: Oxidized, Reduced and 2'-phospho-5'-AMP Bound States. J Mol Biol 247 (1):125–145

Buckley C, Hunter D, Hore P, McLauchlan K (1987) Electron Spin Resonance of Spin-Correlated Radical Pairs. Chem Phys Lett 135 (3):307–312

Budil DE, Thurnauer MC (1991) The Chlorophyll Triplet State as a Probe of Structure and Function in Photosynthesis. Biochim Biophys Acta 1057 (1):1–41

Burnell E, De Lange C (1998) Prediction from Molecular Shape of Solute Orientational Order in Liquid Crystals. Chem Rev 98 (6):2359–2388

Carbonera D, Di Valentin M, Corvaja C, Agostini G, Giacometti G, Liddell PA, Kuciauskas D, Moore AL, Moore TA, Gust D (1998) EPR Investigation of Photoinduced Radical Pair Formation and Decay to a Triplet State in a Carotene-Porphyrin-Fullerene Triad. J Am Chem Soc 120 (18):4398–4405

Carmieli R, Mi Q, Ricks AB, Giacobbe EM, Mickley SM, Wasielewski MR (2009) Direct Measurement of Photoinduced Charge Separation Distances in Donor–Acceptor Systems for Artificial Photosynthesis Using OOP-ESEEM. J Am Chem Soc 131 (24):8372–8373

Closs GL, Forbes MD, Norris JR (1987) Spin-Polarized Electron Paramagnetic Resonance Spectra of Radical Pairs in Micelles: Observation of Electron Spin-Spin Interactions. J Phys Chem 91 (13):3592–3599

Colvin MT, Ricks AB, Scott AM, Co DT, Wasielewski MR (2012) Intersystem Crossing Involving Strongly Spin Exchange-Coupled Radical Ion Pairs in Donor–bridge–Acceptor Molecules. J Phys Chem A 116 (8):1923–1930

Colvin MT, Carmieli R, Miura T, Richert S, Gardner DM, Smeigh AL, Dyar SM, Conron SM, Ratner MA, Wasielewski MR (2013) Electron Spin Polarization Transfer from Photogenerated Spin-Correlated Radical Pairs to a Stable Radical Observer Spin. J Phys Chem A 117 (25):5314–5325

Da Ros T, Prato M, Guldi D, Alessio E, Ruzzi M, Pasimeni L (1999) A noncovalently linked, dynamic fullerene porphyrin dyad. Efficient formation of long-lived charge separated states through complex dissociation. Chem Commun (7):635–636

Dance ZE, Mi Q, McCamant DW, Ahrens MJ, Ratner MA, Wasielewski MR (2006) Time-Resolved EPR Studies of Photogenerated Radical Ion Pairs Separated by p-Phenylene Oligomers and of Triplet States Resulting from Charge Recombination. J Phys Chem B 110 (50):25163–25173

Dance ZE, Ahrens MJ, Vega AM, Ricks AB, McCamant DW, Ratner MA, Wasielewski MR (2008a) Direct Observation of the Preference of Hole Transfer over Electron Transfer for Radical Ion Pair Recombination in Donor-Bridge-Acceptor Molecules. J Am Chem Soc 130 (3):830–832

Dance ZE, Mickley SM, Wilson TM, Ricks AB, Scott AM, Ratner MA, Wasielewski MR (2008b) Intersystem Crossing Mediated by Photoinduced Intramolecular Charge Transfer: Julolidine-Anthracene Molecules with Perpendicular π Systems. J Phys Chem A 112 (18):4194–4201

Elger G, Fuhs M, Müller P, Gersdorff J, Wiehe A, Kurreck H, Möbius K (1998) Time-resolved EPR Studies of Photoinduced Electron Transfer Reactions in Photosynthetic Model Porphyrin Quinone Triads. Mol Phys 95 (6):1309–1323

Forbes MDE, Jarocha LE, Sim S, Tarasov VF (2013) Time-Resolved Electron Paramagnetic Resonance Spectroscopy: History, Technique, and Application to Supramolecular and Macromolecular Chemistry. In: Ian HW, Nicholas HW (eds) Advances in Physical Organic Chemistry, vol Volume 47. Academic Press, pp 1–83

Fuhs M, Elger G, Mobius K, Osintsev A, Popov A, Kurreck H (2000) Multifrequency Time-Resolved EPR (9.5 GHz and 95GHz) on Covalently Linked Porphyrin-Quinone Model Systems for Photosynthetic Electron Transfer: Effect of Molecular Dynamics on Electron Spin Polarization. Mol Phys 98 (15):1025–1040

Giacobbe EM, Mi Q, Colvin MT, Cohen B, Ramanan C, Scott AM, Yeganeh S, Marks TJ, Ratner MA, Wasielewski MR (2009) Ultrafast Intersystem Crossing and Spin Dynamics of Photoexcited Perylene-3, 4: 9, 10-bis (dicarboximide) Covalently Linked to a Nitroxide Radical at Fixed Distances. J Am Chem Soc 131 (10):3700–3712

Gust D, Moore TA, Moore AL (2001) Mimicking Photosynthetic Solar Energy Transduction. Acc Chem Res 34 (1):40–48

Hasharoni K, Levanon H (1995) Attenuation of Intramolecular Electron Transfer Rates in Liquid Crystals. J Phys Chem 99 (14):4875–4878

Hasharoni K, Levanon H, von Gersdorff J, Kurreck H, Möbius K (1993) Photo-Induced Electron Transfer in Covalently Linked Donor–Acceptor Assemblies in Liquid Crystals. Time-Resolved Electron Paramagnetic Resonance. J Chem Phys 98 (4):2916–2926

Hasharoni K, Levanon H, Greenfield SR, Gosztola DJ, Svec WA, Wasielewski MR (1995) Mimicry of the Radical Pair and Triplet States in Photosynthetic Reaction Centers with a Synthetic Model. J Am Chem Soc 117 (30):8055–8056

Hasharoni K, Levanon H, Greenfield SR, Gosztola D, J, Svec WA, Wasielewski MR (1996) Radical Pair and Triplet State Dynamics of a Photosynthetic Reaction-Center Model Embedded in Isotropic Media and Liquid Crystals. J Am Chem Soc 118 (42):10228–10235

Heinen U, Berthold T, Kothe G, Stavitski E, Galili T, Levanon H, Wiederrecht G, Wasielewski MR (2002) High Time Resolution Q-band EPR Study of Sequential Electron Transfer in a Triad Oriented in a Liquid Crystal. J Phys Chem A 106 (10):1933–1937

Hore P (1996) Research Note Transfer of Spin Correlation Between Radical Pairs in the Initial Steps of Photosynthetic Energy Conversion. Mol Phys 89 (4):1195–1202

Hore P, Hunter D, McKie C, Hoff A (1987) Electron Paramagnetic Resonance of Spin-Correlated Radical Pairs in Photosynthetic Reactions. Chem Phys Lett 137 (6):495–500

Jakob M, Berg A, Stavitski E, Chernick ET, Weiss EA, Wasielewski MR, Levanon H (2006) Photoinduced Electron Transfer Through Hydrogen Bonds in a Rod-Like Donor–Acceptor Molecule: A Time-Resolved EPR Study. Chem Phys 324 (1):63–71

Jordan P, Fromme P, Witt HT, Klukas O, Saenger W, Krauss N (2001) Three-Dimensional Structure of Cyanobacterial Photosystem I at 2.5 Ångstrom Resolution. Nature 411 (6840):909–917

Kamlowski A, Zech SG, Fromme P, Bittl R, Lubitz W, Witt HT, Stehlik D (1998) The Radical Pair State in Photosystem I Single Crystals: Orientation Dependence of the Transient Spin-Polarized EPR Spectra. J Phys Chem B 102 (42):8266–8277

Kandrashkin Y, van der Est A (2001) A New Approach to Determining the Geometry of Weakly Coupled Radical Pairs From Their Electron Spin Polarization Patterns. Spectrochim Acta C 57 (8):1697–1709

Kandrashkin YE, van der Est A (2007) Time-resolved EPR spectroscopy of Photosynthetic Reaction Centers: from Theory to Experiment. Appl Magn Reson 31 (1–2):105–122

Kandrashkin YE, Salikhov K, van der Est A, Stehlik D (1998) Electron Spin Polarization in Consecutive Spin-Correlated Radical Pairs: Application to Short-Lived and Long-Lived Precursors in Type 1 Photosynthetic Reaction Centres. Appl Magn Reson 15 (3–4):417–447

Kandrashkin YE, Vollmann W, Stehlik D, Salikhov K, Van der Est A (2002) The Magnetic Field Dependence of the Electron Spin Polarization in Consecutive Spin Correlated Radical Pairs in Type I Photosynthetic Reaction Centres. Mol Phys 100 (9):1431–1443

Kandrashkin YE, Asano MS, van der Est A (2006a) Light-Induced Electron Spin Polarization in Vanadyl Octaethylporphyrin: I. Characterization of the Excited Quartet State. J Phys Chem A 110 (31):9607–9616

Kandrashkin Y, Poddutoori P, Van Der Est A (2006b) Novel Intramolecular Electron Transfer in Axial bis (terpyridoxy) phosphorus (V) Porphyrin Studied by Time-Resolved EPR Spectroscopy. Appl Magn Reson 30 (3–4):605–618

Kay C, Kurreck H, Batchelor S, Tian P, Schlüpmann J, Möbius K (1995) Photochemistry of a Butylene Linked Porphyrin-Quinone Donor-Acceptor System Studied by Time-Resolved and Steady-State EPR Spectroscopy. Appl Magn Reson 9 (4):459–480

Kobori Y, Noji R, Tsuganezawa S (2013) Initial Molecular Photocurrent: Nanostructure and Motion of Weakly Bound Charge-Separated State in Organic Photovoltaic Interface. J Phys Chem C 117 (4):1589–1599

Kobr L, Gardner DM, Smeigh AL, Dyar SM, Karlen SD, Carmieli R, Wasielewski MR (2012) Fast Photodriven Electron Spin Coherence Transfer: A Quantum Gate Based on a Spin Exchange J-Jump. J Am Chem Soc 134 (30):12430–12433

Koradi R, Billeter M, Wuthrich K (1996) MOLMOL: A Program for Display and Analysis of Macromolecular Structures. J Mol Graphics 14 (1):51–55

Kraffert F, Steyrleuthner R, Albrecht S, Neher D, Scharber MC, Bittl R, Behrends J (2014) Charge Separation in PCPDTBT: PCBM Blends from an EPR Perspective. J Phys Chem C 118 (49):28482–28493

Krzyaniak MD, Kobr L, Rugg BK, Phelan BT, Margulies EA, Nelson JN, Young RM, Wasielewski MR (2015) Fast Photo-Driven Electron Spin Coherence Transfer: the Effect of Electron-Nuclear Hyperfine Coupling on Coherence Dephasing. J Mat Chem C

Kurisu G, Zhang HM, Smith JL, Cramer WA (2003) Structure of the Cytochrome b(6)f Complex of Oxygenic Photosynthesis: Tuning the Cavity. Science 302 (5647):1009–1014

Laukenmann K, Weber S, Kothe G, Oesterle C, Angerhofer A, Wasielewski MR, Svec WA, Norris JR (1995) Quantum Beats of the Radical Pair State in Photosynthetic Models Observed by Transient Electron Paramagnetic Resonance. J Phys Chem 99 (12):4324–4329

Lendzian F, Schlüpmann J, von Gersdorff J, Möbius K, Kurreck H (1991) Investigation of the Light-Induced Charge Transfer between Covalently Linked Porphyrin and Quinone Units by Time-Resolved EPR Spectroscopy. Angew Chem Int Edit 30 (11):1461–1463

Levanon H (1987) Spin Polarized Triplets Oriented in Liquid Crystals. Res Chem Intermed 8 (3):287–320

Levanon H, Möbius K (1997) Advanced EPR Spectroscopy on Electron Transfer Processes in Photosynthesis and Biomimetic Model Systems. Annu Rev Bioph Biom 26 (1):495–540

Levanon H, Galili T, Regev A, Wiederrecht GP, Svec WA, Wasielewski MR (1998) Determination of The Energy Levels of Radical Pair States in Photosynthetic Models Oriented in Liquid Crystals with Time-

Resolved Electron Paramagnetic Resonance. J Am Chem Soc 120 (25):6366–6373

Liddell PA, Sumida JP, Macpherson AN, Noss L, Seely GR, Clark KN, Moore AL, Moore TA, Gust D (1994) Preparation and Photophysical Studies of Porphyrin-C60 Dyads. Photochem Photobiol 60 (6):537–541

Loll B, Kern J, Saenger W, Zouni A, Biesiadka J (2005) Towards Complete Cofactor Arrangement in the 3.0 Ångstrom Resolution Structure of Photosystem II. Nature 438 (7070):1040–1044

Lukina EA, Popov AA, Uvarov MN, Kulik LV (2015) Out-of-Phase Electron Spin Echo Studies of Light-Induced Charge-Transfer States in P3HT/PCBM Composite. J Phys Chem B

Mi Q, Ratner MA, Wasielewski MR (2009) Time-Resolved EPR Spectra of Spin-Correlated Radical Pairs: Spectral and Kinetic Modulation Resulting from Electron− Nuclear Hyperfine Interactions. J Phys Chem A 114 (1):162–171

Miura T, Wasielewski MR (2011) Manipulating Photogenerated Radical Ion Pair Lifetimes in Wirelike Molecules Using Microwave Pulses: Molecular Spintronic Gates. J Am Chem Soc 133 (9):2844–2847

Miura T, Carmieli R, Wasielewski MR (2010) Time-Resolved EPR Studies of Charge Recombination and Triplet-State Formation within Donor− Bridge− Acceptor Molecules Having Wire-Like Oligofluorene Bridges. J Phys Chem A 114 (18):5769–5778

Miura T, Aikawa M, Kobori Y (2013) Time-Resolved EPR Study of Electron–Hole Dissociations Influenced by Alkyl Side Chains at the Photovoltaic Polyalkylthiophene: PCBM Interface. J Phys Chem Lett 5 (1):30–35

Niklas J, Beaupré S, Leclerc M, Xu T, Yu L, Sperlich A, Dyakonov V, Poluektov OG (2015) Photoinduced Dynamics of Charge Separation: From Photosynthesis to Polymer–Fullerene Bulk Heterojunctions. J Phys Chem B 119 (24):7407–7416

Norris J, Morris A, Thurnauer M, Tang J (1990) A General Model of Electron Spin Polarization Arising from the Interactions within Radical Pairs. J Chem Phys 92 (7):4239–4249

Pasimeni L, Franco L, Ruzzi M, Mucci A, Schenetti L, Luo C, Guldi DM, Kordatos K, Prato M (2001a) Evidence of High Charge Mobility in Photoirradiated polythiophene–Fullerene Composites. J Mater Chem 11 (4):981–983

Pasimeni L, Ruzzi M, Prato M, Da Ros T, Barbarella G, Zambianchi M (2001b) Spin Correlated Radical Ion Pairs Generated by Photoinduced Electron Transfer in Composites of Sexithiophene/Fullerene Derivatives: A Transient EPR Study. Chem Phys 263 (1):83–94

Poddutoori PK, Zarrabi N, Moiseev AG, Gumbau-Brisa R, Vassiliev S, van der Est A (2013) Long-Lived Charge Separation in Novel Axial Donor–Porphyrin–Acceptor Triads Based on Tetrathiafulvalene, Aluminum (III) Porphyrin and Naphthalenediimide. Chem Eur J 19 (9):3148–3161

Poddutoori PK, Lim GN, Sandanayaka AS, Karr PA, Ito O, D'Souza F, Pilkington M, van der Est A (2015) Axially Assembled Photosynthetic Reaction Center Mimics Composed of Tetrathiafulvalene, Aluminum (iii) Porphyrin and Fullerene Entities. Nanoscale 7 (28):12151–12165

Salikhov KM, Molin YN, Sagdeev R, Buchachenko A (1984) Spin Polarization and Magnetic Effects in Radical Reactions.

Salikhov K, Bock C, Stehlik D (1990) Time Development of Electron Spin Polarization in Magnetically Coupled, Spin Correlated Radical Pairs. Appl Magn Reson 1 (2):195–211

Salikhov K, Kandrashkin YE, Salikhov A (1992) Peculiarities of Free Induction and Primary Spin Echo Signals for Spin-Correlated Radical Pairs. Appl Magn Reson 3 (1):199–216

Sariciftci N, Smilowitz L, Heeger AJ, Wudl F (1992) Photoinduced Electron Transfer from a Conducting Polymer to Buckminsterfullerene. Science 258 (5087):1474–1476

Savitsky A, Moebius K (2006) Photochemical Reactions and Photoinduced Electron-Transfer Processes in Liquids, Frozen Solutions, and Proteins as Studied by Multifrequency Time-Resolved EPR Spectroscopy. Helv Chem Acta 89 (10):2544–2589

Savitsky A, Dubinskii A, Flores M, Lubitz W, Möbius K (2007) Orientation-Resolving Pulsed Electron Dipolar High-Field EPR Spectroscopy on Disordered Solids: I. Structure of Spin-Correlated Radical Pairs in Bacterial Photosynthetic Reaction Centers. J Phys Chem B 111 (22):6245–6262

Savitsky A, Niklas J, Golbeck J, Mobius K, Lubitz W (2013) Orientation Resolving Dipolar High-Field EPR Spectroscopy on Disordered Solids: II. Structure of Spin-Correlated Radical Pairs in Photosystem I. J Phys Chem B 117 (38):11184–11199

Schubert C, Margraf J, Clark T, Guldi D (2015) Molecular Wires–Impact of π-Conjugation and Implementation of Molecular Bottlenecks. Chem Soc Rev

Schweiger A, Jeschke G (2001) Principles of Pulse Electron Paramagnetic Resonance. Oxford University Press, Oxford

Scott AM, Miura T, Ricks AB, Dance ZE, Giacobbe EM, Colvin MT, Wasielewski MR (2009) Spin-Selective Charge Transport Pathways through p-Oligophenylene-Linked Donor− Bridge− Acceptor Molecules. J Am Chem Soc 131 (48):17655–17666

Shaakov S, Galili T, Stavitski E, Levanon H, Lukas A, Wasielewski MR (2003) Using Spin Dynamics of Covalently Linked Radical Ion Pairs to Probe the Impact of Structural and Energetic Changes on Charge Recombination. J Am Chem Soc 125 (21):6563–6572

Stehlik D (2006) Transient EPR Spectroscopy as Applied to Light-Induced Functional Intermediates Along the Electron Transfer Pathway in Photosystem I. In: Golbeck JH (ed) Photosystem I: The Light-Driven Plastocyanin: Ferredoxin Oxidoreductase. Advances in Photosynthesis and Respiration, vol 24. Springer, Dordrecht, The Netherlands, pp 361–386

Stehlik D, Möbius K (1997) New EPR Methods for Investigating Photoprocesses with Paramagnetic Intermediates. Annu Rev Phys Chem 48 (1):745–784

Stehlik D, Bock CH, Petersen J (1989) Anisotropic Electron Spin Polarization of Correlated Spin Pairs in Photosynthetic Reaction Centers. J Phys Chem 93 (4):1612–1619

Tang J, Thurnauer MC, Norris JR (1994) Electron Spin Echo Envelope Modulation Due to Exchange and Dipolar Interactions in a Spin-Correlated Radical Pair. Chem Phys Lett 219 (3):283–290

Tang J, Bondeson S, Thurnauer MC (1996) Effects of Sequential Electron Transfer on Electron Spin Polarized Transient EPR Spectra at High Fields. Chem Phys Lett 253 (3):293–298

Tauber MJ, Kelley RF, Giaimo JM, Rybtchinski B, Wasielewski MR (2006) Electron Hopping in π-Stacked Covalent and Self-Assembled Perylene Diimides Observed by ENDOR Spectroscopy. J Am Chem Soc 128 (6):1782–1783

Thurnauer MC (1979) ESR Study of the Photoexcited Triplet State in Photosynthetic Bacteria. Res Chem Intermed 3 (1):197–230

Thurnauer MC, Clark C (1984) Electron Spin Echo Envelope Modulation of The Transient EPR Signals Observed in Photosynthetic Algae and Chloroplasts. Photochem Photobiol 40 (3):381–386

Thurnauer M, Bowman M, Norris J (1979) Time-Resolved Electron Spin Echo Spectroscopy Applied to the Study of Photosynthesis. FEBS Lett 100 (2):309–312

Thurnauer M, Rutherford A, Norris J (1982) The Effect of Ambient Redox Potential on the Transient Electron Spin Echo Signals Observed in Chloroplasts and Photosynthetic Algae. Biochim Biophys Acta 682 (3):332–338

Thurnauer MC, Poluektov OG, Kothe G (2004) Time-Resolved High-Frequency and Multifrequency EPR Studies of Spin-Correlated Radical Pairs in Photosynthetic Reaction Center Proteins. In: Grinberg O, Berliner L (eds) Very High Frequency (VHF) ESR/EPR. Biological Magnetic Resonance, vol 24. Springer, New York, pp 165–206

Thurnaur MC, Norris JR (1980) An Electron Spin Echo Phase Shift Observed in Photosynthetic Algae: Possible Evidence for Dynamic Radical Pair Interactions. Chem Phys Lett 76 (3):557–561

Tsukihara T, Fukuyama K, Mizushima M, Harioka T, Kusunoki M, Katsube Y, Hase T, Matsubara H (1990) Structure of the [2Fe-2S] Ferredoxin I from the Blue-Green Alga *Aphanothece sacrum* at 2·2 Å Resolution. J Mol Biol 216 (2):399–410

van der Est A (2001) Light-Induced Spin Polarization in Type I Photosynthetic Reaction Centres. Biochim Biophys Acta 1507 (1):212–225

van der Est A (2009) Transient EPR: Using Spin Polarization in Sequential Radical Pairs to Study Electron Transfer in Photosynthesis. Photosynth Res 102 (2–3):335–347

van der Est A, Poddutoori PK (2013) Light-Induced Spin Polarization in Porphyrin-Based Donor–Acceptor Dyads and Triads. Appl Magn Reson 44 (1–2):301–318

van der Est A, Kok M, Burnell E (1987) Size and Shape Effects on the Orientation of Rigid Molecules in Nematic Liquid Crystals. Mol Phys 60 (2):397–413

van der Est A, Fuechsle G, Stehlik D, Wasielewski M (1996) X-and K-band Transient EPR of the Light Induced Radical Ion Pairs in Photosynthetic Model Systems. Berich Bunsen Gesell 100 (12):2081–2085

Verhoeven JW (2006) On the Role of Spin Correlation in the Formation, Decay, and Detection of Long-Lived, Intramolecular Charge-Transfer States. J Photoch Photobio C 7 (1):40–60

Wang Z, Tang J, Norris JR (1992) The Time Development of the Magnetic Moment of Correlated Radical Pairs. J Magn Reson 97 (2):322–334

Wasielewski MR (1992) Photoinduced Electron Transfer in Supramolecular Systems for Artificial Photosynthesis. Chem Rev 92 (3):435–461

Wasielewski MR (2006) Energy, Charge, and Spin Transport in Molecules and Self-Assembled Nanostructures Inspired by Photosynthesis. J Org Chem 71 (14):5051–5066

Wasielewski MR, Johnson DG, Svec WA, Kersey KM, Minsek DW (1988) Achieving High Quantum Yield Charge Separation in Porphyrin-Containing Donor-Acceptor Molecules at 10 K. J Am Chem Soc 110 (21):7219–7221

Wasielewski MR, Gaines III GL, O'Neil MP, Svec WA, Niemczyk MP (1990) Photoinduced Spin-Polarized Radical Ion Pair Formation in a Fixed-Distance Photosynthetic Model System at 5 K. J Am Chem Soc 112 (11):4559–4560

Wasielewski MR, Gaines III GL, O'neil MP, Svec WA, Niemczyk MP (1991) Spin-Polarized Radical Ion Pair Formation Resulting from Two-Step Electron Transfer from the Lowest Excited Singlet State of a Fixed-Distance Photosynthetic Model System at 5 K. Mol Cryst Liq Cryst 194 (1):201–207

Wasielewski MR, Gaines III GL, Wiederrecht GP, Svec WA, Niemczyk MP (1993) Biomimetic Modeling oF Photosynthetic Reaction Center Function: Long-Lived, Spin-Polarized Radical Ion Pair Formation in Chlorophyll-Porphyrin-Quinone Triads. J Am Chem Soc 115 (22):10442–10443

Wasielewski MR, Wiederrecht GP, Svec WA, Niemczyk MP (1995) Chlorin-Based Supramolecular Assemblies for Artificial Photosynthesis. Sol Energ Mat Sol C 38 (1):127–134

Wiederrecht GP, Svec WA, Wasielewski MR (1997) Differential Control of Intramolecular Charge Separation and Recombination Rates Using Nematic Liquid Crystal Solvents. J Am Chem Soc 119 (26):6199–6200

Wiederrecht GP, Svec WA, Wasielewski MR (1999a) Controlling the Adiabaticity of Electron-Transfer Reactions Using Nematic Liquid-Crystal Solvents. J Phys Chem B 103 (9):1386–1389

Wiederrecht GP, Svec WA, Wasielewski MR, Galili T, Levanon H (1999b) Triplet States with Unusual Spin Polarization Resulting from Radical Ion Pair Recombination at Short Distances. J Am Chem Soc 121 (33):7726–7727

Wiehe A, Senge MO, Schäfer A, Speck M, Tannert S, Kurreck H, Röder B (2001) Electron Donor–Acceptor Compounds: Exploiting the Triptycene Geometry for the Synthesis of Porphyrin Quinone Diads, Triads, and a Tetrad. Tetrahedron 57 (51):10089–10110

Wilson TM, Zeidan TA, Hariharan M, Lewis FD, Wasielewski MR (2010) Electron Hopping among Cofacially Stacked Perylenediimides Assembled by Using DNA Hairpins. Angew Chem Int Edit 49 (13):2385–2388

Zech SG, Lubitz W, Bittl R (1996) Pulsed EPR Experiments on Radical Pairs in Photosynthesis: Comparison of the Donor-Acceptor Distances In Photosystem I and Bacterial Reaction Centers. Berich Bunsen Gesell 100 (12):2041–2044

Artificial Photosynthesis Based on 1,10-Phenanthroline Complexes

19

Babak Pashaei and Hashem Shahroosvand

Summary

Natural photosynthesis has been studied recently with the aim of learning fundamental principles and transferring the technology to artificial devices such as dye-sensitized solar cells (DSSCs). In DSSCs, as the leaves of trees, a dye converts sunlight into electricity. Among the several photovoltaic devices such as thin films-based solar cells and single-crystalline Si-based solar cells, DSSC which exhibits efficiency exceed of 13%, is considered to represent low-cost alternatives to conventional inorganic solar cells.

This chapter systematically presents important roles of transition metal complexes based on 1,10-phenanthroline ligand in DSSC. 1,10-phenanthroline can be used as both sensitizer and electrolyte in DSSC which the most efficiency is obtained from cobalt phenanthroline complexes as redox shuttle. It covers not only the most frequently reported and in depth investigated complexes, but also describes some conventional complexes that have led to promising results so far. Moreover, this chapter will survey an introduction about exclusive position of transition metal complex electrolyte as a good candidate to replace with I^-/I_3^- conventional couple redox, to present the interest of readers in this field. Therefore, the main idea is to inspire readers to explore new avenues in the design of new transition metal complex sensitizers and electrolytes to improve the efficiency of DSSC, in particular when compatible sensitizer and electrolyte are selected.

Keywords

Artificial photosynthesis • Phenanthroline • Dye • Sensitizer • Solar cell • Complex • Water oxidation • Water splitting • Photovoltaic • Cobalt • Photochemical reaction

Contents

19.1	Introduction	389
19.1.1	Aim and Scope	389
19.1.2	Dyes in DSSCs	392
19.1.3	Electrolytes in DSSCs	393
19.2	**1,10-Phenanthroline-Based Electrolytes in DSSCs**	395
19.3	**Concluding Remarks**	401
References		403

B. Pashaei • H. Shahroosvand (✉)
Department of Chemistry, University of Zanjan, University Blvd, Zanjan, Iran
e-mail: shahroosvand@znu.ac.ir

19.1 Introduction

19.1.1 Aim and Scope

Faced with the prospect of depleting oil supplies and the certainty of global climate change (IPCC

2007), we are compelled to seek alternative sources to supply our growing energy demands. Several clean energy technologies will play an important role in this challenge, including wind, geothermal, biomass, hydroelectric, and nuclear. However, none of these technologies have the scalable capacity to meet the whole of our global energy demands. Moreover, burning fossil fuels raises carbon dioxide level in the atmosphere. Owing to growing energy demands, exhaustion of oil resources, and global warming issues, there is a need for clean and renewable energy technologies. Photovoltaic technology employing solar energy is regarded as the most efficient technology among all the sustainable energy technologies such as tidal power, solar thermal, hydropower, and biomass. Only the sun, with its virtually limitless supply of fusion energy, can meet our energy needs. Most people around the world live in areas with insolation levels of 150–300 watts per square meter or 3.5–7.0 kWh m^{-2} per day. The world is consuming about 14 terawatts (TW) of energy and will rise to about 40 TW by 2050. Today, in just one hour, the sun provides enough power to supply our energy needs for an entire year (Lewis and Nocera 2006). The Earth receives 174,000 TW of incoming solar radiation at the upper atmosphere. Approximately 30% is reflected back to space while the rest is absorbed by clouds, oceans and land masses. The spectrum of solar light at the Earth's surface is mostly spread across the visible and near-infrared ranges with a small part in the near-ultraviolet. Accessing and utilizing this vast quantity of energy represents a grand challenge in scientific research and engineering (Lewis et al. 2005).

Multi-junction solar cells harvest sunlight by dividing the solar spectrum into portions that are absorbed by a material with a band gap tuned to a specific wavelength range. Combining materials with optimal band gaps is critical for high efficiency.

As shown in Fig. 19.1, the Energy Department's National Renewable Energy Laboratory has announced the demonstration of a 45.7% conversion efficiency for a three-junction solar cell at 234 suns concentration. This achievement represents one of the highest photovoltaic research cell efficiencies achieved across all types of solar cells. Current silicon technologies have thus far experienced limited deployment, primarily due to materials' costs associated with processing of the high quality crystalline silicon used in these devices. Developing cost-effective methods of efficiently capturing solar energy is urgently required.

Among various types of photovoltaics (PV), dye-sensitized solar cells (DSSCs) are an efficient type of devices that have attracted the attention of many researchers because of low fabrication cost, high efficiency, ability to work at low light and mechanical robustness (Hagfeldt et al. 2010; Bomben et al. 2012).

The main components of a DSSC are a cathode, sensitizer dye, photoanode and an electrolyte solution containing a redox couple. The photoanode is a porous nanocrystalline semiconductor such as TiO_2, ZnO layer coated with a monolayer of chemisorbed sensitizers, supported on a conductive glass substrate (SnO_2:F) (Clifford et al. 2011; Argazzi et al. 2004; Nazeeruddin et al. 2004; Colombo et al. 2014).

In DSSCs, as the leaves of trees, a dye converts sunlight into electricity. Dye sensitizer has the same role of chlorophyll in photosynthesis. Chlorophyll is an extremely important biomolecule, critical in photosynthesis, which allows plants to absorb energy from the light (Fig. 19.2b). The basic structure of a chlorophyll molecule is a porphyrin ring, which is coordinated with a central atom. The porphyrin molecules arrange layer by layer together in chloroplast as shown in Fig. 19.2c. As shown in Fig. 19.2d, the porphyrin like structure can modifies by several atoms and group substitution. The π-extended electron in porphyrin lets to harvest the sun light. In artificial photosynthesis, a sensitizer in DSSC designs according to chlorophyll pattern to enhance the molar absorbance (Shahroosvand et al. 2015b).

Figure 19.2a illustrates the primary reaction processes in DSSCs, the balance of which is important to achieve high performance (Katoh et al. 2004; Shahroosvand et al. 2014). First, the dye catches photons of incoming light (sunlight

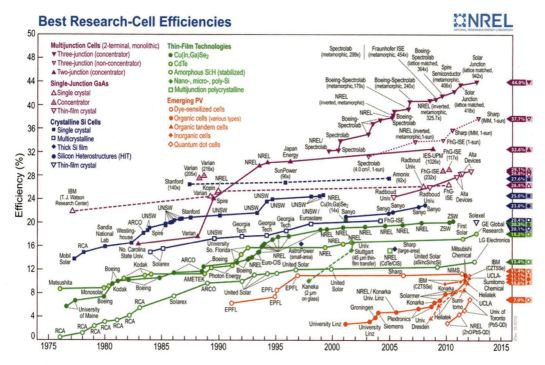

Fig. 19.1 Trends in solar cell efficiencies by technology and year (Courtesy of NREL). Multijunction cells including: three-junction (concentrator): 44%, three-junction (non-concentrator): 37.7%, two-junction (concentrator): 32.6%. Single-junction GaAs including: Single crystal: 26.4%, concentrator: 29.1%, thin film crystal: 28.8%. Crystalline Si cell including: single crystal: 27.6%, multicrystalline: 20.4%, thick Si film: 25.0%, silicon heterostructures: 23.0%, thin-film crystal: 20.1%. Thin-film technologies including: Cu(In,Ga)Se$_2$: 20.3%, CdTe: 18.3%, amorphous Si:H (stabilized): 13.4%. Dye-sensitized solar cells (11.4%), organic cells: 11.1%, organic tandem cells: 10.0%, inorganic cells: 11.1, quantum dot cells: 7.0%

and ambient artificial light) and uses their energy to excite electrons, behaving like chlorophyll in photosynthesis (*Photoexcitation*; step 1). The dye injects this excited electron into the TiO$_2$ (*Electron injection*; step 2) and the electron is conducted away by nanocrystalline titanium dioxide. The electrons can be transported in the semiconductor film as the conducting electrons. Consequently, the counter electrode (CE) collects the electron from the external circuit, and catalyzes the reduction of tri-iodide electrolyte ions (*CE catalysis*; step 3). A chemical electrolyte in the cell then closes the circuit so that the electrons are returned back to the dye (*Regeneration*; step 4). Undesirably, the injected electrons recombine with the oxidized sensitizer dyes (*Recombination*; step 5). This recombination process competes with the regeneration of the oxidized sensitizer dyes by the redox mediator molecules. Finally, the conducting electrons can react with the redox mediator molecules in the solution during transport, before reaching the back contact electrode (*Leak reaction*; step 6). The rate and efficiency of these processes depend strongly on the nature of the sensitizer dye and semiconductor film. Understanding the relationship between the molecule structure of the sensitizing dye and photovoltaic performance is one of the most important tasks for developing a high performance solar cell (Hara et al. 2002; Shahroosvand et al. 2015a).

As seen in Fig. 19.3, this field is growing fast and there is an exponential growth in publications dealing with the role of complexes in DSSCs. This chapter concentrates on the employment of different metal complexes based on 1,10-phenanthroline for engineering of new DSSCs.

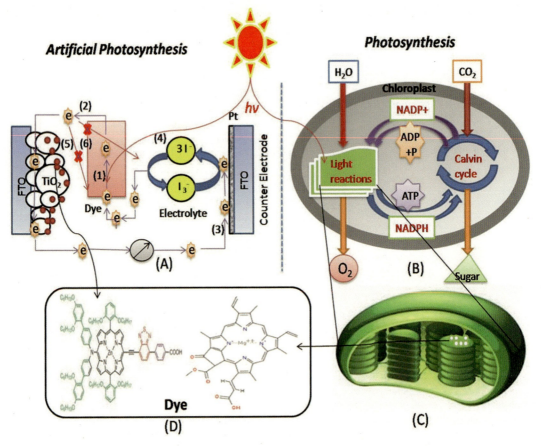

Fig. 19.2 (**a**) comparative schematic of the work principal between artificial photosynthesis (DSSC), and (**b**) natural photosynthesis. Part (**a**): photoexcitation; step 1, electron injection; step 2, CE catalysis; step 3, regeneration; step 4, recombination; step 5, leak reaction; step 6. FTO: fluorine tin oxide. Part (**b**): the occurred reactions in a natural photosynthesis. Part (**c**): chloroplast as light reactor. Part (**d**): the molecular structure of porphyrin which works in a chlorophyll (*right*) and in a DSSC (*left*)

In following, two important components in a DSSC including dye sensitizer and redox electrolyte will be detailed.

19.1.2 Dyes in DSSCs

DSSCs have attracted increasing interest in recent years due to the economical and ecological device fabrication and practically high power conversion efficiency (Li and Diau 2013; Hagfeldt et al. 2010). It is well-known that dyes play an important role in DSSCs, in which they catch photons and inject electrons into the conduction band of the TiO_2 semiconductor, followed by regeneration of the sensitizer by a reversible redox mediator in the electrolyte. It has been realized that the light harvesting efficiency is the key process in the DSSC device. Hence, many efforts have been devoted to increase the absorptivity and stability of the dyes (Robertson 2006). Developing new dye molecules is one approach to extend the light response to cover a broad solar spectrum. Ru polypyridine dyes reveal significant performance owing to their strong absorption in the visible region and long-lived charge-separated excited states (Ardo and Meyer 2009). Two well-known Ru-based dyes are coded **N3** ([Ru(H_2dcbpy)$_2$(NCS)$_2$]; dcbpy =4,4′-dicarboxylato-2,2′-bipyridine) (Nazeeruddin et al. 1993), **N719**

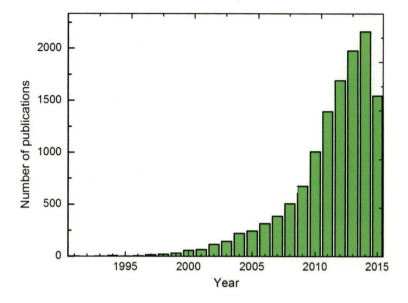

Fig. 19.3 Number of publications annually dealing with DSSCs showing the near exponential growth in interest in recent years (Data collected from Web of Science August 10, 2015)

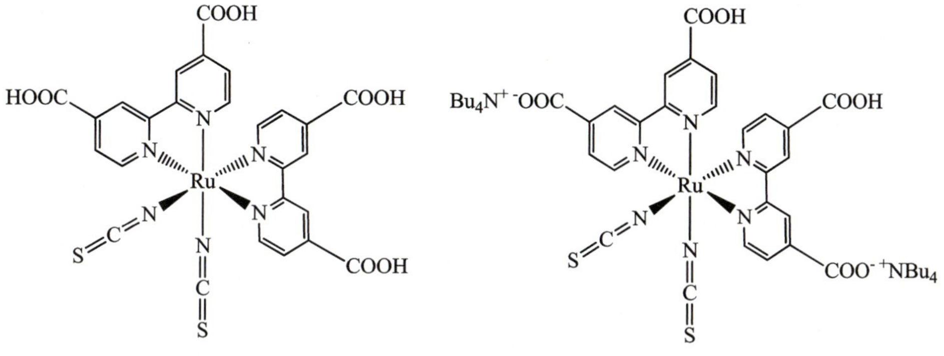

Fig. 19.4 Chemical structure of Ru(II) sensitizers N3 and N719

([Ru(Hdcbpy)$_2$(NCS)$_2$](Bu$_4$N)$_2$) (Nazeeruddin et al. 1999), yielding up to 10% (Fig. 19.4).

Most of modifications in the design of a sensitizer dye for DSSC contain the suitable choice of an ancillary ligand through the tuning the ground and excited-state properties for attaining better light harvesting. In this regard, several approaches have been successfully endeavored to improve the light harvesting abilities of ancillary ligands, which include extending π-conjugation through incorporation of aromatic rings, introducing electron donating/accepting groups in the ancillary ligand, and the use of aliphatic long chains, *etc*. (Aranyos et al. 2003; Wang et al. 2004, 2005; Chen et al. 2006; Kuang et al. 2006; Gao et al. 2008, 2009; Matar et al. 2008; Shahroosvand et al. 2016).

19.1.3 Electrolytes in DSSCs

The electrolyte is one of the most crucial components in DSSCs (Gonçalves et al. 2008; Shahroosvand et al. 2015); it is responsible for the inner charge carrier transport between electrodes and continuously regenerates the dye and itself during DSSC operation. The electrolyte has great influence on the light-to-electric conversion efficiency and long-term stability of the devices. All of the three parameters, the open-circuit photovoltage (V_{oc}), the short-circuit photocurrent density (J_{sc}), and the fill factor (FF) in DSSCs are significantly affected by the electrolyte in DSSCs and by the interaction of the electrolyte with the electrode interfaces (Grätzel 2009). The V_{oc} is the difference between the Fermi level of TiO_2 ($E_{F,n}$) and the redox couple potential of the electrolyte ($E_{F,redox}$), i.e. $V_{oc} = (E_{F,n} - E_{F,redox})/e$, where e is the elementary charge, and $E_{F,n}$ is associated with the titania conduction band edge E_c and free electron density n in the semiconductor film. While the light-harvesting ability of the dye molecules is a determinant factor on the final J_{sc}. Additionally, the device quality is reflected by FF, which is affected by the cell's internal resistance. The internal resistance is composed of resistance of electrolytes, electrodes, and different interfaces such as FTO/TiO_2, TiO_2/electrolyte, TiO_2/dye, dye/electrolyte, and electrolyte/counter electrode.

Several features are essential for the electrolytes used in DSSCs (Wu et al. 2008; Ardo and Meyer 2009; Nogueira et al. 2004): (1) The electrolytes must be capable to transport the charge carriers between photoanode and counter electrode. After the dye injects electrons into the conduction band of TiO_2, the oxidized dye must be quickly reduced to its ground state. (2) The electrolytes must guarantee fast diffusion of charge carriers and produce good interfacial contact with the mesoporous semiconductor layer and the counter electrode. (3) The electrolytes must have long-term stabilities, including thermal, chemical, electrochemical, optical, and interfacial stability, and they also must not cause desorption and degradation of the sensitized dye. (4) The electrolytes should not display a significant absorption in the range of visible light.

The I^-/I_3^- redox couple is the most commonly used redox pair. However, it suffers from severe limitations including inherent corrosive nature, distinct absorption features in the visible region, and a prominent charge recombination arising from I_3^- ion pair formation at the electrode surface (Tian et al. 2011). Overcoming these constraints could lead to further improvements in DSSC performance and ease of fabrication.

Before 2010, studies on the electrolyte predominantly focused on the traditional I^-/I_3^- electrolyte. The progress of new redox mediators falls far behind that of the sensitizing dyes and other materials for the DSSC components. In 2001, for the first time, Nusbaumer et al. used $[Co(dbbip)_2]^{2+}$ (dbbip = 2,6-bis(1'-butylbenzimidazol-2'-yl)pyridine), as redox shuttle in DSSCs, for explaining the effects of cobalt electrolyte on recombination process (Nusbaumer et al. 2001). With this new one-electron redox shuttle, DSSCs sensitized by the dye **Z316** have achieved η value of 2.2% under AM1.5 irradiation (100 mW cm^{-2}). The results showed that in the absence of Co^{II}/Co^{III} redox couple, the absorption signal is decreasing as a result of increasing of dynamic recombination process between oxidized dye and conduction band of semiconductor (Nusbaumer et al. 2001). The advantages of this kind of one-electron outersphere transition metal complex are that, electrolytes are nonvolatile, noncorrosive, light-colored, and tunable potential (0.3–0.9 V) through modification of the ligands. It is very noteworthy that in 2011, Yella et al. improved the efficiency of DSSC to 12.3% by using a $[Co(bpy)_3]^{2+/3+}$-based redox electrolyte (bpy = bipyridine) in conjunction with a donor-π-bridge-acceptor zinc porphyrin (**YD2-o-C8**) as a sensitizer (Yella et al. 2011). However, by modifying the **YD2-o-C8** porphyrin core with the bulky bis(2',4'-bis(hexyloxy)-[1,1'-biphenyl]-4-yl)amine donor (**SM371**) and with incorporation of the proquinoidal benzothiadiazole (BTD) unit into **SM371** afforded the dye **SM315** (Mathew et al. 2014). Figure 19.5 illustrates the J–V curve for the two devices measured under AM 1.5G illumination (1000 W m^{-2} at 298 K). Fabrication of DSSCs utilizing the $[Co(bpy)_3]^{2+/3+}$ redox couple and **SM315** presented a high V_{oc} of 0.91 V, J_{sc} of

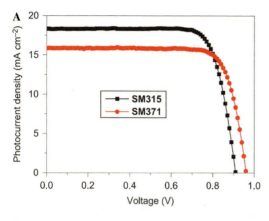 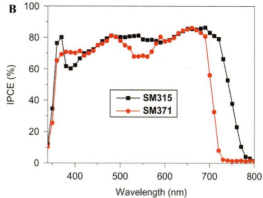

Fig. 19.5 Photocurrent density-voltage characteristic curves (**a**) and incident photon to current efficiency (IPCE) as a function of wavelength (**b**) of various DSSC devices based on SM371 and SM315 (Reprinted with permission from Mathew et al. 2014. Copyright 2014 Nature Publishing Group)

18.1 mA cm^{-2}, FF of 0.78, and efficiency of 13%. At present, Co-based mediators are the most efficient redox couples for DSSCs. By tuning the redox potentials of the Co complex systems and the energy level of the dyes, it is possible to get a DSSC with a higher conversion efficiency in the near future.

Apart from cobalt-based redox shuttles, other transition metal complexes and clusters, such as Ni^{III}/Ni^{IV}, Cu^{I}/Cu^{II}, and ferrocene/ferrocenium (Fc/Fc^{+}), have also been investigated as redox couples (Bai et al. 2011a; Gregg et al. 2001; Hamann et al. 2008; Feldt et al. 2010; Daeneke et al. 2011). In this book chapter, we will investigate transition metal complexes based on 1,10-phenanthroline as electrolytes in DSSCs.

19.2 1,10-Phenanthroline-Based Electrolytes in DSSCs

The first application of 1,10-phenanthroline complex as metal complex electrolyte was reported by Wang et al. who employed different organic sensitizers based on tris(1,10-phenanthroline) cobalt(II/III) ([Co(phen)$_3$]$^{2+/3+}$) redox shuttle in combination with relatively thin TiO$_2$ films of about 6–7 μm, due to the limited electron diffusion length found for the cobalt complexes. They attained the η value of 7.1%, 7.7%, 8.0% and 8.4% with **C230** (Liu et al. 2011), **C233** (Liu et al. 2011), **T3** (Zhang et al. 2011) and **C245** (Xu et al. 2011) dyes, respectively. Even higher efficiencies were achieved using the **C218** dye (Zhou et al. 2011) (9.3%) (Xu et al. 2011) as well as the **C229** dye (9.4%), showing a significant redshift at the absorption edge (Bai et al. 2011b). The molecular structures of these organic photosensitizers are shown in Fig. 19.6.

Compared to the uniped photosensitizers (**C213**, **C218** and **C230**), the corresponding biped **C231–C233** dyes display hyperchromic light absorptions in the visible region upon anchoring on titania nanocrystals. Unfortunately, with respect to the uniped dyes, the utilization of the biped congeners has given rise to noticeable reductions of V$_{oc}$, due to a combined effect of a dye correlated negative shift of titania conduction band edge and a kinetic acceleration of the charge recombination at the titania/electrolyte interface. It is also found that cyclopentadithiophene dyes (**C244–C246**) generally reveal highly efficient dye regeneration, despite the noticeably different ground-state redox potentials of dye molecules. Additionally, impedance study of these cells presents that using dihexyloxy-substituted triphenylamine in a cyclopentadithiophene dye desirably reduces down titania/electrolyte interface charge recombination kinetics, contributing to a high V$_{oc}$.

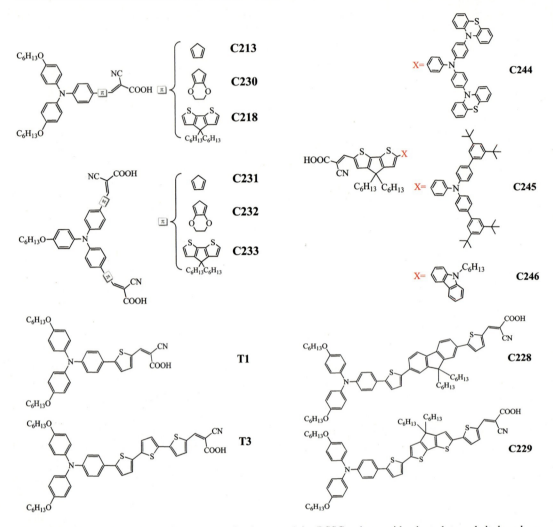

Fig. 19.6 Chemical structure of some organic dyes used in DSSCs along with electrolyte cobalt based on phenanthroline

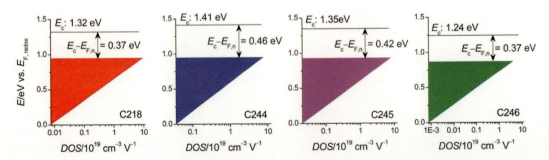

Fig. 19.7 Energy diagrams of dye-coated titania films in contact with cobalt electrolyte. The upper limit of the shaded area represents the $E_{F,n}$ position of a DSSC at the open circuit and simulated AM1.5G conditions. DOS: density of state (Reprinted with permission from Xu et al. 2011. Copyright 2011 Royal Society of Chemistry)

Figure 19.7 shows the energetic diagram of a dye-coated titania film sinking in cobalt electrolyte. It can be obviously realized that the titania conduction band edge E_c level decreases stepwise going **C244** to **C246**, which probably provides a thermodynamic explanation for the gradually decelerated titania/electrolyte interface charge recombination kinetics. On the other hand, it is noteworthy that the E_c can be regarded as the highest energy level that the electron Fermi level may reach, with the result that the energy loss E_c-$E_{F,n}$ decreases from **C244** to **C246**, which agrees well with the increasing free electron density n values.

However, the optimization of DSSCs sensitized with **M14** (Fig. 19.8) in combination with a $[Co(phen)_3]^{2+/3+}$ redox electrolyte yields a DSSC with a η value of 7.2% under AM 1.5 irradiation (100 mW cm^{-2}) (Zong et al. 2012b). The hexapropyltruxene unit retards the rate of interfacial back electron transfer from the conduction band of TiO_2 film to the oxidized electrolyte. This process enables ability of high photovoltages approaching to 0.9 V. The

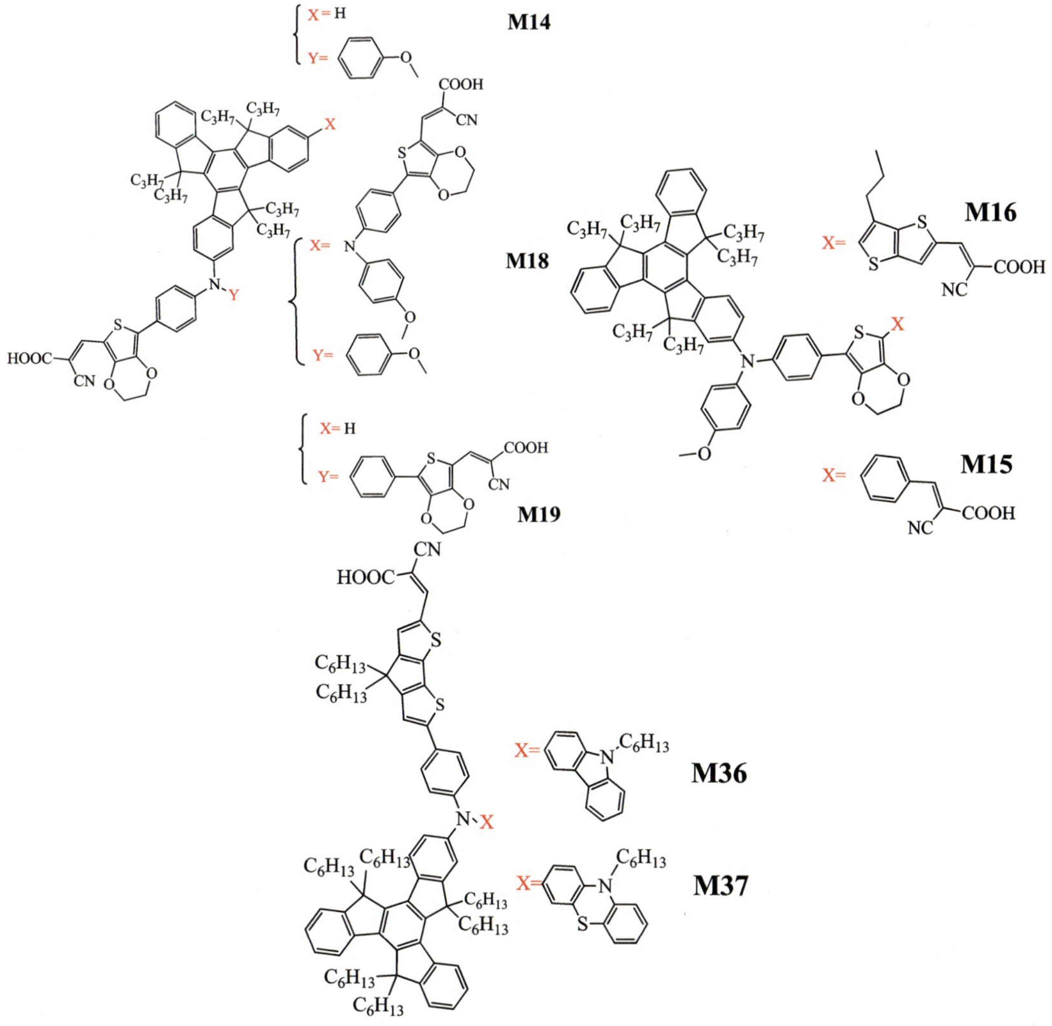

Fig. 19.8 Chemical structure of a series of organic dyes used in DSSCs along with electrolyte cobalt based on phenanthroline

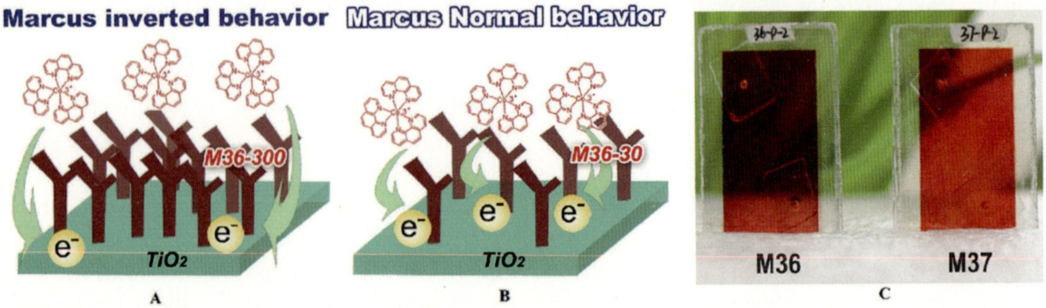

Fig. 19.9 Pictorial illustration for dye layer on titania via (**a**) M36-300, (**b**) M36-30 (the devices prepared from 300 to 30 μM dye solution, i.e. M36-300 and M36-30) and (**c**) a digital photograph of dye-grafted titania films in contact with the Co-phen electrolyte (Reprinted with permission from Gao et al. 2015. Copyright 2015 Elsevier)

measurement of photocurrent transients displayed that the mass transport limitation of the cobalt redox shuttle has been largely removed by using thin TiO$_2$ films. Also, the better efficiency was obtained through the truxene–based organic dye **M16**–sensitized device and [Co(phen)$_3$]$^{2+/3+}$ which displays an efficiency of 7.6% at standard conditions (Zong et al. 2012a).

Additionally, the studies of recombination kinetics of cobalt(III) complexes at titania/dye interface with D-π-A organic dyes, **M36** and **M37**, showed that for **M36** sensitized DSSCs, a Marcus inverted region can be reached for the charge recombination kinetics behavior of cobalt (III) species (Fig. 19.9a, b) (Gao et al. 2015). Marcus theory applies to describe the rate of electron transfer from conduction band to oxidized redox species or oxidized dyes. Benefiting from a Marcus inverted region behavior, the **M36** dye displayed a good compatibility with the [Co(phen)$_3$]$^{2+/3+}$ redox couples. In this region, contrary to all intuition, a further increase in the exergonicity causes a decrease in the reaction rate. Electrochemical impedance spectroscopy (EIS) and intensity modulated photovoltage spectroscopy (IMVS) measurements discovered that the charge recombination kinetics behavior of Co(III) species depends not only on the driving force for recombination, but also the retarding charge recombination ability of the dye layer. A digital photograph of the devices is shown in Fig. 19.9c, where the dye-grafted mesoporous titania films (3-mm-thick) immersed in the Co-phen electrolyte for DSSC fabrication.

Dithieno[3,2-b:2′,3′-d]pyrrole (DTP)-based triphenylamine sensitizers (**XS54–XS57**) (Fig. 19.10) with different hexyloxyphenyl (HOP) substituents were synthesized and applied in DSSCs along with [Co(phen)$_3$]$^{2+/3+}$ redox shuttle (Wang et al. 2013). **XS54** featuring the 4-HOP (hexyloxyphenyl)-DTP spacer, yields a η value of 8.14% in combination with the cobalt electrolyte. Also, electron lifetime studies indicated that charge recombination rates are suggested to be determined by the direction of alkyl chains rather than the number of the alkyl chains. Figure 19.11 indicates the photocurrent density–voltage (J–V) curves and incident photon to current efficiency (IPCE) of the devices employing a [Co(phen)$_3$]$^{2+/3+}$ redox shuttle under AM 1.5 irradiation (100 mW cm^{-2}). The IPCE is the number of collected electrons under short circuit conditions per number of incident photons at a given excitation wavelength. The J$_{sc}$ improvement of **XS54**, mainly benefiting from its broad and high IPCE action area (Fig. 19.11), can be attributed to its stronger molar extinction coefficient (ε) and higher amounts of the dyes absorbed. Unmistakably, the extraordinarily improved J$_{sc}$ of **XS54** significantly contributes to its η values.

Additionally, the photovoltaic performances of [Co(phen)$_3$]$^{2+/3+}$ redox couple were superior to those of I$^-$/I$_3^-$ redox couple for TiO$_2$ film DSSCs sensitized by **XS51–XS52**, representing

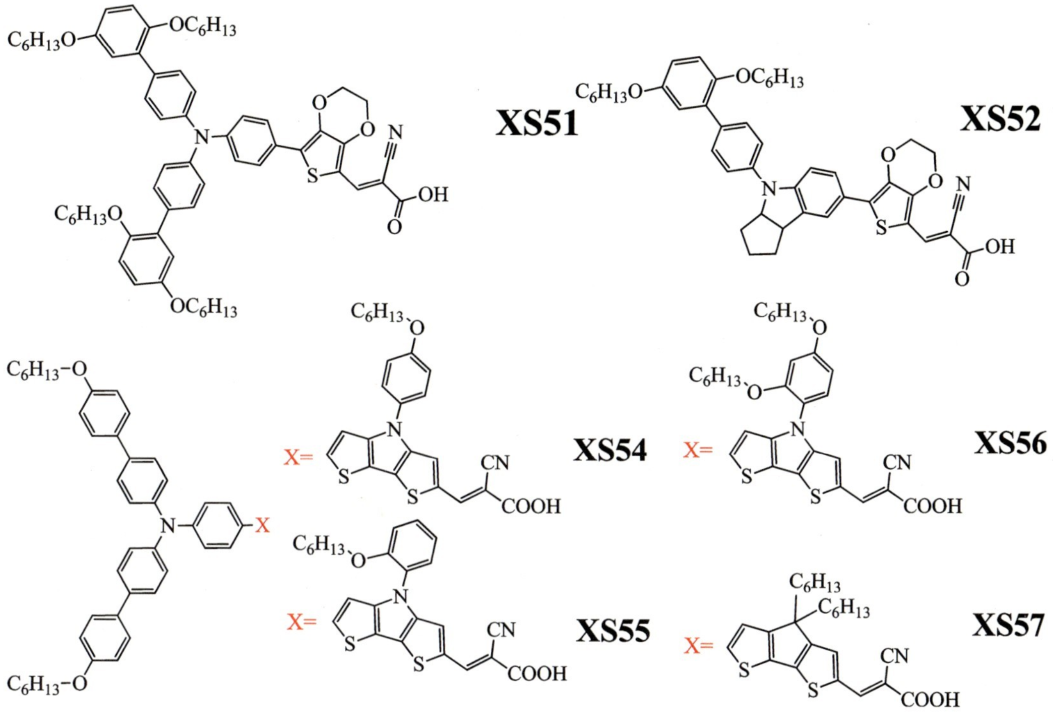

Fig. 19.10 Chemical structure of a series of organic dyes used in DSSCs along with electrolyte cobalt based on phenanthroline

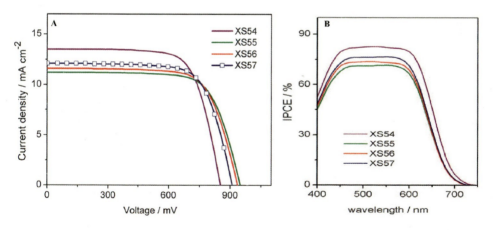

Fig. 19.11 J–V characteristics of DSSCs employing the cobalt electrolyte (**a**), IPCEs action spectra for DSSCs employing the cobalt electrolyte (**b**) (Reprinted with permission from Wang et al. 2013. Copyright 2013 Royal Society of Chemistry)

that rational design of sterically bulky organic dyes is needed for further progress of high-efficiency iodine-free devices (Zhang et al. 2015). The indoline dye **XS52** showed a red-shifted absorption and higher ε than the triarylamine dye. Due to the strong electron-donating capability of indoline unit, **XS52** obtained high J_{sc}.

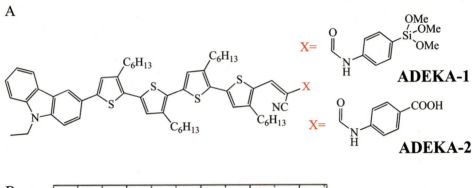

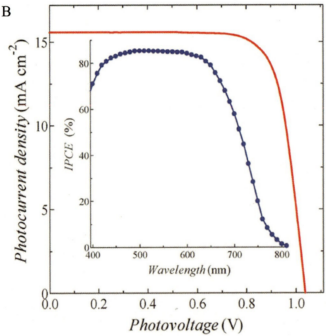

Fig. 19.12 Chemical structure of dyes ADEKA-1 and ADEKA-2 used in DSSCs along with electrolyte cobalt based on phenanthroline (**a**). Typical J–V properties of the ADEKA-1-sensitized solar cell and the IPCE spectrum of the cell (**b**) (Reprinted with permission from Kakiage et al. 2014). Copyright 2014 Royal Society of Chemistry)

DSSCs constructed via a novel metal-free alkoxysilyl carbazole as a sensitizing dye and a $Co^{2+/3+}$-complex redox electrolyte showed a η value of over 12% with V_{oc} higher than 1 V by applying a hierarchical multi-capping treatment to the photoanode (Kakiage et al. 2014). This dye was used with two Co-based mediators, $[Co(Cl-phen)_3]^{2+/3+}$ and $[Co(bpy)_3]^{2+/3+}$. The higher V_{oc} value was obtained via the former one, which has a redox potential of 0.72 V vs. NHE in the **ADEKA-1**-sensitized cell. The J–V curve and IPCE spectrum of the cell are shown in Fig. 19.12.

A mixture of $[Co(phen)_3]^{2+/3+}$ and $[Co(EtPy)_2]^{2+/3+}$ complexes, where EtPy is a terpyridine ligand bearing 3,4-ethylenedioxythiophene (EDOT) substituent, was utilized as redox mediator in **D35** (Fig. 19.13)-sensitized solar cells (Koussi-Daoud et al. 2015). $[Co(EtPy)_2]^{2+}$ complex acts as co-mediator in the presence of $[Co(phen)_3]^{2+}$. The low solubility of this complex prevents its

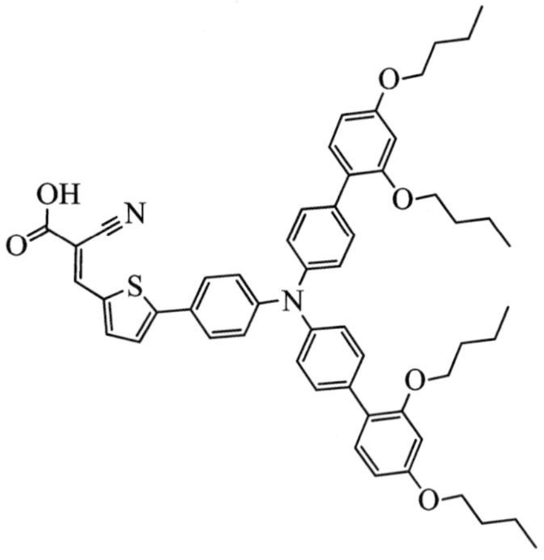

Fig. 19.13 Chemical structure of dye D35

was employed in combination with a bilayer titania thin-film stained with a high-absorption-coefficient organic dye **C218** which generates an impressive power conversion efficiency of 7.0% (Bai et al. 2011a). A broad ~ 87% IPCE plateau in the spectral coverage from 460 to 580 nm was observed for a typical cell with the iodine control electrolyte (Fig. 19.14). However, the copper redox shuttle displayed very low electron transfer rates on several noble metals, carbon black and conducting oxides, resulting in the poor fill factor.

individual applying. The addition of $[Co(EtPy)_2]^{2+}$ to $[Co(phen)_3]^{2+}$ provided an electron cascade which leads to the enhanced cell efficiency. A synergy between the EDOT-functionalized cobalt complex and the PEDOT counter-electrode is responsible for this effect which favors electron transfer and reduces recombination.

For the first time, in 2005, series of blue copper model complexes ($[Cu(SP)(mmt)]^{-/0}$, $[Cu(dmp)_2]^{+/2+}$ and $[Cu(phen)_2]^{+/2+}$) (Table 19.1) were examined for their effectiveness as redox shuttles in DSSCs by Hattori et al. (2005). They afforded a maximum IPCE of 40%, which is thought to stem from a slow regeneration of the **N19** ruthenium photosensitizer owing to the large reorganization energies of copper(I/II) complexes (Ardo and Meyer 2009). The resulting electron self-exchange rate constant decreases in the order of: $[Cu(dmp)_2]^{+/2+} > [Cu(SP)(mmt)]^{-/0} > [Cu(phen)_2]^{+/2+}$. It is in agreement with the order of the smaller structural change between the copper(II) and copper(I) complexes due to the distorted tetragonal geometry. Under the weak solar light irradiation of 20 mW cm^{-2} intensity, the maximum η value was obtained as 2.2% for DSSC using $[Cu(dmp)_2]^{+/2+}$. Whereas, a higher V_{oc} of the cell was attained as compared to that of the conventional I^-/I_3^- couple (Hattori et al. 2005). The $[Cu(dmp)_2]^{+/2+}$ redox shuttle

19.3 Concluding Remarks

Research on dye-sensitized solar cells (DSSC) is progressing at a rapid race. Understanding the structure of various DSSC sensitizers, electrolyte and the search for new molecular structure are critical factors for the development of improved DSSCs. Polypyridyl complexes have gained increasing interest due to their more favorable light harvesting abilities, long-term thermal and chemical stability, compared to other conventional sensitizers and electrolytes for feasible large-scale commercialization of DSSCs. Currently, laboratory cells reach 13% top efficiency by using the combination of a π-extended porphyrin dye, a 1,10-phenanthroline cobalt redox electrolyte, a TiO$_2$ electrode with 20 nm particle size and an additional scattering layer. However, the conventional I^-/I_3^- is highly corrosive, volatile and photoreactive, interacting with common metallic components and materials. Alternative to liquid electrolytes free of the I^-/I_3^- redox couple have been a long term goal in this field, and the ultimate solutions would be all solid state cells, given the inevitable problems of any liquid electrolyte, such as leakage, heavy weight and complex chemistry. Therefore, the sensitizers have to have very high molar extinction coefficient and in this respect optimization of ligands is essential.

Table 19.1 Photovoltaic parameters for DSSCs based on Co-phenanthroline mediators in combination with different dyes

$[Co(phen)_3]^{2+/3+}$, phen = 1,10-phenanthroline

$[Cu(dmp)_2]^{+/2+}$

$[Co(Cl\text{-}phen)_3]^{3+/2+}$, Cl-phen = 5-chloro-1,10-phenanthroline

$[Cu(phen)_2]^{+/2+}$

Electrolyte	Sensitizer	V_{oc} (V)	J_{sc} (mA cm^{-2})	FF	IPCE (%)	η (%)	References
$[Co(phen)_3]^{2+/3+}$	C213	0.94	9.51	0.73	–	6.5	Liu et al. (2011)
	C230	0.90	10.72	0.73	–	7.1	
	C218	0.86	13.26	0.75	–	8.6	
	C231	0.77	10.07	0.74	–	5.8	
	C232	0.79	10.68	0.75	–	6.3	
	C233	0.81	12.24	0.77	–	7.7	
$[Co(phen)_3]^{2+/3+}$	T1	0.82	7.99	0.76	–	5.0	Zhang et al. (2011)
I^-/I_3^-		0.8	7.78	0.75	–	4.6	
$[Co(phen)_3]^{2+/3+}$	T2	0.84	12.98	0.74	–	8.0	
I^-/I_3^-		0.75	12.86	0.71	–	6.9	
$[Co(phen)_3]^{2+/3+}$	C218	0.95	13.30	0.74	–	9.3	Xu et al. (2011)
	C244	0.95	10.54	0.77	–	7.7	
	C245	0.93	12.05	0.75	–	8.4	
	C246	0.87	11.41	0.77	–	7.6	
$[Co(phen)_3]^{2+/3+}$	C228	0.83	7.60	0.74	–	4.7	Bai et al. (2011b)
I^-/I_3^-		0.76	7.78	0.74	–	4.4	
$[Co(phen)_3]^{2+/3+}$	C229	0.85	15.31	0.73	–	9.4	
I^-/I_3^-		0.68	15.20	0.65	–	6.7	
$[Co(phen)_3]^{2+/3+}$	M14	0.83	12.0	0.72	85	7.2	Zong et al. (2012b)
I^-/I_3^-		0.78	11.2	0.69	–	6.0	
$[Co(phen)_3]^{2+/3+}$	M18	0.87	9.0	0.71	–	5.5	
I^-/I_3^-		0.77	9.3	0.68	–	4.9	
$[Co(phen)_3]^{2+/3+}$	M19	0.87	11.2	0.71	–	6.9	
I^-/I_3^-		0.74	11.4	0.68	–	5.7	
$[Co(phen)_3]^{2+/3+}$	M16	0.90	11.9	0.71	–	7.6	Zong et al. (2012a)
I^-/I_3^-		0.68	12.2	0.68	–	5.6	
$[Co(phen)_3]^{2+/3+}$	M15	0.87	9.6	0.70	–	5.8	
I^-/I_3^-		0.76	11.8	0.68	–	6.1	
$[Co(phen)_3]^{2+/3+}$	M36	0.93	15.01	0.69	–	9.58	Gao et al. (2015)
	M37	0.68	11.03	0.83	–	6.19	
$[Co(phen)_3]^{2+/3+}$	XS54	0.85	13.5	0.71	–	8.14	Wang et al. (2013)
	XS55	0.95	11.2	0.70	–	7.45	
	XS56	0.94	11.6	0.69	–	7.48	
	XS57	0.91	12.1	0.70	–	7.68	

(continued)

Table 19.1 (continued)

[Co(phen)$_3$]$^{2+/3+}$	XS51	862	10.5	0.65	78.7	5.88	Zhang et al. (2015)
I$^-$/I$_3^-$		0.75	11.6	0.67	80.6	5.68	
[Co(phen)$_3$]$^{2+/3+}$	XS52	830	11.5	0.68	–	6.58	
I$^-$/I$_3^-$		0.74	12.0	0.69	–	6.13	
[Co(Cl-phen)$_3$]$^{3+/2+}$	ADEKA-1	1.04	15.6	0.77	85	12.5	Kakiage et al. (2014)
	ADEKA-2	–	–	–	–	–	
[Co(phen)$_3$]$^{2+/3+}$/[Co(Etpy)$_2$]$^{2+/3+}$ (EtPy =4′-(2,3-dihydrothieno[3,4-b][1,4]dioxin-5-yl)-2,2′:6′,2″-terpyridine)	D35	0.92	8.4	0.67	–	5.1	(Koussi-Daoud et al. (2015)
[Cu(dmp)$_2$]$^{+/2+}$	N719	0.79	3.2	0.55	–	1.4	Hattori et al. (2005)
[Cu(phen)$_2$]$^{+/2+}$	N719	0.57	0.48	0.43	–	0.12	Hattori et al. (2005)

For comparison the corresponding parameters for DSSCs containing the standard I$^-$/I$_3^-$ redox couple are also reported, when available

V_{oc} open circuit potential, J_{sc} short-circuit current, FF fill factor, $IPCE$ incident photon to current conversion efficiency, η overall power conversion efficiency

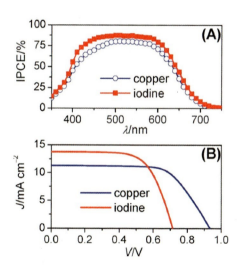

Fig. 19.14 (a) Photocurrent action spectra and (b) J–V characteristics under irradiation of 100 mW cm^{-2} simulated AM1.5G sunlight (Reprinted with permission from Bai et al. 2011a. Copyright 2014 Royal Society of Chemistry)

Acknowledgments The authors thank Dr. M. M. Najafpour and Dr. A. Haddy for their suggestions and critical comments. Authors also acknowledge the University of Zanjan for financial support.

References

Aranyos V, Hjelm J, Hagfeldt A, Grennberg H (2003) Tuning the properties of ruthenium bipyridine dyes for solar cells by substitution on the ligands-characterisation of bis [4,4′-di (2-(3-methoxyphenyl) ethenyl)-2,2′-bipyridine][4,4′-dicarboxy-2,2′-bipyridine] ruthenium(II) dihexafluorophosphate. Dalton Trans:1280–1283

Ardo S, Meyer GJ (2009) Photodriven heterogeneous charge transfer with transition-metal compounds anchored to TiO$_2$ semiconductor surfaces. Chem Soc Rev 38:115–164

Argazzi R, Iha NYM, Zabri H, Odobel F, Bignozzi CA (2004) Design of molecular dyes for application in photoelectrochemical and electrochromic devices based on nanocrystalline metal oxide semiconductors. Coord Chem Rev 248:1299–1316

Bai Y, Yu Q, Cai N, Wang Y, Zhang M, Wang P (2011a) High-efficiency organic dye-sensitized mesoscopic solar cells with a copper redox shuttle. Chem Commun 47:4376–4378

Bai Y, Zhang J, Zhou D, Wang Y, Zhang M, Wang P (2011b) Engineering organic sensitizers for iodine-free dye-sensitized solar cells: red-shifted current response concomitant with attenuated charge recombination. J Am Chem Soc 133:11442–11445

Bomben PG, Robson KC, Koivisto BD, Berlinguette CP (2012) Cyclometalated ruthenium chromophores for the dye-sensitized solar cell. Coord Chem Rev 256:1438–1450

Chen C-Y, Wu S-J, Wu C-G, Chen J-G, Ho K-C (2006) A Ruthenium Complex with Superhigh Light-Harvesting Capacity for Dye-Sensitized Solar Cells. Angew Chem Int Ed 45:5822–5825

Clifford JN, Martínez-Ferrero E, Viterisi A, Palomares E (2011) Sensitizer molecular structure-device efficiency relationship in dye sensitized solar cells. Chem Soc Rev 40:1635–1646

Colombo A, Dragonetti C, Valore A, Coluccini C, Manfredi N, Abbotto A (2014) Thiocyanate-free ruthenium(II) 2,2′-bipyridyl complexes for dye-sensitized solar cells. Polyhedron 82:50–56

Daeneke T, Kwon T-H, Holmes AB, Duffy NW, Bach U, Spiccia L (2011) High-efficiency dye-sensitized solar cells with ferrocene-based electrolytes. Nature Chem 3:211–215

Feldt SM, Cappel UB, Johansson EM, Boschloo G, Hagfeldt A (2010) Characterization of surface passivation by poly (methylsiloxane) for dye-sensitized solar cells employing the ferrocene redox couple. J Phys Chem C 114:10551–10558

Gao F, Wang Y, Shi D, Zhang J, Wang M, Jing X, Humphry-Baker R, Wang P, Zakeeruddin SM, Grätzel M (2008) Enhance the Optical Absorptivity of Nanocrystalline TiO_2 Film with High Molar Extinction Coefficient Ruthenium Sensitizers for High Performance Dye-Sensitized Solar Cells. J Am Chem Soc 130:10720–10728

Gao F, Cheng Y, Yu Q, Liu S, Shi D, Li Y, Wang P (2009) Conjugation of Selenophene with Bipyridine for a High Molar Extinction Coefficient Sensitizer in Dye-Sensitized Solar Cells. Inorg Chem 48:2664–2669

Gao W, Liang M, Tan Y, Wang M, Sun Z, Xue S (2015) New triarylamine sensitizers for high efficiency dye-sensitized solar cells: Recombination kinetics of cobalt(III) complexes at titania/dye interface. J Power Sources 283:260–269

Gonçalves LM, de Zea Bermudez V, Ribeiro HA, Mendes AM (2008) Dye-sensitized solar cells: A safe bet for the future. Energy Environ Sci 1:655–667

Grätzel M (2009) Recent advances in sensitized mesoscopic solar cells. Acc Chem Res 42:1788–1798

Gregg BA, Pichot F, Ferrere S, Fields CL (2001) Interfacial recombination processes in dye-sensitized solar cells and methods to passivate the interfaces. J Phys Chem B 105:1422–1429

Hagfeldt A, Boschloo G, Sun L, Kloo L, Pettersson H (2010) Dye-sensitized solar cells. Chem Rev 110:6595–6663

Hamann TW, Farha OK, Hupp JT (2008) Outer-sphere redox couples as shuttles in dye-sensitized solar cells. Performance enhancement based on photoelectrode modification via atomic layer deposition. J Phys Chem C 112:19756–19764

Hara K, Horiuchi H, Katoh R, Singh LP, Sugihara H, Sayama K, Murata S, Tachiya M, Arakawa H (2002) Effect of the ligand structure on the efficiency of electron injection from excited Ru-phenanthroline complexes to nanocrystalline TiO_2 films. J Phys Chem B 106:374–379

Hattori S, Wada Y, Yanagida S, Fukuzumi S (2005) Blue copper model complexes with distorted tetragonal geometry acting as effective electron-transfer mediators in dye-sensitized solar cells. J Am Chem Soc 127:9648–9654

IPCC A (2007) Intergovernmental panel on climate change. IPCC Secretariat Geneva

Kakiage K, Aoyama Y, Yano T, Otsuka T, Kyomen T, Unno M, Hanaya M (2014) An achievement of over 12 percent efficiency in an organic dye-sensitized solar cell. Chem Commun 50:6379–6381

Katoh R, Furube A, Yoshihara T, Hara K, Fujihashi G, Takano S, Murata S, Arakawa H, Tachiya M (2004) Efficiencies of electron injection from excited N3 dye into nanocrystalline semiconductor (ZrO_2, TiO_2, ZnO, Nb_2O_5, SnO_2, In_2O_3) films. J Phys Chem B 108:4818–4822

Koussi-Daoud S, Schaming D, Fillaud L, Trippé-Allard G, Lafolet F, Polanski E, Nonomura K, Vlachopoulos N, Hagfeldt A, Lacroix J-C (2015) 3,4-Ethylenedioxythiophene-based cobalt complex: an efficient co-mediator in dye-sensitized solar cells with poly (3,4-ethylenedioxythiophene) counter-electrode. Electrochim Acta 179:237–240

Kuang D, Klein C, Snaith HJ, Moser J-E, Humphry-Baker R, Comte P, Zakeeruddin SM, Grätzel M (2006) Ion Coordinating Sensitizer for High Efficiency Mesoscopic Dye-Sensitized Solar Cells: Influence of Lithium Ions on the Photovoltaic Performance of Liquid and Solid-State Cells. Nano Lett 6:769–773

Lewis NS, Nocera DG (2006) Powering the planet: Chemical challenges in solar energy utilization. Proc Natl Acad Sci 103:15729–15735

Lewis N, Crabtree G, Nozik A (2005) Basic energy sciences workshop on solar energy utilization. Office of Science, US Department of Energy, Washington, DC.

Li L-L, Diau EW-G (2013) Porphyrin-sensitized solar cells. Chem Soc Rev 42:291–304

Liu J, Zhang J, Xu M, Zhou D, Jing X, Wang P (2011) Mesoscopic titania solar cells with the tris (1, -10-phenanthroline) cobalt redox shuttle: uniped versus biped organic dyes. Energy Environ Sci 4:3021–3029

Matar F, Ghaddar TH, Walley K, DosSantos T, Durrant JR, O'Regan B (2008) A new ruthenium polypyridyl dye, TG6, whose performance in dye-sensitized solar cells is surprisingly close to that of N719, the 'dye to beat' for 17 years. J Mater Chem 18:4246–4253

Mathew S, Yella A, Gao P, Humphry-Baker R, Curchod BF, Ashari-Astani N, Tavernelli I, Rothlisberger U, Nazeeruddin MK, Grätzel M (2014) Dye-sensitized solar cells with 13% efficiency achieved through the molecular engineering of porphyrin sensitizers. Nature Chem 6:242–247

Nazeeruddin MK, Kay A, Rodicio I, Humphry-Baker R, Müller E, Liska P, Vlachopoulos N, Grätzel M (1993)

Conversion of light to electricity by cis-X_2bis (2,2'-bipyridyl-4,4'-dicarboxylate) ruthenium(II) charge-transfer sensitizers (X = Cl^-, Br^-, I^-, CN^-, and SCN^-) on nanocrystalline titanium dioxide electrodes. J Am Chem Soc 115:6382–6390

Nazeeruddin MK, Zakeeruddin S, Humphry-Baker R, Jirousek M, Liska P, Vlachopoulos N, Shklover V, Fischer C-H, Grätzel M (1999) Acid-base equilibria of (2,2'-bipyridyl-4,4'-dicarboxylic acid) ruthenium (II) complexes and the effect of protonation on charge-transfer sensitization of nanocrystalline titania. Inorg Chem 38:6298–6305

Nazeeruddin MK, Zakeeruddin S, Lagref J-J, Liska P, Comte P, Barolo C, Viscardi G, Schenk K, Grätzel M (2004) Stepwise assembly of amphiphilic ruthenium sensitizers and their applications in dye-sensitized solar cell. Coord Chem Rev 248:1317–1328

Nogueira A, Longo C, De Paoli M-A (2004) Polymers in dye sensitized solar cells: overview and perspectives. Coord Chem Rev 248:1455–1468

Nusbaumer H, Moser J-E, Zakeeruddin SM, Nazeeruddin MK, Grätzel M (2001) $Co^{II}(dbbip)_2^{2+}$ complex rivals tri-iodide/iodide redox mediator in dye-sensitized photovoltaic cells. J Phys Chem B 105:10461–10464

Pashaei B, Shahroosvand H, Abbasi P (2015) Transition metal complex redox shuttles for dye-sensitized solar cells. RSC Adv. 5:94814–94848

Pashaei B, Shahroosvand H, Graetzel M, Nazeeruddin MK (2016) Influence of ancillary ligands in dye-sensitized solar cells. Chem Rev 116(16):9485–9564

Robertson N (2006) Optimizing Dyes for Dye-Sensitized Solar Cells. Angew Chem Int Ed 45:2338–2345

Shahroosvand H, Rezaei S, Abbaspour S (2014) New photosensitizers containing the dipyridoquinoxaline moiety and their use in dye-sensitized solar cells. J Photochem Photobiol B 152:14–25

Shahroosvand H, Najafi L, Khanmirzaei L, Tarighi S (2015a) Artificial photosynthesis based on ruthenium (II) tetrazole-dye-sensitized nanocrystalline TiO_2 solar cells. J Photochem Photobiol B 152:4–13

Shahroosvand H, Zakavi S, Sousaraei A, Eskandari M (2015b) Saddle-shaped porphyrins for dye-sensitized solar cells: new insight into the relationship between nonplanarity and photovoltaic properties. Phys Chem Chem Phys 17:6347–6358

Tian H, Gabrielsson E, Yu Z, Hagfeldt A, Kloo L, Sun L (2011) A thiolate/disulfide ionic liquid electrolyte for organic dye-sensitized solar cells based on Pt-free counter electrodes. Chem Commun 47:10124–10126

Wang P, Zakeeruddin SM, Moser JE, Humphry-Baker R, Comte P, Aranyos V, Hagfeldt A, Nazeeruddin MK, Grätzel M (2004) Stable New Sensitizer with Improved Light Harvesting for Nanocrystalline Dye-Sensitized Solar Cells. Adv Mater 16:1806–1811

Wang P, Klein C, Humphry-Baker R, Zakeeruddin SM, Grätzel M (2005) A High Molar Extinction Coefficient Sensitizer for Stable Dye-Sensitized Solar Cells. J Am Chem Soc 127:808–809

Wang Z, Liang M, Wang L, Hao Y, Wang C, Sun Z, Xue S (2013) New triphenylamine organic dyes containing dithieno [3, 2-b: 2', 3'-d] pyrrole (DTP) units for iodine-free dye-sensitized solar cells. Chem Commun 49:5748–5750

Wu J, Lan Z, Hao S, Li P, Lin J, Huang M, Fang L, Huang Y (2008) Progress on the electrolytes for dye-sensitized solar cells. Pure Appl Chem 80:2241–2258

Xu M, Zhou D, Cai N, Liu J, Li R, Wang P (2011) Electrical and photophysical analyses on the impacts of arylamine electron donors in cyclopentadithiophene dye-sensitized solar cells. Energy Environ Sci 4:4735–4742

Yella A, Lee H-W, Tsao HN, Yi C, Chandiran AK, Nazeeruddin MK, Diau EW-G, Yeh C-Y, Zakeeruddin SM, Grätzel M (2011) Porphyrin-sensitized solar cells with cobalt (II/III)–based redox electrolyte exceed 12 percent efficiency. Science 334:629–634

Zhang M, Liu J, Wang Y, Zhou D, Wang P (2011) Redox couple related influences of π-conjugation extension in organic dye-sensitized mesoscopic solar cells. Chem Sci 2:1401–1406

Zhang Y, Wang Z-h, Hao Y-j, Wu Q-p, Liang M, Xue S (2015) Influence of Triarylamine and Indoline as Donor on Photovoltaic Performance of Dye-Sensitized Solar Cells Employing Cobalt Redox Shuttle. Chin J Chem Phys 28:91–100

Zhou D, Yu Q, Cai N, Bai Y, Wang Y, Wang P (2011) Efficient organic dye-sensitized thin-film solar cells based on the tris (1,10-phenanthroline) cobalt (II/III) redox shuttle. Energy Environ Sci 4:2030–2034

Zong X, Liang M, Chen T, Jia J, Wang L, Sun Z, Xue S (2012a) Efficient iodine-free dye-sensitized solar cells employing truxene-based organic dyes. Chem Commun 48:6645–6647

Zong X, Liang M, Fan C, Tang K, Li G, Sun Z, Xue S (2012b) Design of truxene-based organic dyes for high-efficiency dye-sensitized solar cells employing cobalt redox shuttle. J Phys Chem C 116:11241–11250

Concluding Remarks and Future Perspectives: Looking Back and Moving Forward

Gary F. Moore

Summary

This concluding chapter offers a perspective on photosynthesis research that is both historical and forward-looking. It imagines a future where a significant fraction of our energy demands are supplied by technologies inspired by photosynthesis. However, many scientific, engineering and policy challenges must be addressed for this realization.

Keywords

Artificial photosynthesis • Carbon dioxide • Earth-abundant • Catalyst • Water splitting • Solar energy storage • Electrochemistry • Nanomaterial • Renewable energy

Contents

20.1 Looking Back and Moving Forward 407
References 411

20.1 Looking Back and Moving Forward

While a complete understanding of the photosynthetic process is unavailable, recent advancements provide molecular (Blankenship 2014) as well as global (Archer and Barber 2004) scale details that inspire researchers to not only mimic aspects of this biological process in artificial constructs, but also build systems that, by some metrics, rival those of their biological counterpart (Bard and Fox 1995; Gray 2009; Moore and Brudvig 2011; Tran et al. 2012; Swierk and Mallouk 2013; Walter et al. 2010; Lubitz et al. 2008). Such efforts have been further accelerated by the elucidation of atomic level structure and function of key components of the photosynthetic apparatus, including Photosystem I (PSI) and Photosystem II (PSII) (Ferreira et al. 2004; Loll et al. 2005; Yano et al. 2006; Guskov et al. 2009; Umena et al. 2011). These chlorophyll-based photosystems absorb visible light to initiate a series of electron/hole transfer events that form a wireless current providing the potential needed to oxidize water and power the formation of the biological thermodynamic equivalent of hydrogen, nicotinamide adenine dinucleotide phosphate (NADPH). Concomitant with the generation of electron motive force (emf) across the two photosystems, protons are pumped across a biological membrane. Thus, a fraction of absorbed photonic energy is stored as redox potential in NADPH, and a fraction as proton motive force (pmf) that is used to power the production of adenosine triphosphate (ATP).

G.F. Moore (✉)
School of Molecular Sciences and the Biodesign Institute Center for Applied Structural Discovery (CASD), Arizona State University, Tempe, AZ 85287-1604, USA
e-mail: gary.f.moore@asu.edu

Fig. 20.1 *Make Like a Leaf!* Advances in our understanding of the structure and function of photosynthetic systems inspire the design of future technologies (Photo credit: Brandi Eide)

Together these energy carriers are used to build biological materials from carbon dioxide (CO_2). The products of photosynthetic CO_2 reduction constitute the chemical basis for most life as well as the fossil fuels (ancient biomass) that power our modern societies and economies (Fig. 20.1).

In contrast to the well-oxygenated conditions that prevail on our planet today, early Earth was characterized by the absence of oxygen in the atmosphere and oceans. Atmospheric oxygen concentrations first rose to appreciable levels during the Great Oxidation Event, approximately 2.5–2.3 billion years ago. The evolution of oxygenic photosynthetic organisms is widely accepted as the cause of this rise, although constraining the timing of this evolutionary event has proved difficult (Farquhar et al. 2011). Nonetheless, eons of photosynthetic activity have removed vast quantities of CO_2 from the atmosphere and deposited a significant fraction of carbon in the form of reduced fossil fuels. At present, the human species is rapidly undoing this chemistry and there is general agreement that an anthropogenic increase of CO_2 from the burning of fossil fuels is contributing to global climate change (IPCC 2014). Increasing human population and an associated rate of convenient fossil fuels usage further compound these issues, raising serious environmental and political concerns regarding energy security (Hoffert et al. 1998; Lewis and Nocera 2006). Thus, it is crucial that we devise immediate and longer-term strategies to reduce our dependence on fossil fuels.

Since natural photosynthesis is a solar energy storage process, it is not surprising that its products are widely used for producing non-fossil derived fuels (e.g. bioethanol or biodiesel). There are attempts to use grasses such as *Miscanthus*, which can grow on land that is unsuitable for agricultural farming, to provide biomass for biofuel production (Zhu et al. 2010). There are also efforts to establish algal farms in areas without crops or where traditional crops cannot be grown (Zhu et al. 2010). However, there are limitations to these approaches and it is questionable to what extent such technologies could supplement reliance on fossil fuels. For example, less than 1% of the solar energy absorbed by plants is ultimately stored in the form of chemical bonds (Blankenship et al. 2011; Hammarström et al. 2011). This does not mean parallel solutions should not be pursued; however, in its current form and without the advent of synthetic biology, natural photosynthesis is incapable of satiating the technological energy demands of our modern societies. This is notable in the iconic Keeling Curve graph that shows CO_2 concentration measured over the Mauna Loa Observatory in Hawaii since 1958. While the annual rise and fall of atmospheric CO_2 levels due to seasonal variation in photosynthetic activity are discernible, this is clearly swamped by the overall net rise in CO_2 due to human activities (i.e. the combustion of fossil fuels as unambiguously identified by the unique carbon-14 content of fossil derived carbon).

Transitioning to a sustainable energy future will require alternative energy sources and technologies that replace the burning of fossil fuels. Thus scientists have set to redesign photosynthesis for human technological applications using top-down as well as bottom-up strategies. Artificial photosynthesis, using concepts taken from natural photosynthesis to produce fuels using human engineered systems, has received increasing attention and has been described as a great scientific and moral challenge of our time (Faunce et al. 2013a, b). Among several possibilities, the photoelectrochemical splitting of water to molecular hydrogen and oxygen has been identified as an attractive approach to storing solar energy (Turner 2004). The storage of solar energy as hydrogen, derived from photoelectrochemical water splitting or by electrolysis of water using electricity from other renewable sources, is a relatively clean process and solar driven electrolysis is an existing technology that has great potential for improvement in terms of cost and energy efficiency. Likewise, the continuing decline in the price of conventional photovoltaics as well as the development of new generations of photovoltaic technologies will accelerate widespread use. There are, however, concerns related to the storage, transport, and incorporation of hydrogen fuel into an existing liquid-fuels infrastructure and further improvement in durability, overall efficiency and cost is required for wide-scale deployment. In this context, the development of earth-abundant catalysts capable of replacing the noble metals currently used in many commercial electrolyzers and fuel cells will be beneficial.

An alternative approach is to store solar energy in the chemical bonds of carbon containing molecules derived from the reduction of CO_2 (Turner 2004; Olah et al. 2009; Morris et al. 2009; Kumar et al. 2012; Ronge et al. 2015). This is in some respects more similar to natural photosynthesis where CO_2 absorbed from the atmosphere is converted into reduced forms of carbon that can be used as fuels or chemical feedstock for other applications. However, this approach has proven technologically more challenging especially given the relatively low level of atmospheric CO_2 (> 400 ppm and rising) as well as lack of effective catalysts and CO_2 concentrator technologies. In developing such new materials and technologies we should follow a cautious approach, developing systems that do not spread toxic or highly corrosive materials that are just as or even more polluting than the burning of fossil fuels. In this regard, we can learn a great deal from natural photosynthesis, in particular, the biological strategy of using environmentally safe materials, which are easily accessible and abundant (Najafpour and Govindjee 2011; Scholes et al. 2011).

Electrochemical techniques have been used to examine how enzymes can provide models for renewable energy systems (Armstrong and Hirst 2011). A promising feature of enzymes, and molecular catalysts in general, is their ability to provide discrete three-dimensional environments and coordination spheres for binding a substrate, lowering transition-state energies along a reaction coordinate and releasing a product (Albery and Knowles 1976). Thus, in accordance with the Sabatier principle, a well-designed molecular catalyst can have an exceptionally high activity and, even more importantly, selectivity for catalyzing a desired chemical transformation. These are highly favorable features for catalyzing complex multi-electron and multi-proton reactions such as those associated with solar fuels production. Indeed, bio-inspired hydrogen production catalysts that capture many of the key structural and functional elements of enzymes, while avoiding undesirable features such as size and fragility, have been prepared (Armstrong and Hirst 2011; Helm et al. 2011). Yet, the vast majority of molecular fuel-production catalysts are initially designed and optimized for operation in solution with the use of external chemical agents or an electrode providing the driving force (Artero and Fontecave 2005; Han et al. 2012; Losse et al. 2010; Mulfort et al. 2013) and it is often uncertain how or if a selected complex may function when immobilized at a surface or imbedded in a solar-energy conversion device. Despite these challenges, there has been considerable progress toward the construction and study of systems and subsystems that utilize molecular components attached to mesoporous and nanoparticulate materials (O'Regan and Graetzel 1991; Youngblood et al. 2009; Brimblecombe et al. 2010; Li et al. 2010a, b; Lakadamyali and Reisner 2011; Gardner et al. 2012; Huang et al. 2012; Moore et al. 2011; Zhao et al. 2012; Agiral et al. 2013; Parkinson and Weaver 1984; Badura et al. 2006; Jones et al. 2007; Hambourger et al. 2008; Reisner et al. 2009; Brown et al. 2010; Utschig et al. 2011; Chaudhary et al. 2012; Roy et al. 2012; Berggren et al. 2013; Flory et al. 2014; Wilker et al. 2014), conducting substrates (Andreiadis et al. 2013; Blakemore et al. 2013; Hambourger and Moore 2009; Yao et al. 2012) as well as direct interfaces to light-absorbing semiconducting substrates (McKone et al. 2014; Barton et al. 2008; Kumar et al. 2010; Hou et al. 2011; Moore and Sharp 2013; Krawicz et al. 2013; Cedeno et al. 2014; Krawicz et al. 2014; Seo et al. 2015; Downes and Marinescu 2015).

An additional challenge in the development of catalysts is the difficulty of identifying the active species participating in the catalytic cycle, and perceptions on heterogeneous versus homogeneous materials have blurred with advances in nanomaterials, clusters, metal nanoparticles, and bioinspired hybrid systems (Crabtree 2012; Artero and Fontecave 2013). Interestingly, the active defects in many metal-oxide electrocatalysts have structures reminiscent of the metal-oxo clusters of enzyme reaction centers and the process of building mimics of the catalytic Mn-core of the natural photosynthetic system has been further encouraged with the publication of the cubane-like configuration of the oxygen evolving complex at 1.9 Å resolution (Umena et al. 2011; Young et al. 2012; Kanady et al. 2011). The development of new X-ray spectroscopic techniques for viewing catalyzed reactions at the atomic-scale with increasing time resolution and chemical sensitivity will likely offer additional insights crucial to the design of next-generation catalysts (Neutze et al. 2000; Kirian et al. 2010; Chapman et al. 2006; Chapman et al. 2011; Milathianaki et al. 2013; Redecke et al. 2013).

It is imperative that research efforts are not limited to catalyst only or light-absorber only approaches. The coupling of one-photon and one-electron photochemical charge separation events with multi-electron and multi-proton catalysis (i.e. 4-electron water oxidation, 2-electron hydrogen production and multi-electron CO_2 reduction) imposes formidable chemical challenges (Dempsey et al. 2010; Stubbe et al. 2003; Mayer 2004; Huynh and Meyer 2007; Weinberg et al. 2012; Cyrille et al. 2010). Likewise, mechanisms to store multiple redox equivalents for subsequent electrochemical reactions remain largely unexplored, although biology offers a model of how this can be accomplished. In PSII, a tyrosine residue (TyrZ) functions as a molecular interface and redox mediator between the photo-oxidized primary electron donor ($P680^{\bullet+}$) and the Mn-containing oxygen evolving complex. The oxidation of TyrZ by $P680^{\bullet+}$ occurs with transfer of the phenolic proton to a nearby basic residue (His190). Control over the protonation state of the mediator plays a key role in charge transfer, poising the potential of the tyrosyl/tyrosine redox couple between that of $P680^{\bullet+}$ and the Mn cluster of the OEC while coupling the stepwise electron transfer process with proton activity to effectively remove oxidizing equivalents from the catalytic core. Proton-coupled electron transfer (PCET) is a fundamental aspect of biological energy transduction and will likely be of equal importance to the success of artificial photosynthesis. Further, protons formed during water oxidation in natural or artificial systems need to be efficiently removed from the active site, and subsequently transported to a reducing site, where they can be converted/stored as chemical fuel, or used to generate a pmf across a membrane. Such control over proton transport is especially critical to membrane-based architectures.

Artificial photosynthesis promises a future where solar energy can be utilized not only when it is available, but also stored in chemical bonds as fuels, unleashing a spatial and temporal distribution of renewable energy. To realize this on a global scale, research efforts must be directed to solving current scientific, engineering and social challenges in the field. At present, many components proposed for use in artificial photosynthesis are based on rare and/or toxic compounds that are expensive. Clearly this is unsatisfactory and efforts should be directed toward using environmentally benign materials composed of earth-abundant elements. Materials required for efficient light harvesting, charge separation, water splitting, hydrogen production or CO_2 reduction must be available on a sufficient scale without long-term harmful impacts on the environment and human health.

As the research efforts of artificial photosynthesis have evolved from the study of isolated components to the construction of subsystems and devices, there has been an increasing focus on stability. Thus, it is worth asking what the warranty is on a biological leaf? In the lower limit, the D1 complex which houses the manganese-based water oxidation catalyst lasts for ~30 min before it must be repaired. The ability of biological systems to repair, reproduce and evolve extends this warranty to ~three billion years and counting. From this perspective, we still have much to learn from biology.

Acknowledgment The author acknowledges support from the College of Liberal Arts and Sciences and the Biodesign Institute Center for Applied Structural Discovery (CASD), Arizona State University.

Conflict of Interest The author is not aware of any affiliations, memberships, funding, or financial holdings that might be perceived as affecting the objectivity of this book chapter.

References

Agiral A, Soo HS, Frei H (2013) Visible Light Induced Hole Transport from Sensitizer to Co_3O_4 Water Oxidation Catalyst across Nanoscale Silica Barrier with Embedded Molecular Wires. Chem Mater 25:2264–2273.

Albery WJ, Knowles JR (1976) Evolution of Enzyme Function and the Development of Catalytic Efficiency. Biochemistry 15:5631–5640.

Andreiadis ES, Jacques P-A., Tran PD, Leyris A, Chavarot-Kerlidou M, Jousselme B, et al. (2013) Molecular Engineering of a Cobalt-based Electrocatalytic Nanomaterial for Hydrogen Evolution under Fully Aqueous Conditions. Nat Chem 5:48–53.

Archer MD, Barber J (2004) Molecular to Global Photosynthesis, London: Imperial Coll. Press.

Armstrong FA, Hirst J (2011) Reversibility and Efficiency in Electrocatalytic Energy Conversion and Lessons from Enzymes. Proc Natl Acad. Sci 108:14049–14054.

Artero V, Fontecave M (2005) Some General Principles for Designing Electrocatalysts with Hydrogenase Activity. Coord Chem Rev 249:1518–1535.

Artero V, Fontecave M (2013) Solar Fuels Generation and Molecular Systems: Is It Homogeneous or Heterogeneous Catalysis? Chem Soc Rev 42:2338–2356.

Badura A, Esper B, Ataka K, Grunwald C, Woell C, Kuhlmann J, et al. (2006) Light-driven Water Splitting for (Bio-)Hydrogen Production: Photosystem 2 as the Central Part of a Bioelectrochemical Device. Photochem Photobiol 82:1385–1390.

Bard AJ, Fox MA (1995) Artificial Photosynthesis: Solar Splitting of Water to Hydrogen and Oxygen. Acc Chem Res 28:141–145.

Barton EE, Rampulla DM, Bocarsly AB (2008) Selective Solar-driven Reduction of CO_2 to Methanol Using a Catalyzed p-GaP Based Photoelectrochemical Cell. J Am Chem Soc 130:6342–6344.

Berggren G, Adamska A, Lambertz C, Simmons TR, Esselborn J, Atta M, et al. (2013) Biomimetic Assembly and Activation of [FeFe]-Hydrogenases. Nature 499:66–69.

Blakemore JD, Gupta A, Warren JJ, Brunschwig BS, Gray HB (2013) Noncovalent Immobilization of Electrocatalysts on Carbon Electrodes for Fuel Production. J Am Chem Soc 135:18288–18291.

Blankenship RE (2014) Molecular Mechanisms of Photosynthesis 2nd edition, Blackwell Science.

Blankenship RE, Tiede DM, Barber J, Brudvig GW, Fleming G, Ghirardi M, et al. (2011) Comparing

Photosynthetic and Photovoltaic Efficiencies and Recognizing the Potential for Improvement. Science 332:805–809.

Brimblecombe R, Koo A, Dismukes GC, Swiegers GF, Spiccia L (2010) Solar Driven Water Oxidation by a Bioinspired Manganese Molecular Catalyst. J Am Chem Soc 132:2892–2894.

Brown KA, Dayal S, Ai X, Rumbles G, King PW (2010) Controlled Assembly of Hydrogenase-CdTe Nanocrystal Hybrids for Solar Hydrogen Production. J Am Chem Soc 132:9672–9680.

Cedeno D, Krawicz A, Doak P, Yu M, Neaton JB, Moore, GF (2014) Using Molecular Design to Control the Performance of Hydrogen-producing Polymer-brush-modified Photocathodes. J Phys Chem Lett 5:3222–3226.

Chapman HN, Barty A, Bogan MJ, Boutet S, Frank M, Hau-Riege SP, *et al*. (2006) Femtosecond Diffractive Imaging with a Soft-x-ray free-electron Laser. Nature Phys 2:839–843.

Chapman HN, Fromme P, Barty A, White TA, Kirian RA, Aquila A, *et al*. (2011) Femtosecond X-ray protein nanocrystallography. Nature 470:73–77.

Chaudhary YS, Woolerton TW, Allen CS, Warner JH, Pierce E, Ragsdale SW, et al. (2012) Visible Light-driven CO_2 Reduction by Enzyme Coupled CdS Nanocrystals. Chem Commun 48:58–60.

Crabtree RH (2012) Resolving Heterogeneity Problems and Impurity Artifacts in Operationally Homogeneous Transition Metal Catalysts. Chem Rev 112:1536–1554.

Cyrille C, Robert M, Savéant J-M (2010). Concerted Proton–electron Transfers: Electrochemical and Related Approaches. Acc Chem Res 43:1019–1029.

Dempsey JL, Winkler JR, Gray HB (2010) Proton-coupled Electron Flow in Protein Redox Machines. Chem Rev 110:7024–7039.

Downes CA, Marinescu SC (2015) Efficient Electrochemical and Photoelectrochemical H_2 Production from Water by a Cobalt Dithiolene One-dimensional Metal–organic Surface. J Am Chem Soc 137:13740–13743.

Farquhar J, Zerkle AL, Bekker A (2011) Geological Constraints on the Origin of Oxygenic Photosynthesis. Photosynth Res 107:11–36.

Faunce T, Styring S, Wasielewski MR, Brudvig GW, Rutherford AW, Messinger J, et al. (2013a) Artificial Photosynthesis as a Frontier Technology for Energy Sustainability. Energy Environ Sci 6:1074–1076.

Faunce TA, Lubitz W, Rutherford AW, MacFarlane D, Moore GF, Yang P, et al. (2013b) Energy and Environment Policy Case for a Global Project on Artificial Photosynthesis. Energy Environ Sci 6:695–698.

Ferreira KN, Iverson TM, Maghlaoui K, Barber J, Iwata S (2004). Architecture of the Photosynthetic Oxygen-evolving Center. Science 303:1831–1838.

Flory JD, Simmons CR, Lin S, Johnson T, Andreoni A, Zook J, et al. (2014) Low Temperature Assembly of Functional 3D DNA-PNA-Protein Complexes. J Am Chem Soc 136:8283–8295.

Gardner JM, Beyler M, Karnahl M, Tschierlei S, Ott S, Hammarström L (2012) Light-Driven Electron Transfer between a Photosensitizer and a Proton-Reducing Catalyst Co-adsorbed to NiO. J Am Chem Soc 134:19322–19325.

Gray HB (2009) Powering the Planet with Solar Fuel. Nat. Chem., 1:112.

Guskov A, Kern J, Gabdulkhakov A, Broser M, Zouni A, Saenger W (2009) Cyanobacterial Photosystem II at 2.9 Å Resolution and the Role of Quinones, Lipids, Channels and Chloride. Nat Struct Mol Biol 16:334–342.

Hambourger M, Moore TA (2009) Nailing Down Nickel for Electrocatalysis. Science 326:1355–1356.

Hambourger M, Gervaldo M, Svedruzic D, King PW, Gust D, Ghirardi M, et al. (2008) [FeFe]-hydrogenase-catalyzed H_2 Production in a Photoelectrochemical Biofuel Cell. J Am Chem Soc 130:2015–2022.

Hammarström L, Winkler JR, Gray HB, Styring S (2011) Shedding Light on Solar Fuel Efficiencies. Science 333, 288.

Han Z, Qiu F, Eisenberg R, Holland PL, Krauss TD (2012) Robust Photogeneration of H_2 in Water Using Semiconductor Nanocrystals and a Nickel Catalyst. Science 338:1321–1324.

Helm ML, Stewart MP, Bullock RM, DuBois MR, DuBois DL (2011) A Synthetic Nickel Electrocatalyst with a Turnover Frequency Above 100,000 s^{-1} for H_2 Production. Science 333:863–866.

Hoffert MI, Caldeira K, Jain AK, Haites EF, Harvey LDD, Potter SD, et al. (1998) Energy Implications of Future Stabilization of Atmospheric CO_2 Content. Nature 395:881–884.

Hou Y, Abrams BL, Vesborg PCK, Björketun ME, Herbst K, Bech L, et al. (2011) Bioinspired Molecular Co-catalysts Bonded to a Silicon Photocathode for Solar Hydrogen Evolution. Nat Mater 10:434–438.

Huang J, Mulfort KL, Du P, Chen LX (2012) Photodriven Charge Separation Dynamics in CdSe/ZnS Core/shell Quantum Dot/cobaloxime Hybrid for Efficient Hydrogen Production. J Am Chem Soc 134:16472–16475.

Huynh, MHV, Meyer TJ (2007) Proton-coupled Electron Transfer Chem Rev 107:5004–5064.

IPCC, 2014: Climate Change 2014: Synthesis Report. Contribution of Working Groups I, II and III to the Fifth Assessment Report of the Intergovernmental Panel on Climate Change [Core Writing Team, Pachauri RK, Meyer LA (eds.)]. IPCC, Geneva, Switzerland.

Jones AK, Lichtenstein BR, Dutta A, Gordon G, Dutton PL (2007) Synthetic Hydrogenases: Incorporation of an Iron Carbonyl Thiolate into a Designed Peptide. J Am Chem Soc 129:14844–14845.

Kanady JS, Tsui EY, Day MW, Agapie T (2011) A Synthetic Model of the Mn_3Ca Subsite of the Oxygen-evolving Complex in Photosystem II. Science 333:733–736.

Kirian RA, Wang X, Weierstall U, Schmidt KE, Spence JCH, Hunter M, et al. (2010) Femtosecond Protein Nanocrystallography-data Analysis Methods. Opt Express 18:5713–5723.

Krawicz A, Yang J, Anzenberg E, Yano J, Sharp ID, Moore GF (2013) Photofunctional Construct That Interfaces Molecular Cobalt-based Catalysts for H_2 Production to a Visible-light-absorbing Semiconductor. J Am Chem Soc 135:11861–11868.

Krawicz A, Cedeno D, Moore, GF (2014) Energetics and Efficiency Analysis of a Cobaloxime-modified Semiconductor at Simulated Air Mass 1.5 Illumination. Phys Chem Chem Phys 16:15818–15824.

Kumar B, Smieja JM, Kubiak CP (2010) Photoreduction of CO_2 on p-type Silicon Using Re(bipy-Bu^t)(CO)$_3$Cl: Photovoltages Exceeding 600 mV for the Selective Reduction of CO_2 to CO. J Phys Chem C 114:14220–14223.

Kumar B, Llorente M, Froehlich J, Dang, T, Sathrum A, Kubiak CP (2012) Photochemical and Photoelectrochemical Reduction of CO_2. Annu Rev Phys Chem 63:541–569.

Lakadamyali F, Reisner E (2011) Photocatalytic H_2 Evolution from Neutral Water with a Molecular Cobalt Catalyst on a Dye-sensitised TiO_2 Nanoparticle. Chem Commun 47:1695–1697.

Lewis NS, Nocera DG (2006) Powering the Planet: Chemical Challenges in Solar Energy Utilization. Proc Natl Acad Sci 103:15729–15735.

Li L, Duan L, Xu Y, Gorlov M, Hagfeldt A, Sun L (2010a) A Photoelectrochemical Device for Visible Light Driven Water Splitting by a Molecular Ruthenium Catalyst Assembled on Dye-sensitized Nanostructured TiO_2. Chem Commun 46:7307–7309.

Li G, Sproviero EM, McNamara WR, Snoeberger RC, Crabtree RH, Brudvig GW, et al. (2010b) Reversible Visible-light Photooxidation of an Oxomanganese Water-oxidation Catalyst Covalently Anchored to TiO_2 Nanoparticles. J Phys Chem B 114:14214–14222.

Loll B, Kern J, Saenger W, Zouni A, Biesiadka J (2005). Towards Complete Cofactor Arrangement in the 3.0 Å Resolution Structure of Photosystem II. Nature 438:1040–1044.

Losse S, Vos JG, Rau S (2010) Catalytic Hydrogen Production at Cobalt Centres. Coord Chem Rev 254:2492–2504.

Lubitz W, Reijerse EJ, Messinger J (2008) Solar Water-Splitting into H_2 and O_2: Design Principles of Photosystem II and Hydrogenases. Energy Environ Sci 1:15–31.

Mayer JM (2004) Proton-coupled Electron Transfer: A Reaction Chemist's View. Annu Rev Phys Chem 55:363–390.

McKone JR, Marinescu SC, Brunschwig BS, Winkler JR, Gray, HB (2014) Earth-abundant Hydrogen Evolution Electrocatalysts. Chem Sci 5:865–878.

Milathianaki D, Boutet S, Williams GJ, Higginbotham A, Ratner D, Gleason AE, et al. (2013) Femtosecond Visualization of Lattice Dynamics in Shock-compressed Matter. Science 342:220–223.

Moore GF, Brudvig GW (2011) Energy Conversion in Photosynthesis: a Paradigm for Solar Fuel Production. Annu Rev Condens Matter Phys 2:303–327.

Moore GF, Sharp ID (2013) A Noble-Metal-Free Hydrogen Evolution Catalyst Grafted to Visible Light-absorbing Semiconductors. J. Phys. Chem Lett 4:568–572.

Moore GF, Blakemore JD, Milot RL, Hull JF, Song H-e, Cai L, et al. (2011) A Visible Light Water-splitting Cell with a Photoanode Formed by Codeposition of a High-potential Porphyrin and an Iridium Water-oxidation Catalyst. Energy Environ Sci 4:2389–2392.

Morris AJ, Meyer G.J., Fujita E. (2009) Molecular Approaches to the Photocatalytic Reduction of Carbon Dioxide for Solar Fuels. Acc Chem Res 42:1983–1994.

Mulfort KL, Mukherjee A, Kokhan O, Du P, Tiede DM (2013) Structure-function Analyses of Solar Fuels Catalysts Using in situ X-ray Scattering. Chem Soc Rev 42:2215–2227.

Najafpour MM, Govindjee (2011) Oxygen Evolving Complex in Photosystem II: Better than Excellent. Dalton Trans 40:9076–9084.

Neutze R, Wouts R, van der Spoel D, Weckett E, Hajdu, J (2000) Potential for Biomolecular Imaging with Femtosecond X-ray Pulses. Nature 406:752–757.

O'Regan B, Graetzel M (1991) A Low-cost, High-efficiency Solar Cell Based on Dye-sensitized Colloidal Titanium Dioxide Films. Nature 353:737–740.

Olah GA, Goeppert A, Prakash GKS (2009) Beyond Oil and Gas: The Methanol Economy Wiley-VCH. Weinheim.

Parkinson BA, Weaver PF (1984) Photoelectrochemical Pumping of Enzymic Carbon Dioxide Reduction. Nature 309:148–149.

Redecke L, Nass K, DePonte DP, White TA, Rehders D, Barty A, et al. (2013) Natively Inhibited Trypanosoma Brucei Cathepsin B Structure Determined by Using an X-ray Laser. Science 339:227–230.

Reisner E, Powell DJ, Cavazza C, Fontecilla-Camps JC, Armstrong FA (2009) Visible Light-driven H_2 Production by Hydrogenases Attached to Dye-sensitized TiO_2 Nanoparticles. J Am Chem Soc 131:18457–18466.

Ronge J, Bosserez T, Martel D, Nervi C, Boarino L, Taulelle F, et al. (2015) Monolithic Cells for Solar Fuels. Chem Soc Rev 43, 7963–7981.

Roy A, Madden C, Ghirlanda G (2012) Photo-induced Hydrogen Production in a Helical Peptide Incorporating a [FeFe] Hydrogenase Active Site Mimic. Chem Commun 48:9816–9818.

Scholes GD, Fleming GR, Olaya-Castro A, Grondelle R (2011) Lessons from Nature about Solar Light Harvesting. Nature Chemistry 3:763–774.

Seo J, Pekarek RT, Rose MJ (2015) Photoelectrochemical Operation of a Surface-bound, Nickel-phosphine H_2 Evolution Catalyst on p-Si(111): a Molecular

Semiconductor|Catalyst Construct. Chem Commun 51:13264–13267.

Stubbe J, Nocera DG, Yee CS, Chang MCY (2003) Radical Initiation in the Class I Ribonucleotide Reductase: Long-Range Proton-Coupled Electron Transfer? Chem Rev 103:2167–2201.

Swierk JR, Mallouk TE (2013) Design and Development of Photoanodes for Water-splitting Dye-sensitized Photoelectrochemical Cells. Chem Soc Rev 42, 2357–2387.

Tran PD, Wong LH, Barber J, Loo JSC (2012) Recent Advances in Hybrid Photocatalysts for Solar Fuel Production. Energy Environ Sci 5:5902–5918.

Turner JA (2004) Sustainable Hydrogen Production, Science 305:972–974.

Umena Y, Kawakami K, Shen JR, Kamiya N (2011) Crystal Structure of Oxygen-evolving Photosystem II at a Resolution of 1.9 Å. Nature 473: 55–61.

Utschig LM, Silver SC, Mulfort KL, Tiede DM (2011) Nature-driven Photochemistry for Catalytic Solar Hydrogen Production: a Photosystem I–transition Metal Catalyst Hybrid. J Am Chem Soc 133:16334–16337.

Walter MG, Warren EL, Mckone JR, Boettcher SW, Mi Q, Santori EA, Lewis NS (2010) Solar Water Splitting Cells. Chem Rev 110:6446–6473.

Weinberg DR, Gagliardi CJ, Hull JF, Murphy CF, Kent, CA Westlake BC, et al. (2012). Proton-coupled Electron Transfer Chem Rev 112:4016–4093.

Wilker MB, Shinopoulos KE, Brown, KA, Mulder, DW, King, PW, Dukovic G (2014) Electron Transfer Kinetics in CdS Nanorod-[FeFe] Hydrogenase Complexes and Implications for Photochemical H_2 Generation. J Am Chem Soc 136:4316–4324.

Yano J, Kern J, Sauer K, Latimer MJ, Pushkar Y, Biesiadka J, et al. (2006). Where Water is Oxidized to Dioxygen: Structure of the Photosynthetic Mn_4Ca Cluster. Science 314:821–825.

Yao SA, Ruther RE, Zhang L, Franking RA, Hamers RJ, Berry JF (2012) Covalent Attachment of Catalyst Molecules to Conductive Diamond: CO_2 Reduction Using "Smart" Electrodes. J Am Chem Soc 134:15632–15635.

Young KJ, Martini LA, Milot RL Snoeberger RC, Batista VS, Schmuttenmaer CA, Crabtree RH, Brudvig GW (2012) Light-driven Water Oxidation for Solar Fuels. Coord Chem Rev 256:2503–2520.

Youngblood WJ, Lee S-HA, Kobayashi Y, Hernandez-Pagan EA, Hoertz PG, Moore TA, et al. (2009) Photoassisted Overall Water Splitting in a Visible Light-absorbing Dye-sensitized Photoelectrochemical Cell. J Am Chem Soc 131:926–927.

Zhao Y, Swierk JR, Megiatto JD, Sherman B, Youngblood WJ, Qin D, et al. (2012) Improving the Efficiency of Water Splitting in Dye-sensitized Solar Cells by Using a Biomimetic Electron Transfer Mediator. Proc Natl. Acad Sci 109:15612–15616.

Zhu X, Long SP, Ort DR (2010) Improving Photosynthetic Efficiency for Greater Yield. Annu Rev Plant Biol 61:235–261.

Index

A
Abiotic stress, 186–195, 204, 218–230, 233, 262–270, 307, 313, 315
Activator, 68, 69, 74–81, 86, 88, 89
Aggregation quenching, 39, 40
Alternative electron acceptor, 52–62
Antenna, 2, 6, 12–16, 20, 35, 37, 39, 41, 43–46, 98, 102, 113, 125–130, 135, 136, 147, 150–154, 156, 158, 170–172, 176, 177, 211, 253, 254, 262, 282–284, 290, 344, 360
Artificial photosynthesis, 7, 40, 54, 321–323, 325, 328, 329, 360, 361, 376, 390, 392, 410, 411
Assembly, 20–22, 34, 37, 61, 62, 74, 115–118, 120, 131, 133, 139, 143, 148, 152, 154, 155, 158, 281, 284, 293, 294
ATP synthase, 3, 5, 13, 14, 113, 287

B
Bacteriochlorophyll, 2, 12, 15, 132, 135, 170
Bicarbonate, 52, 57, 58, 60, 206, 309–311

C
Carbon dioxide, 2, 12, 252, 390, 408
Carbon nanotube, 321–324, 326–330
Carotenoid, 12, 14–19, 25, 27, 35, 44, 102, 125, 129, 130, 141, 147, 149, 152, 154, 156, 158, 170, 173, 193, 307, 313
Catalysis, 68, 76, 205, 279, 308, 336, 348, 360, 391, 392, 410
Catalyst, 8, 45, 54, 321, 322, 325, 327, 329, 334–340, 344, 347–349, 351–356, 360, 361, 409–411
Cerium, 322, 337
Chlamydomonas, 42, 116, 130, 131, 134, 136, 137, 139–141, 143–145, 147, 150, 155, 158, 276–296, 313
Chloride, 55, 67–74, 77–79, 86, 89, 100
Chlorophyll, 2, 4, 5, 34–39, 42–44, 46, 52, 53, 56, 69, 98, 99, 113, 114, 125–138, 140, 145, 147–149, 151, 152, 154, 156–158, 169, 192, 208, 210–217, 219, 221, 223–225, 227–234, 251, 262, 281, 284, 288, 290, 307, 308, 314, 376, 390–392, 407
Chlorophyll binding site, 72, 73, 76–79
Chlorophyll fluorescence, 211–217, 219, 221, 223–225, 227–233, 262–270, 310
Chloroplast, 2, 3, 5, 43, 54, 68–70, 100, 112–115, 147, 150, 156, 186, 188, 191, 192, 195, 196, 204, 206, 210, 216, 218, 222, 226, 251–253, 255, 264, 266, 278–281, 286, 287, 292–294, 306–309, 313–315, 390, 392
Cobalt, 57, 336, 394–401
Cold acclimation, 194, 195, 277, 280, 285, 312
Cold adaptation, 280, 286
Complex, 2–6, 8, 12–27, 34–41, 43–45, 52, 53, 55, 57, 69, 74, 79, 83, 87, 98–107, 113–115, 117–120, 124, 125, 127, 130, 131, 133, 137–140, 144–149, 151–155, 157, 158, 176, 191, 206, 207, 214, 218, 222, 224, 226, 227, 229, 230, 250, 252, 254, 263, 269, 277, 282, 287–289, 291–294, 307–309, 311, 313, 315, 336, 347–349, 351–355, 363–365, 367, 368, 378, 379, 394, 395, 400, 401, 409–411
Composite, 39, 325, 329
Crop plant, 204–219, 230, 253, 264, 266, 268
Cyanobacteria, 1, 2, 12, 52, 58, 67, 72–76, 83, 88, 98, 112–115, 121, 125, 128, 132, 137, 138, 141, 144, 146–148, 150, 152–155, 157, 158, 170, 172–174, 176, 181, 206, 252, 276, 334, 344
Cytochrome b559, 61, 226

D
Degradation, 131, 154, 155, 158, 250, 251, 255, 256, 266, 267, 295, 308, 334, 339, 352, 394
Development, 12, 68, 74, 107, 195, 204, 205, 207, 215, 224, 234, 251, 253, 262, 263, 266, 268, 269, 278, 280, 281, 306, 307, 312, 334, 401, 409, 410
Drought, 189–194, 210, 218–225, 227, 262, 264–266, 289, 311, 314
Dye, 20, 21, 44, 45, 54, 390–402

E
Earth-abundant, 409, 411
Electrochemistry, 355
Electrogenic reactions, 102, 106, 107
Electron transfer, 4, 7, 20, 53–58, 60–62, 68, 69, 72, 79, 85, 87, 89, 98–106, 113, 114, 116–118, 121, 124–128, 130–132, 134–137, 139–148, 150, 158, 289, 306, 309, 314, 334, 348, 355, 360, 362, 364, 366, 370, 371, 374, 376–379, 397, 398, 401, 410

Electron transport, 3, 5, 53, 62, 85, 105, 113, 118, 129–131, 144, 152, 192, 195, 206, 207, 211–213, 215–217, 220–222, 224–227, 233, 251–253, 255, 263, 278, 281, 282, 284, 288–290, 293, 295, 306, 309–311, 313–315, 361
Energy transfer, 2, 34–36, 44–46, 125, 128, 135, 137, 150–152, 154, 156, 158, 170–173, 175–177, 179, 181, 281, 282

F
Ferricyanide, 54–56
Fluorescence, 36, 40, 42–44, 69, 80, 139, 170–173, 175, 176, 179, 208, 210–216, 219, 221, 223–225, 227–234, 252, 253, 262–269, 290, 309, 310, 378
Function, 4, 12–25, 27, 41, 43, 52–54, 62, 68, 74, 75, 86, 89, 98, 106, 107, 112, 113, 115, 117, 120, 121, 125, 128–132, 134, 136, 138–140, 142, 144, 146–150, 154, 155, 158, 170–172, 176, 179–181, 186, 187, 191, 210, 212, 213, 215, 221, 222, 226, 229, 264, 266–268, 277, 284, 286, 289, 292–294, 296, 306, 311, 313, 315, 336, 351, 356, 362–364, 371, 373, 381, 395, 407, 408, 410

G
Gas exchange, 192, 207, 208, 211, 212, 216, 219–221, 223, 225, 231–233, 285, 287
Glycine betaine, 186–196
Greening, 307, 308

H
Heat, 37, 170, 190, 211, 213, 224–231, 253, 262, 263, 279, 283, 286, 294, 312
Herbicide, 55, 58, 73, 251, 252, 269, 314–315

I
Inhibitor, 55, 61, 68, 71, 72, 74, 80–83, 85, 86, 88, 89, 121, 131, 193, 284, 288, 290, 292, 307, 314

K
Kinetics, 43, 67–75, 78–89, 99–104, 106, 107, 116, 118, 130, 135, 141, 143, 144, 146, 150, 157, 172, 173, 211, 213–215, 223–225, 227–230, 234, 268, 289, 292, 310, 363, 374, 395, 397, 398

L
Leaf senescence, 224, 266, 267, 306, 308
Light harvesting, 12–23, 25–27, 34–46, 125, 130, 148, 171, 172, 180, 181, 191, 262, 281, 288, 291, 355, 392, 393, 401, 411
Light-induced radical pairs, 364, 380
Lipid, 13, 14, 22, 35, 37, 43, 53, 54, 58–62, 98, 101, 113, 121, 130, 131, 141, 147, 156, 157, 186, 191, 192, 226, 266, 277, 278, 286, 295, 306, 310–315

M
Manganese, 5, 69, 71, 99, 101–107, 226, 308, 337, 344, 349, 351, 352, 354–356, 411
Manganese oxide, 344–355
Mechanism, 5–7, 13, 20, 25, 27, 37, 39, 40, 42–44, 46, 54, 55, 57, 60, 71, 74, 101, 103–107, 131, 135–137, 144, 152, 158, 170–173, 175, 177, 181, 196, 204, 210, 222–224, 226, 227, 230, 251, 255, 256, 262, 264, 276, 277, 280, 282, 287, 290, 311, 312, 327, 329, 336, 337, 340, 344, 346–350, 353, 362, 370, 372, 376, 410
Membrane potential, 103, 104
Mixed site, 38
Molecular switches, 293
Molecular triplet state, 363, 369, 372, 374

N
Nanomaterial, 44, 334, 353, 410
Nanostructure, 322, 323, 325, 328, 329
Nitric oxide, 306–315
Non-photochemical quenching, 39–44, 46, 211, 212, 262, 263, 268, 283–285, 289, 310

O
Oxygen evolution, 6, 54–56, 61, 62, 67–83, 85–89, 98, 107, 293, 308, 313, 322, 329, 347, 353
Oxygen evolving complex, 4, 8, 52–54, 57, 59–60, 69–72, 74, 79, 89, 98, 99, 101–107, 252, 292, 308, 309, 313, 315, 353, 410

P
Phenanthroline, 390–395, 397–399, 401, 403
Phenomics, 263, 264, 268–270
Phenotyping, 230–234, 268, 269
Photoautotrophs, 2, 12, 75, 76, 114, 116, 117, 121, 125, 140, 141, 144–148, 151, 158, 277, 278, 280, 281, 285, 295, 296
Photochemical reaction, 103, 222
Photophosphorylation, 311
Photosynthesis, 1–8, 12–15, 37, 41, 46, 52–55, 61, 68, 112–114, 135, 158, 170, 171, 177, 188–196, 204–211, 213, 215, 216, 218–230, 233, 251–254, 261–264, 266, 268, 269, 279, 280, 282, 286, 293, 306, 309, 311–315, 343, 344, 348, 349, 360, 361, 390–395, 397–401, 403, 408, 409, 411
Photosynthetic apparatus, 68, 130, 186–196, 204, 209, 212, 213, 223, 224, 234, 253, 263, 269, 281, 283, 288, 289, 308, 309, 313–315, 407
Photosynthetic machinery, 192, 204, 208, 213, 221, 264
Photosynthetic performance, 186, 190, 207, 209, 210, 213, 214, 216, 256, 262, 264
Photosynthetic productivity, 207, 208
Photosystem I, 3–5, 113, 170–172, 175–177, 180, 181, 191, 192, 215, 224, 226, 231, 255, 262, 267, 281–284, 288–296, 307, 310, 314, 315, 407
Photosystem II, 2, 3, 5, 6, 35, 39, 52–63, 67–89, 98–107, 113, 114, 121, 129–131, 135–138, 152, 170–172, 175–178, 180, 181, 186, 190–192, 195, 211–213, 215, 217, 221–230, 252–255, 262–265, 267, 268, 281–284, 288–293, 296, 307–310, 313–315, 328, 335, 340, 344, 345, 347–353, 355, 356, 361, 407, 410
Photovoltaic, 44, 46, 58, 390, 391, 398, 402, 409
Phycobilisome, 152, 172–177, 344
Pigment-protein reconstitution, 34–37, 46
Plant growth, 186, 188, 194, 195, 204, 218, 251, 256, 264, 266, 306, 312, 313
Plant hormones, 186–196, 311, 315

Polycyclic aromatic hydrocarbons (PAHs), 249–256
Protein phosphorylation, 296
Proteoliposome, 43, 46, 101, 104
PSI-Cyt b6/f supercomplex, 282, 288, 292–296
Psychrophile, 276–295
PufX protein, 19–21, 23, 24
Purple bacteria, 2, 12–15, 17, 22, 125, 137

Q

Quinone, 13, 14, 19, 25, 27, 52, 54–55, 58–61, 69, 85, 98, 99, 101, 113, 140, 141, 213, 254, 376

R

Reaction center, 2, 5, 35, 68, 98, 113–116, 118, 121, 122, 125, 126, 129, 131–137, 146, 169–171, 176–178, 191, 192, 211, 222, 226, 227, 229, 230, 253, 255, 262, 281, 310, 311, 314, 344, 355, 360, 364, 376, 410
Regulation, 61, 158, 170, 173, 186, 204, 205, 207, 209, 210, 220, 222, 223, 227, 253, 281, 284, 289, 291, 293, 306, 307, 312
Remote sensing, 216, 218, 263, 269, 270
Renewable energy, 390, 409, 411
Rhodobacter sphaeroides, 171

S

Semiconductor, 44, 348–350, 353, 355, 356, 360, 390–392, 394
Senescence, 266, 267, 307, 308
Sensitizer, 390–395, 398, 401

Site-directed mutagenesis, 17, 45, 46, 57, 62
Solar cell, 353, 376, 380, 381, 390, 391, 400, 401
Solar energy storage, 408
Spin polarization, 362, 365, 370, 373–375, 381
S-state cycle, 7, 89
Structure, 2–7, 13, 14, 16–27, 35–38, 41, 42, 44, 52–54, 57, 59, 61, 62, 72–74, 80, 98, 103–105, 113–115, 117–120, 127, 129, 131–133, 135–138, 141, 144–147, 149–154, 156–158, 171, 173, 178–180, 187, 188, 192, 196, 213, 216, 222, 223, 229, 233, 250–252, 255, 277–279, 286, 289, 291, 293, 294, 306, 313, 315, 322, 324, 336, 340, 344, 350, 355, 360, 361, 364, 367, 379–381, 390–393, 395–397, 399–401, 407, 408, 410

T

Thermochromatium tepidum, 25
Time-resolved spectroscopy, 98
Tungsten oxide, 344, 345, 347–356

W

Water oxidation, 4–7, 68, 79, 81, 100–103, 106, 322, 325, 328, 329, 334, 336, 337, 339, 340, 344, 346, 354, 410, 411
Water splitting, 8, 322, 333, 334, 344, 345, 347–356, 409, 411

X

X-ray crystallography, 19, 22

Printed by Printforce, the Netherlands